ADVANCES IN CHRYSOMELIDAE BIOLOGY 1

ADVANCES IN CHRYSOMELIDAE BIOLOGY 1

edited by M.L. Cox

Backhuys Publishers, Leiden

ISBN 90-5782-028-5

PRINTED IN THE NETHERLANDS

TABLE OF CONTENTS

This volume is dedicated to the memory of my beloved wife Parichehr, a collembologist (1936–1998)

MICHAEL L. COX

FOREWORD

The Chrysomelidae or leaf beetles are one of the most species-rich and biologically diverse families of Coleoptera. These beetles are virtually ubiquitous, occurring in nearly any terrestrial habitat on all continents, except Antarctica as well as most islands and some are truly Arctic species. Larvae of the Donaciinae are aquatic, feeding at the roots of their host plants, often in the deep water of rivers and lakes. The chrysomelids successfully colonize hosts on the seashore and on mountains up to the snowline, and with special adaptations are even found in deserts. They include some of the most economically important pests, such as the Colorado potato beetle which not only damage crops directly, but also indirectly by the transmission of plant virus diseases. Some have been successfully employed in the control of weeds in many parts of the world.

They are all phytophagous and feed in the adult and larval stages on any plant organ; and on almost all plants from the primitive cycads, mosses, horsetails, through the Gymnospermae to the advanced Spermatophytes.

This book is the sixth in the series, commencing in 1988 with "Biology of Chrysomelidae" edited by my friend and colleague Pierre Jolivet who has been a great inspiration to me over many years. The 35 chapters are contributed by a total of 57 authors, many world authorities in their particular specialisms. They come from 20 countries, mostly in Europe and the New World, but also from Australasia and the African Continent. The book highlights the continuing world!wide scientific interest in leaf beetles and brings together the results of recent research in a great diversity of topics. These include Palaeontology; Phylogeny; Biogeography; Ecological studies, natural enemies, defences, biological control; Community ecology, Population ecology, conservation; Host plant associations; Reproduction; Biological and Biosystematic studies of individual species or genera.

The larvae and their feeding habits are known for the majority of leaf beetle subfamilies, except the Megascelinae and Aulacoscelinae. The chapter by Cox and Windsor, includes descriptions of the previously undescribed first instar larvae of these subfamilies and uses larval characters to relate the Megascelinae and Aulacoscelinae to the Eumolpinae and Orsodacninae, respectively.

The chapters on mating behaviour in Donaciinae by Bienkowski and chrysomelids in general by Jolivet fill important gaps in our knowledge, mentioned in the foreword to "Novel aspects of the biology of Chrysomelidae".

The majority of chrysomelids show narrow host plant specialization, since they feed on a few, mostly closely related plant species. The relations between one group of chrysomelines and their host plants is the subject of the novel evolutionary study by Hsiao and Pasteels. The chapters by Hespenheide and Dang, Novotny et al., Simmul and de Little increase our knowledge of the host plant relationships of Costa Rican Hispinae; New Guinean chrysomelids and Australian paropsine Chrysomelinae.

Most rain forest canopy studies have been performed in Central and South America, Malaysia, Indonesia or New Guinea. However, the research carried out by Wagner on the arboreal chrysomelid community structure and fauna of different forest types in Central Africa is apparently the first completed on that continent and unusual since it looks at several forest types.

The chapter by Beenen is the first in this series of books to deal directly with the conservation of leaf beetles. It is not only rain forest chrysomelid species which are under threat from deforestation, but several species are already extinct in parts of Europe and others are threatened with extinction because of agricultural activities. This topical chapter highlights the possibilities for conservation and rehabilitation of chrysomelid populations in a cultivated environment.

Anyone interested in the ecology of chrysomelids will find information of value here, because the life cycles of different species in the final, general section of the book are a major key to understanding the ecological strategies of leaf beetles.

I gratefully acknowledge and thank Mr Wil Peters of Backhuys Publishers, Leiden, The Netherlands for his support in publishing this book and continuing this series, of which future volumes are planned. Grateful thanks also go to all of the contributors, for without them such books would not be possible. I also thank Professor Pierre Jolivet for his encouragement, help and guidance to produce another book in this series and to continue his good work started back in 1988. It would also be appropriate here to recognize the great contribution to our knowledge of the Chrysomelidae made by Professor Roy Crowson, who sadly passed away in May. This series is unique and nothing like it is published for any other beetle family or insect group. This volume will be a useful reference resource for agricultural and forestry research workers, university students and museum curators.

Many chrysomelids, especially those in the tropics have yet to be discovered. They provide a vast treasure for naturalists and experimental biologists. In view of the continuing rapid deforestation now occurring in the tropics of South America, Malaysia and Indonesia, there is a urgent need to study the leaf beetles in these new faunas before they vanish from our planet. It is hoped that this book will provide encouragement to my colleagues to carry out these important studies and go some way to fill some of the voids in our knowledge and understanding of the leaf beetles. Several mysteries still remain, including the larval food substrate and position of feeding of the Orsodacninae and Aulacoscelinae. However, it looks increasingly likely from circumstantial evidence that those of the former develop in the apical shoots of oaks, with the recent oviposition in captivity of several Orsodacne females between the brown bud scales. It remains for final instar larvae to be collected in such positions in the wild.

Michael L. Cox

CHRYSOMELIDAE IN FOSSILIZED RESIN: BEHAVIOURAL INFERENCES

GEORGE O. POINAR, JR.
Department of Entomology, Oregon State University, Corvallis, OR 97331, USA,
Tel: 541 752 0917, Fax: 541 737 3643, E-mail: poinarg@bcc.orst.edu

Abstract

A listing of the chrysomelid beetles known from amber deposits is presented, along with the first report of a Mesozoic sagrine chrysomelid from Canadian Cretaceous amber. A discussion of the possible plant hosts of this Cretaceous fossil is presented in light of fossil cycads from that period and location and a possible sedge (Cyperaceae) seed found in the same amber deposits. New findings of chrysomelids in fossilized resin (amber and copal) are presented, along with indications of past behavioural patterns and associations including evidence of predation, gregariousness, feeding behaviour, host plant selection, parasitism, defence mechanisms, mimicry and symbiotic associations.

Introduction

Previous reports of fossil Chrysomelidae have been summarized by Santiago-Blay (1994). All of the Mesozoic and many of the Tertiary fossils are represented by compression fossils in rock layers which are incomplete specimens. Criticisms that most of these are poorly described, rarely show morphological features that provide synapomorphies in cladistic analyses and are debatable as belonging to the Chrysomelidae, are valid. However, identifiable fossils such as those that occur in amber can provide important phylogenetic and biographical information (Reid, 1995).

Chrysomelid fossils occur in a range of amber and copal deposits around the world (Table 1). Copal is defined as immature amber. When resin is no longer pliable by hand, it falls into the category of copal (Poinar, 1992). Over time and through the processes of oxidation and polymerization, semi-fossilized copal acquires the physical characteristics of amber. This is thought to occur within 2-4 million years. But copal can be as young as 50 years and as old as 50,000 years. Some early reports of insects in amber were actually specimens in copal, especially material from Africa. This is why it is important to verify all early reports of amber fossils and most of the specimens described in the 19th Century need to be restudied. A list of the reported genera and species of Chrysomelidae in amber appears in Table 2. No chrysomelids from copal deposits have yet been officially described.

Specimens in fossilized resin have the advantage of being preserved in their entirety, usually without noticeable distortion. Obstructing material can often be polished away, allowing the entire specimen to be viewed from several sides. Unfortunately, characters associated with the reproductive system and around the mouth are usually concealed from view, so only certain external characters can be used for identification purposes.

Advances in Chysomelidae Biology 1, pp. 1–16
Edited by M.L. Cox

Table 1. Deposits of amber and copal containing Chrysomelidae

Deposit	Location	Estimated age	References
Canadian	Alberta, Canada	79 mya	present work
Baltic	Northern Europe	40 mya	Bachofen-Echt, 1949
			Spahr, 1981
			Keilbach, 1982
			Poinar, 1992
			present work
Dominican	Dominican Republic	15-45 mya	Poinar, 1992
			Santiago-Blay et al., 1996
			present work
Mexican	Chiapas, Mexico	22-26 mya	Gressitt, 1963, 1971
			Poinar, 1992
			Santiago-Blay et al, 1996
Colombian	Colombia, S.A.	less than 1000 yrs	present work; Poinar, 1996
Madagascar	Madagascar, Af	less than 1000 yrs	present work
Zanzibar	Zanzibar, Af	untested but probably less than 1000 yrs	Hope, 1836

Table 2. Identified genera and species of Chrysomelidae from amber deposits

Deposit	Taxon	Reference/ author
Baltic	*Altica* Geoffroy 1862	Spahr, 1981
	Anisodera	"
	Cassida L. 1858	"
	Chalepus Thunberg 1805	"
	Chrysomela minutissima	Schaufuss 1891
	C. succini	Giebel 1856
	Colasposoma	Spahr, 1981
	Crioceris(Criocerina) pristina	Germar, 1813
	Cryptocephalus sericeus	Helm 1886
	Donacia Fab. 1875	Spahr, 1981
	Electrolema baltica	Schaufuss, 1891
	Eumolpus Weber 1901	Spahr, 1981
	Galerucella Crotch 1973	"
	Hadroscelus	"
	Haemonia	"
	Hispa	"
	Inclusus	Spahr, 1981
	Lema Fab. 1898	"
	Luperus Geoffroy 1862	"
	Monolepta Erickson 1843	"
	Nodonota Lefevre 1885	"
	Ochrosis	"
	Oposispa scheelei	Uhmann, 1939
	Pachnephorus	Spahr, 1981
	Protanisodera glaesi	Quiel 1910
	Pseudocolaspis	Spahr, 1981
	Sucinagonia javetana	Uhmann, 1939
Mexican	*Profidia nitida*	Gressitt, 1963
		Santiago-Blay et al., 1996
	Asiorestia (Crepidodera) antiqua	Gressitt, 1971
		Santiago-Blay et al., 1996

Table 2. Cont.

Deposit	Taxon	Reference/ author
Dominican	*Alethaxius* Lefevre 1885	Santiago-Blay et al., 1996
	Chalcosicya Blake 1830	”
	Colaspoides Laporte 1833	”
	Cryptocephalus Geoffroy 1862	”
	Diachus LeConte 1880	”
	Glyptoscelis LeConte 1833	”
	Leptonesiotes virkkii	”
	Metachroma LeConte 1858	”
	Sceloenopla ambarensis	”
	Walterianella Bechyne	”
	Cassidinae	present work

First amber fossil of a Mesozoic Chrysomelidae

From the Canadian Cretaceous deposits of Alberta, the first Mesozoic amber chrysomelid has been recovered (Fig. 1). This specimen has been identified as a member of the Sagrinae. Although part of the body is obscured by fungal growth, the enlarged hind femora bearing rows of teeth are clearly visible, a feature characteristic of this subfamily of chrysomelids (Lawrence & Britton, 1991). Fossil sagrines are quite rare and have come from only two sites. *Mesosagrites multipunctataties* Martynov 1935 was described from a single damaged elytron from the Late Jurassic Kara Tau deposits and its placement remains doubtful

Fig. 1. Sagrine chrysomelid in Canadian Cretaceous amber (79 mya). Note thickened and toothed hind femora.

(Santiago-Blay, 1994). More convincing are the specimens of *Eosagra obliquata* Haupt 1950 and *E. subparallela* Haup 1950 from the Eocene "brown coal" deposits near Geiseltal, Germany, which establishes this group in the Palaearctic in the early Tertiary.

The present find places the sagrines in the Nearctic during the Late Cretaceous and shows that the New World distribution of this group has been reduced during the past 79 million years to two relictual populations in South America. Such disjunct distributions of past and present populations of insect taxa are not unkown. Recently a genus of marine water striders was recovered in Dominican amber with the nearest species in the Indo-West Pacific (Andersen & Poinar, 1998). An interesting fossil amber beetle in the family Artematopodidae also demonstrates this point. Some twenty years after Crowson (1973) described the "extinct" genus *Electribius* from Baltic amber, stating that it was most closely related to modern representatives from Central America, two species in the same genus were described from Mexico and El Salvador, respectively (Lawrence, 1995). In all of these cases, the disjunct distribution probably was a result of population reduction caused by extinction events.

It is possible to speculate on the probable host plant of the Canadian fossil sagrine. There are several hypotheses as to the early food plants of sagrines and of chrysomelids in general. One suggests that the early Sagrinae lived on Cycadaceae (Schmitt, 1988) and the extant Australian *Carpophagus banksiae* has been noted visiting *Macrozamia* (Jolivet, 1988), (Table 4). The fossil sagrine could have fed on cycads since fossil evidence has shown that the climate was warm-temperate to subtropical and the cycad, *Nilssonia* has been reported from the Edmonton flora of the Upper Cretaceous of Alberta while *Pseudoctenis* and *Pseudocycas* have been found in the Upper Cretaceous Dunvegan flora of Alberta and British Columbia. Both sites date roughly the same as the amber location (Tidwell, 1998). Thus there is evidence that, as suggested by Jolivet (1988), cycads might have been the ancient hosts for this "stemgroup" of all Chrysomelidae.

Another theory is that the original plant hosts of the sagrines were Angiosperms (Tables 3 & 4) and feeding on monocots is a synapomorphy for all of the known subfamilies, as proposed by Reid (1995). Thus the host plant of the Canadian fossil sagrine could have been some type of reed or rush. By chance, a seed was found in a piece of amber from the same Alberta deposit where the sagrine originated (Fig.2). Botanists have been unable to match this seed with any known plant genus or family, but the morphology closely resembles extant genera of the sedge family (Cyperaceae). This would also agree with the current information known about the depositional site which indicates that the amber was formed in, or adjacent to a fresh-water marsh (Pike, 1995). One might speculate that the spines on the hind femur of the sagrines evolved as a means of grasping stems when feeding or ovipositing on semi-aquatic plants, since falling into water could result in drowning or predation by aquatic organisms.

Chemical analysis of the amber indicated that it was produced by trees of the genus *Agathis* (Araucariaceae) (Lambert *et al.,* 1996). This genus of trees is now restricted to the southern hemisphere where they have a Gondwanian distribution (New Zealand, Australasia and New Caledonia) similar to the Sagrinae which also have presently a nearly Gondwanian distribution with representatives in Australia, India, Africa, South America and southern Asia. These beetles are basically a warm-climate group, most being confined to the tropics or subtropics (Crowson, 1981). The Eocene "brown coal" fossils in Germany and the present find establishes this group in the Northern hemisphere and confirms a past Laurasian dis-

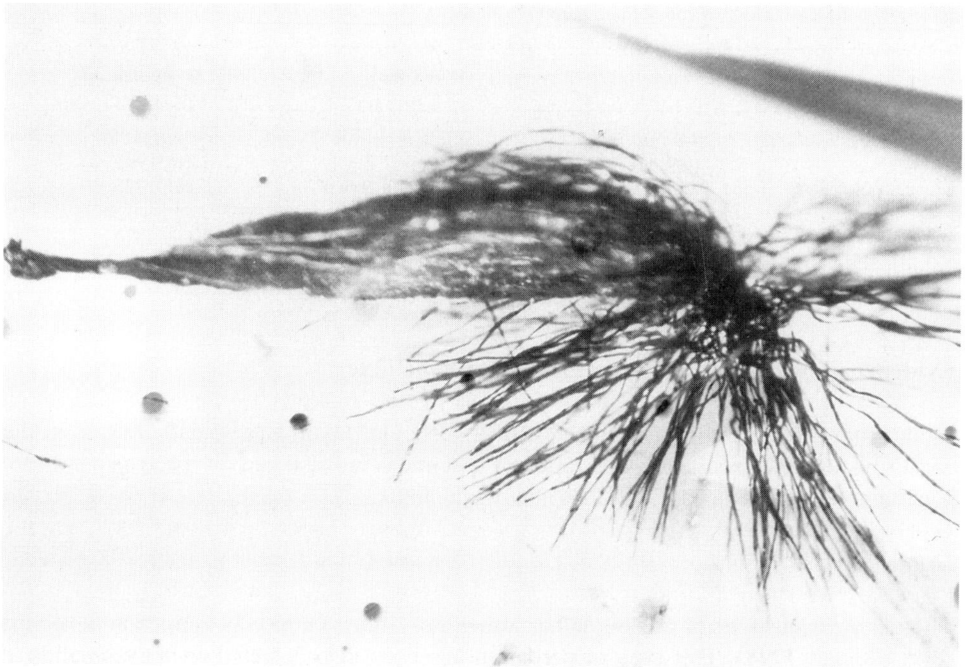

Fig. 2. Seed of a tentative sedge (Cyperaceae) in Canadian Cretaceous amber (79mya) from the same deposit as the sagrine shown in Fig. 1. Length of seed = 3 mm, greatest width = 0.5mm, bristles from the base range from 2.5 – 3.0 mm in length. The basal 40% of the bristles are dark while the upper 60% are hyaline and barely show in the photo.

Table 3. Host plants of larval Sagrinae

Genus	Locality	Plant host	Reference
Atalasis	Neotropical	*Sphaeralcea, Malvastrum, Sida, Gossypium* (Malvaceae)	Jolivet, 1988
Mecynodera	Australia	*Lomandra* (Xanthorrholaceae)	Lawrence & Britton, 1991
Sagra	Pantropical	Leguminosae, Rutaceae Euphorbiaceae, Rhizophoraceae Convolulaceae, Verbenaceae Rubiaceae, Dioscoreaceae Acanthaceae, Meliaceae	Jolivet, 1988; Reid, 1995

tribution of the Sagrinae. It also provides evidence that the subfamily extended back to the time of Pangaea, having radiated over most of the warm areas of the land masses. It appears that past climatic conditions in the Northern hemisphere have been harsher than those in the Southern, explaining why so many more relict species of plants and animals occur in the southern hemisphere today.

Table 4. Plants associated with adult Sagrinae

Taxon	Plant	Association	Reference
Carpophagus banksiae	*Macrozamia* (Cycad)	sitting on leaves	Jolivet, 1988 Reid, 1995
Diaphanops sp.	*Melaleuca* (Myrtaceae)	feeding on pollen	Reid, 1995
Mecynodera sp.	*Acacia* (Mimosaceae)	feeding on pollen	Reid, 1995

Evidence of behaviour in amber fossil Chrysomelidae

Aside from demonstrating the presence of a particular taxon at a specific period in the earth's history, the excellent preservation of specimens in fossilized resin can also indicate something about the behaviour of extinct organisms.

Evidence of predation

There is a curious type of predation that sometimes occurs on insects trapped in fossilized resin (Poinar, 1998). This type of predation has been also detected on chrysomelids in Madagascar copal (Fig.3). It is presumed that the predation occurs after the beetle has

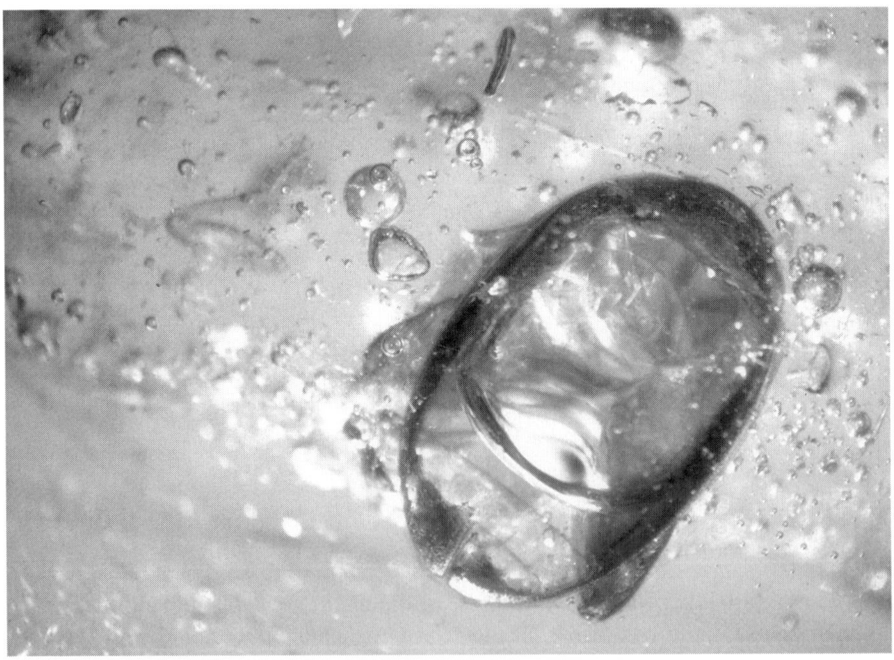

Fig. 3. A chrysomelid in Madagascar copal, with its elytra and internal tissues removed by a predator.

become stuck to the surface of a resin deposit. The predator or predators, which remain unknown at this time, surgically and carefully cut away the cuticle of the prey, which in this case meant removing the elytra of the entrapped leaf beetle. The predator may ingest the elytra or simply cut it away to expose the soft underparts, which are then completely removed with fine precision. Thus the body cavity of the entrapped adult is completely consumed and an air bubble remained in its place when a subsequent resin flow finally completely entombed the specimen. Although the identity of this particular predator remains unknown, it may have left some clues as to its presence since a cluster of elongate scales overlay the air bubble that now fills the empty body cavity of the victimized beetle.

Evidence of gregariousness

Evidence of gregariousness in a population of flea beetles (Alticinae) from the Greater Antilles can be obtained from a single piece of Dominican amber containing eleven adult beetles (Fig.4). The beetles must have been feeding on a leaf that was close to the resin deposit on the Dominican amber tree (*Hymenaea protera*: Poinar, 1991) when they were disturbed. Quite possibly a predator arrived which caused the beetles to jump or fall off the leaf (thanatosis?) into the resin or something might have brushed the leaf into the resin flow.

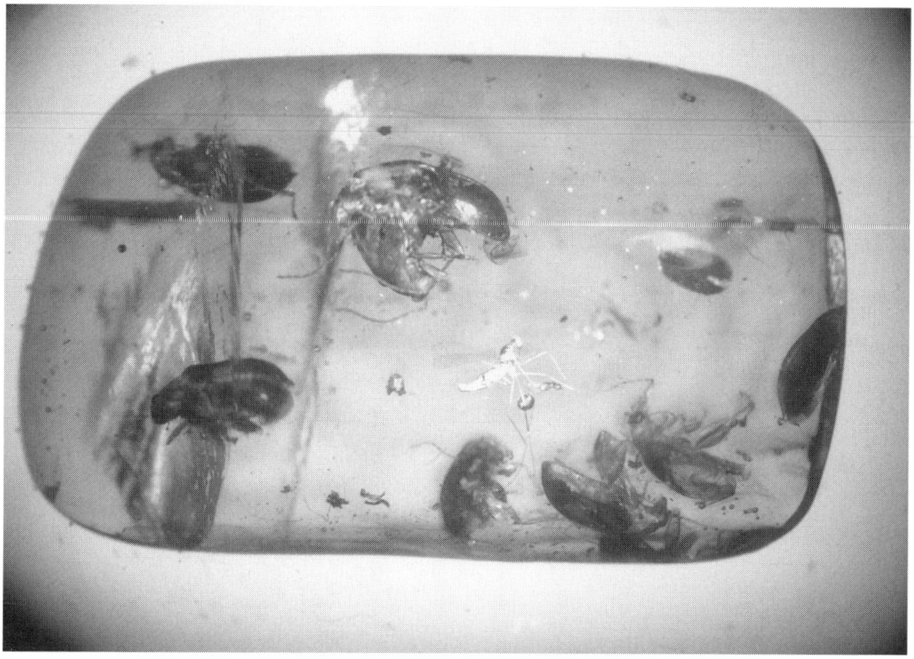

Fig. 4. Eleven adult flea beetles in a piece of Dominican amber provides evidence of gregarious behaviour some 20-40 million years ago.

Evidence of feeding behaviour

Although pieces of fossilized resin containing plant damage along with leaf beetles have not been reported, some leaves in Dominican amber demonstrate feeding damage which is characteristic of that caused by leaf beetles (Fig. 5).

Evidence of host plant selection

Again, unless a recognizable plant part with insect damage is found together with larval or adult chrysomelids, it is impossible to correlate a particular insect with a specific plant to establish host-plant associations. However, many tropical families of plants have characteristic plant hairs and sometimes insects in fossilized resin are associated with them. An example is presented here where a chrysomelid larva (probably a Galerucinae) in Colombian copal has a simple trichome attached to the ventral surface of its body (Figs. 6,7). This type of trichome is characteristic of several plant families, including members of the Combretaceae and Acanthaceae (Gentry, 1993). It would be difficult to specify exactly the plant host without further knowledge of the trichome structure of plants growing in Colombian forests. However, amber insects have been discovered with more characteristic trichomes, like T-shaped hairs characteristic of members of the family Malpighiaceae (Poinar & Poinar, 1999).

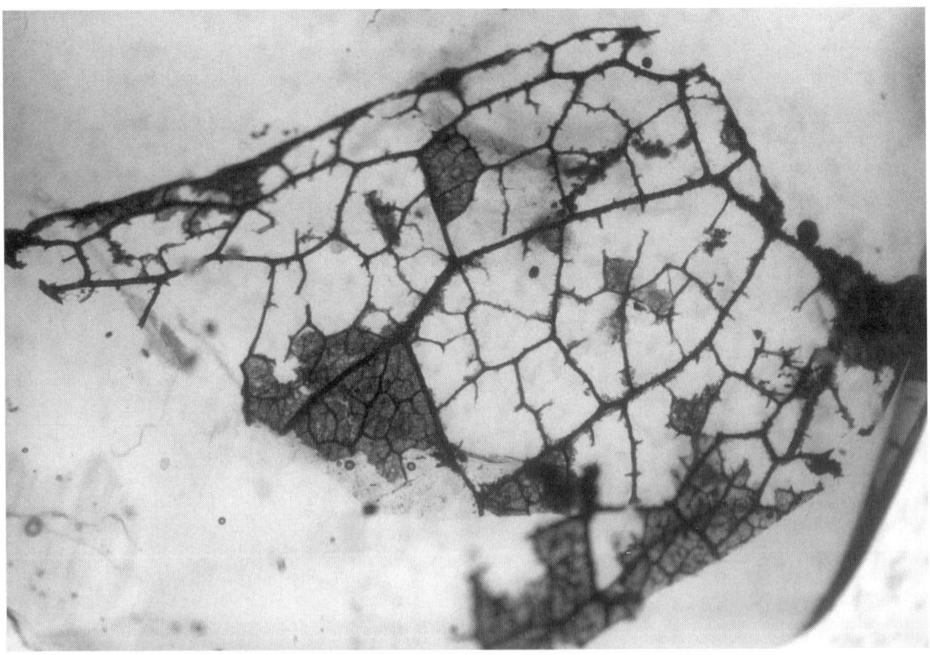

Fig. 5. Feeding damage to a leaf in Dominican amber is characteristic of leaf beetle activity.

Fig. 6. A larval chrysomelid (probably a Galerucinae) in Colombian copal.

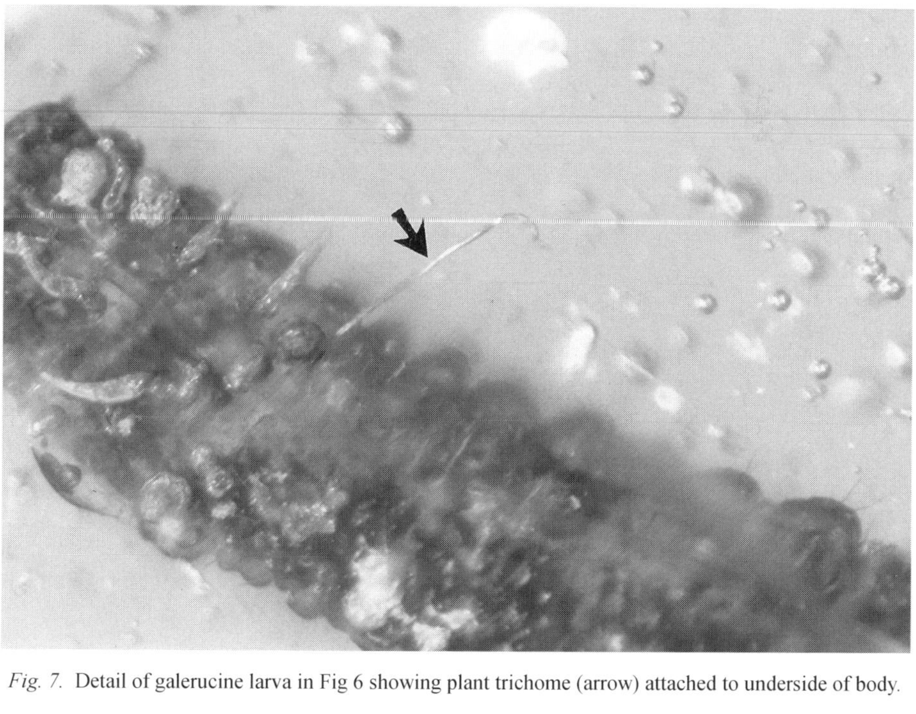

Fig. 7. Detail of galerucine larva in Fig 6 showing plant trichome (arrow) attached to underside of body.

Evidence of parasitism

While leaf beetles with emerging parasites have not been reported in fossilized resin, genera of parasitic wasps with representatives known to parasitize extant chrysomelids have been recovered from amber deposits containing leaf beetles. For instance the oil palm leaf-mining hispid, *Hispolepsis subfasciata* Pic which occurs in South America is parasitized by species of Eulophidae, Encrytidae, and Braconidae (Mariau, 1988), all of which groups are represented in Dominican amber with the fossil hispine, *Sceloenopla ambarensis* Santiago-Blay, Poinar & Craig (1996). More specifically, the prepupa of *H. subfasciata* is parasitized by a species of *Brachymeria* , a genus which occurs in Dominican amber (Fig. 8). This genus of Chalcididae also parasitizes larvae and pupae of tortoise beetles (Buzzi, 1988) which are represented in Dominican amber. Although it is impossible to say that the *Brachymeria* depicted in Fig.8 parasitizes leaf beetles, there is a good possibility that some of the amber ones did.

Defense mechanisms and mimicry

Several Eumolpines in Dominican amber are quite rounded, smooth and shiny. This is especially true with some members of the genus *Metachroma* whose bodies can be characterized as smooth metallic (Fig.9). These modifications are often used by leaf beetles as a protective measure against ants, that find it difficult to grasp the beetle's bodies (Selman, 1988).

Fig. 8. An adult *Brachymeria* (Chalcididae: Hymenoptera) in Dominican amber that could have parasitized leaf beetles.

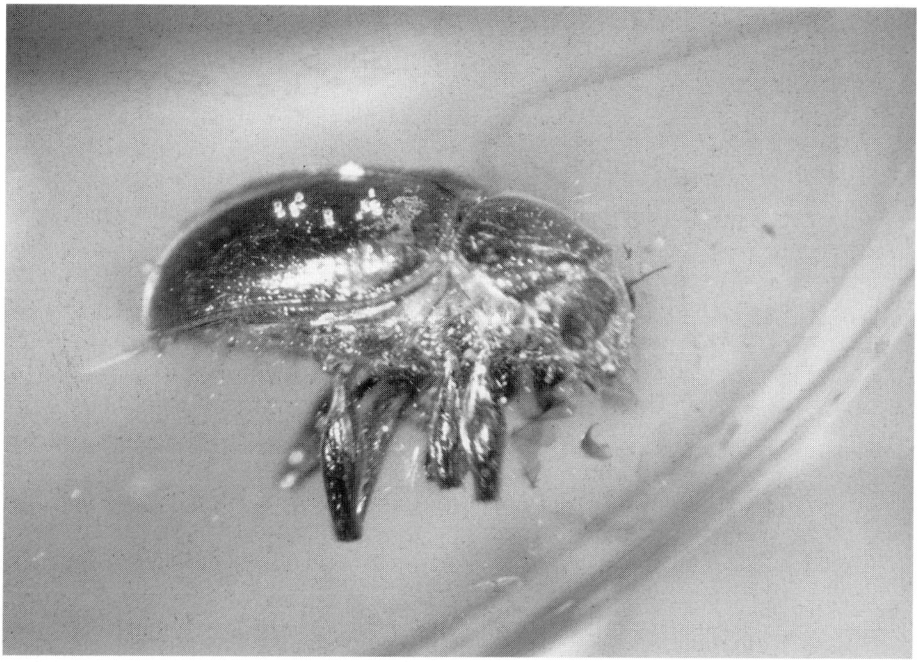

Fig. 9. The smooth, rounded body of this *Metachroma* sp. in Dominican amber may have served as a defence against ants.

Regarding protective coloring or mimicry, eye spots on insects in general are thought to frighten predators by representing the eyes of general, unidentifiable species, although the interpretation of small eye spots is still open to discussion (Balsbaugh, 1988).

The dark spots on the anterior and posterior portions of each elytron on *Leptonesiotes virkkii* Santiago-Blay *et al.,* 1996 in Dominican amber merge into two eyespots when the wings are closed (Fig. 10). These spots could serve as a defense against some types of predators.

Perhaps the clearest evidence of defensive behaviour of chrysomelids in amber is the presence of case-bearing Camptosomata in Dominican amber. A mantle covering the eggs of this group is periodically enlarged upon with faecal material from the growing larvae, forming a protective case which is carried around by the insect. This group abounds in the Neotropics so it is not unexpected that their cases would appear in tropical amber. All of the cases reported to date appear to belong to the Cryptocephalinae (Santiago-Blay *et al.,* 1996) and a recently discovered one in Dominican amber belongs in that same group (Fig.11). This latter case closely resembles extant cases made by members of the genus *Cryptocephalus* (Erber, 1988). It is not clear whether the feeding habits of *Cryptocephalus* larvae include decaying plant material on the forest floor or living leaves on trees and shrubs. The presence of these case-bearing larvae in amber suggests that they did at least crawl on the bark of trees.

Larvae of the Cassidinae do not make cases, but possess caudal and lateral outgrowths which retain excrement, exuviae and debris to conceal themselves (Buzzi, 1988). Two early larval representatives of this group have been found in Dominican (Fig. 12) and

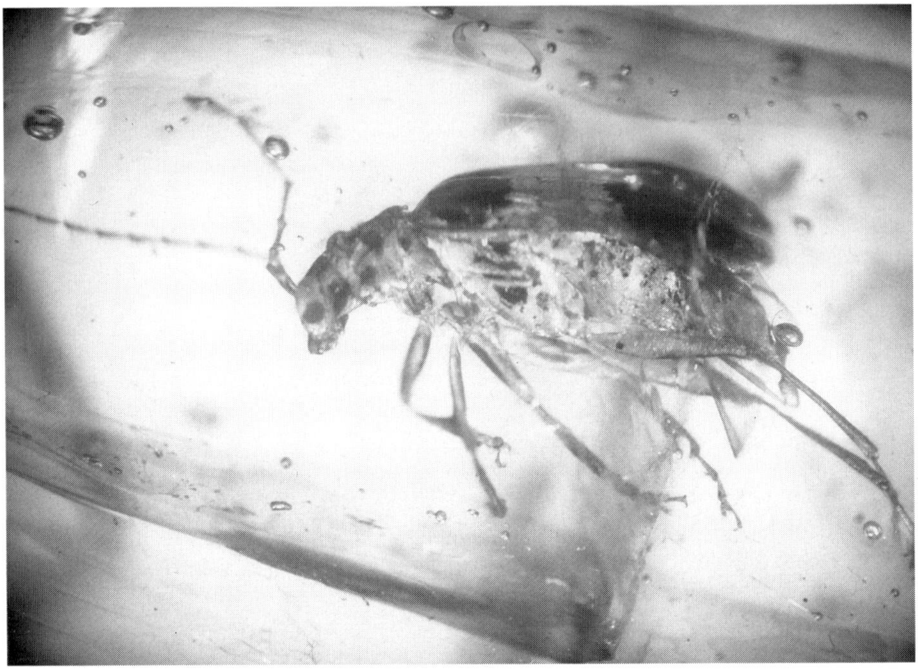

Fig. 10. When the elytra of *Leptonesiotes virkkii* are folded at rest, the color patterns merge to make a pair of eyespots.

Fig. 11. A larval cryptocephaline (probably *Cryptocephalus* sp.) partially protruding from its protective case.

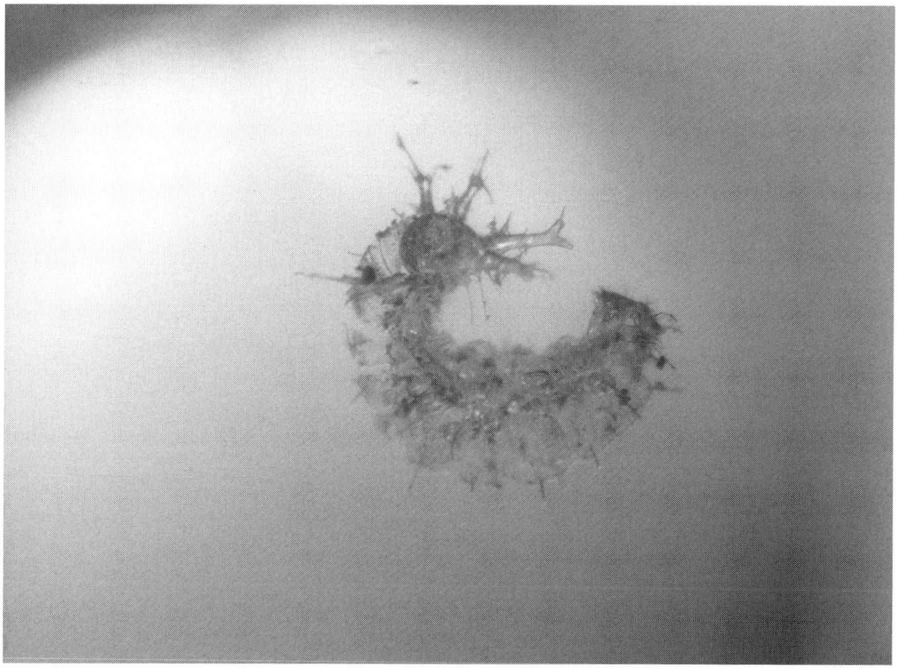

Fig. 12. A larval Cassidinae in Dominican amber.

Baltic amber (Fig. 13). Although debris is absent in these forms (possibly because it came off in the resin) , the caudal spines that serve to hold the material are evident.

Symbiotic associations

Aside from evidence of parasitism by nematodes, fungi , protozoa and viruses, symbiotic associations involving mutualism have rarely been reported involving Chrysomelidae. However the pronotum of *L. virkkii* in Dominican amber (Santiago-Blay *et al.,* 1996) contains a large central cavity (Fig.14), which is probably not an artifact since the remainder of the body shows no sign of distortion. This cavity is lined with a whitish powdery material which may represent some type of microorganism or the product of an organism that has a parasitic or mutualistic association with the adult beetle.

Preservation of internal tissue in amber Chrysomelidae

A study that examined Dominican amber eumolpines with pyrolysis-gas chromatography/ mass spectrometry showed that the volatile components in the resin penetrated both the cuticle and internal tissues of the adult beetles (Stankiewicz *et al.,* 1998). Straight chain hydrocarbons and bicyclic resin products from the fossil tissue were abundant, even in tissues not directly in contact with the amber. It was presumed that resin molecules formed chemical cross links which explained the detailed preservation of the internal tissue, even

Fig. 13. A larval Cassidinae in Baltic amber.

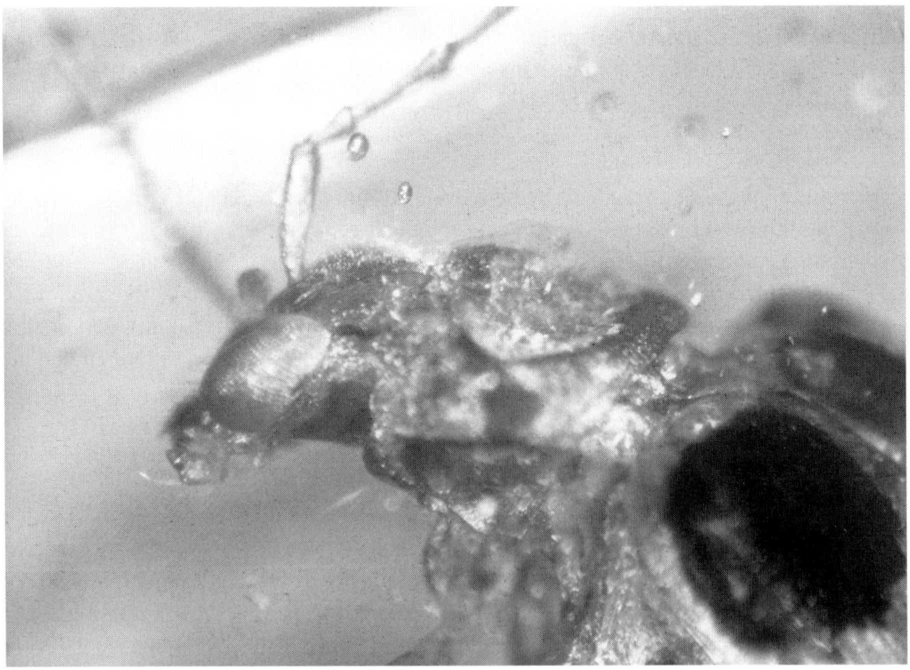

Fig. 14. The pronotum of *L. virkkii* in Dominican amber contains a cavity filled with a powdery substance which could indicate some type of symbiotic or parasitic association.

though the cuticle was degraded. Unfortunately , the study failed to explain how DNA is preserved in amber specimens, although insect DNA sequences, still awaiting analysis, were obtained from a *Metachroma* sp. in Dominican amber (Hendrik Poinar, personal communication).

Conclusions

Aside from establishing the presence of leaf beetles at a particular time and location in the past, amber fossils can also be used to interpret aspects of a wide range of behaviour and interactions that occurred in fossil taxa. Interpretation is dependent in a large part on the habits of extant chrysomelids, since as has been adequately demonstrated by Boucot (1990), the habits of organisms tends to remain steady over time, following the principle of behavioural constancy.

Acknowledgments

The author is indebted to John Doyen and Gary Parsons for assistance in identification of the specimens presented here. Thanks is also extended to Roberta Poinar for comments on the manuscript.

References

Andersen, N. M. & Poinar, Jr., G.O. 1998 A marine water strider (Hemiptera: Veliidae) from Dominican amber. Entomologica scand. 29: 1-9.

Bachofen-Echt, A. 1949 Der Bernstein und seine Einschlusse. Wien: Springer-Verlag. 204 pp.

Balsbaugh, Jr., E.U. 1988 Mimicry and the Chrysomelidae. In: Jolivet, P., Petitpierre, E. and Hsiao, T. H. (eds) Biology of Chrysomelidae. Kluwer Acad. Pub., The Hague pp 261-284.

Boucot, A. 1990 Evolutionary Paleobiology of Behavior and Coevolution. Elsevier, Amsterdam, 725 pp.

Buzzi, Z. J. 1988 Biology of Neotropical Cassidinae. In: Jolivet, P., Petitpierre, E. and Hsiao, T. H. (eds) Biology of Chrysomelidae. Kluwer Acad. Pub., The Hague, pp 559-580.

Crowson, R. A. 1973 On a new superfamily Artematopoidea of polyphagan beetles, with the definition of two new fossil genera from the Baltic amber. J. nat. Hist. 7: 225-238.

Crowson, R. A. 1981 The biology of the Coleoptera. Academic Press, London, 802 pp.

Erber, D. 1988 Biology of Camptosomata Clytrinae-Cryptocephaline-Chlamisinae-Lamprosomatinae. In: Jolivet, P., Petitpierre, E. and Hsiao, T. H. (eds) Biology of Chrysomelidae. Kluwer Acad. Pub., The Hague, pp 513-552.

Gentry, A. 1993 A field guide to the families and genera of woody plants of Northwest South America (Colombia, Ecuador, Peru) with supplementary notes on herbaceous taxa. Univ. Chicago Press, Chicago, 895 pp.

Germar, E. F. 1813 Insecten in Bernstein eingeschlossen, beschrieben aus dem academischen Mineralien-Cabinet zu Halle. Magazin Ent. (Germar). 1: 11-18.

Giebel, C. G. 1856 Fauna der Vorwelt mit steter Berücksichtigung monographisch dargestellt der lebenden Thiere. 2. Gliederthiere 1. Abt. Insecten und Spinnen, Leipzig, 411 pp.

Gressitt, J. L. 1963 A fossil chrysomelid beetle from the amber of Chiapas, Mexico. J. Paleont. 37: 108-109.

Gressitt, J. L. 1971 A second fossil chrysomelid beetle from the amber of Chiapas, Mexico. Univ. Calif. Publs. Ent. 63: 63-64.

Haupt, H. 1950 Die Käfer (Coleoptera) aus der Eozanen Braukohle des Geiseltales. Geologica (Berlin) 6: 1-168.

Helm, O. 1886 Millheilungen uber Bernstein. xiii. Uber die Insecten der Bernsteins. Schr. naturf. Ges. Danzig, N.F. 6: 267-277.

Hope, F. W. 1836 Observations on succinic insects. Trans. R. ent. Soc. London 1: 133-147.

Jolivet, P. 1988 Food habits and food selection of Chrysomelidae. Bionomic and evolutionary perspectives. In:

Jolivet, P., Petitpierre, E. and Hsiao, T. H. (eds) Biology of Chrysomelidae. Kluwer Acad. Pub., The Hague, pp 1-24.
Keilbach, R. 1982 Bibliographie und Liste der Arten tierischer Einschlusse in fossilen Harzen sowie ihrer Aufbewahrungsorte. Dt. ent. Z. N.F. 29: 129-286, 301-491.
Lambert, J. B., Johnson, S. C. and Poinar, Jr., G. O. 1996 Nuclear Magnetic resonance characterization of Cretaceous amber. Archaeometry 38: 325-335.
Lawrence, J. F. 1995 *Electribus* Crowson: alive and well in Mesoamerica, with notes on *Ctesibius* Champion and the classification of Artematopodidae. In: Pakaluk, I. and Slipinski, S. A. (eds.) Biology, phylogeny and classification of Coleoptera: papers celebrating the 80th birthday of Roy A. Crowson. Vol. 1. Muzeum i Instytut Zoologii PAN, Warsaw, pp 411-431.
Lawrence, J. F. & Britton, E. B. 1991 Coleoptera. In: The insects of Australia. Cornell University press, Ithaca, pp 543-683.
Mariau, D. 1988 The parasitoids of Hispinae. In: Jolivet, P., Petitpierre, E. & Hsiao, T. H. (eds) Biology of Chrysomelidae. Kluwer Acad Pub. The Hague, pp 449-461.
Martynov, A. V. 1935 [Note on the fossil insects from Mesozoic deposits in Cheliabinsk district.] Trudy Paleozoologicheskogo Instituta Akademii Nauk SSSR 4: 37-48 (in Russian).
Pike, E. M. 1995 Amber taphonomy and the Grassy Lake, Alberta, amber fauna. Ph.D. thesis. University of Calgary, Alberta, Canada 264 pg.
Poinar, Jr., G. O. 1991 *Hymenaea protera* sp. n. (Leguminoseae: Caesalpinioideae) from Dominican amber has African affinities. Experientia 47: 1075-1082.
Poinar, Jr., G. O 1992. Life in amber. Stanford University press, Stanford, CA, 350 pp.
Poinar, Jr., G. O. 1996 Fossil resins. Lapidary J. 49: 52-56.
Poinar, Jr., G. O 1998 Trace fossils in amber: a new dimension for the ichnologist. Ichnos 6:47-52.
Poinar, Jr., G. O. and Poinar, R. 1999 The amber forest. Princeton University Press (in press)
Quiel, G. 1910 Bemerkungen über Bernstein xiii. Über die Insecten der Bernstein. Schr. naturf. Ges. Danzig, N.F. 6: 267-277.
Reid, C. A. M. 1995 A cladistic analysis of subfamilial relationships in the Chrysomelidae sensu lato (Chrysomeliodea). In: Pakaluk, I. and Slepinski, S. A. (eds.) Biology, Phylogeny and classification of Coleoptera: papers celebrating the 80th birthday of Roy A. Crowson. Vol 2. Muzeum i Instytut Zoologii PAN, Warsaw, pp. 559-631.
Santiago-Blay, J. A. 1994 Paleontology of leaf beetles. In: Jolivet, P. H. , Cox, M. L. and Petitpierre, E. (eds.) Novel aspects of the biology of Chrysomelidae. Kluwer Academic Pub., Dordrecht, pp 1-68.
Santiago-Blay, J. A., Poinar, Jr., G. O. and Craig, P. R. 1996 Dominican and Mexican amber chrysomelids, with descriptions of two new species. In: Jolivet, P.H. A. and Cox, M. L.(eds), Chrysomelidae Biology, vol. 1: The Classification, Phylogeny and Genetics, SPB Academic Pub., Amsterdam, pp 413-424.
Schaufuss, L. W. 1891 Preussens Bernsteinkafer 1. Neue Formen aus der Helm'schen Sammlung im Danziger provinzialmuseum. Berl. ent. Z. 36: 53-64.
Schmitt, M. 1988 The Criocerinae: biology, phylogeny and evolution. In: Jolivet, P. Petitpierre, E. and Hsiao, T. H. (eds), Biology of Chrysomelidae, Kluwer Acad. Pub., The Hague, pp 475-495.
Selman, B. J. 1988 Chrysomelids and ants. In: Jolivet, P., Petitpierre, E. and Hsiao, T.H. (eds), Biology of Chrysomelidae, Kluwer Acad. Pub., The Hague, pp 463-473.
Spahr, U. 1981 Systematischer Katalog der Bernstein-und Kopal-Kafer (Coleoptera). Stuttgarter Beitr. naturk. (Serie B) 80: 107 pp.
Stankiewicz, B. A., Poinar, H. N., Briggs, D. E. G., Evershed, R. P. and Poinar, G. O. 1998 Chemical preservation of plants and insects in natural resins. Proc. R. Soc. B 265: 641-647.
Tidwell, W. D. 1998 Common fossil plants of Western North America. Smithsonain Inst. Press, Washington, D.C., 299 pp.
Uhmann, E. 1939 Hispinin aus baltischem Bernstein. Bernsteinforsch. 4: 19-22.

PRELIMINARY ANALYSES OF CHRYSOMELID PALEODIVERSITY, WITH A NEW RECORD AND A NEW SPECIES FROM DOMINICAN AMBER (EARLY TO MIDDLE MIOCENE)

JORGE A. SANTIAGO-BLAY [1] and PATRICK R. CRAIG [2]
[1]*The Biological Sciences Division, The Biological Sciences Collegiate Division, The Biological Sciences Learning Center, 924 East 57th Street, The University of Chicago, Chicago, IL 60637, USA;* [2]*P. O. Box 545, Monte Rio, CA 95462, USA*

To those who survive major losses
and keep planting seeds of good will.

1. Introduction

Analyses of paleozoodiversity are relatively common for marine biota (for example, blastoids, crinoids, trilobites; Foote, 1996; Raup, 1972; Sepkoski *et al.*, 1981; Valentine, 1969). However, the analysis of terrestrial paleozoodiversity, including chrysomelids, is more problematic because preservation can be more difficult, the localities where large assemblages of fossils (*Lagerstätten*) exist can be very scattered, and many taxa in samples are rare, among others (Signor and Lipps, 1982). Furthermore, many studies of paleozoodiversity rely on counts or identifications at family (for example, Ross and Jarzembowski, 1993) or genus level, assuming that they can be good surrogates of species level biodiversity patterns or of biological phenomena.

The most comprehensive analysis of hexapodan paleodiversity is that of Labandeira and Sepkoski (1993) in which several aspects of the hexapodan fossil record were studied, including, insect biodiversity at the family level, survivorship, and feeding ecology groups through time. In this paper, we study the paleodiversity of the Chrysomelidae, at the subfamily and generic level. The fossil record of the Chrysomelidae consists of a total of 13 subfamilies, approximately 130 genera, and *circa* 360 species (Santiago-Blay, 1994; Santiago-Blay *et al.*, 1996). The Protoscelinae and, at least 24 genera, appear to be extinct or they have not yet been found amongst extant organisms. In addition, we report a new genus for the fossil record (*Acallepitrix*, Alticinae) and describe a new species of *Sceloenopla* (Hispinae).

2. Methods

Analyses of Paleodiversity. A detailed compilation of the fossil record of the Chrysomelidae (Santiago-Blay, 1994) and a recent addition (Santiago-Blay *et al.*, 1996)

Advances in Chysomelidae Biology 1, pp. 17–24
Edited by M.L. Cox

were the data used for the analyses. Also, Labandeira (1994) and about 20 papers cited therein, likely to include records of fossil chrysomelids that Santiago-Blay (1994) may have overlooked, were consulted. Only one record, *Meumolpites* Ponomarenko (Eumolpinae) from the Triassic (Brauckmann and Schlüter 1993), was found and now it is included in the counts. The record of *Colaspoides*? (Eumolpinae) (Santiago-Blay *et al.*, 1996) is a misidentification by author JASB (most specimens have been identified as *Chalcosicya*) and it has not been included in the counts. Counts were made assuming the range-through method, that is, when there is a gap in the fossil record for a taxon, we assumed that the taxon also existed in between the times when the taxon has been recorded.

Specimens. All specimens are reported to have come from the Altamira facies of the El Mamey formation in the Cordillera Septentrional (= northern mountain range) of the Dominican Republic (estimated age, 15-20 million years old, Iturralde-Vincent and MacPhee, 1996). The specimens are to be deposited in the Department of Entomology of the California Academy of Sciences (San Francisco, CA). All measurements are in mm.

3. Preliminary Analyses

Overall biodiversity: subfamilies and genera. Not surprisingly, there is a general increase in diversity with time. Interestingly, there is a relative decline in the Cretaceous and, for the time being, we assume this represents a sampling artifact.

Relationship with angiosperm diversity. When the data of Fig. 1 are plotted in semilog axis (Fig. 2), the fit for the subfamilies, assuming an exponential increase in total taxa with time along with angiosperm expansion (Labandeira and Sepkoski, 1993), is good. However, the fit for the genera is weak, especially for the Cretaceous. The generation of higher level clades, with only a few genera and species representing them, has been reported for many invertebrate groups (Foote 1997). Lower level taxonomic diversity appears later in geologic time. Another complementary possibility is that the appearance (about 138 mya, Valingian) and ascendance of angiosperms (from then to about 100 mya when they became dominant in most plant communities) were "bad news" for the diversification of genera and species of chrysomelids.

Major mouthpart and physiological dietary adaptations to eating broad-leaved plants could just as easily have occurred during the Jurassic, when there were many broad leaved seed ferns, gingkos, cycadophytes and gnetaleans, long before the angiosperms appeared. Thus, higher level chrysomelid clades could have been preadapted to use angiosperm foliage. However, we would have to assume that lower level taxonomic diversity of chrysomelids was tracking angiosperm diversity and, thus, be less abundant. Also, probably there were not that many angiosperms during the Lower Cretaceous, compared to other seed plant taxa, and there may have been lag times before any effect on diversity is detected (Labandeira to Santiago-Blay, pers. comm., Nov. 1998).

However, because of our suspicion that there is a sampling artifact in the Cretaceous, readers should be cautious about the rather interesting possibility that this pattern necessarily means that chrysomelid generic and species diversification may not follow angiosperm biodiversity.

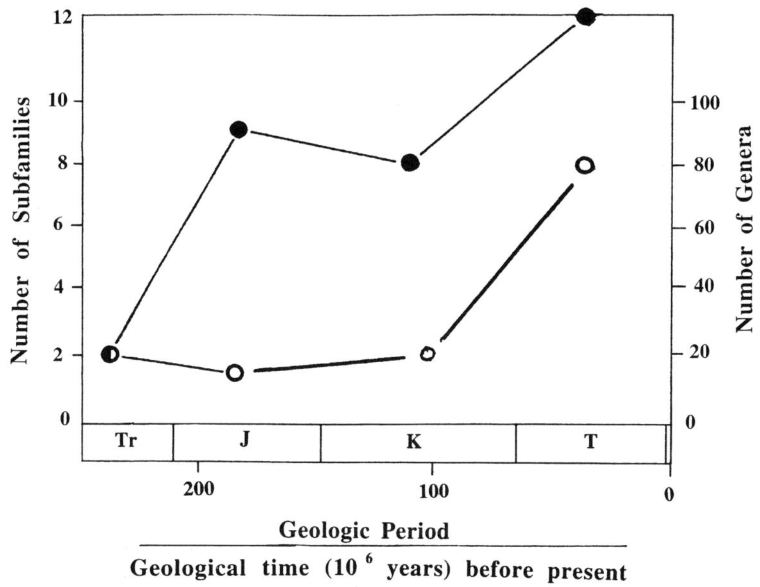

Fig. 1. Diversity of subfamilies (•) and genera (o) for Chrysomelidae, from the Triassic (Mesozoic) to the Tertiary (Cenozoic).

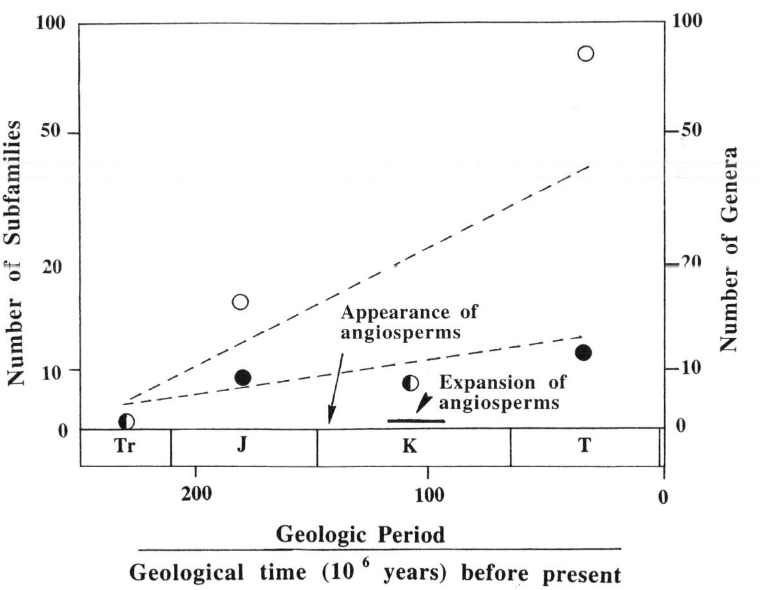

Fig. 2. Diversity of subfamilies (•) and genera (o) for Chrysomelidae, from the Triassic (Mesozoic) to the Tertiary (Cenozoic), on semilogarithmic scale. Dashed lines are possible interpretations of exponential diversification of chrysomelids with angiosperms. Note the relative good match between the dashed line and the data points at the subfamily level and the relative weak match at the generic level. There are 19 subfamilies in the Chrysomelidae (Seeno and Wilcox, 1982) and about 1700 extant genera of chrysomelids. Data on timing of angiosperm evolutionary events from Labandeira and Sepkoski (1993).

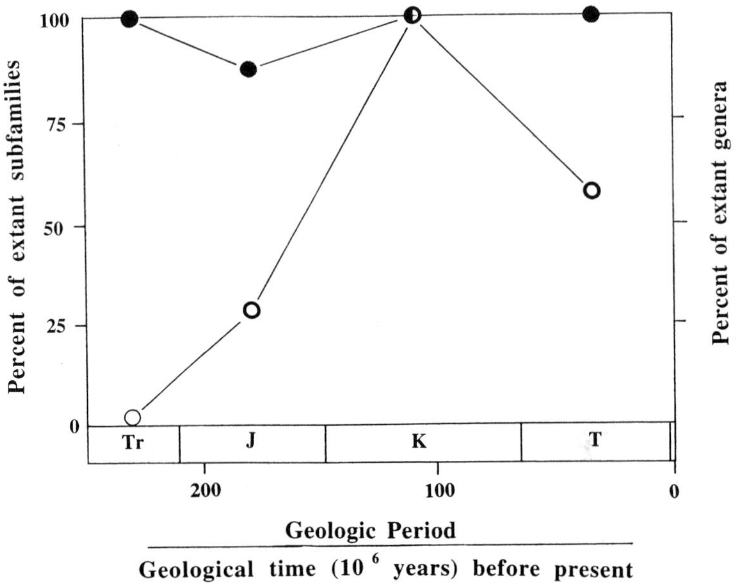

Fig. 3. Lyellian survivorship curves, in percent subfamilies (•) and genera (o) of the extant, for the Chrysomelidae, from the Triassic (Mesozoic) to the Tertiary (Cenozoic).

Survivorship Curves (Fig. 3). At the subfamily level, the extinction rate is very low. Except for the Protoscelinae (reported only for the Jurassic), the other chrysomelid subfamilies reported in the fossil record are extant. This pattern has been documented for the Hexapoda at the family level (Labandeira and Sepkoski, 1993). For the genera, there is considerable variation in the percentage of extant taxa. Many genera reported for the Jurassic (all of those reported for the former Soviet Union) are now extinct. In addition, only two genera have been reported for the Triassic and their generic placement is questionable. About 60% of the genera reported for the Tertiary are still extant.

We consider that the generic level data for the Triassic throughout the Cretaceous are too scarce (two genera for the Triassic, nine for the Jurassic, and nine for the Cretaceous). Therefore, it is premature to make broad generalizations of the patterns of chrysomelid biodiversity and certainly out of the question to attempt reconstruction of ancient paleoecological communities.

4. New genus for the chrysomelid fossil record (Alticinae: *Acallepitrix*) and the description of a new species of *Sceloenopla* (Hispinae)

Subfamily Alticinae

Acallepitrix Bechyné 1959

Figs. 4-7

Diagnosis. Pronotum distinctly punctate, anterior angles bevelled, prebasal transverse impression fades near lateral margins (in this specimen, as in so many other alticines, this

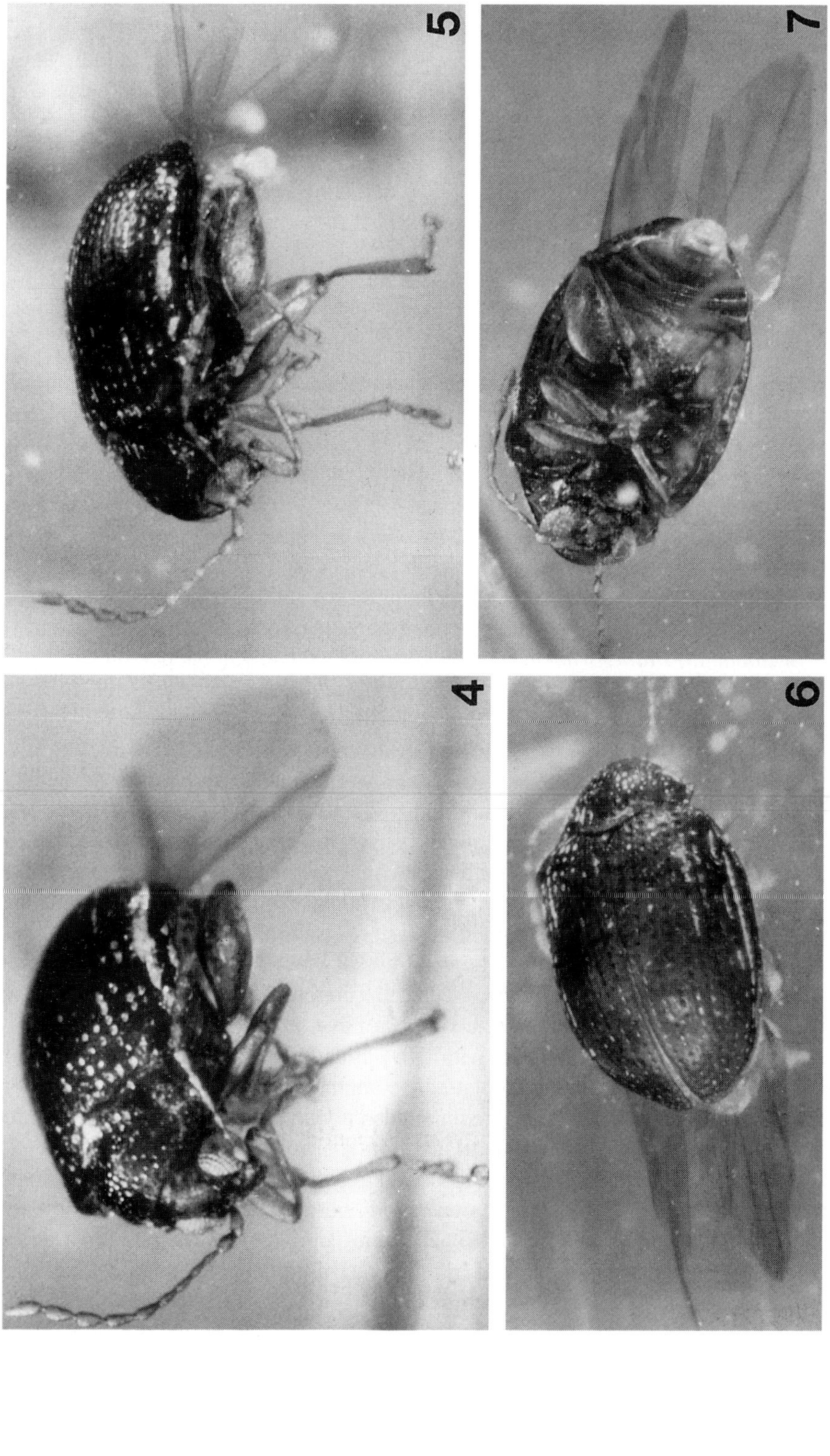

Figs. 4-7. Acallepitrix sp. (Alticinae). 4. Dorsoanterolateral. 5. Lateral. 6. Dorsoretrolateral. 7. Ventral

can be difficult character to see), base doubly sinuous; elytra strongly convex, with nine longitudinal rows of punctures (not counting juxtascutellar and marginal rows), base not wider than pronotal base; the fossil studied is metallic coloured.

Remarks. In the extant fauna, *Acallepitrix* is known from Central and South America (over forty species, Scherer, 1983). This find constitutes a new genus record for the Caribbean and for the chrysomelid fossil record.

Material Examined. One specimen, in Pat Craig's private collection, to be deposited in the California Academy of Sciences, San Francisco, CA.

Subfamily Hispinae

Sceloenopla stainesorum n. sp. Santiago-Blay and Craig

Figs. 8-9

Diagnosis. Antennae cylindrical, without narrow and wide sides; elytra (not head or pronotum) metallic coloured; humeri with one small tooth, not projecting strongly as large tooth or cone; elytra incostate, with narrow carinae; hind elytral angle broadly rounded. *Sceloenopla stainesorum* n. sp., belongs in Group 5a of Uhmann (1937).

According to Staines (pers. comm. to Santiago-Blay, July 1998), *Sceloenopla stainesorum* n. sp. differs from the new species of *Sceloenopla* being described by Samuelson and Staines, as follows: *S. stainesorum* has carina on head, eyes protruding, no transverse basal impression on pronotum, lateral margins of pronotum with notch behind anterolateral margin; elytra with humeral tooth, without costae, margins dark, lateral and apical margins smooth. *Sceloenopla stainesorum* n. sp. differs from *S. mantecada* Sanderson by the number of elytral rows of depressions (only eight in *S. mantecada*), and by their coloration pattern (*S. mantecada* is pale brown, variegated with white, black, and metallic green).

Description. Elongate, dorsally bicolorous (head and prothorax pale brown, elytra very dark metallic bluish green, almost black; 6.7 long, 2.4 maximum width (about three fourths from anterior margin of elytra). Head subtrapezoidal, broader anteriorly, 1.2 wide at eye level, 0.8 at narrowest visible end, posteriorly); vertex broad, with coarse interocular sculpturation and thin, short frontal keel that reaches base of antennae. Antennae filiform, cylindrical (without narrow and wide sides), black, apical portion of distal segments paler, reaching first fourth of elytra, if extended posteriorly; segments 3 and 4 slightly clavate, others subcylindrical. Eyes slightly bulging in dorsal view, obviously bulging in lateral view; smooth band (1/3 to 1/4 eye width) of unsculptured cuticle surrounding eyes, widening and extending posteriorly behind rear margin of eye. Pronotum longer than wide (1.6 maximum width), coarsely sculptured, with two pointed projections anterolaterally (thus forming a sort of chevron) and posterolaterally; laterally slightly narrowing posteriorly; posterior pronotal margin (1.2 wide) distinct, scarcely sculptured, forming deep sulcus. Elytra incostate with ten rows (and a small juxtascutellar row) of coarse, deep depressions (30-40 depressions *per* row), deeper than pronotal depressions, one row of depressions on simple lateral carinae. Humeri with one small tooth, not projecting strongly as large tooth or cone. Width at anterolateral projections, excluding projections, 2.2. Hind elytral angle broadly rounded. Legs pale brown, moderately sized; anterodistal portion of tibiae 1 not modified to receive tarsi. Abdominal sternites brown, with very few setae.

Figs. 8-9. Sceloenopla stainesorum n. sp. (Hispinae) 8. Dorsal. 9. Ventral.

Type Material. Holotype, one specimen, in Pat Craig's private collection, to be deposited in the California Academy of Sciences, San Francisco, CA.

Etymology (Latin, plural). Named after Charles and Susan Staines for their contributions to chrysomedology and for their collegiality to author JASB throughout the years.

Acknowledgments

Drs. David G. Furth (Alticinae) (Department of Entomology, National Museum of Natural History, Washington, D.C.), and R. Wills Flowers (Eumolpinae, Florida A & M, Tallahassee, Florida) greatly assisted author JASB in the identification of the specimens. In addition, Dr. Charles Staines (Hispinae) (Edgewater, MD), and his wife, Susan, read the description of the new species of *Sceloenopla* and provided valuable suggestions. Dr. Conrad C. Labandeira (Department of Paleobiology, National Museum of Natural History, Washington, D.C.) provided valuable input to the section "Relationship with angiosperm diversity". Ms. Molly Tomlinson (Biological Sciences Learning Center, Information Services, University of Chicago) assisted author JASB with manipulating images in Figs. 1–3. Several staff members from the Crerar Library (Univ. Chicago) were very helpful in locating obscure citations by interlibrary loans. All are gratefully acknowledged. Finally, author JASB recognizes the support of relatives, friends, and colleagues in the difficult times he has had to face from mid July 1998 to the present.

References

Brauckmann, C. and Schülter, T. 1993. Neue Insekten aus der Trias von Unter-Franken. Geol. Palaentol. 27: 181-199.
Foote, M. 1996. Perspective: evolutionary patterns in the fossil record. Evolution 50: 1-9.
Foote, M. 1997. The evolution of morphological diversity. Annu. Rev. Ecol. Syst. 28: 129-152.
Iturralde-Vincent, M. A. and MacPhee, R. D. E. 1996. Age and paleogeographical origin of Dominican amber. Science 273: 1850-1852.
Labandeira, C. C. 1994. A compendium of the fossil insect families. Milw. Publ. Mus. Contr. No. 88. 71 pp.
Labandeira, C. C. and Sepkoski Jr. J. J. 1993. Insect biodiversity in the fossil record. Science 261: 310-315.
Raup, D. M. 1972. Taxonomic diversity during the Phanerozoic. Science 177: 1065-1071.
Raup, D. M. 1975. Taxonomic survivorship curves and Van Valen's law. Paleobiology 1: 82-96.
Ross, A. J. and Jarzembowski, E. A. 1993. Arthropoda (Hexapoda; Insecta). pp. 363-426. Chapter 21. *In*: Benton, M. J. (ed.) The Fossil Record 2. Chapman & Hall. London, United Kingdom. 845 pp.
Samuelson, G. A. and Staines, C. L. A new *Sceloenopla* (Coleoptera: Chrysomelidae: Hispinae) from Dominican amber. 6 pp. (unpl. ms.)
Santiago-Blay, J. A. 1994. Paleontology of leaf beetles. *In*: Jolivet, P. Petitpierre, E., and Cox, M. L. (eds.). Novel aspects of the biology of Chrysomelidae (Coleoptera). Series Entomologica (Bari). Kluwer Academic Publ., Dordrecht, The Netherlands. pp. 1-68.
Santiago-Blay, J. A., Poinar, Jr. G. O., and Craig, P. R. 1996. Dominican and Mexican amber chrysomelids, with descriptions of two new species. *In*: Jolivet, P. H. A. and Cox, M. L. (eds.). Chrysomelidae Biology. Vol. 1: The classification, phylogeny, and genetics. SPB Acedemic Publishing bv, Amsterdam, The Netherlands. pp. 413-424.
Scherer, G. 1983. Diagnostic key for the neotropical Alticinae genera. Ent. Arb.Mus. Georg Frey 31/32: 1-89.
Seeno, T. N. and Wilcox, J. A. 1982. Leaf beetle genera. Entomography 1: 1-221.
Sepkoski, Jr., J. J., Bambach, R. K., Raup, D. M., and Valentine, J. W. 1981. Phanerozoic marine diversity and the fossil record. Nature 293: 435-437.
Signor III, P. W. and Lipps, J. H. 1982. Sampling bias, gradual extinction patterns, and catastrophes in the fossil record. Geol. Soc. Amer. Spec. Pap. 190: 291-296.
Uhmann, E. 1937. Amerikanische Hispinen aus dem Zoologisches Museum der Universität Berlin. I. Teil. Subfamilie Amplipalpini und Gattung *Sceloenopla* Chevr. Sond. Mitt. Zool. Mus. (Berlin) 29: 198-213.
Valentine, J. W. 1969. Patterns of taxonomic and ecological structure of the shelf benthos during Phanerozoic time. Palaeontology 12: 684-709.

MITOCHONDRIAL 16S rDNA SEQUENCES AND THEIR USE AS PHYLOGENETIC MARKERS IN LEAF-BEETLES WITH SPECIAL REFERENCE TO THE SUBFAMILY CHRYSOMELINAE

J. GÓMEZ-ZURITA, C. F. GARIN, C. JUAN and E. PETITPIERRE
Lab. Genètica, Dpt. Biologia, Universitat de les Illes Balears and Institut Mediterrani d'Estudis Avançats (IMEDEA), E-07071 Palma de Mallorca, Balearic Islands, Spain

Abstract

The 16S rDNA mitochondrial gene has been sequenced for a fragment of 510 bp in 62 taxa of leaf beetles, mostly in the genera *Timarcha* and *Chrysolina*, and in one species each of Bruchidae and Cerambycidae. These results are discussed with those reported by other authors. The AT content seems to gradually increase from the lowest in the cerambycid *Stenopterus ater* (66.5 %) to the maximum average value of the Galerucinae (78.84 %), in partial agreement with phylogenetic divergences. The 16S gene does not provide a reliable resolution for most of the major lineages at the subfamilial level. Nevertheless, Hispinae + Clytrinae are in a basal positions, Criocerinae and Chlamisinae appear related to *Bruchus* sp., all of them being also basal with respect to the Chrysomelinae. Among Chrysomelinae, monophyletic clades are evident for each of *Timarcha*, *Chrysolina* + *Oreina*, *Entomoscelis* + *Gonioctena*, mostly in agreement with recent taxonomic proposals.

The maximum parsimony tree obtained for *Timarcha* is not resolved enough to establish the relative ancestry of the different clades, but some of them are consistently supported: a) the studied Moroccan species and the south east Spanish relative, *T. espanoli* define a clade, b) the relationships among Iberian endemics with a bifidous mesosternum plus the Balearic *T. balearica* of the same morphological group are unclear, although some of them are closely associated, c) two other species, the European *T. tenebricosa* and the Pyrenean endemic *T. strangulata* do not display significant homology neither between them, nor with regard to any of the other clades, out of the relative affinity of *T. tenebricosa* with *T. balearica*, both sharing similar karyotypes composed of 22 chromosomes, d) the taxa included in the *T. goettingensis* complex are split into at least three different lineages: one covering western and south ranges mostly, another of Central and Cantabrian domains, and the third of the north east Spain mainly, from which the Mideuropean *T. goettingensis* is detached.

The hypothesised molecular phylogeny of *Chrysolina* was used to check its compatibility with plant affiliation and karyological data. The most ancestral clades are those feeding on Lamiaceae and having 2n = 24 chromosomes. At least three clades are in this group composed of the subgenera: a) *Maenadochrysa*, b) *Menthastriella* + *Melasomoptera*, *Erythrochrysa*, and *Chrysolinopsis-Taeniochrysea*, and c) *Chrysomorpha* + *Fastuolina*, and *Centoptera*. In an allied position appear the species of *Chrysolina s.str.*, characterised by a relative polyphagy and 2n = 23 chromosomes in the males. The species of *Stichoptera*, living on Scrophulariaceae, are in the same clade with those of *Paleosticta*, *Ovosoma* and *Colaphodes* which feed on Plantaginaceae, also in partial agreement with the karyological data. Finally the most apomorphous clades are those including *Chalcoidea* and *Allochrysolina*, living on Asteraceae and with 2n = 40 and 42 chromosomes, respectively, one with *Minckia* and *Threnosoma* on Apiaceae, with 2n = 46 or 47 chromosomes, and the last constituted by *Sphaeromela* and *Hypericia*, feeding on Clusiaceae (= Hypericaceae), with 2n = 32 and 38 chromosomes, respectively.

Introduction

Leaf-beetles constitute a large family with more than 35,000 described species (Jolivet, 1988), but with estimations reaching 65,000 considering undescribed taxa (Reid, 1995).

Advances in Chysomelidae Biology 1, pp. 25–38
Edited by M.L. Cox

Cerambycidae is assumed to be the sister family of Chrysomelidae and both are included within the same superfamily of Chrysomeloidea (Crowson, 1981; Lawrence and Britton, 1994; Reid, 1995; Farrell, 1998). However, the basal phylogenetic relationships among Chrysomelidae are controversial, that is Bruchidae and Megalopodidae with Cerambycidae and the other Chrysomelidae (Reid, 1995; Schmitt, 1996). Hsiao (1994a, b) made the first molecular phylogenetic approach to the problem in a preliminary sampling of 13 genera representing the same number of subfamilies.

The leaf-beetles of the subfamily Chrysomelinae, with some 3000 species in 132 genera (Daccordi, 1983, 1994) are one of the best studied taxa from a multidisciplinary phylogenetic approach. The larval morphology allows the separation of them into three main groups, according to the absence *versus* presence of defensive glands and their numbers (Paterson, 1931; Hennig, 1938; Kimoto, 1962; Takizawa, 1976; Cox, 1982). These groups are in part concordant with the chemotaxonomy of their defensive substances (Pasteels, 1993), and their karyological evolution (Petitpierre and Segarra, 1985; Petitpierre, 1988, 1997). Recently, molecular phylogenetic analyses have been performed on a few representatives in several genera of this subfamily (Hsiao, 1994a, 1994b), and some genera much more thoroughly, such as the *Oreina* (Dobler *et al.*, 1996; Hsiao and Pasteels, this volume), *Gonioctena* (Mardulyn *et al.*, 1997) and *Chrysolina* (Garin *et al.*, in press). In addition, another leaf beetle genus belonging to a different subfamily, the galerucine *Ophraella*, also has been carefully analysed from the same standpoint (Futuyma and McCafferty, 1990; Funk *et al.*, 1995).

In the present paper, we attempt to discuss the molecular phylogeny of twenty-three taxa of *Timarcha* and thirty-three of the genus *Chrysolina*, based upon the 16S rDNA sequences, together with those of further chrysomelines, namely *Oreina*, *Entomoscelis*, *Gonioctena* and *Chrysomela*, whose sequences were obtained by ourselves or taken from the data bank and published by previous authors. Also, we have tentatively analysed some subfamilial interrelationships by including further non-Chrysomelinae leaf beetle species of five subfamilies and one representative each of the related families Cerambycidae and Bruchidae to ascertain their genetic homology with Chrysomelidae.

Material and methods

Most of the examined species were collected at different locations in mainland Spain (SP.), others from the Balearic Islands (B.I.), Canary Islands (C.I.), Morocco (MO.), Germany (GE.) and Portugal (PO.). The exact locations for the *Chrysolina* species are reported in Garin *et al.* (in press) and the complete list for all the species can be provided by the authors upon request. The list given in the Appendix provides the country or island of occurrence for the checked species.

The procedures of mtDNA extraction, PCR amplification, cloning and sequencing of approximately the 510 bp of 16S rDNA gene fragment are detailed in Garin *et al.* (submitted). Table 1 shows the accession numbers of the sequences obtained in the present study, plus those published elsewhere.

CLUSTALW v. 1.7 (Thompson *et al.*, 1994) was used for alignment of the sequences. An ambiguous, highly variable region between the positions 240 and 280 was, in some cases, not considered in subsequent analysis. Distance, maximum parsimony and maxi-

Table 1. Accession numbers and references for the 16S rDNA sequences of chrysomelids and allied taxa used in the present study.

Taxa	Accession numbers	Source
Stenopterus	AJ231166	Present paper
Coptocephala	AJ231162	Present paper
Macrolenes	AJ231165	Present paper
Dicladispa	AJ231164	Present paper
Exema	U20678	Funk *et al.*, 1995
Gonioctena	AF014601-AF014619	Mardulyn *et al.*, 1997
Crioceris	AJ231163	Present paper
Chrysolina	AJ008034-AJ008074 AJ008079-AJ008108	Garin *et al.*, in press
Oreina	AJ008075-AJ008076 AJ008109 AJ231167	Garin *et al.*, in press, Present paper
Timarcha	AJ008078 AJ231136-AJ231160	Garin *et al.*, in press, Present paper
Bruchus	AJ231161	Present paper
Entomoscelis	AJ008077	Garin *et al.*, submitted
Chrysomela	AF014621	Mardulyn *et al.*, 1997
Ophraella	U20683, U20685, U20688, U20694, U20700, U20702, U20705, U20710, U20711, U20713, U20715, U20721	Funk *et al.*, 1995
Monoxia	U20682	Funk *et al.*, 1995

Note: As far as we know, the homologous 16S rDNA sequences used in Hsiao (1994a, b), are not available in the data bank and therefore are not included here.

mum likelihood trees were constructed and the support for particular nodes was assessed by bootstrap. All the phylogenetic analyses were performed using PAUP* 4.0.0d64 test version written by David Swofford. MacClade v. 3.04 (Maddison and Maddison, 1992) was used for mapping several characters in the topologies obtained.

Results and discussion

Nucleotide composition

The A+T content of the 16S rDNA fragment in the sampled species of leaf beetles is clearly non-homogeneous as suggested by a Chi-square homogeneity test (not shown). The lowest value of A+T in our screened Coleoptera (Table 2) is found in the cerambycid *Stenopterus ater* (66.51 %), the mean value of the two checked Clytrinae is 70.94 %, the species of Hispinae has slightly more than 72 %, that of Chlamisinae 74.94 %, and the one of Criocerinae 76 %, while the averaged values for Chrysomelinae and Galerucinae are 76.75 % and 78.82 %, respectively. Taking into account that the Cerambycidae occupy an ancestral position within the Chrysomeloidea, and in the Chrysomelidae the most plesiomorphous state of Criocerinae, Hispinae, Chlamisinae and Clytrinae *versus* Chrysomelinae and Galerucinae based on cladistic analyses of morphological data (Lee, 1993; Reid, 1995; Farrell, 1998), the regular increase in the A+T during the evolution of leaf beetles seems quite well-substantiated, except possibly for the Criocerinae, in spite of the small number of the sampled subfamilies. Jermiin and Crozier (1994) observed an evolutionary trend towards increasing A + T content in the hexapod lineage, hypothesised to be caused by

Table 2. Number of sequences, sequence length, minimum and maximum divergences within (with average value in brackets) and among genera and nucleotide composition in the studied chrysomelids and allied taxa.

Taxa	n	Seq. length	Seq. divergence[§]	Nucleotide composition A	C	G	T	A+T
Within genera								
Stenopterus	1	509	-	37.11	22.89	10.60	29.40	66.51
Coptocephala	1	510	-	36.80	19.13	9.93	34.14	70.94
Macrolenes	1	507	-	38.50	18.89	10.17	32.45	70.95
Dicladispa	1	507	-	42.09	18.49	9.49	29.93	72.02
Exema [b]	1	440	-	38.44	18.01	7.06	36.50	74.94
Gonioctena [a]	16	485	0.21-9.38 (6.24)	40.64	15.84	8.16	35.34	75.98
Crioceris	1	511	-	39.95	16.47	7.51	36.08	76.03
Chrysolina	31	509	0.79-11.42 (6.82)	39.54	14.87	9.03	36.55	76.14
Oreina	3	509	1.96-3.34 (2.75)	39.59	14.72	8.83	36.85	76.44
Timarcha	27	512	0.19-5.90 (3.26)	40.25	15.12	8.72	35.92	76.17
Bruchus	1	509	-	39.71	15.50	8.23	36.56	76.27
Entomoscelis	1	510	-	41.02	15.53	7.04	36.41	77.43
Chrysomela [a]	1	471	-	40.78	14.56	7.04	37.62	78.40
Ophraella [b]	11	442	0.23-5.68 (3.56)	41.18	14.79	6.70	37.33	78.51
Monoxia [b]	1	440	-	42.48	14.56	6.31	36.65	79.13
Among genera[+]								
Clytrinae	2	-	17.23	37.65	19.01	10.05	33.29	70.94
Chrysomelinae	79	-	3.14-17.28	40.30	15.11	8.14	36.45	76.75
Galerucinae	12	-	5.26-7.09	41.83	14.68	6.51	36.99	78.82

[a] Mardulyn *et al.*, 1997; [b] Funk *et al.*, 1995; [+] Pairwise comparisons between sequences from different genera; [§] Percentage range of divergence (uncorrected p-distance).

directional mutational pressure. This observation has been confirmed by Frati *et al.* (1997), who found a striking increase in the A+T content of Cytochrome Oxidase II mitochondrial gene in insect orders, from several taxa of Hemimetabola to those of the Holometabola orders. On a much more reduced evolutionary scale, our findings point out seemingly in the same direction.

Nucleotide 16S rDNA divergence and subfamilial phylogenetic relationships in the sampled chrysomelids

The amount of intrageneric uncorrected divergences (Table 2) in *Timarcha* ranges from 0.19 to 5.90 % (in four species in which more than one individual has been sequenced the range is 0 – 0.78 %), in *Oreina* the range is 1.96 – 3.34 %, while in *Chrysolina* it is higher, at 0.79 – 11.42 %. These values are similar to those reported in *Ophraella* 0.23 – 5.68 % (Funk *et al.*, 1995) and *Gonioctena* 0.21 – 9.38 % (Mardulyn *et al.*, 1997), respectively.

Neighbor-Joining trees (NJ) were constructed from the distances calculated using different models of nucleotide substitution. Fig. 1 shows a tree considering all the available sequences of chrysomelids obtained using the Tamura and Nei (1993) distances, similar topologies were obtained using other substitution models. This tree shows high bootstrap support for the monophyly of the sampled genera of Clytrinae (*Coptocephala* + *Macrolenes*) and Galerucinae (*Monoxia* + *Ophraella*). *Crioceris* (Criocerinae) and *Exema* (Chlamisinae)

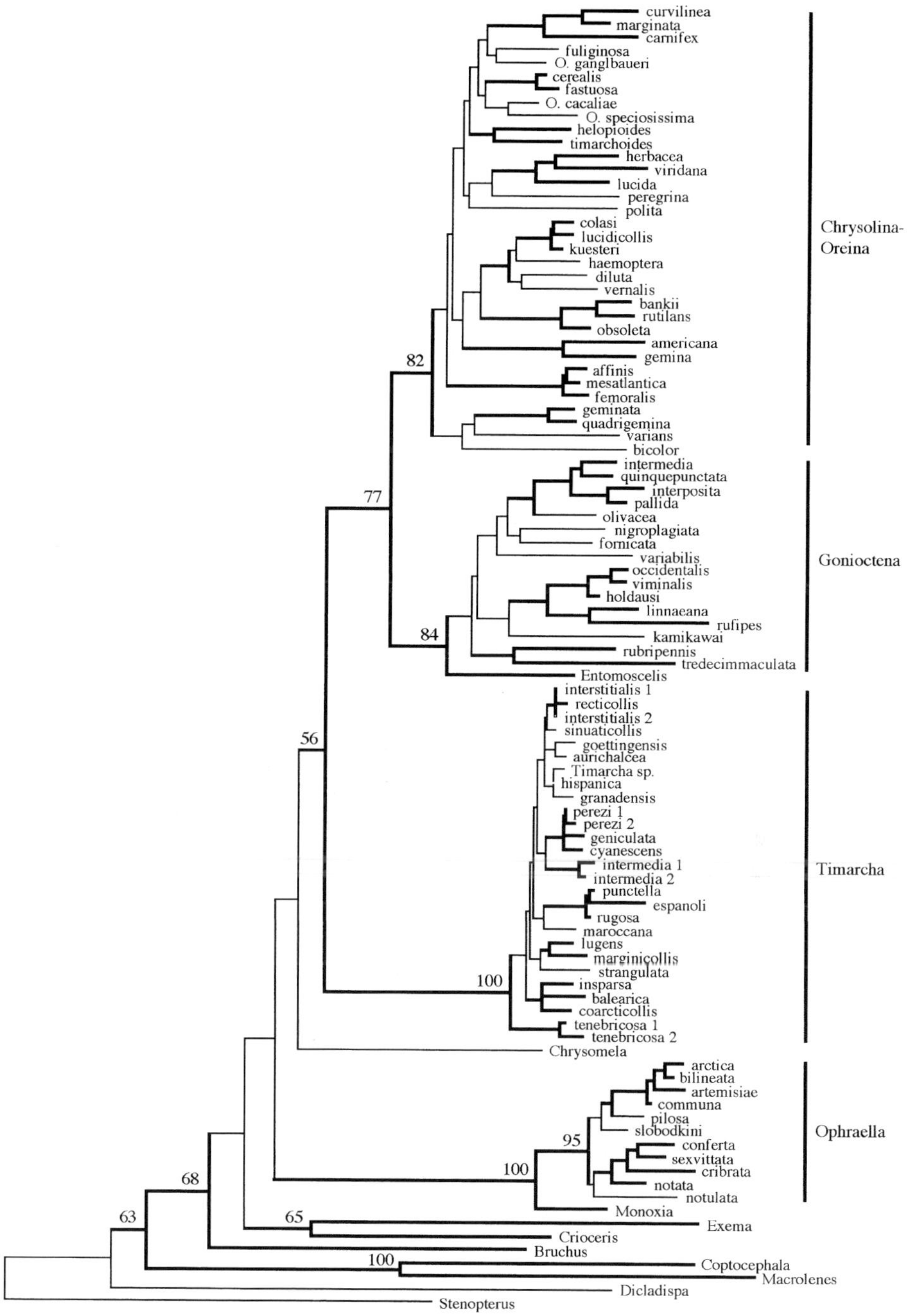

Fig. 1. Neighbor-joining tree showing 16S rDNA-based phylogenetic relationships in chrysomelids assuming a Tamura and Nei (1993) substitution model. The numbers above branches are bootstrap support values for relevant basal nodes after 1000 replications, values lower than 50 % are not shown. For the remaining nodes, no bootstrap values are indicated for simplicity, instead thick branches representing bootstrap values higher than 50 % are shown. The tree is rooted using the *Stenopterus ater* (Cerambycidae) sequence.

appear as sister groups with a bootstrap support of only 65 %. Considering the genera with several species studied, *Ophraella* and *Timarcha* are clearly monophyletic groups, whereas *Gonioctena* and *Chrysolina* appear as sister groups. The sequences of *Entomoscelis* and *Oreina* are included with high support in the *Gonioctena* and *Chrysolina* clades respectively, indicating that *Gonioctena* and *Chrysolina* are not strictly monophyletic based on this marker (see below).

The relationships among the different sampled subfamilies are not well resolved with 16S rDNA sequences. Monophyly of Chrysomelinae is not clearly supported by bootstrap analysis mainly because of the position of *Chrysomela* (Figs.1 and 2), whose sequence is quite divergent from the other Chrysomelinae, therefore showing a problematic and inconsistent position in the tree. On the other hand, in most of the analyses, the Hispinae + Clytrinae form a basal group, whereas the Criocerinae and Chlamisinae represented sequences appear to be related to the one sampled Bruchidae, which in turn are basal with respect to the Chrysomelinae.

The large number of taxa studied above precluded the use of maximum parsimony or likelihood analyses, but a simplified analysis using one representative species chosen at random from each genus gave compatible results with the above topology, confirming the basal position of *Dicladispa* (Hispinae) and the lack of resolution for many of the major lineages. Fig. 2 shows a maximum parsimony tree of the relationships among the different genera in which only the bootstrap values higher than 50 % are given. These results, although with the limitations imposed by the biased and scarce sampling in some of the groups, point out the need for more conserved sequences, such as nuclear ribosomal markers to assess more robustly the subfamilial phylogenetic relationships in a family in which the main lineages are supposed to be of early Tertiary origin at least (>65 Ma) (Crowson, 1981).

16S rDNA phylogenetic relationships within Chrysomelinae genera

Genus *Timarcha*

Figure 3 shows a maximum parsimony 16S rDNA tree produced by successive weighting to minimise the influence of very variable characters in the phylogenetic reconstruction (Farris, 1969). No *a priori* outgroup was constrained in the analysis because of the considerable divergence with any of the other chrysomelid species analysed (Nixon and Carpenter, 1993). As the phylogenetic reconstruction using all available taxa gave evidence of monophyly for the genus, the root is placed between the *Timarcha* ingroup and the remaining outgroup taxa. In this topology, the karyotype of each analysed taxon is also traced as an additional independent character. The polarity of change in the chromosomal evolution is not unambiguously resolved in this tree, but there is a considerable concordance of molecular and cytogenetic data.

In this phylogenetic reconstruction, all of the North African taxa are joined together in a rather consistent clade, including the species *T. maroccana*, *T. rugosa* and *T. punctella*, and the closely related *T. espanoli* from south-eastern Spain. This result agrees with the information provided by morphology, karyology and biogeography which group these taxa as closely related forms, constituting a different lineage from the other studied species.

The Iberian endemics *T. lugens*, *T. marginicollis*, *T. intermedia*, *T. coarcticollis*, *T. in-*

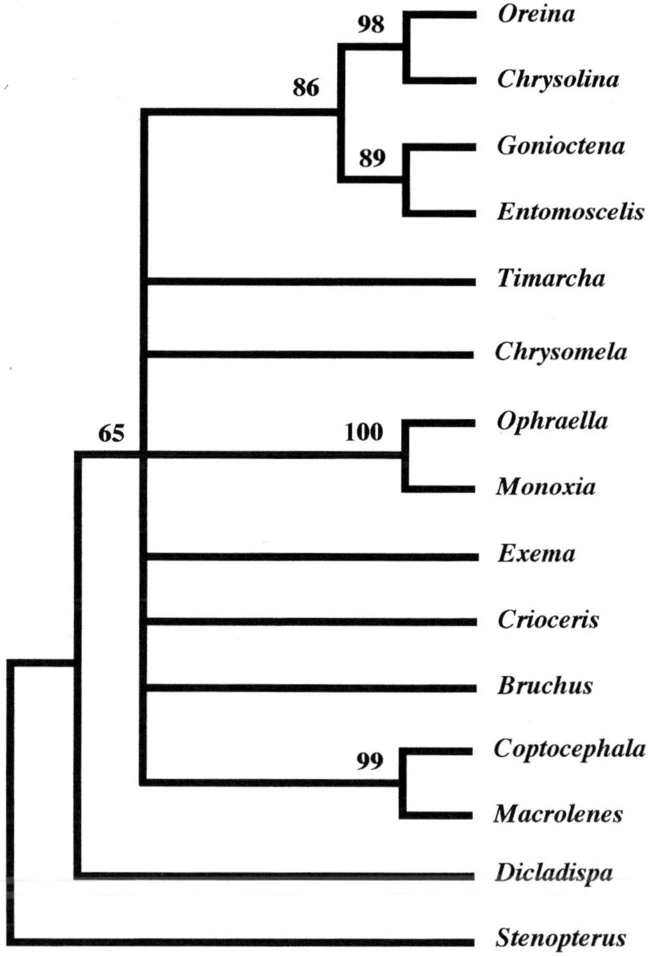

Fig. 2 Maximum parsimony tree depicting phylogenetic relationships among the represented genera. This tree is the resulting consensus of two equally parsimonious trees obtained after an heuristic search with 25 replications with random addition of taxa (ACCTRAN and TBR options in effect). Supported nodes with bootstrap values higher than 50 % after 250 replications are indicated.

sparsa, with 2n = 20 chromosomes, and the Balearic *T. balearica*, with 2n = 22 chromosomes (Petitpierre *et al.*, 1988; Petitpierre, unpubl.) have also been included in the study. All of these species share a bifid mesosternum and a common trend to an obliterate elytral punctuation. Although there are reasons to believe in a possible monophyly of this group, considering karyotypes and morphology, the accurate interrelationships of these taxa are somewhat unclear and loose, detaching all of them from ambiguous positions in the 16S phylogeny. The strong association of *T. coarcticollis* and *T. insparsa* 16S sequences is in accordance with their close morphological similarity and their placement in the same cluster as the Majorcan *T. balearica* can be justified from a biogeographic point of view. The two specimens of *T. tenebricosa*, a European taxon with wide distribution (slightly surpassing the Pyrenees southwards) showed sequences close to *T. balearica*, a species with a karyotype quite similar to that of the former made of 22 chromosomes (Petitpierre, 1970a).

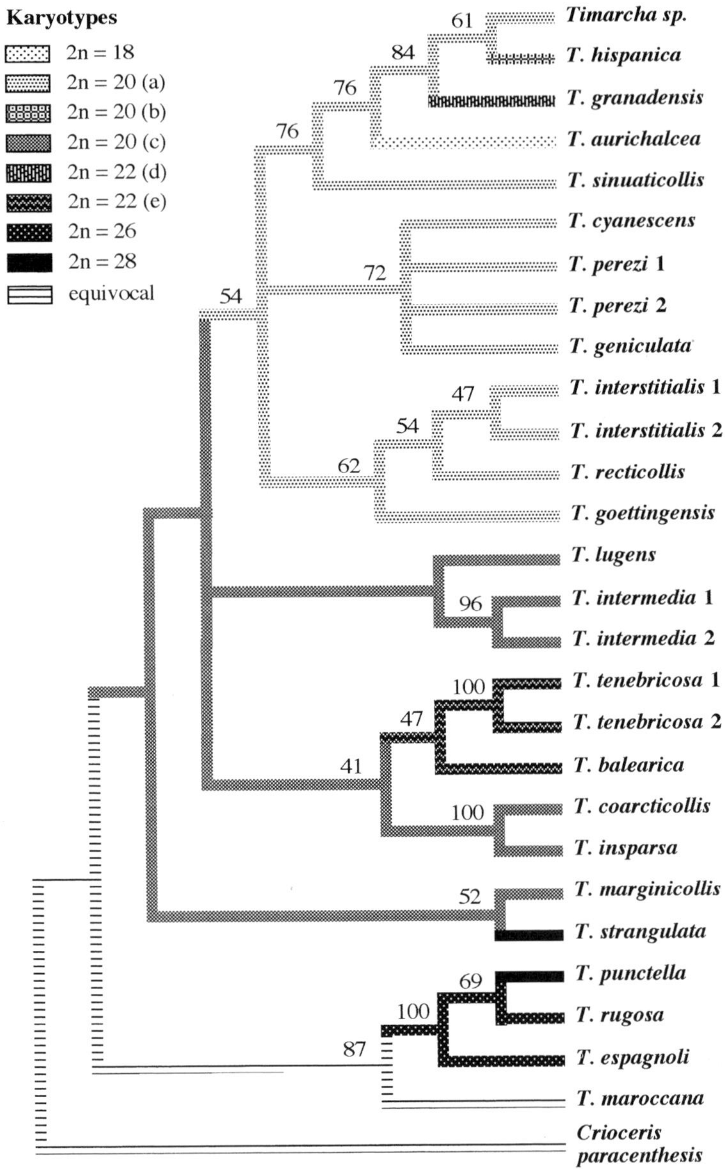

Fig. 3. Strict consensus tree of 11 equally parsimonious trees of 152 steps after an heuristic search (ACCTRAN and TBR options in effect; C.I. = 0.618, R. I. = 0.678) with 10 replications of random addition of *Timarcha* 16S rDNA sequences and one round of weighting of characters proportionally to their rescaled consistency indexes. Bootstrap values above branches were obtained after 100 replications with a heuristic search. On this tree, tracing of the different karyotypes have been performed using McClade v. 3.04 (Maddison and Maddison, 1992) and an ACCTRAN optimisation. There are three different types of karyograms of 2n = 20: asymmetric karyotype with fourth pair being acrocentrics (a); asymmetric karyotype, no acrocentrics (b) and symmetric metacentrics of decreasing size (c). Similarly, two different karyotypes can be distinguished in the species with 2n=22: asymmetric (d) or symmetric karyotypes (e). Patterns for each class are indicated in the Figure. The sequence of *Crioceris paracenthesis* was used as outgroup after establishing the relationships in the ingroup.

The Pyrenean endemic *T. strangulata,* with 2n = 28 chromosomes (Petitpierre, 1970a), belongs to a different morphological group and has no clear relationships with any of the other studied taxa regarding its 16S sequence, so it probably represents a different lineage.

Finally, all of the taxa included in the *T. goettingensis* complex (Petitpierre, 1970b), are in a large cluster separated from all the remaining clades. This complex is composed of taxa characterised by their small or medium size, punctuated elytra and non-bifid mesosternum, and most of them share very similar or identical karyotypes composed of 2n = 20 chromosomes (Petitpierre, 1970a, 1976). This clade also includes the sampled taxa of a different morphological group, represented here by *T. hispanica* from Algarve in South Portugal and *T. granadensis* from Granada in south Spain. There are at least three different lineages within the *T. goettingensis* complex clade: a) the first, including *T. hispanica* and *T. granadensis*, *Timarcha* sp. from León (North West Spain), *T. aurichalcea* from central east Spain and *T. sinuaticollis* from the East Pyrenees, b) the second is composed of taxa covering West and Central Spain, two sampled populations of *T. perezi*, *T. geniculata* and *T. cyanescens*, and c) another group comprising the Catalonian taxa, in north east Spain, two specimens of *T. interstitialis*, and the Central Pyrenean endemic *T. recticollis*. The Mid-European *T. goettingensis* appears to be basal to the latter taxa. These three clades display somewhat different trophic preferences since the taxa of the first two feed on either Rubiaceae or Plantaginaceae while those of the last group and *T. goettingensis* feed on Rubiaceae only (Jolivet and Petitpierre, 1973).

Genera *Gonioctena, Chrysolina* and *Oreina*

Entomoscelis and *Gonioctena* are classified into two different subtribes, Entomoscelina and Paropsina, respectively, in the most recent subfamilial taxonomy (Daccordi, 1994). In spite of this, *Entomoscelis* is clearly positioned within or as a sister group of the *Gonioctena* clade (Figs. 1 and 2) in agreement with the results reported by Hsiao (1994b) who studied only one species of each genus. The larvae of both genera belong to the non-glanduliferous group, characterised by having only one pair of defensive glands (Hennig, 1938; Kimoto, 1962), and the pupae of both genera are much more similar to each other than to those of *Chrysolina* or *Timarcha* (Cox, 1996). The separation of *Entomoscelis* and *Gonioctena* into different subtribes is thereby questionable according to our findings. However, the status of the subtribes Entomoscelina and Paropsina cannot be ascertained until more taxa of the 27 Entomoscelina and the 17 Paropsina genera were sampled.

Chrysolina and *Oreina* are closely related taxa which have been placed in the same genus by Bourdonné and Doguet (1991) and Daccordi (1994), based on non-molecular features. Dobler *et al.* (1996) allozyme data indicate that *Oreina* form a monophyletic clade, but the two *Chrysolina* species studied by them do not allow to establish the sister group status of the latter (see Hsiao and Pasteels, this volume). However, the three of *Oreina* species examined in our study are clearly included in the *Chrysolina* clade in all phylogenetic analyses (Figs. 1 and 4).

The maximum likelihood tree (Fig. 4) is the one providing the best available phylogenetic hypothesis for *Chrysolina*, because it combines the 16S rDNA and Cytochrome Oxidase subunit I preliminary sequence results. The present subgeneric taxonomy of *Chrysolina* is consistent with this molecular phylogeny, since all the species belonging to the same subgenus always appear monophyletic, which is also in agreement with their chromosomal

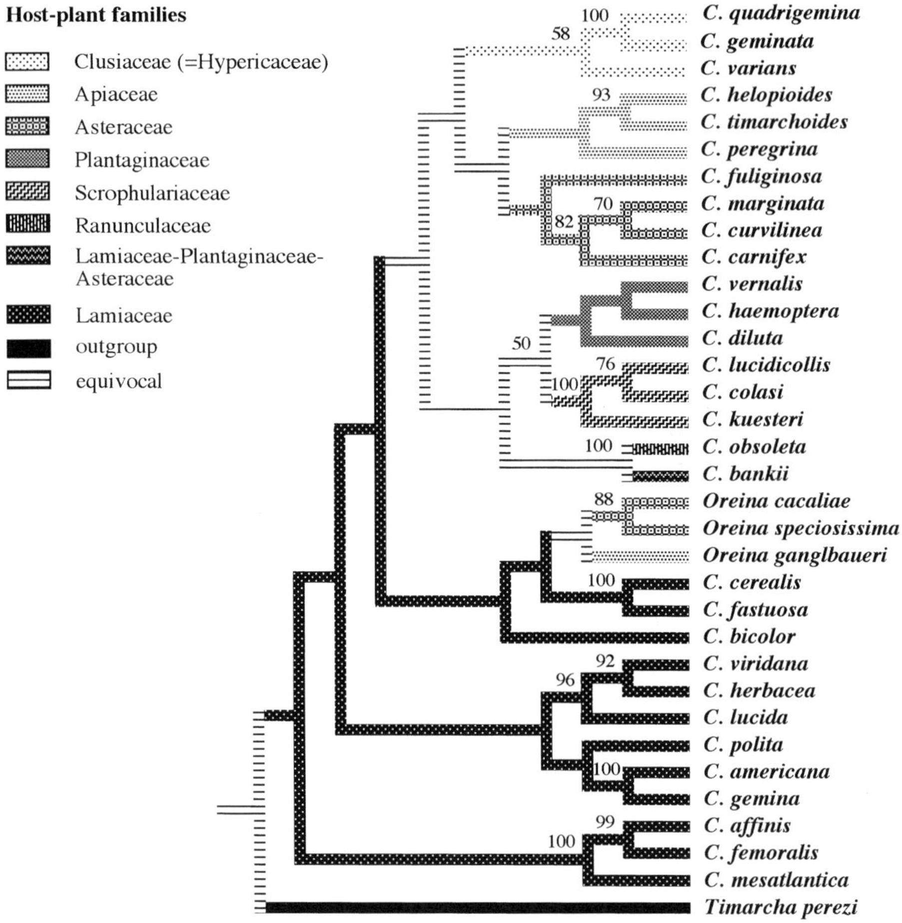

Fig. 4. Maximum likelihood tree of the combined data-set of 16S rDNA and COI sequences in *Chrysolina* and *Oreina*. The numbers above branches are bootstrap proportions higher than 50 % of the corresponding nodes under maximum parsimony and after 100 replications. Patterns on branches represent trophic selections of the different taxa studied which were optimised on the tree assuming ACCTRAN using MacClade v. 3.04 (Maddison and Maddison, 1992). The *Timarcha perezi* sequence has been used as the outgroup.

(Petitpierre, 1975, 1981, 1983) and food-plant choice resemblances (Jolivet and Petitpierre, 1976; Bourdonné and Doguet, 1991). As shown in Fig. 4, in which plant affiliation has been traced on the tree, the most basal clades of *Chrysolina* are those living on Lamiaceae, but the colonisation of these plants by the insects has apparently occurred in three different lineages. First, the species of the subgenus *Maenadochrysa*, *C. affinis*, *C. femoralis* and *C. mesatlantica*, all feeding on thyme (*Thymus* spp.). Second, those of the subgenera *Taeniochrysea* and *Chrysolinopsis*, *C. americana* and *C. gemina*, respectively, the former living on rosemary (*Rosmarinus officinalis*) and lavenders (*Lavandula* spp.), and the latter also on the lavenders. In addition, in the same phyletic branch appear the sister clades feeding on mints (*Mentha* spp.) one including *C. polita* and another with *C. herbacea*, *C. lucida* and *C. viridana*. Third, a further clade feeding on Lamiaceae unites three species each one of a different subgenus, *C. cerealis*, *C. fastuosa* and *C. bicolor*, whose host-plants are quite diverse among the Lamiaceae. All these species of the three clusters share 2n = 24 chromosomes of rather similar sizes and shapes.

The next clade with a supported node close to the previous ones is composed of the *Chrysolina s.str.* subgenus species, *C. bankii* and *C. obsoleta*, having 2n (male) = 23 chromosomes and a quite wide feeding range, at least for adults, on Lamiaceae, Ranunculaceae, Plantaginaceae and Asteraceae. These species are chromosomally derived from the former presumably by the loss of the y-chromosome.

Two additional sister clades supported by a significant bootstrap value are species of the subgenus *Stichoptera*, *C. colasi*, *C. lucidicollis* and *C. kuesteri*, and a much less supported cluster having *C. diluta*, *C. haemoptera* and *C. vernalis*. The species of the first clade are chromosomally heterogeneous (2n = 22, 24, 32) but they share asymmetrical karyotypes with very different size classes, and they feed on *Linaria* spp. (Scrophulariaceae) and show some close morphological characters (Bourdonné and Doguet, 1991; Petitpierre, 1999). Each of the species of the second group are in different subgenera, although they all feed on *Plantago* spp. (Plantaginaceae). Cytogenetically, *C. diluta* has 2n = 36 and *C. haemoptera* and *C. vernalis* display 2n = 40 chromosomes.

A well-established clade unites the species of subgenus *Chalcoidea*, *C. marginata*, *C. carnifex* and *C. curvilinea*, which share karyotypes made of 2n = 40 chromosomes and feeding choices on Asteraceae plants (*Artemisia* and *Achillea* spp.). Another Asteraceae-feeding species, *C. fuliginosa*, belonging to the subgenus *Allochrysolina*, with 2n = 42 chromosomes, seems to be close to the above species despite lacking a supported bootstrap value with regard to them.

Finally, two specialised trophic selections on Apiaceae and Clusiaceae (=Hypericaceae) allow the recognition of further clusters of species. In the first one, *C. helopioides* and *C. timarchoides*, both in the subgenus *Threnosoma* and sharing 2n (male) = 47 chromosomes, are placed together by a consistent bootstrap value, while *C. peregrina* with 2n = 46, is also allied but much more loosely with respect to the former. A similar case is found in the species on Clusiaceae, *C. geminata* and *C. quadrigemina*, both in the subgenus *Hypericia* and sharing 2n = 38 chromosomes, which are in a supported node, while *C. varians*, in the subgenus *Sphaeromela* with 2n = 32 chromosomes, is placed near to them but with a weak support.

Our phylogenetic analyses in *Chrysolina* suggest a minimum of five host plant family shifts. Although some of the shifts are between plant families belonging to the same subclass, such as Lamiaceae to Plantaginaceae and to Scrophulariaceae (subclass Asteridae), others are between plant families of different subclasses, as occur from Lamiaceae to Apiaceae (subclass Rosidae) and from the former to Clusiaceae (subclass Dilleniidae). These plant switches of *Chrysolina* are not correlated with the presumed phylogenetic divergence followed by the host plant families (Cronquist, 1988). A hypothesis of convergent colonisations dependent on the available plants having similar chemical compounds (Mitter and Farrell, 1991) necessary to these leaf beetles, is probably the most plausible explanation to the trophic changes, as found in other Chrysomelidae (Becerra, 1997; Mardulyn *et al.*, 1997).

Acknowledgements

The following colleagues have kindly contributed to the present study by supplying specimens: Gloria Bastazo and José Miguel Vela, Málaga (*C. helopioides*), Javier de Diego, Cantabria (*T. cyanescens*), Frank Fritzlar, Thuringia (*T. goettingensis*), José M. Guzmán, Melilla (*T. punctella, T. rugosa*), Fernando Murria, Zaragoza (*T. perezi* 2), Pedro Oromí, Tenerife (*C. bicolor, C. gemina, C. lucidicollis, C. mesatlantica, C. obsoleta, C. rutilans, T. maroccana*), Antoni Sacarés, Mallorca (*C. viridana*) and Joan and Eduard Vives, Barcelona (*T. interstitialis* 1). This work has been financed by the project number DGICYT PB96-0090 of the Ministry of Education and Culture, Spain. C. G. and J. G-Z. have been supported respectively by FPI post- and predoctoral grants of the Spanish Ministry of Education and Culture.

References

Becerra, J. X. 1997. Insects on plants: macroevolutionary chemical trends in host use. *Science* 276: 253-256.

Bourdonné, J. C. and Doguet, S. 1991. Données sur la biosystématique des *Chrysolina* l. s. (Coleoptera: Chrysomelidae: Chrysomelinae). *Annals Soc. ent. Fr.* (N.S.) 27: 29-64.

Cox, M. L. 1982. Larvae of the British genera of chrysomeline beetles (Coleoptera, Chrysomelidae). *Syst. Ent.* 7: 297-310.

Cox, M. L. 1996. The pupae of Chrysomeloidea. *In:* Jolivet P. H. A. and Cox M. L. (eds.), *Chrysomelidae Biology, vol. 1: The Classification, Phylogeny and Genetics*, Academic Publishers, Amsterdam, The Netherlands, pp. 119-265.

Cronquist, A. 1988. *The evolution and classification of flowering plants*. New York Botanical Garden, New York.

Crowson, R. A. 1981. *The Biology of the Coleoptera*. Academic Press, New York.

Daccordi, M. 1983. Quelques réflexions sur le distribution des Chrysomelinae (Col. Chrysomelidae). *Bull. Soc. ent. Fr.* 88: 448-451.

Daccordi, M. 1994. Notes for phylogenetic study of Chrysomelinae with descriptions on new taxa and a list of all known genera (Coleoptera: Chrysomelidae Chrysomelinae). *In:* Furth D. G. (ed.), *Proc. Third Inter. Symp. Chrysomelidae*, Beijing, 1992, Backhuys Publ., Leyden, pp. 60-84.

Dobler, S., Mardulyn, P., Pasteels, J. M. and Rowell-Rahier, M. 1996. Host-plant switches and the evolution of chemical defence and life history in the leaf beetle genus *Oreina*. *Evolution* 50: 2373-2386.

Farrell, B. D. 1998. "Inordinate fondness" explained: Why are there so many beetles? *Science* 281: 555-559.

Farris, J. S. 1969. A successive approximations approach to character weighting. *Syst. Zool.* 18: 374-385.

Frati, F., Simon, C., Sullivan, J. and Swofford, D. L. 1997. Evolution of the mitochondrial Cytochrome Oxidase II gene in Collembola. *J. Mol. Evol.* 44: 145-158.

Funk, D. J., Futuyma, D. J., Ortí, G. and Meyer, A. 1995. Mitochondrial DNA sequences and multiple data sets: A phylogenetic study of phytophagous beetles (Chrysomelidae: *Ophraella*). *Mol. Biol. Evol.* 12: 627-640.

Futuyma, D. J. and McCafferty, S. 1990. Phylogeny and the evolution of host plant associations in the leaf beetle genus *Ophraella* (Coleoptera: Chrysomelidae). *Evolution* 44: 1885-1913.

Garin, C. F., Juan, C. and Petitpierre, E. 1999. Mitochondrial DNA phylogeny and the evolution of host-plant use in Palearctic *Chrysolina* (Coleoptera, Chrysomelidae) leaf beetles. *J. Mol. Evol.* (in press)

Hennig, W. 1938. Übersicht über die wichtgsten deutschen chrysomelinen. *Arb. physiol. angew. Ent. Berl.* 5: 85-136.

Hsiao, T. H. 1994a. Molecular techniques for studying systematics and phylogeny of Chrysomelidae. *In:* Jolivet P., Cox M. L. and Petitpierre E. (eds.), *Novel Aspects of the Biology of Chrysomelidae*, Kluwer Academic Publishers, Dordrecht, The Netherlands, pp. 237-248.

Hsiao, T. H. 1994b. Molecular phylogeny of chrysomelid beetles inferred from mitochondrial DNA sequence data (Coleoptera: Chrysomelidae). *In:* Furth D. G. (ed.), *Proc. Third Inter. Symp. Chrysomelidae*, Beijing, 1992, Backhuys Publ., Leyden, pp. 9-17.

Hsiao, T. H. and Pasteels, J. M. 1999. Evolution of host-plant affiliation and chemical defense in *Chrysolina-Oreina* leaf beetles as revealed by mtDNA phylogenies. (this volume).

Jermiin, L. S. and Crozier, R. H. 1994. The cytochrome *b* region in the mitochondrial DNA of the ant *Tetraponera rufoniger*: sequence divergence in Hymenoptera may be associated with nucleotide content. *J. Mol. Evol.* 38: 282-294.

Jolivet, P. 1988. Food habits and food selection of Chrysomelidae. Bionomic and evolutionary perspectives. *In:* Jolivet P., Petitpierre E. and Hsiao T. H. (eds.), *Biology of Chrysomelidae*, Kluwer Academic Publishers, Dordrecht, Netherlands, pp. 1-24.

Jolivet, P. and Petitpierre, E. 1973. Plantes-hôtes connues des *Timarcha* Latreille (Col. Chrysomelidae). Quelques considérations sur les raisons possibles du trophisme sélectif. *Bull. Soc. ent. Fr.* 78: 9-25.

Jolivet, P. and Petitpierre, E. 1976. Plantes hôtes connues des *Chrysolina* (Col. Chrysomelidae). Essai sur les types de sélection trophique. *Annals Soc ent. Fr.* 12: 123-149.

Kimoto, S. 1962. A phylogenetic consideration of Chrysomelinae based on immature stages of Japanese species. *J. Fac. Agric. Kyushu Univ.* 12: 67-114.

Lawrence, J. F. and Britton, E. B. 1994. *Australian Beetles*. Melbourne University Press, Melbourne.

Lee, J. E. 1993. Phylogenetic studies on the larvae of the Chrysomelidae (Coleoptera) from Japan. *Jpn. J. Ent.* 61: 409-424.

Maddison, W. P. and Maddison, D. R. 1992. MacClade, Vers. 3.04. Sinauer, Sunderland, Massachusets.

Mardulyn, P., Milinkovich, M. C. and Pasteels, J. M. 1997. Phylogenetic analyses of DNA and allozyme data suggest that *Gonioctena* leaf beetles (Coleoptera; Chrysomelidae) experienced convergent evolution in their history of host-plant family shifts. *Syst. Biol.* 46: 722-747.

Mitter, C. and Farrell, B. 1991. Macroevolutionary aspects of insect-plant relationships. *In:* Bernays E. (ed.), *Insect-plant interactions*, vol. 3, CRC Press, Boca Raton, Florida, pp. 35-78.

Nixon, K. C. and Carpenter, J. M. 1993. On outgroups. *Cladistics* 9: 413-426.

Pasteels, J. M. 1993. The value of defensive compounds as taxonomic characters in the classification of leaf beetles. *Biochem. Syst. Ecol.* 21: 135-142.

Paterson, N. F. 1931. The bionomics and comparative morphology of the early stages of certain Chrysomelidae (Col. Phyt.). *Proc. zool. Soc. Lond.* 4: 879-940.

Petitpierre, E. 1970a. Cytotaxonomy and evolution of *Timarcha* Latr. (Coleoptera, Chrysomelidae). *Genetica Iber.* 22: 67-120.

Petitpierre, E. 1970b. Variaciones morfológicas y de la genitalia en las *Timarcha* Lat. (Col. Chrysomelidae). *Publnes. Inst. Biol. apl. Barcelona* 48: 5-16.

Petitpierre, E. 1975. Notes on chromosomes of ten species of the genus *Chrysolina* Mots. (Coleoptera, Chrysomelidae). *Genetica* 45: 349-354.

Petitpierre, E. 1976. Further cytotaxonomical and evolutionary studies on the genus *Timarcha* Latr. (Coleoptera). *Genetica Iber.* 28: 58-72.

Petitpierre, E. 1981. New data on the cytology of *Chrysolina* (Mots.) and *Oreina* Chevr. (Coleoptera, Chrysomelidae). *Genetica* 54: 265-272.

Petitpierre, E. 1983. Karyometric differences among nine species of the genus *Chrysolina* Mots. (Coleoptera, Chrysomelidae). *Can. J. Genet. Cytol.* 25: 33-39.

Petitpierre, E. 1988. Cytogenetics, cytotaxonomy and genetics of Chrysomelidae. *In:* Jolivet P., Petitpierre E. and Hsiao T. H. (eds.), *Biology of Chrysomelidae*, Kluwer Academic Publishers, Dordrecht, Netherlands, pp. 131-159.

Petitpierre, E. 1997. The value of cytogenetics for the taxonomy and evolution of Leaf Beetles (Coleoptera, Chrysomelidae). *Misc. zool.* 20: 9-18.

Petitpierre, E. 1999. The cytogenetics and cytotaxonomy of the subgenus *Stichoptera* of *Chrysolina* (Coleoptera, Chrysomelidae). *In:* Sobti R. C. and Yadav J. S. (eds.), *Homage book to Prof. H. R. Pajni*, Panjab Univ., Chandigarh, India (in press).

Petitpierre, E. and Segarra, C. 1985. Chromosomal variability and evolution of Chrysomelidae (Coleoptera), particularly that of Chrysomelinae and Palearctic Alticinae. *Entomography* 3: 403-426.

Petitpierre, E., Segarra, C., Yadav, J. S. and Virkki, N. 1988. Chromosome numbers and meioformulae of Chrysomelidae. *In:* Jolivet P., Petitpierre E. and Hsiao T. H. (eds.), *Biology of Chrysomelidae*, Kluwer Academic Publishers, Dordrecht, Netherlands, pp. 161-186.

Reid, C. A. M. 1995. A cladistic analysis of subfamilial relationships in the Chrysomelidae *sensu lato* (Chrysomeloidea). *In:* Pakaluk J. and Slipinski S. A. (eds.), *Biology, phylogeny, and classification of Coleoptera: Papers celebrating the 80th birthday of Roy A. Crowson*, vol. 2, Muzeum i Instytut Zoologii PAN, Warszawa, Poland, pp. 339-632.

Schmitt, M. 1996. The phylogenetic system of the Chrysomelidae – History of ideas and present state of knowledge. *In:* Jolivet P. H. A. and Cox M. L. (eds.), *Chrysomelidae Biology, vol. 1: The Classification, Phylogeny and Genetics*, Academic Publishers, Amsterdam, The Netherlands, pp. 57-96.

Takizawa, H. 1976. Larvae of the genus *Gonioctena* Chevrolat (Coleoptera, Chrysomelidae): Descriptions of Japanese species and the implications of larval characters for the phylogeny. *Kontyû, Tokyo* 44: 444-468.

Tamura, K. and Nei, M. 1993. Estimation of the number of nucleotide substitutions in the control region of mitochondrial DNA in humans and chimpanzees. *Mol. Biol. Evol.* 10: 512-526.

Thompson, J. D., Higgins, D. J. and Gibson, T. J. 1994. Clustal, W.: Improving the sensitivity of progressive multiple sequence alignment through sequence weighting position-specific gap penalties and weight matrix choice. *Nucleic Acids Res.* 22: 4673-4680.

Appendix

Bruchus sp., Mallorca (B.I.); *Chrysolina affinis baetica* (F.), Murcia (SP.); *C. americana* (L.), Mallorca (B.I.); *C. bankii* (F.), Mallorca (B.I.); *C. bicolor* (Ol.), Fuerteventura (C.I.); *C. carnifex* (F.), Barcelona (SP.); *C. cerealis* (L.), Girona (SP.); *C. colasi* (Cobos), Granada (SP.); *C. curvilinea* (Wse.), Murcia (SP.); *C. diluta* (Germ.), Granada (SP.); *C. fastuosa* (Scop.), Lleida (SP.); *C. femoralis* (Ol.), Girona (SP.); *C. fuliginosa* (Ol.), Lleida (SP.); *C. gemina* (Brullé), La Palma (C.I.); *C. geminata* (Payk.), Lleida (SP.); *C. haemoptera* (L.), La Coruña (SP.); *C. helopioides* (Suffr.), Málaga (SP.); *C. herbacea* (Duft.), Teruel (SP.); *C. kuesteri* (Hell.), Girona (SP.); *C. lucida* (Ol.), Almería (SP.); *C. lucidicollis grossepunctata* (Har. Lind.), Lanzarote (C.I.); *C. marginata* (L.), Girona (SP.); *C. mesatlantica* (Kocher), Moyen Atlas (MO.); *C. obsoleta* (Brullé), Tenerife (C.I.); *C. peregrina* (Herr.-Schäff.), Mallorca (B.I.); *C. polita* (L.), Girona (SP.); *C. quadrigemina* (Suffr.), Lleida (SP.); *C. rutilans* (Woll.), Gomera (C.I.); *C. timarchoides* (Bris.), Girona (SP.); *C. varians* (Schäller), Lleida (SP.); *C. vernalis pyrenaica* (Duf.), Lleida (SP.); *C. viridana* (Küst.), Mallorca (B.I.); *Coptocephala scopolina* (L.), Mallorca (B.I.); *Crioceris paracenthesis* (L.), Mallorca (B.I.); *Dicladispa testacea* (L.), Mallorca (B.I.); *Entomoscelis adonidis* (Pall.), Girona (SP.); *Macrolenes dentipes* (Ol.), Mallorca (B.I.); *Oreina cacaliae* (Schrk.), Lleida (SP.); *O. ganglbaueri* (Jakob), Lleida (SP.); *O. speciosissima* (Scop.), Lleida (SP.); *Stenopterus ater* (L.), Mallorca (B.I.); *Timarcha aurichalcea* Bech., Cuenca (SP.); *T. balearica* Gory, Mallorca (B.I.); *T. coarcticollis* Fairm., Cádiz (SP.); *T. cyanescens* Fairm., Cantabria (SP.); *T. espanoli* Bech., Alicante (SP.); *T. geniculata* (Germ.), León (SP.); *T. goettingensis* (L.), Thuringia (GE.); *T. granadensis* Bech., Granada (SP.); *T. hispanica* Herr.-Schäff., Algarve (PO.); *T. insparsa* Rosenh., Granada (SP.); *T. intermedia* Herr.-Schäff. (1), Almería (SP.); *T. intermedia* Herr.-Schäff. (2), Alicante (SP.); *T. interstitialis* Fairm. (1), Barcelona (SP.); *T. interstitialis* Fairm. (2), Barcelona (SP.); *T. lugens* Rosenh., Granada (SP.); *T. marginicollis* Rosenh., Granada (SP.); *T. maroccana* Wse., Moyen Atlas (MO.); *T. perezi* Fairm. (1), Valladolid (SP.); *T. perezi* Fairm. (2), Zaragoza (SP.); *T. punctella* Mars., Nador (MO.); *T. recticollis* Fairm., Lleida (SP.); *T. rugosa* (L.), Nador (MO.); *T. sinuaticollis* Fairm., Girona (SP.); *T. strangulata* Fairm., Lleida (SP.); *T. tenebricosa* (F.) (1), Lleida (SP.); *T. tenebricosa* (F.) (2), Girona (SP.); *Timarcha* sp., León (SP.).

HISTORICAL AND BIOLOGICAL RELATIONSHIPS AMONG HISPINAE INFERRED FROM 12S MTDNA SEQUENCE DATA

TING H. HSIAO[1] and DONALD M. WINDSOR[2]

[1]*Professor Emeritus, Department of Biology, Utah State University, Logan, UT 84322-5305, USA;* [2]*Smithsonian Tropical Research Institute, Apartado 2072, Balboa Ancon, Panama*

Abstract

A preliminary phylogeny to the chrysomelid subfamily, Hispinae (*sensu lato*), was constructed using 12S mtDNA sequence data obtained from 48 focal taxa and an outgroup taxon (Donaciinae). Analysis of the 535 aligned bases in the data set resulted in a cladogram with increasing bootstrap support for recent tribal and generic level divisions within the subfamily. The cladogram isolates all 10 "hispoid" tribes in the basal section of the tree, all 9 "cassidoid" tribes in the derived portion of the tree. The host plants upon which these beetles feed were divided similarly with Monocotyledonae occuring basally and Dicotyledonae apically. One tribe, the Hemisphaerotinae, appear transitional in that they are placed with a clade containing other "hispoid" taxa, occur primarily on monocots yet have exuvial-fecal shields, structures characteristic of "cassidoid" Hispinae. Species exhibiting gregarious larval behavior occurred exclusively among the better-supported, apical tribes. Among the latter taxa, parental guarding of offspring appears to have dual origins, evolving first among Convolvulaceae feeders and later, independently, among taxa feeding on Asteraceae. Several taxonomic rearrangements are suggested based, at least in part, on the better supported portions of the phylogeny.

1. Introduction

While the taxonomic arrangement within the Chrysomelidae has changed considerably over time (Reid, 1995; Schmitt 1996), the close association of the Cassidinae ("Tortoise Beetles"; ~3,000 species) and the Hispinae ("Hispine Beetles"; ~1,500 species) has been reinforced by generations of taxonomists (Seeno and Wilcox, 1982; Jolivet, 1988; Boroweic, 1995). Results of a morphology-based character analysis of selected tribes of Cassidinae by Boroweic (1995) suggested this subfamily was a polyphyletic assemblage that should be merged with the Hispinae. Recent phylogenies based on both 12S mtDNA sequence data from representatives of 11 subfamilies (Hsiao, 1994a) and 18SmtDNA data from representatives of 22 subfamilies of Chrysomelidae (Farrell, 1998) further indicate the close relationship among these taxa and their relatively distant relationship with other chrysomelid subfamilies. Treatments by Crowson (1955), Lawrence and Britton (1991) and Reid (1995) have merged the Cassidinae and Hispinae into a single subfamily, the Hispinae, a convention we follow below.

The hispine-cassidine complex contains species whose larval morphologies and feeding habitats are more diverse than is found in any of the other 18 or so subfamilies of Chrysomelidae (Maulik, 1931, 1932, 1933; Stehr, 1991). Indeed, the radiation and diversification of this group appears largely the result of larval adaptations aimed at overcoming

Advances in Chysomelidae Biology 1, pp. 39–50
Edited by M.L. Cox

the range of environmental conditions, host plants, predators and parasitoids found in exposed, successional habitats. Establishing the phylogeny of this complex is crucial to understanding how larval adaptations and host plant associations may have evolved over time and how particular adaptations may be associated with species diversification rates.

Below, using 12S mtDNA sequence data obtained from 48 species of Hispinae belonging to 18 tribes (Table 1), we construct a preliminary phylogeny for the complex. We then map known larval and adult feeding characteristics and morphological characteristics on to the tips of the phylogeny to determine, 1) whether monocotyledonous or dicotyledonous plants were the initial hosts of species ancestral to this complex, 2) whether endophytic (leaf-mining) or exophytic (surface and axil) feeding is likely to have been the original condition, and 3) whether gregarious larval behaviour and subsocial parental behaviour have evolved more than once within the Hispinae (Windsor and Choe, 1994).

2. Techniques of DNA amplification and sequencing

The detailed procedures used in this study are described in Hsiao (1994a). Briefly, whole insects, either alcohol-preserved or frozen, were used for DNA extraction. The standard extraction involved homogenization of tissues in dry ice, incubation of homogenate in proteinase K, and phenol extraction. This protocol consistently yielded DNA from insects as small as 1 mg. A typical PCR amplification protocol for a 100 fl reaction required 5-20 ng of template DNA, 2 fl of a 20 fM solution of each opposing primer, 8 fl of 10 fM dNTPs, 10 fl of 10X Promega PCR buffer, and 3 fl of 0.1M MgCl2. The mixture was placed in a 0.6 ml microfuge tube, overlaid with mineral oil, preheated in an Ericomp thermocycler at 95 °C for 3 min, and 2 units of Promega Taq polymerase were added. The amplification was carried out in the thermocycler programmed for 35 cycles at 93 °C, 35 sec for denaturing, 55 °C, 35 sec for annealing, and 72 °C, 2 min for extension. A final extension was followed at 72 °C for 5 min.

The 12S rDNA primers are: SR-N-14759 (5'-GACAAAATTCGTGCCAGCAGT-3') and SR-J-14233 (5'-AAGAGCGACGGGCGATGTGT-3'). These primers produced DNA fragments of about 500 base pairs, which cover the more variable region of mitochondrial 12S rRNA gene. The PCR amplified DNA fragment was purified by agarose gel electrophoresis, GeneClean (Bio101, Inc), and precipitated with ethanol. DNA fragment amplified from 100 fl of PCR reaction was sufficient for 2 to 4 sequencing reactions. The procedure of direct double-stranded DNA sequencing of PCR products was identical to that described by Hsiao (1994a). Dideoxy chain termination reaction was carried out with Sequenase version 2.0 kit (United States Biochemical Corporation). Both compliment strands were sequenced twice with one of the amplification primers. Two to three individuals from each species were used to confirm the accuracy of the sequence data.

3. Sequence analysis

DNA sequences from 48 species of Hispinae and an outgroup species of Donaciinae were aligned with the ClustalW program (Higgins et al., 1992) and the GCG multiple alignment program (Wisconsin Sequence Analysis Package, Version 7.0) and adjusted visually for misalignment. Sequence statistics were prepared by Kimura's two-parameter method. The

Table 1. Sequence divergence of Cassidinae, Hispinae and Chrysomelinae in the 12S rDNA gene region.

Level of comparison	Taxon	No. Taxa	Mean % divergence ± S.D.	Range Min	Max
"Cassidoid" Hispinae					
Between species					
	Genus Charidotella	3	14.45 ± 8.71	4.48	20.60
	Genus Stolas	4	15.19 ± 3.51	8.97	18.62
Between genera					
	Tribe Cassidini	11	28.63 ± 4.24	17.68	38.58
	Tribe Stolaini	5	23.59 ± 4.43	13.81	30.26
Between tribes					
	"Cassidoid" Hispinae	11	28.69 ± 5.06	17.50	40.11
"Hispoid" Hispinae					
Between genera					
	Tribe Chalepini	3	37.21 ± 9.80	25.89	43.08
	Tribe Imatidiini	3	37.41 ± 9.59	26.38	43.78
Between tribes					
	"Hispoid" Hispinae	8	37.28 ± 3.55	31.66	42.34
Subfamily Chrysomelinae					
Between species					
	Genus Chrysomela	12	5.86 ± 2.23	0.40	9.40
	Genus Leptinotarsa	12	7.26 ± 2.04	1.80	10.70
Between genera					
	Tribe Chrysomelini	5	12.76 ± 2.95	8.70	17.30
	Tribe Doryphorini	5	10.72 ± 2.34	5.00	13.30
Between tribes					
	Chrysomelinae	8	15.21 ± .86	10.20	22.20

neighbour-joining method was used to reconstruct phylogenetic trees (MEGA, Kumar et al.,1993). The same data sets were also analyzed by PAUP (Phylogenetic Analysis Using Parsimony, Version 3.1.1, Swofford, 1993) to compare congruency of tree topology between neighbour-joining and parsimony methods. Bootstrap values were calculated for support of branching patterns.

Sequencing conducted subsequent to Hsiao (1994b) indicated Donaciinae may be the clade most closely related to the Cassidinae and Hispinae (unpublished data), results consistent with several earlier taxonomic treatments of subfamily relationships within the Chrysomelidae (Crowson, 1994; Jolivet, 1988; however see Reid, 1995). Based on these considerations we chose Donaciinae as the outgroup.

All sequence data are deposited in the Genbank under the accession numbers AF097109-AF097157.

4. Results and discussion

4.1. Sequence divergence among hispine and cassidine taxa

Sequence divergence between "cassidoid" Hispinae species in the genera *Charidotella* and *Stolas* averaged about 15%; approximately twice that occurring between species of

Chrysomelinae (Table 1). The highest sequence divergence (37%) is found among species of "hispoid" Hispinae in the tribes, Uroplatini and Cephaloleiini, more than twice that found in "cassidoid" species and nearly four times that in Chrysomelinae. The same pattern repeats at generic and tribal levels. The divergence values found within the Hispinae are as high as those found between Chrysomelidae subfamilies, suggesting that the "hispoid" taxa are archaic among Chrysomelidae (Hsiao, 1994b; unpublished data).

4.2. Phylogenetic relationships between Hispinae and Cassidinae

The most striking aspect of the phylogeny we obtained is the complete separation of "cassidoid" and "hispoid" groups of Hispinae. All ten "hispoid" tribes are basal on the tree with respect to "cassidoid" tribes (Fig. 1). Further, this phylogeny nearly as cleanly separates taxa by the class of host plant with which they are associated (Table 2). With the exception of some leaf-mining species in the Uroplatini which exploit an array of Dicotyledonae (Jolivet and Hawkeswood, 1995), the basal Hispinae are close associates of Monocotyledonae while the more derived species are predominantly associated with a relatively small number of families of Dicotyledonae (principally Asteraceae, Bignoniaceae, Boraginaceae, Convolvulaceae and Solanaceae). As a group, leaf-mining Hispinae are exceptional in the breadth of both monocot and especially dicot families they exploit (see below). Further, while all larvae of exophagous "hispoid" hispines appear to feed solitarily in Panama (with the exception of a small number of communal leaf-mining species), the "cassidoid" Hispinae include numerous species with gregarious larvae which move en masse among feeding and resting sites (Windsor et al., 1992). The roots of this behaviour may have appeared in ancestral members of the tribe Cassidini. The majority of extant species in this tribe have solitary larvae, however, small numbers of species with gregarious larvae do occur within *Coptocycla*, *Charidotis* and *Hybosa*. Further, our phylogeny places the genus *Chersinellina* at the base of the Stolaini. The many egg, larval and pupal characters this genus shares with members of the Stolaini, rather than the Cassidini, probably justifies its eventual reclassification.

Larval gregariousness seems to have been lost at least once among Stolaini. Larval *Echoma anaglypta* Boheman emerge from clumped egg masses placed in the branching inflorescence heads of *Mikania guaco* H. & B. (Asteraceae) in Panama. The larvae disperse individually over short distances in their scramble to consume developing flowers and buds (Windsor et al., 1995). Other Stolaini and Eugenysini species in the Asteraceae-feeding clade all have highly synchronized and gregarious larvae. The evolutionary loss of gregariousness may additionally have occurred among taxa associated with Boraginaceae, although these instances are less compelling. The larvae of *Eurypedus nigrosignata* Boheman (Ischyrosonychini), a sister taxon in our phylogeny with the highly gregarious *Physonota alutacea* Boheman, emerge from egg masses but wander apart over time. Other Boraginaceae-feeding taxa, such as *Omocera*, *Discomorpha* (Omocerini) and *Orexita* (Cassidini) have solitary larvae, however there is no evidence that this is a derived condition.

Together, the Cephaloleiini (plus Imatidiini) constitute the least derived clade. Although the Imatidiini were once included among the Cassidinae, Monros and Viana (1947) and Boroweic (1995) recognized that larval similarities dictated they be considered within the hispine tribe Cephaloleiini. Our analysis supports that decision. Placed at the base of the

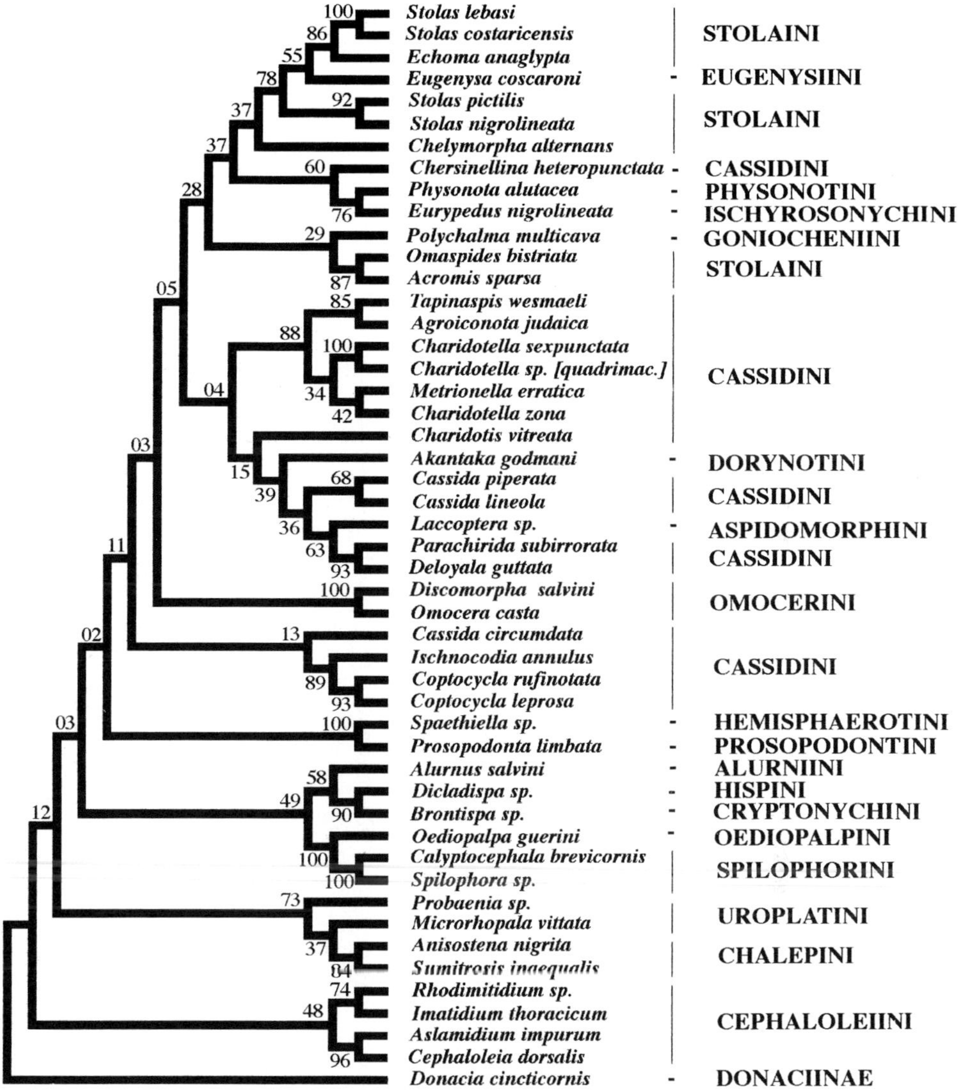

Fig. 1. The neighbour-joining cladogram of Hispinae and Cassidinae generated from 12S mtDNA sequence data (535 aligned bases). The numbers in parentheses are % bootstrap values from 500 replicates.

tree, species in this clade are distinquished by the flattened, "water-penney" larvae which are often found in aquatic and moist refuges provided by herbaceous, monocotyledenous plants in the understorey of tropical and subtropical forests. The proximity of this clade to the Donaciinae, a largely temperate latitude group tied closely to aquatic monocots is suggestive about origins. If anything, these results raise the possibility that the ancestral Hispinae may have diverged from ancestors of Donaciinae to exploit the moist habitats found in leaf axils and rolled leaves of tropical monocots. However, because the boot-strap values we obtained among basal clades are often less than 50%, considerable uncertainty remains

Table 2. Distribution, host plant affiliation, morphology and behaviour of taxa in the order they occur in Fig. 1. Collection data available from the authors.

Species	Tribe	Distri-bution	Host Plant Family	Zone	Terminal Segment	Eggs	Larvae	Par. Care
Stolas lebasi (Boh.)	Stolaini	N	Asteraceae	Leaf	EFA	2	Greg.	-
Stolas costaricensis (Champ.)	Stolaini	N	Asteraceae	Leaf	EFA	2	Greg.	-
Echoma anaglypta (Boh.)	Stolaini	N	Asteraceae	Flower	EA	3	Aggreg.	-
Eugenysa coscaroni Viana	Eugenysini	N	Asteraceae	Leaf	EFA	3	Greg.	+
Stolas pictilis (Boh.)	Stolaini	N	Convolvulaceae	Leaf	EFA	3	Greg.	-
Stolas nigrolineata (Champ.)	Stolaini	N	Convolvulaceae	Leaf	EFA	3	Greg.	-
Chelymorpha alternans Boheman	Stolaini	N	Convolvulaceae	Leaf	EFA	3	Greg.	-
Chersinellina heteropunctata (Boh.)	Cassidini(?)	N	Convolvulaceae	Leaf	EFA	3	Greg.	-
Physonota alutacea Boheman	Physonotini	N	Boraginaceae	Leaf	Fork	3	Greg.	-
Eurypedus nigrosignata Boheman	Ischyrosonychini	N	Boraginaceae	Leaf	EFA	3	Sol.	-
Polychalma multicava (Latr.)	Goniocheniniini	N	Boraginaceae	Leaf	EA	3	Greg.	-
Omaspides bistriata Boheman	Stolaini	N	Convolvulaceae	Leaf	EFA	3	Greg.	+
Acromis sparsa (Boh.)	Stolaini	N	Convolvulaceae	Leaf	EFA	3	Greg.	+
Tapinaspis sp. (?wesmaeli Boh.)	Cassidini	N	Convolvulaceae	Leaf	EFA	1	Sol.	-
Agroiconota judaica (Fab.)	Cassidini	N	Convolvulaceae	Leaf	EFA	1	Sol.	-
Charidotella sexpunctata (Fab.)	Cassidini	N	Convolvulaceae	Leaf	EFA	1	Sol.	-
Charidotella sp. [quadrimaculata Kir.]	Cassidini	N	Convolvulaceae	Leaf	EFA	1	Sol.	-
Metrionella erratica (Boh.)	Cassidini	N	Convolvulaceae	Leaf	EFA	1	Sol.	-
Charidotella zona (Fab.)	Cassidini	N	Convolvulaceae	Leaf	EFA	1	Sol.	-
Charidotis vitreata (Perty)	Cassidini	N	Boraginaceae	Leaf	EFA	1	Sol.	-
Akantaka godmani (Baly)	Dorynotini	N	Bignoniaceae	Leaf	EFA	1	Sol.	-
Cassida piperata Hope	Cassidini	P	Convolvulaceae?	Leaf	EFA	?	Sol.	-
Cassida lineola Creutzer	Cassidini	P	Convolvulaceae?	Leaf	EFA	?	Sol.	-
Laccoptera sp. [?quadrimac.(Thunbg)]	Cassidini	P	Convolvulaceae?	Leaf	EFA	?	?	-
Parachirida subirrorata (Boh.)	Cassidini	N	Convolvulaceae	Leaf	EFA	1	Sol.	-
Deloyala guttata (Oliv.)	Cassidini	N	Convolvulaceae	Leaf	EFA	1	Sol.	-

Discomorpha salvini (Baly)	Omocerini	N	Boraginaceae	Leaf	EFA	3	Sol.	-
Omocera casta (Boh.)	Omocerini	N	Boraginaceae	Leaf	Fork	1	Sol.	-
Cassida (Taiwania) circumdata Herbst	Cassidini	P	Amaranthaceae	Leaf	EFA?	1	Sol.	-
Ischnocodia annulus (Fab.)	Cassidini	N	Boraginaceae	Leaf	EFA	1	Sol.	-
Coptocycla rufonotata (Champ.)	Cassidini	N	Boraginaceae	Leaf	EFA	1	Sol.	-
Coptocycla leprosa (Boh.)	Cassidini	N	Boraginaceae	Leaf	EFA	1	Sol.	-
Spaethiella purpureocincta Spaeth	Hemisphaerotini	N	Arecaceae	Leaf	EFA	1	Sol.	-
Prosopodonta limbata Baly	Prosopodontini	N	Arecaceae	Fold	Spatula	1	Sol.	-
Alurnus salvini Baly	Alurnini	N	Arecaceae	Fold	Spatula	1	Sol.	-
Dicladispa sp. [?armigera Olivier)]	Hispini	P	Poaceae	Mine	none	1	Sol.	-
Brontispa sp. [?longissima (Gestro)]	Cryptonychini	P	Arecaceae	Fold	Spatula	1	Sol.	-
Oediopalpa guerini Baly	Oediopalpini	N	Poaceae	Leaf	EA	1	Sol.	-
Calyptocephala brevicornis Boheman	Spilophorini	N	Arecaceae	Leaf	EA	1	Sol.	-
Spilophora sp.	Spilophorini	N	Arecaceae	Leaf	EA	1	Sol.	-
Probaenia sp.	Uroplatini	N	Asteraceae	Mine	none	1	Sol.	-
Microrhopala vittata (Fab.)	Uroplatini	N	Asteraceae	Mine	none	1	Sol.	-
Anisostena nigrita Olivier	Chalepini	N	Leguminosae	Mine	none	1	Sol.	-
Sumitrosis inaequalis (Weber)	Chalepini	N	Leguminosae	Mine	none	1	Sol.	-
Rhodimitidium sp.	Cephaloleiini	N	Arecaceae	Roll	none	1	Sol.	-
Imatidium thoracicum (Fab.)	Cephaloleiini	N	Heliconiaceae	Roll	none	1	Sol.	-
Aslamidium impurum (Boh.)	Cephaloleiini	N	Marantaceae	Roll	none	1	Sol.	-
Cephaloleia dorsalis Baly	Cephaloleiini	N	Marantaceae	Roll	none	1	Sol.	-
Donacia cincticornis F. [OUTGROUP]	Donaciinae	N	[aquatic monocots]	Stem	none	1	Sol.	-

(N=Neotropic/Nearctic, P=Paleotropic); Larval feeding zone (Leaf=strips leaf surface; Mine=mines leaf interior; Roll=strips interior surface of rolled-leaves; Fold=tunnels through unexpanded, apical leaf); Terminal abdominal segment (EFS=exuvio-fecal shield, ES=exuvial shield; Fork=no shield, eighth tergum forked; Spatula=terminal segments thickened and flattened; none=terminal segments unsclerotized); Oviposition (1=eggs placed solitarily; 2=eggs clumped but not touching; 3=tightly packed); Larvae (Sol.=larvae feed separately; Aggr.=larvae aggregated but do not move synchronously among feeding sites; Greg.=larvae tightly packed and move synchronously among sites); Parental Care ("-" = no maternal care of offspring, "+" = maternal guarding of all immature stages).

concerning the placement of the basal clades relative to the Donaciinae (Fig. 2).

Higher bootstrap values than those reported in this study were obtained by Farrell (1998) for 6 species of Hispinae (4 tribes) in a larger analysis of chrysomeloid subfamily relationships. This analysis isolated Donaciinae and Criocerinae as a sister group to the seed-feeding Bruchinae, the latter complex constituting the out-group to the Hispinae. Considering just the four Hispinae tribes common to both studies, the Alurniini consistently group with the "cassidoid" Hispinae as the more derived clade. While the present study slightly favors the Cephaloleiini as the most basal of the four tribes, the leaf-mining Uroplatini are strongly supported in this position by Farrell's analysis. Both studies suggest an evolutionary progression from endophagous to exophagous larval feeding habits. Considering the aquatic habitats of larval Donaciinae, the endophytic habits of Bruchinae and some monocot-feeding Criocerinae (see Vencl and Morton, this volume; Vencl and Aiello, 1997), one of several evolutionary scenarios has larvae changing from the completely concealed habits of leaf-miners to those within slightly more exposed feeding sites within the axils and leaf scrolls of monocots to exposed leaf-feeding on dicots. If this scenario survives further scrutiny, it will be remarkable in that it appears to have resulted in increased trophic specialization. For example, among leaf-mining Hispinae for which accurate host-plant information has been gathered in Panama, 36 species occur on 18 different families of plants, a ratio of 2 leaf mining species per host plant family (Table 3). Among axil-feeding Hispinae, 4.8 species occur per host plant family. A ratio of 10 species of open, leaf surface feeding "cassidoid" Hispinae, occur on average per host plant family. Remarkably, the entire "cassidoid" assemblage in Panama, roughly 140 species, appears to be supported by only seven host plant families. Thus, the high amount of species diversification which characterizes the apical portion of the Hispinae tree occurs on a small subset of available plant families. We need now to understand why the "cassidoid" hispines have not radiated more broadly among the large number of dicot plant families.

The Hemisphaerotini, typically placed among the Cassidinae, have been considered an intermediate tribe. On our phylogeny, this tribe is distinctly transitional, falling just below the logical boundary between subfamilies. Hemisphaerotini cluster with another enigmatic group, the Prosopodontini. Some species within each of these tribes are associated with Arecaceae (palms), but also include Heliconiaceae, Marantaceae and Cyclanthaceae as host plants. Larval *Prosopodonta* species have a flattened urogomphus while larval *Spaethiella* have a well developed caudal fork upon which they accumulate exuviae and fecal strands – adaptations found widely throughout the cassidoid Hispinae. Our tree suggests that the larval synapomorphies which so well defined the Cassidinae may first have

Table 3. The numbers of Panamanian Hispinae with accurately determined hostplant determinations broadly classified into three larval feeding zones and how they are distributed among plant families.

Trophic Group:	Number of Beetle Species	Number of Plant Families	Species Per Family
Leaf-mining Hispinae:	36	18	2.0 : 1
Rolled and folded leaf-feeding Hispinae:	38	8	4.8 : 1
Open, Surface-feeding Hispinae:	70	7	10.0 : 1

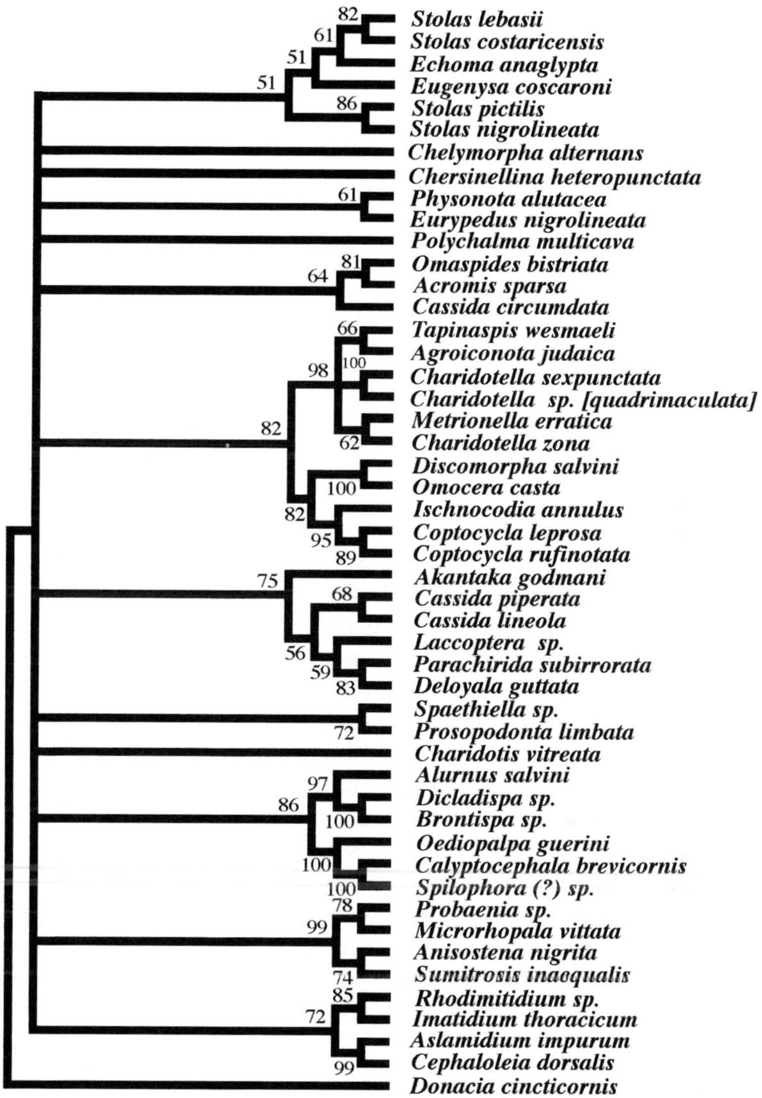

Fig. 2. The PAUP cladogram of Hispinae and Cassidinae generated from 12S mtDNA sequence data (535 aligned bases). Numbers in the graph are % bootstrap values from 100 replicates.

evolved among ancestors associated with monocot host plants. Further, our Fig. 1 contrasts with the trees obtained by Boroweic (1995) in that the Hemisphaerotini appear to be a bridge from the Hispinae to the Cassidinae, rather than fitting subapically on a separate lineage terminating with Cephaloleiini. Comparisons of the two trees is complicated by the many taxa which do not co-occur. Nevertheless, the topologies of the morphology and molecular based trees are quite different.

The exuvial shield of larval Hispinae appears in its simplest form in the tribes Spilophorini and Oediopalpini, two Neotropical "hispoid" tribes which according to larval and pupal morphology and their placement within our phylogeny form a single monophyletic group. However, until there is greater support for the placement of neighbouring basal hispine clades, the homology of the larval *Oediopalpa* and *Spilophora* shields with those of the cassidoid Hispinae remains tentative.

4.3. Hispine host-plant associations

Interestingly, our analysis places the Paleotropical tribe, Hispini, principally containing leaf-mining species on Poaceae, in a separate clade from Neotropical leaf-mining species in the Uroplatini and Chalepini. *Dicladispa* (Hispini) clusters with the Paleotropical, palm-feeding hispine, *Brontispa*. Thus, there is some suggestion in Fig. 1 that the leaf-mining habit may have more than one origin within the Hispinae, one occurring in Neotropical species, the other in Paleotropical species. However, better resolution among basal clades will be needed before this is firmly established. Several morphological features (cylindrical shape, interantennal protuberance in adults and lateral protuberances and a heavily chitinized spatula on the terminal segment of larvae) co-occur in *Brontispa* and other Paleotropical genera and in a single Neotropical genus, *Prosopodonta* (Maulik 1931, 1932, 1933). Rather to our surprize, our phylogeny gives no support to the close ancestry of these New and Old World taxa, suggesting instead that they share prominent larval and adult morphological features by convergence. Alternatively, the morphology of these taxa may have changed little since their mid-Cretaceous separation, while the similarity of their mitochondrial genomes has drifted. The rolled-leaf hispines of the New World are all clustered together at the base of the tree. Rolled-leaf Hispinae and Donacinae have aquatic or semiaquatic larvae. However, caution is necessary in assessing these relationships because bootstrap values in this portion of the tree are very low.

Higher bootstrap values in the upper portion of the tree permit greater confidence in the arrangement of the Cassidinae. Separate origins of maternal care are suggested by the occurrence of one subsocial species (*Eugenysa coscaroni* Viana) within the clade attached to Asteraceae; two subsocial taxa (*Acromis sparsa* Boheman and *Omaspides bistriata* Boheman) in a separate clade attached to Convolvulaceae. Additionally, the tree suggests the earliest cassidines moved from palms and other monocots onto Boraginaceae, subsequently colonizing Convolvulaceae, Bignoniaceae, Solanaceae, and lastly, Asteraceae.

Obtaining a well supported phylogeny for the Hispinae is important if we wish to trace the importance of key innovations in the diversification of this group. Throughout this paper we have cautioned against taking our phylogeny as anything but a preliminary treatment. Further work must seek DNA sequences more conserved over time than those we were able to extract from 12S mtDNA. The excellent resolution Farrell (1998) recently obtained suggests the likely utility of 18S mtDNA in this endeavour. We agree with Reid (1995) when he said that DNA will not replace morphology in phylogenetic reconstruction, but that the two should be considered complementary and used in combination. The combination of morphological and DNA characters could improve the support among basal hispine clades, if problems of weighting such different characters can be solved (Schmitt, 1996).

5. Summary

1. 12S rDNA sequence data were obtained for 48 species of Hispinae (18 tribes) and one outgroup species of Donaciinae.

2. All seven hispoid tribes formed clades basal to the nine cassidoid tribes. However, low resolution among clades in the basal portion of the tree obscure which ancestral monocot-feeding group may have given rise to exophytic, dicot-feeding taxa.

3. The phylogeny provides strong support for the formation of a single tribe encompassing both the Spilophorini and Oediopalpini.

4. The phylogeny supports morphological evidence suggesting the ties of the genus, *Chersinellina* to the Stolaini, rather than the Cassidini.

5. Larval gregariousness which rarely occurs within the tribe Cassidini, occurs throughout the tribe Stolaini, with the exception of at least one species of *Echoma*, where the behaviour apparently has been secondarily lost in connection with highly derived flower feeding habits.

6. Parental care occurs in isolated taxa within the more highly supported "cassidoid" part ot the tree suggesting the behaviour has dual origins, once (*Acromis* and *Omaspides*) in a clade of Convolvulaceae feeding Stolaini and separately and possibly more recently (*Eugenysa*) within the Asteraceae feeding Stolaini.

7. Basal portions of the present phylogeny could be improved by the addition of more highly conserved DNA sequence data, morphological data and the inclusion of additional taxa, especially from the Paleotropics.

Acknowledgements

TH expresses gratitude to Catherine Hsiao for technical assistance, Dr. Jacques Pasteels for providing chrysomelid species for sequencing and Wilford Hanson for assistance in the identification of field collected leaf beetles. We are grateful to Dr. Michael Cox for comments on the manuscript. This research was supported by the Utah Agricultural Experiment Station, Utah State University and the Smithsonian Tropical Research Institute, Panama.

References

Boroweic, L., 1995 Tribal classification of the cassidoid Hispinae (Coleoptera: Chrysomelidae), IN: Pakaluk, J. and Slipinski, S.A. (eds), Biology, Phylogeny, and Classification of Coleoptera. Muzeum i Instytut Zoologii PAN, Warszawa, pp 541-558.

Crowson, R.A., 1955 The natural classification of the families of Coleoptera. Nathaniel Lloyd, London, 1-187 pp.

Crowson, R.A., 1994 Preface. In: Jolivet, P., Cox, M. L. and Petitpierre, E. (eds), Novel Aspects of the Biology of

Chrysomelidae, Kluwer Academic Publishers, Dordrecht, Netherlands, pp 237-248.

Farrell, B.D., 1998 "Inordinate Fondness" explained: why are there so many beetles? *Science* 281: 555-559.

Higgins, D.G., Bleasby, B.J. and Fuchs, R., 1992 CLUSTALV: Improved software for multiple sequence alignment. *CABIOS* 8:189-191.

Hsiao, T.H., 1994a Molecular techniques for studying systematics and phylogeny of Chrysomelidae. In: Jolivet, P., Cox, M. L. and Petitpierre, E. (eds), Novel Aspects of the Biology of Chrysomelidae, Kluwer Academic Publishers, Dordrecht, Netherlands, pp. 237-248.

Hsiao, T.H., 1994b Molecular phylogeny of chrysomelid beetles inferred from mitochondrial DNA sequence data (Coleoptera: Chrysomelidae). In: Furth, D. G. (ed.), Proceedings of the Third International Symposium on the Chrysomelidae, Backhuys Publishers, Leiden, Netherlands, pp 9-17.

Jolivet, P., 1988 Food habits and food selection of Chrysomelidae: Bionomic and evolutionary perspectives. In: Jolivet, P.H., Petitpierre, E. and Hsiao, T. H. (eds), Biology of Chrysomelidae, Kluwer Academic Publishers, Dordrecht, pp 1-24.

Jolivet, P. and Hawkeswood, T., 1995 Host-plants of Chrysomelidae of the World. Backhuys Publishers, Leiden, 281 pp.

Kimoto, S. and Takizawa, H., 1994 Leaf Beetles (Chrysomelidae) of Japan. Tokai Press, Tokyo, 539 pp.

Kumar, S., Tamura, K. and Nei, M., 1993 MEGA: Molecular Evolutionary Genetic Analysis, Version 1.0. The Pennsylvania University, University Park, PA.

Jolivet, P., 1959 Recherches sur L'aille des Chrysomeloidea (Coleoptera). *Mém. Inst. r. Sci. nat. Belg.*, 2(58):1-152.

Lawrence, J.F. and Britton, E.B. 1991 Coleoptera (Beetles). In: CSIRO (ed.), The Insects of Australia. Melbourne University Press, Melbourne, pp. 543-683.

Lawrence, J.F. and Newton Jr., A.F. 1995 Families and subfamilies of Coleoptera (with selected genera, notes, references and data on family-group names), in Pakaluk, J. and Slipinski, S.A. (eds): Biology, Phylogeny, and Classification of Coleoptera. Muzeum i Instytut Zoologii PAN, Warszawa, pp 779-1006.

Maulik, S., 1931 On the structure of larvae of hispine beetles. *Proc. zool. Soc. Lond.* 1931:1137-1162.

Maulik, S., 1932 On the structure of larvae of hispine beetles.-II. *Proc. zool. Soc. Lond.* 1932:293-322.

Maulik, S., 1933 On the structure of larvae of hispine beetles.-III. *Proc. zool. Soc. Lond.* 1933:669-680.

Monros, F. and Viana, M.J. 1947 Revision sistematica de los Hispidae Argentinos (Insecta, Coleop. Chrysomeloid.). *An. Mus. argentino Cienc. nat.* 42:125-324 + 27 pl.

Reid, C.A.M., 1995 A cladistic analysis of subfamilial relationships in the Chrysomelidae senso lato (Chrysomeloidea). IN: Pakaluk, J. and Slipinski, S.A. (eds), Biology, Phylogeny, and Classification of Coleoptera, Muzeum i Instytut Zoologii PAN, Warszawa, pp 559-631.

Schmitt, M., 1996 The phylogenetic system of the Chrysomelidae-history of ideas and present state of knowledge. IN: Jolivet, P.H. and Cox, M.L. (eds), Chrysomelidae Biology, vol.1: The Classification, Phylogeny and Genetics, APB Academic Publishing, Amsterdam, The Netherlands, pp 57-96.

Seeno, T.N., and Wilcox, J.A. 1982 Leaf beetle genera (Coleoptera: Chrysomelidae). *Entomography* 1: 1-221.

Stehr, F.W., 1991 Immature Insects, Vol. 2. Kendall/Hunt Publishing Co. Dubuque, Iowa, 975 pp.

Swofford, D.L., 1993 PAUP: Phylogenetic Analysis Using Parsimony, Version 3.1.1. Illinois Natural History Survey, Champaign.

Vencl, F.V. and Aiello, A. 1997 A new species of leaf-mining *Oulema* from Panama (Coleoptera; Chrysomelidae; Criocerinae). *Jl. N. Y. ent. Soc.* 105:40-44.

Windsor, D.M. and Choe, J.C. 1994 Origins of parental care in chrysomelid beetles. IN: Jolivet, P.H., Cox, M.L. and Petitpierre, E. (eds), Novel Aspects of the Biology of Chrysomelidae, Kluwer Academic Publishers, Dordrecht, Netherlands, pp 111-117.

Windsor, D.M., Riley, E.G. and Stockwell, H.P. 1992 An introduction to the biology and systematics of Panamanian tortoise beetles (Coleoptera: Chrysomelidae: Cassidinae). IN: Quintero, D. and Aiello, A. (eds), Insects of Panama and Mesoamerica., Oxford Univ. Press, pp 372-391.

Windsor, D.M., Valverde de Macedo, M. and De Toledo Siqueira-Campos, A., 1995 Flower feeding by species of *Echoma* Chevrolat (Coleoptera: Chrysomelidae: Cassidinae) on *Mikania* (Asteraceae) in Panama and Brazil. *Coleops Bull.* 49:101-108.

THE FIRST INSTAR LARVA OF *AULACOSCELIS* SP. AND *MEGASCELIS PUELLA* LACORDAIRE (COLEOPTERA: CHRYSOMELIDAE: AULACOSCELINAE, MEGASCELINAE) AND THEIR VALUE IN THE PLACEMENT OF THE AULACOSCELINAE AND MEGASCELINAE

M. L. COX[1] and D. M. WINDSOR[2]
[1]*Museum Associate c / o Department of Entomology, The Natural History Museum, London SW7* 5BD, UK; [2]*Smithsonian Tropical Research Institute, Panama, Unit 0948, APO AA 34002-0948, USA.*

Abstract

Of all of the chrysomelid subfamilies, only the larvae of the Aulacoscelinae and Megascelinae remain undescribed. The first instar larva of *Aulacoscelis* sp. and *Megascelis puella* Lacordaire, both from Panama are described. Their characters are used to relate the Aulacoscelinae and Megascelinae with other chrysomelid subfamilies.

Introduction

Taxonomic History

Aulacoscelinae / *Aulacoscelis*

The name *Aulacoscelis* is probably derived from the greek meaning furrowed leg (Duponchel & Chevrolat, 1842).

The Aulacoscelinae was only established as a separate chrysomelid subfamily by Monrós (1953). It is a small, New World group comprising only 18 described species to date (Suzuki, 1994a). Thirteen of these are presently placed in *Aulacoscelis* Duponchel and Chevrolat, 1842, and five in *Janbechynea* Monrós, 1953. Monrós (1954) divided *Janbechynea* into *Janbechynea sensu stricto* and *Bothroscelis*. According to Arnett (1973) two *Aulacoscelis* species occur in New Mexico, Arizona and California in southwestern USA. However, according to Suzuki (1994a), who reviewed the Aulacoscelinae, only *A. candezei* Chapuis occurs there, but he did not include California in the area of distribution of this genus. In addition, *Janbechynea (Bothroscelis) fulvipes* (Jacoby) occurs in Arizona (Arnett, 1973; Suzuki, 1994a). *A. candezei* and *J. (B.) fulvipes* also occur further south in Central America. In addition, twelve other *Aulacoscelis* spp, plus one other *Janbechynea (Bothroscelis)* sp. occur in Central America. Although Suzuki (1994a: 46) stated that the subgenus *Janbechynea sensu stricto* seems to be limited to South America, he shows in Fig. 1, p. 49 that *J. (J.) elongata* (Jacoby) occurs in Central America. In fact it was described from Playa Vicente

Advances in Chysomelidae Biology 1, pp. 51–70
Edited by M.L. Cox

in Mexico. However, the other two species in *Janbechynea s. s.* only occur in South America as far south as Bolivia, Peru and Brazil.

Apparently, there are undescribed species. J.A. Santiago-Blay is to describe two new species belonging to the subgenus *Bothroscelis* from Mexico and Suzuki (1994a) mentioned that he has also seen undescribed *Aulacoscelis* species in the Pic Collection, Muséum National d'Histoire Naturelle, Paris.

In 1842 Duponchel and Chevrolat briefly described the genus *Aulacoscelis* and the species *melanocera* which they considered close to *Phyllochara* of the Chrysomelinae. In Dejean's catalogue (1837), the genus and species were credited to Chevrolat, but these are nomen nuda and thus the reference is invalid. Stål (1865) described *Aulacoscelis melanocera* without referring to Duponchel and Chevrolat and placed it in the Chrysomelinae. In 1874, Chapuis created the tribe Aulacoscelini (as Aulacoscélites) and transferred *Aulacoscelis* from the Chrysomelinae to the Sagrinae. He also described a second species from Guatemala. Gemminger and Harold (1874) listed the known *Aulacoscelis* species as Sagrinae. Jacoby (1877, 1880, 1888) described and illustrated several *Aulacoscelis* species and listed those species known from Central America. He placed the genus in the Sagridae. In 1903, in the revision of the genera of Sagridae for Wytsman's *Genera Insectorum*, Jacoby excluded *Aulacoscelis* which he referred to the Chrysomelinae. Clavareau (1913) listed the 12 known *Aulacoscelis* species and placed the genus in the tribe Aulacoscelini at the end of the Sagrinae. In the catalogues of Leng (1920) and Blackwelder (1946), *Aulacoscelis* was included in the Aulacoscelini of the Sagrinae. Maulik (1941) briefly stated that *Aulacoscelis* cannot be considered as belonging to the Sagrinae.

In 1946 Crowson revised the genus which he placed in the Aulacoscelini, one of the three Sagrinae tribes. Sagrinae sensu lato was used and included *Orsodacne* and related genera which are now placed in the Orsodacninae, Zeugophorinae and Synetinae; as well as the Hornibiini which are now considered primitive Eumolpinae. Crowson stated that *Aulacoscelis*, though included in the Sagrinae is evidently not nearly related to any of the genera of Sagrini, nor can it easily be included in Orsodacnini, so that he thought that it must constitute a separate tribe. However, he considered *Aulacoscelis* as one of the most anomalous of the entire Chrysomelidae, and from a purely theoretical point of view might even be regarded as meriting a subfamily of its own.

Monrós (1949) excluded the Aulacoscelini from the Sagrinae and suggested a position among the primitive Eumolpinae. Later, in 1953, he established the new subfamily Aulacoscelinae when he described the new genus *Janbechynea* from Bolivia. He made a comprehensive revisional study of the Aulacoscelinae in 1954, but did not elaborate on its phylogenetic or systematic position, merely suggesting that it was among the most ancient of the living Chrysomelidae. Jolivet (1959, 1988), Crowson (1955), Monrós (1955, 1959), Iablokoff-Khnzorian (1966, 1967), Medvedev (1971), Arnett (1973), Jolivet & Petitpierre (1981), Mann & Crowson (1981, 1983a,b, 1984), Seeno & Wilcox (1982), Petitpierre *et al.* (1988), Suzuki (1994a, 1994b, 1996) all recognized the subfamily Aulacoscelinae. However, Suzuki (1974, 1980, 1985, 1988) placed the two tribes Aulacoscelini, plus the Chrysomelini in the subfamily Chrysomelinae. Reid (1995) divided the Chrysomeloidea into 7 families, one of which, the Orsodacnidae comprised the subfamilies Aulacoscelidinae (= Aulacoscelinae) and the Orsodacninae.

Medvedev (1975) described three new *Aulacoscelis* species and a new subspecies of *A. variabilis* Jacoby.

Megascelinae / *Megascelis*

The word Megascelinae is probably derived from the Greek Megas, meaning large, and chele, meaning claw (Nybbakken, 1970). The Megascelinae do have large simple claws which are fused basally, and also an elongate onychium. The claws on the hind legs are the largest.

Fabricius (1801: 477) described *Lema vittata*, which was ascribed to *Megascelis* by Lacordaire (1845: 271), but it was not for a further 25 years that the genus *Megascelis* was established by Sturm (1826: 80). Clavareau (1913) listed approximately 120 species and since that time a further 10 species have been described, mostly by Papp, Monrós, and Pic. *Mariamela*, a monospecific genus, restricted to South America was described by Monrós (1951).

Latreille (1929) included *Megascelis* in the tribe Criocérides of the family Eupodes, whilst De Castelnau (1840) placed *Megascelis* in the tribe Eupodes of the family Chrysomelines. *Megascelis* was included in the tribe Criocérides of the Phytophages by Lacordaire (1845-1848). Chapuis (1874, 1875) included *Megascelis* in the tribe Mégascélides of the section Camptosomes of the Phytophages. *Megascelis* was placed in the family Crioceridae of the Phytophaga by Jacoby (1888-1892), whilst Jacoby and Clavareau (1905) recognised the family Megascelidae. Jacoby (1908) included the Megascelinae in the division Camptosoma of the Chrysomelidae. Clavareau (1913) recognosed the Megascelinae as one of the ten subfamilies comprising the Chrysomelidae. The Megascelinae were included as one of 6 subfamilies comprising the family Donaciidae by Chen (1940). Chûjô (1953-1954) included the Megascelinae as one of 12 subfamilies in the nine divisions forming the 'True Chrysomelidae'. Jolivet (1953-1954, 1959) has variously recognised the family Megascelidae of the superfamily Chrysomeloidea, and also the Megascelinae as one of three subfamilies forming the section Cyclica. The Megascelinae were included in the Chrysomelidae by Monrós (1955), but later in 1959 included with five other subfamilies in the Clytriformes. Medvedev (1971), Arnett (1973) and Mann and Crowson (1983a,b,c,1984) all recognised the Megascelinae. Iablokoff-Khnzorian (1966, 1967) was the first to recognise the Megascelini, as one of the two tribes comprising the Eumolpinae, as did Suzuki (1988); and Reid (1995) who added one new tribe.

Seeno and Wilcox (1982) placed the Megascelinae between the Criocerinae and the Clytrinae, in spite of the publications of Jolivet (1959) and Bechyné and Springlova de Bechyné (1969). Mann and Crowson (1981) placed them between the Megalopodinae and the Lamprosomatinae, which seems more reasonable, but this does not take account of the purely eumolpine characters of these insects (Jolivet, 1997).

Distribution

Aulacoscelinae / *Aulacoscelis*

The Aulacoscelinae are restricted to the Western Hemisphere and are principally Neotropical in distribution. However, two species occur in the southwestern states of the USA. Apparently, *Aulacoscelis hogei* Jacoby, described from Vancouver Island in western Canada is probably a labelling error (Monrós, 1954). The Central American mainland has most

Aulacoscelis species. The subfamily is absent from the Caribbean Islands, as are the Megascelinae, although it occurs at equivalent latitudes on the Central American mainland. Some species have wide distributional ranges (see Monrós, 1954: Fig. 3,page 354), although *A. melanocera*, one of the most widespread, may be a species complex.

Aulacoscelis sp. is widespread in Panama, occurring from Chiriqui Province in the west to Colón Province north of Panama City, east to Panamá Province. Apparently, it does not occur in Costa Rica and Colombia, countries bordering Panama.

This species is described as new in Cox and Windsor (1999 in press).

Megascelinae / *Megascelis*

Only one species, *Megascelis texana* Linell is recorded from North America. However, this is described from Brownsville, Texas, which is in the south east of the State, on the border with Mexico. Approximately 15% of the species are known from Central America from Mexico, south to Panama, whilst some 85% are from South America, with three species common to both regions. Apparently, none occur in the Caribbean. The Megascelinae are principally Neotropical in distribution and are recorded from altitudes of between 5000 and 7000 feet in Mexico. According to Jolivet (1997) *Megascelis* species are thermophilic and hardly exceed >4875 feet in the Panamanian mountains. *Megascelis subtilis* Boheman was described from Honolulu, the Hawaiian Islands, but since specimens of this species are unrepresented in the NHM collections, its status cannot be verified.

Megascelis puella Lacordaire is known from Colombia and Peru (Clavareau, 1913), but this is the first record for Panama. Adults were common at Gamboa, in the Canal Area, 9° 07'N, 79° 40'W, at 35 m altitude on 13 July 1996. Apparently, the adults can appear during May, since they were numerous during the third week of May in 1997 (P. Jolivet, personal communication). In addition, there are specimens in the NHM collected from Isla de Rey, Gulf of Panama on 27 June 1924. The adults may survive in the field for several months, since adults occurred on Taboga Is., Gulf of Panama during September 1924. Apparently, the pupae remain quiescent in the soil until the arrival of heavy rain and adults appear about two weeks after their commencement (P. Jolivet, personal communication).

Aims

Only the larvae of two chrysomelid subfamilies, the Megascelinae and Aulacoscelinae remain undescribed. In addition, the larval feeding habits of these subfamilies are unknown. The aim of this paper is to describe the first instar larva of *Aulacoscelis* sp. and *Megascelis puella* Lacordaire and to use the morphological characters to relate the Aulacoscelinae to the Orsodacninae and the Megascelinae to the Eumolpinae. The first instar larva is the most useful from a phylogenetic standpoint, since it often carries egg bursters which are lost at the first moult and the chaetotaxy is simplified comprising only primary setae which in later instars sometimes becomes obscured by the addition of secondary setae.

The systematic position of the Aulacoscelinae is still much confused which is surprising considering that the adult morphology in this small group is relatively consistent and well studied. Crowson (1946: 85) stated "it would be desirable to obtain larvae of some species of *Aulacoscelis*, as the larval habits and structure should be of great assistance in determining its systematic position".

Materials and methods

Source of Aulacoscelis *first instar larvae*

On 7 May 1993, numerous adults of *Aulacoscelis* sp. were collected from low vegetation at the 1,100 m pass, La Fortuna catchment, 3 km north of Hornitos, Chiriqui, Panama. On 12 May 1997, numerous adults were again collected at this pass as they settled on the vegetation from a migratory swarm. Adults of the same species were observed flying from damaged *Zamia fairchildiana* foliage on 5 May 1991, at Cerro Azul (850 m), 300 km east of the Chiriqui site and 20 km northeast of Panama City. On 25 April 1997, a single adult of the species was observed actively feeding on a *Z. fairchildiana* leaf at the same location. *Aulacoscelis* sp. fed on *Z. fairchildiana* plants throughout the month of May 1997 at this and a second location, Madden Forest (300 m), 20 km north of Panama City. A single oviposition was obtained from one of thirty adults collected from the migratory swarm in Chiriqui, enclosed within a terrarium with soil, cycad and fresh strips of mango for two months. Approximately 40 eggs were clumped without any order, attached to the underside of a piece of cycad stem lying upon the soil. Oviposition occurred on 24 June, eclosion on 6 July, incubation thus requiring about 12 days.

Most of the eggs were fatally attacked by fungi, nematodes and Acari, but 6 larvae successfully ecloded.

Source of Megascelis puella *Lacordaire first instar larvae*

During mid-July 1996 adults of *Megascelis puella* from Gamboa, Panama laid eggs in clusters of about 10 eggs in the leaf folds of *Stizolobium* (= *Mucuna*) *mutisiana* (Leguminosae).

Unfortunately, although the larvae became fully developed, none hatched, possibly because of the growth of fungal hyphae on the eggs. However, eggs containing fully developed larvae, and such larvae dissected from the eggs were mounted in Berlese modified mountant on a microscope slide. Detailed drawings were made of head and body chaetotaxy and mouthpart structure with a Wilde M11 microscope.

Figs 1 & 12 showing head setal arrangement and numbers are adapted from Cox (1976), whilst Figs 9,16 showing body tubercle nomenclature are based on Cox (1976) after Kimoto (1962).

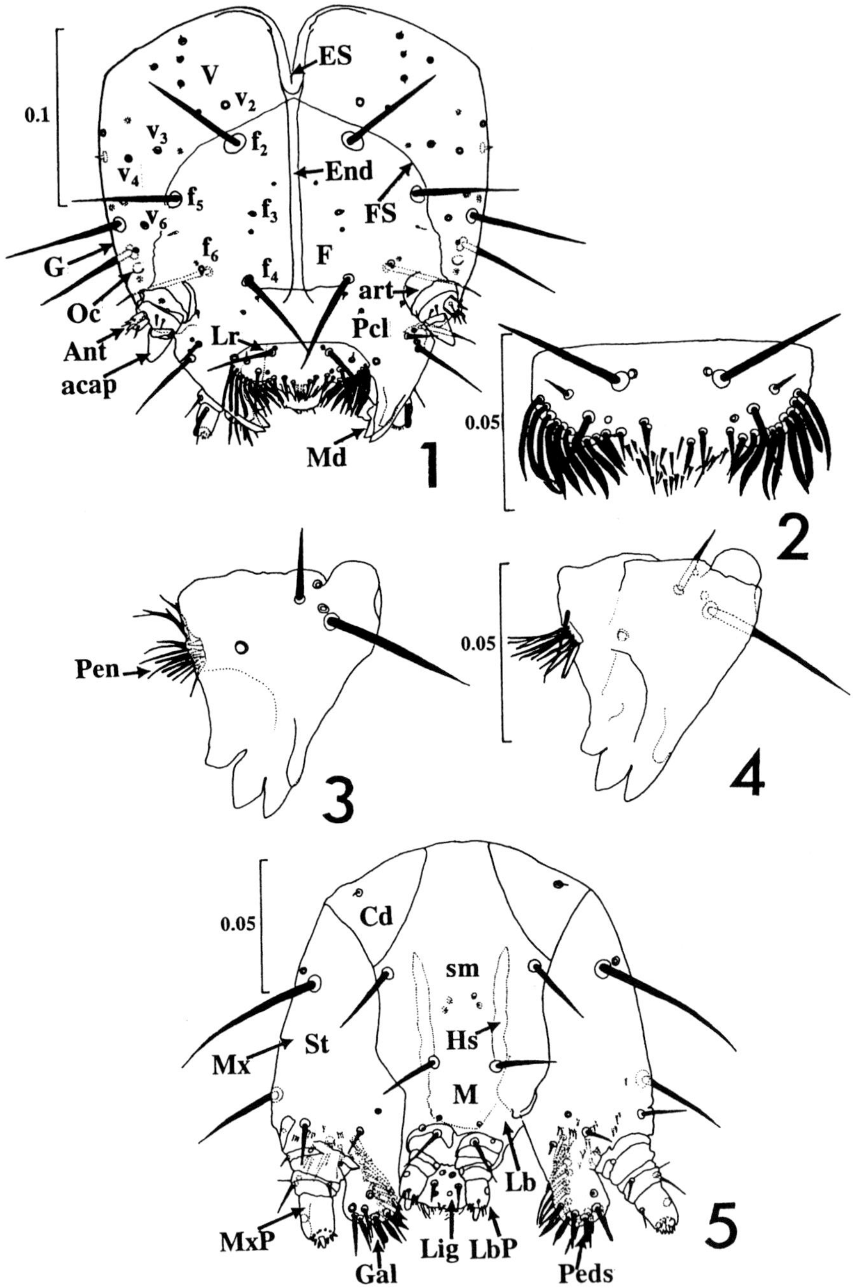

Figs. 1-5 Head parts first instar larva, *Aulacoscelis* sp.: (1) frontal view of head;(2) labrum; (3)left mandible dorsal view; (4) right mandible ventral view; (5) labium and maxillae. Abbreviations: acap, accessory conical antennal process; Ant, antenna; art, basal antennal segment; Cd, cardo; End, endocarina; ES, epicanial suture; F, frons; f_2, f_3, f_4, f_5, f_6, frontal setae 2-6; Gal, galea; Hs, hypopharyngeal sclerite; Lb, labium; LbP, labial palpus; Lig, ligula; Lr, labrum; M, mentum; Md, mandible; Mx, maxilla; MxP, maxillary palpus; Oc, ocellus; Pcl, postclypeus; Peds, pedunculate seta; Pen, penicillus; sm, submentum; St, stipes; V, vertex; v_2, v_3, v_4, v_6, vertex setae 2,3,4 and 6 (represented by sensilla only). Scale line in mm.

Description of first instar larva

***Aulacoscelis* sp.** (Figs 1-11)

Head width : 0.21-0.23 mm, $\bar{x}$ = 0.22 mm (n=5);
Body length: 1.38-1.86 mm, $\bar{x}$ = 1.70 mm (n=5);
Body width: 0.20-0.31 mm, $\bar{x}$ = 0.26 mm (n=5).

General habitus (Figs 6,7). White; head, anal plate pale yellow. Elongate, narrow, approximately same width along entire length; greatest width at about abdominal segments 2-3; length approximately six times greater than width. Tubercles pale yellow, small, indistinct, somewhat darker, more distinct ventrally. Intertubercular plates indistinct, multidentate posteriorly (Fig. 10). Setae well developed, longest on head, anal plate apically, ventrally on abdominal segment 9; ventral body setae generally longer than dorsal setae. Thoracic legs moderately long, prothoracic legs slightly shorter than others.

Head (Figs 1-5) pale yellow, pro- to hypognathous, exserted, slightly elongate-oval; epicranial suture with stem short; frontal sutures narrow, reaching antennae; endocarina (median suture) entire, reaching to indistinct, incomplete, fronto-clypeal suture; frons bearing 3 large setae (f_2, f_4, f_5), 4 campaniform sensilla, f_3, f_6 represented by one sensilla each; vertex bearing no large setae, v_2, v_3, v_4, v_6, represented by sensilla, 4 other sensilla forming triangle dorsal to v_2, gena bearing two large, one very small seta; ocelli, very small, indistinct, one posterior to antennae; clypeal area bearing 4 small setae, plus 2 sensilla; clypeo-labral suture indistinct; labrum (Fig. 2) with anterior margin evenly curved, not incised medially, bearing 2 long, 2 short, plus 2 sensilla dorsally, 6 setae antero-dorsally, 10 or 11 mostly flattened, curved setae along anterior margin on each side, antero-medially bearing numerous elongate spicules; mandibles (Figs 3,4) pale, apical part brown, sub-triangular, apically 4-dentate, dorsally bearing 1 long, 1 shorter seta, plus 3 campaniform sensilla, penicillus well developed, consisting of 12 or 13 densely-packed setae of varying lengths inserted on ventral surface about mid-way along inner margin; antennae short, apparently 2-segmented, anteriorly-directed, attached to head by broad basal articulating membrane, first segment at least two times broader than long, bearing campaniform sensilla laterally, dorsally bearing 4 short setae, segment two, plus an accessory conical process, about two times longer than segment 2, segment 2 pill-box-shaped, apically bearing 4 short setae, a peg-like seta, plus cigar-shaped process, about half length of segment 2; labium (Fig. 5) with submentum bearing 2 long setae, mentum bearing 2 long setae, plus 2 campaniform sensilla anteriorly, labial palpi 2-segmented, apical segment slightly longer, narrower than basal segment, palpiger bearing basally 1 short seta, plus 1 sensilla, segment 1 bearing sensilla intero-laterally, segment 2 with extero-lateral sensilla, apically bearing about 5 very short setae, plus cigar-shaped process, ligula bearing 2 setae, 4 campaniform sensilla, plus numerous, short, curved setae anteriorly; maxillae (Fig. 5) moderately long, cardo bearing very short seta, stipes bearing 1 very long seta basally, with sensilla lateral to it, also shorter seta apically, palpi 3-segmented, basal two segments subequal in length, apical segment two times longer than two basal segments, with about 11 short setae forming ring apically, also cigar-shaped process, palpifer bearing 1 long seta laterally, lacinia not differentiated, apparently fused with galea, galea bearing dorsally 1 sensilla, 3 setae plus pedunculate seta, ventro-laterally bearing fringe of about 13 variable setae, extending ventrally onto apical margin; stipes and basal part of galea ventrally bearing scattered elongate spi-

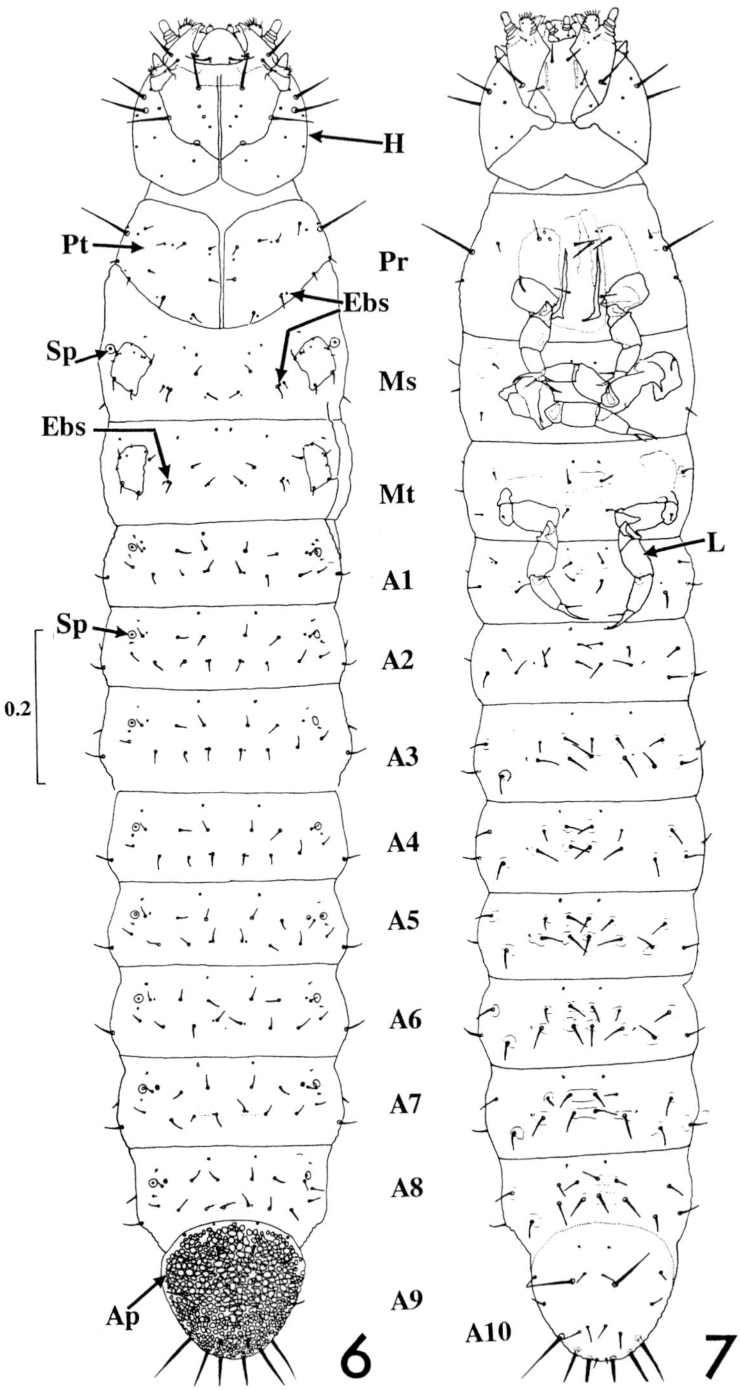

Figs. 6-7 Body general habitus first instar larva, *Aulacoscelis* sp.: (6) dorsal view; (7) ventral view. Abbreviations: A1-10, abdominal segments 1-10; Ap, anal plate; Ebs, egg bursting spine; H, head; L, leg; Ms, mesothorax; Mt, metathorax; Pr, prothorax; Pt, pronotum; Sp, spiracle. Scale line in mm.

cules, in places grouped 2, 3, 4 or 5 together in line; hypopharynx densely setose, with subparallel rods or bracon extending backwards from long mentum setae.

Body (Figs 8-11). Whitish, elongate, subparallel-sided, tubercles indistinct, except for pronotum, tubercles DLae-DLpe on meso- and metathorax, anal plate; small, conical egg bursting spine on pronotum, meso- and metathorax, anterior to seta IV.

Thorax (Figs 9-11). Prothorax with pronotal plate large, bearing 9 or 10 short setae, plus 1 very long antero-lateral seta, setae I-VIII present, setal groups I, III and IV probably bisetate, a conical hatching spine anterior to seta IV, 5 or 6 small campaniform sensilla; tubercles EPa, EPp bearing setae IX, X, respectively; T present, bearing seta XI, plus 2 minute setae anteriorly; P indistinct, seta XII present; As, Ps absent; Es and Ss fused (not fused across mid-ventral line) bearing 2 large setae, XV and XVI. Meso- and metathorax with tubercles Dai, Dpi, Dae, Dpe probably fused, bearing 4 setae, I, II, III, and IV, a small conical egg burster anterior to seta IV; tubercles DLai, DLpi, DLae, DLpe probably fused, bearing 4 setae, V, VI, VII, and VIII, plus 1 or 2 minute setae, and 1 or 2 minute sensilla; annular spiracles present on mesothorax, lateral to DLai-DLpi-DLae-DLpe; EPa, EPp unisetate, bearing seta IX and X, respectively; T bearing 2 minute setae, seta XI missing; tubercles As, Ps absent; Es, Ss unisetate on each side.

Legs (Fig. 11). 3 pairs well developed, moderately long thoracic legs all essentially similar as regards chaetotaxy, prothoracic legs very slightly shorter than others; coxa bearing 4 or 5 long setae, plus about 10 short setae, and several sensilla; trochanters bearing one long seta, 2 or 3 short setae, 4 campaniform sensilla; femora with 3 long setae, 3 short setae; tibiotarsi subequal in length to femora, usually bearing 3 long, 2 slightly shorter setae; tarsunguli elongate, very slightly curved, slightly shorter than tibiotarsi, bearing long seta on inner margin.

Abdomen (Figs 6,7,9). Abdominal segments 1-8 with setae I-XVI present, sometimes borne on indistinct tubercles, setae V, VII, VIII on tubercle DLai-DLae-DLpe, EPa-EPp fused, indistinct tubercle bearing 2 setae, IX, X, tubercles T, P, As, Ps, Es, Ss indistinct, apparently unisetate, except for Ps which is bisetate, bearing seta XIV, also apparently a much reduced seta XI; annular spiracles on tubercles DLai-DLae-DLpe. Abdominal segment 9 dorsally bearing large anal shield, fused across mid-dorsal line, bearing numerous small, variably-shaped pores, probably bearing setae IV-XIV, setae IV, VIII, IX shortest, setae XI, XII, XIII, and XIV long, apically situated; ventrally bearing 2 setae, XV short, XVI long. Abdominal segment 10, ventrally situated, bearing 2 moderately long setae, plus 6 short setae and an anal proleg.

Megascelis puella Lacordaire (Figs 12-17)

Head width 0.27-0.32 mm (n=2).
Body length 0.77-0.81 mm (n=2).

Head (Figs 12-14). Exserted, slightly transverse; epicranial suture moderately long, frontal sutures reaching to antennae, endocarina (Median suture) absent; frons bearing five (f_1, f_2, f_3, f_4, f_5) moderately long to long pointed setae, plus one sensillum ventral to seta f_2, on

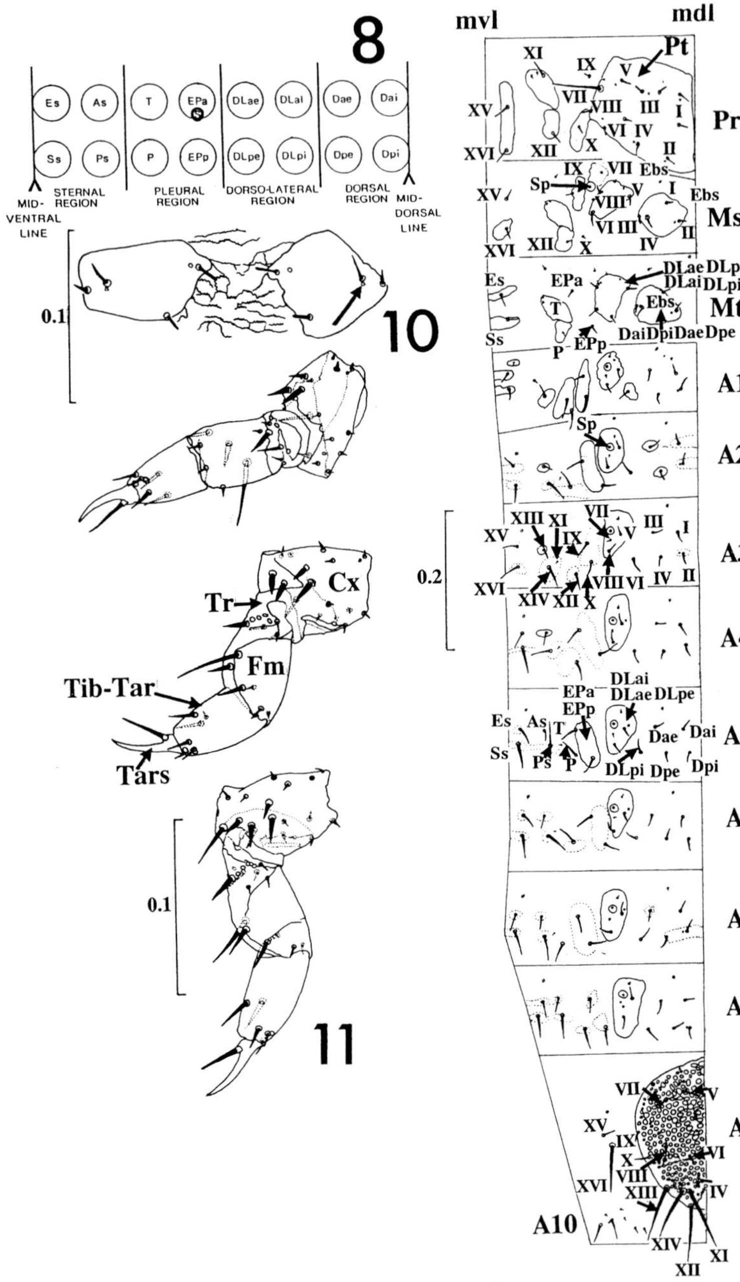

Figs. 8-11 Body and legs first instar larva, *Aulacoscelis* sp.: (8) tubercle nomenclature for typical body segment of generalised chrysomelid (After Cox 1976, modified after Kimoto 1962); (9) left half of the body showing tubercle arrangement and arrangement of setae I-XVI; (10) detail of cuticle between tubercles Dai-Dpi-Dae-Dpe on mesothorax; (11) left, pro-, meso-, metathoracic legs, respectively, outer view. Abbreviations: A1-10, abdominal segments 1-10; Cx, coxa; Ebs, egg bursting spine; Fm, femur; mdl, middorsal line; Ms, mesothorax; Mt, metathorax; mvl, midventral line; Pt, pronotum; Tars, tarsungulus; Tib-Tar, tibiotarsus; Tr, trochanter. Scale line in mm.

each side; vertex bearing six (v_1, v_2, v_3, v_4, v_5, v_6) long slender pointed setae, three or four minute setae, plus sensilla dorsal to v_1, and v_5 on each side; gena bearing six moderately long to long setae on each side; apparently fronto-clypeal, clypeo-labral sutures absent; clypeal area bearing two long setae and one sensillum on each side; labrum with anterior margin evenly curved, not incised medially, bearing three setae on each side; epipharynx bearing two setae on each side; mandibles (Fig. 13) elongate, sub-triangular, apically four-dentate, inner tooth smallest, dorsally bearing one long and one short seta, plus three sensilla, penicillus absent; antennae very short, apparently 2-segmented, somewhat flattened, basal segment bearing two large circular sensilla and one short seta; labium (Fig. 14) with submentum (postmentum) bearing two long, curved setae, plus shorter seta and one sensillum on each side; prementum bearing 1 short seta, plus one sensillum on each side posterior to labial palpi, short 1-segmented palpi apically with one long, plus three short setae, ligula bearing two long, four short setae; maxillae (Fig. 14) short, cardo bearing long curved seta, stipes bearing two long curved setae, plus two sensilla, palpi 2-segmented, apical segment 2 x longer than basal segment with well developed setae apically, palpifer bearing one long seta laterally, lacinia weakly differentiated, bearing two long flattened setae, galea bearing nine shorter setae along inner and apical margin; ocelli absent; hypopharyngeal bracon absent.

Body (Figs 15-17). Whitish, elongate, subparallel-sided, widest at about metathorax; tubercles indistinct, except for pronotum and tubercles DLai-DLpi bearing egg bursting spines dorsal to tubercles DLae-DLpe on meso-, metathorax, dorsal to spiracles on abdominal segments 1-6 or seven; capitate setae dorsally, normal setae ventrally; intertubercular plates present, oblong transverse dorsally, more rounded laterally and ventrally.

Thorax. Prothorax with pronotal plate large, bearing six long setae, two short setae laterally, three sensilla near anterior margin; tubercle EPp absent, tubercle T bisetate, tubercle P, unisetate; tubercle Es-Ss fused, bisetose, also fused across midventral line. Meso- and metathorax bearing nine setae between middorsal line and spiracles; apparently tubercles Dai, Dpi, Dae-Dpe, DLai-DLpi, DLae-DLpe represented; spiracle on mesothorax apparently not on tubercle EPa; tubercle EPp apparently absent; T bisetate; P unisetate; Es-Ss fused, bisetate, also fused across mid-ventral line.

Legs (Fig. 17). Well developed, relatively long, 4-segmented, plus tarsungulus, as long as thoracic segments combined; similar pair on pro-, meso- and metathorax, coxae basally with group of four small setae, at about mid-point laterally with two small setae, apically with two long, pointed setae; trochanters with one long pointed seta and c. five or six campaniform sensilla; femora with about eight pointed setae, two smaller than others; tibiotarsi slightly shorter than femora, usually bearing six or seven pointed setae; tarsunguli elongate, very slightly curved, as long as tibiotarsi, bearing long seta on inner margin.

Abdomen. Abdominal segments 1-7, bearing six setae between middorsal line and spiracle, tubercles Dai, Dpi, Dae, DLai, DLpi present, unisetate, bearing eight or nine setae between spiracle and midventral line with tubercles T, P, As, Ps, Es, Ss present. Es bisetate, apparently fused across mid-ventral line; abdominal segment 8 bearing only three setae between middorsal line and spiracle, bearing six setae between spiracle and midventral line; A9 + A10, bearing a total of four large, plus eight smaller setae.

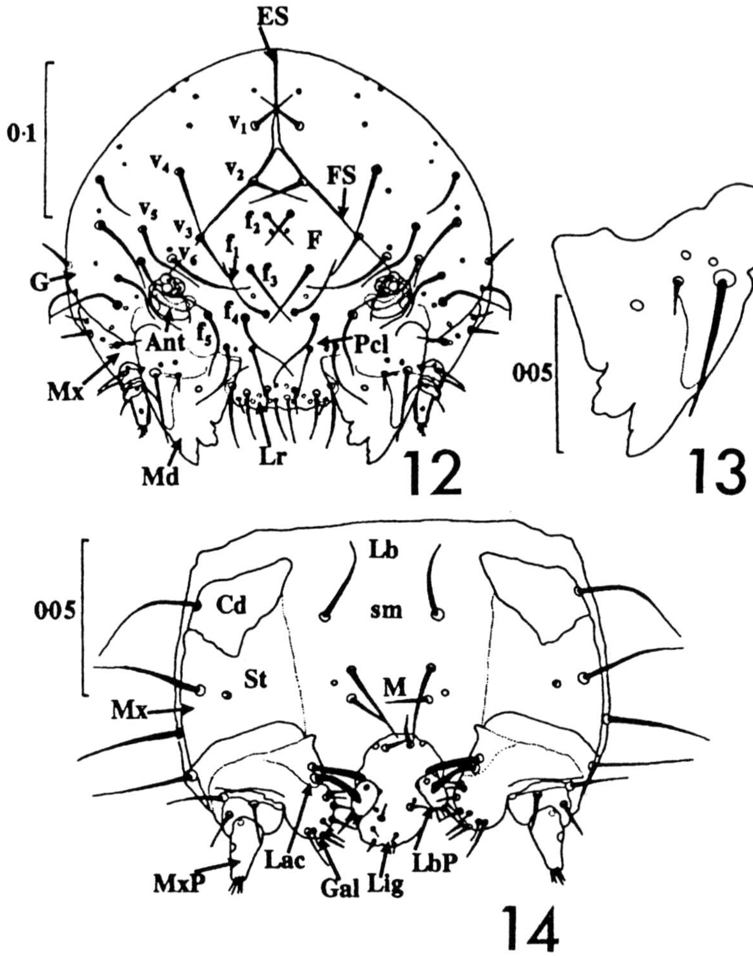

Figs. 12-14 Head parts first instar larva, *Megascelis puella*: (12) frontal view of head; (13) left mandible dorsal view; (14) labium and maxillae. Abbreviations: Ant, antenna; Cd, cardo; Es, epicranial suture; F, frons: f_1, f_2, f_3, f_4, f_5, frontal setae 1-5; Gal, galea; Lb, labium; Lac, lacinia; LbP, labial palpus; Lig, ligula; Lr, labrum; M, mentum; Md, mandible; Mx, maxilla; MxP, maxillary palpus; Pcl, postclypeus; Sm, submentum; St, stipes; v_1, v_2, v_3,v_4, v_5, v_6, vertex setae 1-6. Scale line in mm.

Discussion

Aulacoscelis sp.

The larva of *Aulacoscelis* sp. keys to Orsodacninae at couplet 10 in the key to families and subfamilies of larval Chrysomeloidea by Mann & Crowson (1981). However, it differs from Orsodacninae in the absence of paired urogomphi on abdominal tergite 9.

For a comparison of the first instar larval characters of Aulacoscelinae, Orsodacninae and Chrysomelinae, see Table 1. Characters of relevance and those differing between these subfamilies are as follows.

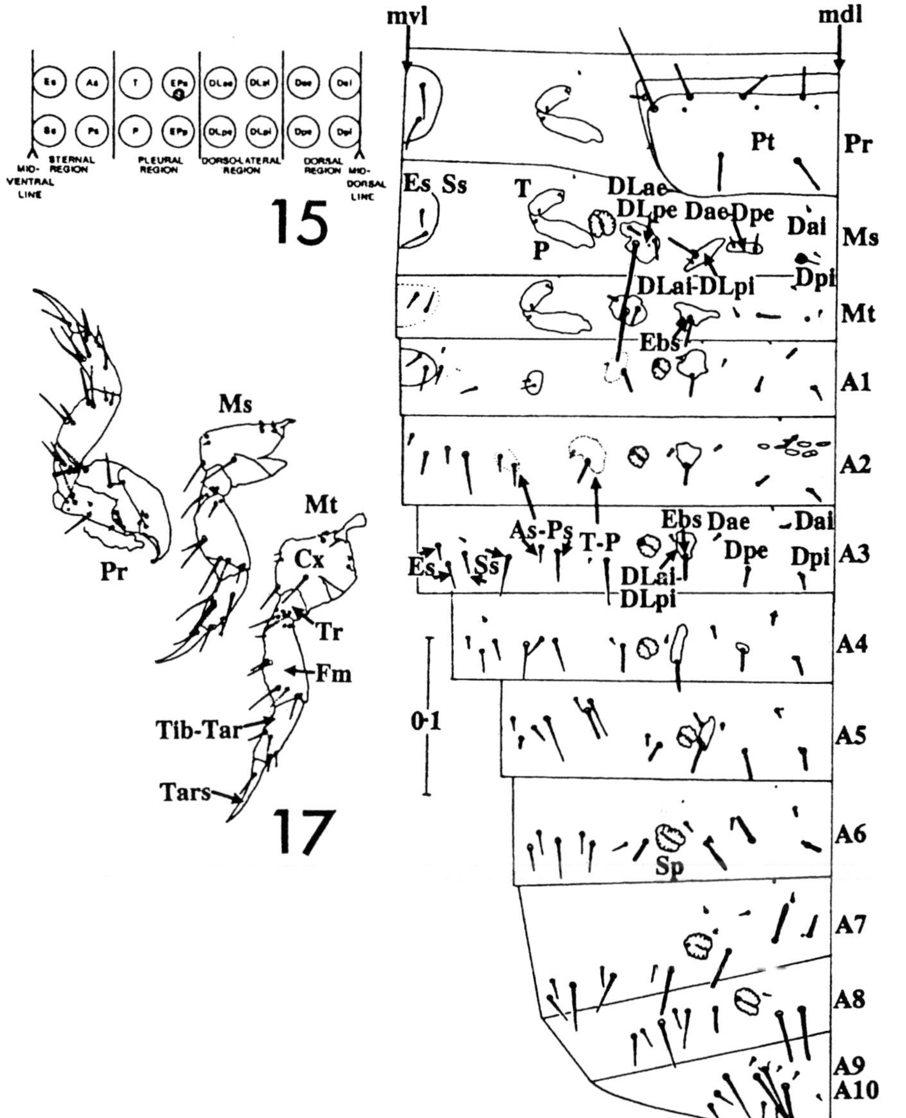

Figs. 15-17 Body and legs first instar larva, *Megascelis puella*: (15) tubercle nomenclature for typical body segment of generalised chrysomelid (after Cox 1976, modified after Kimoto 1962); (16) left half of the body showing tubercle and setal arrangement; (17) left, pro-, meso-, metathoracic legs, outer view. Abbreviations: A1-10, abdominal segments 1-10;Cx, coxa; Ebs, egg bursting spine; Fm, femur; mdl, middorsal line; Ms, mesothorax; Mt, metathorax; mvl, midventral line; Pt, pronotum; Tars, tarsungulus; Tib-Tar, tibio-tarsus; Tr, trochanter.Scale line in mm.

Apparently, the antennae are 2-segmented in Aulacoscelinae and Orsodacninae (incorrectly given as 3-segmented by Cox, 1981) (synapomorphy), whilst in the Chrysomelinae, they are distinctly 3-segmented. In the first two subfamilies, the second antennal segment

Table 1. Comparison of first instar larval characters of Aulacoscelinae, Orsodacninae and Chrysomelinae

Character	Aulacoscelinae	State	Orsodacninae	State	Chrysomelinae	State
Head shape	normal, oval	0	normal, oval	0	normal, oval	0
Head position	protracted	1	protracted	1	protracted	1
Epicranial suture	present	0	present	0	present	0
Frontal suture	present	0	present	0	present	0
Endocarina	present	0	present	0	present	0
Frontal setae	3+ pairs	0	3+ pairs	0	3+ pairs	0
Frontal papillate setae	absent	0	absent	0	absent	0
Fronto-clypeal suture	present	0	present	0	present	0
Antennal segments	2	1	2	1	3	0
Ocelli	1	5	1	5	6	0
Occipital foramen	single	0	single	0	single	0
Hypopharyngeal sclerite	present	0	present	0	present	0
Anterior labral margin	not incised	0	not incised	0	incised	1
Labral setae	> 8	0	> 8	0	> 8	0
Mandibles shape/teeth	triangular/4	0	triangular/6	0	palmate/5,6	1
Penicillus	present	1	present	1	absent (+ rare)	0(1)
Mandibular setae	2	0	2	0	2	0
Cardo	present	0	present	0	present	0
Maxillary palpi	3-segmented	0	3-segmented	0	3-segmented	0
Galea/Lacinia	fused	0	fused	0	fused	0
Postmental setae	4	0	4	0	6	1
Labial palpi	2-segmented	0	2-segmented	0	1 or 2-segmented	0,1
Pronotal type	D-DL	1	D-DL	1	D-DL, D-DL-Ep	1,0
Thoracic egg bursters	Pr, Ms, Mt	0	Pr, Ms, Mt	0	Ms, Mt absent	1 2
Abdominal egg bursters	absent	4	absent	4	I-V, I-II, I absent (rare)	0, 2, 3 4
Spiracles	annular	0	annular biforous	1	annular	0
Dorsal body setae	pointed	0	pointed	0	pointed,capitate	0,1
Dorsal ampullae	absent	0	absent	0	absent	0
Anal plate with numerous pores	present	1	present	1	absent	0
Paired urogomphi	absent	1	present	0	absent	1
Venter of abdomen + long setae	present	1	present	1	absent	0
Tibio-tarsal paronychial appendix	absent	0	absent	0	present	1
claw (pretarsus)	very elongate	0	very elongate	0	short	1
claw seta	present	1	present	1	present	1
pulvillus	absent	0	absent	0	absent	0

0, Plesiomorphic state
1,2,3,4,5, Apomorphic state

is much shorter than the accessory conical process, whilst it is usually much longer in the Chrysomelinae. In the Orsodacninae, the second segment is flat, bearing several long setae, whilst in the Aulacoscelinae, this segment is about half the length of the accessory conical process and bearing only short setae.

In the larva of Aulacoscelinae and Orsodacninae, a small occllus lacking pigmentation is present posterior to the antennae on each side (synapomorphy). The reduction in the number of ocelli probably suggests subterranean or internal feeding habits for these groups.

In contrast, all of the Chrysomelinae including representatives from all of the recognised tribes, have the plesiomorphy of six ocelli on each side of the head. All of these are external foliage feeders.

The mandibles are approximately triangular in the Aulacoscelinae and Orsodacninae with 4 and 6 teeth respectively. They lack a distinct mola in both of these subfamilies, as do the majority of chrysomelid larvae, except those of *Palophagus* which feeds on pollen of *Araucaria* (Kuschel & May, 1990). It is unlikely, therefore, that the larva of *Aulacoscelis* feeds on the pollen in cycad cones. *Aulacoscelis* and *Orsodacne* larvae bear a penicillus consisting of a dense tuft of about 12 or 13, or 10 closely packed setae, respectively (synapomorphy). In the majority of Chrysomelinae, the penicillus is absent, but according to Reid (1995) it is present in *Geomela*. Crowson (1946) defined the larvae of Sagrinae as having mandibles with a single apical tooth and considered that if the larva of *Aulacoscelis* were to have a 2- or 3- toothed mandibular apex, that character alone would warrant the removal of the genus from the Sagrinae, probably to constitute a separate subfamily. The 4-dentate mandibles of *Aulacoscelis* support this removal.

In the pronotum of *Aulacoscelis* apparently tubercle EPp is present, so that the pronotum is of the D-DL type as it is in Orsodacninae (synapomorphy). In the larvae of Chrysomelinae, both the D-DL and D-DL-EP types are both commonly represented.

The egg bursters on the larva of *Aulacoscelis* occur in similar positions on the pronotum, meso- and metathorax to those in the Orsodacninae (Cox, 1981, 1994). They are not situated on strictly homologous tubercles since those of *Aulacoscelis* are on tubercle Dai-Dpi-Dae-Dpe, whilst in Orsodacninae, they occur on tubercle Dae-Dpe. However, they are absent from the abdomen in both groups which is a synapomorphy. In the Chrysomelinae they are rarely absent, they are never on the pronotum, but they may be on the mesothorax; on both the meso- and metathorax; on meso-, metathorax, and abdominal segment 1, sometimes also on abdominal segment 2, rarely also on abdominal segments 2-5. They are usually on the small to large, well sclerotized tubercle DLpi (Cox, 1994).

The spiracles in *Aulacoscelis* are annular, since they are ring-shaped and lack lateral openings or pouches of the atrium. In contrast, those of the Orsodacninae are annular biforous, since they are ring-like, but have two lateral openings or pouches of the atrium. Those of all the Chrysomelinae examined are of the annular type as in the Aulacoscelinae.

In *Aulacoscelis* the ninth abdominal segment bears a well sclerotized dorsal tubercle, the anal plate, apparently with numerous pores, which are probably gland openings, which are also present in the Orsodacninae (Cox, 1981). This is considered a synapomorphy. The position of feeding of the larvae of these two groups is not known with certainty, although Cox (1981) suggested that those of *Orsodacne* may be subterranean root-feeders since the larvae of the Galerucinae, *Luperus* and *Aulacophora* have similar anal plates with similar gland openings and feed at the roots of grasses and Cucurbitaceae, respectively. According to Crowson & Crowson (1996) although the anal plate is well developed, the gland openings are apparently absent in the internal-monocot-bulb-feeding *Exosoma*. The anal plate in Chrysomelinae larvae is usually present, but lacks the numerous pore-like openings.

In this study, the larva of *Aulacoscelis* shares eight synapomorphies with that of Orsodacninae, but only four with that of Chrysomelinae, two of which are only rarely expressed as apomorphies.

The feeding position of *Aulacoscelis* larvae is still unresolved and this must be one of the priorities of future work and to determine if they are cycad-associated.

Megascelis puella *Lacordaire*

The larva of *Megascelis puella* keys to Eumolpinae at couplet 11 in the key to families and subfamilies of larval Chrysomeloidea by Mann and Crowson (1981). At this point they are separated from those of Synetinae by the more or less C-shaped body and abdominal tergite 9, not at all sclerotized.

For a comparison of the first instar larval characters of Megascelinae and Eumolpinae see Table 2. Characters of relevance and those differing between these two subfamilies are as follows.

Apparently, although the antennae of the two subfamilies are 2-segmented, those of

Table 2. Comparison of first instar larval characters of Megascelinae and Eumolpinae.

Character	Megascelinae	State	Eumolpinae	State	Character used by:
Head shape	normal, round	0	normal, round	0	Lee 1993b; Reid 1995
Head position	protracted	1	protracted	1	Lee 1993b
Epicranial suture	present	0	present	0	Reid 1995
Frontal suture	present	0	present	0	Present study
Endocarina	absent	1	absent	1	Lee 1993b
Frontal setae	3 + pairs	0	3 + pairs	0	Lee 1993b
Frontal papillate setae	absent	0	absent	0	Lee 1993b
Fronto-clypeal suture	?absent	?1	present	0	Present study
Clypeo-labral suture	present	0	present	0	Reid 1995
Antennal segments	2	1	2	1	Lee 1993b; Reid 1995
Ocelli	0	6	0, rarely 1	6/5	Reid 1995
Occipital foramen	single	0	single	0	Reid 1995
Hypopharyngeal bracon	absent	1	absent	1	Reid 1995
Anterior margin labrum	not incised	0	not incised	0	Lee 1993b
Labral setae	>8	0	>8	0	Lee 1993b
Mandibles shape/teeth	triangular/4	0	triangular/3,4	0	Lee 1993b; Reid 1995
Penicillus	absent	0	absent	0	Lee 1993b; Reid 1995
Mandibular setae	2	0	2	0	Lee 1993b
Cardo	present	0	present	0	Lee 1993b
Maxillary palpi	2-segmented	1	3-segmented	0	Lee 1993b
Galea/lacinia	not fused	1	?fused	0	Lee 1993b
Postmental setae	3 pairs	1	3 pairs	1	Lee 1993b
Labial palpi	1-segmented	1	1-segmented	1	Lee 1993b; Reid 1995
Pronotal type	D-DL-Ep	0	D-DL	1	Lee 1993b
Thoracic egg bursters	Ms, Mt	1	Ms, Mt	1	Reid 1995
Abdominal egg bursters	A1-6/7	0	A1; A1+2	2	Reid 1995
Spiracles of L1	annular biforous	1	annular biforous	1	Reid 1995
Dorsal body setae	clubbed	1	pointed	0	Present study
Dorsal ampullae	absent	0	absent	0	Reid 1995
Paired urogomphi	absent	1	absent	1	Reid 1995
Venter of abdomen + long setae	present	1	present	1	Reid 1995
Tibio-tarsal paronychial appendix	absent	0	absent	0	Reid 1995
Claw (pretarsus)	very elongate	0	very elongate	0	Lee 1993b; Reid 1995
Claw seta	present	1	present	1	Lee 1993b
Pulvillus	absent	0	absent	0	Lee 1993b

0, Plesiomorphic state; 1,2,5,6, Apomorphic states

Eumolpinae are usually longer, with segments 1 and 2 longer, as well as the conical accessory process being larger. Reid (1995) is incorrect in giving in Table 2, page 589, the Eumolpini as having 1-segmented antennae, This is surprising since earlier (Reid & Storey, 1993) described and figured the larva of *Colasposoma sellatum* Baly as being 2-segmented. Ocelli are usually missing in the Megascelinae, and rarely present in the Eumolpinae. A single, small ocellus is present dorsal to the antennae on each side in the gall-stem feeding *Odontionopa sericea* Gyllenhal. They are missing in the true root-feeding larvae of *Colasposoma fulgidum* Lefèvre, *Euryope terminalis* Baly and *Chrysochus auratus* (F.). Reid (1995) gave Eumolpini as having two ocelli on each side of the head, but in the description of the first instar larva of *C. sellatum* (Reid & Storey, 1993) made no mention of them. Lee (1993b) did not use ocellar number in his cladistical study of Japanese chrysomelid larvae. Crowson (1994) believed that the lack of ocelli in eumolpine larvae suggests originally subterranean or internal-feeding habits. The mandibles are 4-dentate in *Megascelis puella* with the inner tooth the smallest. Apparently, they are 3-dentate in the eumolpines, *C. auratus*, *E. terminalis* and *C. fulgidum*, but 4-dentate in *O. sericea*; the inner tooth may be very small or missing in these species. The maxillary palpi are apparently 2-segmented in the Megascelinae (not including the basal palpifer), whereas they are 3-segmented in the Eumolpinae. The galea and lacinia of the maxillae are apparently separate in the Megascelinae, whereas usually they are fused in the Eumolpinae which agrees with Lee (1993b).

In the pronotum of *Megascelis puella*, apparently the tubercle EPp is missing, or possibly fused with the pronotum, in contrast to the Eumolpinae in which this tubercle is separate. The egg bursters on the larva of *Megascelis puella* are on homologous tubercles (DLai-DLpi) to those of the Eumolpinae, but in this subfamily they are restricted to the mesothorax; meso- and metathorax; meso-, metathorax, abdominal segment 1; meso-, metathorax, abdominal segments 1 and 2; or abdominal segment 1 (see Table 3). In eumolpine larvae the tubercle also bears a total of two setae. The egg bursters are also on most of the body

Table 3. Egg bursters in the Eumolpinae

Species	Ms	Mt	A1	A2	Reference
Hornius grandis (Phil.& Phil.)	-	-	-	-	Jerez 1996
Dictyneis asperatus (Blanchard)	-	-	-	-	Jerez 1996
Stenomela pallida Erichson	-	-	-	-	Jerez 1996
Bromius obscurus (L.)	+	+	-	-	van Emden 1946
Odontionopa sericea Gyllenhal	+	+	+	-	Cox 1994
Colasposoma fulgidum Lefèvre	+	+	+	-	Cox 1994
Colposcelis signata (Mots.)	+	+	+	-	Lee 1993a
Demotina fasciculata Baly	+	+	+	-	Lee 1993a
Demotina modesta Baly	+	+	+	-	Lee 1993a
Hyperaxis fasciata (Baly)	+	+	+	-	Lee 1993a
Lypesthes ater (Mots.)	+	+	+	-	Lee 1993a
Lypesthes lewisii (Baly)	+	+	+	-	Lee 1993a
Scelodonta sauteri Chûjô	+	+	+	-	Lee 1993a
Xanthonia placida Baly	+	+	+	-	Lee 1993a
Chrysochus auratus (F.)	+	+	+	+	Cox 1994
Euryope terminalis Baly	+	+	+	+	Cox 1994
Colasposoma sellatum Baly	-	-	+	-	Reid & Storey 1993

segments in *Zeugophora subspinosa* (F.) of the Zeugophorinae (pro-, meso-, metathorax, A1-6 or 7)(Cox, 1994); in *Temnaspis nankinea* (Pic) (Ms, Mt, A1-8)(Cox, 1994) and *Sphondylia tomentosa* (Lacordaire)(Mt, A1-7)(Megalopodinae)(Schulze, 1996), but the tubercles bearing the egg bursters have a total of three setae, instead of only two.

The simple claws of the known Eumolpinae and *Megascelis puella* larvae suggest the ancestral larvae were never external feeding on foliage (Crowson, 1994).

The spiracles in *Megascelis puella* and the eumolpines *Chrysochus auratus* (Personal observation) and *Colasposoma sellatum* (Reid & Storey, 1993) are annular biforous (bicameral) and have two well developed air tubes or chambers, whereas in the other eumolpines, these are missing and the spiracles are cribriform or septate. Lee (1993b) gave them as annuliform for the Eumolpinae, which is not strictly correct. In addition, in *Megascelis*, the spiracles of the mesothorax and abdominal segments 6 to 8 are largest, subequal in size. However, in the Eumolpinae, the spiracles are largest on the mesothorax, but smaller and subequal in size on abdominal segments 1-8. In addition, Crowson (1994) stated that described eumolpine larvae have annuliform spiracles, but that it would not be surprising if the older bicameral type were found in some of the primitive Eumolpinae. The thoracic and abdominal segments of *Megascelis puella* bearing clubbed setae dorsally, differ from those of Eumolpinae larvae which bear unmodified, long, pointed setae.

According to Crowson (personal communication) the *Megascelis* larva is eumolpine-like in lacking ocelli and endocarina, and having 1-segmented labial palpi, but unusual in seeming to lack a free labrum and having bicameral spiracles. Moreover, it does not appear to have any trace of the venral proleg-like protuberances on the abdominal segments found in some eumolpine larvae. The lack of a free labrum is suggestive of Camptosomata, as are some of the antennal characters.

In conclusion, although Megascelinae and Eumolpinae are closely related on most aspects of larval morphology (they share 12 synapomorphies), there are sufficient significant differences to warrant the recognition of these as separate subfamilies.

References

Arnett, R.H., 1973. The beetles of the United States (A manual for identification). Fourth Printing. Ann Arbor:American Entomological Institute.

Bechyné, J. and Springlova de Bechyné, Y.B., 1969. La posicion sistematica de *Megascelis* Chevrolat (Col. Phytophaga). Revista de la Facultad de Agronomia. Universidad Central de Venezuela, 5(3): 65-76.

Blackwelder, R.E., 1946. Checklist of the Coleopterous insects of Mexico, Central America, the West indies, and South America, Part 4. Bulletin of the United States National Museum, 185: 551-763

Chapuis, F., 1874, in M.Th. Lacordaire and M.F. Chapuis, Histoire Naturelle des Insectes. Genera des Coléoptères ou exposé méthodique et critique de tous les genres proposés jusqu'ici dans cet ordre d'insectes. Volume 10. Famille des Phytophages. Lib. Encyclop. Roret, Paris, iv+455 pp.

Chapuis, F., 1875. Histoire Naturelle des Insectes. Genera des Coléoptères. Voilume 11. Famille des Phytophages. Paris, 420 pp.

Chen, S.H., 1940. An attempt at a new classification of the leaf beetles. Sinensia, 11: 451-481.

Chûjô, M., 1953-1954. A taxonomic study on the Chrysomelidae with special reference to the fauna of Formosa. Technical Bulletin. Kagawa Agricultural College, 5: 19-86; 121-136 (1953); 5: 285-295; 6: 39-51 (1954).

Clavareau, H., 1913. Coleopterorum Catalogus. Pars 51: Chrysomelidae: 1. Sagrinae, 2. Donaciinae, 3. Orsodacninae, 4. Criocerinae, pp 1-103. Berlin: W. Junk.

Cox, M.L., 1976. The taxonomy and biology of the British Chrysomelidae, unpublished Ph.D. thesis, University of Newcastle upon Tyne, UK, 375 pp.

Cox, M.L., 1981. Notes on the biology of *Orsodacne* Latreille with a subfamily key to the larvae of the British Chrysomelidae (Coleoptera). Entomologist's Gazette 32: 123-135.

Cox, M.L., 1994. Egg bursters in the Chrysomelidae, with a review of their occurrence in the Chrysomeloidea

(Coleoptera) in P. Jolivet, M.L. Cox and E. Petitpierre (eds) Novel aspects of the biology of Chrysomelidae, Series Entomologica, Volume 50 (Dordrecht: Kluwer Academic Publishers), pp. 75-110.
Cox, M.L. & Windsor, D.M., in press 1999. The first instar larva of *Aulacoscelis appendiculata* n. sp. (Coleoptera: Chrysomelidae: Aulacoscelinae) and its value in the placement of the Aulacoscelinae. Journal of Natural History, 33.
Crowson, R.A., 1946. A revision of the genera of the chrysomelid group Sagrinae (Coleoptera), Transactions of the Royal Entomological Society of London, **97(4),** 75-115.
Crowson, R.A., 1955. The Natural Classification of the families of Coleoptera. London: Nathaniel Lloyd & Co. [Reprinted 1968 [1967], E.W. Classey, Hampton]
Crowson, R.A., 1994. A long perspective on chrysomelid evolution, in P. Jolivet, M.L. Cox and E. Petitpierre (eds) Novel aspects of the biology of Chrysomelidae, Series Entomologica, Volume 50 (Dordrecht: Kluwer Academic Publishers), pp xix-xxiii.
Crowson, R.A. and Crowson, E.A., 1996. The phylogenetic relations of Galerucinae-Alticinae in, P. Jolivet and M.L. Cox (eds) Chrysomelidae Biology, vol. 1: The Classification, Phylogeny and Genetics (Amsterdam: SPB Academic Publishing bv), pp, 97-118.
Duponchel, P.A.J. and Chevrolat, A., 1842, in D'Orbigny, M.C, Dictionnaire Universel d'Histoire Naturelle, Volume 2, 338.
Emden, von F.I., 1946. Egg bursters in some more families of polyphagous beetles and some general remarks on egg bursters. Proceedings of the Royal Entomological Society of London, (A) 21: 89-97.
Gemminger, M. and Harold, E.F., 1874. Catalogus Coleopterorum hucusque descriptorum synonymicus et systematics. XI. Chrysomelidae (Pars 1), Paris: Sumpta G. Beck. pp. 3233-3478.
Iablokoff-Khnzorian, S.M., 1966. Considérations sur l'édéage des Chrysomélidae et son importance phylogénique. L'Entomologiste, 22: 115-137.
Iablokoff-Khnzorian, S.M., 1967. Considérations sur l'édéage des Chrysomélidae et son importance phylogénique. L'Entomologiste, 23: 65-67.
Jacoby, M., 1877. Descriptions of new species of phytophagous Coleoptera. Proceedings of the Zoological Society of London, 1877: 510-520.
Jacoby, M., 1880, 1888. Biologia Centrali-Americana. Insecta. Coleoptera, VI(1) Phytophaga (part), 1-262; Supplement, 1-374.
Jacoby, M., 1903. Fam. Sagridae, in P.A.G. Wytsman, Genera Insectorum. Coleoptera Phytophaga, Fasc. 14a, 4th edition, pp. 1-11., 1 pl + (addenda) pp. 13-14.
Jacoby, M., 1908. The fauna of British India, including Ceylon and Burma. Coleoptera. Chrysomelidae, Volume 1 . Taylor & Francis: London, UK.
Jacoby, M. and Clavareau, H., 1905. Fam. Megascelidae, in Wytsman P.A.G. ed. Genera Insectorum. Coleoptera Phytophaga, Fasc. 32. 4th edition, 6 pp., 1 pl.
Jerez, V., 1996. Biology and phylogenetic remarks of the sub-Antarctic genera *Hornius*, *Stenomela* and *Dictyneis* (Chrysomelidae: Eumolpinae) in Jolivet, P., Cox, M.L. (eds) Chrysomelidae Biology, vol. 3: General Studies. SPB Academic Publishing bv, Amsterdam, The Netherlands. pp. 239-258.
Jolivet, P., 1953-1954. L'aile des Chrysomeloidea (Coleoptera). Miscellanea Entomologica, 48: 85-101.
Jolivet, P., 1959. Recherche sur l'aile des Chrysomeloidea (Col.). Mémoires de l'Institut Royal des Sciences Naturelles de Belgique, 2(51): 1 180.
Jolivet, P., 1988. Food habits and food selection of Chrysomelidae. Bionomic and evolutionary perspectives in, P. Jolivet, E. Petitpierre, and T.H. Hsiao (eds) Biology of Chrysomelidae, Series Entomologica, Vol. 42 (Dordrecht: Kluwer Academic Publishers) pp. 1-24.
Jolivet, P., 1997. Biologie des Coléoptères Chrysomelides. Société Nouvelle des Editions Boubée: Paris, France.
Jolivet, P. and Petitpierre, E., 1981. Biology of Chrysomelidae (Coleoptera). Bulletin de l'Institut Catalonia Histoire Naturelle, 47 (Sec. Zool., 4): 105-138.
Kimoto, S., 1962. A phylogenetic consideration of Chrysomelinae based on immature stages of Japanese species (Coleoptera). Journal of the Faculty of Agriculture, Kyushu University, 12: 67-116.
Kuschel, G. and May, B.M., 1990. Palophaginae, a new subfamily for leaf-beetles, feeding as adult and larva on Araucarian pollen in Australia (Coleoptera: Megalopodinae). Invertebrate Taxonomy, 3: 697-719.
Lacordaire, J.T., 1845-1848. Monographie des Coléoptères subpentamères de la famille des Phytophages. Volume 2. 4th edition. Bruxelles.
Latreille, P.A., 1829. Les Crustacés, les Arachnides et les Insectes, distribues en Familles naturelles. Paris, Volume 2, 556 pp.
Lee, J.L., 1993a. A systematic study on the larvae of the subfamily Eumolpinae from Japan (Coleoptera: Chrysomelidae). Entomological Research Bulletin, 19: 60-68.
Lee, J.L., 1993b. Phylogenetic studies on the larvae of the Chrysomelidae (Coleoptera) from Japan. Japanese Journal of Entomology, 61(3): 409-424.
Leng, C.W., 1920. Catalogue of the Coleoptera of America, north of Mexico. Mount Vernon, N.Y: John D. Sherman. 470 pp.
Mann, J.S. and Crowson, R.A., 1981. The systematic positions of *Orsodacne* Latr. and *Syneta* Lac. (Coleoptera

Chrysomelidae), in relation to characters of larvae, internal anatomy and tarsal vestiture. Journal of Natural History, 15: 727-749.
Mann, J.S. and Crowson, R.A., 1983a. On the occurrence of mid-gut caeca, and organs of symbiont transmission, in leaf-beetles (Coleoptera: Chrysomelidae). Coleopterists Bulletin, 37: 1-15.
Mann, J.S. and Crowson, R.A., 1983b. On the male reproductive organs and their taxonomic significance in the leaf beetles (Coleoptera: Chrysomelidae). Entomologia Generalis, 9: 75-99.
Mann, J.S. and Crowson, R.A., 1983c. Phylogentic significances of the ventral nerve cord in the Chrysomeloidea (Coleoptera: Phytophaga). Systematic Entomology, 8: 103-119.
Mann, J.S. and Crowson, R.A., 1984. On the digitiform sensilla of adult leaf beetles (Coleoptera: Chrysomelidae). Entomologia Generalis, 9: 121-133.
Maulik, S., 1941. Biology and Morphology of the Sagrinae (Chrysomelidae, Coleoptera). The Annals and Magazine of Natural History, Series 11, 7: 235-254.
Medvedev, L.N., 1971. The ways of evolution and phylogeny of Chrysomelidae (Coleoptera). Proceedings of the 13th International Congress of Entomology (Moscow, 1968), Nauka, Leningrad, 1: 271.
Medvedev, L.N., 1975. Review of the chrysomelid subfamily Aulacoscelinae in entomological collections of the German Democratic Republic. Mitteilungen aus dem Zoologischen Museum in Berlin, 51: 37-40.
Monrós, F., 1949. Sobre la posición sistemática de algunos Eupoda dudosos (Col. Chrysomelidae). Acta Zoologica Lilloana, 7: 545-574.
Monrós, F., 1953. Aulacoscelinae, eine neue Chrysomeliden-Unterfamilie, mit Beschreibung einer neuen bolivianischen Gattung (Col.). Entomologische Arbeiten aus dem Museum Georg Frey, 4: 19-25.
Monrós, F., 1954. Revision of the chrysomelid subfamily Aulacoscelinae. Bulletin of the Museum of Comparative Zoology at Harvard College, 112(4): 321-360.
Monrós, F., 1955. Remarques sur les affinites des familles de Cerambycoidea (Coleoptera). Bulletin de l'Institut Royal des Sciences Naturelles de Belgique, 31: 1-7.
Monrós, F., 1959. Los generos de Chrysomelidae (Coleoptera). Opera Lilloana, 3: 1-337.
Nybakken, O.E., 1970. Greek and Latin in Scientific Terminology. The Iowa State University Press: Ames, Iowa, USA.
Petitpierre, E., Segarra, C., Yadav J.S. and Virkki, N. 1988. Chromosome numbers and meioformulae of Chrysomelidae, in P. Jolivet, E. Petitpierre and T.H. Hsiao (eds) Biology of Chrysomelidae, Series Entomologica, Vol. 42 (Dordrecht: Kluwer Academic Publishers), pp. 161-186.
Reid, C.A.M., 1995. A cladistic analysis of subfamilial relationships in the Chrysomelidae sensu lato (Chrysomeloidea), in J. Pakaluk, and S.A. Slipinski (eds) Biology, Phylogeny and Classification of Coleoptera. Papers celebrating the 80th Birthday of Roy A. Crowson. Volume two. (Warszawa: Muzeum i Instytut Zoologii PAN), pp. 559-631.
Schulze, L., 1996. Life-history and description of early stages of *Sphondylia tomentosa* (Lacordaire)(Coleoptera: Chrysomelidae: Megalopodinae), in P. Jolivet and M.L. Cox (eds) Chrysomelidae Biology, vol. 3: General Studies (Amsterdam: SPB Academic Publishing), pp. 187-199.
Seeno, T.N. and Wilcox, J.A., 1982. Leaf beetle genera. Entomography, 1: 1-221.
Stål, C., 1865. Monographie des Chrysomélides de l'Amérique III. Nova Acta Regiae Societatis Scientiarum Upsaliensis, Third Series 5(II): 177-365.
Sturm, J., 1826. Catalog meiner Insecten-Sammlung. Erster Theil, Kafer. Nurnberg, 207 pp., 4 pls.
Suzuki, K., 1974. Phylogeny of the family Chrysomelidae based on the comparative morphology of the internal reproductive system (Insecta: Coleoptera), unpublished Ph.D. thesis, Tokyo Metropolitan University, Japan, 186 pp.
Suzuki, K., 1980. Phylogeny of the family Chrysomelidae based on the comparative morphology of the internal reproductive system (Insecta: Coleoptera), read at the 16th International Congress of Entomology, Kyoto, Abstr.p. 30.
Suzuki, K., 1985. Phylogenetic relationships among higher taxa of the family Chrysomelidae. Chrysomela, 12: 13-14.
Suzuki, K., 1988. Comparative morphology of the internal reproductive system of the Chrysomelidae (Coleoptera), in P. Jolivet, E. Petitpierre and T.H. Hsiao (eds), Biology of Chrysomelidae, Series Entomologica, Vol. 42 (Dordrecht: Kluwer Academic Publishers), pp. 317-355.
Suzuki, K., 1994a. The systematic position of the subfamily Aulacoscelinae (Coleoptera: Chrysomelidae) in D.G. Furth (ed) Proceedings of the Third International Symposium on the Chrysomelidae, Beijing, 1992 (Leiden: Backhuys Publishers), pp. 45-59.
Suzuki, K., 1994b. Comparative morphology of the hindwing venation of the Chrysomelidae (Coleoptera), in P. Jolivet, M.L. Cox and E. Petitpierre (eds) Novel Aspects of the Biology of Chrysomelidae, Series Entomologica, Vol. 50 (Dordrecht: Kluwer Academic Publishers), pp. 337-354.
Suzuki, K., 1996. Higher classification of the family Chrysomelidae (Coleoptera), in P. Jolivet, M.L. Cox (eds) Chrysomelidae Biology, vol. 1: The Classification, Phylogeny and Genetics (Amsterdam: SPB Academic Publishing bv), pp. 3-54.

INTERNAL STRUCTURE AND PHYLOGENETIC IMPORTANCE OF MALE GENITALIA IN THE EUMOLPINAE

R.W. FLOWERS
Agricultural Research Programs, Florida A&M University, Tallahassee, FL 32307-4100, USA

Abstract

Structures of the male genitalia of 27 taxa of Eumolpinae (Coleoptera: Chrysomelidae) are described and illustrated. Described taxa represent 25 genera distributed across nine of the 14 recognized tribes. Special emphasis is given to descriptions of the endophallus and its modifications. Two major divisions within the Eumolpinae can be identified based on genitalic characteristics: a plesiomorphic group characterized by a relatively simple endophallus, no constriction between the basal hood and the median lobe, a lack of basal spurs and subbasal fenestra on the median lobe, a quadrate tegmen, and relatively distinct ventral sclerites of the eighth and ninth segments; and an apomorphic group characterized by a more complex endophallus which usually contains lobes or sclerotized appendages, a constriction between the basal hood and the median lobe, the presence of a pair of basal spurs and a subbasal fenestra on the median lobe, a triangular tegmen, and reduced or absent ventral sclerites. The function of a complex endophallus during copulation is illustrated based on a pair of *Colaspis pseudofavosa* Riley that were preserved *in copula*. Phylogenetic groups in male structures are briefly discussed.

1. Introduction

Traditionally, morphology of genitalia, especially the male genitalia, has been the "gold standard" for taxonomically separating confusing or closely related insect taxa. The first studies of genitalia in the Coleoptera were conducted on flea beetles (Galerucinae: Alticini [Foudras, 1859, 1860]). The first major comparative survey of the morphology of the male genitalia of Coleoptera was done by Sharp and Muir (1912), which included descriptions of 39 species of Chrysomelidae in 13 subfamilies, and illustrations for many of these. Since Foudras' studies in the mid-1800s, male genitalic structures have been used in differentiating species and genera in the Chrysomelidae subfamilies Cryptocephalinae (Weise 1886, 1889a,b), Donaciinae (Askevold, 1988, 1990, 1991), Criocerinae (White, 1993), Eumolpinae (Flowers, 1997; Reid, 1992), Chrysomelinae (Baly, 1879; Jacques, 1987), and Galerucinae (including alticines) (e.g., Doquet, 1994; Furth, 1992; Heikertinger, 1911, 1913, 1921, 1925; LeSage, 1995; Leonardi, 1978; Wilcox, 1965).

The structure of the male aedeagus with the greatest diversity of characters is the internal sac (Sharp & Muir, 1912), more recently termed the endophallus (Askevold, 1988). This structure is largely or entirely membranous and is withdrawn within the median lobe when at rest in most Coleoptera. (Exceptions can be found in the Lucanidae and Scarabaeidae, where the endophallus is sometimes permanently everted and folded along the median lobe when not in use [Sharp & Muir, 1912].) The endophallus has been used

Advances in Chysomelidae Biology 1, pp. 71–93
Edited by M.L. Cox

sporadically in studies of Scarabaeidae (Woodruff & Beck, 1989), Carabidae (Madge, 1967; Meurgues & Ledoux, 1966), Chrysomelidae (Mann & Crowson, 1996; also for *Timarcha* [Chrysomelinae], Stockmann, 1962, cited in Meurgues & Ledoux, 1966), and Curculionidae (Thompson, 1988). The major drawback to the use of the endophallus is the necessity of everting it in a dead specimen, and the difficulty of doing so for many families of Coleoptera. Several techniques have been used for everting the endophallus and preserving it in its inflated or erect state. These methods include squeezing out the endophallus (Skelley, 1993; works on fresh material only), use of a syringe to inflate the endophallus (Meurgues & Ledoux, 1966), and using pins to pull the endophallus from the median lobe (Sharp & Muir, 1912).

2. Subfamily Eumolpinae

The Eumolpinae is one of the largest subfamilies within the Chrysomelidae, and one of the most taxonomically difficult. Compared to the Galerucinae, Hispinae, and Cryptocephalinae, the Eumolpinae presents itself as a rather homogeneous group of small, compact beetles with generally dark monochromatic coloration. Many genera have been defined on secondary sexual modifications of the males; many other genera are based on gradual characters, such as relative thickness of antennal segments, that lose their discriminatory power as new morphospecies with intermediate measurements are discovered. Within genera, many species were described based on subtle differences in punctation or elytral striation. Again, characters that initially worked well for separating species lost their usefulness with the discovery of additional intermediate forms.

In the Eumolpinae, study of the male genitalia has generally been limited to reporting the form of the apex of the median lobe of the aedeagus (cf. Bechyné, 1950; Blake 1950, 1951, 1967, 1970, 1976). As with external characters, the median lobe's small incremental differences among species, especially within large genera, means that this character is rarely helpful either for identification or for making phylogenetic inferences (an exception is Schultz's 1977 revision of the *Rhabdopterus* north of Mexico). The endophallus in most Eumolpinae is often very difficult to evert enough for study because of its great length. In its retracted state it is much longer than the median lobe; most of this length protrudes through the rear opening of the median lobe and is folded under the basal hood. Thus, methods for inflation that involve building up pressure behind the retracted endophallus inside the median lobe are useless in the Eumolpinae. Adding to the difficulty in working with many taxa are the complex sclerotized structures that fill the median lobe when the endophallus is retracted, thus making it difficult to evert the endophallus by pulling it out through the ostium of the median lobe. Despite these difficulties, Flowers (1995, 1997) illustrated endophallic structures in three genera and showed that study of the endophallus can help resolve taxonomic problems between and within genera.

In the following survey of Eumolpinae male genitalia, terms for the structures of the median lobe and endophallus follow Askevold (1988, 1990, 1991), Askevold & Flowers (1994), and Flowers (1995). This terminology is shown in Figure 17 (reproduced from Flowers 1995). The terms "base" and "apex" are used when describing the position of a structure on the endophallus to describe the relative distance of that structure from the center of the beetle's body, assuming that the median lobe is extruded and the endophallus everted. Thus, during copulation, the "apical sclerite" of the endophallus is indeed apical (posterior) to other sclerites on the endophallus; however, when retracted, these positions are reversed.

At the base of the median lobe is a large hollow basal hood (BH, also called the foramen by Reid [1995]), which lies over the tegmen (TG). The junction between the basal hood the median lobe can vary in different genera, as can the relative lengths of the basal hood and median lobe. The tegmen is lightly sclerotized and Y-shaped or quadrate; in some taxa the arms of the "Y" wrap around the median lobe just ahead of the basal hood. The dorsal lateral lobes, or parameres, of the tegmen, found in many other families of Coleoptera, are lacking in the Eumolpinae. The hollow enclosure formed by the basal hood and the tegmen holds the folded endophallus. The median lobe often has a pair of basal spurs (BS) and a subbasal fenestra (SBF), and is highly variable in length, curvature, and shape of the tip. Associated with the median lobe are up to two additional Y-shaped ventral sclerites lying just above the subgenital plate, corresponding to the eighth and ninth sternites. These are much reduced or missing in many genera.

The endophallus is a long membranous tube which is everted through the ostium at the apex of the median lobe. Most of the endophallus is a simple membrane, but approximately the basal third may have additional sclerotized, folded, or inflated structures. The endophallus shown in Figure 17 illustrates some of these structures. In this example, the endophallic lateral digits (ELD) are associated with a complex basal supporting block (BSB) from which a long flexible sclerite (MEG) protrudes; the ELD's are thus subapical and the endophallus continues distally beyond them, sometimes for a considerable distance. At the base of the endophallus there are additional sclerites, principally another pair of less sclerotized, but movable basal lateral digits (BLD). On the underside of the endophallus between the two sets of digits, there can be a field of fine setae (bb). At the tip of the endophallus are one or more apical sclerites (as). These are associated with a flagellum (often very long), and an elongate, slightly swollen ejaculatory duct, both of which can be either membranous to moderately sclerotized.

3. Survey of Eumolpinae male genitalia

The remainder of this paper is a preliminary survey of the forms of male genitalia found in different tribes and subtribes of the Eumolpinae, emphasizing endophallic structures wherever possible. An attempt was made to include a complete review of genitalic structures for representatives of all the major tribes. However, specimens available to me were primarily New World taxa. This, in turn, somewhat skews the taxonomic coverage, because the eumolpine fauna of the Neotropical region overwhelmingly consists of the tribe Eumolpini. In contrast, Old World faunas are skewed toward two other large tribes, the Typophorini and Adoxini. The classification followed is that of Seeno and Wilcox (1982) with the addition of the tribes Synetini and Megascelidini, which are now considered part of the Eumolpinae (Reid, 1995).

All genitalia were prepared by softening the beetle, removing the abdomen and soaking it for a few minutes in a hot KOH solution. Then the median lobe and associated structures were dissected out and placed for a few minutes longer in the hot KOH until they were sufficiently flexible and transparent. The endophallus was carefully teased out through the apical orifice of the median lobe using hooked insect pins. Preparations were drawn under a dissecting microscope with a camera lucida. In most cases, original drawings were made at a magnification of 70X, but for larger specimens, drawings were made at lower magnifications. In the illustrations that follow, all structures are shown to the same 70X scale, before reduction to page dimensions. Genitalia drawn at a different scale will be so noted. In cases where the endophallus could be only partly everted (due to tearing, or inability to get the entire structure everted), a dotted perpendicular line indicates the extent of the structures actually exposed.

Tribe Hornibiini

Hornibius grandis Philippi [Chile] (Figs.1-3). The basal hood is well sclerotized, continuous with and shorter than the median lobe. No basal spurs or basal swelling are present. There are traces of apodemes laterally. The tegmen is Y-shaped with the lateral arms wrapping around the median lobe. The pattern of sclerotization of the tegmen is unusual for the Eumolpinae in that the outside edges are the most heavily sclerotized. The endophallus is membranous, with the apical area bearing small spicules. The apical sclerite is large and fan-shaped basally, contracting to a twisted apical rod. The flagellum is lightly sclerotized and joins the ejaculatory duct at a right angle. The duct is not swollen, and the seminal

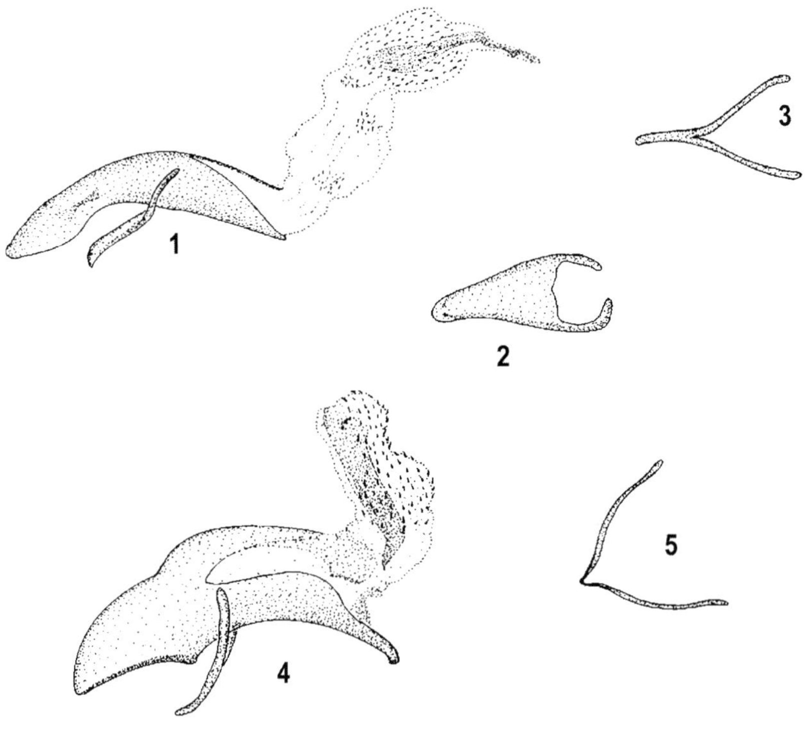

Figs. 1-5. Male genitalia of *Hornibius* and *Syneta*. 1-3, *Hornibius grandis*: 1, lateral view of median lobe and endophallus; 2, tegmen, ventral view; 3, ventral sclerite, ventral view. 4,5, *Syneta carinata*: 4, lateral view of median lobe and endophallus; 5, ventral sclerite, ventral view.

vesicle ducts are visible basally. A well differentiated Y-shaped ventral sclerite lies beneath the median lobe.

This monogeneric tribe is placed at the base of most classifications of the Eumolpinae.

Tribe Synetini

Syneta carinata Mannerheim [USA] (Figs.4-5). The basal hood is well sclerotized, and continuous with and shorter than the median lobe. Basal spurs are lacking and the tegmen is a narrow Y- shaped sclerite with the stem of the "Y" not expanded under the basal hood. The endophallus is relatively short, membranous, and contains regions of small sclerotized spicules. The apical sclerite is large and hook-shaped basally. The ventral sclerite is V-shaped with the two arms constricted at their joining.

Reid (1995) placed the subfamily Synetinae Edwards as a tribe within the Eumolpinae. He considered *Syneta* a plesiomorphic genus with many adult autapomorphies.

Tribe Typophorini

Typophorus sp. [Bolivia] (Figs. 6-9). The basal hood is well sclerotized, continuous with and shorter than the median lobe (Fig. 6). A basal swelling is present but is not developed into basal spurs. The tegmen is quadrate and notched on the rear margin, and does not form a complete ring around the median lobe. The endophallus is entirely membranous and finely spiculate, the apical sclerite is small (Fig. 8). Ventral sclerites (Fig. 9) are well developed.

Metachroma pellucidum Crotch [USA] (Figs. 10-12). Basal hood, median lobe, tegmen, endophallus, and ventral sclerites follow the general pattern seen in *Typophorus*. The middle third of the endophallus (shown in retracted position in Fig. 10) is densely and finely spiculate, and the apical third is sparsely spiculate with spicules fine; due to the small size of *Metachroma*, this can be seen only with a compound microscope at magnification 200X.

Mann & Crowson (1996) illustrated endophalli (uneverted) from two Indian genera in the Typophorini: *Basilepta* and *Nodina*. In both cases the endophalli appear to be mostly membranous, although the *Basilepta* has what seems to be a large apical sclerite and the *Nodina* may have basal sclerites at the ostium.

Tribe Colasposomini

Colasposoma sp. [India] (Figs. 13-15). The basal hood is well sclerotized, continuous with, and subequal to, the median lobe. The base of the median lobe is slightly swollen ventrally. The tegmen (Fig. 14) is quadrate and notched on the rear margin, and does not form a complete ring around the median lobe. The endophallus is membranous except for a weakly sclerotized area where it joins the median lobe (Fig. 13). Two Y-shaped ventral sclerites are present, and the larger, or eighth, sclerite has two small curved auxiliary sclerites at the tip of the arms of the "Y" (Fig. 15). Mann & Crowson (1996) illustrated uneverted endophalli of three Indian species of *Colasposoma*. In all three species, there appeared to be no sclerotized structures present.

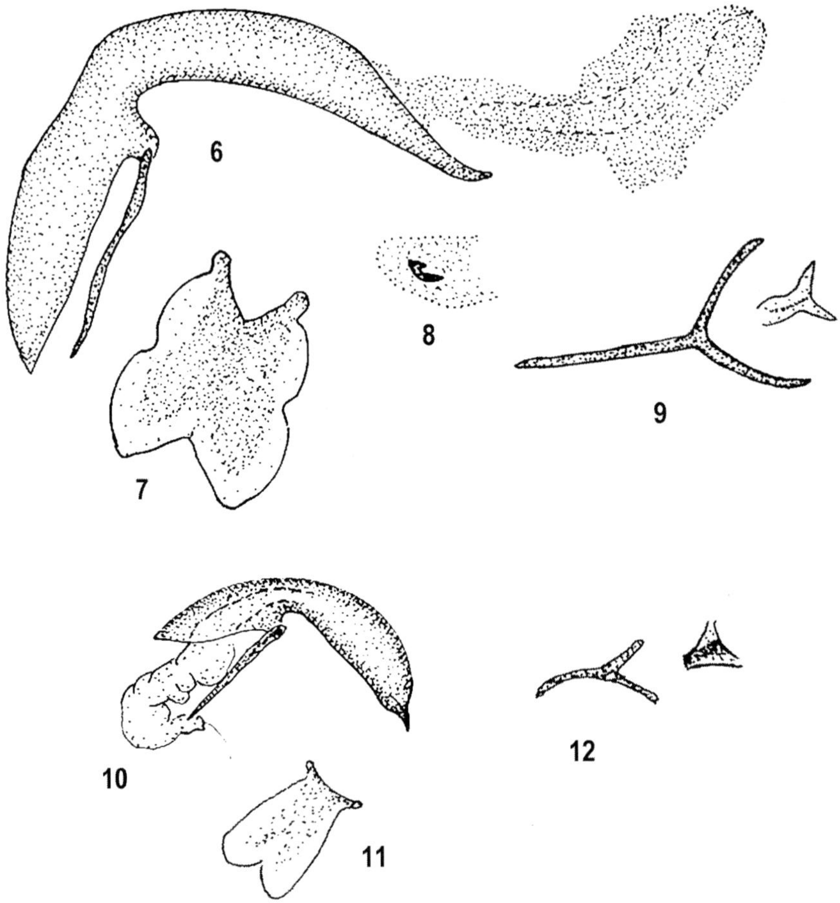

Figs. 6-12. Male genitalia of Typophorini. 6-9, *Typophorus* sp.: 6, lateral view of median lobe and endophallus (dashed line shows uneverted part of endophallus); 7, tegmen, ventral view; 8, apical sclerite; 9, ventral sclerites, ventral view. 10-12, *Metachroma pellucidum*: 10, lateral view of median lobe, showing retracted endophallus; 11, tegmen, ventral view; 12, ventral sclerites, ventral view.

Springlová (1960) established Colasposomini as a tribe, separate from the Eumolpini, based on external characters, and especially because the Colasposomini is primarily an Old World group while the Eumolpini center of diversity is the New World.

Tribe Caryonodini

Caryonoda sp. [Venezuela] (Fig. 16). The basal hood is lightly sclerotized with distinct lateral apodemes, shorter than the median lobe, and with a dorsal constriction at the joint with the median lobe. Basal spurs are present on the median lobe. The tegmen is triangular, and its lateral arms form a ring around the median lobe. The endophallus has small, sclerotized, lobe-shaped BLD's and ELD's. No ventral sclerites were found, but they may have been overlooked due to the very small size of the beetles in this genus.

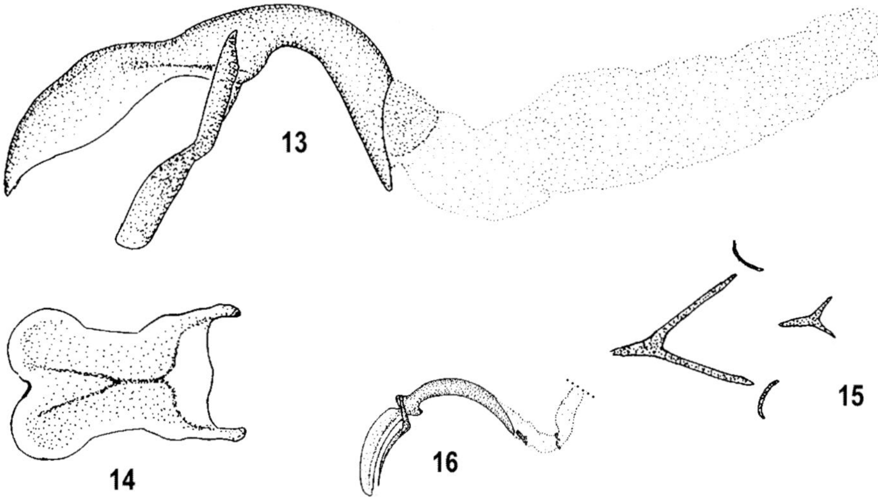

Figs. 13-16. Male genitalia of *Colasposoma* and *Caryonoda*. 13-15, *Colasposoma* sp.: 13, lateral view of median lobe and endophallus; 14, tegmen, ventral view; 15, ventral sclerites, ventral view. 16, *Caryonoda* sp., lateral view of median lobe and endophallus (partially everted).

Bechyné (1951) placed *Caryonoda* in its own tribe, based on the unique structure of its prosternum (see Bechyné & Bechyné, 1961, or Flowers, 1996 for figures). However, the structures of the male genitalia fall within the spectrum of characters exhibited by the Eumolpini.

Tribe Eumolpini

This is a very large tribe with most genera occurring in the Neotropical region. In all genera studied so far, the median lobe bears well differentiated basal spurs, and the junction of the median lobe and the basal hood is constricted. The ventral sclerites below the median lobe are very weak or absent. The endophallus is very diverse, both within and between genera. Chapuis (1874) subdivided the tribe into 21 sections; although a number of these are still recognized, most are considered to be heterogeneous assemblages of genera.

Section Iphimeites

This is the largest and most heterogeneous assemblage within the Eumolpini. The great majority of the 140 genera are Neotropical. The genitalia illustrated here represent only a tiny subset of this fauna.

Hermesia aurata Lefèvre [Brazil] (Fig. 17). The basal hood (BH) is longer than the median lobe, lightly sclerotized, with strong lateral apodemes (AP); a dorsal constriction is present at the joint with the median lobe. Basal spurs (BS) are present on the median lobe and a subbasal fenestra (SBF) is present dorsally. The tegmen (TG) is triangular and weakly sclerotized with lateral arms wrapping around the median lobe. The endophallus has well developed appendages that are also characteristic of many other Eumolpini species: basal

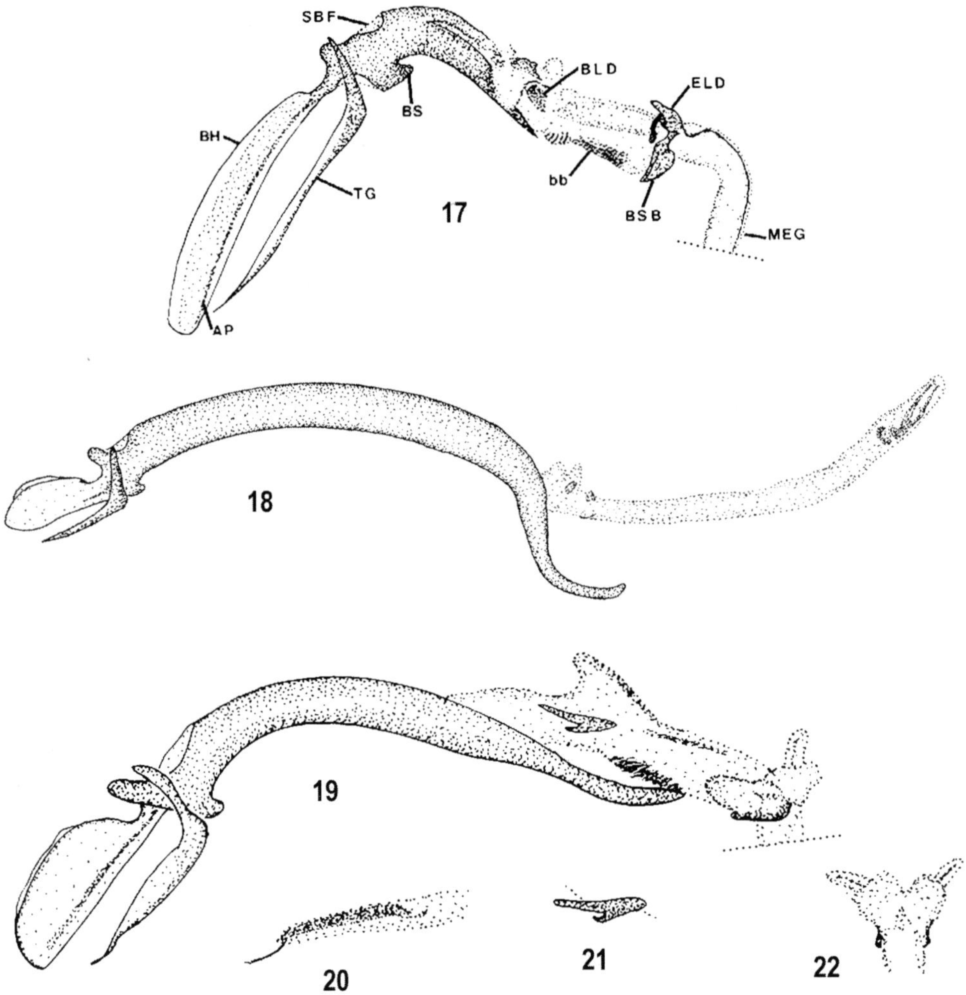

Figs. 17-22. Male genitalia of *Hermesia* and *Adorea*. 17, *Hermesia aurata*, lateral view of median lobe and endophallus (partially everted)(from Flowers 1995). 18-22, *Adorea*: 18, *Adorea cruentata*, lateral view of median lobe and endophallus. 19-22, *Adorea chontalensis*: 19, lateral view of median lobe and endophallus (partially everted); 20, apical sclerite; 21, BLB; 22, apical view of ELDs. Abbreviations: AP = apodeme, bb = basal setal field, BH = basal hood, BLD = basal lateral digit, Bs = basal spur, BSB = basal supporting block; ELD = endophallic lateral digit, MEG = median ejaculatory guide, SBF = subbasal fenestra.

lateral digits (BLD) and endophallic lateral digits (ELD), a basal supporting block (BSB), a median ejaculatory guide (MEG) and basal setal field (bb). For more details on this species and a second *Hermesia* species, see Flowers (1995).

Adorea cruentata Lefèvre [Venezuela] and *A. chontalensis* Jacoby [Costa Rica] (Figs. 18-22) show similar developments of the median lobe, but very different endophalli. In both species, the basal hood is much shorter than the median lobe, lightly sclerotized, with a low dorsal crest and weak lateral apodemes; the tegmen is triangular to spoon-shaped; and the median lobe has well developed basal spurs, a subbasal fenestra, and an extended apex beyond the orifice of the endophallus. In *A. cruentata* (Fig. 18), the endophallic structures are reduced to basal membranous lobes and some tiny sclerites. In *A. chontalensis* (Fig. 19), the BLB's (Fig. 21) are bifid and heavily sclerotized; the ELD's (Fig. 22), although membranous, are well-developed lobes. In both species the apical sclerites (Figs. 18, 20) are large and well sclerotized.

Colaspis sp. [Venezuela] (Figs. 23-26). The basal hood (Fig. 23) is subequal in length to the median lobe, with lateral apodemes, and a constriction is present at the junction with the median lobe; the tegmen is flat and triangular, with lateral arms forming a ring around the base of the median lobe. A subbasal fenestra and basal spurs are present on the median lobe. The endophallus has well developed BLD's, ELD's and bb (dorsal view, Fig. 24); the BLD is largely membranous with a lightly sclerotized central area, and the ELD has a membranous dorsal part and a sclerotized, rod-like, ventral part. The apical sclerite (Fig. 25) is a small, contorted structure. The only evidence of ventral sclerites is a fine, hair-like, Y-shaped sclerite (Fig. 26). In the other genera of Eumolpini studied, ventral sclerites were either not found, or they were very much reduced as in Fig. 26.

Deuteronoda suturalis Lefèvre [Panamá] (Figs. 28-29). The basal hood, median lobe, and endophallus of this species are developed similarly to *Colaspis* (Figs. 23-25). The ELD is membranous with longitudinal sclerotized bars (Fig. 29). In contrast to the ELD of *Colaspis* sp. (Fig. 25), the ELD of *Deuteronoda suturalis* has the dorsal part better developed than the ventral part.

Brachypnoea sp. [Venezuela] (Fig. 27). The basic development of the genitalia of this species follows that of *Colaspis* and *Deuteronoda*. Endophallic structures are reduced but identifiable; this reduction, like that of *Caryonoda* (Fig. 16) is related to the overall small size of the beetles in this genus.

Section Edusites

Tymnes tricolor (Fabricius) [USA] (Figs. 35-37). The lightly sclerotized basal hood bears lateral apodemes and is dorsally constricted at the joint with the median lobe. Basal spurs are present on the median lobe (Fig. 35). The tegmen is triangular with the apical arms partially surrounding the median lobe. The endophallus has both the BLD's and ELD's well developed, and the basal setal field is in a shallow depression with a well-sclerotized margin (Fig. 36). The BLD is narrow and partially sclerotized. A prominent dorsal lobe is present between the BLD's. The ELD's (Fig. 36) have large sclerites on their apical faces. The apical sclerite (Fig. 37) is large and heavily sclerotized.

An interesting modification of the median lobe is the ventral patch of setae just before the apical projection. These setae are associated with what appear to be sensory tubes.

Section Eumolpites

Longeumolpus sp. nr. *amabilis* Springlová [Peru] (Fig. 30), and *Eumolpus* sp. nr. *cupreus* Olivier [Argentina] (Figs. 31-34). The basal hood is lightly sclerotized with lateral apodemes present on each side. A strong constriction is present at the junction of the basal hood and the median lobe. The median lobe is robust and sharply bent in *Eumolpus*, slender and less abruptly curved in *Longeumolpus*; subbasal fenestrae and basal spurs are present in both genera. The endophallus bears well developed BLD's and ELD's in both genera. The apical sclerites (Figs. 30, 34) are relatively large with a tubular basal portion separated from a lobe-like apical portion by spiral grooves.

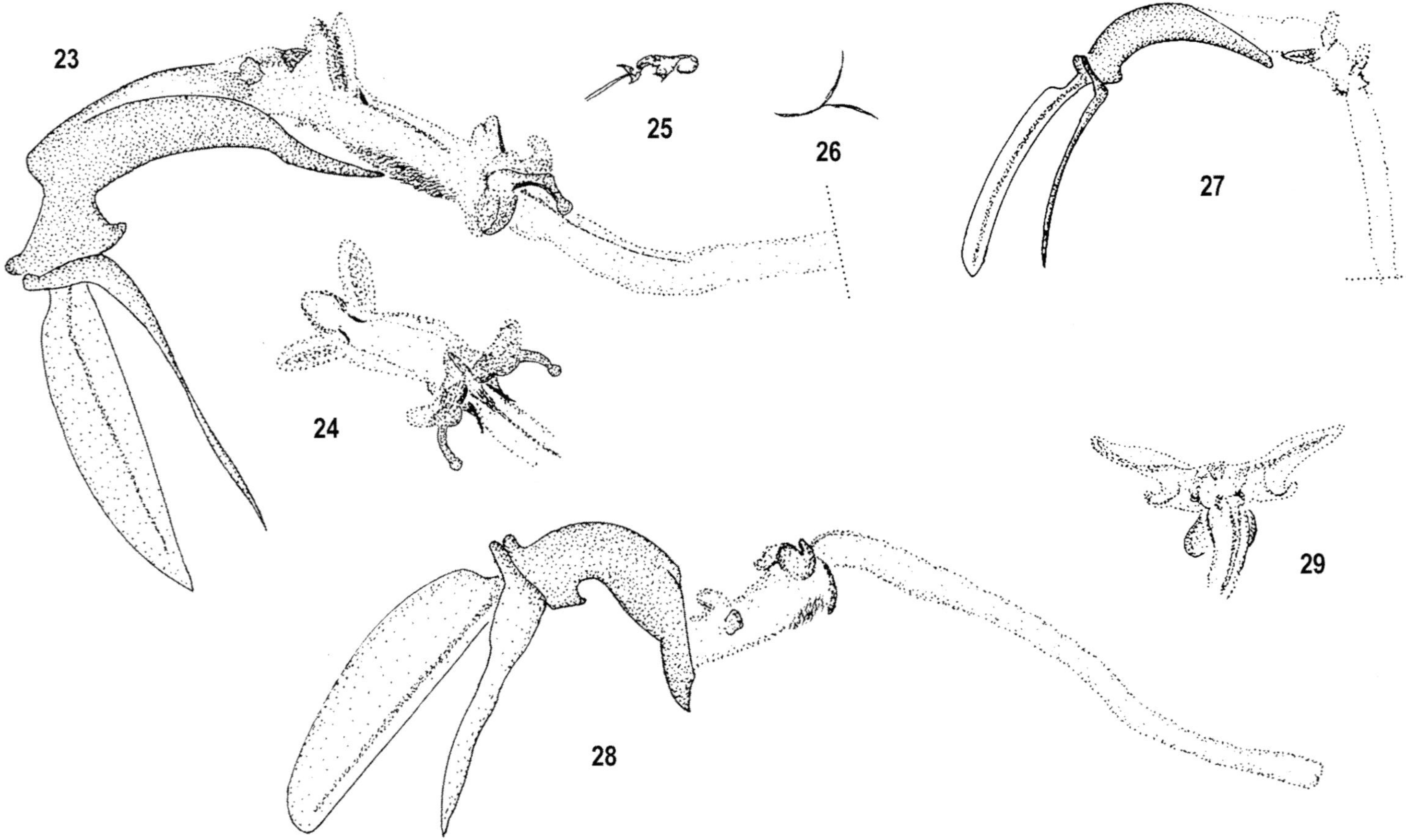

Figs. 23-29. Male genitalia of Iphimeites. 23-26, *Colaspis* sp.: 23, lateral view of median lobe and endophallus (partially everted); 24, dorsal view of endophallus between BLDs and ELDs; 25, apical sclerite; 26, ventral sclerite. 27, *Brachypnoea* sp., lateral view of median lobe and endophallus (partially everted). 28-29, *Deuteronoda suturalis*: 28, lateral view of median lobe and endophallus; 29, apical view of ELDs.

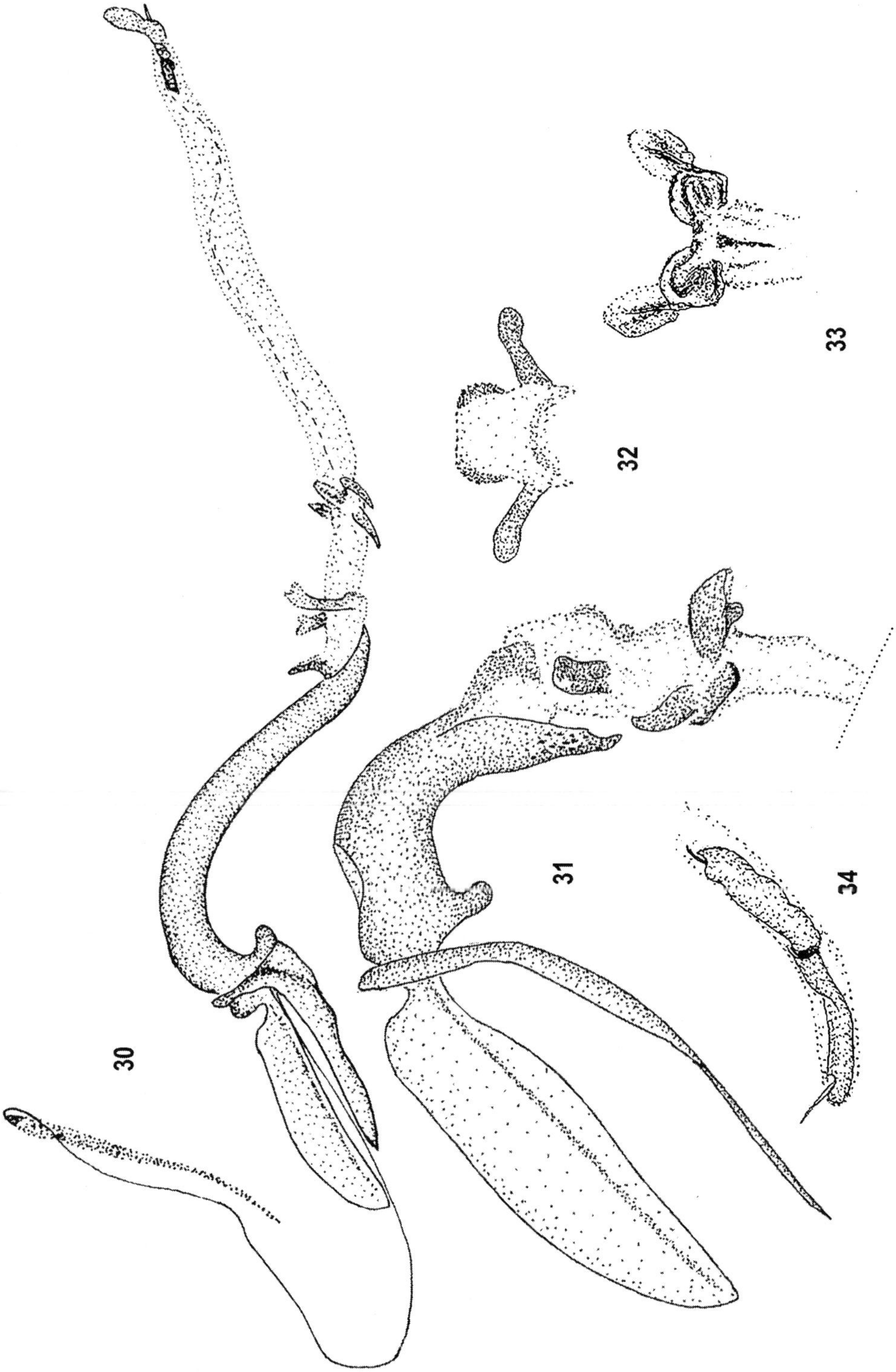

Figs. 30-34. Male genitalia of Eumolpites. 30, *Longeumolpus* sp. nr. *amabilis*, lateral view of spermathecal gland, flagellum, median lobe, and endophallus. 31-34, *Eumolpus* sp. nr. *cupreus*: 31, lateral view of median lobe and endophallus (partially everted); 32, apical view of BLDs; 33, apical view of ELDs; 34, apical sclerite.

The figure of *Longeumolpus* (Fig. 30) also shows the flagellum and lightly sclerotized spermathecal gland. All the Eumolpini studied have structures similar to those found in *Longeumolpus*, although they are illustrated only here and in Fig. 75. In the other tribes of the Eumolpinae, the flagellum and spermathecal gland are unsclerotized, and were often lost during clearing and dissection of the median lobe.

SECTION CORYNODITES

Chrysochus auratus (Fabricius) [USA] (Figs. 38-40). The basal hood is weakly sclerotized and weakly constricted at the junction of the median lobe. No lateral apodeme is present. Basal spurs are present on the median lobe and a subbasal fenestra is present on a strongly flattened dorsal area (Fig. 38). The tegmen (Fig. 39) is triangular with the lateral arms surrounding the median lobe. The endophallus (Fig. 38) is largely membranous, with four membranous projections are present at the base. Immediately apical to this are a pair of weak sclerites that are possibly a reduced BSB. The apical sclerite (Fig. 40) is an elongate twisted structure. Mann & Crowson (1996) illustrated an uneverted endophallus of *Platycorynus peregrinus* (Herbst.), an Old World member of Corynodites. Their illustration shows a complex structure with what may be sclerites on the basal third, and complex internal folding in the apical area.

A general rule in working with eumolpine endophalli is that the larger beetles generally afforded better opportunities of getting the endophallus everted. *Chrysochus*, although a relatively large beetle, was unusually difficult to work with. This was probably due to the sharp bend and compressed base of the median lobe, resulting in an unusually narrow tube through which the endophallic membrane had to be pulled.

SECTION ENDOCEPHALITES

Colaspoides unicolor Jacoby [Costa Rica] (Figs. 41-43), and *Colaspoides* sp. [Brazil] (Fig. 44). The basal hood is lightly sclerotized, with a dorsal constriction at the joint with the median lobe, and no lateral apodemes. Basal spurs and a subbasal fenestra are present on the median lobe. The tegmen is triangular with the lateral arms partially surrounding the median lobe. The endophallic BLD's are reduced and membranous, but a prominent median bilobed structure is present dorsally (Fig. 41). The ELD's are well developed and partially sclerotized. The figure of *C. unicolor* (Fig. 41) shows the ELD's only partially everted; a fully everted set from an unidentified Brazilian *Colaspoides* is shown in Fig. 44. The apical sclerite (Fig. 42) is small and contorted. An unusual feature of both *Colaspoides* studied was the very large spermathecal gland (Fig. 41) which was much longer than the combined median lobe and basal hood (compare with Figs. 28 and 75).

TRIBE HABROPHORINI

Habrophora maculipennis Jacoby [Costa Rica] (Figs. 45-46). The basal hood is weakly sclerotized with lateral apodemes present, and with a constriction present at the junction of the median lobe. A weak subbasal fenestra is present, but basal spurs are absent. The endophallus is largely membranous, with lobes and small sclerotized rods present at the base (Fig. 45). The apical sclerite (Fig. 46) is a small twisted sclerotized structure.

Psathyrocerus fulvipes Blanchard [Chile] (Figs. 47-50). The basal hood is moderately sclerotized and separated from the median lobe, and lacks lateral apodemes. The median

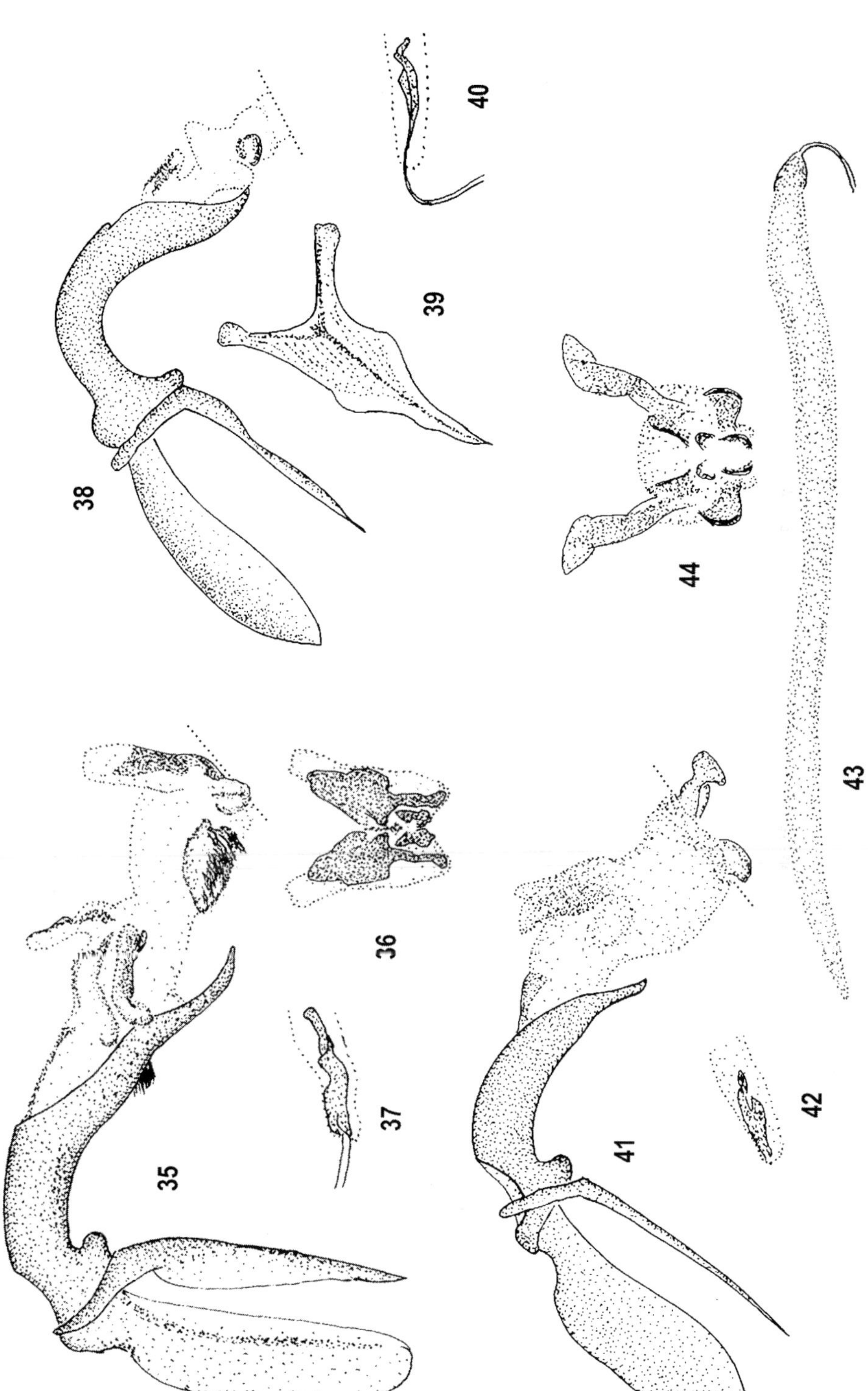

Figs. 35-44. Male genitalia of Edusites, Corynodites, and Endocephalites. 35-37, *Tymnes tricolor*: 35, lateral view of median lobe and endophallus (partially everted); 36, apical view of ELDs; 37, apical sclerite. 38-40, *Chrysochus auratus*: 38, 39, tegmen, ventral view; 40, apical sclerite. 41-44, *Colaspoides*. 41-43, *Colaspoides unicolor*: 41, lateral view of median lobe and endophallus (partially everted); 42, apical sclerite; 43, spermatheal gland. 44. *Colaspoides* sp., apical view of ELDs.

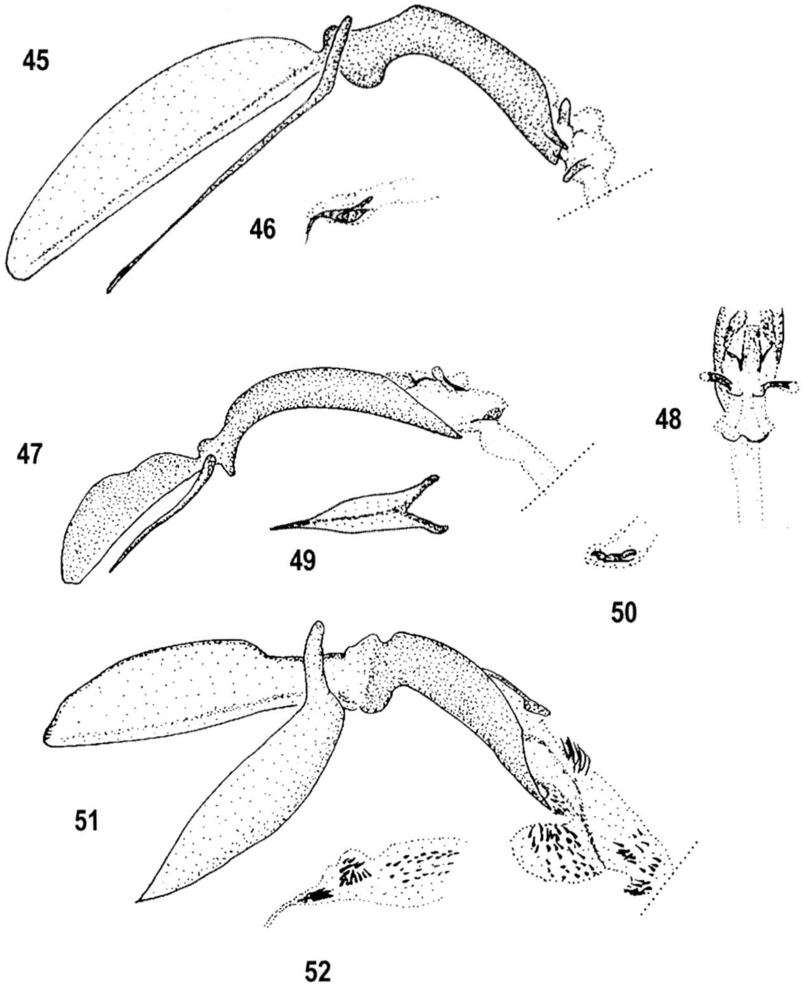

Figs. 45-52. Male genitalia of Habrophorini and Megascelidini. 45-46, *Habrophora maculipennis*: 45, lateral view of median lobe and endophallus (partially everted); 46, apical sclerite. 47-50, *Psathyrocerus fulvipes*: 47, lateral view of median lobe and endophallus (partially everted); 48, dorsal view of base of endophallus; 49, tegmen, ventral view; 50, apical sclerite. 51-52, *Megascelis* sp.: 51, lateral view of median lobe and endophallus (partially everted); 52, apical sclerite.

lobe is shallowly curved with a poorly defined subbasal fenestra, and with basal spurs absent. The tegmen (Fig. 49) is narrowly triangular with the lateral arms not surrounding the median lobe. The endophallus (Fig. 48) has a pair of thin but well-sclerotized lateral arms and a pair of partially sclerotized lobes; between these there are a series of narrow sclerites which are associated with the basal part of the endophallus. A small apical sclerite (Fig. 50) is present.

Tribe Megascelidini

Megascelis sp. [Bolivia] (Figs. 51-52). The basal hood is very weakly sclerotized with dorsal and lateral apodemes. The junction of the basal hood and the median lobe is constricted and partly membranous. The median lobe has a basal swelling ventrally and a subbasal fenestra dorsally. The portion of the tegmen beneath the basal hood is very large and boat-shaped; the lateral arms completely encircle the median lobe. The endophallus has a field of very long dorsal setae, and a large ventral membranous lobe bearing large arrowhead-shaped spicules. A basal supporting block with another field of long heavy curved setae is present on each side of the BSB. A series of narrow lateral sclerites support the ventral lobe. The apical part of the endophallus (Fig. 52) has both small and large arrowhead-shaped spicules, and a small apical sclerite. The flagellum is membranous.

Tribe Adoxini

Section Scelodontites

Graphops marcassita (Crotch) [USA] (Figs.53-55). The basal hood is well sclerotized, except at the very edges, and is continuous with the median lobe. Basal swellings or spurs, and a subbasal fenestra are absent. The tegmen (Fig.55) is quadrate, with an apical V-shaped emargination, and with the lateral arms not curved up around the median lobe. A thin Y-shaped ventral sclerite is present (Fig. 54).

Section Leprotites

Fidia viticida Walsh [USA] (Figs. 56-59). The basal hood is well sclerotized and continuous with the median lobe. A subbasal fenestra and basal spurs are absent, although the venter of the median lobe is expanded. The tegmen (Fig. 57) is quadrate and deeply bilobed at the apex, with the arms not surrounding the median lobe. The endophallus has membranous BLD's with a dorsal lobe, large membranous ELD's with weakly sclerotized ventral lobes, a short MEG, and the BSB reduced to two narrow ventral sclerites. A pair of large dorsal kidney-shaped sclerotized plates between the ELD's and the basal lobes (Fig. 56). The apical sclerite is a simple arrow shaped structure (Fig. 58). The ventral sclerites (Fig. 59) consist of a basal Y-shaped plate with two narrow lateral associated sclerites, and an apical sclerite forming a single bar with a very weakly sclerotized margin.

Xanthonia decimnotata (Say) [USA] (Figs. 60-62). The basal hood is sclerotized and continuous with the median lobe, a subbasal fenestra and basal spurs are absent. The tegmen (Fig. 61) is quadrate, bilobed at the apex, and has short lateral arms. The endophallus is membranous (Fig. 60, shown fully retracted) with a very small apical sclerite. A fine, hair-like ventral sclerite is present (Fig. 62).

Section Bromiites

Bromius obscurus villosulus (Schrank) [USA](Figs. 63-66). The basal hood is moderately sclerotized and continuous with the rest of the median lobe. No distinct basal spurs or subbasal fenestra are present. The tegmen (Fig. 64) is quadrate and bilobed at the posterior margin. The endophallus (Fig. 63) is covered with microspicules and is irregularly folded under the basal hood when retracted. Most of the endophallus appears to be a simple tube – no successful erections were made for this genus – but there appear to be several scler-

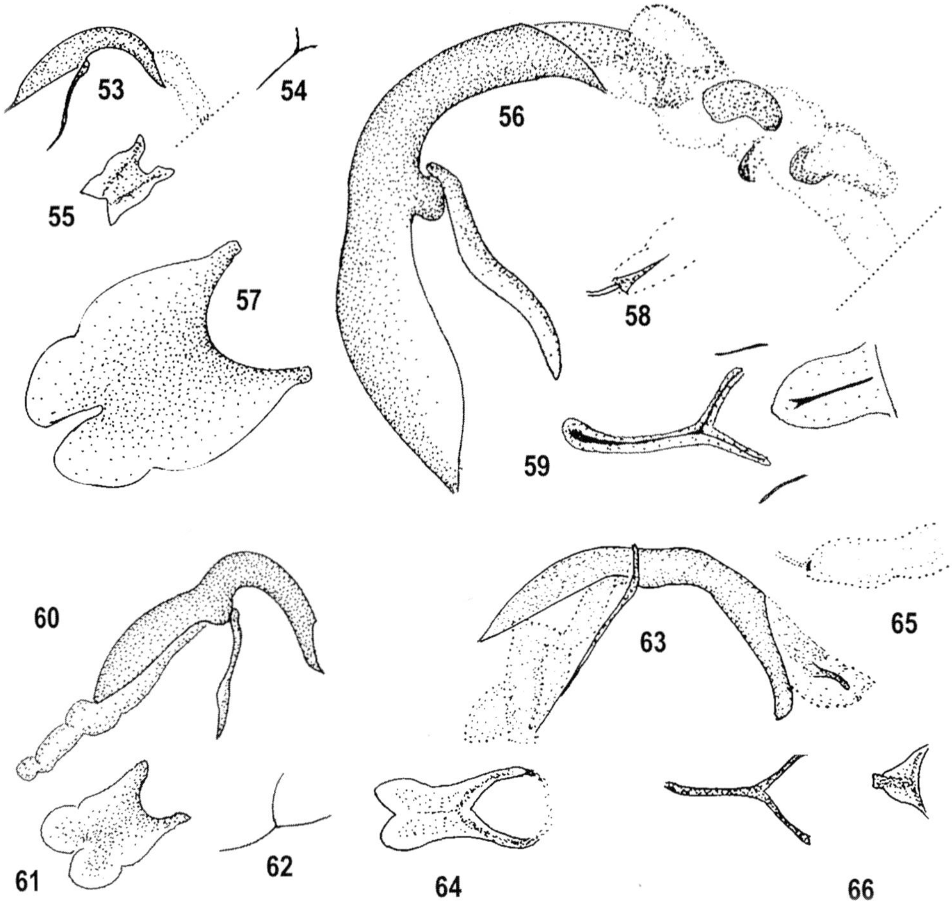

Fig. 53-66. Male genitalia of Adoxini. 53-55, *Graphops marcassita*: 53, lateral view of median lobe and endophallus (partially everted); 54, ventral sclerite; 55, tegmen, ventral view. 56-59, *Fidia viticida*: 56, lateral view of median lobe and endophallus (partially everted); 57, tegmen, ventral view; 58, apical sclerite; 59, ventral sclerites, ventral view. 60-62, *Xanthonia decimnotata*: 60, lateral view of median lobe and endophallus (partially everted); 61, tegmen, ventral view; 62, ventral sclerite,ventral view. 63-66, *Bromius obscurus*: 63, lateral view of median lobe and endophallus (partially everted); 64, tegmen, ventral view; 65, ventral sclerites, ventral view.

ites, including a pair of narrow curved rods, associated with the junction of the base of the endophallus and the median lobe. The apical sclerite (Fig. 65) is tiny, and the flagellum is membranous. Two Y- shaped, well developed, ventral sclerites are present (Fig. 66).

Section Myochroites

Glyptoscelis fascicularis Baly [Venezuela] (Figs. 67-68). The basal hood is sclerotized with a transparent dorsal crest but without lateral apodemes, the hood is constricted at its junction with the median lobe. Basal spurs are present on the median lobe. The lightly sclerotized tegmen is triangular, with broad and almost transparent lateral margins. The endophallus is membranous with a weakly sclerotized basal area and dorsal bars. There is a large ventral lobe near the base, then a pair of smaller ventral flaps and a pair of small

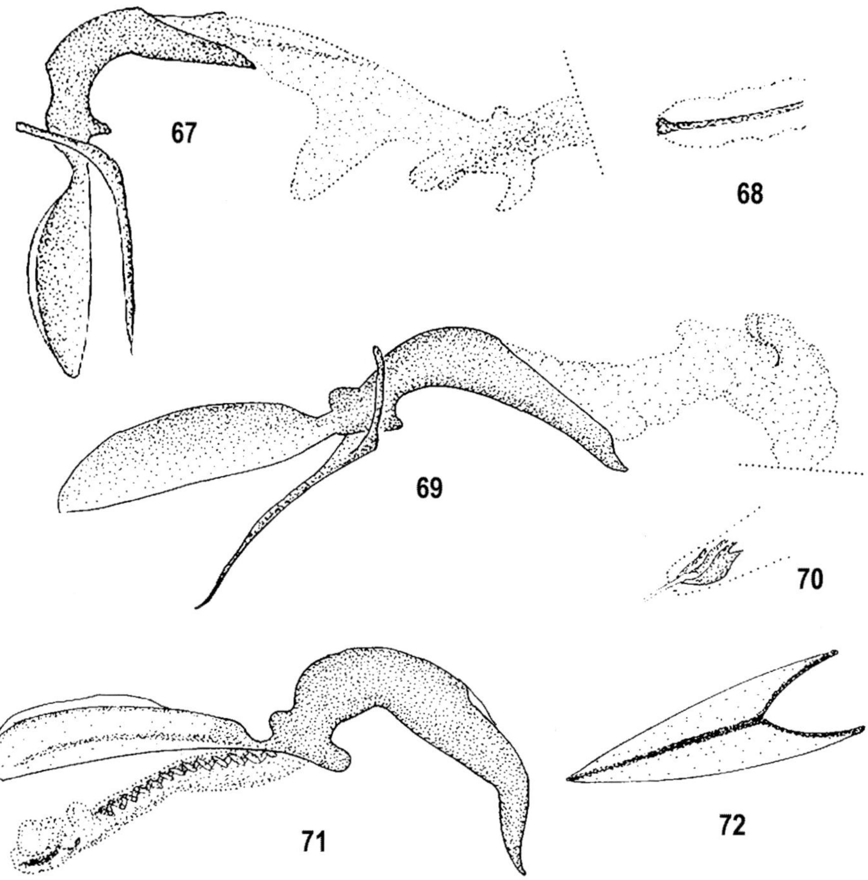

Figs. 67-72. Male genitalia of Myochroites. 67-68, *Glyptoscelis fascicularis*: 67, lateral view of median lobe and endophallus (partially everted); 68, apical sclerite. 69-70, *Glyptoscelis albicans*: 69, lateral view of median lobe and endophallus (partially everted); 70, apical sclerite. 71-72, *Myochrous* sp.: 71, lateral view of median lobe and endophallus (retracted); 72, tegmen, ventral view.

dorsal lobes. What appears to be another, entirely membranous hook-shaped, ventral lobe, covered with fine spicules is also present. The apical lobe (Fig. 68) is straight and narrow.

Glyptoscelis albicans Baly [USA] (Figs. 69, 70). The median lobe, basal hood, and tegmen are similar to those of *G. fasicularis*, except the there is no dorsal crest on the basal hood. The endophallus is entirely membranous with a small median dorsal lobe, and apical to that, a pair of larger dorsal lobes distally. The apical sclerite (Fig. 70) is sclerotized and slightly twisted.

Myochrous sp. [Venezuela] (Figs. 71, 72). The basal hood is lightly sclerotized with a dorsal crest and lateral apodemes. The junction with the median lobe is constricted. Basal spurs and a subbasal fenestra are present on the median lobe. The tegmen (Fig. 72) is triangular. The endophallus (shown fully retracted in Fig. 71) has a spiral structure running along its length. The exact function of this spiral could not be determined, but examination under high magnification indicated that it is a complex invagination of the dorsal surface of the endophallus. The apical sclerite is a small curved structure.

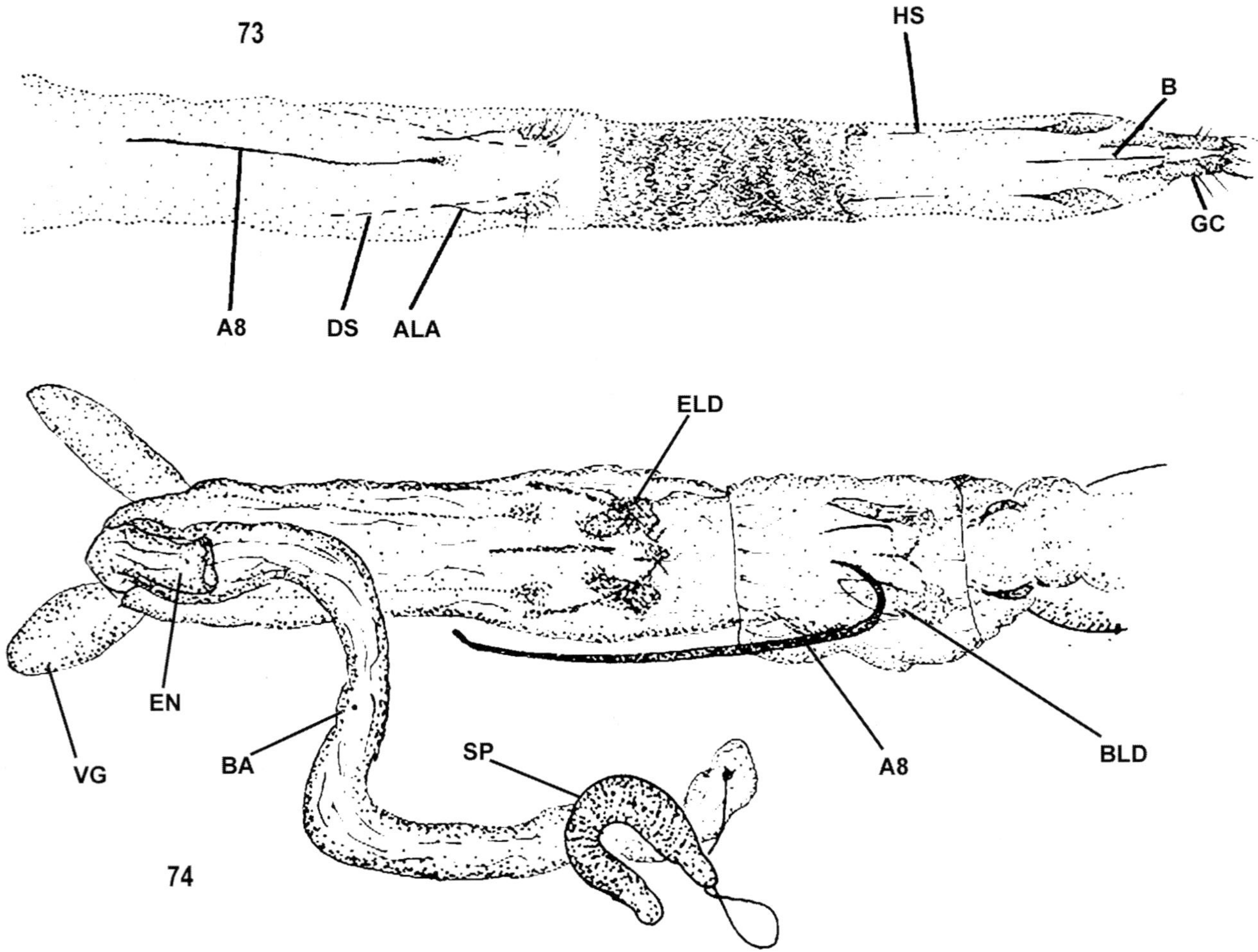

Figs. 73-74. Female characters and genitalia in copula. 73, ovipositor of *Hermesia aurata*, ventral view. 74, *Colaspis pseudofavosa*, female genitalia surrounding male endophallus during copulation. Abbreviations, female structures: ALA = apicolateral arms of sternum VIII, A8 = apodeme of sternum VIII, B = baculum, BA = bursa, DS = dorsal sclerites of segment VIII, GC = gonocoxae, HS = hemisternites of segment XI, VG = vaginal glands. Male structures: BLD = basal lateral digit, ELD = endophallic lateral digit.

4. Function of the endophallus in copulation

The female genitalia in the Eumolpinae, illustrated in their extended condition in Fig. 73 (from Flowers, 1995), usually include segments VIII-XI modified into an elongate ovipositor which telescopes back into itself when at rest. The length of the ovipositor tube and the degree of reduction of the associated sclerites are variable both between and within genera (Flowers, unpublished data). The tube is always capped with a pair of sclerotized gonocoxae, which lie at the mouth of the genital opening of the female in most preserved specimens examined.

Although most copulating eumolpines quickly separate when disturbed, occasionally it is possible to collect a pair that remain joined after death. I was fortunate enough to collect a pair of *Colaspis pseudofavosa* Riley that remained joined after they were dropped in alcohol. Figures 74-76 show several views of the linked male and female structures. In copula, the gonocoxal segment (XI) of the female has been pushed back into the folded ovipositor tube, and the gonocoxae are in contact with the endophallic lateral digits (Figs. 75, 76). Indeed, the apex of the ovipositor tube has been spread so that each gonocoxa lies in a notch between the ELD and the side of the basal supporting block on each side of the endophallus. The ELD's are directed forward over the gonocoxae, while the sides of the basal block appear to be pressed backward (Fig. 76). The female system appears to be under tension, as indicated by the apodeme of sternum VIII, which is rotated to the side and sharply bent inward at the apex (Fig. 74).

Konstantinov (1998) illustrated the copulatory apparati of *Aphthona formosana* Chen (Alticini), also caught *in flagrante delecto*. The endophallus of this system is considerably simpler than in *C. pseudofavosa*, and there is no telescoping ovipositor. The vaginal palpi (analogous to the gonocoxae in Eumolpinae) of the female *Aphthona* are within the vagina, but in this case they are in close contact with the tip of the male median lobe: Konstantinov regarded this contact as an "internal courtship device". The endophallus also bears two small sclerotized structures, but their function is apparently still unknown. Both examples – the Eumolpinae and the Alticini – have in common specific contact of internal sclerotized structures within the female vagina. Konstantinov argued that the vaginal palpi of Alticini are not homologous to the gonocoxae of Eumolpinae and other chrysomelids, but both in appearance and function, they are definitely analogs.

5. Discussion

The current state of our knowledge of the eumolpine male genitalia is too fragmentary to delineate phylogenetic trends at this time. The endophallus shows remarkable variation among the surveyed genera. However, knowledge of the variability of endophallic structures within genera is preliminary. Two examples have been given – *Adorea* and *Glyptoscelis* – which show surprising amounts of variation within relatively small genera. It may be that these characters are uninformative at the generic level.

Notwithstanding our incomplete and uncertain knowledge base, it appears that there are two major generic divisions within the Eumolpinae, based on the morphology of the median lobe and associated sclerites. In Division I, the basal hood is broadly fused to the median lobe, basal spurs are absent or poorly developed, the tegmen is usually quadrate or

75

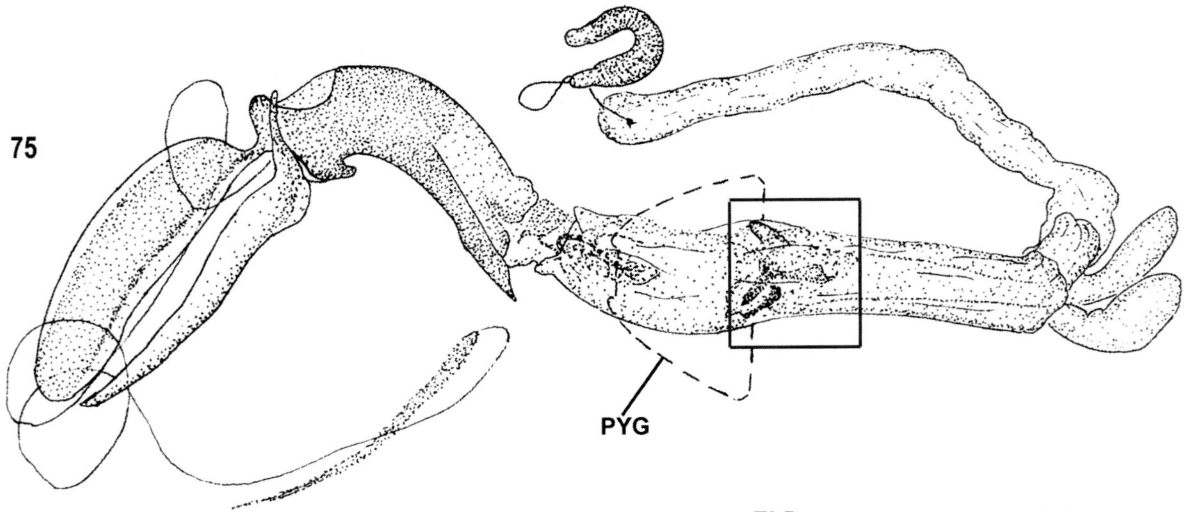

76

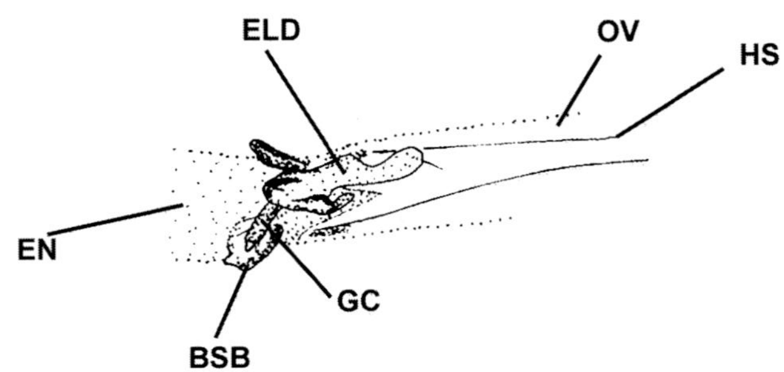

Figs. 75-76. Genitalia of *Colaspis pseudofavosa* in copula. 75, Male and female genitalia, ventral view. Dotted line shows the position of the female pygidium relative to the copulatory organs; square outline shows area shown in Fig. 76. 76, area within square in Fig. 75 with overlying membranes of female ovipositor removed. Abbreviations: BSB = basal supporting block, ELD = endophallic lateral digit, EN = endophallus, GC = gonocoxa, HS = hemisternite, OV = ovipositor.

greatly reduced, and the sternites of segments eight and nine are relatively well sclerotized. Additionally, the endophallus is usually a simple or spiculate membrane, the apical sclerite is small, and the flagellum is membranous. This form of genitalia is found in two tribes considered among the most primitive of the Eumolpinae: Hornibiini and Synetini. It is also characteristic of the Typophorini, and the sections Scelodontites, Leprotites, and Bromiites of the Adoxini.

In Division II, the basal hood is separated from the median lobe by a distinct constriction, the hood is relatively lightly sclerotized and held rigid by apodemes, the median lobe has well-developed basal spurs, or at least a distinct basal swelling, and the subbasal fenestra is usually distinct. The tegmen is usually triangular, and sternites VIII and IX are often not visible, or are reduced to very fine sclerites. The endophalli in this division are often modified, frequently with sclerotized structures, but sometimes only with membranous lobes. In Division II fall the many genera of Eumolpini, the Caryonodini, the Habrophorini, the Megascelidini, and the section Myochroites of the Adoxini.

Combining the above observations with external characters used in the tribal classification of the Eumolpinae, it can be hypothesized that Division II (Eumolpini+Habrophorini+Myochroites) forms an apomorphic group to the remainder of the Eumolpinae. The enlarged basal hood with a constricted base, and the development of lobes and appendages on the endophallus are, as far as known, restricted among Chrysomelidae to this group of Eumolpinae. Within Division I, the Typophorini is also generally considered to be a monophyletic group on the basis of its modified middle and hind tibiae. However, the Adoxini, as currently constituted, is unquestionably polyphyletic. The smaller tribes Caryonodini and Megascelidini are close to Eumolpini-Habrophorini in both internal and external morphology; however, their status as independent tribes needs to be confirmed by further study. As noted above, the male genitalia of *Caryonoda* is no different than what one would expect in a very small Eumolpini.

The inclusion of Megascelidini within the Eumolpinae by Reid (1995) is confirmed by the structure of the median lobe, tegmen, and basal hood. The endophallus of the Bolivian species raises some questions about where in the Eumolpinae this tribe should be placed. In external characters, *Megascelis* is close to the Habrophorini, but the one endophallus successfully everted (although attempts were made with several other *Megascelis* species) most closely resembles the endophallus of *Glyptoscelis fascicularis*.

Reid (1995) also moved the Synetini to the Eumolpinae, arguing that *Syneta* is a plesiomorphic eumolpine with adult apomorphies. The similarities between the endophalli of *Syneta* and *Hornibius* (Figs. 1 and 3) lend support to his view. As noted by Reid, the tegmen is very reduced, and the basal hood is short and broad.

As Reid (1995) commented, the Eumolpinae "are in a mess". At present there can be no assurance that all the currently catalogued genera and species of the three large tribes are properly assigned – this is particularly true of the Eumolpini in the Neotropics – until both genitalic and external characters of each genus are carefully reviewed. Additionally, biodiversity studies throughout the tropics are bringing large numbers of undescribed species to light. To date, reliance on external characters only have given unsatisfactory and contradictory classifications. The study of genitalic characters of both sexes will add a new dimension to the classification of the Eumolpinae and may offer a way out of the taxonomic morass into which this subfamily has fallen.

Acknowledgments

I thank the following individuals and institutions for providing specimens that were used in this study: I. Askevold, Tallahassee, Florida; E. Riley, College Station, Texas; W. Clark, Auburn University Entomological Museum, Auburn, Alabama; J. Liebherr, Cornell University Insect Collection, Ithaca, New York; M. Thomas, Florida State Collection of Arthropods, Gainesville, Florida; and A. Solís, Instituto Nacional de Biodiversidad, Santo Domingo, Costa Rica. I also thank K.E.M. Galley and P. Kovarik for helpful comments on the manuscript. This research was supported in part by a grant (FLAX91005) from the CSREES, USDA, to Florida A&M University, and a National Science Foundation Mid-Career Fellowship (BSR-9003898) to the author.

References

Askevold, I.S. 1988. The genus *Neohaemonia* Szkékessy in North America (Coleoptera: Chrysomelidae: Donaciinae): systematics, reconstructed phylogeny, and geographic history. Trans. Am. ent. Soc. 113:361-430.

—, 1990. Reconstructed phylogeny and reclassification of the genera of Donaciinae (Coleoptera: Chrysomelidae). Quaest. ent. 26:601-664.

—, 1991. Classification, reconstructed phylogeny, and geographic history of the New World members of *Plateumaris* Thomson, 1859 (Coleoptera: Chrysomelidae: Donaciinae). Mem. ent. Soc. Canada 157:1-175.

—, Flowers, R.W. 1994. *Glyptosceloides dentatus*, a genus and species of Eumolpinae new to Chile (Coleoptera: Chrysomelidae). Revta. chil. Ent. 21:69-76.

Baly, J.S. 1879. An attempt to point out the differential characters of some closely-allied species of *Chrysomela*, principally those contained in Suffrian's 11th group: also descriptions of some hitherto uncharacterized forms belonging to the same and other genera of the family. Trans. ent. Soc. London 2:171-179.

Bechyné, J. 1950. Nuovi Eumolpidi Sud-Americani. Memoria Mus. civ. Stor. nat. Verona 2:99-106.

—, 1951. Liste provisioire des Eumolpides de Bolivie et observations diverses sur les espèces de l'Amerique du Sud. (Col. Phytophaga). Ent. Arb. Mus. Georg Frey 2:227-352.

—, de Bechyné, B. S. 1961. Notas sobre Chrysomeloidea Neotropicais II. Bolm. Mus. Para. Emilio Goeldi (Zool.) 37:1-91.

Blake, D.H. 1950. A revision of the beetles of the genus *Myochrous*. Proc. U.S. natn. Mus. 101: 1-64.

—, 1951. A revision of the beetles of the genus *Chalcosicya* Blake (Chrysomelidae) from the West Indies. Bull. Mus. comp. Zool. Harv. 106:287-312.

—, 1967. A revision of the genus *Glyptoscelis* (Coleoptera, Chrysomelidae). Proc. U.S. natn. Mus. 123: 1-53.

—, 1970. A review of the beetles of the genus *Metachroma* Chevrolat (Coleoptera: Chrysomelidae). Smithson. Contr. Zool. 57: 1-111.

—, 1976. The brown semicostate and costate species of *Colaspis* in Mexico and Central America (Coleoptera: Chrysomelidae). Tech. Bull. U.S. Dep. Agric. 1534.

Chapuis, F. 1874. *In* Histoire Naturelle des Insectes. Genera des Coléoptères. T. Lacordaire and F. Chapuis, eds. 10:1-455. Paris.

Doquet, S. 1994. Coléoptères Chrysomelidae, Volume 2 Alticinae. Faune de France 80. Fédération Française des Sociétés de Sciences naturelles, Paris, France.

Flowers, R.W. 1995 *Hermesia* Lefèvre, a resurrected genus of Neotropical Eumolpinae (Coleoptera: Chrysomelidae). Proc. ent. Soc. Washington. 97:35-45.

—, 1996. La subfamilia Eumolpinae (Coleoptera: Chrysomelidae) en América Central. Revta. Biol. trop., Pub. esp. No. 2: 1-60.

—, 1997. Revision of the genus *Cayetunya* Bechyné (Coleoptera: Chrysomelidae: Eumolpinae), with descriptions of three new species. Coleopts Bull. 51:384-399.

Foudras, C. 1859. Alticides (Halticinae). In Mulsant, E. Histoire naturelle des Coléoptères de France. Annls Soc. linn. Lyon (n.s.) 6:138-384.

—, 1860. Alticides (Halticinae). In Mulsant, E. Histoire naturelle des Coléoptères de France. Annls Soc. linn. Lyon (n.s.) 7:17-128.

Furth, D.G. 1992. The New World Blepharida group, with a key to genera and description of a new species (Coleoptera: Chrysomelidae). Jl. N. Y. ent. Soc. 100:399-414.

Heikertinger, F. 1911. Welch Halticinenarten gehören Europa und Nordamerica gemeinsam an? Verh. zool.-bot. Ges. Wien 61:1-20.

—, 1913. *Psylliodes attenuata* Koch, der hopfenoder Hanf-Erdlich. II Teil. Morphologie und Bionomie der Imago. Verh. zool.-bot. Ges. Wien 63:98-136.
—, 1921. Die paläarktischen arten der halticinengattung *Batophila* Foudr. (*Glyptina* Lec.). Koleopt. Rdsch. 9(11-12):89-98.
—, 1925. Monographie der Halticinengattung *Derocrepis* Weise (Coleopt. Chrysomelidae). Weiner ent. Ztg. 42(4-10): 89-98.
Jacques, R.L., Jr. 1987. The Potato Beetles. The genus *Leptinotarsa* in North America (Coleoptera: Chrysomelidae). E.J. Brill. New York. 144p.
Konstantinov, A.S. 1998. On the structure and function of the female genitalia in flea beetles (Coleoptera: Chrysomelidae: Alticinae). Proc. ent. Soc. Wash. 100:353-360.
Leonardi, C. 1978. Studio critíco sulla *Psylliodes picina* (Marsh.) e sulle forme che le sono state attribuite, con particolare geguardo all fauna italana. Atti. Soc. ital. Sci. nat. 119:271-299.
LeSage, L. 1995. Revision of the costate species of *Altica* Müller of North America north of Mexico (Coleoptera: Chrysomelidae). Canadian Ent. 127:295-411.
Madge, R.B. 1967. A revision of the genus *Lebia* Latreille in America north of Mexico (Coleoptera, Carabidae). Quaest. ent. 3:139-242.
Mann, J.S., Crowson, R.A. 1966. Internal sac structure and phylogeny of Chrysomelidae. *In* Jolivet, P.H.A., Cox, M.L., eds. Chrysomelidae Biology. Volume 1: The Classification, Phylogeny and Genetics. SPB Academic Publishing. Amsterdam, The Netherlands. pp. 291-316.
Meurgues, G., Ledoux, G. 1966. Intérêt de l'étude du sac interne dévaginé et en extension. Ann. Soc. ent. Fr.(N.S.) 2:661-668.
Reid, C.A.M. 1992. Revision of the genus *Cheiloxena* Baly (Coleoptera, Chrysomelidae, Eumolpinae) Mem. Mus. Victoria 53:101-114.
—, 1995. A cladistic analysis of subfamilial relationships in the Chrysomelidae sensu lato (Chrysomeloidea), *In* Pakaluk, J., and Slipinski, S.A., eds. Biology, phylogeny, and classification of Coleoptera: papers celebrating the 80th birthday of Roy A. Crowson. Muzeum i Instytut Zoologiii PAN, Warszawa. p. 559-631.
Schultz, W.T. 1977. Review of the genus *Rhabdopterus* (Coleoptera: Chrysomelidae) in America north of Mexico. Ann. ent. Soc. Amer. 70:968-974.
Seeno, T.N., Wilcox, J.A.. 1982. Leaf beetle genera. Entomography 1:1-221.
Sharp, D., Muir, F.A.G.. 1912. The comparative anatomy of the male genital tube in Coleoptera. Trans. ent. Soc. London Part III. (Dec.) 477-642 + 36plates, Ent. Soc. Amer. reprint, 1969.
Skelley, P.E. 1993. A method of genitalia preparation and dry preservation for Coleoptera. Proc. ent. Soc. Wash. 95:131-138.
Springlová – Bechyné, B. 1960. Essai monographique du genre *Eumolpus*. Mém. Inst. r. Sci. nat. Belg. 2nd Ser. Fas. 60. 79 pp.
Stockman, R. 1962. Etude de la variabilité de quelques espèces françaises du genre *Timarcha* (Latreille). Fac. Sc. Paris (thesis).
Weise, J. 1886. Vier neue Pachybrachys-arten. Dt. ent. Z. 30:419-420.
—, 1889a. Forcipes verschiedener Arten von Clytrinae-Gattungen. Dt. ent. Z.. 33:419
—, 1889b. Forcipes verschiedener Arten von Donaccien etc. . Dt. ent. Z. 33:419-120.
White, R.E. 1993. A revision of the subfamily Criocerinae (Chrysomelidae) of North America north of Mexico. Tech. Bull. U.S. Dep. Agric. 1805: 1-158.
Wilcox, J.A. 1965. A synopsis of the North American Galerucinae (Coleoptera: Chrysomelidae).The University of the State of New York; State Educ. Dept., N Y. State Mus. and Sci. Serv., Bull. 400, 226p.
Woodruff, R.E., Beck, B.M.. 1989. Arthropods of Florida and neighboring land areas. Vol. 13: the scarab beetles of Florida (Coleoptera; Scarabaeidae). Part II. The May or June beetles (genus *Phyllophaga*). Florida Dept. Agric., Div. Plant Indust. 225pp.

THE UTILIZATION OF CHORION ULTRASTRUCTURE AND CHORION POLYPEPTIDE ANALYSIS IN RECOGNIZING TAXONOMIC UNITS IN NORTH EUROPEAN GALERUCINI (COL., CHRYSOMELIDAE)

ANNETTE NORDELL-PAAVOLA[1], SEPPO NOKKALA[1], SEPPO KOPONEN[2] and CHRISTINA NOKKALA[1]
[1]*Laboratory of Genetics and* [2]*Zoological Museum, Department of Biology, University of Turku, FIN-20014 Turku, Finland*

1. Introduction

The egg shell or chorion provides a protective shelter for the developing embryo against external physical and biological effects. Although a number of chrysomelid species protect their eggs with excrement or regurgitated food, and even a few cases of parental care are known (see Hilker, 1994; Selman, 1994; Windsor & Choe, 1994), it is evident that the structure of the chorion is of utmost importance to the species survival and is expected to be under strong selection pressures. Variation in the surface structure and coloration of the chorion undoubtedly reflects these pressures, and as pointed out by Petitpierre and Juan (1994) the ultrastructure of chorions can be used to ascertain genetic homologies among allied species. On the other hand, the structure of the chorion may be used as a taxonomic character to separate closely related species (for example Hippa & Koponen, 1986; Selman, 1994). The study of the chorion structure can be extended to the molecular assembly of the chorion, and the analysis and comparison of chorion polypeptides may be used to clarify taxonomic problems (Nokkala & Nokkala, 1994, 1998; Nokkala *et al.*, 1998).

The main object of our study among northern European Galerucini, is the genus *Galerucella* Crotch, which is an interesting and significant genus as it includes both pests of economically important berries (Hippa & Koponen, 1975, 1976) and species that are introduced as biological control agents of weeds (Manguin *et al.*, 1993). Furthermore, the genus has been the focus of taxonomic debate over a hundred years as opinions on what to include in the genus coupled with which name that should be applied to the genus have differed widely (for example Laboissière, 1934; Ogloblin, 1936; Chûjô 1962; Gressitt & Kimoto, 1963; Wilcox, 1965, 1971; Silfverberg, 1974). Also the interpretation of the status of several morphologically very similar forms has divided opinion, especially in the *Galerucella nymphaeae* species complex (see for example Palmén, 1945; Wilcox, 1965; Silfverberg, 1974; Hippa & Koponen, 1986; Nokkala & Nokkala, 1987, 1994, 1998). By reviewing taxonomic literature, it is difficult to define even the approximate number of species belonging to *Galerucella*. In the strict sense, only about ten species are included, whereas in the broad sense up to nearly fifty species may be classified as *Galerucella*.

Advances in Chysomelidae Biology 1, pp. 95–104
Edited by M.L. Cox

Especially difficult seems to be distinguishing between *Galerucella* and the two closely allied genera, *Pyrrhalta* Joannis and *Lochmaea* Weise. It is obvious that a thorough phylogenetic analysis is badly needed to determine the relations within and between these genera.

As a working frame for the present study, we have adopted Silfverberg's (1974) concept of dividing the genus *Galerucella* into two subgenera, *Galerucella* and *Neogalerucella*, and considering *Pyrrhalta* and *Lochmaea* as separate genera. The males of the subgenus *Neogalerucella* carry an apical spur on the middle tibiae, some species also on the front or hind tibiae, while apical spurs are absent from all tibiae in the subgenus *Galerucella*. The subgenus *Neogalerucella* has been further divided into two groups, the smaller *lineola* group, which according to Silfverberg (1974) contains only one species, *G. lineola* (F.), and the *tenella* group including all other species.

In recent papers we have successfully utilized chorion polypeptide analysis to identify separate species in the disputed *Galerucella nymphaeae* species complex (Nokkala & Nokkala, 1994, 1998; Nokkala *et al.*, 1998). We have been able to show that the complex includes two distinct species, *G. nymphaeae* (L.) and *G. sagittariae* (Gyllenhal), and that the *G. nymphaeae* from northern Europe and the *G. nymphaeae* from North America are not conspecific, but the latter resembles more closely the European *G. sagittariae*. These chorion polypeptide analyses confirmed previous results based on morphology of imagos and larvae (for example Hippa & Koponen, 1986).

In the present study, we have compared egg chorion ultrastructure and chorion polypeptide patterns in several North European species of Galerucini, in all six species of *Galerucella* encountered in Finland, *G. nymphaeae* and *G. sagittariae* belonging to the subgenus *Galerucella*, *G. lineola*, *G. calmariensis* (L.), *G. pusilla* (Duftschmid), and *G. tenella* (L.) belonging to the subgenus *Neogalerucella*, and two representatives of closely allied genera, *Lochmaea suturalis* (Thomson) and *Pyrrhalta viburni* (Paykull) to determine whether the structural and chemical characteristics of chorions can be of use in recognizing taxonomic units and revealing phylogenetic relationships in these beetles.

2. Chorion ultrastructure

For studying chorion ultrastructure, beetles were collected at Parainen, Mustfinn, southwest Finland (all *Galerucella* species), and Kaarina, Kuusisto, southwest Finland (*Pyrrhalta viburni*). One pair each of *G. nymphaeae*, *G. sagittariae*, *G. lineola*, *G. calmariensis*, *G. pusilla*, *G. tenella*, and *P. viburni* was placed individually in a small petri dish provided with the appropriate food plant, and the female was allowed to lay eggs for 24 hrs. After this period the adults were removed and the eggs were killed with diethylether vapour.

Chorion ultrastructure was studied with scanning electron microscopy (SEM). Gold coated eggs were studied by JEOL (JSM-5200) microscope using 15 kV and 750 x magnification. The dorsal apex areas of eggs were compared.

The egg shell or chorion of the studied Galerucini species is characterized by raised ridges forming polygonal reticles, with a variably structured (pores or granules) inner space (Fig. 1). The chorion ridges are usually smooth and without pores. This sculptured layer of the chorion is often designated as extrachorion (see Selman, 1994).

On the basis of chorion ultrastructure, three groups were found in the present material:

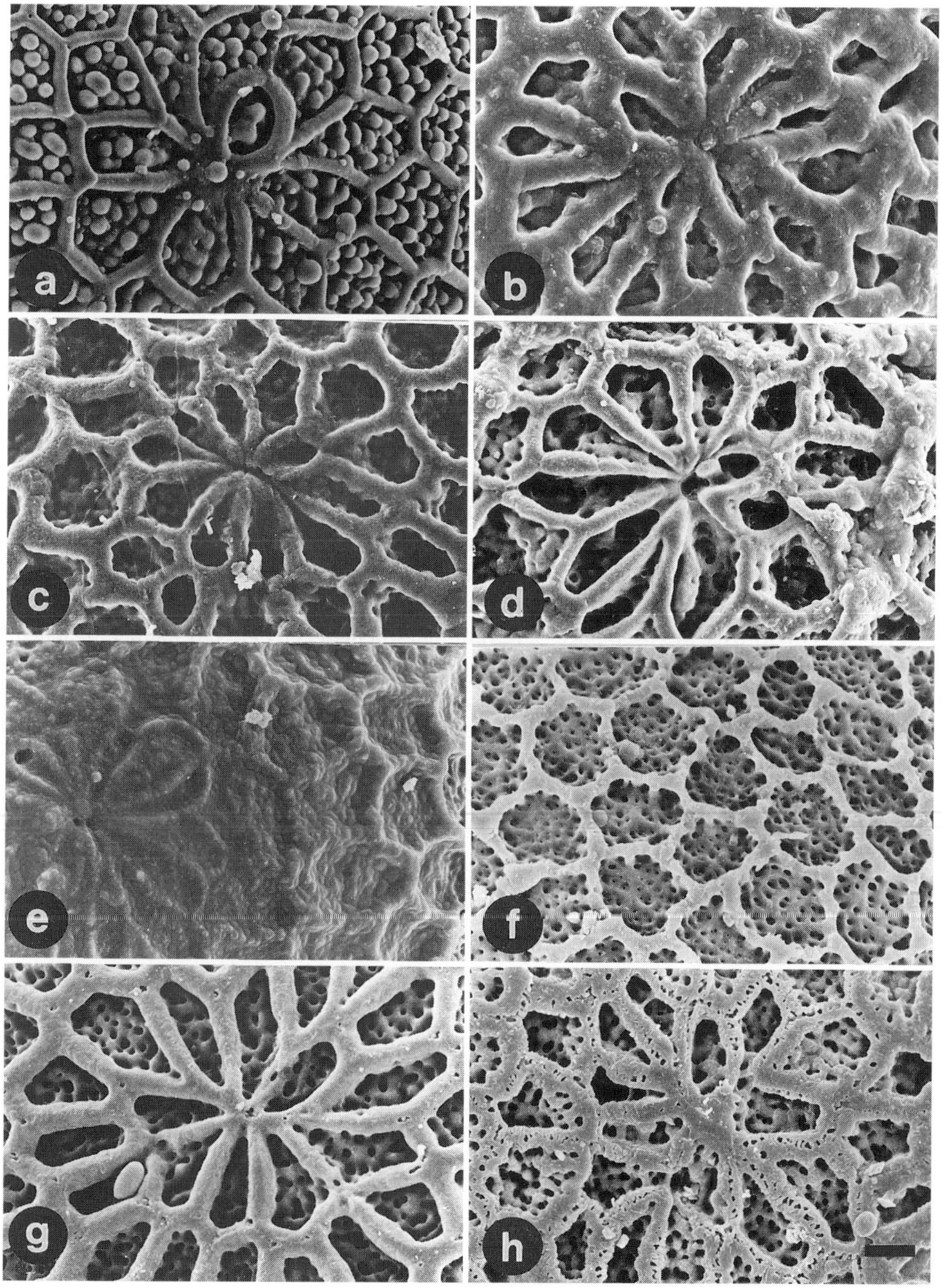

Fig. 1. Chorion ultrastructure around dorsal apex area of eggs in Galerucini. (a) *P. viburni*, (b) *G. lineola*, (c) *G. calmariensis*, (d) *G. tenella*, (e) *G. pusilla*, (f) *G. pusilla*, area adjacent to that shown in e, (g) *G. sagittariae*, (h) *G. nymphaeae*. Bar = 10 μm.

(1) Ultrastructure of the chorion in *P. viburni* with large polygons and coarse granular inner space clearly differs from the others, that is from *Galerucella* species. (2) The chorion of *G. lineola* eggs displays quite small-sized polygons bordered by thick ridges, the inner space with small granules and some pores. In *G. calmariensis* the polygons are somewhat larger and the ridges thinner, and *G. calmariensis* seems to be intermediate between *G. lineola* and the other *Galerucella* species. (3). The remaining species (*G. tenella*, *G. pusilla*, *G. sagittariae* and *G. nymphaeae*) have egg chorions with large(r) polygons with a porous inner space. Even the raised ridges in G. *nymphaeae* are porous and rough, in contrast to all other studied species, probably due to its oviposition on floating leaves of waterlilies (see Discussion).

3. Chorion polypeptides

For chorion polypeptide analysis, samples of *G. nymphaeae*, *G. sagittariae*, *G. tenella*, *G. pusilla*, *G. calmariensis*, *L. suturalis* and *P. viburni* were collected from natural populations in Finland (Table 1).

The method used for chorion polypeptide analysis is described in more detail in Nokkala *et al.* (1998). Briefly, three to five mature eggs from the ovaries of females were dissected in insect Ringer's solution. The eggs were dechorinated and dissolved in 30–40 µl of sample buffer. Samples were heated for 5 min. in a boiling water bath before submitting to SDS-PAGE electrophoresis. We used precast ultrathin gradient gels from Pharmacia (ExelGel™ SDS, gradient 8–18), which were subsequently stained with $AgNO_3$.

In SDS-PAGE electrophoresis molecules migrate in the electric field according to their molecular weights, and are visualized as bands after staining. A band may represent a single polypeptide, that is the product of a single gene locus, or it may be formed by several comigrating polypeptides encoded by different gene loci but sharing similar molecular weight. If a mutation affects the molecular weight of the gene product, it can be observed as a shift in band position on the gel. A locus is revealed if both homozygotes and heterozygotes are detected. Band homology in different species can be inferred from band morphology and colour.

The overall pattern of polypeptide distribution is similar in all species, molecules falling in the range of ca. 10–200 kD (Fig. 2, Table 2). We identified nine major homologous

Table 1. The species and number of populations and individuals studied with chorion polypeptide analysis.

Species	Number of populations	Number of individuals
Galerucella nymphaeae	6	506
G. sagittariae	8	551
G. lineola	3	54
G. tenella	3	186
G. pusilla	2	36
G. calmariensis	2	32
Lochmaea suturalis	1	34
Pyrrhalta viburni	2	17

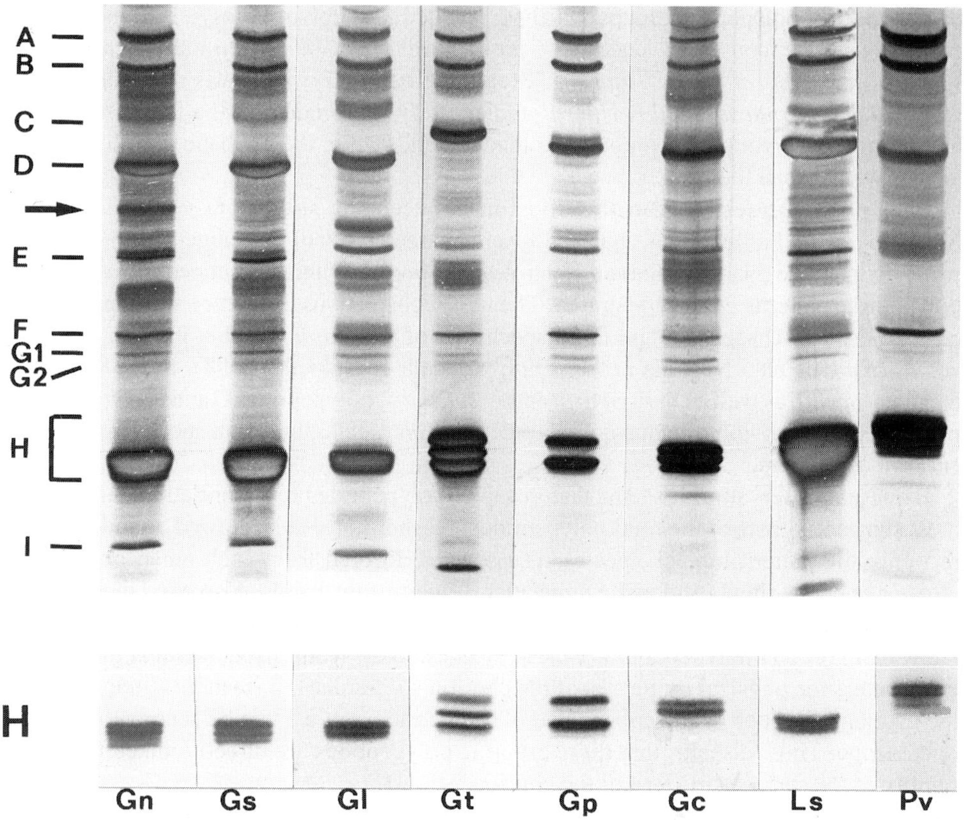

Fig. 2. Chorion polypeptide patterns in Galerucini, Gn = *G. nymphaeae*, Gs = *G. sagittariae*, Gl = *G. lineola*, Gt = *G. tenella*, Gp = *G. pusilla*, Gc = *G. calmariensis*, Ls = *L. suturalis*, Pv = *P. viburni*. Homologous polypeptides A–I are indicated. Polypeptides in the H region are shown in more detail (diluted samples). The 79 kD polypeptide typical for *G. nymphaeae* is indicated by an arrow.

Table 2. Comparison of homologous polypeptides in six species of *Galerucella*, *Lochmaea suturalis* and *Pyrrhalta viburni*. Polypeptides typical for the subgenus *Galerucella* on grey background, the polypeptide E characteristic to the subgenus *Neogalerucella* is boxed.

Species	Polypeptides (kD)									
	A	B	C	D	E	F	G1	G2	H	I
Galerucella nymphaeae	194	172	118	97,4	65	45	40	38	31,29,27,26	16,5
G. sagittariae	194	172	118	97,4	62	45	40	38	31,29,26	16,5
G. lineola	194	172	124	100	72	45	40	38	31,30,29	16
G. tenella	194	172	138	119	72	45	40	38	34,31,28	14
G. pusilla	194	172	132	110	72	45	40	38	32,28	15
G. calmariensis	194	172	126	104	72	45	40	38	31,30,29	15
Lochmaea suturalis	194	172	124	108	68	45	40	–	30,31,29	12
Pyrrhalta viburni	194	172	125	105	65	45	40	–	37,35,33,31	12,5

polypeptides or polypeptide groups, which we labelled A–I. Polypeptides A, B, F and G1+G2 are obviously very conserved, since they were identical in all *Galerucella* and *Neogalerucella* species, in *P. viburni* and *L. suturalis*, except the band G2 which was not found in the specimens of *L. suturalis* and *P. viburni* studied. In *P. viburni* there was a faint band with a higher molecular weight accompanying the G1 band, while the G2 band is of lower molecular weight than the G1.

Four polypeptides marked the division between the subgenera *Galerucella* and *Neogalerucella*. Polypeptides C, D and I were characteristic of the subgenus *Galerucella*, while polypeptide E labelled the subgenus *Neogalerucella*. Within the subgenus *Galerucella*, polypeptide patterns were very similar. The most conspicuous difference was a prominent band in the 79 kD region found in all specimens of *G. nymphaeae*, but in none of the *G. sagittariae* (see also Nokkala *et al.*, 1998; Nokkala & Nokkala, 1998). In the subgenus *Neogalerucella*, as well as in *P. viburni* and *L. suturalis*, polypeptides C and D were heavier in all species, than homologous polypeptides in *Galerucella*, but each species was different from the others.

By visual observation it seems that group H polypeptides represent quantitatively the most substantial part of the total polypeptides. Furthermore, they showed an individually recognizable pattern in each species. The species differed from each other, both in the number of bands shown and in the molecular weights of the bands. Also band morphology and colour showed considerable variation between species. From personal experience we know that group H polypeptides are the very latest to be synthesized, and that their emergence alters the physical properties of the chorion – it hardens, it is easier to remove from the egg, and outlines of the appearance of the chorion surface can be seen with a stereo microscope. Thus, it seems that these group H polypeptides are directly linked to the assembly of the surface structure of the chorion.

The group H polypeptides in *G. nymphaeae* and *G. sagittariae* appeared quite similar at first sight, but inspection of diluted samples revealed four bands in *G. nymphaeae* (the two lighter bands less intensely stained than the two heavier) contrasted to three bands in *G. sagittariae*. Within the subgenus *Neogalerucella*, the bands in the H region were quite different in each species. In *G. lineola* there were three bands very close to each other. The three bands in *G. calmariensis* were distributed over wider range, and the lightest band was considerably fainter than the two others. In *G. pusilla* we observed two strongly stained bands well separated from each other, each accompanied by a faint "shadow" band. *G. tenella* showed three separate strongly stained bands, but also a considerable amount of polymorphism due to heterozygosity was observed in *G. tenella* in this region. On the basis of band morphology, *L. suturalis* with its three bands more closely resembled the species in the subgenus *Galerucella* than those the subgenus *Neogalerucella*. In *P. viburni*, four bands with a characteristic morphology were observed, the two heavier bands being more intensely stained than the two lighter. As a whole, the H region polypeptides were clearly heavier in *P. viburni* than in all the other species studied, as the lightest polypeptides in *P. viburni* were approximately the same size as the heaviest in the others. The presence of four bands in *G. nymphaeae* and *P. viburni*, and the data on heterozygotes (not shown) indicated that polypeptides in this region are encoded by four separate genes.

4. Discussion

The results of the present study demonstrate the usefulness of egg chorion characteristics – structural and chemical – as taxonomic and phylogenetic markers in north European Galerucini. Analysis of chorion ultrastructure and chorion polypeptide patterns are, in fact, analyses of the same subject, only the level of observation is different. We obtained parallel results with both approaches, which confirms the point made by Petitpierre and Juan (1994) of inferring genetic homology on the basis of similar chorion ultrastructure. However, it seems that polypeptide analysis allows more detailed interpretation, as might be expected, since in polypeptide analysis we are studying the components that constitute the visually observable structure.

In general, the chorion ultrastructure of the six species of *Galerucella* and one *Pyrrhalta* was less variable than that in the genus *Chrysolina* (Chrysomelinae) shown by Petitpierre and Juan (1994), indicating that the species studied in the present study are more closely related. However, the ultrastructure of *Pyrrhalta* eggs is distinctly different from that of all studied *Galerucella* species. Within the genus *Galerucella* ultrastructural characteristics of the chorion form a continuum with *G. lineola* at one end and *G. nymphaeae* at the other, without any marked difference between the two subgenera. The properties of the chorion, thickness and the number and density of pores, seem to be correlated with the habitat of each species. As air spaces, the pores have a respiratory function. When pores are numerous, they constitute plastrons which function as lungs when the eggs are submerged. The eggs of *G. lineola* are laid on the leaves of willows and alders where they need to be protected more effectively from desiccation than submergence, while *G. nymphaeae* lays its eggs on the leaves of water lilies. *G. calmariensis*, *G. pusilla*, *G. tenella*, and *G. sagittariae* lay their eggs on the leaves of herbaceous plants, often growing on the shore or on river banks, where they may be occasionally submerged when flooding occurs.

Chorion polypeptide analysis proved to be an effective tool in recognizing taxonomic units. We observed polypeptides identical in all the species studied. Tentatively we might call them Galerucini specific polypeptides. Then we recognized two groups of polypeptides, each typical for one subgenus of *Galerucella*. *L. suturalis* and *P. viburni* were different from both subgenera in respect to these groups. Chorion polypeptide analysis did not support the division of the subgenus *Neogalerucella*, but individual characteristics were found for each species studied. In general, the polypeptide pattern in *L. suturalis* resembles the patterns in the genus *Galerucella* quite closely, whereas the pattern in *P. viburni* was more distinctive.

Also, other characters related to eggs may be valuable in resolving taxonomic and phylogenetic problems. Hilker (1994) analysed feeding deterrents in the eggs of various chrysomelids, including all species analysed in the present study, excepting *G. sagittariae*. In contrast to several other groups, feeding deterrents found in Galerucini are not found in the host plants and thus must be synthesized by the beetles. Hilker (1994) found that the eggs of *G. nymphaeae*, *G. lineola*, *G. calmariensis*, *G. pusilla*, *G. tenella* and *L. suturalis* contained two anthraquinones, chrysophanol and chrysazin, plus dithranol, while the eggs of *P. viburni* contained only chrysophanol. Eggs of another more distantly related Galerucini, *Galeruca tanaceti* (L.), contained chrysophanol and chrysazin. Hilker (1994) suggested that the occurrence of anthraquinones and dithranol is restricted to the Galerucini, whereas their presence could be used as a taxonomic marker at the tribal level. The dissimilarity of

P. viburni compared to the remaining Galerucini is likely to be connected to other discriminating characters of *P. viburni*. It lays eggs during late summer and early autumn, overwinters in an egg stage and protects its eggs well physically (Cox, 1994; Hilker, 1994). Also *G. tanaceti* oviposits in late summer and the overwintering eggs are well protected (see Hilker, 1994). The rest of the species lay eggs during spring and summer, when predators are more frequent and the unprotected eggs benefit more from feeding deterrents.

Still another character related to eggs is the choice of egg-laying site and the pattern of egg laying. From the species in the present study, all species of *Galerucella* lay their eggs on the leaves of their food plant, and *L. suturalis* on moist moss. *P. viburni* gnaws small cavities into the stems of its hostplant for egg-laying, where eggs are laid in batches forming irregular piles (Cox, 1994; Hilker, 1994; personal observation). The exception of *P. viburni* is, again, probably associated with its different life cycle. The eggs of *G. nymphaeae* and *G. sagittariae* are laid in batches of ca. 20 eggs, where the eggs are positioned side by side in one layer (personal observation). The species belonging to the subgenus *Neogalerucella* lay eggs in small batches consisting of 1–4 eggs (personal observation). Also *L. suturalis* lays its egg in batches of 1–5 (Brovdij, 1973).

Petitpierre and Juan (1994) studied and discussed the use of three different and independent characters, chorion ultrastructure, 1C nuclear DNA amount and chromosome numbers, in inferring phylogenetic relations among the Chrysomelinae. Unfortunately we have no information on the 1C DNA contents in the species studied in the present paper, but chromosome numbers are available for each species (Nokkala & Nokkala, 1987). The haploid chromosome number n=17 and a "parachute" sex chromosome system are plesiomorphic characters, as the majority of cytologically inspected Galerucini exhibit these (see also Smith & Virkki, 1978). The only known exceptions are *Galeruca tanaceti* with n=20 (parachute), *G. tenella*, n=13 (parachute) and *G. pusilla*, n=7, which also has a derived sex chromosome system, neoXneoY. Despite similar chromosome number and chromosome size within the n=17 species, changes (inversions) have occurred in the karyotypes, but because of the small size of the chromosomes the analysis of these karyotypes is difficult. The reduction of chromosome number within the subgenus *Neogalerucella* is accompanied by an increase in chromosome size.

The results of the present study as well as all other data considered suggest concordantly that *P. viburni* differs considerably from all *Galerucella* species studied, and from *L. suturalis*. Thus we may quite confidently conclude that *Pyrrhalta* is definitely not congeneric with either *Galerucella* or *Lochmaea*. The absence of *Lochmaea* eggs from chorion ultrastructure analysis is regrettable, since polypeptide analysis placed it closer to *Galerucella* whilst *Pyrrhalta* is more distant to *Galerucella*, although *Pyrrhalta* has been sometimes included in *Galerucella* as a subgenus or *vice versa* (see Silfverberg, 1974). The subdivision of *Galerucella* into two subgenera was supported by chorion polypeptide analysis, but not further subdivision of the subgenus *Neogalerucella*. All *Galerucella* species seemed to be similar enough to be kept in the same genus, and according to our results, *Neogalerucella* should not to be in any case included in *Pyrrhalta* as has been suggested by, for example Gressitt and Kimoto (1963) and Wilcox (1971). Although similar in egg chorion properties, egg-laying characters and life cycle to the remaining *Neogalerucella* species, *G. tenella* and *G. pusilla* are quite distant karyotypically. This serves as a reminder that multiple characters must be studied before defining taxonomic units or constructing reliable phylogenies. It would be very interesting to study the egg chorion properties of the

Nearctic *Tricholochmaea* Laboissière, most species having been described as *Galerucella* (see for example Wilcox, 1965, 1971), and sometimes considered even a separate genus (Silfverberg, 1974), but in most cases placed as a subgenus of either *Pyrrhalta* or *Lochmaea*.

Acknowledgements

We wish to express our thanks to Dr. P. Jolivet for encouraging us to contribute to this volume. Jaakko Puro helped us in collecting the material, Ritva Niemi is thanked for excellent technical assistance in SEM studies. Financial support was received from the University of Turku Foundation, and E.J. Sariola Foundation.

References

Brovdij, V.M., 1973. Galerucini. Fauna Ukraini 19 (17), Kiev, 196 pp. (in Ukrainian)

Chûjô, M., 1962. A taxonomic study on the Chrysomelidae (Insecta: Coleoptera) from Formosa, Part XI. Subfamily Galerucinae. Philipp J. Sci. 91: 1–239.

Cox, M.L., 1994. Diapause in the Chrysomelidae. In: Jolivet P.H., Cox M.L. and Petitpierre E. (eds.). Novel Aspects of the Biology of Chrysomelidae. Kluwer Academic Publishers, Dordrecht, pp. 469–502.

Gressitt, J.L. & Kimoto, S., 1963. The Chrysomelidae (Coleoptera) of China and Korea. Part 2. Pacif. Insects Monogr. 1: 1–108.

Hilker, M., 1994. Egg deposition and protection of eggs in Chrysomelidae. In: Jolivet P.H., Cox M.L. and Petitpierre E. (eds.), Novel Aspects of the Biology of Chrysomelidae. Kluwer Academic Publishers, Dordrecht, pp. 263–276.

Hippa, H. & Koponen, S., 1975. On the damage caused by the species of *Galerucella* (Col., Chrysomelidae) on cloudberry (*Rubus chamaemorus* L.) in Finland and northern Norway. Rep. Kevo Subarctic Res. Stat. 12: 54–59.

Hippa, H. & Koponen, S., 1976. Distribution of the species of *Galerucella* (Col., Chrysomelidae) on cloudberry in Fennoscandia. Rep. Kevo Subarctic Res. Stat. 13: 40–43.

Hippa, H. & Koponen, S., 1986. Morphological, cytological and ethological evidence of reproductive isolation between *Galerucella nymphaeae* (L.) and *G. sagittariae* (Gyll.) (Coleoptera, Chrysomelidae) in Fennoscandia. Ann. Entomol. Fenn. 52: 49–62.

Laboissière, V., 1934. Galerucinae de la faune franHaise (Coléoptères). Ann. Soc. Ent. France 103: 1–108.

Manguin, S., White, R., Blossey, B. & Hight, S.D., 1993. Genetics, taxonomy and ecology of certain species of *Galerucella* (Col.: Chrysomelidae). Ann. ent. Soc. Am. 86: 397–410.

Nokkala, C. & Nokkala, S., 1994. Sympatric speciation in *Galerucella*. In: Jolivet P.H., Cox M.L. and Petitpierre E. (eds.). Novel Aspects of the Biology of Chrysomelidae. Kluwer Academic Publishers, Dordrecht, pp. 259–262.

Nokkala, C. & Nokkala, S., 1998. Species and habitat races in the chrysomelid *Galerucella nymphaeae* species complex in northern Europe. Entomol. Exp. et Applic. 89: 1–13.

Nokkala, C., Nokkala, S. & Nordell-Paavola, A., 1998. European and North American populations of *Galerucella nymphaeae* (Col.: Chrysomelidae): Two separate species revealed by chorion polypeptide analysis. Eur. J. Entomol. 95: 269–274.

Nokkala, S. & Nokkala, C., 1987. Chromosome numbers and chromosomal polymorphism in Finnish species of *Galerucella* Crotch (Chrysomelidae, Coleoptera). Hereditas 106: 51–58.

Ogloblin, D.A., 1936. Chrysomelidae, Galerucinae. Fauna USSR 24 (1): I–XIV + 1–455. (In Russian)

Palmén, E., 1945. Zur Systematik finnischer Chrysomeliden. 1. Gattung *Galerucella* Crotch. Ann. Entomol. Fenn. 11: 140–147.

Petitpierre, E. & Juan, C., 1994. Geneome size, chromosomes, and egg-chorion ultrastructure in the evolution of Chrysomelidae. In: Jolivet P.H., Cox M.L. and Petitpierre E. (eds.), Novel Aspects of the Biology of Chrysomelidae. Kluwer Academic Publishers, Dordrecht, pp. 213–225.

Selman, B.J., 1994. Eggs and oviposition in chrysomelid beetles. In: Jolivet P.H., Cox M.L. and Petitpierre E. (eds), Novel aspects of the biology of Chrysomelidae. Kluwer Academic Publishers, Dordrecht, 69–74.

Silfverberg, H., 1974. The west Palearctic species of *Galerucella* Crotch and related genera (Col., Chrysomelidae). Notul. ent. 54: 1–11.

Smith, S.G. & Virkki, N. 1978. Animal Cytogenetics 3. Insecta 5. Coleoptera. Gebrüder Borntraeger, Berlin, Stuttgart. 366 pp.

Wilcox, J.A., 1965. A synopsis of the North American Galerucinae (Col.: Chrysomelidae). Bull. N. Y. St. Mus. Sci. Serv. 400: 1–221.

Wilcox, J.A., 1971. Coleopterum Catalogus, Supplementa. 78. Chrysomelidae: Galerucinae. Odini, Galerucini, Metacyclini, Sermylini. W. Junk, 's-Gravenhage, pp. 1–220.

Windsor, D.M. & Choe J.C., 1994. Origins of parental care in chrysomelid beetles. In: Jolivet P.H., Cox M.L. and Petitpierre E. (eds.). Novel aspects of the biology of Chrysomelidae. Kluwer Academic Publishers, Dordrecht, pp. 111–117.

A PRELIMINARY CLADISTIC ANALYSIS OF THE SUBTRIBE DISONYCHINA WITH SPECIAL EMPHASIS ON THE SERIES *PARALACTICA* (CHRYSOMELIDAE: GALERUCINAE: ALTICINI)

CATHERINE N. DUCKETT
Universidad de Puerto Rico, P.O. Box 23360, San Juan Puerto Rico, 00931-3360,
E-mail: catherineduckett@hotmail.com

Abstract

Disonychina *sensu* Bechyné and Bechyné is a large morphologically diverse subtribe of Neotropical flea beetles composed of 19 genera that included four generic series. Disonychina is cladistically analyzed and existing phylogenetic series, *Pedilia*, *Disonycha*, *Monomacra* and *Paralactica,* are examined. The taxonomic history of Disonychina and characteristics of the series are discussed.

Disonychina is a subtribe redefined by the use of tergum eight as the pygidium rather than tergum seven, as the latter state is plesiomorphic for Alticini. This condition is recognized for the first time and its importance in the evolution of ovipositors in the flea beetles discussed. Disonychina is also defined by female gonapophyses, when present, external and asetose, gonocoxae flattened, and female sternite VIII shortened. Based on the preceding characters *Enneomacra* Bechyné and *Cuyabasa* Bechyné are removed from the Disonychina. The genera *Etapocanga* and *Ptocadica* are placed in the subtribe Disonychina and in the Series *Monomacra*. The status of *Balzanica* as a junior synonym of *Disonycha* is cladistically supported.

Monophyly of the series *Pedilia* and *Disonycha* is supported, as is the monophyly of the series *Paralactica* minus *Lacpactica*. However, series *Monomacra* is paraphyletic to *Paralactica* series, which suggests the series *Paralactica* is not useful without the creation of a cumbersome number of additional series. Therefore, the series *Paralactica* is eliminated and taxa formerly placed there are placed in the series *Monomacra*.

Extensive reduction of the ovipositor and loss of gonapophyses and gonocoxae is characteristic of most genera, except *Pedilia*.

Diagnoses and phylogenetic characterizations are presented for the apical clade of the Disonychina comprising *Rosalactica*, *Paralacticoides*, *Paralactica*, *Lacpacticoides*, and *Maturacaita*.

Introduction

The phylogeny of the Chrysomelidae has elicited much interest because of their associations with the Angiosperms (Farrell, 1998) and their diversity (Reid, 1995). The Disonychina is a morphologically and taxonomically diverse subtribe (Bechyné & Springlova de Bechyné, 1977, here abreviated Bechyné & Bechyné) of New World flea beetles (Chrysomelidae: Galerucinae: Alticini), most common in the tropics. This group is of interest to taxonomists because of its diversity (both at the generic and specific levels) and to ecologists and agriculturalists because of its close and interesting associations with host plants (Garman, 1889; Chittenden, 1899; Thomas, 1987; Duckett, 1989; Dodge *et al.*, 1990). The Disonychina contains at least 20 genera representing over 600 described species. These are relatively large (4-13 mm in length) usually brightly colored flea beetles (Fig. 1). They

Advances in Chysomelidae Biology 1, pp. 105–136
Edited by M.L. Cox

Fig. 1. Etapocanga pysuna, dorsal habitus, length 7mm.

have attracted the attention of collectors and scientists because of their possibly aposematic coloration, and in some instances, their close associations with specific host plants, possible mimicry complexes and their utility in biological control (Vogt *et al.*, 1979; Thomas,1987; Balsbaugh, 1988; Duckett, 1989).

The disonychine genera have the most developed taxonomic structure of any group among the Alticini because of the hypotheses presented by Bechyné and Bechyné (1966, 1975, 1977). The Bechyné and Bechyné taxonomy (summarized in Table 1.) presents the "Disonychini", as a tribe of the Oedionychinae (Alticidae) containing two subtribes, each of which contains two series. Subtribe Disonychina contains series *Disonycha*, which includes *Disonycha* Chevrolat and *Phenrica Bechyné*, and series *Pedilia* containing *Nephrica* Harold, *Sophraenella* Jacoby, and *Balzanica* Bechyné. Subtribe Monomacrina includes series *Monomacra* (including *Enneomacra* Bechyné & Bechyné, *Heikertingeria* Csiki, *Utingaltica* Bechyné & Bechyné, *Monomacra* Dejean, *Parchicola* Bechyné & Bechyné, *Coroicona* Bechyné, *Dinaltica* Bechyné, *Vilhenaltica* Bechyné & Bechyné, and *Archilactica* Bechyné & Bechyné) and series *Paralactica*, (including *Lacpactica* Bechyné & Bechyné, *Lacpacticoides* Bechyné & Bechyné, *Paralacticoides* Bechyné & Bechyné, *Maturacaita*

Table 1. Taxonomic structure of the Disonychini *sensu* Bechyné and Bechyné 1977

Subtribe *Disonychina*	Subtribe *Monomacrina*
Series ***Pedilia***	Series ***Monomacra***
Pedilia, Clark 1865	*Enneomacra,* Bechyné and Bechyné 1961
Nephrica, Harold 1877	*Heikertingeria,* Ciski 1939
Sophraenella, Jacoby 1905	*Utingaltica,* Bechyné and Bechyné 1961
Balzanica, Bechyné 1959	*Monomacra* Chevrolat 1837
	Parchicola, Bechyné and Bechyné 1975
Series ***Disonycha****	*Coroicona,* Bechyné 1956
Disonycha, Chevrolat 1837	*Dinaltica,* Bechyné 1956
Phenrica, Bechyné 1959	*Vilhenaltica,* Bechyné and Bechyné 1964
	Archilactica, Bechyné and Bechyné 1977
	Series **Paralactica**
	Maturacaita, Bechyné and Bechyné 1977
	Lacpacticoides, Bechyné and Bechyné 1977
	Paralactica, Bechyné and Bechyné 1961
	Paralacticoides, Bechyné and Bechyné 1977
	Rosalactica, Bechyné and Bechyné 1977
	Lacpactica, Bechyné and Bechyné 1977

**Cuyabasa* was included in series Disonychina (Bechyné and Springlova de Bechyné 1966), although it was not mentioned in 1977.

Bechyné & Bechyné, and *Rosalactica* Bechyné & Bechyné) (See Table 1). However, Bechyné and Bechyné proposed taxonomic ranks that are incongruous with most researchers' concept of the Disonychina and Alticinae/Alticini, used characters that are difficult to interpret, and espoused Lamarckian evolutionary philosophy (Bechyné, 1980); consequently their ideas have not been generally accepted (Seeno and Wilcox, 1982; Furth 1985-1988; Reid 1992, 1995).

Although Disonychina has not been generally accepted as a tribe/subtribe, neither have many other groups of flea beetles. There is no currently accepted suprageneric classification within the flea beetles (Seeno & Wilcox, 1982; Furth 1985; Reid 1992). Because of this paucity of accepted groupings and the over 500 genera of alticines, there is a great need for suprageneric studies.

My goals in this chapter include a cladistic analysis of some representative species of the genera of the Disonychina, primarily as a test of the monophyly of the group, but also as a test of the validity of the Bechyné and Bechyné series within the Disonychina. Because of my interest in the genus *Pedilia*, I have analyzed more species of the genera in that series. Moreover, because of a search for the appropriate outgroup for a taxonomic revision of *Pedilia* (Duckett, 1993 and in prep), I have also focused extra attention on genera in the series *Paralactica,* which initially looked like an appropriate outgroup because of its emarginate eyes (compare Fig. 2). Additionally, I hope to establish morphological characters diagnostic of the Disonychina, to delimit the morphological diversity of the group and to test the phylogenetic placement of 2 genera (*Etapocanga* Duckett and *Ptocadica* Harold) which have been suggested as being part of the Disonychina (See below). In addition to these goals, I present a taxonomic history of the Disonychina. I also present diagnoses of genera in series *Paralactica* to help other workers in this poorly known group.

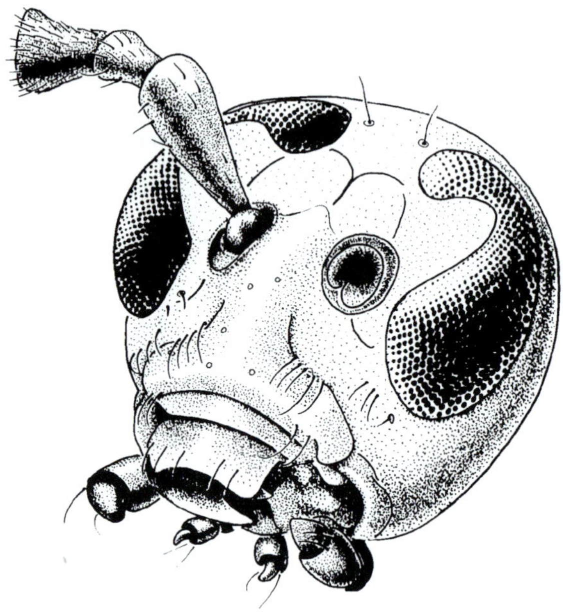

Fig. 2. Head of *Sophreanella fulva*, antenna removed, length 3 mm.

Taxonomic history

The taxonomic history of the Disonychina and the *Paralactica* series is confusing. This is partially because it is characterized by many changes in taxonomic rank and by the creation of new genera based on species originally placed in closely related genera, which in turn are based on, or diagnosed by, poorly defined characters. I will first summarize the taxonomic history of the subtribe as well as the generic series and the characters on which the decisons were based. I will then outline the most confusing taxonomic elements of each generic series.

The tribe Disonychini was first used by Leng (1920) and included *Disonycha* and *Hemiphrynus*, with no character-based definition for the group. Blake (1933, 1955) revised the North American species and the vittate species south of the U.S., commenting that *Disonycha* has "few outstanding characters". However, she recognized the importance of the interantennal carina, the lack of the prothoracic prebasal impression and the obliquely truncate hind angles of the prothorax. Bechyné and Bechyné (1966) later defined the group Disonychina as one of two subtribes of the "Oedionychinae", including seven nominal genera (*Pedilia*, *Nephrica*, *Sophraenella*, *Disonycha*, *Phenrica*, *Balzanica* and *Cuyabasa* Bechyné). *Cuyabasa* was later tacitly removed from Disonychina (Bechyné and Bechyné, 1977), although Bechyné (1983) has not removed *Cuyabasa* from *in litteris* phylogenies. The characters upon which the designation of the Disonychina is based include the lack of an enlarged ultimate metatarsal segment, the lateral portion of the base of the prothorax projecting forward, and the "thoracopleura entire and elongate" (Bechyné & Bechyné, 1975). Except for the unenlarged ultimate metatarsal segment, which is plesiomorphic and shared with the Galerucini and most other Alticini, these characters are subjective and difficult to interpret. "Disonychini" was also distinguished by having a sixth abdominal segment fused to the fifth abdominal sternal segment and articulating under the pygidium (Bechyné & Bechyné, 1975) (see Fig. 4). This description is misleading, not only because the fifth visible ventral sternum is actually sternum VII, but also because the character is present in other groups where it is not easily visible. Later, however, Virkki (1988,1989) summarized the cytological justification of the subtribe Disonychina on the basis of the following characters: 32 sperm per bundle, an even series of metacentric autosomes and supernumerary X chromosomes using primarily species from the genera *Disonycha* and *Phenrica*.

Bechyné and Bechyné (1975) proposed the subtribe Monomacrina of the "Disonychini", at the same time as proposing two new series, series *Paralactica* and *Monomacra*, and clarifying the definitions of the two subtribes. Subtribe "Monomacrina " was distinguished by the presence of a distinct prebasal impression, and "Disonychina " by the absence of the prebasal impression (Bechyné & Bechyné, 1975). (See Table 1 for generic composition of Disonychini *sensu* Bechyné & Bechyné, 1966, 1975, 1977.) This subtribe, Monomacrina, was proposed entirely based on species removed from *Monomacra* (=*Lactica* Erichson). However, the series *Monomacra* and *Paralactica*, proposed within Monomacrina, were characterized using characters showing large amounts of overlap, such as genal length, width of frons, robustness of legs and tarsi, and relative distinctiveness of the ocular sulcus. Series *Monomacra* is ostensibly more robust with longer genae and a more distinctive ocular sulcus. No genitalic characters were given. One of the genera, *Dinaltica*, was suggested to be polyphyletic. A key to the genera in series *Monomacra* was given but is difficult to use.

The Disonychina became the best characterized supra-generic group in the Alticinae in 1977 (Bechyné & Bechyné) with publication of a definition and key to series *Paralactica*. The series was distinguished by the following characters: frons narrow, frequently more so than diameter of eyes; ocular sulci replaced by impressed "points"; gena very short, 1/10 – 1/8 length of eyes; antennae, at least in females, filiform; legs long, and comparatively delicate; little sexual dimorphism of tibiae and first tarsomere. This key is easier to use than that for the series *Monomacra* (Bechyné, 1975) because of the copious illustrations; however, because many characters are subjective (delicate legs, ocular "points," etc.), undescribed species can be difficult to place.

Some of the genera in the *Paralactica* group are well-characterized and easy to identify (for example, *Maturacaita*, and *Lacpacticoides*). These two genera are respectively distinguished by the "sexual dimorphism of the male" (modified pygidium, emargination of sternum VII, apical metafemoral tubercles) and the male sternum VII, (fifth ventrite) very long. However, others such as *Paralacticoides* are distinguished by ambiguous characters such as a post-humeral elytral longitudinal ridge (present only in *some* females), robust morphology, relative density of pubescence, clypeus not bent downwards, and relatively broad interprocoxal plate. Again, there is no discussion of genitalic characters. Because of the vague characterization of genera, these genera are difficult to identify.

The series *Pedilia* is characterized by very strongly emarginate eyes (except *Balzanica*) (compare Fig 2). There is some confusion also about the specific composition of *Pedilia*, although species are easy to place in this series. Bechyné (1959) synonymized *Nephrica* with *Pedilia*, creating the subgenus *Nephrica*, but later Bechyné and Bechyné (1966) ignored this synonymy. *Sophraenella* differs from *Nephrica* and *Pedilia* only in antennal morphology. *Pedilia* is recognized by its round rather than oval dorsal habitus; the oval habitus is characteristic of species assigned to *Nephrica* and *Sophraenella* (Blake, 1951; Bechyné, 1959). I hypothesize that *Sophraenella* and *Nephrica* are synonymous with *Pedilia*. This study aims to test this hypothesis, among others. In the following pages reference to the genus and subgenus *Nephrica* includes species with the following epithets: *didyma* Illiger, *inornata* Jacoby, *basalis* Baly, *clytroides* Bechyné, *cycloptica* Bechyné, *media* Bechyné. *Sophraenella* includes one species *S. fulva* Jacoby. Duckett (1995) synonomized *Balzanica* with *Disonycha* when she concluded *Balzanica cacoscelina* Bechyné, 1959 is conspecific with *Disonycha ornata* Baly, 1876. This study will also test her hypothesis that *Balzanica* should not be classified as a separate genus.

There is also much confusion in the status of *Disonycha, Phenrica* and *Nephrica* (Furth & Savini, 1998). Bechyné (1956) described several new species and made several new combinations under the generic name *Phenrica* (of species formerly described in the genus *Disonycha*). However, the genus was not formally described at that time, nor was a type species designated. Bechyné (1959) redressed these discrepancies, formally describing *Phenrica* and designating *Disonycha austriaca* Schaufuss, a combination from the original paper, as type species. Bechyné and Bechyné (1966) revised *Phenrica*, recognizing two subgenera. *Phenrica austriaca* was designated the type species of the nominate subgenus, and *Nephrica quadrimaculata* (Clark) the type of subgenus *Orthophenrica*. Members of the subgenus *Orthophenrica* are often confused with *Disonycha* or *Nephrica,* as they are larger, with fewer punctations of the vertex and not usually black.

Methods

Choice of taxa. In order to test the monophyly of the series *Paralactica*, all available species of the series were examined (See Table 3). In every case the material examined included at least one sex of the type species of each genus. Specimens of a recently described genus, *Etapocanga*, were also examined. All genera in the Monomacrina were examined except *Vilhenaltica* and *Archilactica*, which were unavailable. *Phenrica* and *Disonycha* were included to address the Bechyné and Bechyné statements of their close relationship to *Pedilia*. All species of *Pedilia*, *Nephrica*, *Sophraenella*, and *Balzanica* , where both sexes are known were studied. Because some species did not differ in the characters used in the analysis, *Sophraenella* and *Balzanica* each resulted in one terminal; *Nephrica* in three, representing six species; and *Pedilia* in four terminals (see Table 3 for specifics).

Three outgroups were selected: *Cuyabasa*, *Macrohaltica* Bechyné, and *Crimissa* Stal. *Cuyabasa* was selected because it had been originally included in the Disonychina (Bechyné & Bechyné, 1966) and in the series *Disonycha*, B. Bechyné *in litteris*; however, it was not mentioned in the last published delineation of the series *Disonycha* (Bechyné & Bechyné, 1977). Seeno and Wilcox (1982) included it in the Oedionychini group. I chose *Cuyabasa*, based on preliminary analysis, as a representative of the putative sister group, Oedionychini. *Crimissa* was selected as a putatively primitive Alticini (Seeno & Wilcox, 1982) and *Macrohaltica* as a completely unplaced outgroup. In total, 46 species were examined (see Table 3.)

Table 2. Character states used in the analysis (character states are ordered unless stated otherwise)

0. Eyes 0 =Eyes, oval	1 = with symmetrical median shaped emargination 2 = asymmetrical with dorsal lobe smaller (see Fig. 2)
1. Transverse clypeal carina UNORDERED	0 = not curving around mandibular bases, not convex directly over labrum 1 = convex, pronounced directly over labrum, curving around mandibular bases (Figs 2 & 8) 2 = ridge-like, not convex and rounded
2. Maxillary palpi 0 = robust and incrassate (Figs. 2&8)	1 = gracile
3. Labrum 0 = transverse (see Fig. 2)	1 = subquadrate, ventral edges beveled
4. Prosternum 0 = with collar-like horizontal ridge to receive head (Fig. 8)	1 = smooth, parallel with axis of beetle
5. Lateral prebasal impression of pronotum 0 = present (Fig. 1)	1 = absent
6. Transverse prebasal impression of pronotum UNORDERED	0 = impression absent 1 = impression oblique and not continuing in center of disc 2 = impression transverse, straight anterior edge not clearly delimited 3 = transverse impression straight, anterior edge clearly delimited (see Fig. 1)
Female characters	
7. Pygidium formed from 0 = from tergum VII	1 = from tergum VIII
8. Epiproct 0 = present	1 = absent
9. Gonapophyses 0 = divided, internal	1 = flattened, divided (2) (Fig. 5b) 2 = flattened, single (Fig. 6a) 3 = entire, "A shaped" 4 = absent
10. Gonocoxae 0 = three dimensional, setose (See Figs 5, 6f-h)	1 = flattened, asetose 2 = absent
11. Sternite VIII 0 = apex as wide as posterior edge of	1 = apex half or 1/3 as wide as the posterior edge of sternum VII 2 = apex less than 1/3 as wide as posterior edge of sternum VII
12. Sternite VIII base 0 = Narrow (Fig 5a)	1 = wide (Figs. 6c-e)
13. Sternite VIII base	

Table 2. Cont.

	0 = rounded	1 = flat 2 = emarginate
14.	Insertion of intersegmental membrane 0 = pointed (Fig. 6c)	1 = straight (Fig. 6e)
15.	Spermatheca with pronounced neck 0 = absent	1 = present (Fig. 6b)
16.	Basal portion of proximal spermathecal duct expanded 0 = expanded (Fig. 6b)	1 = not expanded
17.	Spermathecal pump enlarged 0 = enlarged	1 = parallel sided or smaller (Fig. 6b)
18.	Spermatheca 0 = without flange on pump (Fig. 6b)	1 = with flange on pump
	Male characters	
19.	Apex of sternum VII which articulates with the pygidium UNORDERED	0 = auriculate (Fig. 4a) 1 = transverse and entire (Fig. 4b) 2 = absent (Fig. 4c)
	Median lobe of aedeagus Median dorsal process (20 - 21)	
20.	0 = not recurved (see Figs 3a,b,e,f)	1 = recurved (see Fig. 3c,d)
21.	0 = as long as lateral processes (Fig. 3a)	1 = significantly shorter than lateral processes (seeFig.7) 2 = plate- or bowl-like
22.	Lateral dorsal process of median lobe 0 = not recurved (Fig. 3 a,b, e,f)	1 = recurved inward (see Figs. 3 c-d) 2 = absent
23.	Orientation of lateral dorsal process 0 = oriented in horizontal plane when viewed from dorsal side, broadest aspect when viewed dorsally (Fig. 3 a, e)	1 = oriented in vertical or oblique plane when viewed from dorsal side, broadest when viewed laterally (Fig 3. d) ? = lateral dorsal process absent
24.	Shape of ventral lip of median lobe UNORDERED	0 = round or triangular (Fig. 3 a-d) 1 = produced and elongated apically (Fig 3 e-f) 2 = apically emarginate or bifid (Fig. 7)
25.	Curvature of apex of ventral lip 0 = apex not curving downward sharply (Fig. 3 b,d,f)	1 = apex curving sharply downward

Etapocanga Duckett and *Ptocadica* Harold were also included on the basis of external similarity to the Disonychina and preliminary observations which suggested that they might be disonychines (Duckett, 1995 and Duckett and Moyá in press). See Table 3 for a list of taxa used in this analysis.

Table.3. Taxa used in analysis[1]

Name on Cladogram	Taxa used
Balzanica	*Balzanica cacoscelina* Bechyné *(= Disonycha ornata)*
Coroicona	*Coroicona bilineata* (Jacoby)
Crimissa	*Crimissa curalis* Stäl
Cuyabasa	*Cuyabasa minor* Bechyné
Dinaltica ingea	*Dinaltica ingea* Bechyné
Dinaltica aneipennis	*Dinaltica aeneipennis* (Jacoby)
Disonycha trifasciata	*Disonycha trifasciata* (Clark)
	Phenrica austriaca (Schaufuss)
Disonycha quinquelineata	*Disonycha quinquelineata* (Latreille)
Enneomacra	*Enneomacra acutangulata* (Weise)
Etapocanga	*Etapocanga pysuna,* Duckett
	Etapocanga consejera, Duckett
Heikertingeria	*Heikertingeria weiseana* Bowditch
	Heikertingeria nigrolineata Bryant
	Heikertingeria seminigra Jacoby
Lacpactica	*Lacpactica macula* (F.)
	L. bifasciata (Jacoby)
Lacpacticoides	*Lacpacticoides nicodema* Bechyné
Macrohaltica	*Macrohaltica amethestina* (Olivier)
Maturacaita	*Maturacaita teratopoda* Bechyné and Bechyné
	Maturacaita sp. A.
Monomacra	*Monomacra inermis* Klug
N.didyma	*Nephrica didyma* (Illiger)
N.clytroides	*Nephrica clytroides* (Bechyné)
N.media	*Nephrica media* Bechyné
P.cycloptica	*Pedilia cycloptica* Bechyné
Paralactica	*Paralactica capitata* (Illiger)
	Paralactica helia Bechyné and Bechyné
Paralactica rubinii	*Paralactica rubinii* Bechyné and Bechyné
Paralacticoides	*Paralacticoides bicarinata* Bechyné and Bechyné
Parchicola	*Parchicola yena* Bechyné
P.basalis	*Pedilia basalis* (Baly)
P.inornata	*Pedilia inornata* (Jacoby)
P.sp. A.	*Pedilia sp. A,* undescribed species*
P. sp Q.	*Pedilia sp. Q,* undescribed species
P.globosa	*Pedilia globosa* (F.)
P. sp .P.	*Pedilia sp. P,* undescribed species
Ptocadica	*Ptocadica straminea* Harold
	Ptocadica sp. A. undescribed species
	Ptocadica bifasciata Jacoby
Rosalactica	*Rosalactica maculicollis* (Jacoby)
	R. kermesina Bechyné and Bechyné
S.fulva	*Sophraenella fulva* Jacoby
Utingaltica	*Utingaltica calcarata* (Illiger)
	Utingaltica sellata (Baly)
	Ultingaltica tingomariana Bechyné

* See Duckett 1989, for ecological notes on this species.
[1]Taxa coded as one terminal are grouped together.

Choice of Characters. Morphological characters of the exoskeleton and reproductive systems visible with light microscopy were used. The characters described below were used to analyze relationships in the disonychine genera (Fig.9). The numbers in parentheses in the following paragraphs correspond to the numbers in Table 2 and above the hatch marks on the cladograms in Figures 9 &10. Multistate characters are reported in the text (character #.state#) for example, (21.2) = character 21, state 2. Missing values or characters of uncertain homology are coded (?).

Preparation of genitalia. The whole beetle was softened by heating in distilled water; the abdomen was removed and treated with warm, dilute potassium hydroxide solution (≈10%). Male genitalia were prepared according to the protocol of Smith (1979). Terminology describing male genitalia generally follows Sharp and Muir (1912); I use endophallus instead of internal sac in an effort to draw attention to homology with other orders. I also describe the ventral portion of the median lobe beyond the median orifice as the ventral lip rather than the ventral edge. Other terms were coined to accord with general terms in the German and Spanish language usage of Bechyné and Bechyné (1966, 1975). The lateral dorsal processes and the median dorsal process (see Fig.3) together equal the *Orifizialplatte* of Bechyné and Bechyné, which is also equivalent to the dorsal edge of Sharp and Muir (1912).

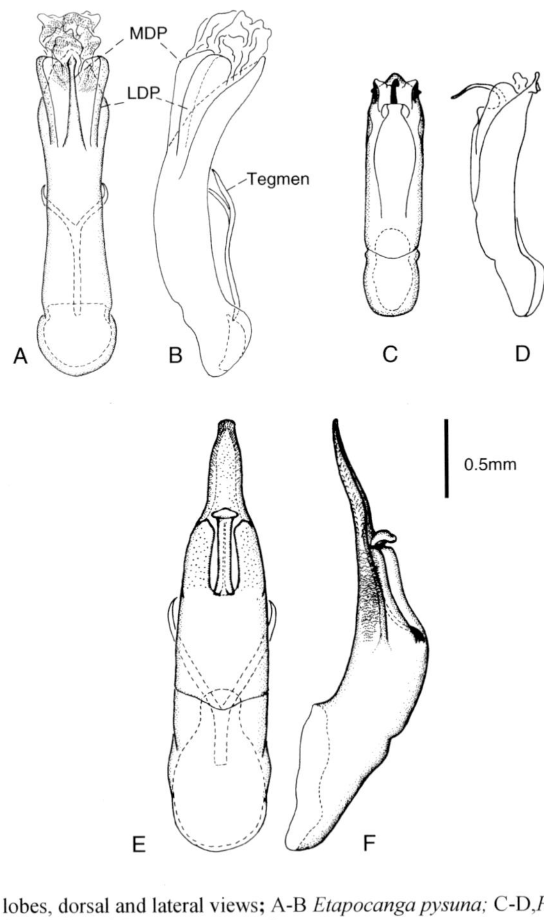

Fig. 3. Male median lobes, dorsal and lateral views; A-B *Etapocanga pysuna;* C-D,*Paralacticoides bicarinata;* E-F,*Pedilia basalis.* Figs. C-D and F tegmen removed.

In females, the spermathecae and ovipositor, including sternite VIII, gonocoxae (vaginal palpi of Konstantinov, 1998), and paraprocts were removed, and prepared with potassium hydroxide using the above protocol. The parts were examined using a compound microscope. Where possible, sternite VIII was detached from the dorsal surface of sternum VII while maintaining its connection to the gonocoxae and the bursa copulatrix. Other terminology follows Mann and Crowson (1983). After initial examination, the genital capsule was separated from the genital system. Abdomens and genitalia were stored in glycerine in vials, with the pinned specimen.

The terminology for the description of spermathecae follows Samuelson (1973), Smith (1985) and Suzuki (1988, 1989), but differs slightly from Suzuki in that the terminology he used is more general to Chrysomelidae and less useful for Alticini. I have coined the term "proximal spermathecal duct" for the thickened melanized portion of the spermathecal duct (see Fig. 6b).

Illustrations were prepared using a Wild M5-C dissecting microscope and Leitz Dialux 20 compound microscope, each fitted with a camera lucida and the latter with differential interference contrast illumination.

Characters analyzed.

GENERAL

0. *Eye shape.* Eye shape is a well-known character for defining certain genera in the Alticini (Scherer, 1962, 1983); eyes strongly emarginate defines *Pedilia*, *Nephrica* and *Sophraenella*, *Etapocanga*, and *Paralactica*. However, it also is used as a key character for *Heikertingeria* and *Phenrica* (Bechyné & Bechyné, 1966; Scherer, 1962) without careful definition of the states. In *Pedilia*, *Nephrica* and *Sophraenella*, and *Paralactica* "strongly emarginate" indicates an eye with two lobes, the dorsal one being smaller. This state is formed by a smooth asymmetrical emargination as in *Etapocanga* and *Pedilia*, *Nephrica* and *Sophraenella* (0.2) (Fig. 2). In *Phenrica* and some *Heikertingeria* (0.1), the eyes are symmetrically shaped and the emargination, although pronounced, does not exceed 2/3 of the depth of emargination in *Etapocanga*. Eyes oval (0.0) is characteristic of the outgroups and *Enneomacra*. In *Ptocadica* most individuals possess oval eyes; however, because some individuals appear to have very slight emarginations, *Ptocadica* was coded with a question mark.

1. *Transverse clypeal carina.* A convex transverse clypeal carina is an obvious feature of *Pedilia*, *Etapocanga*, *Balzanica*, and *Disonycha* (see Fig. 2); in these taxa it is highly convex and curves downward around the labrum to the mandibular articulation (1.1). This carina, if present in the Monomacrina, is not highly convex; it usually slopes outward, grading into the frontal longitudinal carina, and does not curve around the base of the labrum (1.0). In the outgroups, the transverse carina is usually ridge-like and not contiguous with the base of the labrum; if it curves around the mandibular bases, the mandibles are very large and the carina correspondingly wide (1.2). This character clearly separates the genera of the Disonychina and Monomacrina *sensu* Bechyné. This character was coded as unordered.

2. *Maxillary palpi. Pedilia* was defined as having the penultimate segment of the maxillary palpi incrassate (Clark, 1865) (2.0) (Figs. 2 & 8). Within the Disonychina genera this character is shared by the *Balzanica*, *Disonycha*, and *Etopocanga*. The Monomacrina all have very slender gracile maxillary palpi (2.1).

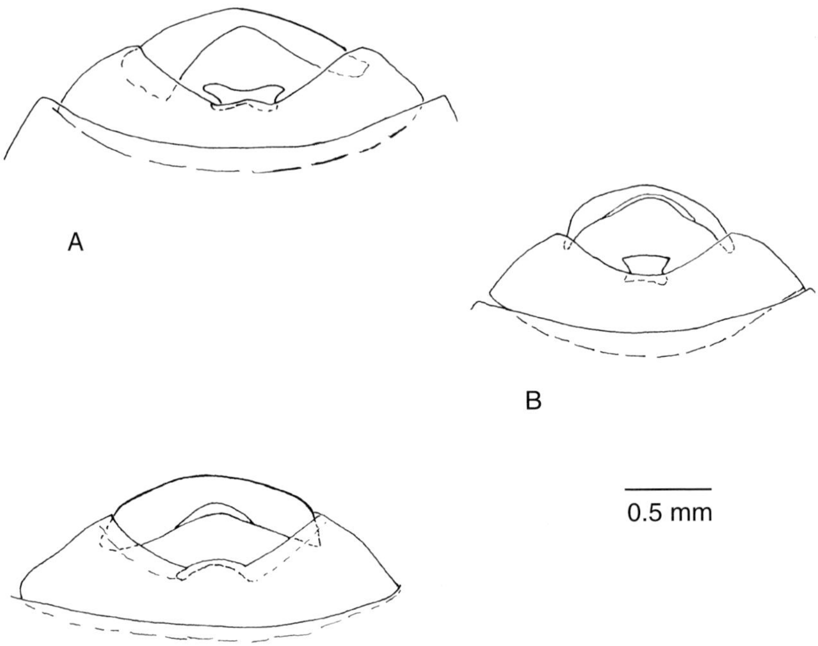

Fig. 4. A-C. Male"pygidia" (tergum VIII) and sternum VII, ventral view; A, *Lacpacticoides*; B, *Pedilia sp. A.*; C, *Nephrica clytroides*

3. *Labrum.* The labrum may be transverse (3.0) (see Fig. 2) or subquadrate, with rounded apical corners (3.1) (compare figures of Bechyné & Bechyné, 1977; Figs. 1, 7, 8, 15).
4. *Prosternum.* The Disonychina genera, with the exception of the rounded species of *Pedilia*, have the anterior portion of the prosternum between the head and the procoxae smooth and flat (4.1). In round *Pedilia* the prosternum is ridged, forming a collar-like shelf when the head is depressed (4.0)(Fig. 8)
5, 6. *Prebasal pronotal impression.* The prebasal pronotal impression is one of the most obvious characteristics of the disonychine genera and the Alticini. It is often used as a key character (Scherer, 1962; Bechyné & Bechyné, 1966, 1975) and was used to separate the Disonychina and Monomacrina (Bechyné and Bechyné 1975), although in the same paper the character is designated "ambiguous."

The prebasal pronotal impression is actually two characters: lateral impressions (5) and a transverse impression (6). The absence of both impressions (5.1, 6.0) is characteristic of *Sophraenella*, *Nephrica*, and one species each of *Pedilia*, *Balzanica*, *Disonycha* and *Cuyabasa*. Lateral pre-basal impressions only (5.0, 6.0) are found only in some *Pedilia*). Other *Pedilia* also have a transverse oblique incomplete impression (6.1) as well. All other genera from the Monomacrina have some type of transverse and lateral, prebasal impressions. However, I perceive two more states of the transverse impression: transverse impression deep, anterior edge not clearly delimited by a sharp line (6.2), and anterior edge of

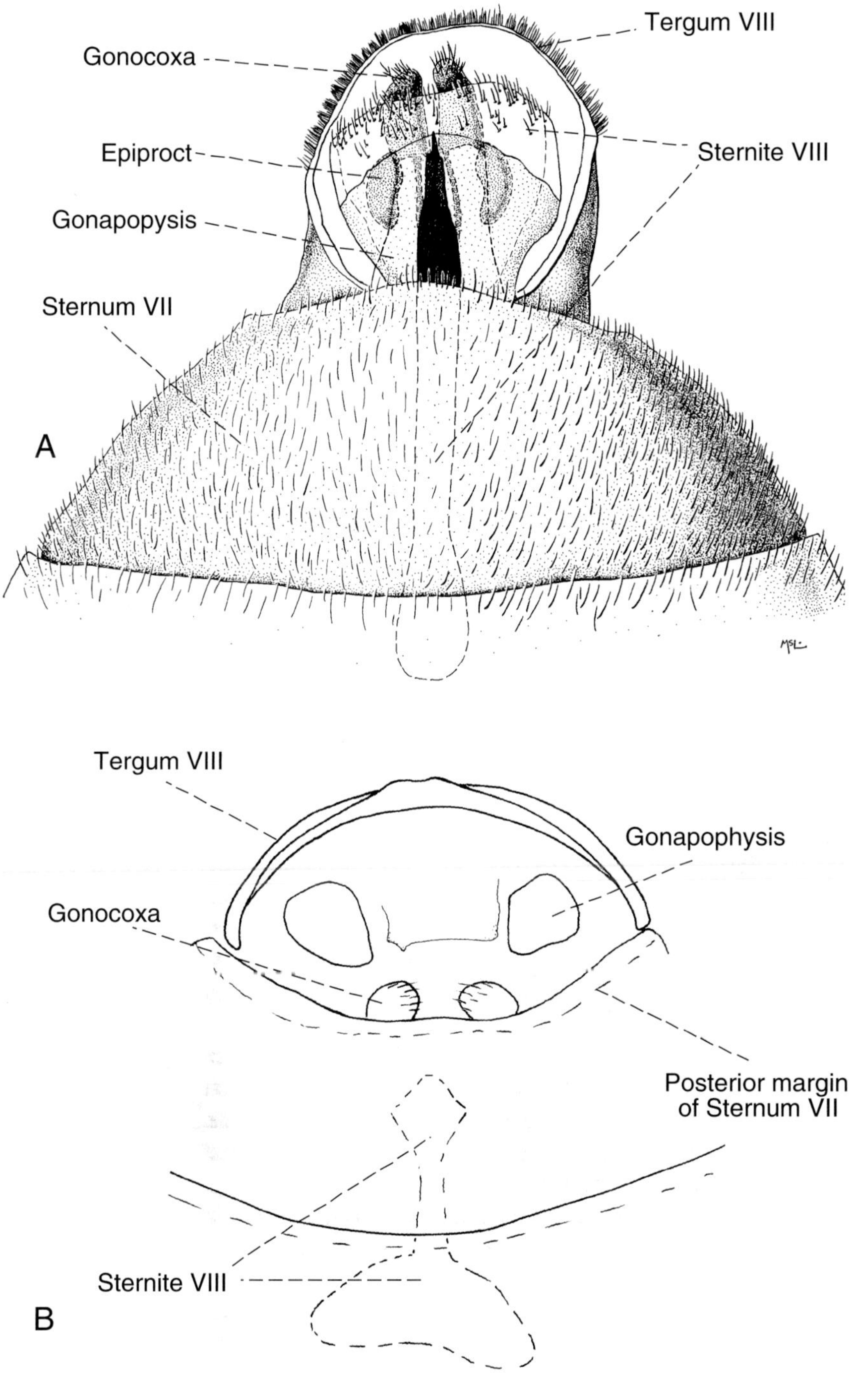

Fig. 5. **A.** Female terminalia, ventral view. A, *Macrohaltica*; B**,** *Pedilia didyma*, basal lateral edges not shown, with sclerotization of membrane between gonapophyses.

transverse impression clearly delimited by a sharp line (6.3) (Fig.1). The character states of character 6 are coded as unordered.

FEMALES

8. *Epiproct*. Absence of an epiproct (8.1) characterizes the Disonychina (except *Enneomacra*) and again appears to be a part of the reduction from a tubular ovipositor. This state is also shared by *Cuyabasa* and the other Oedionychine genera I have dissected (*Alagoasa*, *Walterianella*, *Wanderbiltiana*, *Aspicela*).

9. *Gonapophyses*. Most Alticini having a tubular ovipositor have rod-like internal gonapophyses (9.0) (Fig.5a). The disonychine genera show a variety of different states. *Nephrica* and *Sophraenella* have paired, flattened ventrally oriented gonapophyses (9.1) (Fig.5b). In *Pedilia sensu strictu* the gonapophyses are median and fused (9.2) (Fig.6a) or absent (9.4). *Paralactica capitata* and *Maturacaita* sp., have gonapophyses forming a semi-circular A-shaped structure (9.3). The gonapophyses are absent (9.4) in all other disonychine taxa examined. This character was ordered as a morphocline (Maslin, 1952).

10. *Gonocoxae*. In the Alticini three-dimensional, setose gonocoxae are apparently plesiomorphic (10.0); in the disonychine genera this state is present only in *Pedilia*, *Nephrica* and *Sophraenella* (See Figs 6f-h for diversity) Flattened, asetose gonocoxae (10.1) are present in the other genera of the series *Disonycha*, *Lacpactica*, and *Parchicola*. Gonocoxae are absent (10.2) in genera of the series *Paralactica* (excepting *Lacpactica*), in *Etapocanga*, *Monomacra*, and *Coroicona*.

11, 12, 13, 14. *Sternite VIII*. The shape of the eighth sternite is not highly variable in taxa with a tubular ovipositor, the character states in the outgroups being very similar to those reported for Cerambycidae (Tanner, 1927; Fragoso, 1985). These states include apex almost as wide as the posterior edge of sternum VII (11.0) (Fig.5a), base narrow (12.0) and rounded (13.0) (Fig.5a). In the disonychine genera, however, the shape of sternite VIII is highly variable though always dorsoventrally flattened. In *Pedilia* the base is always wide (12.1) and usually emarginate (13.2) (see Figs 6 c-e); although the variation in the apex is great, varying from 1/2 -1/3 as wide as sternum VII (11.1) or less than 1/3 as wide as sternum VII (11.2); it is never as wide as in the outgroups. *Balzanica*, most species of *Disonycha*, *Monomacra*, *Parchicola*, and *Lacpactica* have a wide apex (11.0) and narrow base. *Coroicona* is unique in having a wide apex and wide, flat base (11.0, 12.1, 13.1).

The shape of the insertion point of the intersegmental membrane between sternite VIII and sternum VII is also a source of variation (14). The sternite is hollow; if one grasps the apex with a set of forceps and the intersegmental membrane with another, the sternite is split. The cavity is indicated by dashed lines (Fig.6c-e). I judge the point of insertion of the intersegmental membrane to be the point where the cavity meets the edge of the sternite. The insertion point may be pointed (14.0)(See Fig. 3c) or rounded (14.1) (Fig.6d).

15, 16, 17, 18. *Spermatheca*. The spermathecae of Alticini often have a pronounced neck (Samuelson, 1973) or collum (Döberl, 1986). However, in the outgroups I examined, the area between the pump and the receptacle was not narrower, but simply bent. All the Disonychina have a pronounced narrowing between the pump and the receptacle (See Fig. 6b) (15.1) except *Enneomacra*.. Other aspects of spermathecal morphology, include a relative expansion of the base of the distal spermathecal duct (16.1) and the spermathecal pump enlarged (17.0) or parallel-sided (17.1). A flange on the pump (for additional insertion surface of the spermathecal muscle) (18.1) is present in *Coroicona*, *Rosalactica*, and *Paralacticoides*, as well as in *Balzanica*, *Cuyabasa*, and one species of *Pedilia*.

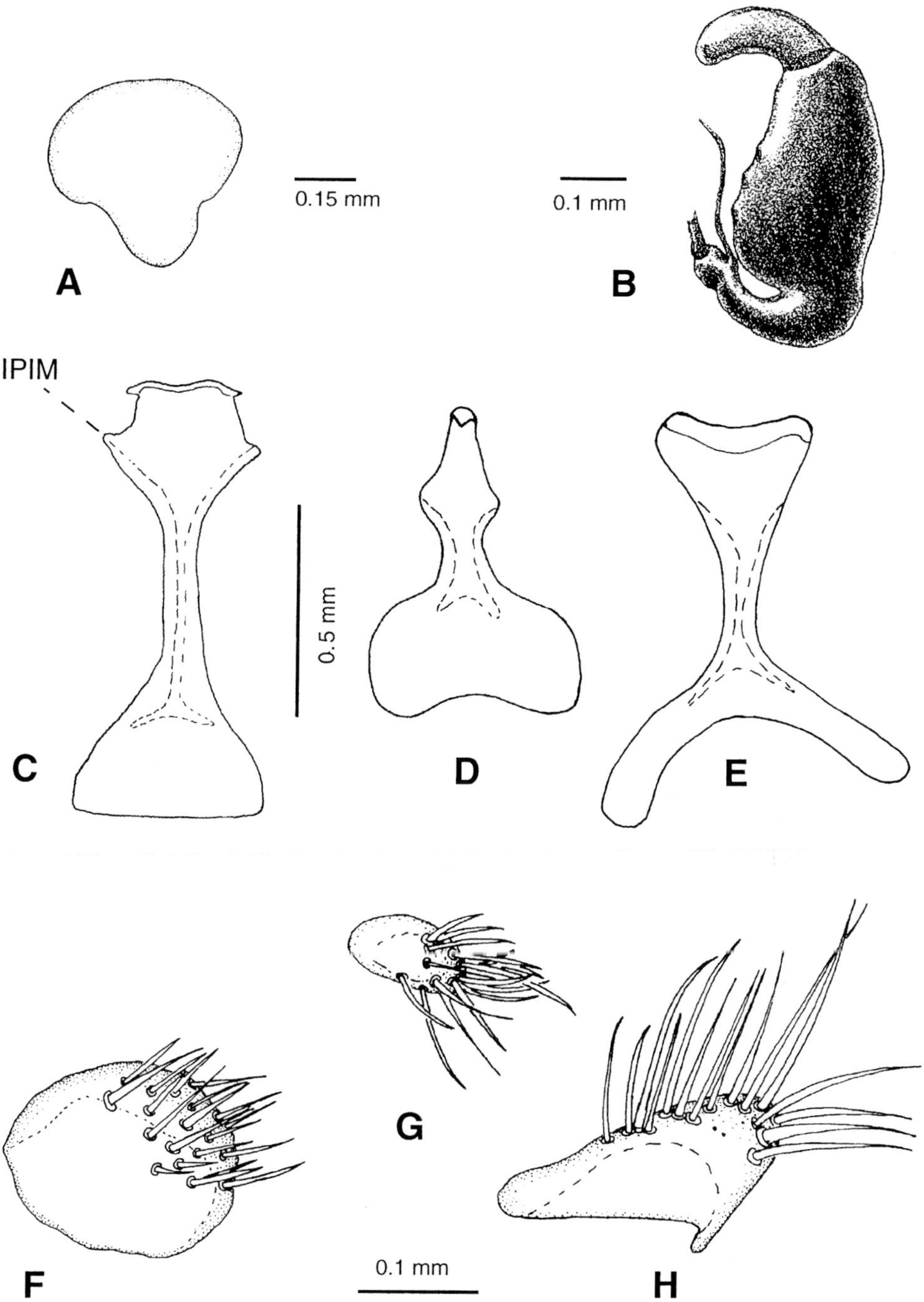

Fig. 6. Parts of female reproductive systems. A, gonopohysis *Pedilia sp. A.*; B, spermatheca, *Etapocanga pysuna;* C, D,E, Sternite VIII, ventral view, C, *E. pysuna,* D, *Pedilia. sp. Q.*; E, *P. globosa;* F-H, left gonocoxae; F. *Sophreanella fulva*; G. *Pedilia. sp. P.* ; H. *Pedilia. sp. Q.*. IPIM = insertion point of intersegmental membrane.

MALES

19. *Apical flange of male sternum VII.* Bechyné and Bechyné (1975) identified this character as the sixth abdominal sternite and a defining character of the Disonychina. The character was indicated in their description of *Paralacticoides* (1977) as (loosely translated) "the sixth abdominal segment of the male is freely visible from underneath in the form of a small transverse plate between the sinus of the fifth abdominal segment and the tip of the pygidium" (see Fig. 4). This is not a sixth ventrite, nor sternite VIII, but a projection of sternum VII, and is present (though not clearly visible) in most taxa dissected. The shape is variable in the series *Paralactica* and series *Pedilia*

In *Enneomacra, Disonycha, Parchicola, Coroicona, Utingaltica, Ptocadica, Lacpacticoides*, and in Series *Pedilia* the flange is auriculate (19.0) (see Fig. 4a). In *Rosalactica, Paralactica, Paralacticoides, Heikertingeria, Dinaltica aeneipennis* Jacoby, and *Pedilia* sp A. the flange is transverse (19.1) (Fig.4b); and in *Maturacaita, Cuyabasa* and *Macrohaltica* the flange appears absent (19.2)(compare Fig. 5a). This character was ordered as a morphocline (Maslin, 1952).

20. *Dorsal curvature median dorsal process of median lobe of aedeagus.* In *Paralacticoides* and *Rosalactica* (Fig.3d) the median dorsal process is recurved like a barb, to point more than 90° behind its usual position (20.1).

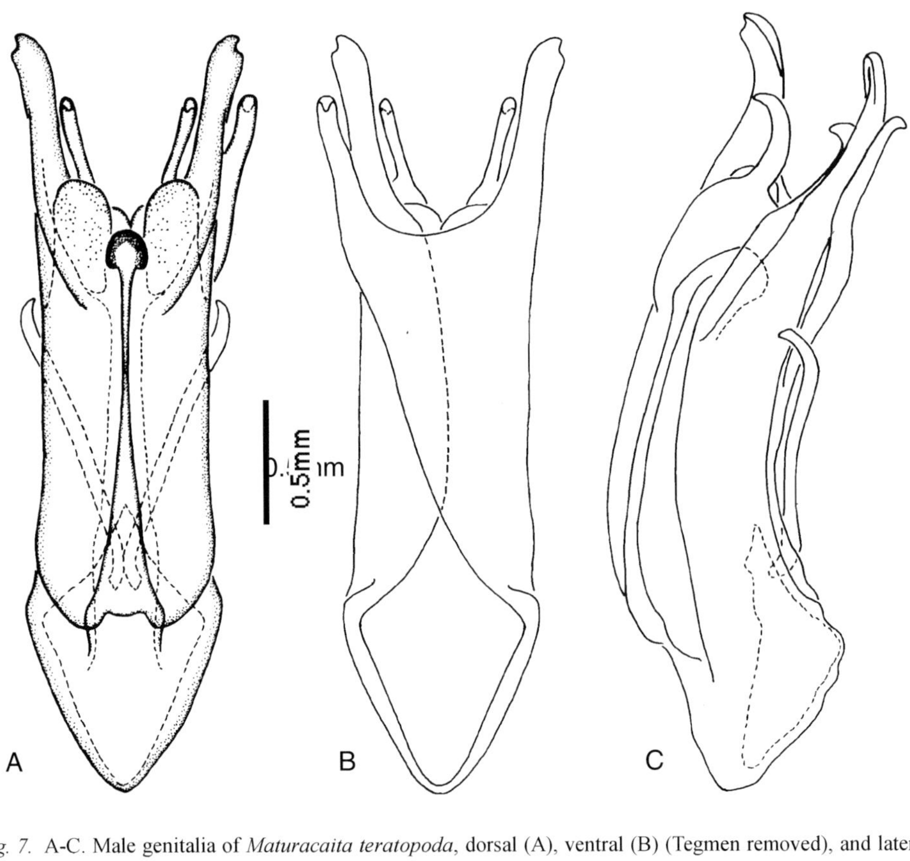

Fig. 7. A-C. Male genitalia of *Maturacaita teratopoda*, dorsal (A), ventral (B) (Tegmen removed), and lateral (C) views.

21. *Length or shape of median dorsal process*. In most taxa the median dorsal process is approximately as long as the lateral processes (21.0)(Fig. 3); however, in some (for example, *Maturacaita,* see Fig. 7) the median process is much shorter (21.1). *Balzanica, Disonycha*, and *Phenrica* may lack the lateral processes; the median structure is much broader, being plate-like or bowl-shaped (21.2)(See Bechyné & Bechyné, 1966). This character is treated as unordered.
22, 23. *Lateral dorsal process of median lobe*. The lateral dorsal process of the median lobe is usually a flat, plate-like structure appearing to lie flat on the surface of the median lobe (22.0 and 23.0)(Fig. 3 a,b). This structure also can be curved inward (22.1), absent (22.2 and 23.? (where homology is uncertain)) or oriented obliquely, rather than lying flat on the dorsal surface of the median lobe (23.1)(Fig. 3c,d).
24. *Outline of ventral lip of median lobe*. Although the outline (general shape) of the ventral lip or edge of the apex of the median lobe is most commonly triangular or rounded (24.0)(Fig. 3a), highly modified ventral edges are found that vary from highly dissected (24.2) (for example, *Maturacaita*, Fig. 7) to greatly elongate (24.1)(Fig. 3.e,f) (for example *Rosalactica*, *Pedilia basalis*). This character is coded as unordered.
25. *Dorso-ventral curvature of ventral apex of median lobe*. In *Lacpacticoides maculicollis*

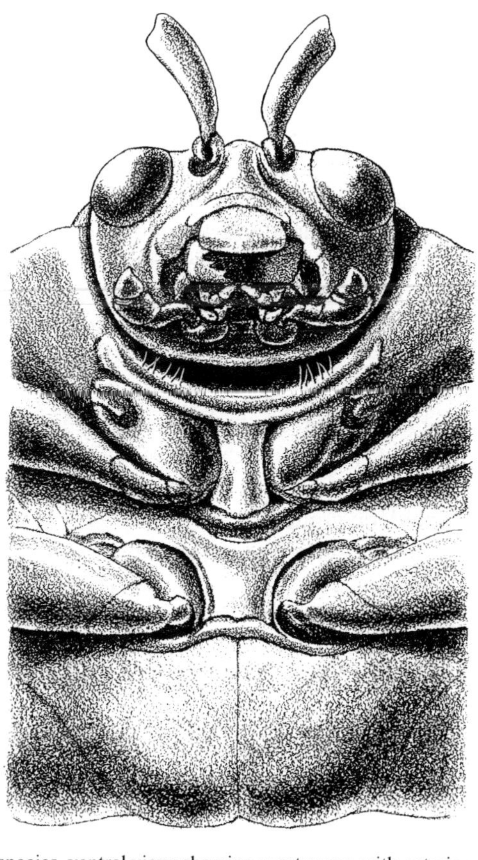

Fig. 8. Pedilia undescribed species, ventral view; showing prosternum with anterior ridge continuous at midline.

and *Paralactica rubinii* the apex of the ventral lip shows a strong ventral curvature, the apex pointing 90° or more from the axis of the median lobe (25.1).

Cladistic Analysis. Cladistic analyses were conducted using Hennig86 version 1.5 (Farris, 1988) using tread, h, h*, mh, mh* and applying bb* to the trees supplied by each command. This protocol is recommended for finding all islands of most parsimonious trees (Maddison, 1991). Twenty-six characters (0 – 25) were coded in 32 terminals, some terminals representing more than one taxon (see Table 3). When possible, transformation series of multistate characters were established by hypothesis of character adjacency, or coded as unordered. Character polarity was established using outgroup comparison (Watrous & Wheeler, 1981). Trees were examined and characters optimized using Clados version 1.2 (Nixon, 1992).

Results of Cladistic Analysis

I discuss relationships of the genera studied in order to clarify the proposed hypotheses of relationship, how they are supported and where further study is needed. Phylogenetic conclusions for the groups are hampered by the paucity of available material, both in number of specimens and in species representation. Because of the small number of species sampled for each genus (except *Pedilia, Nephrica* and *Sophraenella*), this study does not intend to fully characterize incomplete results. However, the diversity within the disonychine genera as a whole, and especially within the series *Paralactica* is accurately characterized.

A cladistic analysis of the 32 terminal taxa using the characters discussed above resulted in 130 most parsimonious cladograms each with 112 steps, Consistency index = 0.39, Retention index = 0.74. The consistency and retention indices are used as measures of homoplasy, relative to the number of possible steps in a cladogram (Farris, 1988). Fig. 9 depicts a strict consensus tree of those cladograms. All of the variation in these cladograms is between species of *Pedilia, Nephrica* and *Sophraenella,* and in the two most apical nodes. In the strict consensus tree (Fig. 9), the relationships among species of *Nephrica* and *Sophraenella* are unresolved as are the relationships between *Maturacaita, Lacpacticoides* and *Paralactica rubinii* Bechyné and Bechyné.

A representative cladogram (Fig. 10) clearly shows support for the series *Pedilia* minus *Balzanica* as a monophyletic group. Support is also present for the series *Paralactica* minus *Lacpactica* as a monophyletic group.

The *Pedilia* clade is supported by eye shape asymmetrically emarginate (2.2) and the morphology of sternite VIII,with the apex 1/3 or less as wide as the posterior edge of sternum VII (11.1, 11.2), base wide (12) and flat or emarginate (13.1, 13.2). *Pedilia sensu stricto*, the species with a round habitus, also appears to be a monophyletic group supported by fused or absent gonapophyses (9.2, 9.4), prosternum modified to receive the head (4.0)(Fig. 8), pronotum with lateral prebasal impressions (5.0). However, the species of the genus and subgenus *Nephrica* and *Sophraenella fulva* Jacoby appear to represent a grade paraphyletic to *Pedilia sensu stricto* in all cladograms. I conclude that *Nephrica* and *Sophraenella* should be accommodated in *Pedilia*; formal synonomy will be presented in a later publication (Duckett in prep). In the discussion below I refer to this clade as series *Pedilia*, which clearly does not include *Balzanica*.

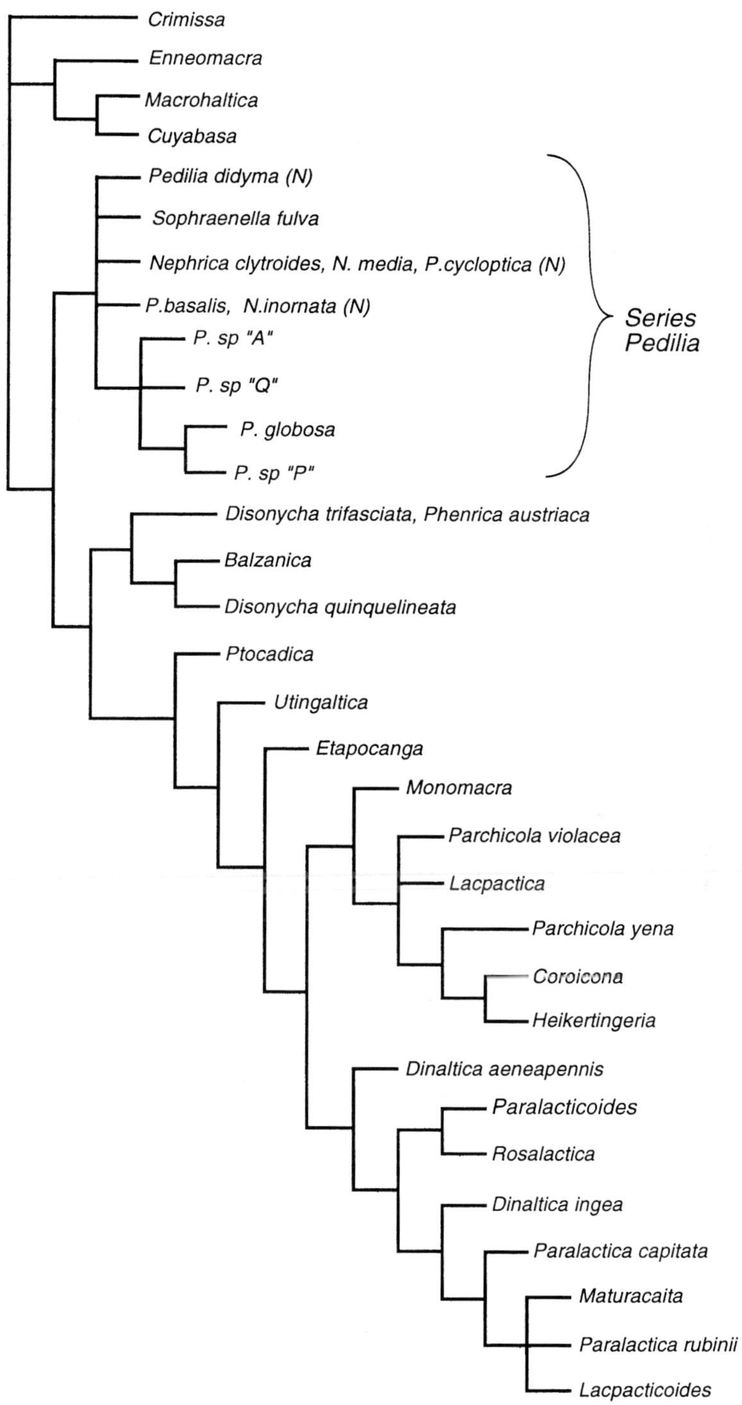

Fig. 9. Strict consensus cladogram. *Pedilia* clade shown with bracket. Species originally described in genus *Nephrica* designated (N).

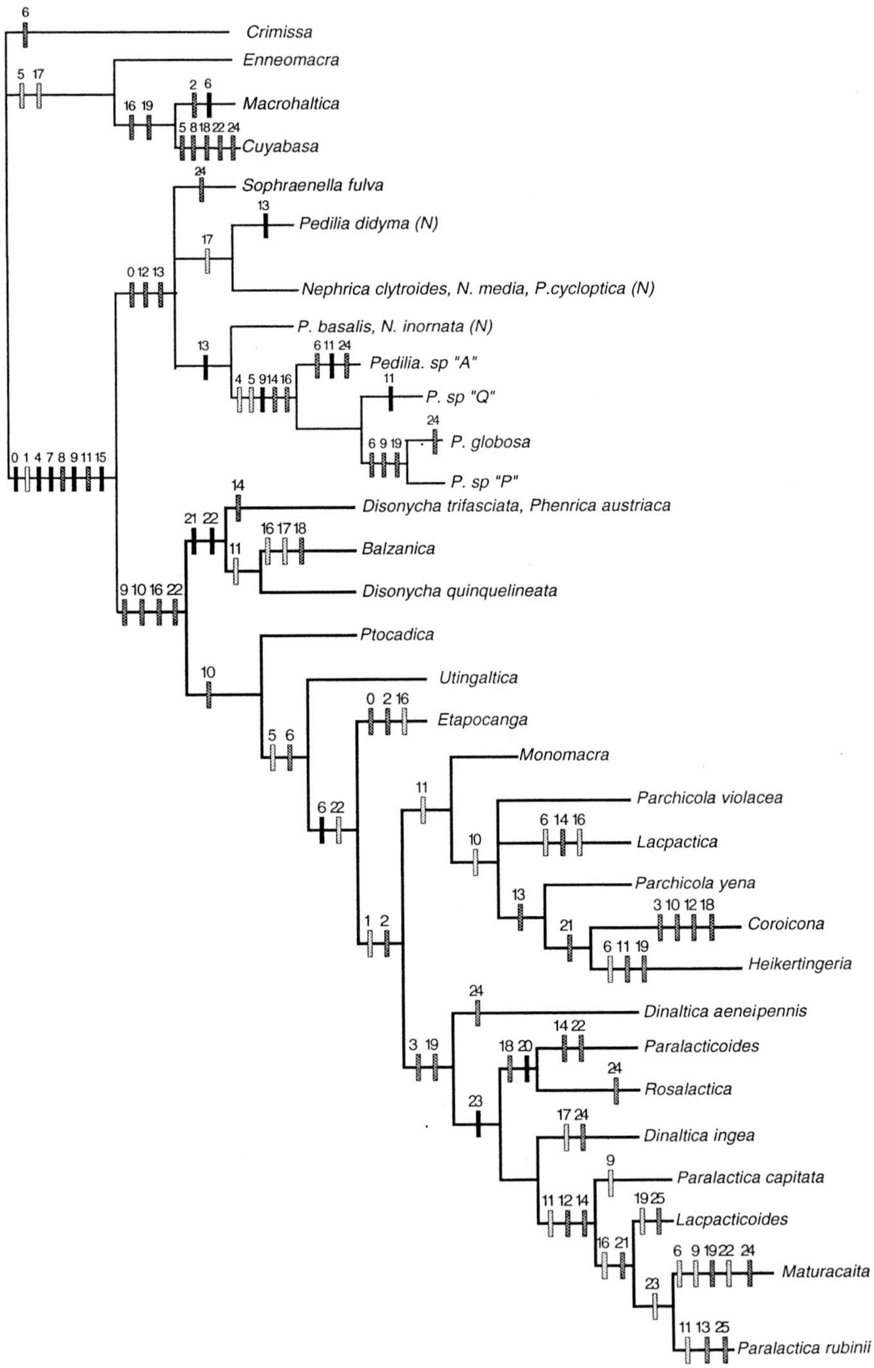

Fig. 10. One of 130 cladograms of the disonychine genera, showing character state changes (See text). Terminals with all species having been originally described in genus *Nephrica* designated (N)

Series *Paralactica* minus *Lacpactica* is supported by a transverse and entire apical flange on male the sternum VII (19.1) (Fig. 4b) and by the lateral dorsal processes of the median lobe oriented obliquely (23.1)(Fig. 3 c-d), the apex of the female sternite VIII is not as wide as sternum VII (11.1, 11.2) (compare Fig. 6d). Some of these characters show reversals near the apex of the cladogram, *Maturacaita* and *Paralactica rubinii* with the lateral dorsal processes oriented horizontally (23), and *Lacpacticoides* with an auriculate flange on the male sternum VII (19.0). In *Maturacaita* this flange appears to be absent (19.2). The apex of sternite VIII (11.2) in its narrowest state supports the clade containing *Paralactica capitata* Illiger, *Lacpacticoides*, and *Maturacaita*, but is reversed in *Paralactica rubinii* (11.1).

The clade containing *Balzanica*, *Disonycha*, and *Phenrica* is supported by the character state "dorsal processes fused to form a plate of the median lobe" (21.2 and 22.2), a structure where the homology with the lateral and median dorsal processes may be somewhat forced. This clade, however, constitutes the subtribe Disonychina *sensu* Bechyné and Bechyné (1966) minus the series *Pedilia*..

The most significant finding of this study is the support for a disonychine clade, that includes all of the disonychine genera of Bechyné and Bechyné (1977), except *Enneomacra*; this being, of course, the formation of the pygidium from tergum VIII (7.1), and less importantly, the loss of the epiproct (8.1).

Character evolution. Several characters in the Disonychina show interesting changes during the evolution of the clade. The most notable are characters of the ovipositor, including gonocoxae (or vaginal palpi (Konstantinov, 1998), gonapophyses and sternite VIII.

The genera of the Disonychina have a reduced ovipositor compared with other Alticini as a result of the loss of the tubular form and the epiproct. However, within the Disonychina the trend toward reduction continues. In the series *Pedilia*, the gonocoxae are three dimensional and setose (9.0)(Figs 6 f-h), though relatively flattened with respect to the outgroups. In *Disonycha*, *Phenrica*, *Balzanica*, as well as *Lacpactica* and *Parchicola*, the gonocoxae are flattened and asetose (9.1), often reduced to amorphous bits of sclerite (very similar to the gonophyses of *Nephrica* and *Sophraenella* compare Fig. 5b). In other more derived disonychine the genera gonocoxae are lost. Within the genus *Pedilia* there is also a trend toward reduction of the gonocoxae (see Duckett, 1993,1995, and in prep) (Fig. 6g).

The gonapophyses also have been reduced and lost in the Disonychina and in the series *Pedilia*. The most plesiomorphic groups, *Nephrica* and *Sophraenella*, have paired flattened gonopophyses (9.1), which are slightly amorphous (see Figs. 5b, 6f), *Pedilia sp. Q* and *P. sp.A* . have median fused gonapophyses (9.2)(Fig. 6a) where as other species of *Pedilia* and most other disonychine genera have lost them altogether (9.4). *Maturacaita* and *Paralactica* appear to have independently evolved a median, fused, recurved gonapophysis (9.3).

Sternite VIII is not reduced in complexity in Disonychina compared to its state in other Alticini, but it is reduced in relative size. In other Alticini, sternite VIII is wide caudally and narrow anteriorly, most frequently with the anterior portion long and narrow, like a handle (Fig.5a). In most Disonychina, the length of the anterior portion relative to the beetle's abdomen is reduced, as would be expected given the loss of overall extension in a non-tubular ovipositor (see Fig. 5b). However, there is increased variation in the width and shape of the anterior (basal) and caudal (apical) areas (compare Figs. 6c-e). In the series *Pedilia* the base is always wide.

Another structure with interesting evolutionary variation is the male median lobe. All Disonychina except the series *Disonycha* have a median lobe with three dorsal processes (see Fig 3). In the series *Disonycha* these processes appear to be fused together (22.2, 23.?), often forming a bowl-like concavity at the top of the median lobe (compare Figs. 1-14, Bechyné & Bechyné, 1966).

In the series *Monomacra* the most derived genera, *Paralactica*, *Paralacticoides*, *Lacpacticoides* and *Rosalactica*, all have the lateral dorsal processes oriented obliquely (Figs.3c,d), rather than lying flat on the dorsal surface of the median lobe. This clade is notable for the extreme variation of the median lobe. The median dorsal process can be recurved as in *Paralacticoides* (Figs 3c,d), or practically absent as in *Maturacaita* (Fig 7). The ventral lip also can show extreme ventral curvature (25.1) as seen in *Lacpacticoides* and *Paralactica rubinii*. This variation exceeds the apparent variation in other clades of Disonychina.

However, it is obvious from the discussion above that many of the genitalic characters used are very plastic and subject to evolutionary reversals. As many of these genera are probably participating in mimicry complexes, with their external appearances maintained somewhat convergently by natural selection, this selection will also confound evolutionary analysis. The reduction of the ovipositor, seen in the Disonychina has caused homoplasy which by definition, is hard to interpret correctly. I judge that this cladogram should be interpreted qualitatively (that is to indicate which groups are in more urgent need of study) or on the large scale, in terms of generic composition of the Disonychina and the series with in it.

Taxonomy

Disonychina

The above analysis justifies the formal removal of *Enneomacra* and *Cuyabasa* from the Disonychina. The strong unequivocal support of the disonychine clade is significant, given the paucity of suprageneric groups in the Alticini and the subjective characters previously used to delimit this group. I recognize the other genera analyzed as members of the Disonychina, including *Ptocadica* and *Etapocanga* (see Table 4). All have tergum VIII as the "pygidium" and do not have epiprocts. I recognize three generic series in the Disonychina. These series are: series *Pedilia*, including *Pedilia* which is probably synonymous with *Sophraenella* and *Nephrica*; series *Disonycha*, including *Disonycha* and *Phenrica*; and series *Monomacra*, including *Ptocadica*, *Utingaltica*, *Etapocanga*, *Monomacra*, *Lacpactica*, *Parchicola*, *Coroicona*, *Heikertingeria*, *Dinaltica*, *Paralacticoides*, *Rosalactica*, *Paralactica*, *Maturacaita*, and *Lacpacticoides*. I have not examined *Vilhenaltica* or *Archilactica* (as they were not available on loan), but I include them in the *Monomacra* series where they were originally placed (see Table 4) because I have no data on which to exclude them. Given the small number of species analyzed from each genus and the paucity of material of many of those species, I do not feel that more formal supra-generic groupings are appropriate at this time.

This analysis suggests the synonymy of *Sophraenella* and *Nephrica* with *Pedilia*, which will be discussed in depth in a revision of all known species, described and undescribed

Table 4. Proposed taxonomic structure of the Disonychina

Series	***Pedilia***	Series	***Monomacra***
	Pedilia		*Ptocadica*
	Nephrica		*Utingaltica*
	Sophraenella		*Etapocanga*
			Monomacra
			Lacpactica
Series	***Disonycha***		*Parchicola*
	Disonycha		*Coroicona*
	Phenrica		*Heikertingeria*
			Dinaltica
			Paralacticoides
			Rosalactica
			Paralactica
			Maturacaita
			Lacpacticoides
			*Vilhenaltica**
			*Archilactica**

*Taxon not examined.

(Duckett in prep). This analysis also suggests that the series *Paralactica* is monophyletic if *Lacpactica* is removed. However, the rest of the series *Monomacra* is paraphyletic to series *Paralactica*. Therefore, the series *Paralactica* is not a taxonomically useful group, unless remaining parts of the series *Monomacra* are further divided.

The position of *Balzanica* inside the *Disonycha* clade indicates that *Balzanica* should be accommodated within *Disonycha* as has been done previously (Duckett, 1995).

Further taxonomic changes are beyond the scope of this study. The results of this analysis indicate the need for further study of the genera in the series *Monomacra* and *Disonycha*. This study also suggests that the genus *Dinaltica* is as Bechyné and Bechyné asserted (1975), in fact polyphyletic, and in urgent need of revision. The same is probably true for the genus *Paralactica*. Without examining more taxa of all the genera in the series *Monomacra* it is premature to suggest further taxonomic changes. However, I am pursuaded that Bechyné and Bechyné may have been premature in splitting up the genus *Monomacra*.

This study also suggests that *Phenrica* may be a junior synonym of *Disonycha*. However, as that genus is so large and morphologically diverse, it is impossible to form a definitive conclusion, based on the one species analyzed. A revision of *Disonycha* and *Phenrica* is needed.

Also, the addition of the characters of the ovipositor to the definition of the Disonychina group suggests re-analysis of some of the taxa Bechyné and Bechyné (1975) moved from the genus *Monomacra* into other genera. Genera such as *Resistenciana* Bechyné and Bechyné and *Strabala* Chevrolat may be misplaced outside the Disonychina. Data presented by Vogt, *et al.* (1979) suggested *Agasicles* Jacoby or some species of *Agasicles* are probably also members of the Disonychina.

Clade Paralactica

Diagnoses and phylogenetic characterizations of genera in the *Paralactica* clade are presented below. As noted above, these characterizations are based on few taxa and should be treated as evidence of the breadth of the series *Paralactica* , rather than reviews of intrageneric variation. Phylogenetic characterizations include both apomorphies and plesiomorphies.

Rosalactica Bechyné and Bechyné 1977:63

TYPE SPECIES. *Rosalactica maculicollis* (Weise), by original designation.
DIAGNOSIS. Elytral base with rufous macula. *Rosalactica* are small, delicate beetles with large eyes and ommatidia and distinguished from other genera of the series *Paralactica* by the rosy-pink basal humeral maculae and by their red body fluids (Bechyné & Bechyné, 1977). Males are distinguished from those of *Paralacticoides* by the less melanized, more robust and apically straightened form of the recurved modified median dorsal process of the median lobe; females by the long, thin basal portion of sternite VIII.
PHYLOGENETIC CHARACTERIZATION. Eyes with symmetrical median emargination. Transverse frontal carina not curving around mandibular bases, not convex. Maxillary palpi gracile. Labrum subquadrate, edges beveled. Pronotum with lateral, prebasal impression, straight anterior edge of the transverse prebasal impression clearly delimited. Pygidium formed from tergum VIII. Epiproct absent. Gonapophyses absent. Gonocoxae absent. Sternite VIII as wide as posterior edge of sternum VII, base wide, rounded; insertion point of intersegmental membrane pointed. Spermatheca with pronounced neck; basal portion of proximal spermathecal duct not expanded. Spermathecal pump parallel-sided with flange.

Apical flange of male sternum VII transverse and entire. Median lobe of aedeagus with barb on median dorsal process; median dorsal process as long as lateral processes. Lateral dorsal process not curved inward; lateral process oriented in oblique plane when viewed from dorsal side. Ventral lip of median lobe elongate, otherwise unmodified.
MATERIAL EXAMINED. *Rosalactica maculicollis basipennis* (Weise). BRAZIL: Amazonas, Borba, Río Madeira, v.1945, B. Pohl, (MZSP); 4 ex., Amazonas, Itacoatiara, vi.1965, Dirings., (MZSP); Pará, Itaituba, Río Tapajos, i.1961. Dirings., (MZSP). PERU: Jacoby col (BMNH). *Rosalactica kermesina* Bechyné and Bechyné, ♂, BRAZIL: Pará Castanhal Sta. Isabel 40 Km. NO Belém, 28-29.x.1962, C. Lindemann, (ZSBS). 16 ex. VENEZUELA: Territorio Federal Amazonas, Culebra, Río Cunucunuma, 3-5.vi.1990, C.N. Duckett and C.J.R. Thompson, (CNDC). *Rosalactica sp.* 14 ex., PERU: Madre de Díos Río Tambopata Res, 30 Km (air) sw Pto. Maldonado, 290m 12° 50' S 69° 20 W (USNM).
REMARKS. According to the Bechyné and Bechyné key (1977) *Rosalactica* is distinguished from *Paralacticoides* by a keel-like prosternal process, rather than a roll-like, or roof-like prosternal process. The authors also identified a difference in pubescence, which I did not perceive. When faded, *Rosalactica* is distinguishable from *Paralacticoides* by its smaller size and genitalia.

Paralacticoides Bechyné and Bechyné 1977 (Fig. 3c,d)

TYPE SPECIES. *Paralacticoides bicarinata* Bechyné and Bechyné, by original designation.

DIAGNOSIS. *Paralacticoides* includes delicate fulvous beetles with bulging eyes and large ommatidia. *Paralacticoides* is distinguished from *Paralactica* by a wider vertex, eyes with a slight emargination rather than a notch, shorter antennae, and the absence of gonapophyses. Both sexes of *Paralacticoides* can be distinguished from other genera by the long finger-like prosternal process which exceeds the coxae; males are distinguished by the highly recurved, strongly melanized modification of the median dorsal process of the median lobe (Figs 3c,d).

PHYLOGENETIC CHARACTERIZATION. Eyes with symmetrical median emargination. Transverse frontal carina not curving around mandibular bases, not convex. Maxillary palpi gracile. Labrum subquadrate, edges beveled. Pronotum with lateral prebasal impression, transverse prebasal impression straight, anterior edge clearly delimited. Pygidium formed from tergum VIII. Epiproct absent. Gonapophyses absent. Gonocoxae absent. Sternite VIII one-half as wide as posterior edge of sternum VII, base wide, rounded; insertion point of intersegmental membrane straight. Spermatheca with pronounced neck; basal portion of proximal spermathecal duct not expanded. Spermathecal pump parallel-sided with flange.

Apical flange of male sternum VII transverse and entire. Median lobe of aedeagus with barb-like recurvature of median dorsal process; median dorsal process as long as lateral processes. Lateral dorsal process curved inward; lateral process oriented in oblique plane when viewed from dorsal side. Ventral lip of median lobe triangular, curving sharply downward.

MATERIAL EXAMINED. *Paralacticoides bicarinata acrocostata* Bechyné and Bechyné, Paratypes: ♂ BRAZIL: Pará, Obidos, vi.1960, Dirings., (MZSP); ♀ same date (MZSP); two ♀ ♀ same location and collector, ii.1958, xi.1955, (MZSP); ♀ Pará, Granja St. Hurt. E.F. Bargança, xi.1956, (MZSP); ♀ Itaituba, (Río Tapajoz), i.1961, Dirings., (MZSP); ♀ Pará, Oriximiná, xi.1960, Dirings., (MZSP); Pará, Mocajuba, Mangabeira, v.1953, Orlando Rego, (QBUM); *Paralacticoides stephania* Bechyné and Bechyné, ♀, VENEZUELA: Bolívar, Km 107, El Dorado-Sta. Elena, 520m, 19-VIII-57, F. Fernandes Y.and C. Rosales, (MIZA).

REMARKS. Bechyné and Bechyné (1977) distinguished *Paralacticoides* from other genera by the thickness and length of the ventral pubescence, the shape or lack of shape of the longitudinal frontal carina, and a longitudinal elytral ridge extending backwards from the humeral callus. All of these characteristics are difficult to distinguish and therefore practically useless for identification purposes. I detected no difference in pubescence between *Paralacticoides* and *Rosalactica* as described in the text (Bechyné & Bechyné, 1977). The longitudinal frontal carina does broaden as Bechyné and Bechyné indicated; however, this can be easily confused with the transverse carina which is stated to be absent. The trait is also somewhat variable and similar to the state in *Rosalactica*.

The longitudinal elytral carina is present though difficult to see in some females of *P. bicarinata*, but it is not visible in males, nor in females of *P. stephania* Bechyné and Bechyné.

Paralactica Bechyné and Bechyné 1961

TYPE SPECIES. *Paralactica capitata* (Illiger), by original designation.
DIAGNOSIS. *Paralactica* includes fulvous beetles with large eyes, having a conspicuous ≈60° notch on the mesal margin of the eye near the antennal socket; antennae very long; prosternal process flat, recurved dorsally, following the contour of coxa. *Paralactica capitata* includes beetles with dark vertices. Females can be distinguished from other genera by the notched eye; the presence of a single A-shaped gonapophysis; and sternite VIII with a very narrow apex, continuing narrowed and widening abruptly at the base. Males can be distinguished by the very thick ventral lip of the median lobe and by the form of the lateral dorsal processes of the median lobe which curve inward, meeting medially and forming a ring over the ventral lip.
PHYLOGENETIC CHARACTERIZATION. Eyes with symmetrical median emargination. Transverse frontal carina not curving around mandibular bases, not convex. Maxillary palpi gracile. Labrum subquadrate, edges beveled. Anterior portion of prosternum smooth. Pronotum with lateral prebasal impression, transverse prebasal impression straight, anterior edge clearly delimited. Pygidium formed from tergum VIII. Epiproct absent. Gonapophyses of female entire, A-shaped. Gonocoxae absent. Sternite VIII narrow apically, base wide, flattened, with insertion of intersegmental membrane straight. Spermatheca with pronounced neck, basal portion of proximal spermathecal duct not expanded. Spermathecal pump parallel-sided and without flange.

Apical flange of male sternum VII transverse and entire. Median lobe of aedeagus without barb on median dorsal process; median dorsal process as long as lateral processes. Lateral dorsal process of median lobe recurved inward; lateral process oriented in oblique plane when viewed from dorsal side. Ventral lip of median lobe triangle-shaped, apex unmodified.
MATERIAL EXAMINED. *Paralactica capitata* (Illiger) 14 ex. BRAZIL: Pará, Marajó, P. Pedras, 12.iii.1978, M.F. Torres (MPEG); Pará, Santarewsinho, Mun. de Itaituba, Río Tapajoz, viii.61, Dirings., (MZSP); Canidé (Río Gurupí, x.1964, B. Malkin, (MZSP). *Paralactica rubinii* ♀ and ♂ Amazonas, Uaupes, 30.i.1964, C. Lindemann, (ZSBS); ♂ VENEZUELA: Territorio Amazonas. Upper Río Cunucunuma, Tapara, 20.iv.1950, J. Maldonado, (USNM). Estado Bolívar, Km 72, near San Isidro, 5.i.1990, C.N. Duckett, (CNDC). *Paralactica sp. A* ♀. BRAZIL: Amapá, Serra Lombard, Olemargu 10-vii 1961, J. and B. Bechyné, (MIZA). *Paralactica sp. B* 3 ex. BRAZIL: Amazonas, Manaus, vi.1972. F. Oliveira, (DZUP); 1 ex. SURINAME: Brokopando district, Brownsberg Natuurpark, Mazaroni Plateau, 400-500 m, 23 Aug 1982, W.E. Steiner, (USNM).
REMARKS. *Paralactica rubinii* Bechyné & Bechyné is very similar to *P. helia* Bechyné & Bechyné and *P. capitata* externally, however, *Paralactica rubinii* differs greatly from *P. capitata* in the form of the genitalia. Apomorphies of *P. rubinii* include: on the median lobe, the apex of the ventral lip curves sharply downward, the median dorsal process is shorter than the lateral processes; the base of female sternite VIII is also emarginate. Other differences of *P. rubinii* include gonapophyses absent, apex of sternite VIII 1/3 the width of sternum VII and lateral processes of median lobe horizontally oriented.

Maturacaita Bechyné and Bechyné 1977
(Fig.7a-c)

TYPE SPECIES. *Maturacaita teratopoda* Bechyné and Bechyné 1977:101 by original designation.

DIAGNOSIS. *Maturacaita* includes fulvous beetles distinguished from other genera by a very narrow prosternal process with a raised longitudinal lamella; very long, filiform antennaewith segments IV - VIII piceous; large eyes with prominent median notch around base of antennae; frontal longitudinal carina narrow (cf. Bechyné & Bechyné 1977, Fig. 1, p. 102).

Males with hemispherical emargination of sternum VII; median lobe deeply emarginate, dorsal and ventral lips appearing horn-like, asymmetrical (see Figs 7a, b, c). Females with gonapophyses forming a three dimensional semicircular structure, A-shaped in profile.

PHYLOGENETIC CHARACTERIZATION. Eyes with symmetrical median emargination. Transverse frontal carina narrow, produced over labrum. Labrum longer than wide. Maxillary palpi long and gracile. Anterior portion of prosternum smooth and parallel with the axis of the beetle. Pronotum with lateral prebasal transverse impressions, anterior edge of transverse impression not clearly delimited. Pygidium formed by tergum VIII.

Female genitalia with gonapophyses fused, semicircular, A-shaped in cross section; gonocoxae absent; sternite VIII narrow apically, base wide with posterior edge flat; insertion point of intersegmental membrane straight.

Spermatheca with pronounced neck between receptacle and pump; basal portion of proximal spermathecal duct expanded; spermathecal pump parallel-sided, flange absent.

Male sternum VIII with internal apical flange absent.

Median lobe of aedeagus without barbed structure or median dorsal process; lateral dorsal process oriented in a vertical plane, not curved inward; ventral lip of median lobe bifid, not curving sharply downward.

MATERIAL EXAMINED. *Maturacaita teratopoda*, Bechyné and Bechyné. Holotype, ♂, BRAZIL: Amazonas, Maturacá, 23.1.1963, C. Lindeman, (ZSBS); Paratype, ♀ VENEZUELA: Bolívar, Guayaraca, Auyantepui, 1100 m, 25.iv.1956, F. Fernandez Y. and C.J. Rosales, (MIZA); ♂, Territorio Federal Amazonas, Occamo, 14.iv.1965, F. Fernandez Y., (MIZA); *Maturacaita sp.* A., ♀, PERU: Upper Río Maranon, x.14.1924, "F6103" H. Bassler, (AMNH).

Lacpacticoides Bechyné and Bechyné 1977:60

TYPE SPECIES: *Lacpacticoides nicodema* (Bechyné) 1957, by original designation.

DIAGNOSIS. *Lacpacticoides* includes small, delicate beetles, with a smooth prosternal surface and prosternal processes and with the elytral margin slightly rounded as if forming a bead. Males of *Lacpacticoides* can be separated from those in all other genera by the length and shape of sternum VII. Sternum VII is almost as long as segments IV - VI together. This segment is additionally modified internally and externally; the anterior internal edge appears scalloped, each side of the internal portion being strongly rounded (visible in a cleared specimen). The median portion of the posterior edge is produced into a small rounded lobe which corresponds to a similar emargination on the pygidium.

Females can be distinguished by the almost trapezoidal shape of sternite VIII with a narrow, flattened apex and a flat base almost twice the width of the apex.
PHYLOGENETIC CHARACTERIZATION. Eyes with symmetrical median emargination. Transverse frontal carina not curving around mandibular bases, not convex. Maxillary palpi gracile. Labrum subquadrate, edges beveled. Prosternum smooth. Pronotum with lateral prebasal impression, transverse prebasal impression straight, anterior edge clearly delimited. Pygidium formed from tergum VIII. Epiproct absent. Gonapophyses absent. Gonocoxae absent. Sternite VIII narrow apically; base wide, flat; insertion of intersegmental membrane straight. Spermatheca with pronounced neck; basal portion of proximal spemathecal duct expanded. Spermathecal pump parallel-sided, without flange.

Apical flange of male sternum VII auriculate. Median lobe of aedeagus with barb; median lobe of aedeagus significantly shorter than lateral process. Lateral dorsal process of median lobe curved inward; lateral process oriented in oblique plane when viewed from dorsal side. Ventral lip of median lobe triangle-shaped, apex curving sharply downward.
MATERIAL EXAMINED. *Lacpacticoides nicodema* (Bechyné) 1957, (all determined by J. Bechyné 1967). ♂, BRAZIL: Pará, Canta Galo, i. 1957, Dirings., (MZUP); 2 ♀♀ Amapá, Maturacá, 23.i. 1963, C. Lindemann, (ZSBS); VENEZUELA: Rió Siapó, T.F. Amazonas, 11.xii.1972, 140m, C. Lindemann (ZSBS).
REMARKS. There is a second species, *Lacpacticoides tepuicola* Bechyné and Bechyné 1977, which is known from four females. This species was not available for study.

Acknowlegements

I am pleased to thank R. Wyttenbach, Q.D. Wheeler, J. V. McHugh, S.R. Mandry, M. Luckow, J.K. Liebherr, T. Kaplan, N. Jacobson, and P. Fraissinet for improving comments on earlier drafts of the manucript. My study especially benefited from discussions with Vilma Savini, José Santisteban, C.A.M. Reid, J. V. McHugh, D.G.Furth, R. W. Flowers and C. Breidenbach. I thank Frances Fawcett for spermathecal and aedeagal drawings, and Mike Linkenhocker for the *Etapocanga* habitus and the *Macrohaltica* ovipositor. R. Wyttenbach was also instrumental in the construction of many of the plates.

I thank the following curators and institutions for their kindness in visiting their institutions and for loans of material: L.H. Herman, The American Museum of Natural History, New York, (AMNH); M. Monné, Brazilian National Museum, Río De Janiero (QBUM); Biologische Zentralanstalt, Berlin (BZB); S. L. Shute British Museum (Natural History) London (BMNH); D. Kavannaugh and R. Brett, California Academy of Sciences (CAS); R. Anderson, Canadian Museum of Nature (CMN); C. Bordón, Colección Carlos Bordón, Maracay, Venezuela (CCB); J. K. Liebherr and R. Hoebeke, Cornell University (CUIC); F. Hieke, Musum für Naturkunde de Humbolt— Universität zu Berlin (MNHU); Museum George Frey (Frey); D. G. Furth, Museum of Comparative Zoology, Harvard University, Boston (MCZC); N. Bertí, Museum National d'Histoire Naturelle, Paris (MNHN); T. Pimentel Chaves, Museu de Zoología Paraense Emilio Goeldi, Belém do Pará (MPEG); C. Costa, Museu de Zoología da Universidade de São Paulo, São Paulo (MZSP); L. J. Joly and V. Savini, Instituto de Zoologia Agricola, Universidad Central de Venezuela, Maracay (MIZA); A. Harada and C. R. Fonseca, Invertebrate Collection of INPA, Manaus (INPA); A. Solís, Museo Nacional de Costa Rica, San José (MNCR); P. Lindskög, Naturhistoriska

Riksmuseet, Stockholm (NHRS); E. G. Riley, College Station, Texas (EGRC) and Texas A& M University (TAMU); R.E. White, United States National Museum (Natural History), Washington (USNM); M. A. Alameida, Universidade Federal do Paraná, Curitiba (DZUP); G. Scherer, Zoologische Statssammlung des Bayerischen Staates, Munich, (ZSBS); O. Martin, Zoologisk Museum University of Copenhagen, Copenhagen (ZMUC).

Financial support was provided in part by the following grants: a Fulbright-Hayes grant for study in Venezuela; a Theodore Roosevelt Memorial Fund grant (American Museum of Natural History) and an H. E. Moore Fund grant (Cornell) for study in Costa Rica; two Ernst Mayr grants (Harvard) for museum work in London, Munich and Berlin. I would like to thank Q. D. Wheeler for making museum study in Paris possible. Other financial support was provided by Hatch grants to Q. D. Wheeler. Parts of this manuscript were typed by Karen Herzog, with funds provided by the Cornell University office of disabilty. Ron Hoy kindly made graphics design software and equipment in his lab available for my use. All support is gratefully acknowledged. Salvador Mandry provided additional computer consultation. Finally, I would like to thank Mike Cox for inviting me to contribute to this volume and for his kind and thorough editorial advice.

References

Baly, J.S. 1879. Descriptions of new genera and of uncharacterized species of Halticinae. Trans. ent. Soc. London, 1879:157-184, 283-323.

Balsbaugh, E.U. 1988. Mimicry and the Chrysomelidae. in: Jolivet, P., Petitpierre E., &. Hsiao T.H. (Eds.) Biology of Chrysomelidae. Kluwer Academic Publishers, Dordrecht, The Netherlands . pp. 261-284.

Bechyné, J. 1956. Beitrage zur Kenntnis der neotropischen Alticiden und Galeruciden. Ent. Arb. Mus. George Frey, 7:965-1071.

Bechyné, J. 1957. Voyage de M. le Dr. A Roman au Bresil (1914-1915) Eumolpides, Galerucides et Alticides (Col: phytophaga). Ark. Zool., 11 (10):133-152

Bechyné, J. 1959. Beitrage zur Kenntnis der Alticdenfauna Boliviens (Coleop. Phytoph.) Beitr. neotrop. Fauna, 1:269-381.

Bechyné, J. 1980. El Jeannelismo y la evolucion:concepto de las leyes organicas sin excepcion. Grafindustral Aragua, Maracay, Venezuela. 178 pages.

Bechyné, J. & B.Springlova de Bechyné. 1961. Notas sobre Chrysomeloidea neotropicais. Bol. Mus. Pará Emilio Goeldi (Zool)., no. 33:1-50, 7 figs.

Bechyne, J. & B.Sprınglova de Bechyné. 1964. Notes sur quelques Chrysomeloidea neotropicaux (Coleoptera: Phytophaga). Rev. Fac. Agron. Univ. Cent. Venez., 3(3):69-123, 27 Figs.

Bechyné, J. & B.Springlova de Bechyné. 1966. Eividenz der bisher bekannten Phenrica-Arten (Col.Phytophaga, Alticidae). Ent. Tidskr., 87:142-170, 16 figs.

Bechyné, J. & B.Springlova de Bechyné. 1975. Notas sobre la serie filetica de *Monomacra* y sus formas convergentes (Col. Phytophaga, Alticidae). Revta Fac. Agron. Univ. cent. Venez. 8:25-140.

Bechyné, J. & B.Springlova de Bechyné. 1977. On the phylogenesis of some neotropical alticids (Col. Phytophaga). Stud. neotrop. Fauna Envir., 12:81-146. (in German).

Bitsch, J. and R. Matsuda. 1979. Morphologie abdominale des insects. Traité de Zoologie, Anatomie, Systématique, Biologie, Libraires de L'Académie de Médecine, Paris, 463-490, illus.

Blake, D.H. 1933. Revision of the beetles of the genus *Disonycha* occurring in America north of Mexico. Proc. U.S. natn. Mus., 82:1-66. 8 figs.

Blake, D.H. 1951. Synonymies and new species of flea beetles (Coleoptera: Chrysomelidae). Proc. ent. Soc. Wash. 53:138-147.

Blake, D.H.1955. Revision of the vittate species of the chrysomelid beetle genus *Disonycha* from the Americas south of the United States. Proc. U.S. natn. Mus., 104:1-86.

Chevrolat , L. A. A. 1837. *In:* Dejean, Catalogue des coleopteres de la collection de M. le comte Dejean, livr. 5, pp 361-443. Paris. Second Editon. [not 1835].

Chittenden, M. 1899. The spinach flea beetle. Bull. U.S. Bur. Ent. 19, p.80-85.

Clark, H. 1865. An examination of the Halticidae of South America. J. Ent., 2:375-412.

Csiki, E. 1939. in: Heikertinger and Csiki. 1939. Coleopterorum Catalogus, Chrysomelidae: Halticinae, I, pars 166:1-336. [Vol. 25.] Junk publishers, Gravenhage.

Dejean, P.F.M.A. 1837. Catalogue des coleopteres de la collection de M. le comte Dejean, livr. 5, pp 361-443. Paris. Second Editon. [not 1835].
Döberl, M. (1986). Die Spermathek als Bestimmungshilfe bei den Alticinen. Ent. Bl. Biol. Syst. Käfer, 82, (1-2), 3-14.
Dodge, K.L., P.W. Price, J. Kettunen & J. Tavanaien. 1990. Preference and performance of the leaf beetle *Disonycha pluriligata* (Coleoptera: Chrysomelidae) in Arizona and comparisons with beetles in Finland. Envir. Ent.19:905-910.
Duckett, C.N. 1989. Natural history of *Pedilia* sp. A and its interactions with other herbivores of *Passiflora pittieri*. Entomography 6:381-389.
Duckett, C.N. 1993. A revison of genus *Pedilia* Clark and a cladistic analysis of series *Paralactica* and the Disonychini (Coleoptera: Chrysomelidae:Alticinae), unpublished Ph. D. Dissertation, Cornell Univ. Ithaca, NY.
Duckett, C.N. 1994 (1995) *Etapocanga*, a new genus and two new species of Neotropical flea beetle (Coleoptera:Chrysomelidae:Alticinae). J. N.Y. ent. Soc. 102:344-349.
Duckett, C.N. 1995. Variation in Reproductive Structures. Chrysomela 29:4-5.
Duckett, C.N. and Moyá, S. 1999. A new species of *Ptocadica* Harold (Chrysomelidae: Alticinae) from Costa Rica. Coleopts. Bull. *in press*.
Farrell, B.D. 1998. "Inordinate Fondness" Explained: Why are there so many beetles? Science 281:555-559.
Farris, J.S., 1988. Hennig86 documentation. Farris, Stony Brook, N.Y.
Feuillet, C., 1989. Diversity and distribution of Guianan Passifloraceae. in: Tropical Forests, Academic Press, New York, pp. 311-318.
Fragoso, S.A. 1985. The terminalia as a basis for the classification of Cerambycidae (Coleoptera) subfamilies. Revta. bras. Ent. 29:125-134.
Furth, D.G. 1985. Relationships of Paleartic and Neartic Genera of Alticinae. Entomography 3:375-392.
Furth, D.G. 1988. The jumping apparatus of flea beetles (Alticinae)- the metafemoral spring. in: Jolivet, P., Petitpierre E., & Hsiao (Eds.) Biology of Chrysomelidae Kluwer Academic Publishers, Dordrecht, The Netherlands pp.285-299.
Furth, D. G. and V. Savini. 1998. Corrections, clarifications, and additions to the 1996 checklist of the Alticinae of Central America, including Mexico (Coleoptera: Chrysomelidae). Insecta Mundi. 12:133-138.
Garman, H. 1889. The pigweed flea-beetle. Second Ann. Rep. Kentucky Agr. Exp. Sta. 17-30.
Harold, E. von. 1877. Beitrage zur Kenntnis dur Peruanischen Kaferfauna (Halticinae). Dt. ent. Z., 21:129-152
Heikertinger, F. and C.Csiki. 1939. Coleopterorum Catalogus, Chrysomelidae: Halticinae, I, pars 166:1-336. [Vol. 25.] Junk publishers, Gravenhage.
Heikertinger, F. and C. Csiki. 1940. Coleopterorum Catalogus. Chrysomelidae: Halticinae, II, pars 169:337-635. [Vol. 25.] Junk publishers, Gravenhage.
Hemenway, R. and W.H. Whitcomb. 1968. The life history of *Disonycha glabrata* (Coleoptera: Chrysomelidae). J. Kans. ent. Soc. 41:174-178.
Illiger, J.C.W. 1807. Verzeichniss der Arten der Flohkäfer, *Halticae*, in der Hellwig – Hoffmanseggischen Sammlung, mit Beschreibung der neuen, und Bezeichnung der übrigen Arten. Magazin für Insektenkunde, 6:81-188.
Jacoby, M. 1891. Biologia Centrali-Americana, Insecta, Coleoptera, Supplement to Phytophaga. 6(1):1-349.
Jacoby, M. 1902. Descriptions of new species of Coleoptera of the family Halticidae from South and Central America. Proc. zool. Soc. Lond., 1902, 1:171-204, illus.
Jacoby, M. 1904 (1905). Descriptions of thirty-two new species of Halticinae (phytophagous Coleoptera) from South and Central America. Proc. zool. Soc. Lond. 1904. 2:396-413.
Konstantinov, A.S. 1987. Morphological structures used to identify female leaf beetles of the genus *Altica*. (Coleoptera: Chrysomelidae).Zool. Zh. Vol. LXVI, No.1, pp. 42-50.
Konstantinov, A.S. 1998. On the structure and function of the femal genitalia in flea beetles (Coleoptera: Chrysomelidae: Alticinae). Proc. ent. Soc. Wash. 100:353-360.
Leng, C.W., 1920. Catalogue of the Coleoptera of America, north of Mexico, 470 pp. Mount Vernon, N.Y.
Maddison, D.R. 1991. The Discovery and Importance of Multiple Islands of Most-Parsimonious Trees. Syst. Zool. 40:315-328.
Mann, J.S. & Crowson, R.A. (1983). On the internal reproductive organs and their taxonomic significance in the leaf beetles (Coleoptera: Chrysomelidae). Entomologia gen., 9, 75-99.
Maslin, T. P., 1952. Morphological criteria of phyletic relationships. Syst. Zool. 1:49-70.
Nixon, K.C. Clados Version 1.2, c. 1992. Kevin C. Nixon, P.O. Box 270, Trumansburg, NY 14886.
Reid, C.A.M. 1992. The leaf-beetle genus *Microdonacia* Blackburn (Coleoptera: Chrysomelidae: Galerucinae): revision and systematic placement. Syst. Ent. 17, 359-387.
Reid, C.A.M. 1995. A cladistic analysis of subfamilial relationships in the Chrysomelidae *sensu lato* (Chrysomeloidea). In: Biology and Classification of Coleoptera: Papers celebrating the 80th Birthday of Crowson Roy A.,. Pakaluk J. and Slipinski S.A., Eds. (Museum I Instytut Zoologi PAN, Warsaw, 1995. pp. 559-631.
Samuelson, G.A. 1973. Alticinae of Oceania (Coleoptera: Chrysomelidae). Pacif. Insects Monogr. 30:1-165.

Scherer, G. 1962. Bestimmungsschlüssel der neotropischen Alticinen-Genera (Coleoptera: Chrysomelidae: Alticinae). Ent. Arb. Mus. George Frey, 13, 497-607.

Scherer, G. 1983. Diagnostic key for the Neotropical Alticinae genera (Coleoptera: Chrysomelidae: Alticinae). Ent. Arb. Mus. Georg Frey 31/32:1-89.

Scherer, G. 1988. The origins of the Alticinae. in: Jolivet, P., E. Petitpierre, & T.H. Hsiao (Eds.) Biology of Chrysomelidae Kluwer Academic Publishers, Dordrecht, The Netherlands, pp. 115-130.

Seeno, T.N., & J.A. Wilcox. 1982. Leaf beetle genera (Coleoptera: Chrysomelidae). Entomography, 1:1-222.

Sharp, D., & F.A.G. Muir. 1912. The comparative anatomy of the male genital tube in Coleoptera. Trans. ent. Soc. London, 1912: 477-642; illus.

Smith, E.H. 1979. Techniques for the dissection and mounting of the male (aedeagus) and female (spermathecal) genitalia of the Chrysomelidae (Coleoptera). Coleopts Bull., 33:93-103.

Smith, E.H. 1985. Revision of the genus *Phyllotreta* Chevrolat of America north of Mexico part I. The maculate species (Coleoptera: Chrysomelidae, Alticinae). Fieldiana, new series, No. 28.

Suziki, K. 1988. Comparative Morphology of the Internal Reproductive System of the Chrysomelidae (Coleoptera). in: Jolivet, P., Petitpierre E., & Hsiao T.H. (Eds.) Biology of Chrysomelidae Kluwer Academic Publishers, Dordrecht, The Netherlands., pp. 317-355.

Suzuki, K. 1989. Character correlation in internal reproductive system of the family Chrysomelidae (Coleoptera). Entomography 6:511-518.

Tanner, V.M. 1927. The female genitalia of Coleoptera. Trans Am. ent. Soc. 53:3-50.

Thomas, C.D. 1987. Behavioral determination of diet breadth in insect herbivores:the effect of leaf age on choice of host species by beetles feeding of Passiflora vines. Oikos 48:211-216.

Verma, K.K. Male Reproductive organs as taxonomic characters for a broad classification of Chrysomelidae. Entomography. 3:485-488.

Virkki, N. 1988. Cytotaxonomy of Alticinae. in: Jolivet, P., Petitpierre E., & Hsiao T.H. (Eds.) Biology of Chrysomelidae Kluwer Academic Publishers, Dordrecht, The Netherlands, pp. 187-203.

Virkki, N. 1989. On the cytological justification of the flea beetle subtribes Oedionychina and Disonychina (Bechyné & Springlova de Bechyné, 1966). Entomography 6: 545-550.

Vogt, G.B. & J.U. McGuire Jr. & A.D. Cushman. 1979. Probable evolution and morphological variation in South American Disonychine flea beetles (Coleoptera: Chrysomelidae) and their Amaranthaceous hosts. Tech. Bull. U.S. Dep. Agric. No. 1593. 148 pages.

Watrous, L.E., & Q.D. Wheeler. 1981. The out-group comparison method of character analysis. Syst. Zool. 30:1-11.

Weise, J. 1921. Wissenschaftliche Ergebnisse der schwedischen entomologischen Reise des Herrn Dr. A. Roman in Amazonas 1914-1915. 6. Chrysomelidae. Ark. Zool., 14:1-205.

Appendix A

Disonychina Character Matrix

Crimissa	02 000 110 000 000 000 10? 0 0 00 0 0
Macrohaltica	02 100 040 000 000 001 002 0 0 00 0 0
Enneomacra	02 000 000 000 000 000 000 0 0 00 0 0
Cuyabasa	02 000 100 100 000 001 012 0 0 10 1 0
Balzanica	11 001 1?1 141 000 010 01? ? 2 2? 0 0
Disonycha quinquelinata	11 001 101 141 000 011 100 ? 2 2? 0 0
Disonycha trifasciata (+)	11 001 101 141 100 111 100 ? 2 2? 0 0
Monomacra	10 101 031 142 000 011 100 0 0 00 0 0
Parchicola yena	10 101 031 141 001 011 100 0 0 00 0 0
Parchicola violacea	10 101 031 141 000 011 100 0 0 00 0 0
Coroicona	10 111 031 142 011 011 110 0 1 00 0 0
Lacpactica	10 101 021 141 000 110 100 0 0 00 0 0
Maturacaita	10 111 021 132 211 110 102 0 1 00 2 0
Paralactica capitata	10 111 031 132 211 111 101 0 0 11 0 0
Paralactica rubinii	10 111 031 142 112 110 101 0 1 10 0 1
Paralacticoides	10 111 031 142 100 111 111 1 0 11 0 0
Lacpacticoides	10 111 031 142 211 110 100 0 1 11 0 1
Rosalactica	10 111 031 142 100 011 111 1 0 01 1 0
Etapocanga	21 001 031 142 110 010 10? 0 0 00 0 0
Pedilia globosa	21 000 011 140 112 111 101 0 0 00 1 0
P. sp. A.	21 000 011 120 212 111 100 0 0 00 2 0
P. didyma	21 001 101 110 112 010 000 0 0 00 0 0
P. sp. Q.	21 000 001 120 212 111 100 0 0 00 0 0
P. sp. P	21 000 011 140 112 111 101 0 0 00 0 0
Sophreanella fulva	21 001 101 110 111 010 100 0 0 00 1 0
P. cycloptica (+)	21 001 101 110 111 010 000 0 0 00 0 0
Nephrica inornata (+)	21 001 101 110 112 010 100 0 0 00 0 0
Dinaltica aeneipennis	10 111 031 142 100 011 101 0 0 00 2 0
Dinatica ingea	10 111 031 142 101 011 00? 0 0 11 2 0
Utingaltica	11 001 021 142 100 011 100 0 0 10 0 0
Protocadica	?1 001 101 142 100 011 100 0 0 10 0 0
Heikertingeria	10 101 011 141 101 011 101 0 1 00 0 0

ECOLOGICAL AND BIOGEOGRAPHICAL ASPECTS OF THE ANDALUSIAN LEAF BEETLE ENDEMISMS

JOSÉ M. VELA and GLORIA BASTAZO
Coronel Osuna, 18, 4-E, 29006 Málaga, Spain

Abstract

Thirty species and six subspecies of leaf beetles have been reported as endemic to Andalusia. We provide precise (1 km x 1 km) location data, as well as information on areal type, altitudinal and pluviosity ranges, bioclimatic level, lithology, phenology and host plant data. Also, an assessment of the conservational status of the taxa is made. Besides, the influences of all the variables are commented and discussed. The assemblages of taxa are defined as a function of the total variables. Furthermore, the Andalusian endemic fauna is quantitatively compared to that of the Iberian Peninsula, Italy and Bulgaria; subfamilial structure in these countries is also compared. Some historical remarks on the statement of the total and endemic diversity are made.

Introduction

An endemism is a taxon that lives in a specific locality or region, regardless of the area that it occupies (Rapoport, 1982). So, we can speak about Holarctic, Palaearctic, European or Sicilian endemisms. There can be distinguished between neoendemisms, for which the actual distribution and place of origin is the same and paleoendemisms, which occupy a given area after the extinction of the rest of the populations (Margalef, 1974). Endemisms are a valuable part of biodiversity for two reasons: i) from the conservational point of view, such taxa are more prone to extinction than non endemic ones and ii) they can inform us about the history of faunas and areas.

Andalusia, with an area of 87,268 km^2, lies between 38°44' and 39°00' N and 7°31' and 1°38' W. Situated in meridional Spain, it occupies the greatest part of the southern Iberian Peninsula. The Straits of Gibraltar are its most remarkable coastal phenomenon, in that they delimit the Atlantic waters to the west and Mediterranean waters to the east, and separate Andalusia from the African continent. Andalusia is divided into eight provinces, which from east to west are: Almería, Granada, Jaén, Málaga, Córdoba, Sevilla, Cádiz and Huelva.

From the tectonic point of view, Andalusia is a boundary area between plates and is disturbed by horizontal and vertical movements. During the Miocene, almost all of Andalusia was submerged under the Tethys Sea, the predecessor of the Mediterranean. The former sea was limited to the north by the Iberian massif, and to the south by the massif Bethic-Riffian. The erosion of both blocks produced materials that piled up in the sea; after, the

Advances in Chysomelidae Biology 1, pp. 137–158
Edited by M.L. Cox

collision of the massifs resulted in the Bethic range rising, involving a part of the Bethic-Riffian block. Remains of the old sea at the north of the new Bethic range became filled with alluvial sediments giving rise to the Guadalquivir Valley. Southwards, the straits of Gibraltar opened, resulting in the present day situation. The tectonic septentrional block, namely the Iberian Plateau, became folded and subsided, forming a clear lithologic frontier with the Guadalquivir Valley.

Andalusia presents three main structural units (Díaz del Olmo, 1987): i) the Sierra Morena, that is the base of the Iberian Central Plateau, which belongs to the hercinian orogeny; ii) the Guadalquivir Depression, filled with materials eroded both from the northern Sierra Morena as from the southern Bethic Ranges; iii) the Bethic Range, produced in the Alpine orogeny.

The highest altitude in Andalusia is also the highest in the Iberian Peninsula: Mulhacén peak (3,478 m) in the Sierra Nevada (Granada). The Andalusian zones with an altitude higher than 2,000 m amount to about 200 km^2, spread in some sierras in the provinces of Granada, Jaén and Almería. The greater part of Andalusia has an altitude of between 200 and 600 m.

The Andalusian region is wholly included in the Mediterranean climate. Although, this climate presents some variations in accordance with the topography and orography (Gómez-Moreno, 1987), its general tendencies are: i) increase of rainfall westwards: in western Andalusia there is the locality with the highest pluviometry in Spain: 2000 litres/m^2/year in Grazalema (Cádiz); in the eastern part is the most arid zone of Spain or even Europe, Cabo de Gata (Almería), with 135 litres/m^2/year; ii) increase of rainfall with altitude; iii) very irregular rainfall, normally ranging between 200 – 800 litres/m^2/year, with sudden and strong precipitations in a few hours; iv) different atmospheric dynamics between the western part, with Atlantic influence, and the eastern zone, with Mediterranean influences; v) the continental nature of the prevailing temperature in the subsident areas and a temperate nature in the littoral ones; vi) a long dry summer; vii) nearly 2800 hours of sunshine per year, the highest in the Iberian Peninsula.

Vegetal communities in Andalusia are valuable both in the quantitative as in the qualitative aspects in that they contain endemisms in high numbers and high diversity (Ortega *et al.*, 1987). The altitudinal and latitudinal variation of the environmental conditions interact with the physiological features of the plants giving place to different bioclimatic levels (Rivas Martínez, 1987). The Mediterranean region in the Iberian Peninsula accounts for 6 bioclimatic levels (Rivas Martínez, 1984), of which 5 are found in Andalusia (Ortega *et al.*, 1987): thermomediterranean, mesomediterranean, supramediterranean, oromediterranean and crioromediterranean. In the Andalusian landscapes, the dominant vegetation is scrub, because of degradation of the soil. Woody species prevail over herbaceous ones; evergreens over deciduous; also, small, hard leaved or sclerophyllous species are frequent. In general, growth is slow, because of the long periods of drought.

In this paper, we will list for the first time the census of leaf beetles endemic to Andalusia; together we will report their corology and discuss some of the biological and abiotic variables of their habitats. The endemism diversity of Andalusia and Iberian Peninsula will be quantitatively compared to that of other south European countries.

Material and methods

Locality, phenology and host plant data for specimens were extracted from three sources: i) collections made by authors; ii) collections made by colleagues, who gave us either the specimens or their data; iii) a research of literature undertaken to find records on Andalusian endemisms; these relevant papers were: Rosenhauer (1856), Heyden (1870), Allard (1870), Kiesenwetter (1873), Medina (1875), Weise (1891), Daniel & Daniel (1904), Escalera (1931), Bechyné (1948, 1950), Jolivet (1954), Cobos (1952, 1953, 1954a, 1954b, 1954c, 1958, 1987), Codina (1960, 1963), Burlini (1968), Jolivet & Petitpierre (1973), Daccordi & Petitpierre (1977), Warchalowski (1985, 1991), Doguet & Petitpierre (1986), Doguet *et al.* (1996), Bastazo & García-Raso (1985), Bastazo & Vela (1985), Bastazo *et al.* (1993), Vela & Bastazo (1985, 1993), Döberl (1994), Gruev & Döberl (1997) and Bastazo (1998). The different concerned variables were recorded from a variety of sources:

1. The locality data were assigned by means of UTM (Universal and Transversal of Mercator) coordinates, at the present time recommended for cartographic representation of the distribution areas (Cartan, 1978; Palomo & Antúnez, 1992). All the mentioned UTM coordinates in Table 1 must be referred to zone S and to spindles 29 (which includes 100 x 100 square km grids: PC, PB and compensating zones QC, QB, QA) and 30 (which includes 100 x 100 square km grids: UH, UG, UF, VH, VG, VF, WH, WG, WF, XG and compensating zones TH, TG, TF, TE). The map of Andalusia with the superimposed 10 x 10 square km grid is shown in Figure 1. The locality data, in table 1, are mainly provided with a precision of 1 x 1 square km, that is, with four digits; to translate them to the 10 x 10 square km reticle, it is necessary to read only the first and third digits. The location of a given locality in the UTM coordinate system was carried out based on the series of Military Maps of Spain of the Geographic Service of Army, scale 1: 50,000.

2. The considered distributional types of Andalusian endemisms were: i) macroareal (abridged as ma): many or several (mainly more than 4) populations in distant places; ii) mesoareal (me): a few populations (less than 4) in moderately distant places; iii) microareal (mi): species having one only population or a few very close.

3. The altitude data were obtained in two forms: i) when recorded by authors, an altimeter was used; ii) otherwise, by means of contour levels on the Military Maps of Spain (scale 1:50,000).

4. Pluviosity data were obtained from pluviometric maps by Capel Molina (1987) and Junta de Andalusía (1987a, 1997). Pluviosity data were categorized (Rivas Martínez, 1987) in following ombroclimates (expressed in litres/m^2/year): i) Arid (A): < 200; ii) Semiarid (SA): 200 – 350; iii) Dry (D): 350 – 600; iv) Subhumid (SH): 600 – 1000; v) Humid (H): 1000 – 1600; vi) Hyperhumid (HH): > 1600.

5. A bioclimatic level is each one of the thermoclimatic types that are disposed in a altitudinal or latitudinal series, with a typical phytocenosis. The bioclimatic level is but a part of the universal phenomena of altitudinal or latitudinal zonation. The Mediterranean bioclimatic levels are defined (Rivas Martínez, 1982, 1987) in accordance with the annual average temperature (T), the average of the lowest temperatures of the coldest month (m) and the average of the highest temperatures of the coldest month (M) (all in degrees Celsius). For table 1, five levels were considered: i) crioromediterranean: T < 4°, m < -7°, M < 0°; ii) oromediterranean: T = 4° to 8°, m = -7° to -4°, M = 0° to 2°; iii) supramediterranean: T = 8° to 13°, m = -4° to -1°, M = 2° to 9°; iv) mesomediterranean: T = 13° to 17°, m = -1° to

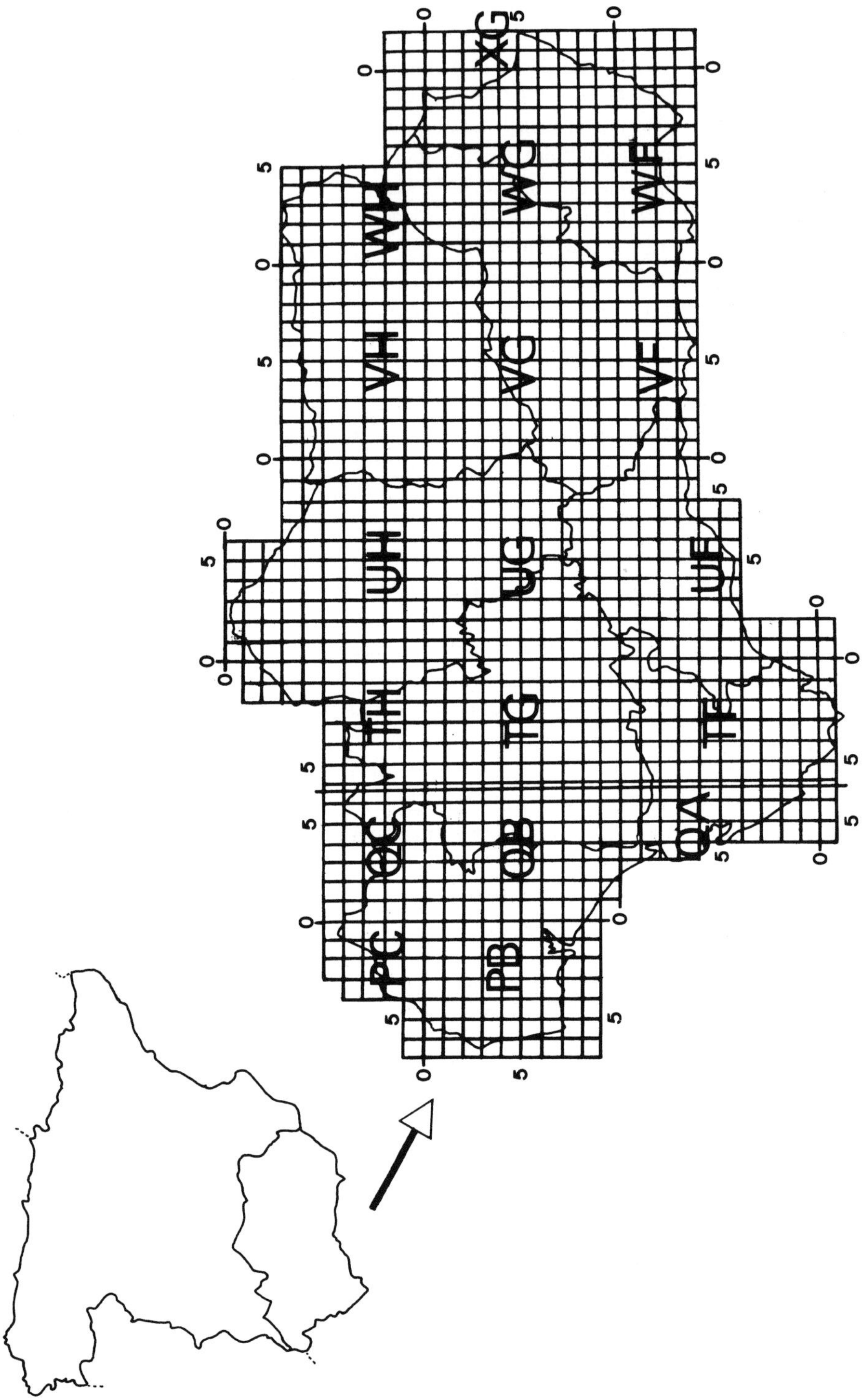

Fig. 1. Situation of Andalusia in the Iberian Peninsula and UTM coordinates with overimposed 10 x 10 square km grid.

4 °, M = 9° to 14°; v) thermomediterranean: T = 17° to 19°, m = 4° to 10°, M = 14° to 18°. The assignment of a level corresponding to a specific locality was carried out by superimposition on the Map of the Vegetation series of Spain, S = 1: 200,000 (Rivas Martínez, 1985).

6. The lithological data were obtained from the Minero-Geologic Map of Andalusia (S = 1:400,000) (Junta de Andalusía, 1985).

7. Conservational status was assessed as recommended by the World Conservation Union document (IUCN, 1994). The following categories were recognized for the Andalusian leaf-beetle endemisms: i) Endangered (EN); ii) Vulnerable (VU); iii) Lower Risk (LR), with two subcategories: Near Threatened (nt) and Least Concern (lc); iv) Data Deficient (DD): information to make a direct or indirect assesment of the risk of extinction is inadequate. The criteria used for categories EN, VU and LR are those proposed in IUCN (1994).

8. A clustering analysis was performed using SPSS for MS WINDOWS release 6.1, in which the dissimilarity coefficient matrix was obtained from binary squared Euclidean distances. The dendrogram was done using average linkage (between groups) method. Presence/absence (1/0) matrix was obtained using the following 31 variables: 0-1000m, 1000-2000m, > 2000m, A, SA, D, SH, H, H-HH, thermo, meso, supra, oro, crioro, limestone, dolomite, schist/micaschist, marble, marl, clay, gypsum, sandstone, peridotite, late autumn, winter, late winter-early spring, late spring, spring-summer, late summer, greater part of the year, greater part of the year not in winter.

9. χ^2 tests were developed to compare faunas. When 20 % of the frequencies are below 5 or one of them is below 1, this analysis is not recommended (Cochran's rule); for this reason, subfamilies showing a small number of species have not been considered in these statistical analyses. In each χ^2 table, expected frequencies are shown in brackets placed to the right of observed frequencies, and χ^2 values for each cell below them; the keys for statistical levels of significance are: * ($p < 0.05$), ** ($p < 0.01$), *** ($p < 0.001$). Probabilities accepted to reject the nulle hypothesis were in no case below 95 %.

In this paper, the concept of endemism has been applied in a strict sense; therefore, the considered Andalusian endemisms are species/subspecies that have not been found anywhere out of Andalusia, until the present.

Results and discussion

Up to the present state of knowledge, 30 species and 6 subspecies were recorded as endemic to Andalusia. Table 1 shows their distribution, phenology and host plant data; some physical and biological features of the habitat were extracted from the primary distributional data, such as area extent, range of altitude and pluviosity, bioclimatic level, lithology and the assessment of their conservational status.

Taxonomically, the greater proportion of Andalusian endemisms is found in the subfamily Chrysomelinae, with 14 species and 4 subspecies. Lower numbers are found in the Alticinae, with 6 spp., the Clytrinae with 4 spp. and 2 sspp., and similar numbers in the Galerucinae, with 4 spp. The subfamilies Eumolpinae and Cryptocephalinae are represented only by 1 sp. each.

Table 1. Endemic Chrysomelidae from Andalusia

species	UTM coordinates	distributional type	altitude	pluviosity	bioclimatic level	substratum	phenology	host plants	conservational status
Labidostomis bolivari garciaolmoi	UF1962,VF0784,VF7396	me	1700-2700	H	supra-oro	limestone,marble,micaschist	VI-VII	Quercus faginea Lam., Berberis hispanica Boiss. et Reuter	EN
L. nevadensis	VG2854,VG9106,VG9708,VF6999,VF7199	me	1000-3100	SA-H	meso-crioro	micaschist,clay, marl	III-VIII	?	EN
Coptocephala scopolina alticola	VG6108,VG8905,VG6406,VF7199	mi	2000-3100	SH-H	oro-crioro	micaschist	VI-VII	?	EN
Smaragdina cobosi	UG9037,UF1962,UF8296,VH80,VG5703, VG9512,WG0697,WG0895,WG1498	ma	900-1800	SH-H	meso-supra	limestone,dolomite,micaschist	V-VII	Quercus faginea Lam., Q. rotundifolia Lam.	LR: nt
Otiothraea filabrensis	UF3385,WG2854,WG5222,WF6483	me	200-2000	A-D	termo-supra	gypsum	III-V	?	EN
O. avilai	UG9159,VG5223,VG6108	me	500-2000	D-SH	termo-supra	gypsum	II-VI	?	EN
Pachybrachis rondanus	UF06	mi?	??	SH?	meso?	limestone?	?	?	DD
Macrocoma cylindrica	WF6385,WF7565	mi	0-300	A	termo	marl,volcanic soils	V-VI	Asteraceae flowers	VU
Chrysolina affinis nevadensis	VG9106,VG9708	mi	2000-2400	SH	oro	micaschist	V-VII	Thymus sp. ?	EN
C. fuliginosa coriacea	TF9890,UF4877,UF6864,UF9164,VF0266, VF0867VF0967,WG0391	ma	0-1200	D-SH	termo-meso	micaschist,clay, marl	II-IV, XII	Mantisalca salmantica (L.) Briq. et Cavillier	VU
C. helopioides	QB6241,QA5637,QA6124,QA6917,TH5703, TG04,TF5638,TF7800,TF8511,TF9890, TE8195, UF06, UF5156,UF6276,UF7466	ma	0-700	D-H	termo	limestone,marl, marble,clay,sandstone,grawvac	IX-V	Ferula communis L.	LR: nt
C. janbechynei	VG2854,VG8935,WG7117	me	600	SA-D	meso	gypsum,sandstone?	IV-V	Artemisia barrelieri Besser	LR: nt
C. lucida suarezi	WF0678,WF1094,WF1293,WF1770,WF2351, WF3890	mi	50-900	D	termo-meso	marl, alluvial soils, metapelitic shale	II-IX	Mentha rotundifolia (L.) Hudson	VU
C. colasi	VG6601,VG9106,VG9708,VF6999,VF7199	mi	2100-3300	SH-H	oro-crioro	micaschist	V-VIII	Sideritis glacialis Boiss.	EN
Cyrtonus almeriensis	WF4175,WF4275	mi	50	SA	termo	limestone	XII-II	?	EN
C. contractus	VG5804,VF7092	mi	1700-2300	SH	supra-oro	limestone,micaschist	VII-X	Santolina ?	EN
C. fairmairei	UF1962,UF2662,UF2365	mi	1100-1700	SH-H	meso-supra	limestone	I-XII	?	EN
C. mateui	VG4605,VG5703,VG9708	mi	1800-2000	SH	supra-oro	limestone,micaschist	V-X	?	EN
C. minor	TE7093,UF1359,UF1863,UF1962	me	300-1700	SH-H	termo-supra	limestone,marl	III-IV,VII-XI	?	VU
C. pardoi	WG1093,WG1498,WG7169	me	1300-1700	D-SH	supra	limestone,dolomite	IV-V	?	VU
C. plumbeus gadorensis	WF4175, WF59	mi	0-700	SA	termo	limestone	I-II	Artemisia sp.	EN
Timarcha granadensis	VF4713,WH4612,WG1436,WG1587	me	700-2000	D-SH	meso-oro	limestone,marl	VII-IX	?	VU
T. insparsa	VG5804,VG6509,VG6604,VG7001, VG7101,VG9708,VF7199	mi	1200-3100	SH-H	meso-crioro	micaschist	II-IV,VII-X	Plantago nivalis Boiss.	VU
T. lugens	VG6406,VG6601,VG6604,VG7200, VF6999,VF7199	mi	2000-3480	SH-H	oro-crioro	micaschist	V-IX	Veronica fruticulosa L.	VU

Table 1. Cont.

species	UTM coordinates	distributional type	altitude	pluviosity	bioclimatic level	substratum	phenology	host plants	conservational status
Timarcha marginicollis	UG5727,VG5504,VG5608,VG5804,VG6604, VG7305,VG9708,VF6788,WF0500, WG0595,WG1093,WG1498	me	950-3000	SH-H	meso-crioro	limestone,dolomite, micaschist	V-IX	Asperula sp.,Galium sp.	VU
T. parvicollis	PB8123,QB1597,QB6132,QA5237,QA5436, TG3644,TF4871,TF6749,TF7605,TF8510, TF8663,TF8774,TE7093,TE7095,UG72, UF0239,UF06,UF1762,UF1863,UF2436 UF2566,UF2662,UF2867,UF4876,UF4956, UF5445,UF6599,UF6767,UF7061,UF7093, UF7526,UF7566,UF7676,UF7765,UF3564 UF8664,UF8694,UF9164,UF9174,VG3723, VG6128,VF0090, VF0767, VF0775 VF2468, WG1498	ma	0-1500	D-HH	termo-supra	limestone,dolomite, marl,clay,micaschist	I-XII	Galium sp.	LR: nt
Galeruca baetica	PB8421,PB9113,QB0809,QA1798QA2966, QA53	mi	0-50	D-SH	termo	sand,calcareous sandstone	IV-VII	Brassicaceae	VU
Calomicrus foveolatus	TF9144,TF9156,TF9949,UF0239,UF1661, UF1946,UF1974,UF3048,UF4055,VG3511	ma	350-1650	SH-HH	termo-meso	limestone,dolomite, marble,peridotite,micaschist	V-VII	Quercus suber L., Q. rotundifolia Lam.	LR: lc
C. sordidus	TF9850,UF0239,UF0246,UF0563,UF1946, UF2043,UF4055,VG3180	ma	300-1200	SH-H	termo-meso	marl,schist,peridotite	IV-V	Quercus suber L., Q. rotundifolia Lam.	LR: nt
Luperus abdominalis	TF9008,UG9037,UF0239,UF1974,UF2571, VG3180,VG6809,WG0387,WG0895	ma	200-1500	SH-HH	termo-supra	limestone,dolomite, micaschist,peridotite	V-VII	Quercus rotundifolia Lam., Q. faginea Lam.	LR: lc
Anthobiodes heydeni	TF9153,UF0067,UF2174	me	450-1000	SH-H	termo-meso	limestone,marl	V	?	VU
Aphthona maldesi	VG6505,WG0385,WG0387,WG1436	ma	1400-2500	H	supra-oro	limestone,dolomite, micaschist	VI-X	Helianthemum hirtum (L.) Miller	VU
Longitarsus andalusicus	VF1287	mi?	1150	SH	meso	schist	VIII	Thymus mastichina L. ?	DD
L. petitpierrei	TE7093,TE8195	mi	70-250	SH	termo	marl	XI-XII	Plantago lanceolata L.	EN
L. seticollis	UF0246,UF1359,UF1661,UF1742,UF1762, UF1863UF1962,UF1974UF2365,UF2368, UF2466,UF3048	mi	500-1700	SH-H	meso-supra	limestone,marble,peridotite	XI-VII	Phlomis crinita Cav., P. purpurea L.	VU
L. velai	TF8774,TF9965	mi	500-1250	H-HH	meso	limestone,marl	XII-III	Echium albicans Lag. et Rodr.	EN

Types of area sizes

The endemisms can be classified according to the degree of eurychory-stenochory of their ranges. Considering the size and number of the populations, as well as the distance between them, we can observe that there exist 8 species/subspecies with a macroareal-type distribution, 10 mesoareal and 18 microareal (table 1). The last two types are prevalent, so that we can say that most endemics exist in few localities, not widely separated from each other.

Altitude

The distribution in the different altitude levels (Figure 2) shows 20 spp./sspp. living below 1000m, half of them exclusively in this zone. Between 1000 – 2000 m are found 21 spp./sspp., 5 of them exclusively at this altitude. Above 2000 m, 10 taxa can be found, although only 4 are typical of this altitude. If we compare these data with the surface occupied by the different altitudes in Andalusia (data taken from Junta de Andalusía, 1987b), we can see that 86.2 % of the Andalusian land surface is below 1000 m, 13.1 % between 1000-2000 m and 0.7 % is above 2000 m. As predicted by a mere area/species relationship, the first altitudinal zone would be expected to hold a higher number of endemisms, and the highest zone a much smaller number. Several factors give account for these differences:

i) In Andalusia, lowlands have been used by man for centuries in intensive and widespread agriculture: 54.1 % of Andalusian soil is cultivated (Junta de Andalusía, 1988), and this soil is mainly situated in lowlands.

ii) In recent times, reafforestation policies have resulted in considerable expansion; of the 23.5 % of woodlands, 11.2 % are plantations of *Eucalyptus*, 31.8 % of *Pinus* (*P. pinea*, *P. pinaster*, *P. halepensis*, principally) and 57 % are *Quercus* and other autochthonous tree

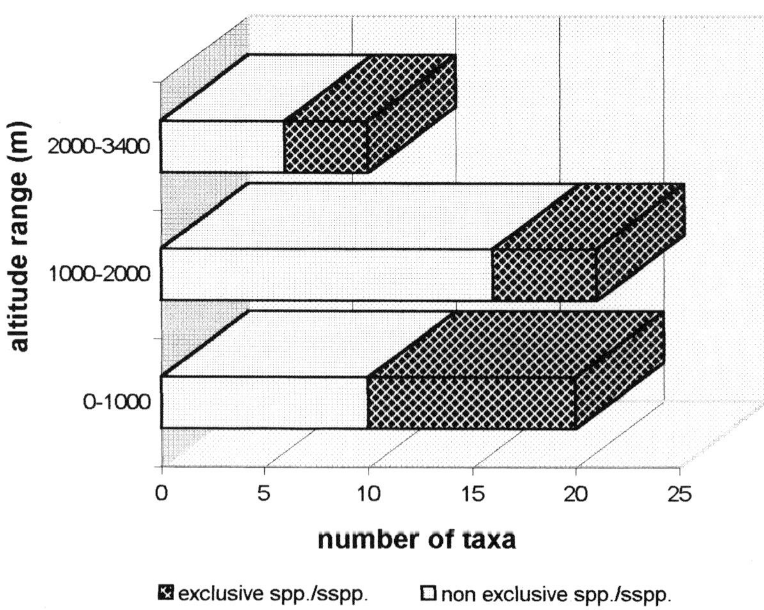

Fig. 2. Number of taxa exclusive and non exclusive to each altitudinal level.

formations (Junta de Andalusía, 1988); therefore, only 13.4 % of the Andalusian landscape is composed of natural woodlands, more or less dispersed, principally *Quercus*. Furthermore, only 11.7 % of the total surface are scrublands and 4.2 % are grasslands. It is worthwhile mentioning that the greater part of this natural vegetation is best represented in the mountains, rather than in lowlands.

So, it is very possible that fragmentation and destruction of habitat has produced local extinctions of endemic populations in the lowlands.

The existence of more endemic species at higher altitudes can be a result of the insularity of these environments, which causes reproductive isolation of populations, scarce genetic flow and local divergence, thus promoting that speciation takes place more likely in high mountains than at mid- levels or on lowlands (Mani, 1962).

Pluviosity

The classification according to pluviometry indicates that subhumid (SH) localities are occupied by most species/subspecies (that is, 27). In humid (H) and hyperhumid (HH) localities 23 spp./sspp. can be found, while in dry (D) localities only 11. In the arid (A) and semiarid (SA) localities 7 spp./sspp. subsist; of these, 6 are exclusive to these ambients of low pluviometry: *Otiothraea filabrensis*, *Macrocoma cylindrica*, *Chrysolina janbechynei*, *C. lucida suarezi*, *Cyrtonus almeriensis* and *C. plumbeus gadorensis*.

Species living in arid, semiarid or dry climates show some peculiar adaptations in order to maintain their body fluids and to obtain water. Some of these are nocturnal, such as *Cyrtonus almeriensis*, *C. plumbeus gadorensis* and *Chrysolina janbechynei*; other are flower eaters, such as *Macrocoma cylindrica*; others live on plants associated with watercourses, such as *Chrysolina lucida suarezi,* or the adult activity takes place in late winter or early spring, such as *Otiothraea filabrensis*.

Bioclimatic levels

Relative to the occupancy of the bioclimatic levels, Figure 3 shows the distribution of the endemic taxa. We can see that 6 taxa are exclusive to the thermomediterranean, although 16 can live at this level. In the mesomediterranean we can find 21 taxa, 17 in the supramediterranean, 12 in the oromediterranean and 6 in the crioromediterranean. Four species/subspecies are exclusive to the oro-crioromediterranean: *Coptocephala scopolina alticola*, *Chrysolina affinis nevadensis*, *C. colasi* and *Timarcha lugens*, all of them endemics of the Sierra Nevada. Other Sierra Nevada endemisms such as *T. insparsa* can live at the mesomediterranean level.

Influences of substratum

Table 2 shows the number of species in each type of substratum. The greater number of species (21) can be found on limestone, but also a high number can live on micaschists/schists. 12 species live on marl. Fewer species are to be found on marble (4), peridotite (4), gypsum (3), sandstones (3) and other types of substrata (4).

Some type of stenoic relationship can be remarkable such as that of *Otiothraea filabrensis*, *O. avilai* and *Chrysolina janbechynei* on gypsum, that of *Longitarsus petitpierrei* on marls

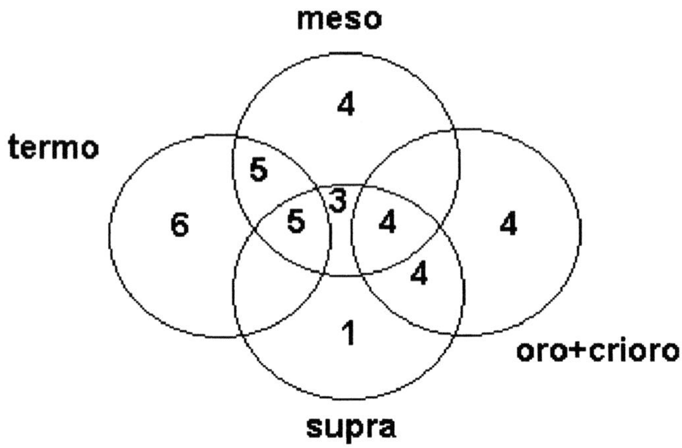

Fig. 3. Endemism diversity along the bioclimatic levels.

Table 2. Number of taxa in different lithological substrata (not exclusives)

Lithology	Number of spp./sspp.
limestone/dolomite	21
marl	12
marble	4
micaschist/schist	17
peridotite	4
gypsum	3
sandstone	3
other	4

and that of *Coptocephala scopolina alticola*, *Chrysolina affinis nevadensis*, *C. colasi*, *Timarcha insparsa* and *T. lugens* on micaschist, but the latter relationship may be coincidental, because these five taxa are endemic to the Sierra Nevada, which is in its major part a micaschistic massif. On the other hand, *Cyrtonus almeriensis*, *C. fairmairei*, *C. plumbeus gadorensis* and *Longitarsus velai* are found exclusively on limestone.

Phenology

According to the temporal appearance of adults, the species can be divided as in Table 3. Eighteen are spring or spring-summer taxa. Late autumn, winter and late winter-early spring taxa total 9. Five spp./sspp. extend their adult activity throughout the greater part of the year. It is interesting to note that 6 species can be found in winter: *Cyrtonus almeriensis*, *C. fairmairei*, *C. plumbeus gadorensis*, *Timarcha parvicollis*, *Longitarsus velai*, *L. seticollis*: of these, *C. almeriensis*, *C. plumbeus gadorensis* and *L. velai* exclusively in the colder months. One species is typical of late autumn: *Longitarsus petitpierrei*. Three species live as adults only in late summer: *Cyrtonus contractus*, *Timarcha granadensis* and *Longitarsus andalusicus*.

Table 3. Seasonal occurrence of taxa.

Period	Taxa
Late autumn	*Longitarsus petitpierrei*
Winter	*Cyrtonus almeriensis, C. plumbeus gadorensis, L. velai*
Late winter-early spring	*Otiothraea filabrensis, O. avilai, Chrysolina fuliginosa coriacea, C. helopioides, Cyrtonus pardoi*
Late spring	*Labidostomis bol. garciaolmoi, L. nevadensis, Coptocephala scopolina alticola, Smaragdina cobosi, Macrocoma cylindrica, Chrysolina affinis nevadensis, C. janbechynei, Calomicrus foveolatus, C. sordidus, Luperus abdominalis, Anthobiodes heydeni*
Spring-summer	*Chrysolina colasi, Cyrtonus mateui, Timarcha insparsa, T. lugens, T. marginicollis, Galeruca baetica, Aphthona maldesi*
Late summer	*Cyrtonus contractus, Timarcha granadensis, Longitarsus andalusicus*
Greater part of the year, including winter	*Cyrtonus fairmairei, Timarcha parvicollis, Longitarsus seticollis*
Greater part of the year, not in winter	*Chrysolina lucida suarezi, Cyrtonus minor*

Ecological affinities

A clustering analysis has been performed taking together all the studied variables. This analysis has served to establish species/subspecies assemblages in accordance with their ecological affinities. Figure 4 shows two main groups of taxa. One of them includes the following spp./sspp.: *Labidostomis bolivari garciaolmoi*, *L. nevadensis*, *Coptocephala scopolina alticola*, *Chrysolina affinis nevadensis*, *C. colasi*, *Cyrtonus contractus*, *C. mateui*, *C. pardoi*, *Timarcha granadensis*, *T. insparsa*, *T. lugens*, *T. marginicollis*, *Aphthona maldesi* and *Longitarsus andalusicus*. This group consists of species living in habitats with the following ecological features: i) high or medium mountains; ii) above the mesomediterranean bioclimatic level; iii) climate subhumid to humid; iv) appearing from late spring to late summer; v) many of them live obligatorily or facultatively on schists/micaschists. These species mainly spread in mountains of the provinces of Jaén and Granada.

The other main group in Figure 4 is composed of two sub-groups; the first of these includes the following taxa: *Otiothraea filabrensis*, *O. avilai*, *Pachybrachis rondanus* (little available information), *Macrocoma cylindrica*, *Chrysolina lucida suarezi*, *C. janbechynei*, *Cyrtonus almeriensis*, *C. plumbeus gadorensis*, *Galeruca baetica* and *Longitarsus petitpierrei*. These are species mainly living: i) in low or mid-levels; ii) amongst vegetation of thermo to mesomediterranean-type, mainly grasses or scrub; iii) in areas with pluviosity from arid to dry (except *Longitarsus petitpierrei*); iv) on marl, gypsum, limestone or sands; v) with adult appearance periods from late autumn to late spring. Some of these species live in semi-arid steppe areas of Almeria or Granada and others live in coastal prairies from Cádiz and Huelva.

The second sub-group includes the following spp./sspp.: *Smaragdina cobosi*, *Chrysolina fuliginosa coriacea*, *C. helopioides*, *Cyrtonus fairmairei*, *C. minor*, *Timarcha parvicollis*, *Calomicrus foveolatus*, *C. sordidus*, *Luperus abdominalis*, *Anthobiodes heydeni*, *Longitarsus seticollis* and *L. velai*. Ecological features of these taxa are: i) mainly mid-level altitudes; ii) from thermo (only *C. helopioides*) to supramediterranean-type vegetation; iii) pluviosity

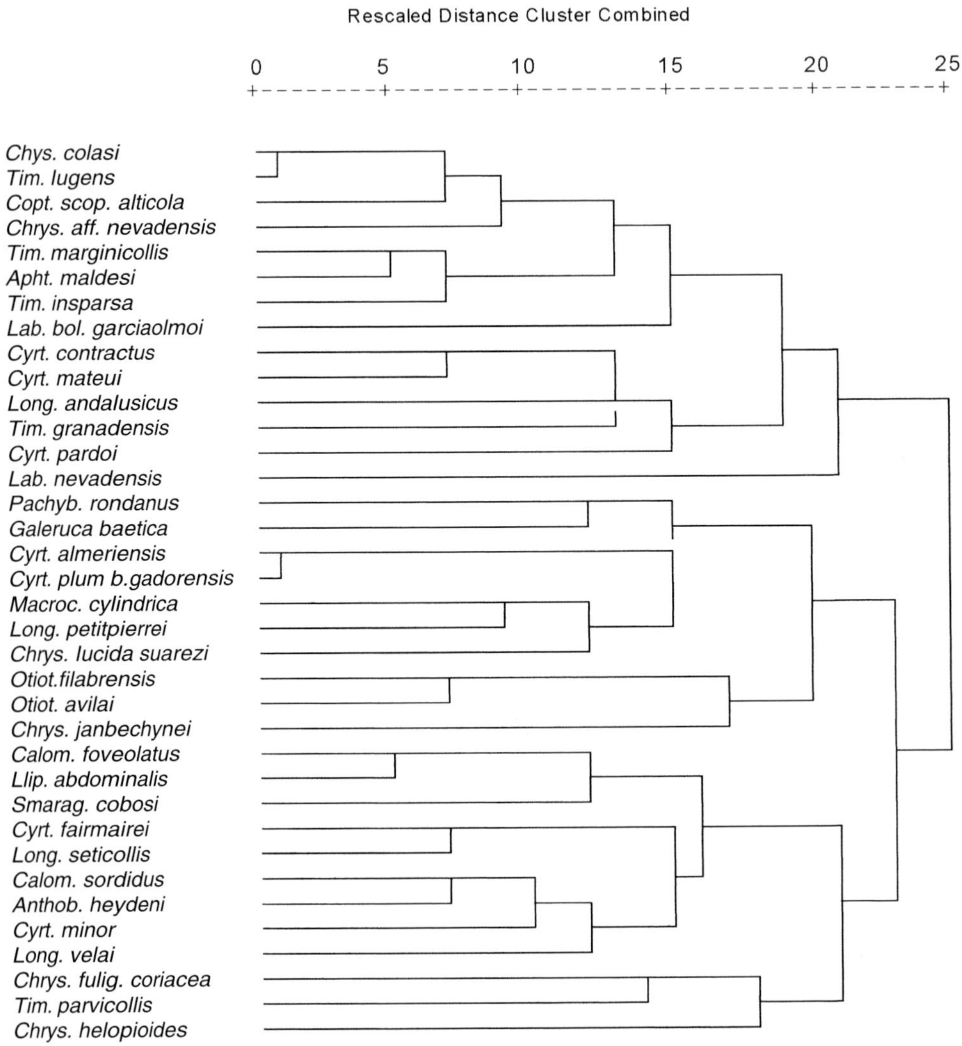

Fig. 4. Cluster dendrogram of the endemic taxa according to the ecological variables and phenology.

from dry to hyperhumid; iv) mainly on limestone; v) adult period in winter (*L. velai*), early spring, late spring or even the greater part of year. These species mainly live in scrub or in the montane forests of Málaga, Cádiz, Córdoba or Jaén sierras.

Host Plants

Host plants were identified for 22 out of the 36 taxa considered in this study. Clytrinae as *Labidostomis bolivari garciaolmoi*, *Smaragdina cobosi*, and Galerucinae Luperini as *Calomicrus foveolatus*, *C. sordidus* and *Luperus abdominalis* feed on flowers and young, tender leaves of *Quercus* (Fagaceae), but the first of them also likes flowers of *Berberis* (Berberidaceae). The flowers of an Asteraceae species are eaten by *Macrocoma cylindrica* and the leaves of this vegetal family by *Chrysolina janbechynei*, *C. fuliginosa coriacea*, *Cyrtonus plumbeus gadorensis* and perhaps *C. contractus*. The Lamiaceae are selected by *Chrysolina affinis nevadensis*, *C. lucida suarezi*, *C. colasi*, *Longitarsus andalusicus* and *L. seticollis*. The family Apiaceae is chosen by *Chrysolina helopioides*; Plantaginaceae by *Timarcha insparsa* and *Longitarsus petitpierrei*; Rubiaceae by *Timarcha marginicollis* and *T. parvicollis*; Scrophulariaceae by *Timarcha lugens*; Brassicaceae by *Galeruca baetica* and Boraginaceae by *Longitarsus velai*.

Indeed, the present information indicates that at a plant generic level, the endemic fauna is monophagous. The only exceptions are *Labidostomis bolivari garciaolmoi*, which shows preference not for a vegetal family but for the available floral parts of Berberidaceae and Fagaceae, and *Timarcha marginicollis*, which feeds on *Asperula* and *Galium* (Rubiaceae).

Conservational status

This conservational status assessment is an essay to evaluate the endemic leaf beetles for conservation policy aims. Figure 5 shows the number of taxa in each category. We have considered 14 taxa as endangered. Of them, 10 are known to live in one unique locality (microareal species) and 4 live in at least two separate localities (mesoareal): *Labidostomis bolivari garciaolmoi*, *L. nevadensis*, *Otiothraea avilai* and *O. filabrensis*. We have made the assessment of endangered because of the small population sizes and the scarcity of known populations. A similar number of species (13) is considered vulnerable. Of these, 6 are microareal, 5 mesoareal and 2 macroareal; these two taxa are *Chrysolina fuliginosa coriacea* and *Aphthona maldesi*. Seven species are at a Lower Risk, 5 of these near threat-

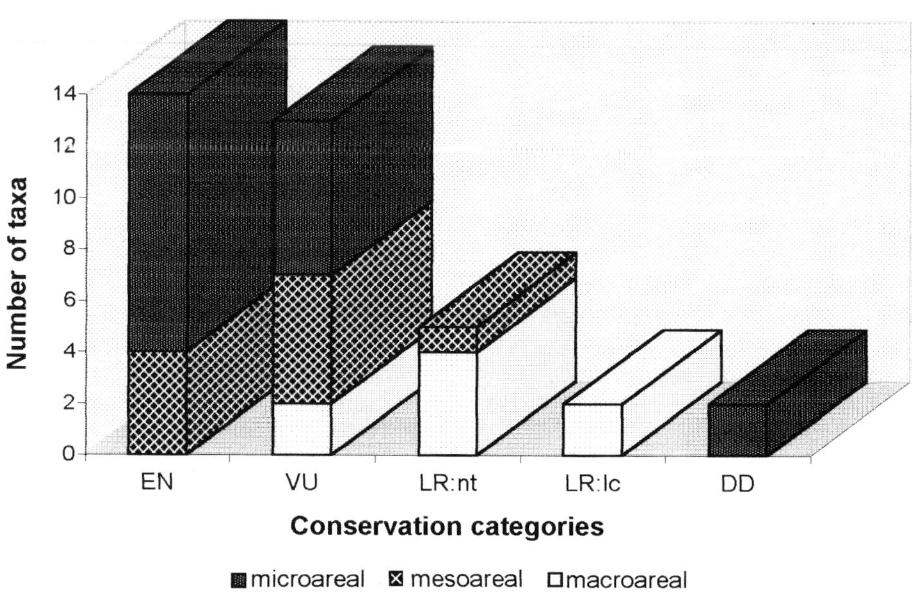

Fig. 5. Repartition of taxa according to their conservational status and type of distribution.

ened (nt) and 2 are of least concern (lc). As yet we do not have sufficient data to assess the status of *Pachybrachis rondanus* or *Longitarsus andalusicus*.

Biogeographic remarks

Table 4 presents the results obtained in a comparison of the taxonomic composition of Chrysomelidae in three Southern European countries: Iberian Peninsula, Italy and Bulgaria, detailing total faunal diversity, endemic fauna and the endemic diversity in Andalusia. It can be observed that the Italian Peninsula (including Sicily and excluding Sardinia), with a geographic area of 276,934 km^2, holds a high diversity: 790 spp., of which 50 (6.3 %) are Italian endemisms (Biondi *et al.*, 1994). Bulgaria, with 110,910 km^2, has 526 species; but of them, only 6 (1.1 %) are endemisms (Gruev & Tomov, 1998). The Iberian Peninsula has an area of 596,732 km^2 and its leaf beetle diversity is 630 spp., 122 of which (19.4 %) are Iberian endemisms. Some relevant comments can be outlined:

1. Considering the total and endemic diversity of the Iberian Peninsula, Italy and Bulgaria (Table 5): i) the total diversity of the Iberian Peninsula is lower than that which would be expected in relation to its total area (see above) and to the expected number in relation to the diversity of the other countries; it is necessary to emphasize that some sampling deficiences exist in Iberian fauna, although they cannot justify the striking difference in regard to the Italian fauna; ii) the total diversities in Italy and Bulgaria are no more different than would be expected; iii) the endemic fauna in the Iberian Peninsula is significantly higher than that expected in relation to total species; iv) the Italian and Bulgaria endemisms are significantly lower in number.

The different orography, climatic patterns and biogeographic history of these countries can offer some possible explanations: i) From the Alpine orogeny in the Miocene, the Pyrenean chain became a rather impermeable barrier both northwards from the Iberian shield and southwards from France; this phenomenon was strengthened by the absence of deep valleys that could act as corridors and could make dispersal easier. Probably, the last glacial event produced a southerly movement in the European species, many of which could not pass through the Pyrenees, although a few of them reached the Iberian Peninsula. ii) Mountain ranges in the Iberian Peninsula are oriented in an east-west direction, so this made difficult, if not impossible, the latitudinal movements of the biota in accordance to the cycles of advance and retreat of glaciers and cold ambients. When cold was in retreat, and Mediterranean vegetation colonized the warm lands, most of the European species could not survive and became extinct (Quezel *et al.*, 1980). iii) However, North Italian Alps functioned as permeable barriers because of the existence of north-south oriented valleys that served as corridors for dispersal. Furthermore, Italian ranges (Apennines) are mainly oriented north-south, making easy the dispersal and colonization of fauna. iv) Bulgaria has no natural impenetrable external frontiers, and so, its leaf-beetle endemic diversity is low. v) Moreover, ecological factors may also be invoked. The Italian Peninsula, as well as Bulgaria, has a larger area with an Eurosiberian climate, and only a smaller part has a Mediterranean climate (Adams and Faure, 1997); so they have many Eurosiberian species together with transmediterranean ones. The Iberian Peninsula has only a narrow northern zone, from Galicia to the Pyrenees with an Eurosiberian climate (Ortega *et al.*, 1987); the rest of Spain and Portugal have Mediterranean-type vegetation. The Iberian Peninsula lacks a great number of Eurosiberian species because of a simple species/area effect. vi) Other

Table 4. Comparative analysis of total and endemic species in three different Southern European countries (data taken form Biondi *et al.*, 1994 for Italy and from Gruev and Tomov, 1998 for Bulgaria) in relation to Andalusian endemisms.

	Iberian fauna		Andalusian	Italian fauna		Bulgarian fauna	
	Total fauna	Endemic fauna	endemic fauna	Total fauna	Endemic fauna	Total fauna	Endemic fauna
Donaciinae	17 (2.70 %)	2 (1.64 %)	0	29 (3.67 %)	0	18 (3.42 %)	0
Orsodacninae	2 (0.32 %)	0	0	2 (0.25 %)	0	2 (0.38 %)	0
Zeugophorinae	1 (0.16 %)	0	0	3 (0.38 %)	0	2 (0.38 %)	0
Criocerinae	12 (1.90 %)	0	0	19 (2.40 %)	1 (2 %)	12 (2.28 %)	0
Clytrinae	45 (7.14 %)	8 (6.56 %)	4 (13.33 %) + 2 ssp.	53 (6.71 %)	0	35 (6.65 %)	0
Cryptocephalinae	97 (15.40 %)	26 (21.31 %)	1 (3.33 %)	123 (15.57 %)	7 (14 %)	67 (12.74 %)	0
Lamprosomatinae	1 (0.16 %)	0	0	1 (0.13 %)	0	1 (0.19 %)	0
Eumolpinae	8 (1.27 %)	1 (0.82 %)	1 (3.33 %)	13 (1.65 %)	0	11 (2.09 %)	1 (14.67 %)
Chrysomelinae	136 (21.59 %)	48 (39.34 %)	14 (46.67 %) + 4 ssp.	135 (17.09 %)	11 (22 %)	85 (16.16 %)	1 (14.67 %)
Galerucinae	42 (6.67 %)	14 (11.47 %)	4 (13.33 %)	56 (7.09 %)	12 (24 %)	32 (6.08 %)	1 (14.67 %)
Alticinae	241 (38.25 %)	23 (18.85 %)	6 (20 %)	320 (40.51 %)	19 (38 %)	231 (43.92 %)	3 (50.00 %)
Cassidinae	24 (3.81 %)	0	0	32 (4.05 %)	0	28 (5.32 %)	0
Hispinae	4 (0.63 %)	0	0	4 (0.51 %)	0	2 (0.38 %)	0
Total	630	122	30 (+ 6ssp.)	790	50	526	6

Table 5. χ^2 analysis of raw frequencies for total and number of endemic species living in each country. Expected frequencies are shown in brackets, and χ^2 values for each comparison below both frequencies. Degrees of freedom (d.f.): 1 for individual cells, 2 for row subtotals (that is, distribution of total spp. and endemisms per country, respectively) and 1 for column subtotals (that is, country distributions according to total species and endemisms). χ^2 total value for the whole contingency table: 103.2 (2 d.f., $p < 0.001$). Levels of statistical significance: -: $p > 0.05$; *:$p < 0.05$; **: $p < 0.01$; ***: $p < 0.001$.

	Iberian pen.	Italy	Bulgaria	Subtotal
Total species	630 (689.0) / 5.05 (*)	790 (769.6) / 0.54 (-)	526 (487.4) / 3.06 (-)	1946 / 8.65 (*)
Endemisms	122 (63.0) / 55.2 (***)	50 (70.4) / 5.91 (*)	6 (44.6) / 33.4 (***)	178 / 94.5 (***)
Subtotal	752 / 60.25 (***)	840 / 6.45 (*)	532 / 36.46 (***)	

recent features of the Mediterranean ecosystem in Spain and Portugal can be recorded as factors that impoverish diversity: fires, early human occupation, extensive agriculture, urbanizations, prevalence of serial degradation steps, soil erosion and large scale reafforestation involving monospecific woodlands, using allocthonous species of genera *Pinus* and *Eucalyptus*.

In relation to the endemic fauna, the high diversity recorded in the Iberian Peninsula is remarkable. Again, several historical and ecological explanations can be proposed: i) The historical explanation is supported by the different paleogeographic origins of the Iberian Peninsula (Bons, 1973; Cano-García, 1987). In the Upper Tertiary (-20 m.y.), Western Europe was isolated from Africa by the old Tethys sea, that covered all Andalusia. With the Alpine orogeny, the Bethic, Rif and Atlas Ranges were formed, accompanied by a sea level fall of the Mediterranean. In the upper Miocene-early Pliocene (5-6 m.y. ago), the Bethic range was connected to northwest African mountain ranges. It gave rise to a typical Bethic-Maghrebinian fauna, that experienced vicariance phenomena when the Straits of Gibraltar progressively subsided until the final separation of Europe from Africa at the beginning of the Pleistocene (1.65 m.y. ago) (Quezel *et al.*, 1980). ii) As stated above, the Pyrenees are an almost impenetrable barrier, but other ranges have also acted as barriers: the Cantabric, Iberic, Central, Subbetic and Penibetic mountains. Together, the complexity of the Iberian relief and the climatic, lithologic and vegetation patchiness, produced a high degree of isolation in the leaf-beetle populations. Many places of the Mediterranean area did not suffer the effects of glaciation so that tertiary Mediterranean species could experience reproductive isolation in small populations and underwent speciation (Pons, 1975; Oosterbroek & Arntzen, 1992). iii) The small number of endemisms in the Italian fauna is well correlated with its area, but moreover with the much smaller area of the Mediterranean-type vegetation area in this country in relation to Iberian Peninsula. iv) The small number of endemisms in Bulgaria may be explained because it has few of its own mountainous ranges: the Sredna Gora range, Rhodopi mountains, Pirin mountain. Its greatest range, the Stara Planina, reaches Serbia. On the other hand, no great barriers separate it from surroundings countries. Therefore, Bulgaria has a good number of Balkan endemisms, but only 6 of them exclusively.

Finally, Andalusia, with 87,268 km^2, has 30 endemic species and 6 subspecies. To this number we can add 37 Iberian endemisms that are also found in this region, resulting in a total of 67 Iberian endemisms living in Andalusia. Arguments such as those advanced above can serve to justify this high figure.

Table 6. χ^2 analysis of total species per subfamily living in each country. Degrees of freedom (d.f.): 1 for individual cells, 2 for row subtotals (that is, distribution of subfamilial structure per country, respectively), 8 for column subtotals (that is, country distributions according the subfamilial structure). χ^2 total value for the whole contingency table: 13.8 (16 d.f., $p > 0.5$). Key for significance levels as in table 5.

	Iberian pen.	Italy	Bulgaria	Subtotal
Donaciinae	17 (20.7) / 0.66 (-)	29 (26) / 0.35 (-)	18 (17.3) / 0.04 (-)	64 / 1.05 (-)
Criocerinae	12 (10.4) / 0.21 (-)	19 (17.5) / 0.13 (-)	12 (11.6) / 0.01 (-)	43 / 0.35 (-)
Clytrinae	45 (43.1) / 0.19 (-)	53 (54.0) / 0.02 (-)	35 (35.9) / 0.02 (-)	133 / 0.23 (-)
Cryptocephalinae	97 (92.9) / 0.18 (-)	123 (116.5) / 0.36 (-)	67 (77.5) / 1.42 (-)	287 / 1.96 (-)
Eumolpinae	8 (10.4) / 0.55 (-)	13 (13.0) / 0.00 (-)	11 (8.6) / 0.67 (-)	32 / 1.22 (-)
Chrysomelinae	136 (115.3) / 3.72 (-)	135 (144.5) / 0.62 (-)	85 (96.2) / 1.30 (-)	356 / 5.64 (-)
Galerucinae	42 (42.1) / 0.00 (-)	56 (52.8) / 0.06 (-)	32 (35.1) / 0.27 (-)	130 / 0.33 (-)
Alticinae	241 (256.4) / 0.92 (-)	320 (321.6) / 0.01 (-)	231 (214) / 1.35 (-)	792 / 2.28 (-)
Cassidinae	24 (27.2) / 0.38 (-)	32 (34.1) / 0.13 (-)	28 (22.7) / 0.23 (-)	84 / 0.74 (-)
Subtotal	622 / 6.81 (-)	780 / 1.68 (-)	519 / 5.31 (-)	

Table 7. χ^2 analysis of endemic species per subfamily living in each country. Degrees of freedom (d.f.): 1 for individual cells, 1 for row subtotals (that is, distribution of endemism subfamilial structure per country, respectively), 4 for column subtotals (that is, country distributions according the subfamilial structure of endemisms). χ^2 total value for the whole contingency table: 16.16 (4 d.f., $p < 0.01$). Key for significance levels as in table 5.

	Iberian endemisms	Italian endemisms	Subtotal
Clytrinae	8 (5.7) / 0.96 (-)	0 (2.3) / 2.33 (-)	8 / 3.29 (-)
Cryptocephalinae	26 (23.4) / 0.11 (-)	7 (9.6) / 0.71 (-)	33 / 0.82 (-)
Chrysomelinae	48 (41.8) / 0.92 (-)	11 (17.2) / 2.24 (-)	59 / 3.16 (-)
Galerucinae	14 (18.4) / 1.06 (-)	12 (7.6) / 2.58 (-)	26 / 3.64 (-)
Alticinae	23 (29.7) / 1.53 (-)	19 (12.2) / 3.72 (-)	42 / 5.25 (*)
Subtotal	119 / 4.58 (-)	49 / 11.58 (*)	

2. In relation to the taxonomic structure, that is, the number of total species per subfamily, it is interesting to note that this is not significantly different among the Iberian Peninsula, Italian and Bulgarian faunas (Table 6). This means that the differences in the number of species in any one given subfamily occurring in these three countries may be due to chance and not to significant differences. If the endemic subfamilial structure is compared between the Iberian and Italian (including Sicily) Peninsulas, only a small difference is significant in relation to the number of Alticinae species (Table 7), no other subfamily having significant differences. The endemic structure in Bulgaria has not been compared because the very small number of species makes the χ^2 analysis inadequate.

3. Table 8 compares the taxonomic structure among Iberian leaf-beetle fauna (excluding endemic species to avoid redundant data in the analysis), Iberian endemisms (omitting those exclusive to Andalusia, for the same reasons) and Andalusian endemisms. Only five subfamilies have been considered because of the low numbers of the others. The results indicate the following differences from those expected: i) slightly higher number of Iberian endemic Cryptocephalinae; ii) higher number of endemic Chrysomelinae both in the

Table 8. χ^2 analysis of Iberian species, Iberian endemisms and Andalusian endemisms per subfamily. Degrees of freedom (d.f.): 1 for individual cells, 2 for row subtotals (that is, distribution of subfamilial structure per faunal type, respectively), 4 for column subtotals (that is, faunal type distribution according the subfamilial structure). χ^2 total value for the whole contingency table: 52.67 (8 d.f., $p < 0.001$). Key for significance levels as in table 5.

	Iberian fauna (excluding endemisms)	Iberian endemisms (excluding Andalusian ones)	Andalusian endemisms	Subtotal
Clytrinae	37 (35.4) / 0.04 (-)	4 (7.2) / 1.44 (-)	4 (2.3) / 1.20 (-)	45 / 2.68 (-)
Cryptocephalinae	71 (75.4) / 0.38 (-)	25 (15.6) / 5.73 (*)	1 (5.01) / 3.21(-)	97 / 9.32 (**)
Chrysomelinae	88 (107.1) / 3.42 (-)	34 (21.8) / 6.80 (**)	14 (7.0) / 6.91 (**)	136 / 17.13 (***)
Galerucinae	28 (33.1) / 0.78 (-)	10 (6.7) / 1.58 (-)	4 (2.2) / 1.54 (-)	42 / 3.90 (-)
Alticinae	218 (189.9) / 4.16 (*)	17 (38.7) / 12.13 (***)	6 (12.5) / 3.35 (-)	241 / 19.64 (***)
Subtotal	442 / 8.78 (-)	90 / 27.68 (***)	29 / 16.21 (**)	

Iberian Peninsula and in Andalusia; iii) in Alticinae, a slightly higher number of Iberian species and a much lower number of Iberian endemisms.

4. A similar comparison in respect of the Italian fauna, shown in Table 9, indicates that there is a much higher number of endemic Galerucinae than expected, but the endemism diversity structure of the rest is not different to the total diversity structure in each subfamily. Only four subfamilies have been considered in this analysis in that Criocerinae has only one endemic species and the rest none.

Conclusions

Our study has revealed 30 species and 6 subspecies endemic to Andalusia. In regard to the prevalence of subfamilies, Andalusian endemic Chrysomelinae are high in number if compared with the total Iberian Chrysomelinae. Endemic species in other subfamilies are not different in number to those expected in relation to subfamilial structure, both in the Iberian total fauna and Iberian endemisms.

The ecological features considered in this paper have influenced the present day distribution and location of populations, but the biogeographic history of the region is also an important factor in the origin and distribution of populations of endemic species.

Much is as yet unknown on the corology, ecology and biogeography of the Andalusian endemic leaf beetles. Likely, the census presented here will be modified by two types of changes: i) possibly some of the taxa will be found outside of Andalusia, mainly those spread in the semiarid steppes of eastern Granada and Almeria, which do not have natural frontiers to the steppes of Murcia and even Albacete; ii) in the future, some new species will be described from Andalusia; of the considered taxa, 16 have been described after 1950, and four of them: *Otiothraea avilai*, *Aphthona maldesi*, *Longitarsus petitpierrei* and *Longitarsus velai*, very recently.

Table 9. χ^2 analysis of total and endemic Italian fauna per subfamily. Degrees of freedom (d.f.): 1 for individual cells, 1 for row subtotals (that is, distribution of subfamilial structure per faunal type, respectively), 3 for column subtotals (that is, faunal type distribution according the subfamilial structure). χ^2 total value for the whole contingency table: 16.91 (3 d.f., $p < 0.01$). Key for significance levels as in table 5.

	Italian fauna (excluding endemisms)	Italian endemisms	Subtotal
Cryptocephalinae	116 (113.5) / 0.06 (-)	7 (9.5) / 0.66 (-)	123 / 0.72(-)
Chrysomelinae	124 (124.6) / 0.00 (-)	11 (10.4) / 0.03 (-)	135 / 0.03 (-)
Galerucinae	44 (51.7) / 1.15 (-)	12 (4.3) / 13.59 (***)	56 / 14.74 (***)
Alticinae	301 (295.3) / 0.11 (-)	19 (24.7) / 1.31 (-)	320 / 1.42 (-)
Subtotal	585 / 1.32 (-)	49 / 15.59 (**)	

Acknowledgements

To our collegues and friends Miguel A. Alonso Zarazaga (Madrid), José Miguel Ávila (Granada), Manuel Baena (Córdoba), Pablo Barranco (Almería), Pedro Coello (Cádiz), Juan de Ferrer (Algeciras), Frank Fritzlar (Jena), Manuel Huertas (Huelva), Francisco Sánchez-Piñero (Granada) and Andrzej Warchalowski (Wroclaw) for sending us specimens or their locality data. To Raymond J. Connor (Cómpeta, Málaga), who kindly spent much of his time in improving our English usage and wording. Critical comments and statistical advice of Paul Palmqvist (Málaga) strongly contributed to improving the manuscript. The kindness and patience of Michael L. Cox (London) have made possible the publication of this paper.

References

Adams, J.M. and Faure, H. 1997. Preliminary vegetation maps of the world since the last glacium maximum: an aid to archaeological understanding. *J. archaeol. Sci.*, 24: 623-647.

Allard, F. 1870. In Heyden, L., Entomologische Reise nach dem Südlichen Spanien, der Sierra Guadarrama und Sierra Morena, Portugal und den Cantabrischen Gebirgen. *Berl. ent. Z., Berlín*, 14: 1-176.

Bastazo, G. 1998. Two new species of *Longitarsus* Berthold, 1827 from Southern Spain (Coleoptera: Chrysomelidae: Alticinae). *Elytron*, 11: 31-38.

Bastazo, G. and García-Raso, J.E. 1985. *Longitarsus caroli* n. sp. from the South of the Iberian Peninsula. *Atti Soc. ital. Sci. nat.*, 126: 94-100.

Bastazo, G. and Vela, J.M. 1985. Notas corológicas, tróficas y sistemáticas sobre *Longitarsus* (Col. Chrys. Alticinae) del sur de la Península Ibérica. Actas II Congr. Ib. Ent., Lisboa: 151-163.

Bastazo, G., Vela, J.M. and Petitpierre, E. 1993. Datos faunísticos sobre Alticinae ibéricos (Col. Chrysomelidae). *Boln. Asoc. esp. Entomología*, 17: 379-394.

Bechyné, J. 1948. Contribution à la connaissance du genre *Timarcha* (Col. Phytophaga, Chrys.). Latr. 12: Ètudes phylogénétiques et zoogéographiques. *Sb. nar. mus. Praze*, 4B (2):1-63.

Bechyné J. 1950. 7ª contribution à la connaisance du genre *Chrysolina* Motsch. *Ent. Arb. Mus.Georg Frey*, 1: 47-185.

Biondi, M., Daccordi, M. Regalin, R. and Zampetti, M. 1994. Coleoptera Polyphaga XV (Chrysomelidae, Bruchidae). In Minelli, A., Ruffo, S. and La Posta, S. (eds.). Checklist delle specie della fauna italiana, 60. Calderini, Bologna, 34 pp.

Bons, J. 1973. Herpetologie Marocaine II. Origines, évolution et particularités du peuplement herpétologique du Maroc. *Bull. Soc. Sci. nat. phys. Maroc*, 53 (1-2): 63-109.

Burlini, M. 1968. Revisione delle specie italiani e della maggior parte delle specie europee del genere *Pachybrachis* Chevr. *Memorie Soc. ent. Ital.*, 47: 11-116.

Cano García, G. 1987. Andalusía, un espacio diferenciado. In Cano García, G. (ed.). Geografía de Andalusía. Ediciones Tartessos, Sevilla, vol. I, pp. 11-48,
Capel Molina, J.J. 1987. El clima de Andalusía. In Cano García G. (ed.) Geografía de Andalusía. Ediciones Tartessos, Sevilla, vol. II, pp. 99-185.
Cartan, M. 1978. Inventaires et cartographies de repartitions d´especes. Faune et flore. Editions du Centre National de la Recherche Scientifique, París, 127 + XIX pp.
Cobos, A. 1952. Nuevas especies de coleópteros Phytophagoidea de la provincia de Almería (Sudeste Español). *Boll. Assoc. romana Ent.*, 7(1): 1-8.
Cobos, A. 1953. Especies nuevas de Coleópteros de la provincia de Almería. *Archos. Inst. Aclim. Almería*, 1: 127-134.
Cobos, A. 1954a. Coleópteros de Sierra Nevada (Familias Cerambycidae y Chrysomelidae). *Archos. Inst. Aclim., Almería*, 2: 139-156.
Cobos, A. 1954b. Una breve campaña entomológica por las sierras de Tejeda y Almijara (provincia de Málaga). *Archos. Inst. Aclim., Almería*, 3: 29-39.
Cobos, A. 1954c. Nota sobre el género *Cyrtonus* Latr.(Col. Chrys.) y diagnosis de formas nuevas ibero-marroquíes. *Tamuda*, 2: 318-324.
Cobos, A. 1958. Datos para el catálogo de los coleópteros de España. Especies de los alrededores de Málaga (adiciones y correcciones a la lista de 1949). *Graellsia*, 16: 23-28.
Cobos, A. 1987. La coleopterofauna endémica almeriense. *Graellsia*, 43: 3-17.
Codina, F. 1960. Apuntes sobre Chrysomelidae ibero-marroquíes (Coleoptera). *Graellsia*, 18: 37-50.
Codina, F. 1963. Apuntes sobre Chrysomelidae ibero-marroquíes (Coleoptera). *Graellsia*, 20: 69-77.
Daccordi, M. and Petitpierre, E. 1977. Coleópteros crisomélidos de la Sierra de Cazorla (Jaén) y descripción de una nueva especie de *Clytra* Laich. *Miscnea. ent.*, 4(1): 125-136.
Daniel, K. and Daniel, J. 1904. Neue palärktische Koleopteren. *Münch. Koleopt. Z.*, 2(1): 76-93.
Díaz del Olmo, F. 1987. El relieve de Andalusía. In Cano García, G. Geografía de Andalusía. Ediciones Tartessos, Sevilla, vol. II, pp. 11-98.
Döberl M. 1994. Bemerkenswerte Alticinenfundee aus Westereuropa (col. Chrys.). *Ent. Nachr. Berl.*, 38(3): 179-182.
Doguet, S. and Petitpierre, E. 1986. Deux nouvelles espèces d´*Aphthona* d´Espagne (Col. Chrysomelidae). *Revue. fr. Ent.*, 8(4): 145-148.
Doguet, S. Bastazo G. Bergeal M. and Vela J.M. 1996. Contribution à l´etude des Chrysomelidae d´Andalousie (Col.). *Nouv. Revue Ent.*, 13(4): 315-323.
Escalera, M. 1931. Los *Labidostomis* metálicos del Mediterráneo occidental (Col. Chrys.). Eos, 7: 107-130.
Gómez Moreno, M.L. 1987. Apuntes de Geografía Física de Andalusía. Imprenta Universidad de Málaga, Málaga, 142 pp.
Gruev, B. and Döberl, M. 1997. General distribution of the flea beetles in the Palaeartic Subregion (Col. Chrys. Alticinae). *Scopolia*, 37:1-496.
Gruev, B. and Tomov, V. 1998. Catalogus fauna Bulgaricae. Coleoptera Chrysomelidae. Pensoft Publishers, Sofía-Moscow, 160 pp.
Heyden, L. von 1870. Entomologische Reise nach dem Südlichen Spanien, der Sierra Guadarrama un Sierra Morena, Portugal und den Cantabrischen Gebirgen. *Berl. ent. Z., Berlín*, 14:1-176.
IUCN, 1994. IUCN Red List Categories. IUCN, Gland, 21 pp.
Jolivet, P. 1954. Notes écologiques sur les Chrysomeloidea de la Sierra Nevada. *Bull. Annls. Soc. r. ent. Belg.*, 90(3): 69-72.
Jolivet, P. and Petitpierre, E. 1973. Plantes hôtes connues des *Timarcha* Lat. (Col. Chrys.). Quelques considérations sur les raissons possibles du trophisme sélectif. *Bull. Soc. ent. Fr.*, 78(1-2): 9-25.
Junta de Andalusía, 1985. Mapa geológico-minero de Andalusía. Consejería de Economía e Industria, Sevilla.
Junta de Andalusía, 1987a. Plan Especial de protección del medio físico y catálogo de la provincia de Málaga. Publ. Cons. Obras Públicas y Transportes, Sevilla, 437 pp.
Junta de Andalusía, 1987b. Informe general del medio ambiente en Andalusía 1987. Publ. Agencia de Medio Ambiente, Sevilla, 452 pp.
Junta de Andalusía, 1988. Medio ambiente en Andalusía. Publ. Agencia de Medio Ambiente, 367 pp.
Junta de Andalusía, 1997. CD-Rom La información ambiental de Andalusía. Consejería de Medio Ambiente.
Kiesenwetter H.V. 1873. Bemerkungen zur bearbeitung der *Luperus*-Arten in der monographie des Gallerucides von Joannis in Marseul´s Abeille T. III Ao. 1866 p. 1 flg. *Berl. Ent. Zeitsch.*, 17(1/2): 23-31.
Mani, M.S. 1962. Introduction to high altitude entomology. Methuen, London. 302 pp.
Margalef, R. 1974. Ecología. Ediciones Omega, Barcelona, 951 pp.
Medina, M. 1875. Coleópteros de Andalusía del Museo de Historia Natural de la Universidad de Sevilla. *Anales Soc. esp. Hist. nat.*, 24: 271-272.
Oosterbroek, P. and Arntzen, J.W. 1992. Area-cladograms of circum-Mediterranean taxa in relation to Mediterranean palaeography. *J. Biogeogr.*, 19: 3-20.

Ortega, F. Molero, J. and García, E. 1987. La vegetación de Andalusía. In Cano, G. (ed.). Geografía de Andalusía, vol. II. Ediciones Tartessos, Sevilla, pp. 187-226.
Palomo, L.J. and Antúnez, A. 1992. Los atlas de distribución de especies. In Vargas J.M., Real R. and Antúnez A. (eds.). Objetivos y métodos biogeográficos. Aplicaciones en Herpetología. *Monografías de Herpetología Asoc. Herp. Esp.*, vol 2, pp. 39-50.
Pons, A. 1975. Introduction. In Pons, A. Reille, M. Triat, H. Coûteaux, M. Jalut, G. Öner, S. Planchais, N. and Vernet, J.L. Les données historiques et l´étude de la flore méditerranéenne. Colloques Internationaux du CNRS, nº 235: La flore du bassin Méditerranéen: Essai de Systématique synthétique. Montpellier, pp. 305-326.
Quézel, P. Gamisans, J. and Gruber, M. 1980. Biogéographie et mise en place des flores méditerranéens. Nat. Monsp., nº hors série: 41-51.
Rapoport, E. 1982. Areography. Geographical strategies of species. Pergamon Press, Exeter, 269 pp.
Rivas Martínez, S. 1982. Etages bioclimatiques, secteurs chorologiques et séries de végétation de l´Espagne mediterranéenne. *Ecol. mediterr.*, 8 (172): 275-288.
Rivas Martínez, S. 1984. Pisos bioclimáticos de España. *Lazaroa*, 5: 33-43.
Rivas Martínez, S. 1985. Mapa de las series de vegetación de España. Serv. Publ. Minist. Agricultura, Pesca y Alimentación, Madrid.
Rivas Martínez, S. 1987. Nociones sobre Fitosociología, Biogeografía y Bioclimatología. In Peinado-Lorca, M. and Rivas Martínez, S. (Eds.). La vegetación de España. Serv. Publ. Univ. Alcalá de Henares, 544 pp.
Rosenhauer, W.G. 1856. Die Thiere Andalusiens, nach dem Resultate einer Reise zusammengestellt. Erlangen, 307-345 (Chrysomelidae).
Vela, J.M. and Bastazo, G. 1985. Trofismo y fenología de las taxocenosis de Crisomélidos (Coleoptera) de la Sierra de las Nieves (Málaga, sur de España). Actas II Congr.Ib. Ent., Lisboa: 165-175.
Vela, J.M. and Bastazo, G. 1993. El género *Otiothraea* Warchalowski, 1990 (Col. Chrys. Clytrinae) en la Península Ibérica: descripción de una especie nueva y asignación de otra previamente conocida. *Zool. Baetica*, 4: 31-37.
Warchalowski, A. 1985. Revision der gattung *Labidostomis* Germar, 1824 (Coleoptera, Chrysomelidae, Clytrinae). *Polskie Pismo ent.*, 55:621-765.
Warchalowski, A. 1991. Über die rot und schwarz gefleckten Arten der Untergattung *Coptocephala* s.str. *Genus*, 2(3): 229-279.
Weise, J. 1891. Bekannte und neue Chrysomeliden aus Spanien. *Dt. ent. Z., Berlin*, 35(2): 373.

Appendix

Complete names of the considered species/subspecies.

Labidostomis bolivari Escalera, 1931 ssp. *garciaolmoi,* Cobos, 1954
L. nevadensis Daniel & Daniel, 1904
Coptocephala scopolina (Linnaeus, 1767) ssp. *alticola* Cobos, 1954
Smaragdina cobosi (Codina, 1963)
Otiothraea filabrensis (Cobos, 1957)
O. avilai Vela et Bastazo, 1993
Pachybrachis rondanus Burlini, 1968
Macrocoma cylindrica (Küster, 1846)
Chrysolina (Maenadochrysa) affinis (Fabricius, 1787) ssp. *nevadensis* Cobos, 1952
C. (Allochrysolina) fuliginosa (Olivier, 1790) ssp. *coriacea* (Suffrian, 1851)
C. (Threnosoma) helopioides (Suffrian, 1851)
C. (Chalcoidea) janbechynei Cobos, 1953
C. (Melasomoptera) lucida (Olivier, 1807) ssp. *suarezi,* Cobos, 1952
C. (Cobosorina) colasi (Cobos, 1952)
Cyrtonus almeriensis Cobos, 1953
C. contractus Fairmaire, 1882
C. fairmairei Rosenhauer, 1856
C. mateui Cobos, 1954
C. minor Fairmaire, 1883
C. pardoi Cobos, 1953
C. plumbeus gadorensis Cobos, 1954
Timarcha granadensis Bechyné, 1948
T. insparsa Rosenhauer, 1856
T. lugens Rosenhauer, 1856
T. marginicollis Rosenhauer, 1856
T. parvicollis Rosenhauer, 1856
Galeruca baetica Weise, 1891
Calomicrus foveolatus Rosenhauer, 1856
C. sordidus Kiesenwetter, 1873
Luperus abdominalis Rosenhauer, 1856
Anthobiodes heydeni Allard, 1870
Aphthona maldesi Doguet et Petitpierre, 1986
Longitarsus andalusicus Gruev, 1972
L. petitpierrei Bastazo, 1998
L. seticollis Mohr, 1962
L. vela, Bastazo, 1998

BIOLOGY AND BIOGEOGRAPHY OF DESERT LEAF BEETLES OF CENTRAL ASIA

IGOR K. LOPATIN
Department of Zoology, Byelorussian State University, Prospekt Skoriny 4, 220080 Minsk, Belarus

1. Introduction

Desert insects of the world, despite their taxonomic differences, possess similar adaptations to an arid climate. The adaptations are primarily related to water acquisition and retention. They have been studied mostly in the Tenebrionidae, which are a characteristic component of desert communities. Similar adaptations in the family Chrysomelidae have been less well studied, probably because this group consists mostly of mesophilous forms. However, several works (Lopatin, 1975, 1977; Lopatin & Kulenova, 1986; Medvedev, 1982) contain information on biology and distribution of leaf beetles in Central Asian desert environments.

2. Materials

Specimens for the present study were collected from 1956 to 1969 in Turkmenistan, Uzbekistan and Tajikistan, and in the deserts of Kazakhstan during the 1985 expedition. During the project, the biology of many leaf beetle species was also investigated, including their life cycle, host plants and behaviour.

3. Physical environment and plant communities

Asian deserts form part of the great belt of Old World, Northern Hemisphere deserts. They occur from Northern Africa to North Western China. This paper covers only those parts of deserts that occur in Turkmenistan, Uzbekistan, Tajikistan, Kazakhstan, Mongolia, and Xinjiang (North Western China).[1]

The distinctive attributes of the deserts of this region include high summer temperature, low precipitation (from 100-200mm), absence of surface water deposits, sandy soils and saline ground waters. Together these factors influence the local plant communities.

Lowland deserts of Middle Asia have their maximum (up to 70%) rainfall in the spring, followed by late autumn and winter. In summer, from mid-day to the beginning of October,

[1] In Russian literature it is common to divide Central Asia on its western (Middle Asia) and eastern parts (Central Asia). The latter includes Mongolia, North Western China and Tibet.

Advances in Chysomelidae Biology 1, pp. 159–168
Edited by M.L. Cox

there is no appreciable precipitation. Snow does not occur there. Summer is hot —40–44 °C in the afternoon in full shade. Winter may be cold – sometimes the temperature drops below -30 °C – but commonly it is mild with average January temperatures of +1.4 °C in Karakum and –4.1 °C in Kizylkum.

Deserts of Central Asia have greater geographic relief, colder winters and the maximum precipitation in the summer. Different regions have different average temperatures: +2.5 °C in Tsaidom, +11.6 °C in Kashgar. Winter is cold: the average January temperature in the Mongolian Gobi is–18.0 °C and –7.2 °C in Kashgar. Precipitation varies from 100mm to 219mm in Ashlan.

It is possible to distinguish several different types of desert based on soil composition: sand, sand and rubble, rubble and gypsum, clay and loess, solonchak, and stone. Plant communities are represented by several different types, influenced by soil composition. Several assemblages of trees favor sandy soils. They consist of *Haloxylon persicum*, *H. aphyllum*, some species of *Calligonum*, *Ammodendron connolyi*, *Salsola richteri*, *S. rigida*, *Ephedra strobilacea*, *Hammada leptoclada*, *Artemisia herba-alba*, *Tournefortia sogdiana*, *Aristida pennata* and *Carex physodes* etc.

Rubble deserts on tertiary plateaus contain several species of *Artemisia*, *Annabasis salsa*, *Salsola rigida* and *S. gemmascens*. Clay and loess deserts are characterized by *Calligonum*, *Artemisia*, *Salsola richteri*, *Halocnemum strobilaceum*, *Eurotia*, *Poa*, *Stipa* etc. Finally, solonchak is occupied by *Halocnemum strobilaceum*, several species from the genus *Salsola* (*arbuscula*, *laricifolia*, *rigida*, *gemmascens*), *Eurotia ceratoides*, *Annabasis salsa*, *Kalidium caspicum*, *Hammada leptoclada*, *Atriplex cana*, *Limonium gmelini*, *L. caspicum* and also *Tamarix ramosossimus*, *T. gracilis*, and *T. laxa*.

There are also several types of desert in Central Asia. Sand deserts of Jungaria contain *Haloxylon persicum*, *H. ammodendron*, *Tamarix*, *Reaumuria songorica*, *Ephedra przevalskii*, *Calligonum mongolicum*. *Nitraria sphaerocarpa*, *Caragana leucophloea*, *Atraphaxis compacta*, *Sueda* and *Kochia* are common on compacted clay and stone soils. *Atriplex cana*, *Halocnemum strobilaceum*, *Kalidium caspicum*, *K. gracile*, and *Lycium ruthenicum* are common on solonchak.

The river beds of the Central Asian deserts have a unique plant community. In the Russian literature it is called Tugai. The Tugai is formed by a combination of mesotherm trees and bushes. Typically these are poplar forests inhabited by *Populus diversifolia*, *P. pruinosa*, *P. ariana* which are accompanied by *Salix wilhelmsiana*, *Elaeagnus angustifolia*, *E. orientalis*, *Tamarix hispida*, *T. ramosissimus*, and *Halimodendron halodendron*. Trees are covered with vines *Clematis orientalis*, *Cynanchum sibiricum*, and *Calystegia sepium*. *Calamogrostis pseudophragmites* is common on the forest borders.

The trees *Ulmus pumila*, *Populus diversifolia*, *Salix caspia*, *Elaeagnus moorcrofti* and *Sophora alopecuroides* are distributed in the eastern part of this region. Forests in the river beds function as dispersal pathways by which mesophilous insects can penetrate desert environments.

4. Composition and structure of the leaf beetle fauna.

Of the 800 species of leaf beetles of Central Asia, 233 species occur in deserts, accounting for 29% of the fauna. These 233 species belong to 10 subfamilies, however the majority are Cryptocephalinae (72 species – 30.9%), Alticinae (42 species – 18.0%), and Eumolpinae (30 species – 12.8%). Altogether these three subfamilies constitute 62% of the desert fauna. The other subfamilies are represented as follows: Donaciinae – 4.2%, Criocerinae – 3.8%, Clytrinae – 7.2%, Chrysomelinae – 5.1%, Galerucinae – 8.1%, Hispinae – 0.8%, and Cassidinae – 5.5%. In the Palaearctic region as a whole the three most speciose subfamilies are the Alticinae, Chrysomelinae, and Cryptocephalinae. The species composition of deserts differs significantly. Cryptocephalinae and Eumolpinae which dominate in this fauna are better adapted to xeric environments because their larvae are soil-dwelling or case-bearing.

As was noted by Kryzhanovsky (1965) the desert fauna has a high percentage endemism. There are 158 (67.8%) endemic species in Central Asia. Some genera are also endemic: *Jaxartiolus*, *Aphilenia*, *Parnops*, *Dermestops*, *Andosia*, *Atomyria*, *Anidania*, *Adoxinia*, *Crosita*, *Theone*, *Nyctiphantus*, *Tschitsherinula*, and *Eremocassis*. Other genera are subendemic, with only a few species represented beyond this region: *Thelyterotarsus*, *Chloropterus*, *Chrysocharis*, *Callipta*, *Ischyronota*, and *Macromonycha*. This indicates a fauna of considerable age.

5. Regional characteristics of Central Asian desert fauna

Central Asian deserts can be divided into two types (Appendix). The first type includes the deserts of Turan (Karakum, Kizylkum, and Muyunkum) and a few small areas in Turkmenia, Uzbekistan, Tadzhikistan and western Kazakhstan. The second type includes the deserts of eastern Kazakhstan, Mongolia, and Xinjiang.

Examining this Appendix reveals that the leaf beetle fauna of the western subregion is richer than the eastern subregion (217 species and 73 species, respectively). The other characteristic of the western subregion is the elevated hierarchical level at which the endemism occurs. The genera *Aphilenia, Dermestops, Andosia, Atomyria, Anidania, Callipta, Tshitsherinula* and endemic species of *Cheilotoma, Antipa, Bedelia, Macrocoma, Malegia, Acmenychus, Chiridula*, and *Macromonycha* are restricted to this region. The eastern subregion does not have endemic genera and species endemism is much lower there. 217 species are known from the western subregion, 142 of them are endemic (65.4%). The eastern subregion has 73 species, 12 of which (16.4%) are endemic. These facts are indicative of the relative ages of the two desert regions and support the commonly accepted view that the deserts of Turan are older than the deserts of Mongolia andXinjiang. Undoubtedly, the main source of the leaf beetle fauna of Turan is the xeric eastern Mediterranean subregion.

6. Adaptations to desert life

Desert Chrysomelidae are very diverse morphologically, horologically, and phenologically, and have a broad range of host plants. True xerophilous species inhabit the sand and solonchak deserts. They feed on *Haloxylon*, *Calligonum*, *Ammodendron connolyi* and *Hammada leptoclada*. Many species are light colored with hirsute venters. They are active not only in spring but also in summer. This ecological group contains several species of *Antipa*, *Aphilenia*, *Cryptocephalus* (*C. semiargenteus*, *C. curtissimus*, *C. kiritschenkiellus*, *C.*

rufofasciatus, *C. rubi*, *C. rubi properus*), *Labidostomis curtus*, *Clytra rufina*, *Pachybrachis kaplini*, *Thelyterotarsus minimus*, *T. similis*, *T. baeckmanni*, *T. regeli*, *T bergi*, *Adoxinaa spinipes*, *Nyctiphantus* sp. sp., *Tschitsherinula paradoxocara*, *Calomicrus deserticola*, *Phyllotreta parfentjevi*, *Ph. mollis*, *Longitarsus turcomanorum*, and *Ischyronota conicicollis*.

Despite the presence of Chrysomelidae in deserts, they are mesophilous insects. This paradoxical situation can be resolved by the selection of moist habitats, such as riverbeds, or by timing their development so that they are short lived and actively reproducing only in early spring. Tugai leaf beetle species complexes consist of mesophilous species associated with *Populus diversifolia*, *P. pruinosa*, *Salix wilhelmsiana*, *Elaeagnus angustifolia*, *Tamarix hispida*, *Clematis orientalis*, and *Glycyrrhiza glabra*. Salinous parts of Tugai are inhabited by *Halimodendron halodendron*, *Salsola*, and *Limonium*. Among common species of Chrysomelidae in Tugai of Central Asia are *Crioceris duodecimpunctata hypopsila*, *C. iliensis*, *Smaragdina unipunctata*, *Jaxartiolus baekmannianus*, *Cryptocephalus tschimganensis*, *C. tataricus*, *C. turangae*, *C. tamaricis*, *Pachybrachis skopini*, *P. glycerrhizae*, *Stylosomus major*, *Parnops glasunovi*, *Chloropterus unguiculatus*, *C. grandis*, *C. lefevrei*, *Bedelia kokanica*, *Chrysochares punctatus*, *C. asiaticus orientalis*, *Diorhabda eleongata*, *Calomicrus gussakovskii*, *Monolepta russica*, *Phyllotreta latevittata*, *Longitarsus rectelineatus*, *L. rubiginosus*, *Altica balassogloi*, *Epitrix caucasica*, *Chaetocnema conducta*, *C. sinuata*, *Psylliodes dilutella*, *Acmenychus inermis*, *Eremocassis weisei*, and *Cassida turcmenica*.

Another mesophilous complex consists of species with spring activity. The adults disappear at the onset of summer, but larvae live eating the plant roots or mining stems or leaves, or protected by cases. Exposed surface-feeding larvae are active at night. *Crioceris oshanini*, *Cheilotoma ivanovi*, *Thelyterotarsus iliensis*, *Phyllotreta beschkentica*, *Longitarsus primaeveris*, *L. picicollis*, and *L. excisipennis* belong to this group. Several species of the genera *Macroplea*, *Donacia*, and *Plateumaris* feed on aquatic plants of ponds and rivers.

7. Host plant selection

Despite similar ecological conditions, the deserts of the western and eastern subregions are very different floristically. The family Chenopodiaceae is the dominant taxon in Turan. It is followed by Asteraceae, Fabaceae, Brassicaceae, Liliaceae, Poaceae, Polygonaceae, Apiaceae, Zygophyllaceae, and Plunbaginaceae. Asteraceae is the dominant group in Mongolia, followed by Fabaceae, Poaceae, and Chenopodiaceae. The family Polygonaceae is very poorly represented (Lavrenko, 1962). Representatives of the Liliaceae and Apiaceae are common among ephemeras.

Is the abundance of different plant families correlated with the abundance of their leaf beetle herbivores? Many chrysomelid species consume Chenopodiaceae. For example, *Cryptocephalus rubi properus*, *Thelyterotarsus similis*, *Aphilenia ornata*, *Adoxinia spinipes*, and *Phyllotreta parfentjevi* feed on *Haloxylon persicum* and *H. aphyllum*. *Pachybrachis arnoldii* feeds on *Nanophyton*. All Turanian species of *Antipa*, *Pachybrachis kazakhstanicus*, *Andosia reitteri*, and *Phyllotreta ustulata* feed on *Salsola rigida*, *Salicornia* and *Anabasis salsa*. *Cryptocephalus rubi properus* feeds on *Halocnemum strobilaceum* whilst *Cryptocephalus coronatus* feeds on *Camphorosma lessingi*. All of these species are Turanian endemics.

Many Chrysomelidae of the Central Asian deserts are associated with two genera of polygonaceous plants. For example *Clytra rufina, Cryptocephalus monilis, C. kiritschenkiellus, C. curtissimus, C. rufofasciatus, C. semiargenteus, Thelyterotarsus baekmanni, T. regeli, T. bergi, Aphilenia hauseri*, and *A, parvula* feed on *Calligonum*.

Consortium of another plant genus *Atraphaxis* is very characteristic for Central Asia. Most of its herbivores (several dozen beetle species from different families) are concentrated in xeric habitats of the foothills. Also in deserts *Cheilotoma ivanovi, Clytra atraphaxidis, Thelyterotarsus iliensis*, and *T. dzhungarus* feed on *Atraphaxis*. Special studies are needed to understand the host plant/herbivore relationships of *Atraphaxis*.

The family Asteraceae is represented by several species of *Artemisia* in the deserts of Central Asia (*A. terrae-albe, A. badghyzi, A. sublessingiana, A. turanica*). *Labidostomis curtus, Cryptocephalus bohemius, Anidania luctuosa, Chrysolina graminis artemisiae, Crosita bogutensis, C. altaica faldermanni*, all known species of *Theone* and many species of *Galeruca* feed on these plants.

Fabaceae are also a common food plant for many leaf beetles. *Cryptocephalus balassogloi, C. bitaeniatus, C. prasolovi, Pachybrachis nigropunctatus, Macrocoma sarvadensis*, and *Diorhabda persica* feed on *Alhagi camelorum, A. kighizorum*, and *A. persarum*.

Spring ephemera feed on the leaves of Brassicaceae (especially on *Malcolmia bungei*). These are *Colaphellus apicalis, C. sophiae hoefti*, many species of *Phyllotreta* (*P. beschkentica, P. banghaasi* etc.), and *Psylliodes sophiae. Entomoscelis erythrocnema* feeds on *Adonis*.

A special group consists of species feeding on poplar and willow. *Cryptocephalus turangae, C. bodungeni, C. malaxanthus flaviventris, Pachybrachis skopini, P. rufescens, Parnops glasunovi, Atomyria sarafschanica*, and *Bedelia kokanica*, as well as the widespread *Chrysomela saliceti* and *Crepidodera plutus* belong to this group. *Altica balassogloi* feeds on *Elaeagnus angustifolia* and *E. orientalis* (Elaeagnaceae). Vines *Cynanchus sibiricum* and *Calystegia sepium* attract *Chrysochares punctatus* and *C. asiaticus orientalis*.

The trophic specialization of desert leaf beetles is not well known, but the scant data indicate a close affinity between endemic leaf beetles and common desert plants.

8. Conclusion

The desert fauna of the world has a long history of adaptation to the specific conditions of the desert environment. An arid climate is the most important of these conditions. As a result, there has been a great deal of taxonomic diversification, although there are a limited number of life forms. These life forms evolved under the pressure of a limited water supply, widely fluctuating daily temperatures and other conditions which are absent in humid regions. Desert environments and fauna are under great pressure due to over grazing, road construction and oil and gas mining. Therefore, the study of desert fauna and the ecology of its components are of the utmost urgency.

Acknowledgments

I extent my warmest thanks to my colleagues N. Vandenberg and A. Konstantinov of the USDA, Systematic Entomology Laboratory, Washington, D.C. for help with the preparation and translation of this manuscript.

References

Kryzhanovsky, O. L. 1965. The structure and origin of terrestrial fauna of Central Asia. Nauka. Moscow-Leningrad. 419 pp. (In Russian).

Lavrenko, E. M. 1962. The basic traits of the botanical geography of deserts of Eurasia and North Africa. Acad. Sci. Leningrad. 168 pp. (In Russian).

Lopatin, I. K. 1975. The leaf beetles (Coleoptera, Chrysomelidae) of the Mongolian People's Republic. In: Insects of Mongolia, 3. Nauka, Leningrad: 191-231. (In Russian).

Lopatin, I. K. 1977. The leaf beetles (Coleoptera, Chrysomelidae) of Central Asia and Kazakhstan. Nauka, Leningrad. 268 pp. (In Russian).

Lopatin, I. K. and K. Z. Kulenova, 1986. The leaf beetles (Coleoptera, Chrysomelidae) of Kazakhstan. Nauka, Alma-Ata. 200 pp. (In Russian).

Medvedev, L. N. 1982. The key of Chrysomelidae of Mongolia. Nauka. Moscow. 302 pp. (In Russian).

Appendix

Distribution of Chrysomelidae in deserts of Central and Middle Asia.

Name of species	Subreg.1	Subreg.2
I. DONACIINAE		
Macroplea mutica F.	+	+
M.appendiculata Panz.	+	
Donacia clavipes F.	+	+
D. vulgaris Zschach	+	
D. v.issykensis Jcbs.	+	
D. bactriana Wse.	+	
D. b.sahlbergi Jcbs.	+	
D. marginata Hop.	+	
D. impressa Payk.	+	
D. cinerea Hbst.	+	
II. CRIOCERINAE		
Oulema melanopus L.	+	
Lema quadrimaculata Gebl.	+	
L. decempunctata Gebl.	+	+
Crioceris quatuordecimpunctata Scop.	+	+
C. duodecimpunctata L.	+	+
C. d.hypopsila Jcbs.	+	
C. bicruciata Sahlb.	+	
C. oshanini Dobrn	+	+
C. iliensis Wse.	+	+
III. CLYTRINAE		
Labidostomis curtus Lop.	+	
L. senicula Krtz.	+	+
L. s.deserta Medv.	+	
L. metallica centrisculpta Pic	+	
L. bipunctata Mnnh.		+
Cheilotoma ivanovi Jcbs.	+	
Coptocephala apicalis Lac.	+	
C. gebleri Gebl.	+	+
C. unifasciata deserta Medv.	+	
C. australis Medv.	+	+
Clytra rufina Sols.	+	

Appendix. Cont.

Name of species	Subreg.1	Subreg.2
C. atraphaxis punctata Wse.	+	
C. a.maculifrons Zubk.	+	
Smaragdina unipunctata Oliv.	+	
S. discolor Sols.	+	
Antipa silensis Wse.	+	
A. nigriventris Lef.	+	
A. puncticollis Medv.	+	
IV. CRYPTOCEPHALINAE		
Jaxartiolus baeckmannianus Jcbs.	+	
J. reductosignatus Lop.	+	
J. kozlovi Lop.		+
Cryptocephalus anticus Suffr.	+	
C. melanocephalus flaviventris Lop.	+	
C. bodungeni Jcbs.	+	
C. b.tadzhika Jcbs.	+	
C. semiargenteus Rtt.	+	
C. undulatus Suffr.	+	+
C. monilis Wse.	+	+
C. curtissimus Pic	+	
C. rufofasciatus Sols.	+	
C. kiritschenkiellus Lop.	+	
C. tamaricis Sols.	+	+
C. wittmeri Lop.	+	
C. kozlovi Lop.		+
C. coronatus Suffr.	+	
C. sarafschanensis iliensis Wse.	+	
C. s.betpakdalensis Lop.	+	
C. borochorensis Pic	+	
C. balassogloi Jcbs.	+	
C. prasolovi Lop.	+	
C. bohemius Drap.	+	
C. bitaeniatus Sols.	+	
C. rubi properus Wse.	+	
C. r.glasunovi Jcbs.	+	
C. impressipygus Ogl.	+	
C. tataricus Gebl.	+	
C. globulus Lop.	+	+
C. agnus Wse.		+
C. crux Gebl.	+	+
C. hammadae Lop.	+	
C. dilutellus Jcbs.	+	+
C. tschimganensis Wse.	+	
C. turangae Lop.	+	
Pachybrachis kazakhstanicus Lop.	+	
P. arnoldii Lop.	+	+
P. gobicus Lop.		+
P. nigropunctatus Suffr.	+	
P. pistor Brt.	+	+
P. mitjaevi Lop. et Kul.	+	
P. auliensis Brt.	+	+
P. skopini Lop.	+	
P. cribricollis Pic	+	
P. kaplini Lop.	+	
P. rufescens Lop.	+	

Appendix. Cont.

Name of species	Subreg.1	Subreg.2
P. ghilarovi Lop.	+	
P. jastschenkoi Lop.		+
P. sinkianensis Lop.		+
P. abditus Lop.	+	
P. jacobsoni Lop.	+	
P. heptapotamicus Lop.	+	
P. marki Lop.	+	
P. kirghizicus salsolae Lop.	+	
Thelyterotarsus iliensis Lop.	+	
T. minimus Jcbs.	+	
T. similis Lop.	+	
T. margaritae Lop.	+	
T. limbatus Lop.	+	
T. issykensis Lop.	+	
T. inopinatus Lop.	+	
T. sogdianua Lop.	+	
T. mogoltavicus Lop.	+	
T. przevalskii Lop.	+	
T. mongolicus Lop.		+
T. baeckmanni Jcbs.	+	
T. regeli Jcbs.	+	+
T. r.bergi Jcbs.	+	
T. dzhungarus Lop.	+	
T. gobicus Medv.et Vor.		+
Stylosomus major Brt.	+	+
S. weberi Reitt..	+	
S. tamaricis Herr.-Sch.		+
V. EUMOLPINAE		
Aphilenia hauseri Wee.	+	
A. ornata Reitt.	+	
A. parvula Wse.		+
A interrupta Wse.	+	
A. i.gobica Lop.		+
Parnops glasunovi Jcbs.	+	
P. atriceps Pic		+
Dermestops ahngeri Jcbs.	+	
Chloropterus grandis Wse.	+	+
Ch. lefevrei Reitt.	+	
Ch. versicolor Mor.	+	
Ch. unguiculatus Lop.	+	
Bedelia kokanica Sols.	+	
Pachnephorus tesselatus Duft.	+	+
P. jacobsoni Lop.	+	
P. turcomanicus Medv.	+	
Andosia reitteri Wse.	+	
Atomyria sarafschanica Sols.		+
Anidania luctuosa Sols.	+	
Adoxinia spinipes Reitt.	+	
Macrocoma sarvadensis Sols.	+	
Malegia turkestanica Reitt.	+	
M. jacobsoni Sum.	+	
Chrysochares punctatus Gebl.	+	+
Ch. asiaticus orientalis Lop.	+	+
Callipta borealis Lop.	+	
C. fausti balchana Lop.	+	

Appendix. Cont.

Name of species	Subreg.1	Subreg.2
C. f.badghyza Lop.	+	
C. f.murgabica Lop.	+	
C. f.palvanica Lop.	+	
Chrysochus chinensis Baly		+
VI. CHRYSOMELINAE		
Chrysolina grata Fald.	+	
Ch. turanica Reitt.	+	
Cli. graminis artemisiae Motsch.	+	+
Ch. aeruginosa centralasiae Lop.		+
Ch. sacarum Wse,	+	
Ch. s.embiensis Lop.	+	
Ambrostoma quadriimpressum Motsch.		+
Crosita heptapotamica Jcbs.	+	
C. bogutensis Lop.	+	
C. altaica faldermanni Kryn.	+	+
Colaphellus apicalis Men.	+	
C. sophiae hoefti Men.	+	
VII. GALERUCINAE		
Galeruca nigrolineata Mnnh.	+	
G. interrupta armeniaca Wse.	+	
Theone costipennis Kirsch.	+	
Th. margelanica Krtz.	+	
Th. silphoides Dalm.	+	?
Nyctiphantus hirtus Wse.	+	+
N. h.nocturnus Sem.	+	
N. custos Sem.	+	
N. bergi Sem.	+	
Diorhabda persica Fald.	+	
D. elongata Brull.	+	+
D. rybakovi Wse.	+	+
Xanthogaleruca luteola Miill.	+	
Galerucella calmariensis L.	+	+
G. nymphaeae L.	+	+
Tschitsherinula paradoxocara Jcbs.	+	
Monolepta russica Gmel.	+	
Calomicrus deserticola Ogl.	+	
C. gussakovskii Ogl.	+	
VIII. ALTICINAE		
Phyllotreta ustulata Lop.	+	
Ph. parfentjevi Shap.	+	
Ph. pallidipennis Reitt.	+	+
Ph. weiseana Jcbs.	+	
Ph. konevi Lop.	+	
Ph. mollis Konst.	+	
Ph. beschkentica Lop.	+	
Ph. banghaasi Heik.	+	
Ph. nemorum L.		+
Aphthona gracilis Fald.	+	
A. weiseana Konst.	+	
A. nonstriata Goeze	+	

Appendix. Cont.

Name of species	Subreg.1	Subreg.2
A. flaviceps All.	+	
A. hauseri Wse.	+	
A. maculata All.	+	
Longitarsus rectelineatus Foudr.	+	
L. acuticollis Lop.	+	
L. desertorum Heik.	+	+
L. turcomanorum Lop.	+	
L. primaeveris Lop.	+	
L. tmetopterus Jcbs.	+	
L. pellucidus Foudr.	+	+
L. tournefortii Medv.et Vor.		+
Altica balassogloi Jcbs.	+	
A. deserticola Wse.	+	+
A. tscharynensis Ogl.	+	
A. palustris Wse.	+	
Crepidodera plutus Latr.	+	+
Epitrix abeillei Baud.	+	+
E. caucasica Heik.	+	
Hermaeophaga ruficollis Luc.	+	
Chaetocnema semicoerulea Koch	+	+
Ch. splendens ljudmilae Lop.	+	
Ch. breviuscula Fald.	+	
Ch. hortensis Geoffr.	+	+
Ch. sinuata Wse.	+	+
Ch. nebulosa Wse.	+	+
Ch. orientalis Baud.	+	
Ch. conducta Motsch.	+	
Ch. psylloides Pic	+	
Psylliodes reitteri parallela Wse.	+	+
P. dilutella Heik.	+	
P. sophiae Heik.	+	
P. macella Wse.	+	+
IX. HISPINAE		
Acmenychus inermis Zubk.	+	+
Hispa atra L.	+	+
X.CASSIDINAE		
Ischyronota desertorum Gebl.	+	+
I. conicicollis Wse.	+	+
I. elevata Reitt.	+	
I. spaethi Reitt.	+	
I. schusteri Spth.	+	+
Chiridula semenovi Wse.	+	
Macromonycha apicalis Gebl.	+	
Eremocassis weisei Jcbs.	+	
Cassida berolinensis Suffr.	+	+
C. sareptana Krtz.	+	
C. palaestina Rche.	+	
C. turcmenica Wse.	+	+
C. undecimnotata Gebl.	+	+

BIOLOGICAL CONTROL OF COLORADO POTATO BEETLE, *LEPTINOTARSA DECEMLINEATA* (SAY) (CHRYSOMELIDAE: CHRYSOMELINAE)

ALISON F. MOLDENKE and RALPH E. BERRY
Department of Entomology, 2046 Cordley Hall, Oregon State University, Corvallis, Oregon 97331-2907, USA

Abstract

Resistance to chemical insecticides increasingly is thwarting efforts to control Colorado potato beetle (*Leptinotarsa decemlineata* (Say)), a serious, sometimes devastating, pest of solanaceous crops such as potato, eggplant and tomato. Biological control, used alone or in conjunction with traditional pesticides, is therefore becoming more and more important to effective management of this pest. This article reviews recent research on the use of a variety of organisms as control agents against Colorado potato beetle, including *Bacillus thuringiensis*, spiroplasmas, and other bacteria; the fungus *Beauveria bassiana*; entomopathogenic nematodes; and arthropod predators and parasitoids.

1. Introduction

The Colorado potato beetle (CPB), *Leptinotarsa decemlineata* (Say) (Coleoptera: Chrysomelidae), is a major pest on solanaceous crops throughout much of the United States, Mexico, Europe and Asia (Weber and Ferro, 1994; Foster and Flood, 1995). Economic damage caused by this insect is most important in potato (*Solanum tuberosum*), but is also significant in tomato (*Lycopersicon esculentum*), eggplant (*Solanum melongena*) and other solanaceous crops (Foster and Flood, 1995; Hamilton and Lashomb, 1996; Hamilton *et al.*, 1996). Because of the stress of defoliation, plant resources are diverted from fruit or tuber formation, lessening yields by as much as 50% or more, depending on the time of defoliation (Trumble *et al.*, 1993; Mahr *et al.*, 1995; Nault and Kennedy, 1996; Grafius, 1997). In extreme cases, damage may be so severe that no crop is produced. The significance of this is evident when one considers that the commercial value of potato production in the Pacific Northwest, which accounts for nearly half the production in the United States, was estimated at over $1 billion in 1996. In Michigan, control costs and crop losses (~11%) cost the potato industry over $14 million in 1994 (Grafius, 1997).

CPB has rapidly developed resistance to most pesticides (Weber and Ferro, 1994). Furthermore, some pesticides used to control CPB, such as aldicarb (Temik), have produced significant undesirable environmental effects because of their persistent contamination of both soil and groundwater (Munster *et al.*, 1996). Successfully controlling CPB without adverse environmental effects thus presents a continuing challenge throughout much of the world. Biological control, particularly in combination with integrated pest management and expert systems, is being looked to increasingly to provide at least part of the solution.

Advances in Chysomelidae Biology 1, pp. 169–184
Edited by M.L. Cox

Ferro (1994) reviewed the status of biological control efforts against CPB; our review will emphasize the research reported since 1994.The organisms being investigated as control agents are much the same today as at the time of Ferro's review: microbial organisms, entomopathogenic nematodes, and arthropod predators and parasitoids.

2. Microbial control agents

Microbial control agents that have been tested against CPB include bacteria, especially *Bacillus thuringiensis* subsp. *tenebrionis* (*Btt*) and its δ–endotoxins and spiroplasmas; microsporidians; and fungi. We found no studies of microsporidians since Ferro's (1994) review; spiroplasmas, on the other hand, have excited a flurry of research. Studies on bacteria and fungi have concentrated on *B. thuringiensis* and the fungus *Beauveria bassiana.*

2.1. Bacteria

2.1.1. *Bacillus thuringiensis*

Bacillus thuringiensis, particularly *B. thuringiensis* subsp. *tenebrionis* (also known as *B. thuringiensis* var. *san diego*) and the crystalline protein δ–endotoxins contained in its spore, continue to be the subject of intensive research in control of CPB. Use of *Btt* can reduce defoliation and CPB numbers markedly (Hough-Goldstein and Whalen, 1993; Cloutier and Bauduin, 1995; Hough-Goldstein *et al.*, 1996; Schrod *et al.,* 1996). Most attention has focused on the CryIIIA δ–endotoxin, which is effective specifically against CPB and other coleopterans. Martinez-Ramirez and Real (1996) found that this toxin can be proteolytically digested by CPB gut contents or chymotrypsin and binds to brush-border membrane vesicles from its midgut. They could not confirm earlier reports that this toxin acted in its intact (nonproteolytically processed) form. Carroll *et al.* (1997) demonstrated that chymotrypsin-treated CryIIIA protein was more soluble at the neutral pH characteristic of CPB gut than was the native toxin. The digested protein bound to CPB midgut membranes and retained significant insecticidal activity. Novillo *et al.* (1997) subsequently characterized a chymotrypsin-like protease and other proteases from the midgut of Colorado potato beetle. These likely are the activators of the CryIIIA endotoxin and may interfere with control strategies involving the use of protease inhibitors as well.

Lambert *et al.* (1992) described a CryIIIC toxin active against CPB but not against several lepidopterans, and Bradley *et al.* (1995) isolated a CryIB crystal protein from *Btt* and an undescribed strain that is active against both coleopteran chrysomelid larvae (including CPB) and lepidopteran larvae. Both the CryIIIC and the CryIB toxins are less toxic to CPB than CryIIIA; both require solubilization and proteolytic digestion before application to exert toxicity maximally (CryIB) or at all (CryIIIC). Such proteins may, however, be useful in a transgenic application.

In order for *Btt* to be effective, it must be consumed. Antifeedants have been proposed as control measures for CPB (Murray *et al.*, 1993; Alford, 1994; Ghidiu *et al.*, 1996), and feeding stimulants have been proposed to enhance *Btt* consumption (Hough-Goldstein *et al.*, 1991). Interactions between these compounds and the effects of *Btt*, as well as antifeedant effects of *Btt* itself, are therefore of interest. Researchers in several laboratories (Hough-

Goldstein *et al.*, 1991; Arpaia and Ricchiuto, 1993; Altre *et al.*, 1996; Ghidiu *et al.*, 1996) have detected no antifeedant or repellent effects of *Btt* in either small or large larvae or adults, although feeding inhibition attributable to gut disruption occurred in second instars after several hours (Altre *et al.*, 1996). Hough-Goldstein *et al.* (1991) also found that feeding stimulants used in conjunction with *Btt* did not increase consumption of foliage sufficiently to increase *Btt*-induced mortality. Murray *et al.* (1993) tested the use of citrus limonoid antifeedants as a stress inducer in early instars before application of *Btt*. Limonoid application did not affect mortality relative to that observed in larvae not pretreated with limonoid. *Btt* and limonoid, singly or in combination, did slow larval development relative to that of the controls.

Mulch, which has been used in control programs for CPB (Stoner, 1993,1997; Stoner *et al.*, 1996; Brust 1994, 1996), may enhance the effectiveness of *Btt* application. Brust (1996) found that the combination of straw mulch and *Btt* treatment reduced defoliation and increased yield to levels equivalent to those achieved with esfenvalerate, a synthetic pyrethroid, and was significantly more effective than either mulch or *Btt* alone. Natural predators were enhanced by the use of mulch, but were not affected by the presence of *Btt*.

Although formulations of *Btt* are effective against CPB, their application is logistically difficult (Grafius, 1997) and should be timed to maximize exposure of the earlier larval stages, which are most susceptible (Bystrak *et al.*, 1994; Ghidiu *et al.*, 1994). Mortality depends on spray coverage, temperature, and time between application and rainfall; mortality is higher at higher temperatures and when rain is delayed after application (Bystrak *et al.*, 1994; Ferro *et al.*, 1997). Timing of application relative to peak egg hatch also is important. Application within 4 days of peak (30%) egg hatch in the spring appears to control larvae adequately to prevent yield loss; at the same time, it minimizes the selecting for resistance that may occur with the additional treatments required by earlier application (Ghidiu and Zehnder, 1993).

The development of *Btt* transgenic crops, including potato, tomato, and eggplant, to control CPB (Perlak *et al.*, 1993; van der Salm *et al.*, 1994; Chen *et al.*, 1995; Arpaia *et al.*, 1997a,b, 1998; Innacone *et al.*, 1997; Jelenkovic *et al.*, 1998) is a significant new approach. The use of transgenic plants simplifies application logistics, presents little or no threat to natural enemies [although Hilbeck *et al.* (1998) have recently reported a third-trophic-level effect on *Chrysoperla carnea*], and may provide more lasting control (Perlak *et al.*, 1993; Roush, 1994). The likelihood that CPB will develop resistance to *Btt*, however, is considered a major constraint to its use, particularly in transgenic plants (Arpaia and Ricchiuto, 1993; Ferro, 1993; Whalon *et al.*, 1993, 1994; Gould *et al.*, 1994; Whalon and Wierenga, 1994; Wierenga *et al.*, 1996; Grafius, 1997; Arpaia *et al.*, 1998). Low levels of selection and the presence of refugia under current conditions of *Btt* application (relative immunity of late instars and adults, poor persistence in the field, and temporal and spatial refugia) may minimize the development of resistance so long as those conditions are maintained (Ferro, 1993). Incorporation of the toxin into plants, however, could increase selection pressure and the likelihood that resistance would develop. On the other hand, fitness costs associated with resistance (Whalon *et al.*, 1993; Trisyono and Whalon, 1997) may slow its development. Although Altre *et al.* (1996) were unable to demonstrate such costs in terms of beetle development, their sample sizes were too small to analyze statistically. Furthermore, as they point out, costs may lie in other aspects of the life history and behaviour. Roush (1994) also presented evidence that a laboratory strain of CPB resistant to sprayed *Bt* does not develop on transgenic plants, indicating that resistance mechanisms to sprays and plants may be different. Studies of resistance development are further complicated by temporal and spatial natural variation in CPB populations (Weisz *et al.*, 1994; Robertson *et al.*, 1995; Midgarden *et al.*, 1997).

Development of resistance could be postponed or lessened by several tactics, including developing transgenic plants that express the toxin at low levels, which may delay development but not kill susceptible individuals; developing plants that express toxin at very high levels, so as to render resistance genes functionally recessive; management techniques such as crop rotation and use of other biological control agents; mixing transgenic with nontransformed plants; and using transgenic plants where immigration of susceptible insects into the population is probable (Ferro, 1993; Gould *et al.*, 1994; Hoy and Head, 1995; Arpaia *et al.*, 1998). Rahardja and Whalon (1995), on the basis of their finding that resistance to CryIIIA δ–endotoxin is incompletely dominant and may involve more than one locus, suggested that resistance management may be possible if begun before resistance factors are fixed in the population. All the strategies of resistance management mentioned above, however, present practical problems in terms of implementation. Furthermore, understanding of their potential usefulness is limited by lack of knowledge of the behaviour and biology of Colorado potato beetle, particularly in the presence of transgenic plants (Gould *et al.*, 1994). Hoy and Head (1995), for example, found that using mixtures of transgenic and nontransgenic plants may result in indirect selection for resistance because of a tendency of more resistant larvae to move onto nontransgenic plants.

2.1.2. Spiroplasmas

Spiroplasmas are helical, walless microorganisms related to bacteria (*Prokaryotae*: Class Mollicute) that can cause diseases in plants, insects and mammals (Tanada and Kaya, 1993). Hackett *et al.* (1996) have described a species, *Spiroplasma leptinotarsa,* that appears specific to the genus *Leptinotarsa*. It is widely distributed throughout North America (Hackett *et al.*, 1992) and has been identified in Europe as well (Hackett *et al.*, 1997). This species appears to be a commensal, rather than a disease-causing, organism, but has been proposed for use in biocontrol as a genetically engineered vector for insect-killing toxins, specifically the δ–endotoxin from *Btt* (Hackett *et al.*, 1992). The spiroplasmas are thought to be spread from insect to insect by contact with regurgitated and fecal material (Clark, 1982). Thus, genetically engineered spiroplasmas, once applied, would spread naturally through the CPB population and would not require frequent reapplication, as does *Btt* (Hackett *et al.*, 1992).

Spiroplasma leptinotarsa, first isolated from CPB gut fluid in 1982 (Clark, 1982), is found in association with the midgut microvilli of the beetle and does not grow in the hemocoel (Clark, 1982; Hackett *et al.*, 1996). Difficulties in culturing the organism *in vitro* have restricted its use in genetic transformation and biological control. Recently, however, new methods have been developed to facilitate the culture of *S. leptinotarsa* (Konai *et al.*, 1996) and to assess intraspecific variation (Hackett *et al.*, 1997). Interest in capitalizing on its commensal relationship with CPB to facilitate its use as a genetically engineered vector has revived. Hackett *et al.* (1997), who identified distinct serological and genomic variants from different areas of North America and Poland, cautioned that such differences should be considered in developing strains for use in biocontrol. This area is now poised for major advances in genetic transformation and development of application methodology.

2.1.3. Other bacteria

A novel approach to the biocontrol of CPB is the use of ice-nucleating bacteria such as *Pseudomonas syringae* to reduce the supercooling capacity of overwintering insects (Lee *et al.*, 1994, 1996). Either topical application of *P. syringae* near a spiracle or exposure to the bacteria in the soil decreases the supercooling capacity of CPB by some 3 to 4 °C. Lee *et al.* (1996) proposed the use of such bacteria to enhance the effectiveness of mulch treatments in inducing temperature-caused mortality.

2.2. Fungi: Beauveria bassiana

Beauveria bassiana (Balsamo) Vuillemin is effective against many insects. In CPB, it has generally been used against foliar-feeding stages. Long *et al.* (1998) found that *B. bassiana* did not affect mortality of CPB eggs or the first instars hatching from them. They therefore recommended that the fungus be applied only after most eggs have hatched. The earlier stages appear to be more susceptible; because of this and because the fourth instar continues to feed at rates as high as or higher than those of untreated larvae (Fargues *et al.*, 1994), delaying treatment until the CPB have reached the later larval stages seems inadvisable. Poprawski *et al.* (1997) found unformulated conidia to be even more effective than insecticide treatments in reducing populations of larval CPB. Couteaudier *et al.* (1996) increased virulence of *B. bassiana* against CPB by hybridization with *B. sulfurescens*, a species that produces an insecticidal toxin but was not pathogenic to CPB in their tests; they suggest that protoplast fusion may be useful in improving biocontrol by entomopathogenic fungi.

Because *B. bassiana* does infect so many insects, care must be taken to minimize impact on predators when using the fungus as part of an integrated management program against CPB. Todorova *et al.* (1994) found that the predator *Coleomegilla maculata* De Geer (the 12-spotted ladybird beetle) was very sensitive to one strain of *B. bassiana*, but not to another; both strains caused high mortality in CPB. The strain less pathogenic to *C. maculata* slows development, possibly prolonging and increasing its predatory potential (Todorova *et al.*, 1996). Thus, careful selection of strain may permit safe use of the fungus in the same management program with this predator. Poprawski *et al.* (1997) were unable to determine whether *B. bassiana* infected the predaceous hemipteran *Perillus bioculatus* (F.) when released in the same plot. Additional studies of the susceptibility of these and other predators to *B. bassiana* are needed.

3. Entomopathogenic nematodes

Entomopathogenic nematodes belonging to the families Steinernematidae and Heterorhabditidae present great potential for successfully controlling insects, such as CPB, that pass at least part of the life cycle in the soil. These nematodes and their associated symbiotic bacteria reproduce inside insect hosts, killing the hosts by septicemia and bacterial toxins. Natural inhabitants of the soil, they appear to pose little or no threat in their natural genetic state to plants, vertebrates, or other noninsect invertebrates and do not require registration for use in the United States (Poinar, 1991; Wouts, 1991; Kaya and Gaugler, 1993; Smart, 1995). Their often broad host range, their rapid lethal effect, and the ease of their application add to their attractiveness as control agents for soil-dwelling organisms (Kaya and Gaugler, 1993; Gaugler *et al.*, 1997).

CPB has acquired the reputation of being somewhat resistant to entomopathogenic nematodes, *Steinernema* spp. in particular (Wouts, 1991; Thurston *et al.,* 1994). Nevertheless, several investigators have recently achieved encouraging results with both *Steinernema* and *Heterorhabditis* species.

Nickle *et al.* (1994) reported that *S. carpocapsae* delivered in pelletized pesta (a protective pelletized formulation based on wheat flour) reduced adult emergence of CPB by 91 to 100%. Saringer *et al.* (1996) tested *Steinernema carpocapsae* (All strain) and *Heterorhabditis* (HP88 strain) against CPB under laboratory and field conditions. In both instances, *S. carpocapsae* was more effective, resulting in 89% mortality under optimal field conditions, compared with 60% with the *Heterorhabditis* strain.

Berry *et al.* (1997) compared pathogenicity of five strains of *Heterorhabditis* (*H. bacteriophora, H. indicus, H. marelatus*, OH23, and OH 95) and 2 species of *Steinernema* (*S. oregonensis* and *S. riobravis*) against CPB in a laboratory bioassay. *Heterorhabditis* strains exotic to the region from which the beetles were collected gave highest mortality (up to 100%), with *H. marelatus* being particularly effective. Exotic *Heterorhabditis* spp. also were more effective than exotic *Steinernema*, in contrast to the work of Saringer *et al.* (1996). Berry *et al.* (1997) attributed the differences they found to differences in foraging strategy between the genera. All strains persisted in the field for 14 weeks, but none reproduced in CPB under the experimental conditions. However, A. F. Moldenke (unpublished observations) has found that *H. marelatus* and a heterorhabditid strain closely related to OH95 can reproduce in Colorado potato beetle. Difference in species and strain behaviour, as well as in the immunological response of the beetles to these strains (Thurston *et al.*, 1994; Berry *et al.*, 1997) may account for the discrepancies among these studies.

Development of a strain able to reproduce in CPB, while retaining long-term field persistence, could provide a valuable aid in biological control of this pest. Clearly further work is needed to determine the strains most effective both for insect control and for their persistence and reproductive success. Furthermore, many factors seem to be involved in regulating host susceptibility to entomopathogenic nematodes, including nematode strain, habitat and behaviour; host defensive behaviour and grooming, attractiveness to the nematodes, morphological characteristics, host plant and immunological factors; and abiotic conditions, such as temperature, soil texture and relative humidity. Few of these have been examined with respect to improving the efficacy of entomopathogenic nematodes against Colorado potato beetle. Such studies would greatly enhance the usefulness of these organisms in control of Colorado potato beetle.

4. Arthropod parasitoids and predators

Heimpel and Hough-Goldstein (1992), Hough-Goldstein *et al.* (1993), Ferro (1994), Mena-Covarrubias *et al.* (1996), Hilbeck and Kennedy (1996), and Hilbeck *et al.* (1997) discussed predation and surveyed several groups of arthropod natural enemies that, properly managed, may contribute more to control than they have to date. Known predators include mites; phalangids; some 30 species of spiders; carabids, such as *Calledia decora* (F.), *Lebia grandis* (Hentz) and *L. analis* (Dejean), and *Pterostichus chalcites* Say; coccinellids, such as *Coleomegilla maculata* (De Geer) and *Hippodamia convergens* Guérin-Méneville; hemipterans such as *Oplomus spp.*, *Perillus bioculatus* (F.), *Podisus maculiventris* (Say) and *Geocoris punctipes* (Say); and the neuropterans (chrysopids) *Chrysopa* spp. and *Chrysoperla carnea* (Stevens). Parasites seem to be confined to two parasitoid genera: the tachinid fly *Myiopharus* spp., and the hymenopteran egg parasitoid *Edovum puttleri*.

Brust (1994, 1996) has found that mulching enhances the presence of natural predators and, alone or with *Btt*, decreases defoliation and increases yields. Stoner *et al.* (1996) also found that mulching decreased beetle numbers and defoliation, although yield was not affected; predators may have been influential here also, although they were not studied directly. Natural enemies may be used more effectively if such studies are expanded.

The effective predator groups may change within a season and over the years in a given system, and nearby alternative host plants may provide critical refugia both for predators and for CPB susceptible to insecticides and genetically engineered crop plants (Mena-Covarrubias *et al.*, 1996). Nevertheless, little is known about the biology and, especially, the interactions among predators of CPB.

Many natural enemies of CPB are tropical and do not prosper in temperate regions where CPB are pests, and natural enemies by themselves generally do not reduce damage below economic thresholds. Perhaps because of these factors and the manipulations required to circumvent them, the effectiveness of their use has been viewed with some pessimism (Hough-Goldstein *et al.*, 1993), even as the need for decreased dependence on synthetic insecticides has increased. Only a few studies of arthropod natural predators have been published recently, and work on many of the genera studied earlier has not been pursued. A few genera, including *Myiopharus*, *Edovum*, *Podisus*, *Perillus* and *Coleomegilla*, have been studied recently; the principal results of these are described below.

4.1. Myiopharus spp *(Diptera: Tachinidae)*

The tachinid fly genus *Myiopharus* has been identified as the principal and only known indigenous parasitoid of the CPB in the United States (Ferro, 1994; Lopez *et al.*, 1997a,b). The biology of *Myiopharus*, particularly *M.* (= *Doryphorophaga) doryphorae* and *M.* (=*Adoryphorophaga) aberrans* (Gollands *et al.*, 1991; Lopez *et al.*, 1997a,b), has been fairly well studied in the northeastern United States. Here, differences in the biology of the two species and relationship with the host may be exploitable in control strategies (Gollands *et al.*, 1991; Lopez *et al.*, 1997b). However, less is known about the biology of this genus in other areas of the United States or for other parts of the world, where different climatic conditions, particularly temperature (Lopez *et al.*, 1997a), may affect its phenology and synchronization with its host.

The solitary parasitoid is larviposited as a first instar into larval CPB. It remains dormant during host larval development and develops in the prepupa or pupa after the host has burrowed into the soil. *M. aberrans* may also larviposit into pre- or post-diapause adults (Lopez *et al.*, 1997b), although it is not clear whether the parasitoid larvae develop under these circumstances. Superparasitism is rare in nature, even when suitable hosts are limited, because of discrimination against already parasitized hosts by ovipositing females (Lopez *et al.*, 1995) and guarding of hosts after oviposition (Lopez *et al.*, 1997b). The parasitoid emerges as an adult at the same time that the CPB adults are emerging from the soil (Lopez *et al.*, 1992). Both *Myiopharus* species overwinter in the host and can be important in limiting survival of overwintering CPB adults; they also cause high mortality in late season (second generation) CPB (Gollands *et al.*, 1991; Lopez *et al.*, 1992; Ferro, 1994).

The basic biology of *Myiopharus* spp. and their interactions with each other, their host, and other biocontrol agents require considerably more study if the potential of the organism as a biocontrol agent is to be realized. Lopez and Ferro (1995) have explored the compatibility of *Myiopharus* spp. and *Btt.* Larvae oviposited into *Btt*-intoxicated hosts could not develop if the host died. If the host survived *Btt* treatment, however, the parasitoid usually developed successfully; low, slightly intoxicating doses of *Btt* may even enhance exposure of the host to the tachinid (Lopez and Ferro, 1995). Furthermore, *Myiopharus* tended not to oviposit into severely intoxicated, unresponsive larvae. Thus, careful timing of *Btt* application relative to the active period of the fly could enhance compatibility of these control measures.

Interaction of *Myiopharus* spp. with other internal parasites, such as nematodes, has scarcely been examined. Application of nematodes, especially more pathogenic exotic nematodes, for control of CPB (Berry *et al.*, 1997) may affect population and parasitization dynamics of the tachinid. Steinernematids can attack tachinids (Kaya, 1984), but the ability of heterorhabditids to do so apparently has not been investigated. In view of the promise that both the nematodes and the tachinid hold for biological control of CPB (Ferro, 1994; Berry *et al.*, 1997), information on these interrelationships will be extremely valuable in developing new strategies for biological control and integrated pest management of CPB.

4.2. Edovum puttleri (Hymenoptera: Eulophidae)

The biology and behaviour of *Edovum puttleri* Grissell, an egg parasitoid of CPB, was the subject of more than 30 studies in the decade before 1994. In his review of that literature, Ferro (1994) was very pessimistic as to the usefulness of this parasitoid in control of Colorado potato beetle. Apparently *E. puttleri* is more effective against CPB in eggplant than in potato, for reasons that remain unclear (Vasquez *et al.*, 1997). We found only two other studies of this parasitoid (Hamilton and Lashomb, 1996; Hamilton *et al.*, 1996) published since 1994. In these studies, Hamilton and coworkers were particularly interested in the compatibility of this parasitoid with insecticides (esfenvalerate, oxamyl, and rotenone) used to control CPB in eggplant fields. All of these compounds were detrimental to *E. puttleri*, whether applied in the food, through contact with leaves, or on CPB egg masses parasitized by the wasp. The authors recommended use of alternative insecticides where *E. puttleri* is to be released and cited unpublished data that *Bt* had little effect and might be suitable for such use (Hamilton *et al.*, 1996). Growers releasing *E. puttleri* reduced insecticide use and lowered costs while increasing both yield and profits relative to those using conventional pesticide programs (Hamilton and Lashomb, 1996).

4.3. Podisus maculiventris and Perillus bioculatus (Hemiptera: Pentatomidae)

The spined soldier bug (*Podisus maculiventris)* and the two-spotted stink bug (*Perillus bioculatus*) are generalist predators that often prey on CPB. The host range of *P. maculiventris* is broader and includes beneficials such as *Coleomegilla* spp. (Hough-Goldstein *et al.*, 1996). *Podisus maculiventris* can consume as many CPB as can *P. bioculatus* (more in some experimental circumstances); they also remain in plots longer (Hough-Goldstein and McPherson, 1996). The data of Hough-Goldstein and McPherson (1996) corroborated earlier work indicating that *P. bioculatus* prefers earlier life stages of CPB than does *P. maculiventris*. Effectiveness of one or the other in an augmentative release therefore would be affected by the predominant stage of CPB in the field, and the two may be used at different times in the season for maximum predatory effectiveness.

Podisus maculiventris shows promise for control of CPB in Europe (Stamopolous and Chloridis, 1994; Colazza *et al.*, 1995) as well as in its native North America, even though CPB does not seem to be an optimal prey. Life span, body weight, and egg production of *P. maculiventris* were significantly lower when they were fed CPB reared on potato or tomato than when they were fed mealworms (Valicente and O'Neil, 1995); the CPB host plant had no effect on either parameter. Bugs given prey only occasionally had a longer life span and laid far fewer eggs, with delayed first oviposition, than when prey was available daily. Having five, rather than one, CPB available per feeding shortened lifespan but decreased age at first oviposition. Trends were similar when prey were fed on tomato or potato. Although CPB may be a relatively poor host, availability of alternative prey in the field may maintain *P. maculiventris* reproductive potential and populations while contributing to CPB predation.

Perillus bioculatus also has shown promise for control of CPB in both Europe and North America (Hough-Goldstein and Whalen, 1993; Cloutier and Bauduin, 1995; Hough-Goldstein *et al.*, 1996; Saint-Cyr and Cloutier, 1996). Although commonly considered a generalist, this predator seems to prefer CPB; acceptability of alternative prey is influenced by maternal diet (Saint-Cyr and Cloutier, 1996).

Hough-Goldstein and co-workers (Hough-Goldstein and Whalen, 1993; Hough-Goldstein *et al.*, 1996) obtained marked reduction in CPB egg masses and small and large larvae after release of *P. bioculatus* as either nymphs or eggs. The eggs and small nymphs, however, were susceptible to predation and would benefit from protective shelters until they became larger (Hough-Goldstein *et al.*, 1996). Survival of small nymphs also may be enhanced by releasing them with a few larger nymphs; communal feeding in this situation appears to benefit survival of the smaller stadia (Cloutier, 1997).

Release as early instar nymphs simplifies storage and transportation of *P. bioculatus* (Hough-Goldstein *et al.*, 1996). On the other hand, early stages are less likely to disperse from the release point (Lachance and Cloutier, 1997). Dispersal is enhanced by lower predator density, higher temperature (above a 19 °C threshold), and predator hunger (Lachance and Cloutier, 1997), and by appropriate placement of pheromone sources in the field, particularly in the presence of prey (Sant'Ana *et al.*, 1997).

Perillus bioculatus and *Btt* together are quite effective, even at a reduced frequency of *Btt* application (Hough-Goldstein and Whalen, 1993). Certain insecticides also are compatible with release of *P. bioculatus* (Hough-Goldstein and Whalen, 1993). Mass rearing techniques are needed if this predator is to reach its potential in biological control (Hough-

Goldstein and Whalen, 1993; Lachance and Cloutier, 1997). Colazza *et al.* (1995) have developed techniques that could be used in mass rearing of *P. bioculatus* for augmentative release.

4.4. Coleomegilla maculata (Coleoptera:Coccinellidae)

Coleomegilla maculata De Geer, the 12-spotted ladybird beetle, is a generalist predator that consumes soft-bodied insects and eggs in both its larval and adult stages. CPB eggs and early instars are frequent prey; older larvae and adult *C. maculata* can attack third- and fourth-instar CPB as well (Giroux *et al.*, 1995). Even adult predators seem to prefer smaller prey; when small size is the result of poor adaptation of the prey to its host plant, differential selection by predators for prey size could affect the genetic mechanisms driving host plant adaptation in the pest (Lu *et al.*, 1996).

A diet limited to CPB eggs can slow development and increase mortality of *C. maculata*, especially in early instars and in nonadapted populations; it also can increase the preoviposition period, but fecundity is not reduced (Munyaneza and Obrycki, 1997a). If alternative food, such as pollen and pea aphids, is available to early instars, as would be likely in the field, such effects need not preclude the use of *C. maculata* against CPB (Munyaneza and Obrycki, 1998). Prey consumption increases with predator age and, in all stadia, with temperature from 10 to 30 °C (Giroux *et al.*, 1995); search efficiency is inversely related to prey density (Munyaneza and Obrycki, 1997b).

Because *C. maculata* would be exposed to any synthetic pesticide used against CPB in the field, questions have been raised about the relative susceptibility of predator and prey to insecticides and development of resistance in the predator. Coll *et al.* (1994) found that adult *C. maculata* were quite mobile--much more so than CPB--and gene flow among populations was high. Estimated gene flow between treated and untreated fields was not high enough, however, to slow development of resistance in *C. maculata* in response to selection pressure in treated fields, enhancing its usefulness in integrated pest management programs. Because of the high mobility of the adults and the effective predation by fourth instars, Giroux *et al.* (1995) suggested that inundative release of third instars would maximize effectiveness by maximizing predation by fourth instars while minimizing loss of predators through adult flight. Such a practice would select for resistance in *C. maculata* released in insecticide-treated fields and could enhance its subsequent dispersal through the area by the surviving adults. On the other hand, if the CPB population is susceptible and no alternative food is available, the larval predators likely would not survive to reproduce or disperse.

As a coleopteran, *C. maculata* may be susceptible to *Btt* and therefore incompatible with it in IPM programs. Giroux *et al.* (1994b), however, found that *Btt* did not increase mortality to *C. maculata* larvae exposed to it in a pollen diet, although developmental time was increased by exposure. This complemented their earlier work (Giroux *et al.*, 1994a) showing that *Btt* in a pollen diet did not increase mortality in adults, although it did diminish their predation rate for a time when applied to CPB eggs.

5. Conclusion

As resistance of CPB to chemical pesticides increases and spreads, development of strategies leading to implementation of effective and economical biological control becomes increasingly urgent. In developing such strategies, several factors must be taken into account, including

– interactions among biocontrol organisms used simultaneously, such as tachinid parasitoids and entomopathogenic nematodes, predators that compete for prey or prey on each other, and *Btt* used in conjunction with coleopteran predators

– potential of cultural practices, such as mulching, as enhancers of natural enemy habitat

– natural variation in susceptibility of insects to biocontrol (for example, Robertson *et al.*, 1995 and discussion by Weisz *et al.*, 1994).

– effects of resistant strains of plants and different mechanisms of plant resistance on arthropod natural enemies and enemy/pest interactions (Barbour *et al.*, 1997).

In addition to such biological factors, economical and efficient methods to produce enough biocontrol organisms for use on a wide scale are not available for most promising candidates. Clearly, biological control of CPB remains a complex and challenging endeavour requiring extensive, intensive, and immediate attention.

Acknowledgments

We thank Christine Armer for helpful comments on the manuscript.

References

Alford, R., 1994. Deployment strategies for antifeedants in management of Colorado potato beetle. In: Zehnder, G.W., Powelson, M.L., Jansson, R.K. and Raman, K.V. (eds) Advances in Potato Pest Biology and Management. American Phytopathological Society Press, St. Paul, Minnesota, pp 342-357.

Altre, J.A., Grafius, E.J. and Whalon, M.E., 1996. Feeding behavior of cryIIIA-resistant and susceptible Colorado potato beetle (Coleoptera: Chrysomelidae) larvae on *Bacillus thuringiensis tenebrionis*-transgenic CryIIIA-treated and untreated potato foliage. J. econ. Ent. 89: 311-317.

Arpaia, S. and Ricchiuto, B., 1993. Effects of *Bacillus thuringiensis* toxin extracts on feeding behavior and development of Colorado potato beetle (Coleoptera: Chrysomelidae) larvae. Envir. Ent. 22: 334-338.

Arpaia, S., Chiriatti, K. and Giorio, G, 1998. Predicting the adaptation of Colorado potato beetle (Coleoptera: Chrysomelidae) to transgenic eggplants expressing CryIII toxin: the role of gene dominance, migration, and fitness costs. J. econ. Ent. 91:21-29.

Arpaia, S., Gould, F. and Kennedy, G.G., 1997a. Potential impact of *Coleomegilla maculata* DeGeer (Coleoptera: Coccinellidae) predation on adaptation of *Leptinotarsa decemlineata* Say (Coleoptera:Chrysomelidae) to Bt-transgenic potatoes. Entomologia exp. appl. 82: 91-100.

Arpaia, S., Mennella, G., Onofaro, V., Perri, E.F. and Rotino, G.L., 1997b. Production of transgenic eggplant (*Solanum melongena* L.) resistant to Colorado potato beetle (*Leptinotarsa decemlineata* Say). Theor. appl. Gen. 95: 329-334.

Barbour, J.D., Farrar, R.R., Jr. and Kennedy, G.G., 1997. Populations of predaceous natural enemies developing on insect-resistant and susceptible tomato in North Carolina. Biol. Control 9: 173-184.

Berry, R.E., Liu, J. and Reed, G., 1997. Comparison of endemic and exotic entomopathogenic nematode species for control of Colorado potato beetle (Coleoptera: Chrysomelidae). J. econ. Ent. 90: 1528-1533.

Bradley, D., Harkey, M.A., Kim, M.-K., Biever, K.D. and Bauer, L.S., 1995. The insecticidal CryIB crystal protein of *Bacillus thuringiensis* ssp. *thuringiensis* has dual specificity to coleopteran and lepidopteran larvae. J. Invert. Path. 65: 162-173.

Brust, G.E., 1994. Natural enemies in straw-mulch reduce Colorado potato beetle populations and damage in potato. Biol. Control 4(2): 163-169.

Brust, G.E., 1996. Interaction of mulch and *Bacillus thuringiensis* subsp. *tenebrionis* on Colorado potato beetle (Coleoptera:Chrysomelidae) populations and damage in potato. J. econ. Ent. 89: 467-474

Bystrak, P.G., Sanborn, S. and G. Zehnder, G., 1994. Methods for optimizing field performance of *Bacillus thuringiensis* endotoxins against Colorado potato beetle. In: Zehnder, G.W., Powelson, M.L., Jansson, R.K. and Raman, K.V. (eds) Advances in Potato Pest Biology and Management. American Phytopathological Society Press, St. Paul, Minnesota, pp 386-402.

Carroll, J., Convents, C., Van Damme, J., Boets, A., Van Rie, J. and Ellar, D.J., 1997. Intramolecular proteolytic cleavage of *Bacillus thuringiensis* Cry3A δ–endotoxin may facilitate its coleopteran toxicity. J. Invert. Path. 70: 41-49.

Chen, Q., Jelenkovic, G., Chin, C.K., Billings, S., Eberhardt, J., Goffreda, J.C. and Day, P., 1995. Transfer and transcriptional expression of coleopteran cryIIIB endotoxin gene of *Bacillus thuringiensis* in eggplant. J. Am. Soc. hort. Sci. 120: 921-927.

Clark, T. B., 1982. Spiroplasmas: diversity of arthropod reservoirs and host-parasite relationships. Science 217: 57-59.

Cloutier, C., 1997. Facilitated predation through interaction between life stages in the stinkbug predator *Perillus bioculatus* (Hemiptera: Pentatomidae). J. Insect Behav. 10: 581-598.

Cloutier, C. and Bauduin, F., 1995. Biological control of the Colorado potato beetle *Leptinotarsa decemlineata* (Coleoptera: Chrysomelidae) in Quebec by augmentative releases of the two-spotted stinkbug *Perillus bioculatus* (Hemiptera: Pentatomidae). Can. Ent. 127: 195-212.

Colazza, S., Czepak, C. and Isidorao, N., 1995. Introduction of two American predators to Italy, *Podisus maculiventris* (Say) and *P. connexivus* Bergroth, for the biocontrol of nonindigenous pests (Heteroptera: Pentatomidae). Redia 78: 379-388. (In Italian; English abstract).

Coll, M., de Mendoza, G. and Roderick, G.K., 1994. Population structure of a predatory beetle: the importance of gene flow for intertrophic level interactions. Heredity 73(3): 228-236.

Couteaudier, Y., Viaud, M. and Riba, G., 1996. Genetic nature, stability, and improved virulence of hybrids from protoplast fusion in *Beauveria*. Microb. Ecol. 32: 1-10.

Fargues, J., Delmas, J.C. and Lebrun, R.A., 1994. Leaf consumption by larvae of the Colorado potato beetle (Coleoptera: Chrysomelidae) infected with entomopathogen, *Beauveria bassiana*. J. econ. Ent. 87: 67-71.

Ferro, D.N., 1993. Potential for resistance to *Bacillus thuringiensis*: Colorado potato beetle (Coleoptera: Chrysomelidae)--a model system. Am. Entomologist 38: 38-44.

Ferro, D.N., 1994. Biological control of the Colorado potato beetle. In: Zehnder, G.W., Powelson, M.L., Jansson, R.K. and Raman, K.V. (eds) Advances in Potato Pest Biology and Management. American Phytopathological Society Press, St. Paul, Minnesota, pp 357-375.

Ferro, D.N., Slocombe, A.C., Mercier, C.T. and Christ, T., 1997. Colorado potato beetle (Coleoptera: Chrysomelidae): residual mortality and artificial weathering of formulated *Bacillus thuringiensis* subsp. *tenebrionis*. J. econ. Ent. 90:574-578.

Foster, R. and Flood, B. (eds.), 1995. Vegetable Insect Management (with emphasis on the Midwest). Meister Publishing Company, Willoughby, Ohio. 206 pp.

Gaugler, R., Lewis, E. and Stuart, R. J., 1997. Ecology in the service of biological control: the case of entomopathogenic nematodes. Oecologia 109: 483-489.

Ghidiu, G.M. and Zehnder, G.W., 1993. Timing of the initial spray application of *Bacillus thuringiensis* for control of the Colorado potato beetle (Coleoptera: Chrysomelidae) in potatoes. Biol. Control 3: 348-352.

Ghidiu, G.M., Collins, D.E. and Krifman, G.W., 1996. Laboratory and field studies of *Bacillus thuringiensis* subsp. *tenebrionis* as a feeding deterrent to Colorado potato beetle (Coleoptera: Chrysomelidae). J. Agric. Ent. 13: 3493-57.

Ghidiu, G.M., Hamilton, G.C. and Krifman, G.W., 1994. *Bacillus thuringiensis* ssp. *tenebrionis* sprays affect hatching of Colorado potato beetle (Coleoptera: Chrysomelidae) eggs and subsequent larval dispersal. J. econ. Ent. 87: 1298-3101.

Giroux, S., Coderre, D., Vincent, C. and Coté, J.C., 1994a. Effects of *Bacillus thuringiensis* var. *san diego* on predation, effectiveness, development, and mortality of *Coleomegilla maculata lengi* (Col: Coccinellidae) larvae. Entomophaga 39: 61-69.

Giroux, S., Côté, J.-C., Vincent, C., Martel, P. and Coderre, D., 1994b. Bacteriological insecticide M-One® effects on predation efficiency and mortality of adult *Coleomegilla maculata lengi* (Coleoptera: Coccinellidae). J. econ. Ent. 87: 39-43

Giroux, S., Duchesne, R.M. and Coderre, D., 1995. Predation of *Leptinotarsa decemlineata* (Coleoptera: Chrysomelidae) by *Coleomegilla maculata* (Coleoptera: Coccinellidae): comparative effectiveness of predator developmental stages and effect of temperature. Envir. Ent. 24: 748-754.

Gollands, B., Tauber, M..J. and Tauber, C.A., 1991. Seasonal cycles of *Myiopharus aberrans* and *M. doryphorae* (Diptera:Tachinidae) parasitizing Colorado potato beetle in upstate New York. Biol. Control 1: 153-163.

Gould, F., Follett, P., Nault, B. and Kennedy, G.G., 1994. Resistance management strategies for transgenic potato plants. In: Zehnder, G.W., Powelson, M.L., Jansson, R.K. and Raman, K.V. (eds) Advances in Potato Pest Biology and Management. American Phytopathological Society Press, St. Paul, Minnesota, pp. 255-277.

Grafius, E., 1997. Economic impact of insecticide resistance in the Colorado potato beetle (Coleoptera:Chrysomelidae) on the Michigan potato industry. J. econ. Ent. 90: 1144-1151.

Hackett, K.J., Henegar, R.B., Whitcomb, R.F., Lynn, D.E., Konai, M., Schroder, R.F. Gasparich, G.E., Vaught, J.L. and Cantelo, W.W., 1992. Distribution and biological control significance of Colorado potato beetle spiroplasmas in North America. Biol. Control 2(3): 218-225.

Hackett, K.J., Lipa, J.J., Gasparich, G.E., Lynn, D.E., Konai, M., Camp, M. and Whitcom, R.F., 1997. The spiroplasma motility inhibition test, a new method for determining intraspecific variation among Colorado potato beetle spiroplasmas. Int. J. syst. Bact. 47: 33-37.

Hackett, K.J., Whitcomb, R.F., Clark, T.B., Henegar, R.B., Lynn, D.E., Wagner, A.G., Tully, J. G., Gasparich, G.E., Rose, D.L. and Carle, P., 1996. *Spiroplasma leptinotarsae* sp. nov., a mollicute uniquely adapted to its host, the Colorado potato beetle, *Leptinotarsa decemlineata* (Coleoptera:Chrysomelidae). Int. J. syst. Bact. 46: 906-911.

Hamilton, G.C. and Lashomb, J., 1996. Comparison of conventional and biological control intensive pest management programs on eggplant in New Jersey. Fla. Ent. 79: 488-496.

Hamilton, G.C., Lashomb, J.H. and Patt, J.M., 1996. Impact of selected insecticides used in eggplant production on *Edovum puttleri* Grissell (Hymenoptera: Eulophidae). J. ent. Sci. 31: 442-439.

Heimpel, G.E. and Hough-Goldstein, J.A., 1992. A survey of arthropod predators of *Leptinotarsa decemlineata* (Say) in Delaware potato fields. J. agric. Ent. 9: 137-142.

Hilbeck, A. and Kennedy, G.G., 1996. Predators feeding on the Colorado potato beetle in insecticide-free plots and insecticide-treated commercial potato fields in eastern North Carolina. Biol. Control 6: 273-283.

Hilbeck, A., Baumgartner, M., Fried, R.M. and Rigler, F., 1998. Effects of transgenic *Bacillus thuringiensis* corn-fed prey on mortality and development time of immature *Chrysoperla carnea* (Neuroptera: Chrysopidae). Envir. Ent. 27: 480-487.

Hilbeck, A., Eckel, C. and Kennedy, G.G., 1997. Predation on Colorado potato beetle eggs by generalist predators in research and commercial potato plantings. Biol. Control 8: 191-196.

Hough-Goldstein, J. and McPherson, D., 1996. Comparison of *Perillus bioculatus* and *Podisus maculiventris* (Hemiptera: Pentatomidae) as potential control agents of Colorado potato beetle (Coleoptera: Chrysomelidae). J. econ. Ent. 89: 1116-1123.

Hough-Goldstein, J. and Whalen, J., 1993. Inundative release of predatory stink bugs for control of Colorado potato beetle. Biol. Control. 3(4): 343-347.

Hough-Goldstein, J.A., Heimpel, G.E., Bechmann, H.E. and Mason, C.E., 1993. Arthropod natural enemies of the Colorado potato beetle. Crop Protection 12(5): 324-334.

Hough-Goldstein, J., Janis, J.A. and Ellers, C.D., 1996. Release methods of *Perillus bioculatus* (F.), a predator of the Colorado potato beetle. Biol. Control 6: 114-122.

Hough-Goldstein, J., Tisler, A.M., Zehnder, G. W. and Uyeda, K. A., 1991. Colorado potato beetle (Coleoptera: Chrysomelidae) consumption of foliage treated with *Bacillus thuringiensis* var. *san diego* and various feeding stimulants. J. econ. Ent. 84: 87-93.

Hoy, C.W. and Head, H.G., 1995 Correlation between behavioral and physiological responses to transgenic potatoes containing *Bacillus thuringiensis* delta-endotoxin in *Leptinotarsa decemlineata* (Coleoptera: Chrysomelidae). J. econ. Ent. 99: 480-486.

Innacone, R., Grieco, P.D. and Cellini, F., 1997. Specific sequence modifications of a cry3B endotoxin result in high levels of expression and insect resistance. Plant mol. Biol. 34:485-496.

Jelenkovoc, G., Billings, S., Chen, Q., Lashomb, J., Hamilton, G. and Ghidiu, G., 1998. Transformation of eggplant with synthetic cryIIIA gene produces a high level of resistance to the Colorado potato beetle. J. Amer. Soc. hort. Sci. 123: 19-25.

Kaya, H. K., 1984. Effect of the entomogenous nematode *Neoplectana carpocapsae* on the tachinid parasite *Compsilura concinnata* (Diptera: Tachinidae). J. Nematol. 16: 9-13.

Kaya, H. K. and Gaugler, R., 1993. Entomopathogenic nematodes. A. Rev. Ent. 38: 181-206.

Konai, M., Hackett, K.J., Williamson, D.L., Lipa, J.J., Pollack, J.D., Gasparich, G.E., Clark, E.Z., Vacek, D.C. and Whitcomb, R.F., 1996. Improved cultivation systems for isolation of the Colorado potato beetle spiroplasma. Appl. envir. Microbiol. 62: 3453-3458.

Lachance, S. and Cloutier C., 1997. Factors affecting dispersal of *Perillus bioculatus* (Hemiptera: Pentatomidae), a predator of the Colorado potato beetle (Coleoptera: Chrysomelidae). Envir. Ent. 26: 946-954.

Lambert, B., Hofte, H., Annys, K., Jansens, S., Soetart, P. and Peferoen, M., 1992. Novel *Bacillus thuringiensis* insecticidal crystal protein with a silent activity against coleopteran larvae. Appl. envir. Microbiol. 58: 2536-2542.

Lee, R.E., Jr., Costanzo, J. P., Kaufman, P. E., Lee, M.R. and Wyman, J.A., 1994. Ice-nucleating active bacteria reduce the cold-hardiness of the freeze-intolerant Colorado potato beetle (Coleoptera: Chrysomelidae). J. econ. Ent. 87: 377-381.

Lee, R.E., Jr., Steigerwald, K.A., Wyman, J.A., Costanzo, J. P. and Lee, M.R., 1996. Anatomic site of application of ice-nucleating active bacteria affects supercooling in the Colorado potato beetle (Coleoptera: Chrysomelidae). Envir. Ent. 25: 465-469.

Long, D.W., Drummond, F.A. and Groden, E., 1998. Susceptibility of Colorado potato beetle (*Leptinotarsa decemlineata*) eggs to *Beauveria bassiana*. J. Invert. Path. 71: 192-183.

Lopez, E.R. and Ferro, D.N., 1995. Larviposition response of *Myiopharus doryphorae* (Diptera:Tachinidae) to Colorado potato beetle (Coleoptera:Chrysomelidae) larvae treated with lethal and sublethal doses of *Bacillus thuringiensis* Berliner subsp. *tenebrionis*. J. econ. Ent. 88: 870-874.

Lopez, R., Ferro, D.N. and Van Driesche, R. G., 1992. Overwintering biology of *Myiopharus aberrans* and *Myiopharus doryphorae* (Dipt.: Tachinidae) larval parasitoids of the Colorado potato beetle (Col.: Chrysomelidae). Entomophaga 37: 311-315.

Lopez, R., Ferro, D.N. and Van Driesche, R.G., 1995. Two tachinid species discriminate between parasitized and non-parasitized hosts. Entomologia exp. appl. 74: 37-45.

Lopez, R., Ferro, D.N. and Elkinton, J.S., 1997a. Temperature-dependent development rate of *Myiopharus doryphorae* (Diptera: Tachinidae) within its host, the Colorado potato beetle (Coleoptera: Chrysomelidae). Envir. Ent. 26: 655-660.

Lopez, E.R., Roth, L.C., Ferro, D.N., Hosmer, D. and Mafra-Neto, A., 1997b. Behavioral ecology of *Myiopharus doryphorae* (Riley) and *M. aberrans* (Townsend), tachinid parasitoids of the Colorado potato beetle. J. Insect Behav. 10: 49-78.

Lu, W.H., Kennedy, G. and Gould, F., 1996. Differential predation by *Coleomegilla maculata* on Colorado potato beetle strains that vary in growth on tomato. Entomologia exp. appl. 81:7-14.

Mahr, S.,. Wyman, J., Radcliffe, E., Hoy, C. and Ragsdale, D., 1995. Potatoes. In: R. Foster and B. Flood (eds), Vegetable Insect Management (with emphasis on the Midwest). Meister Publishing Company, Willoughby, Ohio, Chap. 5. 206 pp.

Martínez-Ramírez, A.C. and Real, M.D., 1996. Proteolytic processing of *Bacillus thuringiensis* CryIIIA toxin and specific binding to brush-border membrane vesicles of *Leptinotarsa decemlineata* (Colorado potato beetle). Pestic. Biochem. Physiol. 54: 115-122.

Mena-Covarrubias, J., Drummond, F. A. and Haynes, D. L., 1996. Population dynamics of the Colorado potato beetle (Coleoptera: Chrysomelidae) on horsenettle in Michigan. Envir. Ent. 25: 68-77.

Midgarden, D., Fleischer, S.J., Weisz, R. and Smilowitz, Z., 1997. Site-specific integrated pest management impact on development of esfenvalerate resistance in Colorado potato beetle (Coleoptera: Chrysomelidae) and on densities of natural enemies. J. econ. Ent. 90: 855-867.

Munster, C.L., Skaggs, R.W. and Pemmireddy, V.R., 1996. Effect of water table management on the fate of the pesticide aldicarb. Trans. Amer.Soc. agric. Engin. 39: 55-66.

Munyaneza, J. and Obrycki, J.J., 1997a. Reproductive response of *Coleomegilla maculata* (Coleoptera: Coccinellidae) to Colorado potato beetle (Coleoptera: Chrysomelidae) eggs. Envir. Ent. 26: 1270-1275.

Munyaneza, J. and Obrycki, J.J., 1997b. Functional response of *Coleomegilla maculata* (Coleoptera: Coccinellidae) to Colorado potato beetle eggs (Coleoptera: Chrysomelidae). Biol. Control 8: 215-224.

Munyaneza, J. and Obrycki, J.J. 1998. Development of three populations of *Coleomegilla maculata* (Coleoptera: Coccinellidae) feeding on eggs of Colorado potato beetle (Coleoptera: Chrysomelidae). Envir. Ent. 27: 117-122.

Murray, K.D., Alford, A.R., Groden, E., Drummond, F.A., Storch, R.H., Bentley, M.D. and Sugathapala, P.M., 1993. Interactive effects of an antifeedant used with *Bacillus thuringiensis* var. *san diego* delta-endotoxin on Colorado potato beetle (Coleoptera: Chrysomelidae). J. econ. Ent. 86: 1793-1801.

Nault, B.A. and Kennedy, G.G., 1996. Evaluation of Colorado potato beetle (Coleoptera: Chrysomelidae) defoliation with concomitant European corn borer (Lepidoptera: Pyralidae) damage on potato yield. J. econ. Ent. 89: 475-480.

Nickle, W.R., Connick, W.J., Jr. and Cantelo, W.W., 1994. Effects of pesta-pelletized *Steinernema carpocapsae* (All) on Western corn rootworms and Colorado potato beetles. J. Nematol. 26: 249-250.

Novillo, C., Castañero, P. and Ortego, F., 1997. Characterization and distribution of chymotrypsin-like and other digestive proteases in Colorado potato beetle larvae. Arch. Insect Biochem. Physiol. 36: 181-201.

Perlak, F.J., Stone, T.B., Muskopf, Y.M., Petersen, L.J., Parker, G.B., McPherson, S.A., Wyman, J., Love, S., Reed, G., Biever, S.D. and Fischoff, D.A., 1993. Genetically improved potatoes: protection from damage by Colorado potato beetles. Plant mol. Biol. 22: 313-321.

Poinar, G. O., Jr., 1991. Genetic engineering of nematodes for pest control. In: K. Maramoresch (ed.) Biotechnology for Biological Control of Pests and Vectors. CRC Press, Boca Raton, FL, pp 77-93.

Poprawski, T.J., Carruthers, R.I., Speese, J. III, Vacek, D.C. and Wendel, L.E., 1997. Early season applications of the fungus *Beauveria bassiana* and introduction of the hemipteran predator *Perillus bioculatus* for control of Colorado potato beetle. Biol. Control 10: 48-57.

Rahardja, U. and Whalon, M. E. 1995. Inheritance of resistance to *Bacillus thuringiensis* subsp *tenebrionis* cryIIA delta-endotoxin in Colorado potato beetle (Coleoptera: Chrysomelidae). J. econ. Ent. 88: 21-26.

Robertson, J.L., Preisler, H.K., Ng, S.S., Hickle, L.A. and Gelernter, W.D., 1995. Natural variation: a complicating factor in bioassays with chemical and microbial pesticides. J. econ. Ent. 88: 1-10.

Roush, R.T., 1994. Managing pests and their resistance to *Bacillus thuringiensis*: can transgenic crops be better than sprays? Biocontrol. Sci. Technol. 4:501-506.

Saint-Cyr, J.F. and Cloutier C., 1996. Prey preference by the stinkbug *Perilllus bioculatus*, a predator of the Colorado potato beetle. Biol. Control 7: 251-258.

Sant'Ana J., Bruni, R., Abdul Baki, A.A. and Aldrich, J.R., 1997. Pheromone-induced movement of nymphs of the predator, *Podisus maculiventris* (Heteroptera: Pentatomidae). Biol. Control 10: 123-128.

Saringer, G., Fodor, A., Nadasy, M., Lucskai, A. and Georgis, R., 1996. Possibilities of biological control using entomopathogenic nematodes against *Leptinotarsa decemlineata* L. and *Athalia rosae* L. larvae. Mededelingen Faculteit Landbouwkundige en Toegepaste Biologische Wetenschappen Universiteit Gent 61(3B): 961-966.

Schrod, J., Basedow, Th. and Langenbruch, G.A., 1996. Studies on bionomy and biological control of the Colorado potato beetle *Leptinotarsa decemlineata* Say, (Col.: Chrysomelidae) in Hesse (FRG) (in German). J. Appl. Ent. 120: 619-626.

Smart, G.C., Jr. 1995. Entomopathogenic nematodes for the biological control of insects. J. Nematol. 27: 529-533.

Stamopolous, D.C. and Chloridis, A., 1994. Predation rates, survivorship and development of *Podisus maculiventris* (Het.: Pentatomidae) on larvae of *Leptinotarsa decemlineata* (Col.: Chrysomelidae) and *Pieris brassica* (Lep.: Pieridae) under field conditions. Entomophaga 39: 3-9.

Stoner, K.A., 1993. Effects of straw and leaf mulches and trickle irrigation on the abundance of Colorado potato beetles (Coleoptera: Chrysomelidae) on potato in Connecticut. J. ent. Sci. 28: 393-403.

Stoner, K..A., 1997. Influence of mulches on the colonization by adults and survival of larvae of the Colorado potato beetle (Coleoptera: Chrysomelidae) in eggplant. J. ent. Sci. 32: 7-16.

Stoner, K.A., Ferrandino, F.J., Gent, M.P.N., Elmer, W.H. and Lamondia, J.A., 1996. Effects of straw mulch, spent mushroom compost, and fumigation on the density of Colorado potato beetle (Coleoptera: Chrysomelidae) in potatoes. J. econ. Ent. 89: 1267-1280.

Tanada, Y. and Kaya, H. K., 1993. Other bacterial infections. In: Tanada, Y. and Kaya, H. K. (eds) Insect Pathology. Academic Press, New York. pp 147-170.

Thurston, G.S., Yule, W.N. and Dunphy, G.B., 1994. Explanations for the low susceptibility of *Leptinotarsa decemlineata* to *Steinernema carpocapsae*. Biol. Control 4: 53-58.

Todorova, S.I., Cote, J.C. and Coderre, D., 1996. Evaluation of the effects of two *Beauveria bassiana* (Balsamo) Vuillemin strains on the development of *Coleomegilla maculata lengi* Timberlake (Col: Coccinellidae). J. appl. Ent. 120: 159-163.

Todorova, S.I., Cote, J.C., Martel, P. and Coderre, D., 1994. Heterogeneity of 2 *Beauveria bassiana* strains revealed by biochemical tests, protein profiles and bioassays on *Leptinotarsa decemlineata* (Col: Chrysomelidae) and *Coleomegilla maculata* Lengi (Col., Coccinellidae) larvae. Entomophaga 39: 159-169.

Trisyono, A. and Whalon, M.E., 1997. Fitness costs of resistance to *Bacillus thuringiensis* in Colorado potato beetle (Coleoptera: Chrysomelidae). J. econ. Ent. 90: 267-271.

Trumble, J. T., Kolodny-Hirsch, D. M. and Ting, I. P., 1993. Plant compensation for arthropod herbivory. A. Rev. Ent. 38:93-119.

Valicente, F.H. and O'Neil, R.J., 1995. Effects of host plants and feeding regimes on selected life-history characteristics of *Podisus maculiventris* (Say) (Heteroptera, Pentatomidae). Biol. Control 5: 449-461.

van der Salm, T., Bosch, D., Honee, G., Feng, L., Munsterman, E., Bakker, P., Stiekema, W.J. and Visser, B., 1994. Insect resistance of transgenic plants that express modified *Bacillus thuringiensis cryIA(b)* and *CryIC* genes: a resistance management strategy. Plant mol. Biol. 26: 51-59.

Vasquez, B., Lashomb. J.H., and Hamilton, G., 1997. Alightment of *Edovum puttleri* Grissell (Hymentopera: Eulophidae) in response to colors mimicking three hosts of the Colorado potato beetle (Coleoptera: Chrysomelidae). J. ent. Sci. 32: 386-397.

Weber, D.C. and Ferro, D.N., 1994. Colorado potato beetle: diverse life history poses challenge to management. In: Zehnder, G.W., Powelson, M. L., Jansson, R. K. and Raman, K. V. (eds) Advances in Potato Pest Biology and Management. American Phytopathological Society, St. Paul, Minn, pp 54-70.

Weisz, R., Saunders, M., Smilowitz, Z., Hua, H.W. and Christ, B., 1994. Knowledge-based reasoning in integrated resistance management--the Colorado potato beetle (Coleoptera: Chrysomelidae). J. econ. Ent. 87: 1384-1399.

Whalon, M.E. and Wierenga, J.M., 1994. *Bacillus thuringiensis* resistant Colorado potato beetle and transgenic plants: some operational and ecological implications for deployment. Biocontrol Sci. Technol. 4: 555-561.

Whalon, M.E., Miller, D.L., Hollingworth, R.M., Grafius, E.J. and Miller, J.R., 1993. Selection of a Colorado potato beetle (Coleoptera:Chrysomelidae) strain resistant to *Bacillus thuringiensis*. J. econ. Ent. 86: 226-233.

Whalon, M. E., Rahardja, U., and Verakalasa, P., 1994. Selection and management of *Bacillus thuringiensis*-resistant Colorado potato beetle. In: Zehnder, G.W., Powelson, M.L., Jansson, R.K. and Raman, K.V. (eds) Advances in Potato Pest Biology and Management. American Phytopathological Society, St. Paul, Minn, pp 309-321.

Wierenga, J.M., Norris, D.L. and Whalon, M.E., 1996. Stage-specific mortality of Colorado potato beetle (Coleoptera: Chrysomelidae) feeding on transgenic potatoes. J. econ. Ent. 89: 1047-1052.

Wouts, W. M., 1991. *Steinernema* (*Neoaplectana*) and *Heterorhabditis* species. In: W.R. Nickle (ed) Manual of Agricultural Nematology. Marcel Dekker, New York, pp 855-898.

WOLBACHIA INFECTIONS IN CHRYSOMELIDS

ROSANNA GIORDANO[1] and JAN J. JACKSON[2]
[1]*Department of Biology, University of Vermont, Burlington, VT, USA;* [2]*Pioneer Hi-Bred International, Windfall Research Center, IN, USA*

Abstract

Wolbachia bacteria are a commonly occurring infection of arthropods. We present the effects that this selfish cytoplasmic organism can have on its hosts, and focus on available information regarding *Wolbachia* infection(s) in chrysomelids. The only studied case is that involving the subspecies *Diabrotica virgifera virgifera* LeConte.

1. Introduction

Rickettsial bacteria in the genus *Wolbachia* are of interest as a result of the ways in which they manipulate their hosts reproductive abilities for the sake of furthering their own dissemination and the effects that this in turn can have on their hosts. Disruption of reproductive patterns can significantly alter the genetic structure of a population as well as affect life history parameters of the host.

Effects of *Wolbachia* infections in invertebrates vary depending on the identity of the host and strain of bacteria. *Wolbachia* have been shown to cause the conversion of genetic males into fully reproductive females in several species of isopods. *Wolbachia* infections are common in isopods and females of infected populations have been shown to be composed exclusively of genetically male individuals (Rousset *et al* ., 1992a; Martin *et al.*, 1994; reviewed by Rigaud, 1997). Bouchon *et al.*, (1998) recently tested 80 crustaceans representing 5 orders and found that all the infected species belonged to the Isopoda. Moreover, with the exception of two estuarine species, all others were terrestrial. This study, as well as a negative screen for *Wolbachia* infections in mollusks (Schilthuizen & Gittenberger, 1998) suggests that *Wolbachia* may be restricted to terrestrial environments.

It has been also demonstrated that the presence of *Wolbachia* induces parthenogenesis in several haplo/diploid species of hymenopteran wasps (Stouthamer *et al.*, 1993; Schilthuizen & Stouthamer, 1997; reviewed by Stouthamer, 1997). However, the most common phenomenon expressed in diploid insects is that of cytoplasmic incompatibility. This can occur when a male of a *Wolbachia*-infected population mates with a female that is not infected or when matings occur between individuals infected with different strains of *Wolbachia*. Cytoplasmic incompatibility typically results in the production of non viable eggs.

We could predict a degree of variation to the expression of a phenotype, as well as a higher rate of speciation for an organism which can exist in the extremely variable and isolated internal biotic environment of myriads of hosts. Indeed, it has been shown that

Advances in Chysomelidae Biology 1, pp. 185–196
Edited by M.L. Cox

levels of cytoplasmic incompatibility can vary from being very severe, resulting in 100% mortality of offspring as in *Tribolium confusum* Duval (Wade & Stevens, 1985) to complete viability as in *Drosophila mauritiana* Bigot (Giordano *et al.*, 1995) and some populations of *D. simulans* Sturtevant (Hoffman *et al.*, 1996). Multiple strains of *Wolbachia* have been identified. This has been determined by observing bidirectional incompatibility between infected populations of the same species as well as DNA sequencing of portions of several genes, such as the 16S rRNA (O'Neill *et al.*, 1992), cell division, *fts*Z (Werren *et al.*, 1995), and cell membrane, *wsp* (Zhou *et al.*, 1998) genes.

The polymerase chain reaction has rendered what was once thought to be a rare infection, into one of the more commonly recognized intracellular microorganisms infecting arthropods (for a list of infected organisms see review in Werren & O'Neill, 1997). A recent report indicated that *Wolbachia* infections may be widespread in ants (Wenseleers *et al.*, 1998), this together with a previous finding of infection in several species of termites (Bandi *et al.*, 1997) raises questions as to the possible role that *Wolbachia* may play in the structure of insect societies (Hurst, 1997; Frank, 1997). Moreover, its distribution has been shown to extend beyond arthropods to include several species of filarial nematodes (Genchi *et al.*, 1998; Bandi *et al.*, 1998).

The genera of rickettsia most closely related to *Wolbachia* include pathogenic organisms such as *Anaplasma marginale*, *Ehrlichia canis*, *E. equii* and *Cowdria ruminantium*. Together with the more distantly related *Rickettsia tsutsugamushi* and *R. prowazekii*, they are gram-negative, alpha-proteobacteria that can parasitize mammals and are vectored by arthropods. Thus far, two reports have shown that *Wolbachia* can cause disease: Juchault *et al.* (1974) observed nervous tissue necrosis after injecting *Wolbachia* harbored by *Armadillidium* in *Porcellio dilatatus* Brandt and Ratzenburg, while Min and Benzer (1997) observed muscle and nerve degeneration and early death in infected *D. melanogaster* Meigen. In both cases, disease was subsequent to *Wolbachia* proliferation in tissues. If we consider that the pathogenic effect occurred in *P. dilatatus* as a result of being exposed to a new *Wolbachia* strain, we may infer from this that the pathogenic infection in *D. melanogaster* has been recently acquired. Despite the existence of these two pathogenic strains, up to this point no reports have been made of *Wolbachia* in vertebrates.

Published laboratory results testing the effects of *Wolbachia* on host fitness in *D. simulans* from Riverside, CA indicate a slight disadvantage to the infected state; however no difference was seen in the number of eggs laid by field collected, infected and uninfected flies (Hoffmann & Turelli, 1988; Turelli & Hoffmann, 1995). It remains to be determined whether there is a disadvatage to *Wolbachia* infections in the wild. Laboratory tests on other *Drosophila* species have not shown evidence of negative effects on fitness (Giordano *et al.*, 1995; Hoffmann *et al.*, 1996; Hoffmann & Turelli, 1997). In *D. melanogaster* infection persist in the wild despite low levels of cytoplasmic incompatibility and imperfect maternal transmission, leading to the conclusion that *Wolbachia* infection may in some cases provide beneficial effects to the host (Hoffmann & Turelli, 1997). In *T. confusum* sperm from males infected with *Wolbachia* has a fertility advantage over sperm of uninfected males, even when competing with sperm from three individuals within a single female. This phenomenon would increase the speed at which a *Wolbachia* infection would spread through a population. However, infected females had fewer offspring than uninfected females (Wade & Chang, 1995).

Nothing is known regarding the molecular mechanism by which *Wolbachia* manipulates the reproductive capabilities of its host. What has been observed thus far in three *Trichogramma* species is that haploid unfertilized eggs from thelytokous females become diploid due to the failure of the two sets of chromosomes to segregate in the first mitotic anaphase (Stouthamer & Kazmer, 1994). A delay in the division of nuclei containing paternal chromatin was seen in the incompatible eggs of *Nasonia vitripennis* (Walker) (Reed & Werren, 1995). In *D. simulans* a delay in paternal chromatin condensation and in the movement of the male chromosomes during anaphase has been observed, resulting errors lead to the formation of embryos with aneuploid or haploid nuclei and eventual death (Callaini *et al.*, 1994, 1996, 1997).

Portions of the 16S rRNA gene were first utilized to phylogenetically characterize the relationship between strains of *Wolbachia* (O'Neill *et al.*, 1992; Rousset *et al.*, 1992b). However this gene is too conserved to allow discrimination between putative species within the genus *Wolbachia.* Subsequently the cell division *ftsZ* gene was used by Werren *et al.*, (1995), this gene divided strains of *Wolbachia* tested in two groups, A and B. A high degree of variation was noted among members of group B, while variation among members of group A was so small that their relationship was largely unresolved using this gene. Recently, Bandi *et al.*, (1998), using the *ftsZ* gene proposed the existence of two new groups of *Wolbachia* strains harbored by filarial nematodes. Group C includes *Wolbachia* strains harbored by two species in the genus *Dirofilaria* and three species in the genus *Onchocerca.* Group D includes two species in the genus *Brugia* and one species each in the genera *Wuchereria* and *Litomosoides*. Moreover they found that the phylogeny of the filarial hosts was congruent with that of the *Wolbachia* strains, indicating that there has not been any recent horizontal transmission of *Wolbachia* strains between host species.

Because of the high degree of similarity between members of group A *Wolbachia* it is important that a more variable gene be used to resolve the relationships between its members. For this purpose, Zhou *et al.* (1998) recently used the *wsp* gene which codes for an outer surface protein of *Wolbachia*. A phylogenetic analysis of 28 taxa using the *wsp* gene was able to discern eight groups within group A previously described using *ftsZ*. This may be a more promising gene for future phylogenetic studies, especially if they involve strains of *Wolbachia* belonging to group A.

Arthropods infected with two strains of *Wolbachia* are common. Werren *et al.*, (1995) found that nine of 26 infected species were doubly infected. Double infections have also been demonstrated in *N. vitripennis* (Breeuwer, *et al.*, 1992), *Aedes albopictus* (Skuse) (Sinkins *et al.*, 1995), *Gryllus pennsylvanicus* Burmeister and *G. integer* Scudder (Giordano *et al.*, 1997). Moreover double infections have been created in the laboratory by microinjecting *Wolbachia*-infected egg cytoplasm directly into the egg (Sinkins, *et al.*, 1995). Doubly infected females are compatible with either strain of singly infected males, whereas doubly infected males are unidirectionally incompatible with singly infected females, this serves to drive the double infection through a population (Sinkins *et al.*, 1995). Bacterial strains in *D. simulans* with double infections have been shown to segregate if the larvae are kept in crowded rearing conditions (Sinkins *et al.*, 1995), and in *N. vitripennis* if exposed to a prolonged larval diapause (Perrot-Minnot *et al.*, 1996). This separation of strains could result in the development of bidirectional incompatibility within a population or species.

Two strains of *Wolbachia* found in *D. mauritiana* and *D. simulans* have been described, which while unable to modify sperm and cause incompatibility can however rescue the incompatibility effect caused by a closely related strain (Mercot & Poinsot, 1998; Bourtzis *et al.*, 1998). Such a strain and the evolution of *Wolbachia* towards neutrality has in fact been theoretically predicted as long as the frequency of maternal transmission is increased (Prout, 1994; Turelli, 1994; Hurst & McVean, 1996). The rescue positive *Wolbachia* strain in *D. simulans* was found in a population which originated on Mount Kilimanjaro, Tanzania. If a strain which cannot cause, but can rescue incompatibility is derived from an incompatibility causing strain, then the Kilimanjaro *Wolbachia* strain may be one of the oldest found in *D. simulans* (Mercot & Poinsot, 1998). The ability of a *Wolbachia* strain to rescue sperm modification, can explain how a strain which cannot cause incompatibility can be maintained and spread in an infected population. This also suggests that the molecular mechanisms involved in causing and rescuing incompatibility can be indepedent of each other.

Wolbachia strains have been successfully transferred, by means of embryonic microinjection techniques, between closely related strains of *Drosophila*, *Trichogramma* wasps and isopods (Boyle *et al.*, 1993; Giordano *et al.*, 1995; Grenier *et al.*, 1998; Rigaud & Juchault, 1995) as well as between more distantly related insects, such as the transfer of a *Wolbachia* strain from *A. albopictus* to *D. simulans* (Braig *et al.*, 1994). These transfers have shown that while expression of cytoplasmic incompatibility is largely *Wolbachia* dependent, there are host-parasite interactions that can affect its severity.

It has been proposed that *Wolbachia* can promote speciation in populations that harbor strains that are bidirectionally incompatible. The *Nasonia* sibling species complex is composed of 3 species that are reproductively incompatible as a result of *Wolbachia* infections (Breeuwer & Werren, 1990). Incompatibility types also exist between different populations of *Culex pipiens* and *Gryllus* spp. in North America, while these have not all been fully explored, preliminary evidence indicates that they may be the result of *Wolbachia* infections (Yen & Barr, 1971; Giordano *et al.*, 1997).

Phylogenetic analysis of *Wolbachia* strains from distantly related hosts, show a high degree of similarity, this indicates that infections are being transmitted horizontally as well as vertically (O'Neill *et al.*,1992; Rousset *et al.*, 1992b; Werren *et al.*, 1995). Transmission of *Wolbachia* has been shown to occur in isopods via hemolymph contact with an open wound (Rigaud & Juchault, 1995). No other patterns have been discerned as to how the bacteria are vectored, however it is possible that parasites or parasitoids that achieve an intimate relationship with their hosts, such as some Hymenoptera, Diptera, mites or nematodes, could be implicated in the transmission.

2. *Wolbachia* in chrysomelids:

Infections of *Wolbachia* in chrysomelids have not been widely reported. This is likely because very few species have been tested. Given that chrysomelids are one of the most species-rich and widespread of Coleoptera families and that *Wolbachia* infections have been shown to be common in arthropods (our study of temperate insects found 16% of 615 taxa to be infected, R. Giordano and H. M. Robertson unpublished, while Werren *et al.*, 1995 found that 16.9% of 157 tropical species of arthropods were infected with *Wolbachia*), we predict that one simply has to study the chrysomelids to uncover species that harbor this infection.

Thus far the only species that have been documented as infected are as follows: *Chelymorpha alternans* (Boheman) and *Chersinella heteropunctata* (Boheman) in the tribe Stolaini, *Acromis sparsa* (Boheman) in the tribe Cassidini, both in the subfamily Cassidinae (Borowiec, 1955; Werren *et al*., 1995), and *Diabrotica virgifera virgifera* LeConte tribe Luperini, subfamily Galerucinae (Krysan & Smith, 1987; Giordano *et al*., 1997).

Using a PCR assay based on *ftsZ* primers which can specifically amplify group A or B *Wolbachia*, Werren *et al.* (1995) determined that both *C. alternans* and *C. heteropunctata* are infected with a strain of *Wolbachia* belonging to group B, while *A. sparsa* was found to be doubly infected with strains belonging to groups A and B. There is no additional information regarding the identity of the actual *ftsZ* gene sequence for the above three species. The species of chrysomelid that up to now has been more thoroughly investigated is *D. v. virgifera*. A phylogenetic analysis of *Wolbachia* strains using the *ftsZ* gene, shows that the strain infecting populations of *D. v. virgifera* in the corn belt of the United States belongs to group A. The *Wolbachia* strain found in *D. v. virgifera* is the sister group to all the other strains in group A (Figure 1), which includes bacterial strains from *Drosophila*, parasitic Hymenoptera, Lepidoptera and a curculionid, with the exception of that found in the parasitic wasps *Tricopria drosophilae* (sensu Kieffer) and *Melittobia* sp. (Giordano *et al.*, 1997).

Because all the in depth information that we currently possess regarding infections of *Wolbachia* in chrysomelids is restricted to that in *D. v. virgifera* we shall focus the rest of this narrative on this species and its uninfected subspecies *D. v. zeae* Krysan and Smith.

D. v. virgifera and *D. v. zeae* are New World diabroticites, the distribution of the former extends from northern Mexico to as far north as southern Canada, while *D.v. zeae* has been recorded from Nicaragua to southern Texas. Recently *D. v. virgifera* was accidentally introduced in eastern Europe (Maceljski & Barcic, 1993).

The first report of a rickettsia-like infection in *D. v. virgifera* was made by Degrugillier *et al.*, (1991). These authors investigated the possible causes for a loss in viability of a colony of *D. v. virgifera*. Using electron microscopy, they determined that a picornavirus might have been the cause of the abnormality in sperm, in the process they also detected rickettsia-like organisms which were present in the testes and cytoplasm of spermathecal epithelial cells of both diseased and healthy individuals. The authors also noted a personal communication from B. Riemann and S. Newman who reported the presence of rickettsia-like organisms in accessory gland cells and the antennae of males (Degrugillier *et al.*, 1991). There have been few reports of *Wolbachia* infections in tissues other than gonadal, but this is probably a result of the lack of such studies. One recent exception is the study of Min and Benzer (1997) who demonstrated the presence of *Wolbachia* in the brain, muscle and retina of *D. melanogaster*.

D. v. zeae has been described as a subspecies of *D. v. virgifera* (Krysan *et al.*, 1980), this was determined on the basis of taxonomical as well as multiple behavioral studies such as mating choice experiments, competition, response to sex pheromone and insemination and sperm activity. Krysan *et al.* (1980) concluded that the two subspecies are morphologically very similar, that there is no ecological or temporal isolation between them and no premating barriers. However, when they crossed *D. v. virgifera* males from South Dakota with female *D. v. zeae* from Mexico they obtained very few to no offspring, while the reciprocal cross was fertile (Krysan & Branson, 1977). These crosses were repeated by Giordano *et al.* (1997), who when crossing *D. v. virgifera* males from South Dakota with *D. v. zeae* females from Southern Texas obtained the same results. Likewise, the reciprocal cross was

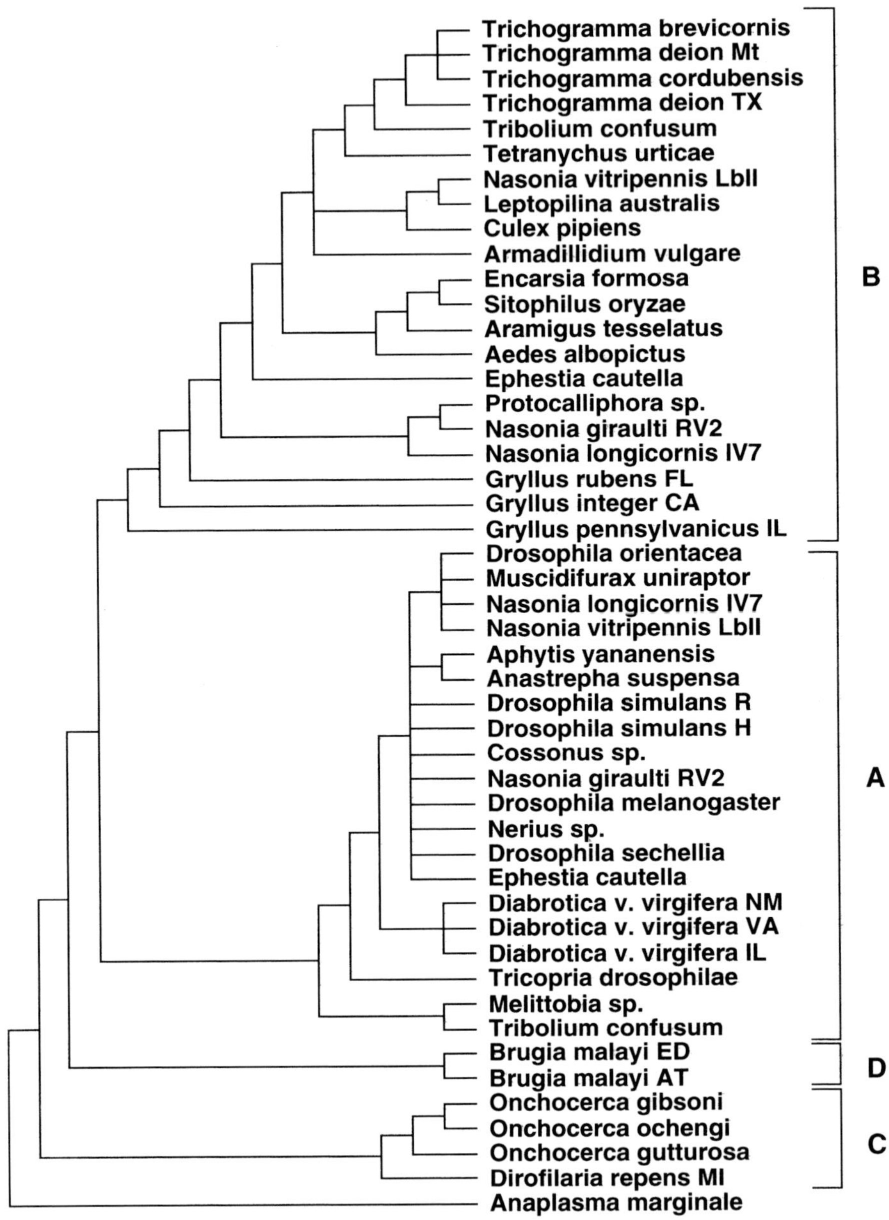

Fig. 1. Strict consensus of 36 most parsimonious trees estimated using partial DNA sequences of the *ftsZ* gene. *Anaplasma marginale* was designated as outgroup. These trees are 948 steps long, with retention index of 0.874. Designation A, B, C and D refer to groups identified by Werren *et al.* 1995, and Bandi *et al.* 1998. *Wolbachia* sequences are represented using the host name.

also fertile. Knowing that *D. v. virgifera* is infected with *Wolbachia*, Giordano *et al.* (1997) tested whether this infection was the cause of the unidirectional incompatibility described by Krysan and Branson (1977). They found that when tetracycline treated and thus uninfected

Table 1. Comparisons of mean percent egg hatch +/- SE, for crosses using infected and tetracycline-treated (T) *D. v. virgifera* and naturally uninfected *D.v. zeae*, using the Mann-Whitney U test. **P<0.0001

Female	Male	Cross	N	Eggs	Hatch (%)	Comparison
D. v. virgifera	x *D. v. virgifera*	*a*	10	4,706	74.9 ± 5.5	a vs b NS
D. v. virgifera	x *D. v. virgifera* T	b	10	5,109	81.1 ± 3.4	
D. v. virgifera T	x *D. v. virgifera* T	c	10	5,187	66.4 ± 6.2	
D. v. virgifera T	x *D. v. virgifera*	d	14	5,627	0.4 ± 0.1	d vs c **
D. v. zeae	x *D. v. zeae*	e	10	4,608	83.4 ± 4.0	
D. v. zeae	x *D. v. virgifera*	f	15	6,433	0.3 ± 0.1	f vs e **
D. v. zeae	x *D. v. virgifera* T	g	15	6,376	74.8 ± 6.5	g vs e NS
D. v. virgifera	x *D. v. zeae*	h	15	7,735	82.0 ± 4.3	h vs a NS
D. v. virgifera T	x *D. v. zeae*	i	15	8,447	75.5 ± 5.3	i vs c NS

Modified from Giordano *et al.*, 1997 PNAS 94: 11439-11444. Reprinted with permission copyright 1997, National Academy of Sciences, U.S.A.

D. v. virgifera males mated with *D. v. zeae*, the incompatibility was eliminated, egg hatch from this cross was not significantly different from its control. This pattern was also seen when uninfected *D. v. virgifera* males were crossed with tetracycline treated and uninfected *D. v. virgifera* females. The strain of *Wolbachia* found in *D. v. virgifera* is capable of causing severe incompatibility. When infected *D. v. virgifera* males were crossed with naturally uninfected *D. v. zeae* females, 0.3% of eggs hatched from a total of 6433 eggs, likewise when infected *D. v. virgifera* males were crossed with tetracycline treated and thus uninfected *D. v. virgifera females*, 0.4% eggs hatched from a total of 5627 eggs (Table1). Giordano *et al.* (1997) also found that infection with *Wolbachia* is widespread in populations of *D. v. virgifera* throughout the corn belt of the United States, and that they are all likely to be infected with the same strain. Sequences of the *ftsZ* gene of *Wolbachia* infecting individuals from New Mexico, Virginia and Illinois are identical (Figure 1). All the major maize growing States north of, and including New Mexico, had populations that were infected with *Wolbachia*. However, two populations sampled in Duncan and Willcox, Arizona were not infected (Figure 2). The authors concluded that the population of *D. v. virgifera*, which has in recent time spread (Krysan & Branson, 1987) throughout most of North America, is likely to be the descendant of a very small infected population or perhaps a single female which became infected in the area including and surrounding northern New Mexico. If this is indeed the case, it would follow that the population of *D. v. virgifera* should reflect a high degree of homogeneity. Preliminary studies of the genetic variation found in populations of *D. v. virgifera* and its closely related species *D. barberi*, indicate that the latter has a higher degree of genetic variation than the former (Szalanski & Powers, 1996). The precise distribution of the uninfected populations of *D. v. virgifera* in Arizona still need to be determined, but they are likely to be the remnant of the original uninfected population. A more unlikely scenario is that these are populations that naturally became uninfected.

3. Possible uses of *Wolbachia* for control of *D. v. virgifera* and *D. v. zeae*.

Studies of *Wolbachia* contribute to our understanding of sex determining mechanisms, cellular events subsequent to fertilization, vertical and horizontal transmission of microor-

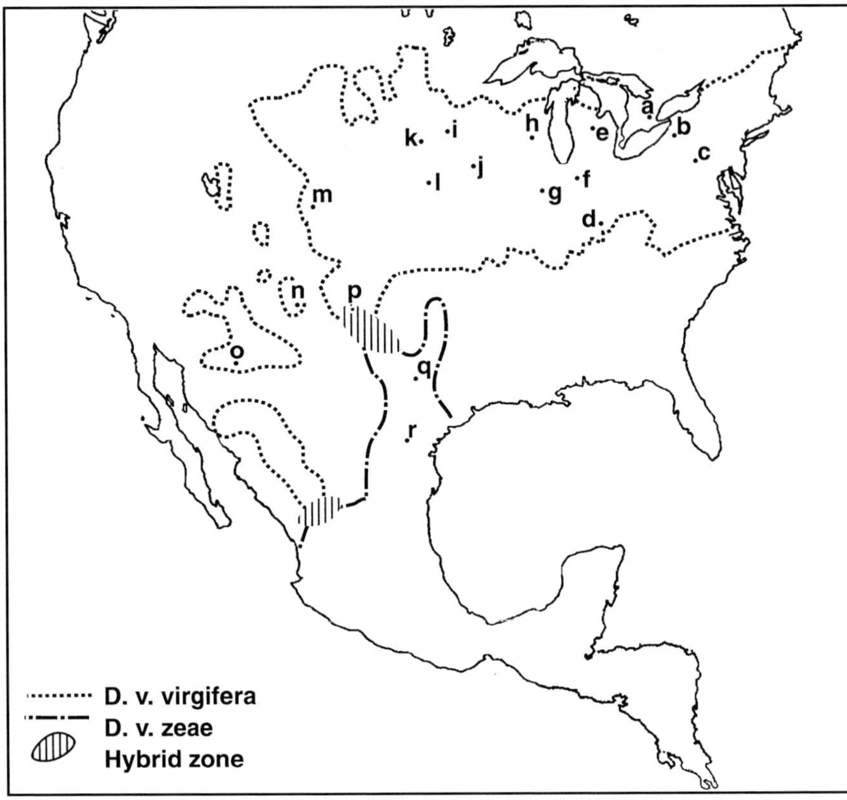

Fig. 2. Distribution and sampling sites of *D. v. virgifera* and *D. v. zeae*. a. Guelph, Ontario, Canada; b. Freeville, N.Y.; c. University Park, PA; d. Louisville, KY; e. East Lansing, MI; f. West Lafayette, IN; g. Urbana, IL; h. Madison, WI; i. Lamberton, MN; j. Ames, IA; k. Brookings, SD; l. Mead, NE; m. Fort Collins, CO; n. Moriarty, NM; o. Duncan and Willcox, AZ; p. 3 sites: 4 miles NE of Edmonson, Swisher, Co.; 6 miles NE of Edmonson, Swisher; 5 miles NE of Hart, Hale Co.; q. Stephenville, TX, r. Uvalde, TX

ganisms, and speciation among others. However, *Wolbachia* also offers new opportunities to manipulate insect pest populations through population suppression and displacement.

D. v. virgifera and *D. v. zeae* are major pests of corn throughout North America and insecticides are commonly used to control their damage. Applications to the soil are designed to protect the corn root system from larval feeding. Aerial applications are used to reduce adult populations, especially gravid females. These applications can have serious off-target effects including the suppression of beneficial insect populations and ground water contamination. If handled improperly, they can pose a health risk to chemical handlers and maize producers. A bait system composed of a feeding arrestant with a reduced amount of insecticide (Chandler, 1998), has helped to reduce off-target effects and human exposure rates. Crop rotation is another means which has been used to offset damage by corn rootworms. However the long term viability of these methods need additional evaluation. Insecticide resistance has been documented in populations of *D. v. virgifera* in Nebraska (Meinke *et al.*, 1998), and preference for adult feeding and oviposition in soybean fields has been reported from the states of Indiana and Illinois (Gray *et al.*, 1998), both of these traits could spread throughout the Corn Belt.

New suppressive methods are needed for corn rootworm species such as *D. v. virgifera* and *D. v. zeae*. Several new options may serve as effective compliments or alternatives to current chemical-based approaches. Insect-parasitic nematodes can suppress rootworm larval populations and provide root protection that effectively protects corn yields (Jackson, 1996; Wright *et al.*, 1993). However, further development of this method is needed in order to improve its economics and consistency. Genetically modified plants, resistant or more tolerant to *Diabrotica* feeding are being developed by several companies. *Wolbachia* could be another means by which to suppress populations of *D. v. zeae* and *D. v. virgifera*. In fact, the model system of infection in *D. v. virgifera* and lack thereof in *D. v. zeae* offers considerable potential to evaluate *Wolbachia* as a suppression method on a major agronomic pest complex. Furthermore, a *Wolbachia* based suppressive mechanism could be rotated to reduce resistance development and compliment other existing methods.

While the theory and practice of using *Wolbachia* to manipulate a pest population remains largely untested, reports of the eradication of a field population of *Culex* in Burma (Laven, 1967) and India (Curtis *et al.*, 1982) as well as the swift and dramatic movement of a *Wolbachia* infection monitored in populations of *D. simulans* in California (Turelli & Hoffmann, 1991) indicated that further evaluation of the potential of *Wolbachia* as a biological means of pest control is justified.

It is possible to capitalize on the absence of *Wolbachia* in *D. v. zeae* populations. The release of *Wolbachia* infected *D. v. virgifera* males into an uninfected population would be akin to a sterile male release without the drawbacks of irradiated or otherwise altered males. Rearing methods for *D. v. virgifera* are well established and it is possible to effectively isolate males (Branson & Jackson, 1988; Branson *et al.*, 1988).

There is also the potential to develop a laboratory population of *D. v. virgifera* infected with a *Wolbachia* strain that is incompatible with the strain present in the field population. Males infected with the incompatible strain could be released for the control of the resident *D. v. virgifera* population. If the introduction of a new strain of *Wolbachia* in *D. v. virgifera* is successful this method could then be utilized to control other *Diabrotica* pests, such as *D. barberi* Smith and Lawrence, *D. balteata* LeConte, *D. longicornis* (Say) and *D. undecimpunctata howardi* Barber. The opportunity to learn about and use *Wolbachia* as a suppressive mechanism within the chrysomelids and especially with *Diabrotica* species looks very promising. An intergrade zone has been reported between *D. v. virgifera* and *D. v. zeae* in northern Texas (Krysan *et al.*, 1980), we presume that there is natural control taking place at this locality where infected *D. v. virgifera* meet uninfected *D. v. zeae* populations. The study of these populations may help us to expand this suppressive interaction beyond the boundaries of the intergrade zone.

Acknowledgements

R.G. dedicates this work to the memory of Ellis Gilmore MacLeod, whose teaching and friendship shall continuously inspire his students.

References

Bandi, C., Sironi, M., Nalepa, C.A., Corona, S., and Sacchi, L., 1997. Phylogenetically distant intracellular symbionts in termites. Parassitologia, 39: 71-75.

Bandi, C., Anderson, T. J. C., Genchi, C., and Blaxter, M.L., 1998. Phylogeny of *Wolbachia* in filarial nematodes. Proceedings of the Royal Society of London Series B, Biological Sciences 265: 2407-2414.

Borowiec, X., 1995. Tribal classification of Cassidoid Hispinae (Coleoptera: Chrysomelidae). In: Pakaluk, J. and Slipinnski, S.A. (eds), Biology, phylogeny and classification of Coleoptera: papers celebrating the 80th birthday of Roy Crowson. Warszawa: Muzeum i Instit. Zool. PAN, pp. 541-558.

Bouchon, D., Rigaud, T., and Juchault, P., 1998. Evidence for widespread *Wolbachia* infection in isopod crustaceans: molecular identification and host feminization. Proceedings of the Royal Society of London Series B, Biological Sciences, 265: 1081-1090.

Bourtzis, K., Dobson, S.L., Braig, H.R., and O'Neill, S.L., 1998. Rescuing *Wolbachia* have been overlooked. Nature, 391: 852-853.

Boyle, L., O'Neill,S.L., Robertson, H.M., and Karr, T.L., 1993. Interspecific and intraspecific horizontal transfer of *Wolbachia* in *Drosophila*. Science, 260: 1796-1799.

Braig, H.R., Guzman, H., Tesh, R.B., and O'Neill, S.L., 1994. Replacement of the natural *Wolbachia* symbiont of *Drosophila simulans* with a mosquito counterpart. Nature, 367: 453-455.

Branson, T.F., and Jackson, J.J., 1988. An improved diet for adult *Diabrotica virgifera virgifera* (Coleoptera: Chrysomelidae). Journal of the Kansas Entomological Society, 61: 353-355.

Branson, T.F., Jackson, J.J., and Sutter, G.R., 1988. Improved method for rearing *Diabrotica virgifera virgifera* (Coleoptera: Chrysomelidae). Journal of Economic Entomology, 81: 410-414.

Breeuwer, J. A. J., Stouthamer, R., Barns, S. M., Pelletier, D. A., Weisburg, W. G., Werren, J.H., 1992. Phylogeny of cytoplasmic incompatibility micro-organisms in the parasitoid wasp genus *Nasonia* (Hymenoptera: Pteromalidae) based on 16S ribosomal DNA sequences. Insect Molecular Biology, 1: 25-36.

Breeuwer, J.A.J., and Werren, J.H., 1990. Microorganisms associated with chromosome destruction and reproductive isolation between two insect species. Nature, 346: 558-560.

Callaini, G., Dallai, R. and Riparbelli, G., 1997. *Wolbachia*-induced delay of paternal chromatin condensation does not prevent maternal chromosomes from entering anaphase in incompatible crosses of *Drosophila simulans*. Journal of Cell Science, 110: 271-280.

Callaini, G., Riparbelli, M.G., and Dallai, R., 1994. The distribution of cytoplasmic bacteria in the early *Drosophila* embryo is mediated by astral microtubules. Journal of Cell Science, 107: 673-682.

Callaini, G., Riparbelli, M.G., Giordano, R., and Dallai, R., 1996. Mitotic effects associated with cytoplasmic incompatibility in *Drosophila simulans*. Journal of Invertebrate Pathology, 67: 55-64.

Chandler, L. D., 1998. Comparison of insecticide-bait aerial application methods for management of corn rootworm (Coleoptera: Chrysomelidae). Southwestern Entomologist, 23:147-159.

Curtis, C.F., Brooks, G.D., Ansari, M.A., Grover, K.K., Krishnamurthy, B.S., Rajagoplalan, P.K., Sharma, L.S., Sharma, V.P., Singh, D., Singh, K.R.P., and Yasuno, M., 1982. A field trial on control of *Culex quinquefasciatus* by release of males of a strain integrating cytoplasmic incompatibility and a translocation. Entomologia Experimentalis et Applicata, 31: 181-190.

Degrugillier, M.E., Degrugillier, S.S., and Jackson, J.J., 1991. Nonoccluded, cytoplamsic virus particles and rickettsia-like organisms in testes and spermathecae of *Diabrotica virgifera*. Journal of Invertebrate Pathology, 57: 50-58.

Frank, S.A., 1997. Cytoplasmic incompatibility and population structure. Journal of Theoretical Biology, 184: 327-330.

Genchi, C., Sacchi, L., Bandi, C., and Venco.L., 1998. Preliminary results on the effect of tetracycline on the embryogenesis and symbiotic bacteria (*Wolbachia*) of *Dirofilaria immitis*. Parassitologia, 40: 72.

Giordano, R., Jackson, J.J., and Robertson, H.M., 1997. The role of *Wolbachia* bacteria in reproductive incompatibilities and hybrid zones of *Diabrotica* beetles and *Gryllus* crickets. Proceedings of the National Academy of Science USA, 94: 11439-11444.

Giordano, R., O'Neill, S. L. and Robertson, H. M., 1995. *Wolbachia* infections and the expression of cytoplasmic incompatibility in *Drosophila sechellia* and *D. mauritiana*. Genetics, 140: 1307-1317.

Grenier, S., Pintureau, B., Heddi, A., Lassabliere, F., Jager, C., Louis, C., and Khatchadourian, C., 1998. Successful horizontal transfer of *Wolbachia* symbionts between *Trichogramma* wasps. Proceedings of the Royal Society of London Series B, Biological Sciences, 265: 1441-1452.

Gray, M. E., Levine, E., O'Neal, M., Spencer, J. L., Isard, S. A., and Steffey, K. L. 1998 Crop rotation and western corn rootworm, *Diabrotica virgifera virgifera* LeConte, management: have we lost a pest management tool in the eastern corn belt? Proceedings of the Sixth Australasian Applied Entomological Research Conference, The University of Queensland, Brisbane, Australia.

Hoffmann, A. A., and Turelli, M., 1988. Unidirectional incompatibility in *Drosophila simulans*: inheritance, geographic variation and fitness effects. Genetics, 119: 435:444.

Hoffmann, A. A., and Turelli, M., 1997. Cytoplasmic incompatibility in insects. In: O'Neill, S., Hoffman, A.A., and Werren, J.H. (eds), Influential passengers: Inherited microorganisms and arthropod reproduction. Oxford University Press, Oxford, pp 42-80.

Hoffmann, A. A., Clancy, D. and Duncan, J., 1996. Naturally-occurring *Wolbachia* infection in *Drosophila simulans* that does not cause cytoplasmic incompatibility. Heredity, 76, 1-8.

Hurst, G.D.D., 1997. *Wolbachia*, cytoplasmic incompatibility, and the evolution of eusociality. Journal of Theoretical Biology, 184: 99-100.

Hurst, L.D., and McVean, G.T., 1996. Clade selection, reversible evolution and the persistence of selfish elements: the evolutionary dynamics of cytoplasmic incompatibility. Proceedings of the Royal Society of London Series B, Biological Sciences, 262: 97-104.

Jackson, J.J., 1996. Field performance of entomopathogenic nematodes for suppression of western corn rootworm (Coleoptera: Chrysomelidae). Biological and Microbial Control, 89: 366-372.

Juchault, P., Legrand, J.J., and Martin, G., 1974. Action interspecifique du facteur feminisant responsable de la thelygenie et de l'intersexualite du Crustace *Armadillidium vulgare* (Isopode Oniscoide). Annales d'Embryologie et de Morphogenese, 7: 265-276.

Krysan, J.L. and Branson, T.F., 1977. Inheritance of diapause intensity in *Diabrotica virgifera*. The Journal of Heredity, 68: 415-417.

Krysan, J.L., and Smith, R., 1987. Systematics of the *virgifera* species group of *Diabrotica* (Coleoptera: Chrysomelidae: Galerucinae). Entomography, 5: 375-484.

Krysan, J.L., Smith, R.F., Branson, T.F., and Guss, P.L., 1980. A new subspecies of *Diabrotica virgifera* (Coleoptera: Chrysomelidae): description, distribution, and sexual compatibility. Annals of the Entomological Society of America, 73: 123-130.

Laven, H., 1967. Eradication of *Culex pipiens fatigans* through cytoplasmic incompatibility. Nature, 216: 383-384.

Maceljski, M., and Barcic, J. I., 1993. *Diabrotica virgifera virgifera* LeConte (Coleoptera: Chrysomelidae) - kukuruzna zlatica. Fragmenta Phytomedica Herbologica, 21: 173-183.

Martin, G., Gruppe, S.G., Laulier, M., Bouchon, D., Rigaud, T., and Juchault, P., 1994. Evidence for *Wolbachia* spp. in the estuarine isopod *Sphaeroma rugicauda* (Crustacea): a likely cytoplasmic sex ratio distorter. Endocytobiosis and Cell Research, 10: 215-225.

Meinke, L.J., B.D. Siegfried, R.J. Wright, and L.D. Chandler, 1998. Adult susceptibility of Nebraska western corn rootworm (Coleoptera: Chrysomelidae) populations to selected insecticides. Journal of Economic Entomology, 91:594-600.

Mercot, H., and Poinsot, D., 1998. And discovered on Mount Kilimanjaro. Nature, 391: 853.

Min, K-T., and Benzer, S., 1997. *Wolbachia*, normally a symbiont of *Drosophila*, can be virulent, causing degeneration and early death. Proceedings of the National Academy of Science USA, 94: 10792-10796.

O'Neill, S.L., Giordano, R., Colbert, A.M.E., Karr, T.L., Robertson, H.M., 1992. 16S rRNA phylogenetic analysis of the bacterial endosymbionts associated with cytoplasmic incompatibility in insects. Proceedings of the National Academy of Science USA, 89: 2699-2702.

Perrot-Minnot, M.J., Guo, L.R., and Werren, J.H., 1996. Single and double infections with *Wolbachia* in the parasitic wasp *Nasonia vitripennis*: effects on compatibility. Genetics, 143: 961-972.

Prout, T., 1994. Some evolutionary possibilities for a microbe that causes incompatibility in its host. Evolution, 48: 909-911.

Reed, K.M., and Werren, J.H., 1995. Induction of paternal genome loss by the paternal-sex-ratio chromosome and cytoplasmic incompatibility bacteria (*Wolbachia*): a comparative study of early embryonic events. Molecular Reproduction and Development, 40: 408-418.

Rigaud, T., 1997. Inherited microoorganisms and sex determination of arthropod hosts. In: O'Neill, S., Hoffman, A.A., and Werren, J.H. (eds), Influential passengers: inherited microorganisms and arthropod reproduction. Oxford University Press, Oxford, pp 81-101.

Rigaud, T., and Juchault, P., 1995. Success and failure of horizontal transfers of feminizing *Wolbachia* endosymbionts in woodlice. Journal of Evolutionary Biology 8: 249-255.

Rousset, F., Bouchon, D., Pintureau, B., Juchault, P., and Solignac, M., 1992a. *Wolbachia* endosymbionts responsible for various alterations of sexuality in arthropods. Proceedings of the Royal Society of London Series B, Biological Sciences, 250: 91-98.

Rousset, F., Vautrin, D., and Solignac, M. 1992b. Molecular identification of *Wolbachia*, the agent of cytoplasmic incompatibility in *Drosophila simulans*, and variability in relation with host mitochondrial types. Proceedings of the National Academy of Science USA, 250: 91-98.

Schilthuizen, M., and Gittenberger, E., 1998. Screening mollusks for *Wolbachia* infection. Journal of Invertebrate Pathology, 71: 268-270.

Schilthuizen, M., and Stouthamer, R., 1997. Horizontal transmission of parthenogenesis-inducing microbes in *Trichogramma* wasps. Proceedings of the Royal Society of London Series B, Biological Sciences, 264: 361-366.

Sinkins, S.P., Braig, H.R., and O'Neill, S.L., 1995. *Wolbachia* superinfections and the expression of cytoplasmic incompatibility. Proceedings of the Royal Society of London Series B, Biological Sciences, 261: 325-330.
Stouthamer, R., 1997. *Wolbachia*-induced parthenogenesis. In: O'Neill, S.L., Hoffman, A.A., and Werren, J.H. (eds), Influential passengers: inherited microorganisms and arthropod reproduction. Oxford University Press, Oxford, pp 102-122.
Stouthamer, R. and Kazmer, D.J., 1994. Cytogenetics of microbe-associated parthenogenesis and its consequences for gene flow in *Trichogramma* wasps. Heredity, 73: 317-327.
Stouthamer, R., Breeuwer, J.A.J., Luck,R.F., and Werren, J.H., 1993. Molecular identification of microorganisms associated with parthenogenesis. Nature, 361: 66-68.
Szalanski, A.L., and Powers, T.O., 1996. Molecular Diagnostics of three Diabrotica (Coleoptera: Chrysomelidae) pest species. Journal of the Kansas Entomological Society, 69: 260-266.
Turelli, M., 1994. Evolution of incompatibility-inducing microbes and their hosts. Evolution, 48: 1500-1513.
Turelli, M., and Hoffmann, A.A., 1991. Rapid spread of an inherited incompatibility factor in California *Drosophila*. Nature, 353: 440-442.
Turelli, M., and Hoffmann, A.A., 1995. Cytoplasmic incompatibility in *Drosophila simulans:* dynamics and parameter estimates from natural populations. Genetics, 140: 1319-1338.
Wade, M.J., and Chang, N.W., 1995. Increased male fertility in *Tribolium confusum* beetles after infection with the intracellular parasite *Wolbachia*. Nature, 373: 72-74.
Wade, M. J., and Stevens, L., 1985. Microorganism mediated reproductive isolation in flour beetles (genus *Tribolium*). Science, 227: 527-528.
Wenseleers, T., Ito, F., Van Borm, S., Huybrechts, R., Volckaert, F., and Billen, J., 1998. Widespread occurrence of the micro-organism *Wolbachia* in ants. Proceedings of the Royal Society of London Series B, Biological Sciences, 265: 1447-1452.
Werren, J.H. and O'Neill, S.L., 1997. The evolution of heritable symbionts. In: O'Neill, S.L., Hoffman, A.A., and Werren, J.H. (eds), Influential passengers: inherited microorganisms and arthropod reproduction. Oxford University Press, Oxford, pp 1-41.
Werren, J.H., Zhang, W., Guo, L.R., 1995. Evolution and phylogeny of *Wolbachia*: reproductive parasites of arthropods. Proceedings of the Royal Society of London Series B, Biological Sciences, 251: 55-71.
Wright, R.J., Witkowski, J.F., Echtenkamp, G., and Georgis, R., 1993. Efficacy and persistence of *Steinernema carpocapsae* (Rhabditida: Steinernematidae) applied through a center-pivot irrigation system against larval corn rootworms (Coleoptera: Chrysomelidae). Journal of Economic Entomology, 86: 1348-1354.
Yen, J.H., and Barr, A.R., 1971. New hypothesis of the cause of cytoplasmic incompatibility in *Culex pipiens*. Nature, 232: 657-58.
Zhou, W., Rousset, F., and O'Neill, S., 1998. Phylogeny and PCR-based classification of *Wolbachia* strains using *wsp* gene sequences. Proceedings of the Royal Society of London Series B, Biological Sciences, 265:509-515.

CHEMICAL CUES MEDIATING INTERACTIONS BETWEEN CHRYSOMELIDS AND PARASITOIDS

MONIKA HILKER and TORSTEN MEINERS
Freie Universität Berlin, Institut für Zoologie, Haderslebener Str. 9, D-12163 Berlin, Germany

Abstract

Empirical evidence of chemically mediated interactions between Chrysomelidae and their parasitoids is reviewed. We considered that chemical cues guiding a parasitoid during host location may be released by each developmental host stage and even by the plant used by the host. Additionally, we took into account that a foraging parasitoid may use different cues for host habitat location, host finding and recognition, and assessment of host suitability. Known sources of chemicals that have been proved to modify foraging behaviour of parasitoids with chrysomelid hosts are described and listed. Possible adaptations of parasitoids to chrysomelid pheromones and allomones are discussed.

Introduction

Chrysomelidae are heavily attacked by parasitoids, although they live concealed or display protective devices against enemies. No developmental stage of Chrysomelidae is spared by parasitoids. Several chrysomelid species, which occur as harmful pests of cultivated plants, have effectively been controlled by release of parasitoids (Colazza & Bin, 1992; Field & Kwong, 1994; Wajnberg & Hassan, 1994). The wide range of hymenopteran and dipteran endo- and ectoparasitoid species of Chrysomelidae, their biological diversity and economic importance have comprehensively been reviewed by Cox (1994). Nevertheless, despite this plethora of knowledge about parasitoids of Chrysomelidae, up to now only a few studies have addressed the question of how a parasitoid locates its chrysomelid host and recognizes its suitability. For numerous parasitoids with hosts other than Chrysomelidae, host search is known to be mediated both by innate and learned responses to physical and chemical cues (Schmidt, 1991; Vinson, 1985, 1991; Weseloh, 1981). The importance of chemical cues for a successful host search has been proved for many parasitoid species. Successful host search of a parasitoid includes the processes of host habitat location, host finding within the habitat, host recognition, acceptance of suitable hosts, and possibly manipulation of the host by anaesthetizing, by modifying synchronization or circumventing the host´s immune reaction (Fig. 1) (Godfray, 1994). A wide range of volatile and non-volatile components from the host habitat, the food plants of the host, the host integument, host secretions, and from host excreta have been identified as cues that guide a parasitoid during these different steps of host approach and indicate the suitability of the host encountered (reviewed in Jones, 1981; Kainoh, 1990; Rutledge, 1996). But what is known

Advances in Chysomelidae Biology 1, pp. 197–216
Edited by M.L. Cox

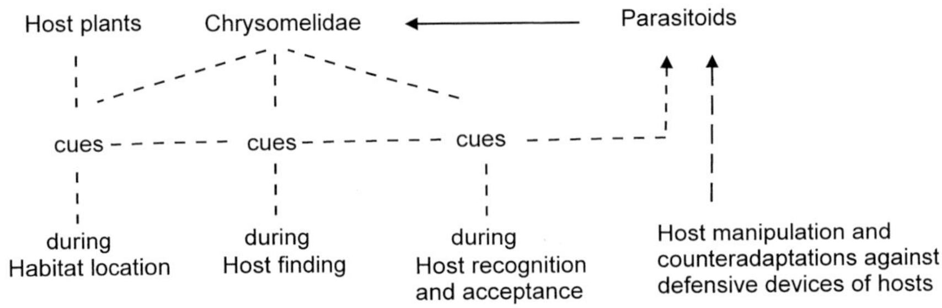

Fig. 1. Chemical interactions during host approach of parasitoids to Chrysomelidae.

about such chemical cues for parasitoids of chrysomelid hosts? We focus on this question with regard to recent studies which strongly suggest or clearly prove chemical mediation of chrysomelid – parasitoid interactions. We especially address investigations of two galerucine species, which are heavily attacked by egg parasitoids: the elm leaf beetle *Pyrrhalta (Xanthogaleruca) luteola* Müller, which is parasitized by *Oomyzus gallerucae* Fonscolombe (Hymenoptera, Eulophidae), and the tansy leaf beetle *Galeruca tanaceti* L., whose eggs are infested by *O. galerucivorus* Hedqvist (Tables 1-3). Furthermore, emphasis is on recent chemoecological studies of parasitoids of Bruchinae. Since the information necessary for successful host location in parasitoids varies during the process of host approach, the parasitoids′ different behavioural steps during host location are considered separately (Fig. 1, Table 4).

Host habitat location

In contrast to generalist parasitoids, specialized parasitoids need to search for the typical habitat of the host stage to be parasitized. The problem of host habitat location may be circumvented by staying in the host′s habitat when once encountered, but wind may transport tiny parasitoids to other places, or hosts may leave a habitat, so that the task of habitat location arises again. Numerous parasitoids with hosts other than Chrysomelidae are known to be attracted from long distances by volatiles emitted from (1) the intact food plant of the host, (2) the damaged food plant, (3) host frass, (4) exocrine glandular secretions of the host which serve the host as defensive devices or as sex pheromones (Agelopoulos *et al.*, 1995; Du *et al.*, 1996; Mattiacci *et al.*, 1993; Rutledge, 1996; Tumlinson *et al.*, 1992). For only a few parasitoids of Chrysomelidae, chemical cues are known to provide information for host habitat location.

Cues from food plants of hosts

The effects of food plants of hosts on parasitoids have recently been studied in the habitat location of two egg parasitoid species of Galerucinae: *Oomyzus gallerucae* (Hymenoptera, Eulophidae), a wasp which heavily attacks eggs of the elm leaf beetle *Pyrrhalta luteola*, and *O. galerucivorus*, a parasitoid of the egg stage of *Galeruca* spp. (Meiners & Hilker, 1997; Meiners *et al.*, 1997). While *P. luteola*, often occurring as a pest on elm, deposits its eggs on the under-surface of elm leaves, *Galeruca tanaceti* – the other investigated host

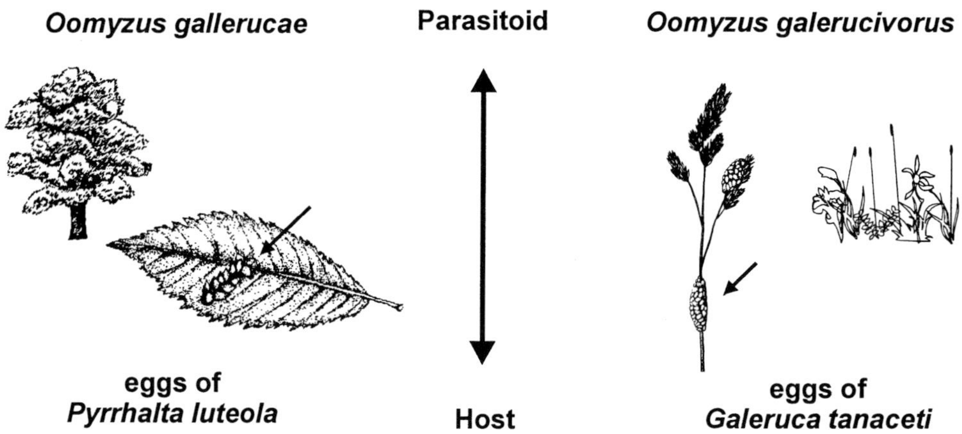

Fig. 2. Host and host habitats of *Oomyzus gallerucae* (elm trees) and of *O. galerucivorus* (herbaceous layer).

Table 1. Response of female *Oomyzus gallerucae* to odours from the elm leaf beetle host, *Pyrrhalta luteola*, and elms (*Ulmus campestris*) offered in a four arm olfactometer[1].

Odour source	Response
mixed volatiles from the plant-host complex	
elm leaves with eggs, female beetles, faeces	*attractive*
host volatiles	
eggs	*not attractive*
female beetles	*not attractive*
larvae	*not attractive*
faeces of adult beetles	*attractive*
faeces of larvae	*attractive*
plant volatiles	
intact elm leaves	*not attractive*
elm leaves damaged with cork borer	*not attractive*
elm leaves damaged with razor blade	*not attractive*
elm leaves damaged by adult beetles	*not attractive*
plant volatiles depending on egg deposition	
elm leaves (damaged by adult beetles) with eggs deposited	*attractive*
elm leaves (damaged by adult beetles) + eggs separately	*not attractive*
elm leaves (damaged by adult beetles) with eggs deposited and removed afterwards	*attractive*
elm leaves (damaged with razor blade) with transplanted eggs	*attractive*
elm leaves (damaged with razor blade) with oviduct secretion applied	*attractive*

[1]Meiners & Hilker 1997

species – lays its eggs preferably onto tips of dead grass culms. Thus, these closely related parasitoids face different problems when searching for a suitable host habitat. While *O. gallerucae* needs to approach the living plant infested by the host eggs, *O. galerucivorus* has to find upright standing (dead plant) substrate with the host eggs on its tips (Fig. 2).

O. gallerucae is attracted by volatiles from elm leaves (*Ulmus campestris* L.) onto which host eggs have been deposited (Table 1). Volatiles from intact elm leaves and from elm leaves damaged by feeding do not attract this parasitoid. Eggs themselves are not attractive from a distance, nor is the parasitoid attracted to an experimental combination of volatiles from host eggs, removed from a leaf, mixed with volatiles from elm leaves that never carried host eggs. Thus, the attractive effect of volatiles from elm leaves with eggs is not due to a synergism of egg volatiles and plant volatiles. Instead, deposition of eggs obviously induces the elm leaf to change its volatile blend in such a way that egg parasitoids are attracted (Meiners & Hilker, 1997). For oviposition, a gravid elm leaf beetle gnaws a small groove into the under-surface of a leaf and glues several rows comprising a total of about 10 to 30 eggs with its oviduct secretion into this groove. When elm leaves were damaged by careful use of a razor blade simulating the action of a female elm leaf beetle, the parasitoid was not attracted. However, when oviduct secretion has been smeared into this wound, the leaves became attractive (unpublished data from Meiners & Hilker). When responding to plant volatiles induced by oviposition of the host, *O. gallerucae* reacts to cues that guarantee the presence of the suitable developmental stage of the host. Up to this study induction of plant volatiles attractive to parasitoids were only known to be elicited by the feeding of herbivores (overview in Karban & Baldwin, 1997). For example, corn infested with feeding larvae of *Helicoverpa zea* Boddie emits volatiles that attract the larval parasitoid *Cotesia marginiventris* Cresson (Turlings *et al.*, 1991, 1995). The study of host habitat location in *O. gallerucae* shows that oviposition of a host onto its food plant induces the emission of plant volatiles which are attractive to the egg parasitoid (Meiners & Hilker, 1997; and unpublished data from Meiners & Hilker). In contrast to *O. gallerucae*, the egg parasitoid *O. galerucivorus* was not attracted to volatiles from food plants of its polyphagous host *G. tanaceti* (Table 2). Leaves of *Tanacetum vulgare* L., *Achillea millefolium* L., and *Brassica pekinensis* Ruprecht were tested and none of these plant species was found attractive, neither when tested with intact leaves nor with leaves damaged by adult *G. tanaceti*. Nor does

Table 2. Response of female *Oomyzus galerucivorus* to odours from the tansy leaf beetle host, *Galeruca tanaceti*, and different host plants (*Tanacetum vulgare*, *Achillea millefolium*, *Brassica pekinensis*) offered in a four arm olfactometer[1].

Odour source	Response
mixed volatiles from the plant-host complex	
tansy leaves, female beetles, faeces	*not attractive*
host volatiles	
female beetles	*not attractive*
faeces	*not attractive*
plant volatiles	
intact tansy leaves	*not attractive*
intact yarrow leaves	*not attractive*
artificially damaged tansy leaves	*not attractive*
tansy leaves damaged by beetles	*not attractive*
yarrow leaves damaged by beetles	*not attractive*
chinese cabbage leaves damaged by beetles	*not attractive*
hay (dried culms of cock´s foot)	*not attractive*

[1]Meiners *et al.* 1997

odour from dead plant material (dead grass culms of *Dactylis glomerata* L., which mostly carry the host eggs on their tips) elicit any positive response in the wasp. Thus, habitat location of *O. galerucivorus* seems not to be mediated by plant volatiles. Neither does this parasitoid seem to rely on common volatiles from the wide range of food plants of its hosts, nor does it learn specific plant odours (Meiners *et al.*, 1997).

Volatiles from the host plant are known to be attractive in several parasitoids of Bruchinae (Table 4). For these parasitoids, the seeds, rather than the vegetative part of the plant emit the attractive volatiles (Cortesero *et al.*, 1993). *Uscana lariophaga* Steffan (Hymenoptera; Trichogrammatidae) is an egg parasitoid of the bruchids *Callosobruchus maculatus* (F.) and *Bruchidius atrolineatus* (Pic). Volatiles, both from uninfested cowpea grains and infested ones attract the egg parasitoid. *Dinarmus basalis* Rondani (Hymenoptera, Pteromalidae) and *Eupelmus vuilleti* (Crawford) (Hymenoptera, Eupelmidae) are ectoparasitoids of several bruchid species developing in seeds of Leguminosae. These parasitoid species are as well attracted by volatiles from both infested and uninfested seeds (Monge & Cortesero, 1996). Females of *E. vuilleti* were shown to be highly sensitive to volatiles from that kind of seeds which they had contacted previously during their early adult stage prior to their emergence. In contrast to this enhanced response in the case of contact experience with seeds, experience just with odour of seeds does not increase the sensitivity towards seed volatiles. Contact experience does not cause a change in the reaction of the ectoparasitoid *D. basalis* to plant volatiles (Cortesero & Monge, 1994).

Cues from adult hosts

A very cunning way to *stay* within the correct habitat was observed in *Emersonella niveipes* Girault (Hymenoptera, Eulophidae), an egg parasitoid of the cassid *Stolas* spp. The adult parasitoid is phoretic on the elytra of host females and is thereby transported not only to its preferred habitat, but even to a later egg batch of the host individual. The wasps were seen parasitizing host eggs before the beetle had finished egg laying. Wasps were only found on female beetles, not on males. Interestingly, cassid females with phoretic wasps lay significantly more egg clutches than those without any parasitoids, whereas the total number of eggs deposited does not differ between females with phoretic wasps and those without. Thus, the presence of a parasitoid influences the spatial oviposition behaviour in these cassidines, not the fecundity (Carroll, 1978). It is unknown how this parasitoid differentiates between cassid males and females, and how it selects for females that are soon to lay eggs. The phoretic scelionid *Telenomus euproctidis* (Wilcox) uses a component of the sex pheromone of its host, the tussock moth *Euproctis taiwana* (Shiraki) (Lepidoptera: Lymantriidae) to find host females (Arakaki *et al.*, 1996).

Host females are also attractive to the parasitoid *Uscana lariophaga* (Hymenoptera, Trichogrammatidae), which attacks eggs of the bruchid *Callosobruchus maculatus* on cowpeas, but is not phoretic as described above (Table 4). Van Huis *et al.* (1994) provided strong evidence that sex pheromones of *C. maculatus* females are attractive to this egg parasitoid. Male bruchids do not elicit a significant response in *U. lariophaga*. Mate finding by sex pheromones is only known in a few chrysomelid species (Chuman *et al.*, 1987; Guss *et al.*, 1982, 1983a, b, 1984, 1985; Krysan *et al.*, 1986), but host sex pheromones are already known to attract several egg parasitoids with hosts other than Chrysomelidae (Arakaki *et al.*, 1996, 1997; Lewis *et al.*, 1982; Noldus, 1988).

Similar to sex pheromones, also aggregation pheromones might play a role in host finding for parasitoids of leaf beetles. Morris *et al.* (1996) reviewed some evidence of communication by aggregation pheromones in a few chrysomelid species. In several other phytophagous insect species, aggregation pheromones are known to be used as kairomones by predators and parasitoids for prey/host habitat location (for example, Borden, 1982, 1985; Mizutani *et al.*, 1997; Wiskerke *et al.*, 1993).

Effects of volatiles from gravid host females were also tested in the above mentioned egg parasitoids of Galerucinae (Table 1, 2). Neither *O. gallerucae* nor *O. galerucivorus* showed any response to them (Meiners & Hilker, 1997; Meiners *et al.*, 1997). Odour from faeces of the hosts did only attract the parasitoid of the elm leaf beetle. Thus, volatiles from the plant and from the host are involved in habitat location in *O. gallerucae*, whereas no attractive long range volatiles have been detected in *O. galerucivorus*. How *O. galerucivorus* locates the host habitat is still unknown.

Cues from host eggs

Volatiles from eggs of the bruchid *Callosobruchus maculatus* were found to be attractive to the trichogrammatid egg parasitoid *Uscana lariophaga* (Van Huis *et al.*, 1994) (Table 4). Ormel *et al.* (1995) speculated on the source and nature of these attractive odours since eggs *per se* probably do not emit any odours (Kaiser *et al.*, 1989). The authors suggested that volatiles from the beetles, their faeces or their pheromones adhere to eggs. Sex pheromones and epideictic pheromones are known to be involved in the intraspecific communication of *C. maculatus* (Cork *et al.*, 1991; Credland & Wright, 1990; Messina *et al.*, 1992; Pajni & Gupta, 1990). The adherence of sex pheromones to eggs and leaves of the host plant was shown by Zaki (1985) and Noldus *et al.* (1991). Thus, parasitoids do not need to rely solely on the sex pheromone producing host, but may perceive the plant- or egg-adhered pheromone also after emission. Also the glue that attaches bruchid eggs to seeds might contain volatiles attractive to the egg parasitoid. For example, the secretion around eggs of the pentatomid bug *Nezara viridula* (L.) contains a kairomone for the egg parasitoid *Trissolcus basalis* (Wollaston) (Hymenoptera, Scelionidae) (Bin *et al.*, 1993).

Table 3. Comparison of specificity of kairomones from the faeces of galerucine beetles eliciting host finding behaviour in two egg parasitoids closely related to each other [1,2,3].

	Oomyzus gallerucae	*Oomyzus galerucivorus*
Species	species-specific (*Pyrrhalta luteola*)	genus-specific (*Galeruca* spp.)
Host stage	unspecific (larva, adult)	unspecific (larva, adult)
Host gender	unspecific male, female	unspecific male, female
Host plant	genus-specific (*Ulmus* spp.)	unspecific (diverse plant families)

[1]Meiners & Hilker, 1997; [2]Meiners *et al.*, 1997; [3]Meiners & Hilker, unpublished data

Cues from host larvae

The ectoparasitoid *Eupelmus vuilleti* (Hymenoptera, Eupelmidae) is strongly attracted by the odour from larval frass of the host *Bruchidius atrolineatus* while larvae themselves elicit only a weak olfactory response (Cortesero & Monge, 1994). *Dinarmus basalis* also lives ectoparasitically upon larvae of *B. atrolineatus*, but is attracted by odour from host larvae and not by host larval frass (Monge & Cortesero, 1996). Thus, these two ectoparasitoids favour different cues for the habitat location of the same host (Table 4). On the other hand, both ectoparasitoid species are not attracted to odours from larvae and faeces of another common host species, *Callosobruchus maculatus*, when previously reared on *B. atrolineatus*. When these ectoparasitoids have contact experience with *C. maculatus* larvae from their early adult stage, also odours from this host are attractive. However contact must occur during the early adult stage before the emergence from a seed, when the ectoparasitoid imago is still within the larval chamber of its host in contact with both the seed and host larval remains. Although adult experience prior to emergence is crucial for the later host habitat location, any pre-adult experience does not influence the response to larval host odours in *E. vuilleti* and *D. basalis*.

Table 4. Known sources of chemical stimuli mediating approach of parasitoids to chrysomelid hosts (including Bruchinae).

Approach / Parasitoid	Chemical stimuli	Chrysomelid host (parasitized stage)
Host habitat location	*Volatiles from:*	
Oomyzus gallerucae	* food plant with eggs * faeces of adult beetles * faeces of beetle larvae	*Pyrrhalta luteola* (egg)[1]
Uscana lariophaga	* uninfested and infested seeds of food plant * female beetles * host eggs	*Callosobruchus maculatus* (egg)[2,3]
Uscana lariophaga	* uninfested and infested seeds of food plant	*Bruchidius atrolineatus* (egg)[2]
Dinarmus basalis	* uninfested and infested seeds of food plant * larvae	*Bruchidius atrolineatus* (larva)[4]
Eupelmus vuilleti	* uninfested and infested seeds of food plant * larvae * pupae * faeces	*Bruchidius atrolineatus* (larva, pupa)[4,5,6]
Eupelmus vuilleti	* larvae	*Callosobruchus maculatus* (larva)[6]
Host finding		
Eupelmus vuilleti	* female beetles	*Bruchidius atrolineatus* (larva)[5]
Edovum puttleri	* eggs	*Leptinotarsa decemlineata* (egg)[7]
	Contact kairomones from:	
Oomyzus gallerucae	* faeces of adults and larvae	*Pyrrhalta luteola* (egg)[1]
Oomyzus galerucivorus	* faeces of adults and larvae	*Galeruca tanaceti* (egg)[7]

[1]Meiners & Hilker, 1997; [2]Van Huis et al., 1994; [3]Van Alebeek & van Huis, 1997; [4]Monge & Cortesero, 1996; [5]Cortesero et al., 1993; [6]Cortesero & Monge, 1994; [7]Meiners *et al.*, 1997

Table 4. Cont.

Approach / Parasitoid	Chemical Stimuli	Chrysomelid host (parasitized stage)
Host recognition	*Contact kairomones from:*	
Oomyzus gallerucae	* egg chorion	*Pyrrhalta luteola* (egg) [1]
Oomyzus galerucivorus	* extrachorion	*Galeruca tanaceti* (egg) [7]
Host suitability		
Edovum puttleri	* egg chorion ? * marker pheromone deposited by parasitoid	*Leptinotarsa decemlineata* (egg)[8,9]
Tetrastichus asparagi	* egg chorion * marker pheromone deposited by parasitoid	*Crioceris asparagi* (egg)[10]
Tetrastichus spec.	* egg chorion * marker pheromone deposited by parasitoid	*Crioceris duodecimpunctata* (egg)[10]
Host manipulation	*Allomones from:*	
Eupelmus orientalis	* poison gland of female parasitoid * saliva of parasitoid larva	*Callosobruchus maculatus* (larvae)[11]

[1]Meiners & Hilker, 1997; [7]Meiners *et al.*, 1997; [8]Lashomb *et al.*, 1987; [9]Gardenghi *et al.*, 1994; [10]Van Alphen, 1980; [11]Doury et al., 1995

Cues from host pupae

No chemical cues from Chrysomelid host pupae are known to be used by parasitoids for habitat location.

Host finding by parasitoids

Once a parasitoid has found the habitat/microhabitat of its host, it needs to encounter the prey (Fig. 1). While some parasitoid species find the host just by physical contact, others are guided by visual and/or chemical cues (Vinson, 1985). The euphorine braconids *Microctonus vittatae* Muesebeck, *M. punctulatae* Loan and Wylie and *Townesilitus psylliodis* (Loan) were observed to follow moving flea beetle adults (several *Phyllotreta* spp.), which they probably respond visually to (Wylie, 1984). Chemical information during the process of host finding is primarily provided by olfaction of close range volatiles from the host or by contact with chemicals produced by the host (Table 4).

Cues from adult hosts

Contact with substrate contaminated with faeces of the adult elm leaf beetle *Pyrrhalta luteola* elicited a positive response in the eulophid egg parasitoid *Oomyzus gallerucae* (Meiners & Hilker, 1997). When encountering a substrate with host faeces, *O. gallerucae* displayed a host finding behaviour which is typical for numerous parasitoid species: it bowed the tips of its antennae to the ground and drummed them with high frequency onto the substrate. The same behaviour has been observed when the egg parasitoid *O. galerucivorus* encounters substrate contaminated with faeces of its galerucine host *Galeruca tanaceti* (Meiners *et al.*, 1997). The parasitoid's response to faeces of adult *G. tanaceti* was independent of the food plant its polyphagous galerucine host had fed on. Faeces of beetles fed on *Tanacetum vulgare*, *Achillea millefolium*, and *Brassica pekinensis* elicited antennal drumming in *O. galerucivorus*. However, both in *O. galerucivorus* and in *O. gallerucae*, the response to faeces was dependent on the producing species. Faeces of the non-host chrysomelid species *Phaedon cochleariae* F. fed on *B. pekinensis* did not elicit any response in *O. galerucivorus*. No arrestment was observed in *O. gallerucae* when it contacted faeces of larvae of *Opisthograptis luteola* L. (Lepidoptera, Geometridae) fed with elm leaves (Meiners, unpublished data) (Table 3).

Olfaction of sex pheromones of the bruchid *Bruchidius atrolineatus* did not attract the larval ectoparasitic wasp *Eupelmus vuilleti* from long range, but increased its locomotory activity within the habitat (Table 4). As host larvae and adults can coexist in the same habitat, sex pheromones could increase the probability that a parasitoid encounters its host larvae (Cortesero *et al.*, 1993).

Cues from host eggs

Eggs of the elm leaf beetle *Pyrrhalta luteola* offered in close proximity to parasitoid females of *Oomzus gallerucae* were not attractive. Even when the parasitoids passed the eggs at only 1 mm distance, no positive response occurred. The same behaviour was observed in *O. galerucivorus*, when eggs of the tansy leaf beetle *Galeruca tanaceti* were offered. Thus, both *Oomyzus* species seem to encounter their hosts by chance, when they are in the correct habitat (Meiners & Hilker, 1997; Meiners *et al.*, 1997).

Eggs of several *Leptinotarsa* spp. are known to be heavily attacked by *Edovum puttleri* Grisell (Hymenoptera, Eulophidae), a wasp which has frequently been studied in relation to the biological control of the Colorado beetle (Colazza & Bin, 1992; Corrigan *et al.*, 1991; Gardenghi *et al.*, 1994; Jansson & Lashomb, 1988; Lashomb *et al.*, 1987; Maini & Nicoli, 1990; Mityakina *et al.*, 1993). Leonard *et al.* (1987) showed that eggs of *Leptinotarsa decemlineata* (Say) emit a volatile, hexane-soluble kairomone that increases searching activity in *E. puttleri* (Table 4). The choice of host species that is preferentially approached by *E. puttleri*, is influenced by the host species which the wasp emerged from and by the plant experienced by the emerging wasp (Jansson & Lashomb, 1988). Very similar results were reported by Hofstetter & Raffa (1997) for the encyrtid parasitoid *Oencyrtus kuvanae* (Howard) which parasitizes eggs of the gypsy moth. Jansson and Lashomb could not decide from their results whether pre-adult and/or early adult experience determines host selection behaviour in *E. puttleri*. When an egg mass has once been encountered, the host choice of *E. puttleri* is no longer affected by the plant species which was previously experienced by the emerging wasp (Jansson & Lashomb, 1988).

Cues from host larvae

The egg parasitoid *Oomyzus galerucivorus* displays a positive contact response to faeces of larvae and adult *Galeruca tanaceti* (as described above) (Table 4). Since larvae and egg-laying adults of the latter do not occur at the same time of the year, the parasitoid's response to the larval faeces is probably of no biological significance for host finding. However, *Oomyzus galerucivorus* might use kairomones from larval faeces for habitat location. Before *G. tanaceti* pupates and enters diapause during summer, *O. galerucivorus* has thereby the chance to locate the habitat where its host very likely occurs in the subsequent autumn. A more proximate explanation would be that food metabolism in larvae and adults of *G. tanaceti* is so similar that both discharge the kairomones eliciting antennal drumming and arrestment behaviour in the parasitoid.

Cues from host pupae

Neither are there any close range volatiles nor any contact chemicals of chrysomelid pupae known to inform pupal parasitoids during the process of host finding.

Host recognition and host acceptance by parasitoids

When a parasitoid encounters a potential host, it needs to recognize the correct species (Fig. 1). Suitable hosts may divulge their identity by physical and chemical cues (Arthur, 1981) (Table 4).

Although the eggs of the tansy leaf beetle *Galeruca tanaceti* are covered by an extrachorion with conspicuous morphological structures (Messner, 1983; Scherf, 1956), chemical rather than physical features help identify them as the correct host for the eulophid parasitoid *O. galerucivorus*. This was proved by a simple experimental trick: the extrachorion is liquid during egg deposition and hardens within about an hour. When the liquid extrachorion is completely taken up with a piece of filter paper, it does not harden as usual and does not develop its typical morphological structures. Nevertheless, *O. galerucivorus* exhibits host recognition behaviour when contacting a filter paper soaked with extrachorion. Thus, the parasitoid recognizes its host from chemicals, even when typical morphological features of the host surface are lacking. The chemical cues emanating from the extrachorion are best soluble in dichloromethane (Meiners *et al.*, 1997). Chemicals from the egg surface are also important as host recognition cues for *O. gallerucae*, an egg parasitoid of the elm leaf beetle (Meiners & Hilker, 1997).

When a parasitoid recognizes that it has encountered its host species, it still needs to check the individual host suitability. Host age and the degree of prior host parasitization are crucial factors for a parasitoid. Several parasitoids are known to assess the suitability of a host before parasitization (Godfray, 1994).

The age of larval hosts may be assessed by "measuring" host size or cues of the host´s integument (Schmidt, 1974). Few eulophid wasps with chrysomelid hosts are known to assess the age of their host. The egg parasitoid *Edovum puttleri* prefers younger eggs of the Colorado beetle to older ones (Lashomb *et al.*, 1987). However, it is unknown by which means *E. puttleri* assesses the age of eggs. Strand and Vinson (1983) showed that *Telenomus heliothidis* Ashmead, an egg parasitoid of the noctuid moth *Heliothis virescens* (F.) has a preference for more spherical eggs. They concluded that measuring shape can be an indirect mechanism to assess host age. *Tetrastichus asparagi* Crawford, which parasitizes advanced embryos and larvae of *Crioceris asparagi* L., parasitizes older eggs with a higher frequency than younger ones. The same result was found for another *Tetrastichus* species attacking

C. duodecimpunctata L. Both *Tetrastichus* species reject younger eggs almost immediately after antennal contact (Van Alphen, 1980). Thus, they are able to assess host age by external cues of the egg chorion. The chemistry of insect egg shells undergoes several changes (for example, oxidation, crosslinking of maternal proteins) during the hardening process (Hinton, 1981; Margaritis, 1985). This process might still be ongoing in young eggs and differences in the chemical composition at different times after oviposition could offer parasitoids a way to evaluate host age.

Those parasitoids which can only successfully develop in an unparasitized host or a host with a relatively low extent of prior parasitizations, should be able to recognize whether a host already harbours parasitoids or whether the degree of parasitization allows successful reproduction and development of the offspring. Several parasitoids with hosts other than Chrysomelidae are known to mark hosts or host areas which they have already visited by marking pheromones (Conti *et al.*, 1997; Greany & Oatman, 1972; Higuchi & Suzuki, 1996; Price, 1970; Salt, 1937; Van Baaren *et al.*, 1994, 1995). The same eulophid wasps, which were mentioned above for being able to assess host age, are also capable of discriminating between parasitized and unparasitized hosts. *E. puttleri* females are supposed to leave a chemical signal on the host egg surface, which deters conspecific females from further oviposition into the marked egg. When experimentally preventing females from perceiving the external chemical marker, they cannot distinguish between parasitized and unparasitized eggs (Gardenghi *et al.*, 1994). Even though *E. puttleri* females are capable of host discrimination, superparasitism frequently occurs. However, no more than a single parasitoid can successfully develop within an egg of the Colorado beetle. Also the *Tetrastichus* spp. which parasitize *Crioceris* eggs are able to distinguish between parasitized and unparasitized hosts mainly by external cues of the eggs. In the majority of encounters *Tetrastichus* spp. reject parasitized eggs already after antennal contact (Van Alphen, 1980).

Host manipulation and counteradaptations against defensive devices

Chrysomelidae are well-known for their impressive diversity of both physical and chemical defensive devices (Blum, 1994; Hilker, 1994; Pasteels *et al.*, 1994, 1996). This rich diversity may serve to escape from adaptations of enemies (Pasteels & Grégoire, 1984). Nevertheless, several defensive devices of insects which protect against generalist enemies are known to be overcome by specialists. These specialists have adapted behaviourally and/or physiologically to their hosts (Fig. 1).

Behaviourally, parasitoids of chrysomelids are known to overcome physical barriers like the faecal covering of egg masses of the hispine *Microrhopala vittata* (F.) which effectively reduces egg mortality resulting from sucking and chewing predators. *Chrysonotomyia* spp. (Hymenoptera: Eulophidae) are able to oviposit through the relatively penetrable leaf surface or the seam between the leaf and the faecal covering (Damman & Cappuccino, 1991). Vehement behavioural defence like waving a faecal shield by larvae of *Cassida rubiginosa* Mueller protects against ants, but not against parasitoids (Eisner *et al.*, 1967).

Physiologically, host defences may be circumvented by (1) chemical mimicry, (2) paralysis, (3) suppression of the host´s immune system and modification of host physiology or (4) the use of detoxifying and special excretory mechanisms:

1) The overcoming of host defences by parasitoids even extends to those host species that live guarded by ants. For example Mutillidae succeed in parasitizing myrmecophilous clytrine larvae (Barbier, 1976; Richards, 1980). A parasitoid's successful attack of a host living together with ants may be possible, when parasitoids are capable of effective defence against the ants or when they are disguised by chemical mimicry as the parasitoid *Lysiphlebus cardui* (Marshall) (Aphidiidae) is by mimicking chemicals of the integument of its host *Aphis fabae cirsiiacanthoidis* Scop. (Dettner & Liepert, 1994). An interesting example of a chemically defended parasitoid species is the hyperparasitoid *Alloxysta brevis* Thomson which attends aphid colonies guarded by ants and produces 6-methyl-5-hepten-2-one in its mandibular glands. This compound was shown to repel the ants attending the aphid colonies (Völkl *et al.*, 1994). The ketone also dominates the exocrine defensive secretion of larvae of the chrysomeline *Gonioctena viminalis* L. (Dettner & Schwinger, 1987).
2) The ovipositing parasitoid female and the parasitizing larva may emit compounds inducing paralysis of its host (Beckage, 1997). Doury *et al.* (1995) showed that the ectoparasitic wasp *Eupelmus orientalis* Crawford (Hymenoptera, Eupelmidae) stings its host, larvae of *Callosobruchus maculatus* (F.), prior to oviposition. During stinging, a venom from the poison gland is injected and induces permanent paralysis and developmental arrest of host larvae. The freshly hatched eupelmid parasitoid larva bites through the integument of the bruchid larva to enter its host. During that process, the parasitoid larva releases a paralysing factor into the host, most probably with its saliva. Older instar larvae of the parasitoid do not emit the paralysing chemical.

An interesting convergence of chemicals from a parasitoids' paralysing venom and the defensive secretion of adult eumolpine leaf beetles is putrescine (1,4-diamino butane): This amine is present both in the defensive secretion of *Chrysochus asclepiadeus* Pallas (as diacetyl-putrescine) (Dobler *et al.*, 1998) and the venom of the bethylid *Goniozus legneri* Gordh, which is a primary external parasitoid of lepidopteran larvae. Prior to oviposition, the parasitoid mother paralyzes host larvae by repeated stinging (Skinner *et al.*, 1990). Polyamines are also known to be present in the paralyzing venoms of various spiders and are believed to block cation channels in nerve or muscle plasma membranes (Quicke, 1988, 1997).
3) Mechanisms of the suppression of the host's immune system and modifications of the host's physiology by parasitoids have intensively been studied during the last decade (Beckage, 1998; Coudron & Puttler, 1988; Piek & Spanjer, 1986; Skinner *et al.*, 1990; Strand, 1986). As yet, only a few of these immunologically suppressive components and those modifying host physiology have been identified in parasitoids with hosts other than Chrysomelidae. The chemical nature of these compounds ranges from large molecular weight proteins to polypeptides and amino acids (Leluk *et al.*, 1989), amino acid derivatives (Shimizu *et al.*, 1993), other amine derivatives (Skinner *et al.*, 1990), ecdysteroids and ecdysteroid-type compounds (for example, Brown *et al.* 1993), and terpenes (for example, juvenile hormone; Beckage, 1991). Even viruses, virus-like particles and teratocytes are known to be injected into the host, in order to prevent encapsulation (Beckage, 1998; Tanaka & Wago, 1990).
4) Chemical defence of chrysomelid hosts has obviously been overcome by numerous parasitoid species but not much is known about the underlying mechanisms. For example, toxins such as cyanogenic compounds present in all stages of the chrysomeline *Paropsis atomaria* Olivier do not prevent parasitoids from successful attacking of their hosts

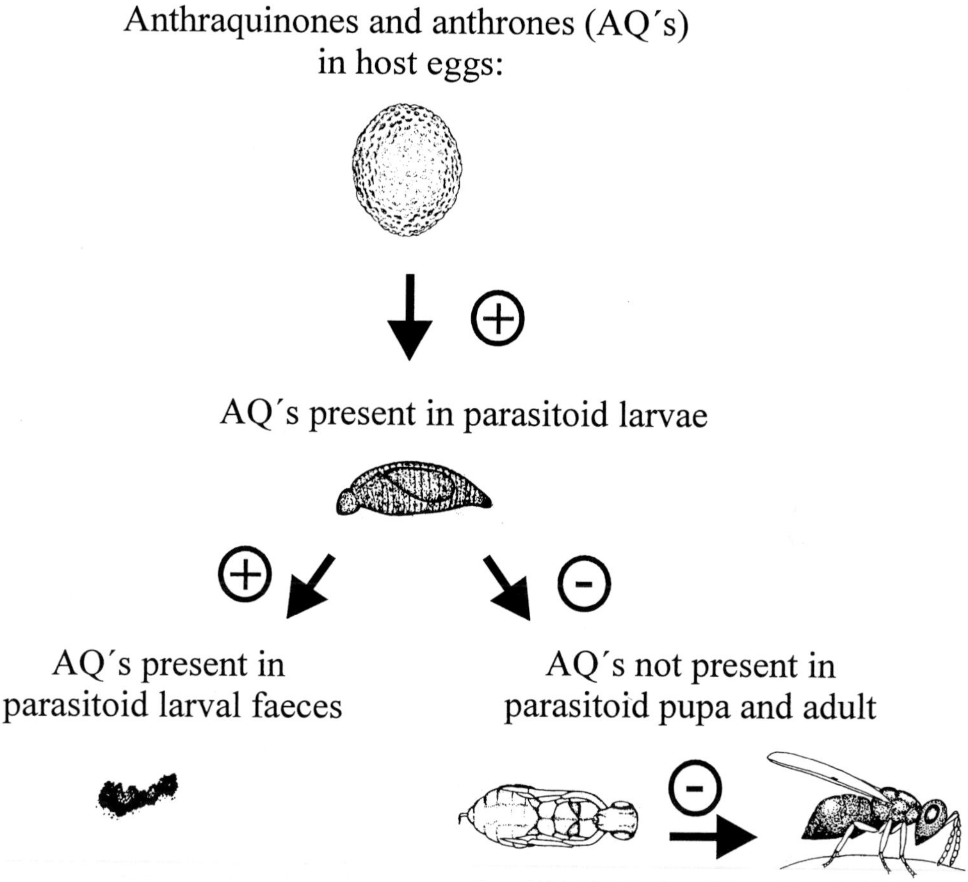

Fig. 3. Fate of anthraquinones and anthrones (=AQ´s) in eggs of *Galeruca tanaceti* which are parasitized by the eulophid wasp *Oomyzus galerucivorus*. +, transfer of AQ´s takes place; -, no transfer of AQ´s takes place.

(Nahrstedt & Davis, 1986; Peck, 1963). These parasitoids need physiological adaptations to detoxify the host toxins. Also the braconid *Apanteles zygaenarum* (Marshall) which parasitizes cyanogenic lepidopteran larvae of *Zygaena*, was shown to cope with cyanides by the enzyme rhodanese, which detoxifies hydrocyanic acid to thiocyanate, which then may be easily excreted (Jones *et al.*, 1962).

Unaltered excretion of host toxins rather than metabolization was shown in the parasitoid *Oomyzus galerucivorus*. This eulophid wasp attacks eggs of *Galeruca* spp., which contain anthrones and anthraquinones known as cytotoxins and repellents against predators (Hilker *et al.*, 1992; Hilker & Köpf, 1994; Howard *et al.*, 1982; Thomson, 1987). Meiners *et al.* (1997) showed that larvae of *O. galerucivorus* obviously excrete these anthraquinones and anthrones unaltered with their faeces, since the parasitoid´s faecal pellet, which is released by the parasitoid larva prior to pupation and is always found next to the pupa within the host eggs, contains anthrones and anthraquinones (Fig. 3). Thus, *O. galerucivorus* does not store defensive components of the host to use them for its own protection. Up to now, very few studies investigated whether host defensive compounds are sequestered by parasitoids (Rowell-Rahier & Pasteels, 1992). Benn *et al.* (1979) reported the transfer of pyrrolizidine alkaloids from the lepidopteran host *Nyctemera annulata* Boisduval to a *Microplitis* spp. The endoparasitoids *Diadegma armillata*, *Herpestomos brunnicornis* and *Triclistus yponomeutae* (Hymenoptera: Ichneumonidae) sequester butenolides of their lepidopteran host *Yponomeuta* spp. (Fung, 1988). The few studies on the transfer of host defensive compounds between the second and third trophic level contrast with the numerous studies of sequestration of protective compounds between plants and herbivores (Blum, 1981, 1994; Harborne, 1993; Pasteels *et al.*, 1994).

Specialist enemies may even turn the tables by not only utilizing host defensive compounds for their own protection, but also by using them as kairomones to locate their prey (for example, Mattiacci *et al.*, 1993). Numerous volatile defensive compounds from exocrine glands are well-known to protect larvae of Chrysomelinae against generalist predators (Blum, 1994; Pasteels *et al.*, 1994). Pupae of several chrysomeline species do not shed off the exuvia of the last larval instar, which contains exocrine glands filled with volatile defensive secretions, but keep the exuvia attached and spend the pupal period hanging exposed from the host plant. Both larvae with defensive glands and openly hanging pupae of Chrysomelinae are attacked by several parasitoid species, albeit reports indicate a low parasitization rate (review in Cox, 1994; Dowden, 1939; Togashi, 1974). While the defensive devices of these chrysomelines are known to be used as kairomones by specialist predators (Köpf *et al.*, 1997), their possible kairomonal effects on parasitoids have not been studied as yet.

Conclusions

The rich biological diversity of Chrysomelidae (Jolivet *et al.*, 1994; Jolivet & Cox, 1996) faces a high behavioural, morphological, and physiological diversity of parasitoids (Godfray, 1994; Quicke, 1997). Thus, interactions between Chrysomelidae and their parasitoids may be expected to be mediated by a multi-faceted repertoire of cues. The present review shows that parasitoids may exploit chemicals of Chrysomelidae and their food plants for successful host attack, as chrysomelids themselves do with secondary plant chemicals. We have just only begun to realize the importance of chemical stimuli necessary for a parasitoid to approach and successfully attack a chrysomelid host. The few chemical signals known to be involved in chrysomelid – parasitoid interactions and the studies that strongly suggest chemical mediation between chrysomelids and parasitoids may give us an idea of the impressive abilities of parasitoids to perceive and decode natural chemicals for their own purposes (Table 4). The significance of host chemicals in the evolution of host specialization and host shifts in phytophagous insects has been emphasized, discussed and statistically tested, but has been hardly outlined in parasitoids (for example, Becerra, 1997; Berenbaum, 1983; Ehrlich & Raven, 1964; Shaw, 1988; Vet & Dicke, 1992). Sequestration of host toxins by phytophagous insects is well known (for example, Pasteels & Rowell-Rahier, 1991), but only a very few studies have investigated whether parasitoids sequester noxious components of phytophagous insects, perhaps to prevent predation or hyperparasitization (see this review). It will be an interesting task to follow the fate of plant toxins sequestered by chrysomelids and to keep track of the numerous *de novo* produced chrysomelid toxins into the third trophic level.

Future studies will hopefully untwist the chain of synomonal, kairomonal and allomonal molecules, which is pulled back and forth between the different trophic levels during evolution.

Acknowledgement

We thank Olaf Kahl and Johannes Steidle for very helpful comments on earlier drafts of the manuscript.

References

Agelopoulos, N.G., Dicke, M., and Posthumus, M.A. 1995. Role of volatile infochemicals emitted by feces of larvae in host-searching behavior of parasitoid *Cotesia rubecola* (Hymenoptera: Braconidae): A behavioral and chemical study. J. Chem. Ecol. 21: 1789-1811.

Arakaki, N., Wakamura, S., and Yasuda, T. 1996. Phoretic egg parasitoid, *Telenomus euproctidis* (Hymenoptera: Scelionidae), uses sex pheromone of tussock moth *Euproctis taiwana* (Lepidoptera: Lymantriidae) as a kairomone. J. Chem. Ecol. 22: 1079-1085.

Arakaki, N., Wakamura, S., Yasuda, T., and Yamagishi, K. 1997. Two regional strains of a phoretic egg parasitoid, *Telenomus euproctidis* (Hymenoptera: Scelionidae), that use different sex pheromones of two allopatric tussock moth species as kairomones. J. Chem. Ecol. 23: 153-161.

Arthur, A.P. 1981. Host acceptance by parasitoids. In: Nordlund, D.A., Jones, R.L., and W.J. Lewis (eds.), Semiochemicals, their role in pest control. John Wiley and Sons, New York: 97-120.

Barbier, J. 1976. Note sur le parasitisme de Clytrinae (Col. Chrysomelidae) par de Mutillidae (Hym.). Entomologiste 32: 163-165.

Becerra, J.X. 1997. Insects on plants: Macroevolutionary chemical trends in host use. Science 276: 253-256.

Beckage, N.E. 1991. Host-parasite hormonal relationships: A common theme? Expl. Parasit. 72: 332-338.

Beckage, N.E. 1997. Parasites and pathogens. Effects on host hormones and behavior. Chapman & Hall, 352 pp.

Beckage, N.E. 1998. Die Geheimwaffen der Schlupfwespen. Spektrum der Wissenschaften 2: 78-85.

Berenbaum, M. 1983. Coumarins and caterpillars: A case for coevolution. Evolution 37: 163-179.

Benn, M., DeGrave, J., Gnanasunderam, C., and Hutchins, R. 1979. Host plant pyrrolizidine alkaloids in *Nyctemera annulata* Boisduval: Their persistence through the life-cycle and transfer to a parasitoid. Experientia 35: 731-732.

Bin, F., Vinson, S.B., Strand, M.R., Colazza, S., and Jones, W.A. 1993. Source of an egg kairomone for *Trissolcus basalis*, a parasitoid of *Nezara viridula*. Physiol. Ent. 18: 7-15.

Blum, M.S. 1981. Chemical defences of arthropods. Academic Press, New York, 562 pp.

Blum, M.S. 1994. Antipredator devices in larvae of the Chrysomelidae: a unified synthesis for defensive eclecticism. In: Jolivet, P.H., Cox, M.L., and Petitpierre, E. (eds.), Novel aspects of the biology of Chrysomelidae. Kluwer Academic Publishers, Dordrecht: 277-288.

Borden, J.H. 1982. Aggregation pheromones. In: Mitton, J.B. and Sturgeon, K.B. (eds.), Bark beetles in North American conifers: ecology and evolution. University of Texas Press, Austin: 74-139.

Borden, J.H. 1985. Aggregation pheromones. In: Kerkut, G.A. and Gilbert, L.I. (eds.), Comprehensive insect physiology, biochemistry and pharmacology. Vol. 9. Pergamon Press, New York: 257-285.

Brown, J.J., Kiuchi, M., Kainoh, J., and Takeda, S. 1993. In vitro release of ecdysteroids by an endoparasitoid, *Ascogaster reticulatus* Watanabe. J. Insect Physiol. 39: 229-234.

Carroll, C.R. 1978. Beetles, parasitoids and tropical morning glories: a study in host discrimination. Ecol. Ent. 3: 79-85.

Chuman, T., Guss, P.L., Doolittle, R.E., McLaughlin, J.R., Krysan, J.R., Schalk, J.M., and Tumlinson, J.H. 1987. Identification of female produced sex pheromone from banded cucumber beetle, *Diabrotica balteata* LeConte (Coleoptera: Chrysomelidae). J. Chem. Ecol. 13: 1601-1616.

Colazza, S. and Bin, F. 1992. Introduction of the oophage *Edovum puttleri* Griss. (Hymenoptera: Eulophidae) in Italy for the biological control of Colorado potato beetle. In: Bin, F. (ed.), Proc. 4th Europ. workshop, Insect parasitoids, (Perugia, April 3-5, 1991). Redia 74: 203-225.

Conti, E., Jones, W.A., Bin, F., and Vinson, S.B. 1997. Oviposition behavior of *Anaphes iole*, an egg parasitoid of *Lygus hesperus* (Hymenoptera: Mymaridae; Heteroptera: Miridae). Ann. ent. Soc. Am. 90: 91-101.

Cork, A., Hall, D.R., Blaney, W.M., and Simmonds, M.S.J. 1991. Identification of a component of the female sex pheromone of *Callosobruchus analis* (Coleoptera: Bruchidae). Tetrahedron Letters 32: 129-132.

Corrigan, J., Lashomb, J., May, M., and Schal, C. 1991. Oviposition behavior of *Edovum puttleri,* reared on two hosts, *Leptinotarsa decemlineata* and *Leptinotarsa texana*. Entomologia exp. appl. 61: 139-148.

Cortesero, A.M. and Monge, J.P. 1994. Influence of pre-emergence experience on response to host and host plant odours in the larval parasitoid *Eupelmus vuilleti*. Entomologia exp. appl. 72: 281-288.

Cortesero, A.M., Monge, J.P., and Huignard, J. 1993. Response of the parasitoid *Eupelmus vuilleti* to the odours of the phytophagous host and the host plant in an olfactometer. Entomologia exp. appl. 69: 109-116.

Coudron, T.A. and Puttler, B. 1988. Response of natural and factitious hosts to the ectoparasite *Euplectrus plathypenae* (Hymenoptera: Eulophidae). Ann. ent. Soc. Am. 81: 931-937.

Cox, M.L. 1994. The Hymenoptera and Diptera parasitoids of Chrysomelidae. In: Jolivet, P.H., Cox, M.L., and Petitpierre, E. (eds.), Novel aspects of the biology of Chrysomelidae. Kluwer Academic Publishers, Dordrecht: 419-468.

Credland, P.F. and Wright, A.W. 1990. Oviposition deterrents of *Callosobruchus maculatus* (Coleoptera: Bruchidae). Physiol. Ent. 15: 285-298.

Damman, H. and Cappuccino, N. 1991. Two forms of egg defence in a chrysomelid beetle: egg clumping and excrement cover. Ecol. Ent. 16: 163-167.

Dettner, K. and Liepert, C. 1994. Chemical mimicry and camouflage. A. Rev. Ent. 39: 129-154.

Dettner, K. and Schwinger, G. 1987. Chemical defense in the larvae of the leaf beetle *Gonioctena viminalis* L. (Coleoptera: Chrysomelidae). Experientia 43: 458-460.

Dobler, S., Daloze, D., and Pasteels, J.M. 1998. Sequestration of plant compounds in a leaf beetle´s defensive secretion: cardenolides in *Chrysochus*. Chemoecology 8: 111-118.

Doury, G., Rojas-Rousse, D., and Periquet, G. 1995. Ability of *Eupelmus orientalis* ectoparasitoid larvae to develop on unparalysed host in absence of female stinging behaviour. J. Insect Physiol. 41: 287-296.

Dowden, P.B. 1939. *Schizonotus sieboldi,* an important parasite of imported willow leaf beetle (*Plagiodera versicolora*). J. agric. Res. 58: 581-592.

Du, Y.-J., Poppy, M., and Powell, W. 1996. Relative importance of semiochemicals from first and second trophic level in host foraging behavior of *Aphidius ervi*. J. Chem. Ecol. 22: 1591-1605.

Ehrlich, P.R. and Raven, P.H. 1964. Butterflies and plants: A study in coevolution. Evolution 18: 586-608.

Eisner, T., Van Tassel, E., and Carrel, J.E. 1967. Defensive use of a "fecal shield" by a beetle larva. Science 158: 1471-1473.

Field, R.P. and Kwong, R.M. 1994. Biological control of the elm leaf beetle. Plant Prot. Quart. 9: 47-49.

Fung, S.Y. 1988. Butenolides in parasitoids and adults of small ermine moths, *Yponomeuta spp.* (Lepidoptera: Yponomeutidae). Proc. K. ned. Akad. Wet. 91: 363-367.

Gardenghi, G., Mazzoli, G.-L., and Tommasini, M.-G. 1994. Behavior and superparasitization of *Edovum puttleri* Grissell (Hym. Eulophidae), an egg parasitoid of *Leptinotarsa decemlineata* (Say) (Col. Chrysomelidae). Bollettino dell' Istituto di Entomologia "Guido Grandi" della Universita degli Studi di Bologna 48: 137-154.

Godfray, H.C.J. 1994. Parasitoids. Behavioral and evolutionary ecology. Princeton University Press, Princeton, 473 pp.

Greany, P.D. and Oatman, E.R. 1972. Analysis of host discrimination in the parasite *Orgilus lepidus*. Ann. ent. Soc. Am. 65: 377-383.

Guss, P.L., Camey, R.L., Sonnet, P.E., and Tumlinson, J.H. 1983a. Stereospecific sex attractant for *Diabrotica cristata* Harris (Coleoptera: Chrysomelidae). Envir. Ent. 12: 1296-1297.

Guss, P.L., Tumlinson, J.H., Sonnet, P.E., and McLaughlin, J.R. 1983b. Identification of a female produced sex pheromone from the southern corn rootworm *Diabrotica undecimpunctata howardi* Barber. J. Chem. Ecol. 9: 1363-1375.

Guss, P.L., Tumlinson, J.H., Sonnet, P.E., and Proveaux, A.T. 1982. Identification of female produced sex-pheromone of the western corn rootworm. J. Chem. Ecol. 8: 545-556.

Guss, P.L., Sonnet, P.E., Camey, R.L., Branson, T.F., and Tumlinson, J.H. 1984. Response of *Diabrotica virgifera virgifera, D. v. zea,* and *D. porracea* to stereoisomers of 8-methyl-2-decyl propanoate. J. Chem. Ecol. 10: 1123-1132.

Guss, P.L., Sonnet, P.E., Camey, R.L., Tumlinson, J.H., and Wilkin, P.H. 1985. Response of northern corn rootworm *Diabrotica barberi* Smith and Lawrence, to stereoisomeres of 8-methyl-2-decyl propanoate. J. Chem. Ecol. 11: 21-26.

Harborne, J.B. 1993. Introduction to ecological biochemistry. Academic Press Limited, 383 pp.

Higuchi, H. and Suzuki, Y. 1996. Host handling behavior of the egg parasitoid *Telenomus triptus* to the egg mass of the stink bug *Piezodorus hybneri*. Entomologia exp. appl. 80: 475-479.

Hilker, M. 1994. Egg deposition and protection of eggs in Chrysomelidae. In: Jolivet, P.H., Cox, M.L., and Petitpierre, E. (eds.), Novel aspects of the biology of Chrysomelidae. Kluwer Academic Publishers, Dordrecht: 263-276.

Hilker, M., Eschbach, U., and Dettner, K. 1992. Occurrence of anthraquinones in eggs and larvae of several Galerucinae (Coleoptera: Chrysomelidae). Naturwiss. 79: 271-274.

Hilker, M. and Köpf, F. 1994. Evaluation of the palatability of chrysomelid larvae containing anthraquinones to birds. Oecologia 100: 421-429.

Hinton, H.E. 1981. Biology of insect eggs. Vol. I-III. Pergamon Press, Oxford.

Hofstetter, R.W. and Raffa, K.F. 1997. Effects of host diet on the orientation, development, and subsequent generations of the gypsy moth (Lepidoptera: Lymantriidae) egg parasitoid *Ooencyrtus kuvanae* (Hymenoptera: Encyrtidae). Envir. Ent. 26: 1276-1282.

Howard, D.F., Phillips, D.W., Jones, T.H., and Blum, M.S. 1982. Anthraquinones and anthrones: occurrence and defensive function in a chrysomelid beetle. Naturwissenschaften 69: 91-92.

Jansson, R.K. and Lashomb, J.H. 1988. Host-habitat conditioning in the eulophid parasitoid, *Edovum puttleri:* effects of herbivore's host and its food plant. Entomologia exp. appl. 46: 173-180.

Jolivet, P.H., Cox, M.L., and Petitpierre, E. (eds.) 1994. Novel aspects of the biology of Chrysomelidae. Kluwer Academic Publishers, Dordrecht, 582 pp.

Jolivet, P.H. and Cox, M.L. (eds.) 1996. Chrysomelidae biology, (Vol. I-III). SPB Academic Publishing bv, Amsterdam.

Jones, D., Parsons, J., and Rothschild, M. 1962. Release of hydrocyanic acid from crushed tissues of all stages in the life-cycle of species of the Zygaeninae (Lepidoptera). Nature 153: 52-53.

Jones, R.L. 1981. Chemistry of semiochemicals involved in parasitoid-host and predator-prey relationship. In Nordlund, D.A., Jones, R.L., and Lewis, W.J. (eds.), Semiochemicals, their role in pest control. John Wiley and Sons, New York: 921-927.

Kainoh, Y. 1990. Chemicals controlling the behaviour of parasitoids and predators. In: Bay-Petersen, J. (ed.) The use of natural enemies to control agricultural pests. FFTC Book Series 40: 212-221.

Kaiser, L., Pham-Delegue, M.H., Backchine, E., and Masson, C. 1989. Olfactory response of *Trichogramma madidis* Pint. et Voeg.; effect of chemical cues and behavioural plasticity. J. Ins. Behav. 2: 701-710.

Karban, R. and Baldwin, I.T. (1997). Induced responses to herbivory. The University of Chicago Press. 319 pp.

Köpf, A., Rank, N.E., Roininen, H., and Tahvanainen, J. 1997. Defensive larval secretions of leaf beetles attract a specialist predator *Parasyrphus nigritarsis.* Ecol. Ent. 22: 176-183.

Krysan, J.R., Wilkin, P.H., Tumlinson, J.H., Sonnet, P.E., Carney, R.L., and Guss, P.L. 1986. Responses of *Diabrotica lemniscata* and *D. longicornis* to stereoisomers of 8-methyl-2-decyl propanoate and studies on the pheromone of *D. longicornis.* Ann. ent. Soc. Am. 79: 742-766.

Lashomb, J., Krainacker, D., Jansson, R.K., Ng, Y.S., and Chianese, R. 1987. Parasitism of *Leptinotarsa decemlineata* (Say) eggs by *Edovum puttleri* Grisell (Hymenoptera: Eulophidae): effects of host age, parasitoid age, and temperature. Can. Ent. 119: 75-82.

Leluk, J., Schmidt, J., and Jones, D. 1989. Comparative studies on the protein composition of hymenopteran venom reservoirs. Toxicon 27:105.

Leonard, D.E., Wu, Z.-X., and Ferro, D.N. 1987. Responses of parasite *Edovum puttleri* to kairomone from eggs of Colorado potato beetle, *Leptinotarsa decemlineata.* J. Chem. Ecol. 13: 335-344.

Lewis, W.J., Nordlund, D.A., Gueldner, R.C., Teal, P.E.A., and Tumlinson, J.H. 1982. Kairomones and their use for management of entomophagous insects. XIII. Kairomonal activity for *Trichogramma* spp. of abdominal tips, excretion and a synthetic sex pheromone blend of *Heliothis zea* (Boddie) moths. J. Chem. Ecol. 8: 1323-1331.

Maini, S. and Nicoli, G. 1990. *Edovum puttleri* (Hymenoptera: Eulophidae): Biological activity and responses to normal and frozen eggs of *Leptinotarsa decemlineata* (Coleoptera: Chrysomelidae). Entomophaga 35: 185-194.

Margaritis, L.H. 1985. Structure and physiology of the eggshell. In: Kerkut, G.A. and Gilbert, L.I. (eds.), Comprehensive insect physiology, biochemistry and pharmacology. Vol. 1. Pergamon Press, New York: 151-230.

Mattiacci, L., Vinson, S.-B., Williams, H.-J., Aldrich, J.-R., and Bin, F. 1993. A long-range attractant kairomone for egg parasitoid *Trissolcus basalis,* isolated from defensive secretion of its host, *Nezara viridula.* J. Chem. Ecol. 19: 1167-1181.

Meiners, T. and Hilker, M. 1997. Host location in *Oomyzus gallerucae* (Hymenoptera: Eulophidae), an egg parasitoid of the elm leaf beetle *Xanthogaleruca luteola* (Coleoptera: Chrysomelidae). Oecologia 112: 87-93.

Meiners, T., Koepf, A., Stein, C., and Hilker, M. 1997. Chemical signals mediating interactions between *Galeruca tanaceti* L. (Coleoptera, Chrysomelidae) and its egg parasitoid *Oomyzus galerucivorus* (Hedqvist) (Hymenoptera, Eulophidae). J. Ins. Behav. 10: 523-539.

Messina, F.J., Kemp, J.L., and Dickson, J.A. 1992. Plasticity in the egg-spacing behavior of a seed beetle: effect of host deprivation and seed patchiness (Coleoptera: Bruchidae). J. Ins. Behav. 5: 609-621.

Messner, B. 1983. Dopa-Oxidase-gehärtete Sekrete schützen das Eigelege von *Galeruca tanaceti* L. (Coleptera: Chrysomelidae). Entomol. Nachrichten u. Berichte (Leibzig) 27: 221-223.

Mityakina, O.-N., Maksimov, V.-N., and Klimenko, A.-A. 1993. Diurnal timing of oviposition of *Edovum puttleri* (Hymenoptera, Eulophidae), an egg parasite of Colorado potato beetle. Zool. Zh. 72: 71-76.

Mizutani, N., Wada, T., Higuchi, H., Ono, M., and Leal, W.S. 1997. A component of a synthetic aggregation pheromone of *Riptortus clavatus* (Thunberg) (Heteroptera: Alydidae), that attracts an egg parasitoid, *Ooencyrtus nezarae* Ishii (Hymenoptera: Encyrtidae). Appl. Ent. Zool. Tokyo 32: 504-507.

Monge, J.P. and Cortesero, A.M. 1996. Tritrophic interactions among larval parasitoids, bruchids and Leguminosae seeds; influence of pre-and post-emergence learning on parasitoids response to host and host-plant cues. Entomologia exp. appl. 80: 293-296.

Morris, W., Grevstad, F., and Herzig, A. 1996. Mechanisms and ecological functions of spatial aggregation in chrysomelid beetles. In: Jolivet, P.H. and Cox, M.L. (eds.), Chrysomelidae biology, SPB Academic Publishing bv, Amsterdam: 303-322.

Nahrstedt, A. and Davis, R.H. 1986. (R)-Mandelonitrile and prunasin, the sources of HCN in all stages of *Paropsis atomaria* (Coleoptera: Chrysomelidae). Z. Naturf. C. 41: 928-934.

Noldus, L.P.J.J. 1988. Response of the egg parasitoid *Trichogramma pretiosum* to the sex pheromone of its host *Heliothis zea*. Entomologia exp. appl. 48: 293-300.

Noldus, L.P.J.J., Potting, R.P.J., and Barendregt, H.E. 1991. Moth sex pheromone adsorption to leaf surface – Bridge in time for chemical spies. Physiol. Ent. 16: 329-344.

Ormel, G.-J., Gort, G., and Van Alebeek, F.A.N. 1995. Analysing host location in *Uscana lariophaga* (Hymenoptera: Trichogrammatidae), an egg parasitoid of bruchids (Coleoptera: Bruchidae), using Cox's proportional hazards model. Bull. ent. Res. 85: 113-123.

Pajni, H.R. and Gupta, S. 1990. Role of antennal sensillae in perception of sex pheromone in bruchids, *Callosobruchus maculatus* (F.) and *C. chinensis* (L.). Indian J. Exp. Biol. 28: 174-178.

Pasteels, J.M. and Grégoire, J.-C. 1984. Selective predation on chemically defended chrysomelid larvae. A conditioning process. J. Chem. Ecol. 10: 1693-1700.

Pasteels, J.M. and Rowell-Rahier, M. 1991. Proximate and ultimate causes for host plant influence on chemical defense of leaf beetles (Coleoptera: Chrysomelidae). Entomologia gen. 15: 227-235.

Pasteels, J.M., Rowell-Rahier, M., Braekman, J.-C., and Daloze, D. 1994. Chemical defence of adult leaf beetles updated. In: Jolivet, P.H., Cox, M.L., and Petitpierre, E. (eds.), Novel aspects of the biology of Chrysomelidae. Kluwer Academic Publishers, Dordrecht: 289-302.

Pasteels, J.M., Rowell-Rahier, M., Ehmke, A., and Hartmann, T. 1996. Host-derived pyrrolizidine alkaloids in *Oreina* leaf beetles: physiological, ecological and evolutionary aspects. In: Jolivet, P.H. and Cox, M.L. (eds.), Chrysomelidae biology, SPB Academic Publishing bv, Amsterdam: 213-226.

Peck, O. 1963. A catalogue of the nearctic Chalcidoidea (Insecta: Hymenoptera). Can. Entomol. 30: 1-1092.

Piek, T. and Spanjer, W. 1986. Chemistry and pharmacology of solitary wasp venoms. In: T. Piek (ed.), Venoms of the hymenoptera. Academic Press, London: 161-307.

Price, P.W. 1970. Trail odours: recognition by insect parasitoids in cocoons. Science 170: 546-547.

Quicke, D.L.J. 1988. Spider venoms and the prospect of safer insecticides. New Scient. 1640: 38-41.

Quicke, D.L.J. 1997. Parasitic wasps. Chapman and Hall, London. 470 pp.

Richards, O.W. 1980. Scolioidea, Vespoidea, and Sphecoidea. Hymenoptera, Aculeata. Handbook for the Identification of British insects. Royal Entomological Society of London, London, 118 pp.

Rowell-Rahier, M. and Pasteels, J.M. 1992. Third trophic level influences of plant allelochemicals. In: Rosenthal, G.A. and Berenbaum, M.R. (eds.), Herbivores. Their interactions wirh secondary plant metabolites, Vol. II, Academic Press, San Diego: 243-277.

Rutledge, C.E. 1996. A survey of identified kairomones and synomones used by insect parasitoids to locate and accept their hosts. Chemoecology 7: 121-131.

Salt, G. 1937. The sense used by *Trichogramma* to distinguish between parasitized and unparasitized hosts. Proc. R. Soc. (B) 122: 57-75.

Scherf, H. 1956. Zum feineren Bau der Eigelege von *Galeruca tanaceti* L. (Col., Chrysom.). Zool. Anz. 157: 124-130.

Schmidt, G.T. 1974. Host acceptance behaviour of *Campoletis sonorensis* (Hym., Ichneumonidae). Ann. ent. Soc. Am. 67: 835-844.

Schmidt, J.M. 1991. The role of physical factors used in tritrophic interactions. In: Bin, F. (ed.), Proceedings of the 4th European workshop, Insect parasitoids, (Perugia, April 3-5, 1991). Redia 74: 43-94.

Shaw, S.R. 1988. Euphorine phylogeny: the evolution of diversity in host utilization by parasitoid wasps (Hymenoptera: Braconidae). Ecol. Ent. 13: 323-335.

Shimizu, T., Tatsuki, T., and Takeda, N. 1993. Aromatic amino acids in the venom of the braconid parasitoid *Apanteles kariyai*. Z. Naturforsch. C. 48:108-109.

Skinner, W.S., Dennis, P.A., and Quistad, G.B. 1990. Partial characterization of toxins from *Goniozus legneri* (Hymenoptera: Bethylidae). J. econ. Entomol. 83: 733-736.

Strand, M.R. 1986. The physiological interactions of parasitoids with their hosts and their influence on reproductive strategies. In: Waage, J. and Greathead, D. (eds.), Insect parasitoids. Academic press, London: 97-136.

Strand, M.R. and Vinson, S.B. 1983. Factors affecting host recognition and acceptance in the egg parasitoid *Telenomus heliothidis* (Hymenoptera: Scelionidae). Envir. Ent. 12: 1114-1119.

Tanaka, T. and Wago, H. 1990. Ultrastructural and functional maturation of teratocytes of *Apanteles kariyai*. Arch. Ins. Biochem. Physiol. 13: 187-197.

Thomson, R.H. 1987. Naturally occurring quinones III. Recent advances. Chapman and Hall, London, 732 pp.

Togashi, I. 1974. The hymenopterous parasites of *Oulema oryzae* (Kuwayama) (Coleoptera, Chrysomelidae) in Ishikawa Prefecture. Mushi 48: 7-13.

Tumlinson, J.H., Turlings, T.C.J., and Lewis, W.J. 1992. The semiochemical complexes that mediate insect parasitoid foraging. Agric. Zool. Rev. 5: 221-245.

Turlings, T.C.J., Loughrin, J.H., McCall, P.J., Röse, U.S.R., Lewis, W.J., and Tumlinson, J.H. 1995. How caterpillar-damaged plants protect themselves by attracting parasitic wasps. Proc. natn. Acad. Sci. U.S.A. 92: 4169-4174.

Turlings, T.C.J., Tumlinson, J.H., Eller, F.J., and Lewis, W.J. 1991. Larval-damaged plants: Source of volatile synomones that guide the parasitoid *Cotesia marginiventris* to the microhabitat of its host. Entomologia exp. appl. 58: 72-85.

Van Alebeek, F.A.N. and Van Huis, A. 1997. Host location in stored cowpea by the egg parasitoid *Uscana lariophaga* Steffan (Hym., Trichogrammatidae). J.Appl.Entomol. 121:399-405.

Van Alphen, J.J.M. 1980. Aspects of the foraging behaviour of *Tetrastichus asparagi* Crawford and *Tetrastichus* spec. (Eulophidae), gregarious egg parasitoids of the asparagus beetle *Crioceris asparagi* L. and *C. duodecimpunctata* L. (Chrysomelidae). I. Host species selection, host stage selection and host discrimination. Neth. J. Zool. 30: 307-325.

Van Baaren, J., Boivin, G., and Nenon, J.-P. 1994. Intra- and interspecific host discrimination in two closely related egg parasitoids. Oecologia 100: 325-330.

Van Baaren, J., Nenon, J.-P., and Boivin, G. 1995. Comparison of oviposition behavior of a solitary and a gregarious parasitoid (Hymenoptera: Mymaridae). J. Ins. Behav. 8: 671-686.

Van Huis, A., Schütte, C., Cools, M.H., Fanget, P., Van der Hoek, H., and Piquet, S.P. 1994. The role of semiochemicals in host location by *Uscana lariophaga,* egg parasitoid of *Callosobruchus maculatus.* In: Proc. 6th internat. working conf. on stored-product protection Vol. 2: 1158-1164.

Vet L.E.M. and Dicke, M. 1992. Ecology of infochemical use by natural enemies in a tritrophic context. A. Rev. Ent. 37: 141-172.

Vinson, S.B. 1985. The behaviour of parasitoids. In: Kerkut, G.A. and Gilbert, L.I. (eds.), Comprehensive insect physiology, biochemistry and pharmacology. Vol. 9. Pergamon Press, New York: 417-469.

Vinson, S.B. 1991. Chemical signals used by parasitoids. In: Bin, F. (ed.), Proc. 4th Europ. workshop, Insect parasitoids, (Perugia, April 3-5, 1991). Redia 74: 15-42.

Völkl, W., Hübner, G., and Dettner, K. 1994. Interactions between *Alloxysta brevis* (Hymenoptera, Cynipoidea, Alloxystidae) and honeydew-collecting ants: how an aphid hyperparasitoid overcomes ant aggression by chemical defence. J. Chem. Ecol. 20: 2901-2915.

Wajnberg, E. and Hassan, S.A. 1994. Biological control with egg parasitoids. Oxford University Press. 304 pp.

Weseloh, R.M. 1981. Host location by parasitoids. In Nordlund, D.A., Jones, R.L., and Lewis, W.J. (eds.), Semiochemicals, their role in pest control. John Wiley and Sons, New York: 79-96.

Wiskerke, J.S.C., Dicke, M., and Vet, L.E.M. 1993. Larval parasitoid uses aggregation pheromone of adult hosts in foraging behaviour: A solution to the reliabilty – detectability problem. Oecologia 93: 145-148.

Wylie, H.G. 1984. Oviposition and survival of three nearctic euphorine braconids in crucifer-infesting flea beetles (Coleoptera: Chrysomelidae). Can. Ent. 116: 1-4.

Zaki, F.N. 1985. Reactions of the egg parasitoid *Trichogramma evanescens* Westw. to certain insect pheromones. Z. angew. Ent. 99: 448-453.

MACROEVOLUTIONARY ASPECTS OF LARVAL SHIELD DEFENCES

FREDRIC V. VENCL [1,2] and TIMOTHY C. MORTON[3]
[1] *Department of Neurobiology and Behavior, The State University of New York at Stony Brook, Stony Brook, NY 11794-5230, USA;* [2] *The Smithsonian Tropical Research Institute;* [3] *Department of Ecology and Evolution, The University of Chicago, Chicago, IL 60637-1573, USA*

Abstract

Chief among the numerous barriers to colonization of the leaf-surface adaptive zone by herbivorous insects are plant chemical defences and natural enemies. Although the larval stage is a particularly vulnerable one to both factors, a restricted number of Chrysomelidae lineages with grazing larvae have successfully diversified on leaf surfaces. The majority of them however, have waste-based structures, such as fecal shields or parasols. There are direct chemical links between larval diets and predator defences. Necessary for survival, the fecal shields deployed by larval Alticinae, Criocerinae, and cassidoid Hispinae are chemical defences derived from both primary and secondary host plant metabolites. Larvae transform particular host precursors before their incorporation into the shields, where they act to repel or deter invertebrate predators. A chemical comparison of the unrelated species, *Lema trilinea* White and *Plagiometriona clavata* (F.), in the same trophic niche, revealed striking convergence in defensive chemistry. We used a combination of phylogenetic inference, chemical analysis, and bioassays to compare the efficacy of shield defences that differed in chemical complexity. The shields of the derived *Lema* lineage, containing 'novel' steroidal alkaloids from historically younger solanaceous hosts, are more effective than those of the sister taxon, *Neolema,* whose shields are derived from chemically simpler commelinaceous hosts. We hypothesize that after *Lema* colonized the Solanaceae, their shields became more potent, broad spectrum defences that enhanced their diversification rate relative to the *Neolema* clade that retained the ancestral, Commelinaceae. Chemical commonality probably facilitated host shifts. Moreover, the predators able to circumvent the derived *Lema* shield defence containing novel compounds are descendants of a subset of the enemies that attack the ancestral defence, and thus, the *Lema* clade shed most of its former enemies. These findings support the nasty-host hypothesis which posits that host specialization is a selective response to natural enemies through dietary dependence on chemically well-endowed hosts for defensive compounds. Predation is implicated as a major determiner driving host shifts through chemical enhancement of shield defences. An escape-and-diversify scenario based on shield defences may have also occurred in the radiation of the New World cassidoid Hispinae. Fecal shields as chemical defences are 'key innovations' that likely fostered the adaptive radiation of the New World Criocerinae and of the Hispinae on leaf surfaces.

1. Introduction

Members of the Chrysomelidae exhibit a remarkable variety of morphological and behavioral attributes for defence against predation. Among strategems such as thanatosis, saltation, maternal care, crypsis, stridulation, and mimicry, to name just a few, chemical defences have received the most attention. Over the past twenty years, considerable study has been devoted to defensive mechanisms that sequester or synthesize chemicals that fortify hemolymph, reflexively bled exudates, and exocrine gland secretions (Ferguson *et al.*,

Advances in Chysomelidae Biology 1, pp. 217–238
Edited by M.L. Cox

1985; Pasteels *et al.*, 1994; and references therein). Although perhaps more banal, larval waste-retention is nevertheless, a very widespread trait within the Chrysomelidae (Crowson, 1981; Blum, 1994; Olmstead, 1994). Despite their ubiquity, waste-based structures have been neglected by chemical ecologists; little is known about their effectiveness, either as physical, or as chemical barriers to attacking enemies (Pasteels *et al.*, 1988; Blum, 1994; Olmstead, 1994). The available evidence supports the hypothesis that many fecal shields, scatoshells, fecal cases, and other waste-embellished characters function primarily as non-chemical, barrier-type defences (Eisner *et al.*, 1967; Root and Messina, 1983; Erber, 1988; Olmstead and Denno, 1993). Nonetheless, we believe there is an important chemical link between waste-retention and the remarkably successful radiation of the Chrysomelidae.

The observations reported herein summarize approximately seven years of field and laboratory studies investigating the chemical ecology of the fecal shields in members of the Alticinae, Criocerinae, and cassidoid Hispinae. We begin by outlining the ecological context of leaf-grazing and by tracing historical patterns in larval defences and trophic habits within the Chrysomelidae. It has long been supposed that larval herbivorous insects that feed on plants rich in secondary chemicals might have readily available dietary sources of prefabricated defensive compounds. This expectation turns out to be incomplete. We present data that support the idea that fecal shields[1], rather than being inadvertent repositories, or 'waste-dumps' for toxic metabolites, are instead, complex chemical adaptations for defence against predation whose compositions are determined by the selective elimination of particular, transformed secondary, as well as primary host plant metabolites that act to modify predator behavior. We compare the shield systems of two unrelated species occupying the same trophic niche in an effort to identify ecological factors responsible for shield chemical convergence. We then explore the phylogenetic implications of shield dependence on dietary sources of defensive chemicals, the evolution of host specialization, and the role of shield defences in beetle diversification. We conclude with a discussion of the evolution of the fecal shields as key innovations in the radiation of the Criocerinae and of the Hispinae in the leaf surface adaptive zone.

2. The leaf surface-larval herbivore interface

In terms of diversity and abundance, phytophagous beetles must be counted among the most important primary consumers. Plant-feeding, however, and especially surface folivery, evidently represent an insurmountable ecological 'hurdle' for most other insect groups (Southwood, 1973; Strong *et al.*, 1984). While serious 'bottom-up' (that is, plant-imposed) obstacles, such as low leaf nutrient quality, attachment, mechanical, and chemical defences, restrict plant exploitation, there are equally severe, 'top-down' (that is, ecological) barriers that include desiccation, and natural enemies. Additionally, there may even be the 'sideways' barrier imposed by competition, to overcome (Denno *et al.*, 1995). Despite these formidable obstacles, a restricted number of insect taxa have formed intimate and prolonged relationships with their host plants, in which all aspects of the life cycle occur on

[1] In this chapter, we will use the term 'shield' in reference to both fixed shields and mobile parasols, both being waste-laced structures that lie on or above larvae. This is an intentionally broad definition because both shield types, due to their dorsal location, are potentially of strategic importance as either physical, or as chemical barriers, to thwart enemy attacks.

the host's leaves. Although a minority of insect groups, including Chrysomelidae, feed on plants, they account for the majority of insect species. Moreover, herbivorous insects are consistently more species-rich than are their corresponding non-herbivorous sister-groups (Mitter *et al.*, 1988; Farrell, 1998), indicating that colonization of the unsaturated leaf surface 'adaptive zone' (*sensu* Simpson, 1944) may well have fostered herbivore cladogenesis (Ehrlich and Raven, 1964; Southwood, 1973; Farrell, 1998).

The larval stage of the herbivore's life cycle is probably the most vulnerable one, especially for grazers (ectophytes). Not only are they physically exposed on the leaf surface, ectophytic larvae are flightless and lack the hardened cuticle of the adult. Because of their high apparancy and low vagility, larval ectophytes are reliable targets, and thus the risk of discovery by natural enemies is much greater for them than it is for internal feeders (Cox, 1994). Hence, predators and parasitoids may be the most important selective barriers to the evolution of the surface-feeding habit. Our research subjects for example, immature cassidoid-hispines and criocerines, are targets of between 65 and 70% of the known enemies for all the life stages combined (Schmitt, 1988; Cox, 1994, 1996; Olmstead, 1996). Furthermore, the majority of described chrysomelid species (>60%) have larvae with internal-feeding habits such as gallicolous, aquatic, stem, root, or leaf-mining; the widespread occurrence of endophytic trophic behaviors further supports the hypothesis that natural enemies have been generally limiting to the colonization of leaf surfaces.

3. Phyletic ordering of larval trophic habits and distribution of defensive syndromes

It is thought that larval surface folivory (ectophylly) is a derived trophic mode within the Chrysomelidae. Ectophylly is necessarily predicated on the evolution of some kind of protective mechanism (Crowson, 1981). The internal-feeding mode would seem to obviate the need for defence; basal subfamilies are characterized by subterranean, aquatic, or endophytic larval stages that presumably lack physical or chemical defences. Among chrysomelid subfamilies with ectophytic larvae, there is evidence for the phyletic ordering of various defensive strategems. The vast majority of species with ectophytic larvae, comprising nearly 20% of all described chrysomelids, exhibit some type of chemical defence, either through glandular secretions, sequestered allelochemicals, regurgitants, autohemoragic responses, or waste-retention. The latter category represents the majority of putatively protective strategems.

Exocrine glands and fortified hemolymph defences are restricted to the larval members of the derived Galerucinae and Chrysomelinae (Pasteels *et al*, 1984). In contrast, the occurrence of defensive syndromes based on fore-and-aft discharges, such as fecal shields and regurgitants, appears to be evenly represented among many subfamilies, most notably the members of the Camptosomata, Alticinae, cassidoid Hispinae (*sensu* Reid, 1995), and the Criocerinae. Chemical screening of a large sample of waste-based structures would likely reveal many to have antibiotic, repellent, or deterrent properties. Thus, we have directed our attention to these much neglected and very curious larval characteristics.

4. Fecal shields of shining leaf beetles

The most notable and least understood characteristic of the subfamily Criocerinae, the shining leaf beetles, is the larval habit of waste-retention; feces accumulate on the larval dorsum to form viscous shields (Fig. 1A, B). Shining leaf beetles are also notable for their dietary eclecticism: they have colonized members of both mono- and dicotyledonous plant families. In the New World for example, members of the genus *Lema* feed almost entirely on plants in the Solanaceae, a family recognized for many toxic secondary compounds. The closely related members of *Neolema* feed exclusively on plants in the Commelinaceae, a chemically unremarkable monocotelydonous family (Monrós, 1960; Schmitt, 1988; White,1993).

We suspected *Lema's* shield to be a chemically-based defence because *Solanum dulcamara* contains highly toxic steroidal glycoalkaloids (SGAs) and steroidal saponins (Mathé and Mathé, 1979; Tingey, 1984; Roddick, 1986; Hegnauer, 1992). In contrast, we expected *Neolema's* shield to function primarily as a physical barrier. We employed a generalist predatory ant in a bioassay to measure shield effectiveness as a defence. As background for the present study, we report briefly on methodology and findings of previous work comparing the shield defences of *L. trilinea* White and *N. sexpunctata* (Olivier).

4.1. Bioassays and diet modification experiments

We employed an ecologically relevant generalist predatory ant as a bioassay to test the hypothesis that fecal shields function as a defence. Justifying their use in bioassays, predatory ants are acknowledged to be important sources of larval mortality (Selman, 1988; Hölldobler and Wilson, 1990; Cox, 1996). Long-recognized as reliable detectors of chemically active substances (Eisner and Meinwald, 1966), ants proved helpful in our search for repellent or deterrent compounds.

Larvae of both *L. trilinea* and of *N. sexpunctata* with host-derived fecal shields suffered few losses when exposed to ant attack in the assays. The behavior of the ants indicated that both species' shields functioned as deterrent, rather than repellent chemical barriers that thwarted ant attacks after contact. Larvae lacking fecal shields, however, suffered nearly 100% mortality. If restricted to a substitute diet, larvae of both species suffered high mortality, even though they were covered by normal-looking shields. Deterrence was fully restored to larvae reared on the modified diet by the mechanical removal of their modified-diet shields and replacement with shields transferred from larvae feeding on the natural host. This last finding supported the idea that the larvae of both species were incapable of synthesizing any defensive of their own, and thus they are entirely dependent on their hosts for defensive substances.

4.1.2. Chemical screening of host plants and beetle shields

The bioassay-guided extraction and fractionation protocol used in our studies was specifically designed to partition natural compounds quickly and without degradation. It effectively removed all non-volatile organic compounds of moderate to high polarity, deactivated enzymes, and selectively separated the extract into non-polar, alkaline, and polar fractions (Fig. 2). After completion of the extraction/fractionation phase, fractions were

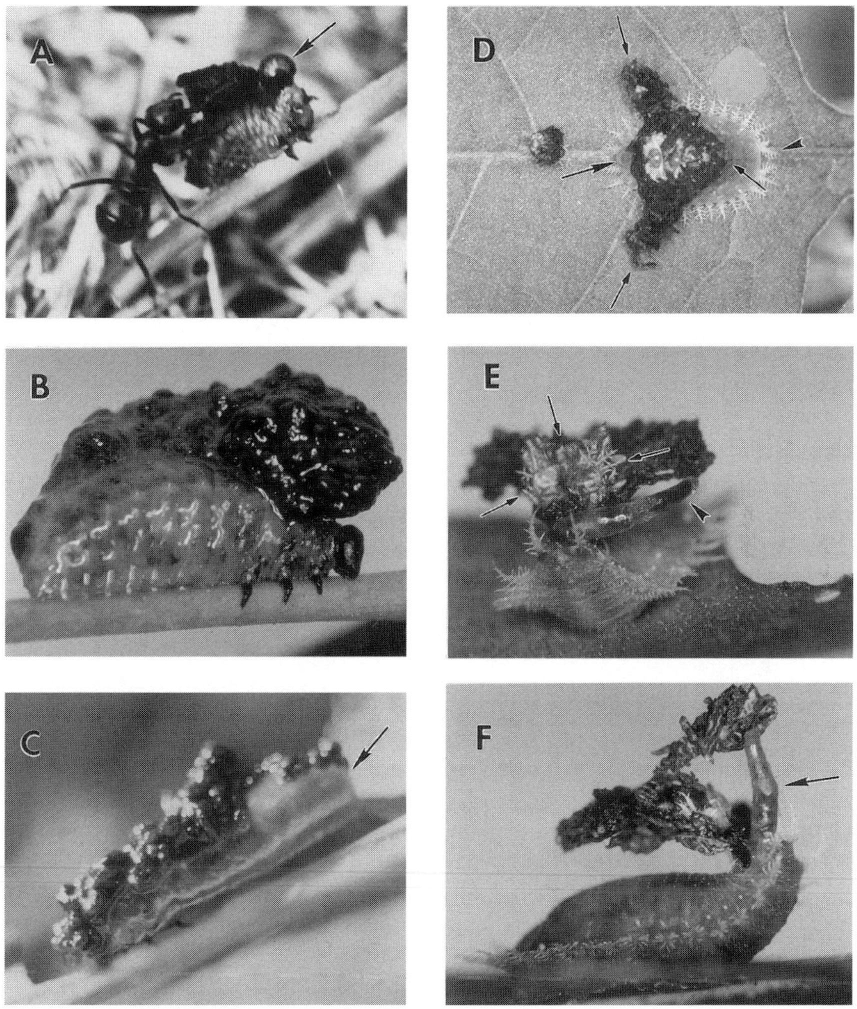

Fig. 1. A) Larva of the leaf-beetle, *Lema trilinea* under attack by the predatory ant, *Formica subsericea.* The larva has arched its head backward and regurgitated a droplet of buccal fluid (arrow). After contacting the fecal shield, (arrow-the dark mass on top of the larva), the ant hastily withdrew, vigorously groomed its antennae, and wiped its mouth parts against the substrate. Deposition of feces on top of the larva is due to a dorsally opening anus (from Vencl and Morton, 1999). B) Shield-bearing larva of *Neolema sexpunctata.* C) Due to a dorsal anus (arrow), feces of *Blepharida rhois* accumulate to form a shield, evident as the glistening mass on top of the larva (from Vencl and Morton, 1998a). D) The fecal shield of a fifth instar, *Plagiometriona clavata* larva (dorsal view) is attached to a highly movable appendage, the furca, which emanates from the 8th abdominal segment. Shown at rest in a level position, the shield is held like a parasol above the larva and is visible as the deltoid-shaped, dark structure (bounded by small arrows); it can be lifted, tilted, and rapidly flagged in the direction of a physical disturbance. The larva has fully retracted its telescoping anus (large arrow). A smaller (4 mm), second instar larva (left) is shown for comparison. Larvae are fringed with mechano-receptive scoli (arrowhead). E) Shown from behind, the parasol of *P. clavata* is a composite structure formed by the accumulation of exuviae (light material bounded by small arrows is a freshly cast exuvium), and of feces (dark material in background) deposited by a telescoping anus. Formed by the evagination of the 9th and 10th abdominal segments, the anus (arrowhead) is here extended to a length of 4.5 mm and is about to apply a bolus of feces (dark material visible within at the distal end) to the parasol's right flank. F) The telescoping anus (arrow), shown loading feces onto a dorsal projection, can deliver fecal material containing defensive compounds to any part of the parasol-like shield.

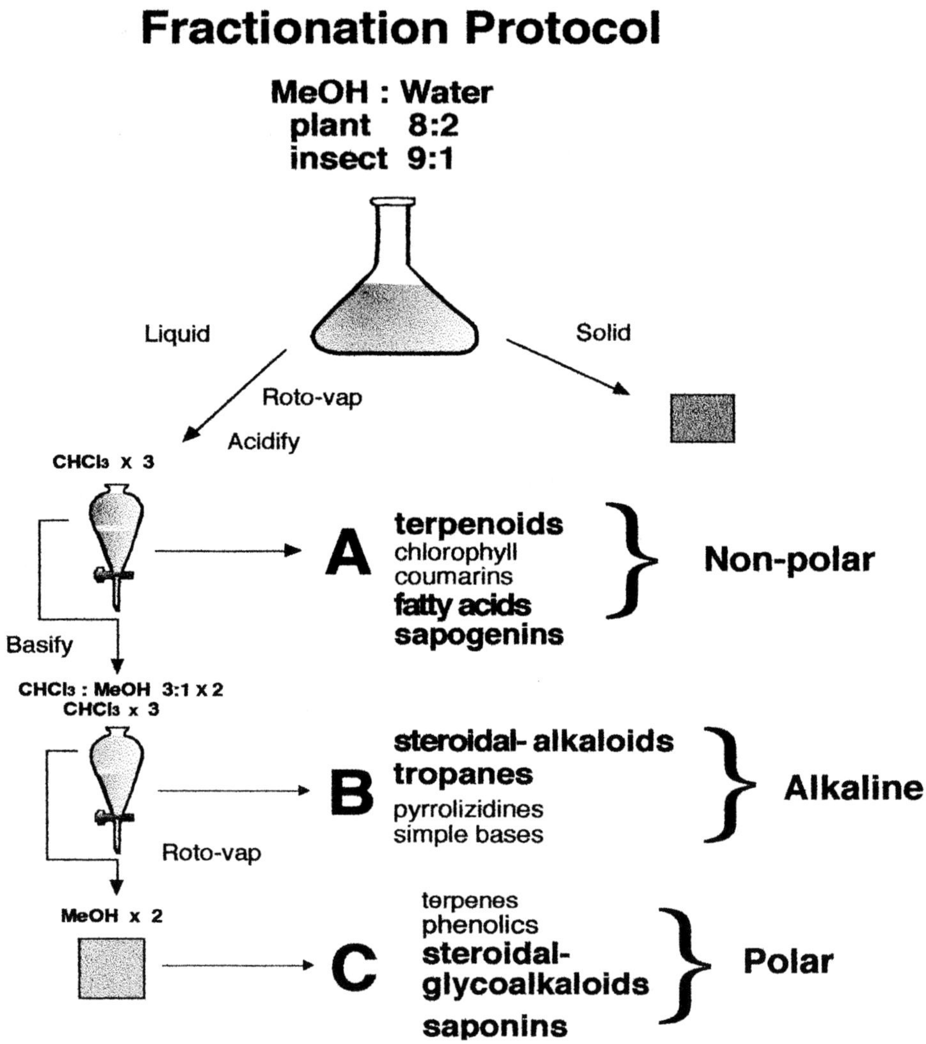

Fig. 2. The extraction/fractional protocol used to isolate compounds in plant and insect samples. As the dielectric constants of the solvent systems increased from A through C, compounds with increasing polarity were isolated. Fresh leaves or feces were weighed and extracted by vigorous maceration (leaves) or sonication (feces) in methanol (MeOH). Extracts were filtered, evaporated under reduced pressure, acidified, and then fractionated with chloroform ($CHCl_3$), filtered, and evaporated to yield the non-polar, non-alkaloid A fraction. The remaining aqueous phase was made strongly basic and was extracted with $CHCl_3$: MeOH twice, and then two times with $CHCl_3$. This second organic phase was reduced in volume to yield the moderately polar, alkaline B fraction. The remaining aqueous solution was evaporated and redissolved to yield the highly polar C fraction. Compounds in bold were later identified from fecal shields using GC-MS, HPLC, and FAB-MS.

bioassayed for repellence or deterrence. Active fractions were chemically analyzed and putative defensive compounds were isolated and characterized. Finally, pure standards of identified compounds or their closest available analogs were assayed to confirm their anti-predator activity.

Table 1. Chemical composition, compound source, and hos: associations of fecal shield-bearing Chrysomelidae species.

beetle subfamily / taxon[a]	host family	shield type[b]	shield chemical composition[c] compound name	origin[d]	mode of action[e]	reference
Alticinae						
Blepharida rhois	Anacardiaceae	FS	phytol	1°	D	Vencl and Morton, 1998
			hexadecanoic acid methyl ester	1°	D	
			octadecatrienoic acid methyl ester	1°	D	
			methyl gallate	2°	D	
			gallic acid	2°	D	
			1-,4-, and 5-galloyl glucoses	2°	D	
Hispinae						
Plagiometriona clavata	Solanaceae	P	phytol derivatives	1°	D	Vencl *et al.*, 1999
			hexadecanoic acidderivatives	1°	D	
			octadecenoic acid derivatives	1°	D	
			diosgenin	2°	D	
			diosgenone metabolites	2°	D	
			solasodine	2°	D	
			hydroxysolasodine-chacotriose	2°	?	
			hydroxysolasodine-solatriose	2°	?	
Eurypedus nigrosignata	Boraginaceae	P	hexadecanoic acid	1°	D?	Gómez, 1997
			octadecanoic acid	1°	D?	
			behenic acid	1°	?	
			α-pinene	2°	R	
			β-terpinene	2°	R	
			sabinene	2°	R	
			germacrene D	2°	R/D	
			caryophyllene	2°	R/D	
			β-sitosterol	2°	?	
Coptocycla leprosa	Boraginaceae	P	phytol	1°	D?	Gómez, 1997
			hexadecanoic acid methyl ester	1°	D?	
			octadecadienoic acid methyl ester	1°	D?	
			octadecanoic acid methyl ester	1°	D?	
			octadecenoic acid methyl ester	1°	D?	
			N-acetyltyramine	2°	D?	

Table 1. Cont.

beetle subfamily / taxon[a]	host family	shield type[b]	shield chemical composition[c] compound name	origin[d]	mode of action[e]	reference
Criocerinae						
Lema trilinea	Solanaceae	FS	hexadecanoic acid	1°	D	Morton and Vencl, 1998
			diosgenone metabolites	1°	D	
			solasodine	2°	D	
			hydroxy-solasodine chacotriose	2°	D	
			hydroxy-solasodine solatriose	2°	D	
Lema trivittata	Solanaceae	FS	octadecatrienoic acid methyl ester	1°	D	Vencl and Morton, unpubl.
			octadecadienoic acid methyl ester	1°	D	
			octadecanoic acid methyl ester	1°	D	
			octadecenoic acid methyl ester	1°	D	
			tropane derivatives	2°	D	
Neolema sexpunctata	Commelinaceae	FS	phytol derivatives	1°	D	Morton and Vencl, 1998
			hexadecanoic acid	1°	D	

[a] decreasingly derived taxa from top to bottom; [b] FS=fecal shield, P=parasol; [c] relative abundance greater than 5% of integrated GC peak area; [d] L=synthesized by larva, 1°=primary plant metabolism, 2°=secondary plant metabolism; [e] D=deterrent, R=repellent.

4.1.3. Chemical comparison of shields derived from noxious and benign host plants

The larval defence of *N. sexpunctata* appears to be based primarily on derivatives of the diterpene alcohol, phytol, and on the fatty acid (FA), palmitic (hexadecanoic) acid (Table 1). The chemical data indicate that larvae remove two carbons from phytol, but then stop the reaction. We suspect that the limited transformation of phytol may be significant; it may facilitate the compartmentalization and/or the elimination of this potential toxin, perhaps by increasing its polarity, at the same time it may enhance phytol's deterrent activity. In any case, these plant metabolites are evidently more valuable in the feces than they are as an energy source. This may be due in part to the paucity of other defensive candidates in the diet, and to the apparent inability of larvae to synthesize their own defensive compounds. The fact that phytol itself exhibits significant deterrence against ants is perhaps the best explanation for its retention in feces.

In contrast to the phytol-based defence of *N. sexpunctata, L. trilinea's* non-polar, alkaline, and polar shield fractions all exhibited deterrence (Table 1). The major component of the plant non-polar fraction, phytol, was absent in feces, thus indicating that *L. trilinea* can metabolize it completely. However, *L.trilinea's* feces contained two aglycone derivatives of the host's saponin, dioscin. The insect deglycosylated the saponin precursor and then added an additional double bond before its incorporation into the shield defence. The alkaloid fraction of *L. trilinea* feces was also deterrent in the bioassays. The results of chemical analysis (FAB-MS) indicate that fecal shields contain a mixture of alkaloidal glycosides and aglycones, both of dietary origin, but significantly transformed by larvae (Fig. 3).

These findings demonstrated that the shields of *N. sexpunctata* and *L.trilinea* functioned as chemical defences, the efficacy of which depended on transformed primary and secondary host precursors. Remarkably, even the innocuous monocot host provided equally effective precursors for shield defences.

5. The fecal shield of *Blepharida rhois*

Instead of discarding their feces, larvae of the sumac flea beetle, *Blepharida rhois* (Forster), retain it on their backs to form fecal shields (Fig. 1C). Both adults and larvae of *B. rhois*, are dietary specialists that feed on the smooth sumac, *Rhus glabra,* of Long Island, NY. We observed that ants attacking shielded larvae retreated and groomed vigorously, indicating the possible presence of chemical deterrents. To examine whether shields were chemical, rather than physical barriers against predation, we employed the bioassay, diet modification experiments, and chemical screening protocol described above.

Shields of larvae reared on *R. glabra* thwarted ants, while larvae that had their shields removed were readily seized. Larvae reared on a substitute diet were defenceless. However, mortality was reduced after their diet-modified shields were replaced with ones obtained from larvae reared on the host plant. The shield defence was found to be a mixture of three fatty acids, a suite of tannins, their metabolites and phytol (Table 1). All shield compounds, or their precursors, were obtained entirely from the host plant. The *B. rhois* shield thus represents another instance of a combinatorial strategy, one using metabolites from both primary and secondary plant pathways for chemical defence (Vencl and Morton, 1998).

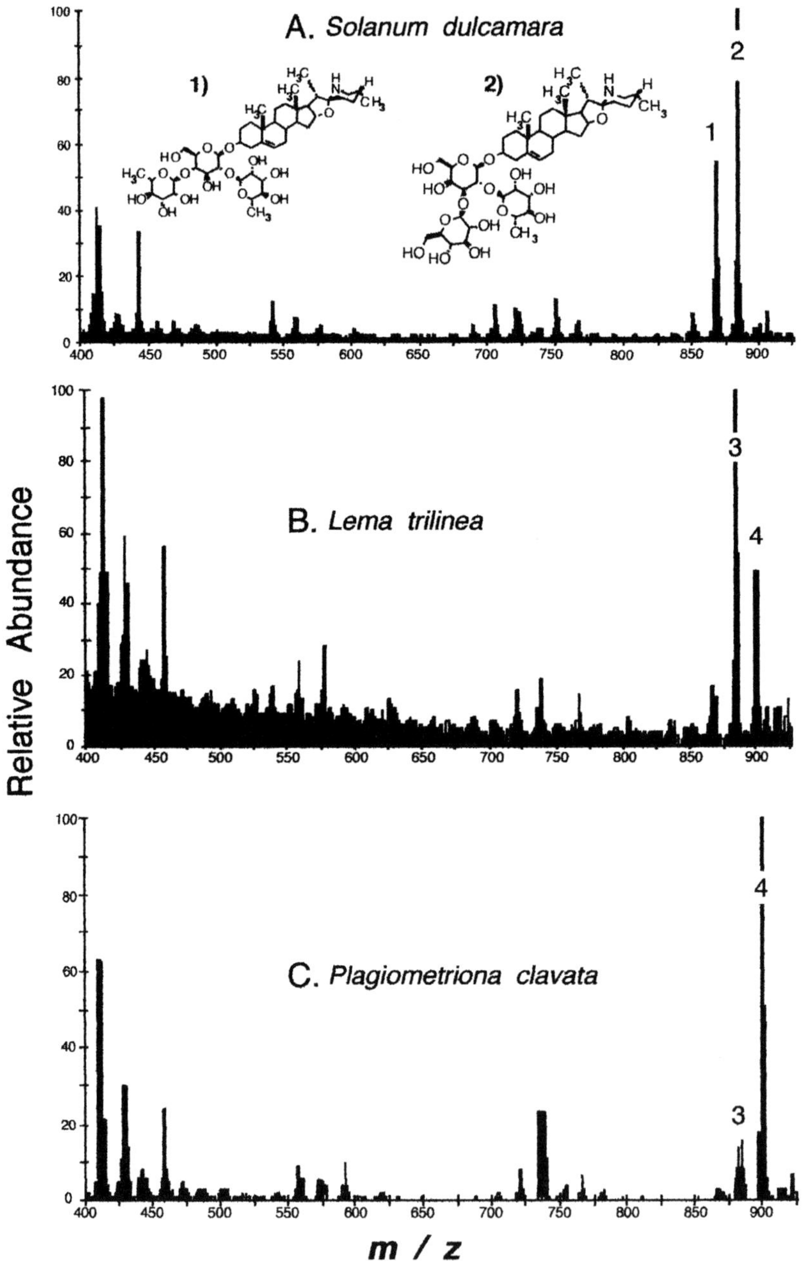

Fig. 3. (Opposite) Positive ion fast-atom bombardment mass spectrometry of the polar C fractions of the host plant, (A) *Solanum dulcamara*, the shield feces of *L. trilinea* (B), and of *P. clavata* (C). The two major plant peaks, (1) and (2), are the steroidal glycoalkaloids (SGAs) solasodine chacotriose and solasodine solatriose (structures l and 2, respectively). Both beetle species are capable of significant transformations of the hosts SGAs. In addition to complete deglycosilation, the fragmentation patterns for shield feces are consistent with two hydroxylated solasodine glycosides (peaks 3 and 4); each fragment is 16 mass units greater (*m/z* 884 and *m/z* 900) than the parent plant compounds (peaks 1 and 2), thus indicating the addition of one hydroxyl, probably on carbon 24, although the exact location has yet to be determined. The majority of the plant SGA, due to their lower polarity, appear in the B fraction as aglycones.

6. Chemical convergence in shield defences

Whenever similar character states evolve independently in similar environments, one can ask if they are adaptations that benefit their bearers. Comparisons of convergent characters, especially if they are complex, can be useful for the identification of particular agents of contemporary selection that maintain them as functionally beneficial adaptations. We compared the larval fecal shields of two common but unrelated North American taxa, the tortoise beetle, *Plagiometriona clavata* (F.), and the shining leaf beetle, *L. trilinea*. Both species occupy the same trophic niche, to wit, they feed on the leaves of woody nightshade, *S. dulcamara* (Solanaceae), often on the same host plant. Their shield systems differ morphologically: *L.trilinea's* shield is a sessile, fecal mound created by a dorsal anus (Fig. 1A). In contrast, the tortoise beetle's shield is held aloft, like a parasol, above the larva, and can be directionally aimed at attacking enemies. Parosol-type shields are composite structures, formed from the accumulation of feces and exuviae. A remarkable telescoping anus, formed by periodic eversion of the invaginated 9th and 10th abdominal segments (Stehr, 1987), deposits feces onto the parosol-type shield (Fig. 1D to F).

Using a combination of diet modification and solvent leaching treatments to remove the chemical component, *P. clavata's* shields were assayed and found to function, like *L. trilinea's*, as an effective chemical defence, necessary for larval survival (Vencl *et al.*, 1999). *Plagiometriona clavata's* shield owed its effectiveness to four major deterrent components: (1) SGAs, (2) a suite of metabolites derived from the steroidal saponin, dioscin, (3) FAs, and (4) derivatives of phytol. All shield metabolites, except perhaps a few FAs, had host plant precursors (Table 1; Fig. 3).

There are remarkable chemical parallels between the shield systems of both species. The profile of FAs, especially the large amount of hexadecanoic acid, is similar. Although there is more oleic acid in the tortoise beetle's feces, the suite of dioscin derivatives is virtually identical in quantity for both species as well as in their MS fragmentation patterns. The most interesting similarity between shield systems is how the two beetle species transformed and used the SGAs and saponins, potentially the most dangerous allelochemicals they encounter; both are considered extremely toxic because they can bind with membrane-bound sterols, alter membrane permeability, and cause loss of cell integrity (Mitchell and Harrison, 1985; Keukens *et al.*, 1996). They only differ from one another by the substitution of nitrogen with oxygen in their respective F rings and both produce their toxic effects by insertion of the apolar portions of the F ring into the target membrane. To compartmentalize and eliminate allelochemicals from the organism, metabolic detoxification systems typically increase allelochemical polarity (that is, aqueous solubility) by using some combination of hydrolysis, oxidation, conjugation with a polar moiety, or reduction. (Dauterman and Hodgson, 1978; Duffey, 1980). Astonishingly, both beetles appear to use a three-step hydrolysis-reduction strategy to 'sequester' SGAs in their shield defences; they first hydroxylate, deglycosylate, and then finally add extra double bonds and additional hydroxyls to both SGA and to the diosgenin residues (Fig. 3). Hydroxylation of the apolar portions of saponins and SGAs might well reduce their membrane binding tendency, at the same time it increases their polarity, making them more readily confinable within the gut and facilitating their elimination in feces. SGA-laced feces, not accidentally we think, become strategically positioned, on or above the larvae of both species, to act as deterrents in shield defences.

7. Phylogenetic aspects of shields and diet evolution-the nasty host hypothesis

The host plant associations of the Criocerinae and of the cassidoid Hispinae are characterized by two striking features: narrow host range, superimposed on a polyphagous background at higher taxon levels. Members of both subfamilies have not only colonized mono- and dicotyledonous plant families, they have each colonized more than a dozen families. The radiation of these groups appears puzzling; micro- and macroevolutionary processes seem decoupled. For instance, although the New World members of the genus *Lema* are predominantly dietary specialists, host shifts have occurred between taxonomically unrelated plant families (Dioscoreaceae and Solanaceae).

Although the reasons for its recurrent evolution have been extensively debated, dietary specialization in herbivorous insects, particularly in the leaf beetles, is thought to be favoured by two major selective factors—plant defences and natural enemies. The long-entertained 'biochemical-barrier' explanation maintains that the diet breadth of an insect clade evolves primarily as an adaptation to specific toxic secondary compounds that characterize the chemical defences of its host plants (Ehrlich and Raven, 1964; Feeny, 1976; Berenbaum, 1990)—secondary chemicals are feeding barriers to all but a few herbivore species that are adapted to them (Fraenkel, 1959; Feeny, 1976). However, the sufficiency of this hypothesis has been questioned (Bernays and Graham, 1988; Futuyma and Keese, 1992). Natural enemies, thought to be important determiners of plant suitability, may limit herbivores to only those plants that act as refugia where predators inflict the lowest mortality (Brower, 1958; Price *et al.,* 1980; Jeffries and Lawton, 1984). Most specialized herbivores, even in the absence of their hosts, are less vulnerable to predation than their generalist counterparts (Damman, 1987; Denno *et al.,* 1990; Cornelius and Bernays, 1995; Dyer, 1995); specialists, perhaps by the assimilation of their hosts' secondary compounds, are rendered unpalatable to predators (Smiley *et al.,* 1985; Pasteels *et al.,* 1988; Jones *et al.,* 1989).

The relative importance of plant chemistry and of natural enemies in the evolution of diet breadth is still much debated (Jaenike, 1990; Futuyma and Keese, 1992; Rausher, 1992); diet breadth is likely to be a response to both factors, but exactly how and why, remains unresolved. The central question then, is to determine whether an herbivore evolves a narrower diet in response to chemical dissimilarities that present higher metabolic hurdles for transfer between host taxa, or whether an herbivore colonizes plants that provide higher levels of chemical protection from its predators.

Using fecal shields as a model of chemically linked, tripartite trophic interactions, we tested the idea that generalist predators play a determining role in diet evolution, and ultimately, in the diversification of beetle species with ectophytic larvae. If host chemical complexity is positively correlated with shield composition, as Table 1 suggests, then the question arises whether shields of varying chemical complexity differ in effectiveness. In the limit, we wanted to know if predation pressure has had a predictable impact on dietary evolution, perhaps reflected in correlated patterns of differential susceptibility to predation and in quantitative or qualitative chemical escalation in shield defences. The 'nasty' host hypothesis (NHH) states that herbivores evolve restricted diets principally because they are rendered less vulnerable to natural enemies through the incorporation, either by active or passive means, of the inimical compounds available in their diets (Smiley *et al.,* 1985; Denno *et al.,* 1990; Gauld *et al,* 1992; Rank *et al.,* 1996). Potential hosts would be scarce because the vast majority of plants lack such noxious compounds, and hence, would be unsuitable as chemical donors for shield defences. The evolutionary outcome of this process would likely be a reduction in diet breadth. In our interpretation of the hypothesis, 'novel' (that is, derived) compounds are thought to improve shield effectiveness to the extent that the lineage bearing the 'better' defence escapes its former enemies long enough to diversify. A test of this 'escape-and-radiate' scenario (*sensu* Thompson, 1989), and its relation to NHH requires, a synthesis of phylogenetic and ecological information and must demonstrate that: (1) a compound is 'novel', (2) relative to closely related beetles that have retained the ancestral shield defence, the lineage with the novel defensive chemistry supports fewer predatory species and/or suffers less predation, and (3) the clade with the novel defence is more diverse than its equally old, sister taxon that lacks it.

7.1. Methods

7.1.1. Phylogenetic analysis

To recover the history of host plant-beetle interaction, we analyzed larval and adult morphological characters using maximum parsimony (MP). To confirm beetle identifications, field-collected and borrowed specimens from the United States National Museum were compared to the Jacoby types in the Bowditch collection of the Museum of Comparative Zoology. Only taxa (n=55) were considered for inclusion in the analysis that had known host associations, a series with a minimum of 6 specimens, and for which male genitalic characters could be examined.

Morphological characters were examined under incident light with a dissecting microscope equipped with an ocular reticule at 10X and 50X. Male aedeagi were cleared in warm KOH. The search for characters was limited to those that could be reliably categorized into two or more discrete states. Continuous characters (for example, proportions) were expressed modally (> or <) for each species. Some morphological characters were previously described by Lacordaire (1885), Monrós (1960), and White (1993) and we retained much of their terminology (Table 2). Our analysis found 20 of 34 characters to be informative. We further culled taxa by excluding all species with seriously incomplete data (>3 missing values). Of the 55 taxa at the beginning of the study, 22 species had complete information on all characters. Nine redundant, and thus uninformative taxa were deleted, leaving 13 in the analysis.

We inferred the higher phylogenetic relationships among New World criocerine taxa with PAUP (version 3.1.1; Swofford, 1993). Using an exact method, the BRANCH-and-BOUND algorithm, we found all most parsimonious trees. Confidence in nodes was estimated using the BOOTSTRAP option with 500 replicates. Unweighted characters were polarized *a posteriori* using the outgroup method. *Crioceris* was designated as the outgroup (Schmitt, 1988). We tried *Lilioceris* as a more distant out-group but lost all phylogenetic signal within the ingroup as a result.

7.1.2. Predator-choice bioassays

We used an array of ecologically relevant generalist predators in both laboratory and field choice assays to compare the relative effectiveness of shield defences differing in chemical

Table 2. Description of characters of criocerine species and outgroup taxa. Only informative characters used in the analysis are presented.

#	character	states
2	vertex produced	(0) yes; (1) no
3	pronotum arcuate	(0) yes; (1) no
4	claws	(0) separate; (1) connate
7	aedeagal lateral lobes	(0) convergent; (1) parallel; (2) medial only
9	pronotal constriction	(0) absent; (1) submedial to basal; (2) medial
10	antennal tubercles	(0) widely separated; (1) close
12	antennal length	(0) <1/2 body length; (1) >1/2 body length
13	apical antennal segment elongated	(0) <2x length; (1) >2x length
14	eye notch	(0) shallow (<50% diameter); (1) moderate (=50% diameter); (3) deep (>1/2 diameter)
15	disc color pattern	(0) solid; (1) marginal stripe; (2) stripes+spots; (3) stripes; (4) spots
16	post-ocular constriction	(0) weak, <1mm; (1) moderate; (2) deep, >1mm
17	9th elytral stria	(0) incomplete; (1) complete
19	metafemora in male swollen	(0) no, width<1/2 length; (1) yes, width>1/2 length
24	clypeal punctation: location	(0) marginal; (1) scattered
25	clypeal punctation: density	(0) sparse<2/mm^2; (1) moderate; (2) dense, >4/mm^2
27	pronotal width at constriction	(0) >85% of apex; (1) <85% of apex
28	medial pit	(0)absent to weak; (1) present
29	pronotal width/length ratio	(0) >1; (1) <1
30	elytral depression	(0) none; (1) feeble; (3) deep
34	iridescence	(0) no; (1) yes

composition. Predator-choice assays were performed at The Smithsonian Tropical Research Institute at Barro Colorado Island, Panamá, and in Crystal Brook Park, Long Island, New York during 1996-97. Each assay consisted of the simultaneous presentation of three larvae, representing each of the two shield types: shields produced either on a diet of *Commelina* sp. or those produced on a diet of *Solanum* sp. Laboratory assays with coleopterous or asopine pentatomid predators, conducted in 9 cm petri dishes, were monitored for three hours, or until two prey items were taken. Field assays with hymenopterous predators were conducted near active foraging trails or nests for 30 minutes or until two prey items were taken.

7.2. Results

7.2.1. Phylogenetic analysis

The MP analysis of the morpho-characters produced a strict concensus tree of 62 steps. Despite the fact that no monophyletic groups were found with greater than Felsentein's (1985) recommended 95% significance level, three taxon-clusters were defined at >70%, the level found previously to provide sufficient branch support (Hillis and Bull, 1993). Although we did not find a great many useful synapomorphies, the lateral lobes of the aedeagus, depth of the eye-notch, punctation of the 9th stria, antennal length, and form of the pronotal constriction proved to be reliable indicators of three natural groupings at the generic level (Fig. 4). The results here, despite the small number of useful morphological synapomorphies at our disposal, support the monophyly of the New World genus *Lema*, its close relationship to *Neolema*, and more distantly, to *Oulema*.

7.2.2. Host associations and terminal diversities

According to the available literature and our field work, of 265 described *Lema* species, the majority of species with known hosts (30) feed on members of the Solanaceae. An additional seven species are recorded from the chemically similar, but taxonomically disparate Dioscoreaceae (Windsor pers com.; Flowers, pers com.; Vencl, unpub.). The sister taxon to *Lema* revealed by our study is *Neolema*. It contains 136 species for which only 12 hosts are presently known; all are members of the Commelinaceae. As for indigenous New World *Oulema*, we presently know 25 species, all but 4 of which have their hosts in the Commelinaceae.

Using the assumption of maximun parsimony (that is, fewest evolutionary changes), we attempted to recover the history of host-use by mapping the distribution of host plant families onto the beetle phylogeny (Fig. 4). We obtained three results: (1) the ancestral host is evidently the Commelinaceae; (2) the Solanaceae is the derived host of the *Lema* clade, and; (3) comparison of the terminal diversities between the sister taxa, *Lema* and *Neolema*, which by definition, are of equal evolutionary age (Mitter *et al.*, 1988), reveals the *Lema* clade to be significantly more species-rich (X^2=40.96; $P < 0.001$).

7.2.3. Predator-choice bioassays

Predator assays, pitting shields differing in chemical complexity against one another, demonstrated that the more complex *Solanum*-based shields were, on average, more effective against a wider array of invertebrate generalist predators than were the *Commelina*-based shields (Fig. 5).

7.3. Discussion

To evaluate the role of shield defences and the relevance of the nasty host hypothesis in the adaptive radiation of the Criocerinae, we sought three pieces of information: (1) a beetle phylogeny which incorporates host-use data, (2) a comparison of the inferred host shifts with the distribution among hosts of the chemical characters that are utilized in shield defences and, (3) a demonstration which reveals that, relative to closely related beetles which have retained the ancestral defence, the lineage with the novel (derived) defensive chemistry supports fewer predatory species and/or suffers less predation.

With few reliable morphological apomorphies, we anticipated a substantial amount of homoplasy, reduced resolution, and soft polytomies. However, since our goal was an estimation of higher phyletic relationships within the New World members of the *Lema*-group, we hoped to recover, through the use of morphological data alone, a stable topology at that level, and to compare the observed groupings for agreement with the current taxonomy. The results of the analysis, although preliminary, indicate that a clear relationship exists in the New World: the *Oulema* clade is ancestral to the two sister taxa, *Lema* and *Neolema*. This finding supports the idea that *Oulema* retains an old and specialized relationship with its commelinaceous host plants which probably predates the separation of Gondwana some

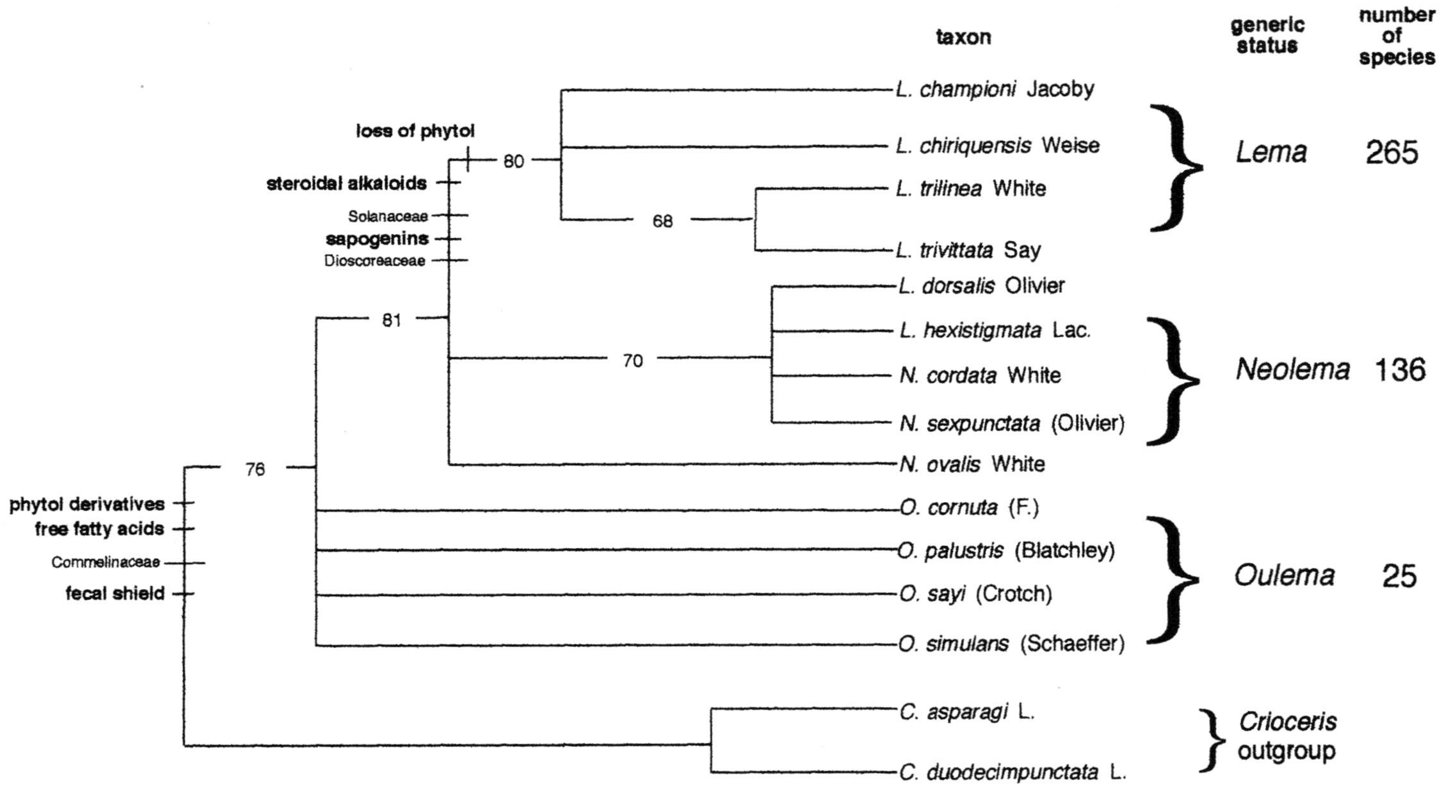

Fig. 4. Shortest majority rule consensus tree of criocerine taxa with 62 steps found from maximum parsimony analysis with an exact search based on the morphology data set (Table 1). Host plant families and shield compounds are mapped onto the tree. The inferred history of the incorporation of host-derived allelochemicals into shield defences is based on the inferred historical reconstruction of host use. Bootstrap proportions for 500 replicates supporting each node are given. R.I. = 0.72; C.I. and H.I. = 0.5 (after Vencl and Morton, 1999).

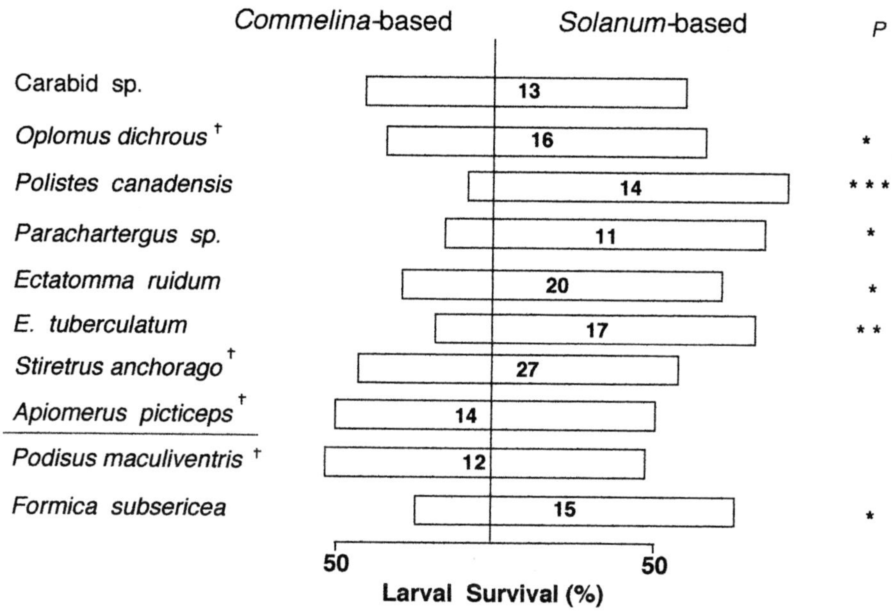

Fig. 5. Predatory choice bioassays comparing the effectiveness of leaf beetle larval fecal shields in Panamá (above line) and New York (below line). The shields differ in chemistry: *Solanum*-based shields of *Lema* species shields (right) contain complex plant alkaloids in addition to some primary host metabolites, while *Commelina*-derived shields of *Neolema* and *Lema* (*Oulema*) species (left) contain only primary host metabolites. Each bar equals the total number of larvae that survived in each bioassay; the proportion of each shield type is represented by the right or left shift of the bar. † = hemipterous, members of the pentatomid subfamily Asopinae; the remainder are hymenopterous. A choice assay consisted of the presentation of three larvae with each type of shield to a predator in a 9 cm Petri dish (carabid and hemipterous predators) for three hours. Assays with hymenopterous predators were conducted for 30 mins along foraging trails or near nests. Probabilities calculated by the binomial test: $*P < 0.05$; $**P < 0.01$; $***P < 0.001$. n = total number of larvae taken; number of replicates for each predator assay = 5–18.

70 mya (Cloud, 1978). In contrast, *Lema* has a more recent relationship with its solanaceous hosts. The results of our MP analysis are in basic agreement with Schmitt's (1988) hypothesis that the sequence of host shifts within the Criocerinae was from mono- to dicotyledonous plant families and concordant with their respective appearances (Crowson, 1981).

7.3.1. Shield efficacy and beetle diversification-a provisional scenario

The limitations of the present study notwithstanding, we propose the following provisional scenario (Fig. 6):

After splitting from the ancestral commelinaceous hosts, the *Lema* clade likely colonized the phytol- and saponin-containing Dioscoreaceae. Reports of *Dioscorea*-feeding Old World *Lema* support the antiquity of the association (Chûjô, 1951). The shift to the steroidal alkaloid-rich Solanaceae, when these plants appeared in the New World sometime in the Oligocene (Schmitt, 1988), was initially *enabled* by a chemical exaptation, or 'bridge' (*sensu* Ehrlich and Raven, 1964), that is, the ability to transform steroidal molecules. The bioassay and terminal diversity data are in support of an 'escape-and-radiate'

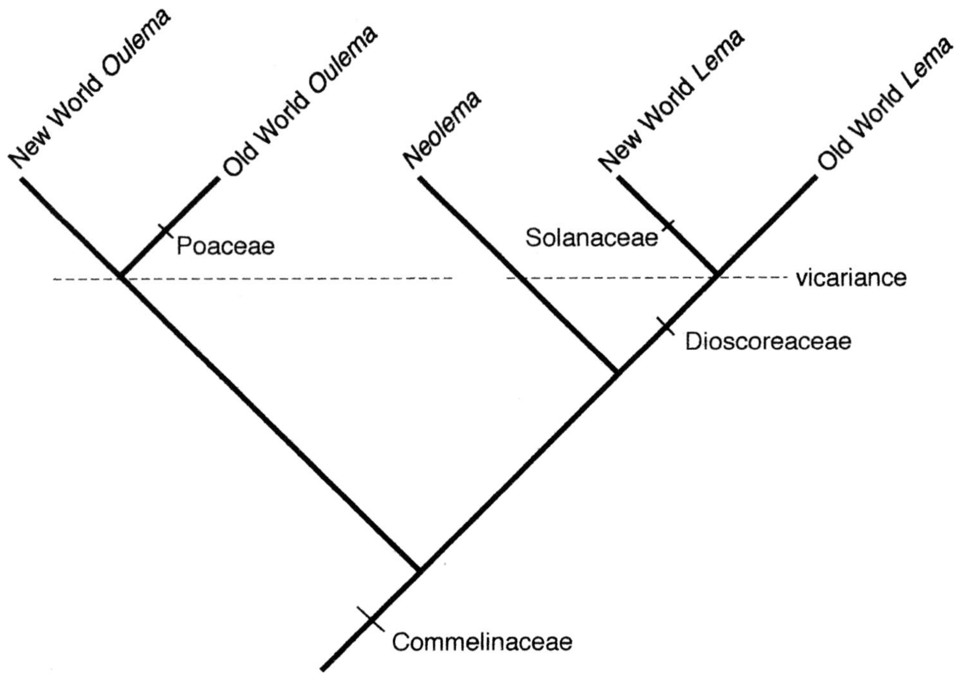

Fig. 6. Hypothetical reconstruction of host shifts in the *Lema*-group, inferred from phytochemical data and host records (after Vencl and Morton, 1998b).

scenario. *Lema's* defence appears to be better than the ancestral defences based solely on host plant primary metabolites. We hypothesize that the numerical superiority of the *Lema* clade is a consequence of the incorporation of the novel and more potent deterrent steroidal glycoalkaloids and steroidal saponins into shield defences. Despite their present diversity and abundance, the predators now able to circumvent *Lema's* shield defence, the members of the Asopinae (Hemiptera: Pentatomidae), are descended from a subset of the enemies attacking the antecedent shield defences. Thus, the *Lema* lineage evidently has 'shed' most of its erstwhile enemies. We surmise that the qualitatively superior defence allowed the *Lema* clade to enter a new adaptive zone that precluded its former enemies, ultimately enhancing its rate of diversification by reducing losses to predation. If so, predation may well have *driven* host shifts at the same time it narrowed *Lema's* diet by making the former hosts relatively unsafe in regard to reduced fitness. We predict a similar scenario for tortoise beetles once their shield chemical ecology is elucidated.

8. Evolutionary transformations in the shields of tortoise beetles

The phylogenetic analysis of the Neotropical Hispinae by Hsiao and Windsor (this volume) revealed three striking evolutionary patterns in diet and in shield morphology that further support the nasty-host hypothesis. First, the older hispinoid taxa, with endophytic larvae that lack shields, feed on members of older monocotyledonous families, the palms and members of the Zingerberales. Second, the more derived hispinoids possess surface-feeding larvae with relatively simple shield structures, such as wedges and spatulas, that lack a fecal (chemical) component. In contrast, the more derived tortoise beetles, all with ectophytic larvae, feed on younger (ca. Oligocene) dicotyledonous taxa, such as the forget-me-nots (Boraginaceae), morning glories (Convolvulaceae), asters (Asteraceae), and nightshades (Solanaceae). In support of the idea that they are important defences, shields are reduced and fecal retention is limited or very temporary in species that exhibit maternal larval care, and removal of the guardian significantly increases loses to predation (Windsor, 1987; Windsor and Choe, 1994). The inferred transformation series in shield morphology is as follows:

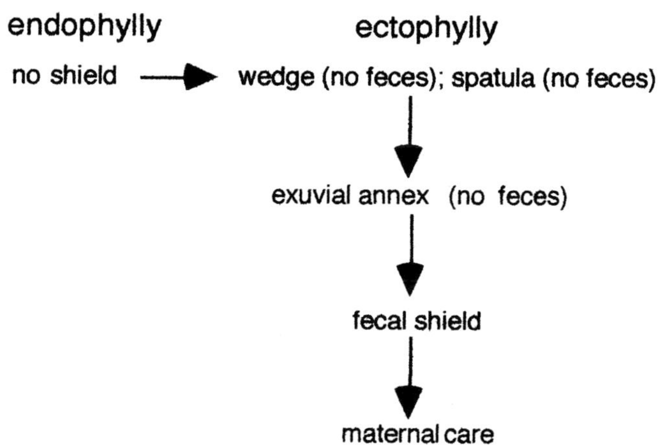

We suspect that the radiation of the Hispinae can be largely explained by the nasty-host hypothesis. The shield defence may have had its origin as a simple physical barrier in basal Hispinae. It then underwent a morphological elaboration that was subsequently accompanied by the addition of digestive wastes inadvertently bearing a chemical component. Initially, wastes contained only primary plant metabolites. Plants not known for complex secondary chemistry, may nonetheless, be chemically well defended through the up-regulation of primary pathways (Berenbaum, 1995). Primary metabolites, as our work suggests, also have deterrent or repellent properties when included in larval shields. The further chemical elaboration of tortoise beetle shields coincided with the appearance of dicotyledonous plants known for qualitative defences based on secondary metabolites. As tortoise beetles colonized these plants, the availability of diverse and noxious allelochemicals increased, and thus, the shields containing the novel, and more toxic host-derived compounds, likely became more broadly effective as chemical defences.

9. Fecal shields as key chemical innovations for colonization of the leaf surface

With the evidence now at our disposal, it appears that host shifts in both the Criocerinae, and perhaps also in the cassidoid Hispinae, occurred as a selective response to predation pressure. Initially, host shifts followed trajectories predicted by plant chemical commonality, but were later constrained by the dietary availability of more potent, novel (that is, derived) chemicals that served to bolster shield efficacy. At least for the two leaf beetle groups for which we have information about chemical shield ecology, it may be that high predation rates for their vulnerable, ectophytic larval stages not only impelled host shifts, but also limited them to only prospective host taxa that possessed suitably effective defensive precursors.

Fecal shields, unlike reflexive discharges, remain 'deployed' as both physical and as chemical barriers, as long as larvae continue to eat. In contrast to eversible glandular secretions, sequestered defensive substances in hemolymph, and regurgitants that are projected in response to immediate threats, shields containing deterrents or repellents may thwart attacking enemies long before they can damage the larval cuticle. The fecal shield, as such, represents a critical adaptation, or 'key innovation', that likely facilitated the successful colonization of the leaf surfaces adaptive zone by a limited handful of chrysomelid lineages. In contrast to those taxa with internal trophic modes, the diversification of the subfamilies with larvae attacking leaf surfaces was necessarily accompanied by the evolution of a chemically fortified fecal component which served to augment the ancestral barrier-type shield defence, or by some other form of chemical defensive strategy. Our understanding about the selective role played by natural enemies in the evolution of plant-acquired defences, diet breadth, and larval trophic habits in the Chrysomelidae is still very fragmentary, especially with respect to traits that rely on waste-retention. To redress the inadequacy of our knowledge, many more studies are needed that combine phylogenetic information about host shifts and the distribution of compounds across taxa, detailed chemical analyses, and experimental verification of the effectiveness of larval defences against predators and parasitoids in natural settings.

Acknowledgments

We thank S. Czybora for many useful improvements to the manuscript. J. C. Schultz, A. D. Carlson, and F. Johnson contributed importantly to this work by providing valuable technical advice and access to facilities. Support by the Andrew W. Mellon Foundation Grant for Exploratory Research to the senior author is gratefully acknowledged.

References

Berenbaum, M.R., 1990 Evolution of specialization in insect-umbellifer associations. Annu. Rev. Entomol. 35: 319-345.

Berenbaum, M.R., 1995 Turnabout is fair play: secondary roles for primary compounds. J. Chem. Ecol. 21: 925-940.

Bernays, E.A. and Graham, M., 1988 On the evolution of host specificity in phytophagous arthropods. Ecology 69: 886-892.

Blum, M.S., 1994 Antipredator devices in larvae of the Chrysomelidae: a unified synthesis for defensive eclecticism. *In* Jolivet, P.H., Cox, M.L., and Petitpierre, E. (eds.), Novel Aspects of the Biology of the Chrysomelidae. Kluwer Academic Publishers, Dordrecht, The Netherlands, pp 277-288.

Brower, L.P., 1958 Bird predation and foodplant specificity in closely related species. Am. Nat. 92: 183-187.

Chûjô, M., 1951 A taxonomic study on the Chrysomelidae (Insecta-Coleoptera) of Formosa. Techn. Bull. Kagawa Ag. Col. 2: 71-120.

Cloud, P., 1978 Cosmos, earth, and man. Yale University Press, NY.

Cornelius, M.L., and Bernays, E.A., 1995 The effect of plant chemistry on the acceptability of caterpillar prey to the Argentine ant *Iridomyrmex humilis* (Hymenoptera: Formicidae). J. Insect Behavior 8: 579-593.

Cox, M.L., 1994 The Hymenoptera and Diptera parasitoids of the Chrysomelidae. *In* Jolivet, P.H., Cox, M.L., and

Petitpierre, E. (eds.), Novel Aspects of the Biology of the Chrysomelidae. Kluwer Academic Publishers, Dordrecht, The Netherlands, pp 419-467.
Cox, M.L., 1996 Insect predators of the Chrysomelidae, *In* Jolivet, P.H.A., and Cox, M.L. (eds.), Chrysomelidae Biology, Vol. 2, Ecological Studies. SPB Academic Publishing, Amsterdam, The Netherlands, pp. 23-91.
Crowson, R.A., 1981 The biology of the Coleoptera. Academic Press, London.
Damman, H., 1987 Leaf quality and enemy avoidance by larvae of a pyralid moth. Ecology 68: 87-97.
Dauterman, W.C., and Hodgson, E., 1978 Detoxication mechanisms in insects. *In* Rockstein, M. (ed.), Biochemistry of Insects. Academic Press, New York, pp 541-577.
Denno, R.F., Larsson, S., and Olmstead, K.L., 1990 Role of enemy-free space and plant quality in host-selection by willow beetles. Ecology 71: 124-137.
Denno, R.F., McClure, M.S., and Ott, J.R., 1995 Interspecific interactions in phytophagous insects: Competition reexamined and resurrected. Annu. Rev. Entomol. 40: 297-331.
Duffey, S.S., 1980 Sequestration of plant natural products by insects. Annu. Rev. Entomol. 25: 447-477.
Dyer, L.A., 1995 Tasty generalists and nasty specialists? A comparative study of anti-predator mechanisms in tropical Lepidoptera. Ecology 76: 1483-1496.
Ehrlich, P.R., and Raven, P.H., 1964 Butterflies and plants: a study in coevolution. Evolution 18: 568-608.
Eisner, T., and Meinwald, J., 1966 Defensive secretions of arthropods. Science 153: 1341-1350.
Eisner, T., Van Tassell, E., and Carrel, J. E., 1967 Defensive use of a "fecal shield" by a beetle larva. *Science* 158: 1471-1473.
Erber, D., 1988 Biology of the Camptosomata Clytrinae-Cryptocephalinae-Chlamisinae-Lamprosomatinae. *In* Jolivet, P.H.A., Petitpierre, E., and Hsiao, T.H. (eds.), The Biology of the Chrysomelidae. Kluwer Academic Publishers, Dordrecht, The Netherlands, pp 541-552.
Farrell, B.D., 1998 "Inordinate Fondness" Explained: Why Are There So Many Beetles? Science 281: 555-559.
Feeny, P.P., 1976 Plant apparancy and chemical defense. *In* Wallace, J.W., and Marshall, R.L. (eds.), Biochemical Interactions between Plants and Insects. v 10. Recent Advances in Phytochemistry. Plenum, NY, pp 1-40.
Felsenstein, J., 1985 Confidence limits on phylogenies: an approach using the bootstrap. A. Rev. Genet. 22: 521-565.
Ferguson, J.E., Metcalf, R.L., and Fisher, D.C., 1985 Disposition and fate of cucurbitacin B in five species of diobroticites. J. Chem. Ecol. 11: 1307-1321.
Fraenkel, G.S., 1959 Râison d'être of plant secondary compounds. Science 129: 1466.
Futuyma, D.J., and Keese, M.C., 1992 Evolution and coevolution of plants and phytophagous arthropods. *In* Rosenthal, G.A. and Berenbaum, M. (eds.), Herbivores: Their interactions with secondary plant metabolites, 2E volume II: Evolutionary and ecological processes. Academic Press, NY, pp 439-475.
Gauld, I.D., Gaston, K.J., and Janzen, D.H., 1992 Plant allelochemicals, tritrophic interactions and the anomolous diversity of tropical parasitoids: the 'nasty' host hypothesis. *Oikos* 65: 353-357.
Gómez, N.E.Q., 1997 The fecal shield of larvae of tortoise beetles (Cassidinae: Chrysomelidae): a role for chemical defense using plant-derived secondary compounds. Ph. D. dissertation, Technischen Universität Carolo-Wilhelmina, Braunsweig.
Hegnauer, R., 1992 Chemotaxonomie der Pflanzen. Birkhäuser Verlag, Basel.
Hillis, D.M., and Bull, J.J., 1993 An empirical test of boostrapping as a method for assessing confidence in phylogenetic analysis. Sys. Biol. 42: 182-192.
Hölldobler, B., and Wilson, E.O., 1990 The Ants. Belknap Press of Harvard University, Cambridge, MA.
Jaenike, J., 1990 Host specialization in phytophagous insects. A. Rev. Ecol. Syst. 21: 243-247.
Jeffries, M.J., and Lawton, J.H., 1984 Enemy free space and the structure of ecological communities. Biol. J. Linn. Soc. 23: 269-286.
Jones, C.G., Whitman, D., Compton, S., Silk, P., and Blum, M.S., 1989 Reduction in diet breadth results in sequestration of plant chemicals and increased efficiency of chemical defense in a generalist grasshopper. J. Chem. Ecol. 15: 1811-1822.
Keukens, E.A., De Vrije, T., Jansen, I.A., De Boer, H, Jassen, M., De Kroon, A.I., Jongen, W.M., and De Kruijff, B., 1996. Glycoalkaloids selectively permeabilize cholesterol containing biomembranes. Biochem. et Biophys. Acta 1279: 243-250.
Lacordaire, J.T., 1885 Monographie des Coléoptères subpentamères de la famille des Phytophages. 1(1). Mém. Soc. r. Sci. Liège 3: 1-740.
Mathé, I., Jr., and Mathé, I., Sr., 1979 Variation in alkaloids in *Solanum dulcamara*. *In* Hawkes, J.G., Lester, R.N., and Skelding, A.D. (eds.), The Biology and Taxonomy of the Solanaceae. Linn. Soc. Symp. 7. Academic Press, London, pp. 211-222.
Mitchell, B.K., and Harrison, G.D., 1985 Effects of *Solanum* glycoalkaloids on chemosensilla in the Colorado potato beetle: A mechanism of feeding deterrence? J. Chem. Ecol. 11: 73-83.
Mitter, C.B., Farrell, B., and Wiegmann, B., 1988 The phylogenetic study of adaptive zones: Has phytophagy promoted insect diversification? Am. Nat. 132: 107-128.
Monrós, F., 1951 Notas sobre Criocerinae. Actá zool. lilloana 11: 467-482.

Monrós, F., 1960 Los Generos de Chrysomelidae (Coleoptera). Op. lilloana III. Univ. Nac. Tucuman: 1-337.
Morton, T.C., and Vencl, F.V., 1998 Larval beetles (Chrysomelidae: Criocerinae) form defense from recycled host compounds discharged as fecal wastes. J. Chem. Ecol. 24: 765-786.
Olmstead, K.S., 1994 Waste products as chrysomelid defenses, *In* Jolivet, P.H., Cox, M.L., and Petitpierre, E. (eds.), Novel Aspects of the Biology of the Chrysomelidae. Kluwer Academic Publishers. The Netherlands, pp 311-318.
Olmstead, K.L., 1996 Cassidine defenses and natural enemies. *In* Jolivet, P.H.A., and Cox, M.L. (eds.), Chrysomelidae Biology, Vol. 2, Ecological Studies. SPB Academic Publishing, Amsterdam, The Netherlands, pp. 3-21.
Olmstead, K.L., and Denno, R.F., 1993 Effectiveness of tortoise beetle larval shields against different predator species. Ecology 74: 1394-1405.
Pasteels, J.M., Rowell-Rahier, M., Braekman, J.C., Daloze, D., 1984 Chemical defenses in Leaf beetles and their larvae: The ecological, evolutionary and taxonomic significance. Biochem. Syst. Ecol. 12: 395-406.
Pasteels, J.M., Braekman, J.C., and Deloze, D., 1988 Chemical defense in the Chrysomelidae. *In* Jolivet, P.H.A., Petitpierre, E., and Hsiao, T.H. (eds.), The Biology of the Chrysomelidae. Kluwer Academic Publishers, Dordrecht, The Netherlands, pp 231-250.
Pasteels, J.M., Rowell-Rahier, M., Braekman, J.C., and Daloze, D., 1994 Chemical defense of adult leaf beetles updated. *In* Jolivet, P.H., Cox, M.L., and Petitpierre, E. (eds.), Novel Aspects of the Biology of the Chrysomelidae. Kluwer Academic Publishers, The Netherlands, pp 289-301.
Price, P., Bouton, C., Gross, P., McPheron, B., Thompson, J.N., and Weise, A. E., 1980 Interactions among three trophic levels: influence of plants on interactions between insect herbivores and natural enemies. A. Rev. Ecol. Syst. 11:41-65.
Rank, N., Smiley, J., and Köpf, A., 1996 Natural enemies and host plant relationships for chrysomelid leaf beetles feeding on Salicaceae. *In* Jolivet, P.H.A., and Cox, M.L. (eds.), Chrysomelidae Biology, Vol. 2, Ecological Studies. SPB Academic Publishing, Amsterdam, The Netherlands, pp 147-171.
Rausher, M.D., 1992 Natural selection and the evolution of insect-plant interactions. *In* Roitberg, B.D., and Isman, M.B. (eds.), Insect Chemical Ecology, Chapman and Hall, NY, pp 20-88.
Reid, C.A.M., 1995 A cladistic analysis of subfamilial relationships in the Chrysomelidae *sensu lato* (Chrysomeloidea). *In* Pakaluk, J., and Slipinski, S.A. (eds.), Biology, Phylogeny, and Classification of Coleoptera: Papers celebrating the 80th birthday of Roy A. Crowson. Muzeum i Instytut Zoologii PAN, Warzawa, pp. 559-631.
Roddick, J.G., 1986 Steroidal alkaloids of the Solanaceae. *In* D'Arcy, W.G. (ed.), Solanaceae: Biology and Systematics. Columbia University Press, New York, pp 201-222.
Root, R.B., and Messina, F.J, 1983 Defensive adaptations and natural enemies of a case-bearing beetle, *Exema canadensis* (Coleoptera: Chrysomelidae). Psyche 90: 67-80.
Schmitt, M., 1988 The Criocerinae: Biology, phylogeny, and evolution. *In* Jolivet, P.H.A., Petitpierre, E., and Hsiao, T.H. (eds.), The Biology of the Chrysomelidae. Kluwer Academic Publishers, Dordrecht, The Netherlands, pp 475-496.
Selman, B.J., 1988 Chrysomelids and ants. *In* Jolivet, P. H., Petitpierre, E., Hsiao, T.H. (eds.), The Biology of the Chrysomelidae. Kluwer Academic Publishers, Dordrecht, The Netherlands, pp 463-473.
Simpson, G.G., 1944 Tempo and mode in evolution. Columbia. U. Press, New York.
Smiley, J.T., Horn, J.M., and Rank, N.E., 1985 Ecological effects of salicin at three trophic levels: new problems from old adaptations. Science 229: 649-651.
Southwood, T.R.E., 1973 The insect/plant relationship – an evolutionary perspective. Symp. R. ent. Soc. Lond. 6: 3-30.
Stehr, F. W., 1987 Immature Insects. Vol 2. Kendel/Hunt, Dubuque, Iowa.
Strong, D.R., Lawton, J.H., and Southwood, T.R.E., 1984 Insects on Plants. Harvard U. Press, Cambridge, MA.
Swofford, D.L., 1993 PAUP: phylogenetic analysis using parsimony. V. 3.1.1, Illinois Natural History Survey, Champaign, Illinois.
Thompson, J.N., 1989 Concepts of coevolution. Trends. Ecol. Evol. 4: 179-183.
Tingey, W.M., 1984 Glycoalkaloids as pest resistance factors. Amer. Potato. J. 61: 157-167.
Vencl, F.V., and Morton, T.C., 1998 The fecal shield defense of the sumac flea beetle, *Blepharida rhois* (Chrysomelidae: Alticinae). Chemoecology 8: 25-32.
Vencl, F.V., and Morton, T.C., 1999 Did chemical change in shield defenses promote diversification of shining leaf beetles (Chrysomelidae: Criocerinae)? Proceedings of the fourth International Symposium on the Chrysomelidae of the XX International Congress of Entomology, Firenze, 1996. Mus. reg. Sci. nat. Torino, pp 1-14.
Vencl, F.V., Morton, T.C., Mumma, R.C., and Shultz, J.C., 1999 Shield defense of a larval tortoise beetle. J. Chem. Ecol. 25(3): 549-566
White, R.E., 1993 A revision of the Subfamily Criocerinae (Chrysomelidae) of North America North of Mexico. Tech. Bull. U. S. Dept. Agri. 1805: 1-158.
Windsor, D.M., 1987 Natural history of a subsocial tortoise beetle, *Acromis sparsa* Boheman (Chrysomelidae: Cassidinae) in Panama. Psyche 94: 127-150.
Windsor, D.M., and Choe, J.C., 1994 Origins of parental care in chrysomelid beetles. *In* Jolivet, P.H., Cox, M.L., and Petitpierre, E. (eds.), Novel Aspects of the Biology of the Chrysomelidae. Kluwer Academic Publishers, The Netherlands, pp. 111-117.

POSSIBLE MIMETIC COMPLEXES IN CENTRAL AMERICAN *CEPHALOLEIA* (COLEOPTERA: CHRYSOMELIDAE: HISPINAE)

C.L. STAINES
3302 Decker Place, Edgewater, MD 21037, USA

Abstract

The genus *Cephaloleia* is speciose and widely distributed in the Neotropics. Life history work has been conducted on very few species. Mimetic complexes are postulated for 17 *Cephaloleia* species from Mesoamerica. The complexes involve beetles in the families Chrysomelidae, Cantharidae, Lampyridae, Cleridae, and Lycidae.

1. Introduction

Mimicry, the resemblance of one animal to another, has long fascinated naturalists. Mimicry increases the resemblance (homoplasy) between species and may affect taxonomy by making it difficult to separate species and possibly obscuring the relationships between species. The description of mimicry complexes are informative to both ecologists and taxonomists.

The theory of mimicry in insects was first proposed by Bates (1862). Since that time there has been a steady flow of literature. Cott (1940) has summarized the early literature. Rettermeyer (1970) reviewed the literature since Cott and recognized four types of mimicry in insects- Batesian, Müllerian, Peckhamian (aggressive), and Wasmannian.

Both Müllerian and Batesian mimicry depend on aposematic (warning) coloration. In Müllerian mimicry, two distasteful species share a color pattern, maximizing the effect for each other by minimizing the number of patterns predators must remember. In Batesian mimicry, an uncommon and relatively edible or harmless species (the mimic) resembles a more common distasteful species (the model), thus deceiving predators which are likely to experience and avoid the model. Because distastefulness is a relative and variable characteristic (Turner, 1984), it is impossible to positively categorize mimetic relationships as Müllerian or Batesian (Huheey, 1976). Therefore the existence of mimicry and its character as Batesian or Müllerian must be considered speculative.

Resemblance among Müllerian mimics does not need to be close since both species are distasteful. Huheey (1984) believed that Müllerian mimicry is an unfortunate term since there is no deception involved.

Batesian mimicry involves close resemblance between the model and mimic since deception of the predator by prey is involved. A Batesian mimic benefits at the expense of both its predator and its model. By encountering and eating pseudaposematic mimics, predator learning is impeded, causing greater predation of the model than would otherwise be

Advances in Chysomelidae Biology 1, pp. 239–246
Edited by M.L. Cox

the case (Huheey, 1964). The Batesian mimic has no deterrent, its sole means of surviving predation is to prevent attack through its resemblance to its model.

Sometime during their life cycle, most species of Chrysomelidae feed on plant tissue. Because many plants produce chemicals to deter predators, chrysomelids have the potential of sequestering these chemicals for their own defense against predators (Pasteels *et al.*, 1988). There is a growing body of literature documenting the presence of distastefulness among a wide range of chrysomelids. This would indicate that many, if not all, chrysomelids are relatively distasteful either by sequestered chemicals or those produced by the beetle (Pasteels *et al.*, 1988, 1994). Schaefer (1996) postulated that asopine pentatomids (Heteroptera: Pentatomidae) may derive their defensive allomones from their chrysomelid prey. Because of this, the Chrysomelidae are excellent candidates for involvement in mimicry complexes either as Müllerian comimics or Batesian models.

Mimicry was first reported in the Chrysomelidae by Gahan (1891). Since then, mimicry has been reported in many chrysomelid subfamilies (Balsbaugh 1988, Hespenheide 1996).

Mimicry was first reported in hispines by Gahan (1891) who reported a mimetic complex of three *Lema* species (Chrysomelidae: Criocerinae) and *Hispa* species in Borneo. Gahan (1913) reported a mimetic complex involving *Erythroplatys corallifer* White (Coleoptera: Cerambycidae: Cerambycinae), *Epiplecta maculata* Maulik (Coleoptera: Tenebrionidae), and *Sceloenopla maculata* (F.) (Hispinae). Shelford (1902, 1916) stated that some *Gonophora* species (hispine) mimicked Lycidae (Coleoptera). Maulik (1919) reported several mimetic complexes involving hispines: *Strongylium rufipenne* Kollar (Tenebrionidae) and the hispine *Lasiochila excavata* Baly; *Estigmenida variabilis* Gahan (Cerambycidae: Lamiinae) and the hispine *Lasiochila cylindrica* (Hope); *Plaxomicrus hispoides* Aurivillius (Cerambycidae: Lamiinae) , *Lema monstrosa* Baly (Criocerinae), and the hispines *Dactylispa cladophora* (Guérin-Méneville) and *D. spinosus* (Weber); and a number of Curculionidae (Coleoptera) which resemble hispines. Lane (1951) remarked on the resemblance of the cerambycine cerambycid *Zelliboria* to hispines. Linsley (1959) suggested another possible mimetic complex involving the hispine *Estigmena chinensis* Hope and the cerambycid *Estigmenida variabilis* Gahan. Menier (1985) illustrated a mimetic complex consisting of an oedemerid beetle, a Tenebrionidae, the lamiine cerambycid *Carterica mimi* Bolon, a clerid beetle, an alticine chrysomelid *Cerichrestus* sp., and a hispine *Chalepus* sp. Mawdsley (1992) reported a large mimicry ring involving species from the beetle families Lycidae, Lampyridae, Melyridae, Cleridae, Elateridae, Buprestidae, Ptilodactylidae, Oedemeridae, Pedilidae, Cerambycidae, and the hispine *Odontota scapularis* (Olivier). Hespenheide (1996) reported mimicry between the mirid *Lopidea robiniae* (Uhler) (Heteroptera: Miridae), the clytrine *Anomoea laticlavia laticlavia* (Forster) (Chrysomelidae: Clytrinae), and the hispine *Odontota dorsalis* (Thunberg). Hespenheide (1996) also mentioned five mimetic complexes involving at least 50 species of hispines at La Selva Biological Station, Costa Rica.

The genus *Cephaloleia* contains 185 described New World species (Uhmann, 1957), 88 which occur north of Panama (Staines, 1996). The genus is widespread in the Neotropics from Mexico to Argentina. Adults and larvae feed in the furled leaves or flowers of various genera of Zingiberaceae, Heliconiaceae, Maranthaceae, Cyclanthaceae, Costaceae, and Arecaceae (Jolivet, 1989). Biological and ecological studies have been conducted on several Mesoamerican species (Strong, 1977a,b, 1981, 1982a,b; Auerbach and Strong, 1981; Morrison and Strong, 1981; Seifert, 1982).

2. Methods

While working on a revision of the Central American species of *Cephaloleia* I was struck by the resemblance of various *Cephaloleia* species to cantharids, clerids, lycids, and other chrysomelids. The following discussion hypothesizes mimetic complexes and does not have experimental or field data to substantiate it. Hopefully, the ideas presented will encourage field work by naturalists in the tropics.

The known aspects of model and mimic species were examined to see if these species could both be encountered by the same predator. The biology of *Cephaloleia* would seem to preclude this but *Cephaloleia* adults are long-lived (Morrison & Strong, 1981) and move to another leaf when the one being fed on unfurls (Strong, 1977a). The zingiberalen host plants occur in discrete clumps in various degrees of isolation from conspecifics. Therefore adult *Cephaloleia* may be encountered by predators when moving from one leaf to another.

Specimens of each species identified as mimetic were examined to determine the degree of overall morphological similarity. The most important potential predators in mimetic associations involving *Cephaloleia* are diurnal birds and lizards. Color vision is well developed in both of these groups so it is probably safe to assume that similarities in morphology and color pattern noted by humans would also be noted by these predators.

3. Hypothesis

Several *Cephaloleia* species (*C. bella* Baly, *C. belti* Baly, *C. championi*, Baly, *C. consanguinea* Baly, *C. luctuosa* Guérin-Méneville, *C. quadrilineata* Baly, *C. suaveola* Baly, and *C. vicina* Baly) have the basic color pattern of dark head, pale pronotum, and dark elytra with yellow vittae. A series of *Cephaloleia belti* (Fig. 1) from Costa Rica contained a specimen of the clerid *Perilypus reventazon* Ekis (Fig. 2). Without the aid of a magnification, the clerid was indistinguishable from the hispine apart from the long, dense pubescence covering the dorsal surface. Mimicry is well documented in clerids (Menier, 1985; Mawdsley, 1994). Most clerids are predaceous as larvae and adults (Arnett, 1973). Ekis (1977) stated that *Perilypus* species mimicked lampyrids, cantharids, lycids, and chrysomelids. He did not give specific information on any complex. Mawdsley (1994) suggested that *Perilypus* mimicked Criocerinae and Galerucinae (Chrysomelidae). Ekis (1977) gave some natural history information on *P. reventazon*; most specimens were collected by beating tree branches intertwined with lianas. This method was employed in collecting *C. belti* and *P. reventazon*. This appears to be Müllerian mimicry.

Cephaloleia irregularis Uhmann resembles a cantharid. Uhmann (1930) remarked on the similarity of this species to cantharids. From my observations it is not uncommon to receive cantharids in box lots of Neotropical hispines. Cantharids are protected by chemicals such as dihydromatricaria acid (Meinwald *et al.*, 1968). Mimicry is not well documented in Cantharidae. Massee (1959) reported that the cantharid *Malthinus flaveolus* (Paykull) mimics the mirid *Campyloneura virgula* (Herrich-Schaeffer). Menier (1985) and Mawdsley (1994) reported several clerid genera that mimic cantharids. Cantharid larvae are predaceous, whereas adults feed on pollen and nectar (Arnett, 1973). This appears to be Müllerian mimicry.

Figs. 1-2. 1. *Cephaloleia belti*. 2. *Perilypus reventazon*.

Cephaloleia instabilis Baly and *C. stenosoma* Baly are variable in coloration and resemble lycids. Mimetic complexes of lycids and cerambycids are well documented (Linsley, 1961). Shelford (1902, 1916) discussed mimicry between the hispine *Gonophora* and lycids in New Guinea. Rettermeyer (1970) stated that lycids appear to be distasteful to most predators. Lycid models are usually large, flat beetles with expanded elytra, short serrate antennae, an elongate head, and aposematic coloration (Linsley, 1961). Lycid larvae and adults are predaceous (Arnett, 1973). This appears to be Müllerian mimicry.

Cephaloleia marginella Uhmann is dark with yellowish pronotal and elytral margins and resembles a lampyrid. I have seen lampyrids with a similar color pattern from Costa Rica. Some lampyrids are distasteful (Sexton, 1960). Lamyprid larvae and some adults are predaceous (Arnett, 1973). This appears to be Müllerian mimicry.

Cephaloleia costaricensis Uhmann (Fig. 3) and *C. exigua* Uhmann have a reddish pronotum and metallic blue elytra; *C. gilvipes* Uhmann is metallic blue. All of these species resemble *Altica* (Fig. 4), *Diphaulaca* (Fig. 5), and other Alticinae. Mimetic complexes involving Alticinae have been well documented (Balsbaugh, 1967, 1988; Lindroth, 1971). *Altica* are associated with various Onagraceae, Ericaceae, Lythraceae, Rosaceae, Vitaceae, and other plant families (Phillips, 1977a,b; Scherer, 1969); *Diphaulaca* are associated with various Fabaceae and Solanaceae (Flowers & Janzen, 1997). This appears to be Müllerian mimicry.

Cephaloleia barroi Uhmann and *C. metallescens* Baly (Fig. 6) are broader than other *Cephaloleia* and resemble various *Plagiodera* (Fig. 7) and *Phaedon* (Fig. 8) (Chrysomelinae). DeRoe & Pasteels (1982) found that adult Palearctic and Nearctic Chrysomelinae were protected by defensive glands and reflex bleeding. The defensive

Figs. 3-5. 3. *Cephaloleia costaricensis*. 4. *Altica* sp. 5. *Diphaulaca* sp.

Figs. 6-8. 6. *Cephaloleia metallescens*. 7. *Plagiodera* sp. 8. *Phaedon* sp.

glands produce a cardenolide (Daloze & Pasteels, 1979). Mimicry involving Chrysomelinae models is well documented (Balsbaugh, 1988). *Plagiodera* is associated with various Flacourtiaceae (Flowers & Janzen, 1997); *Phaedon* have been taken on Asteraceae and Fabaceae (Balsbaugh, 1989). This appears to be Müllerian mimicry.

4. Discussion

Field demonstration of a selective advantage for a species is difficult. Experimental design is difficult: too few mimics and statistics are invalid; too many mimics and the natural system is disturbed. Even controlled experiments are open to other interpretations (see Waldbauer & Sternburg, 1976, 1983). It is probably impossible to design experiments which prove a resemblance between two species is a mimetic relationship (Hespenheide, 1995).

Another problem is to determine which species is the model and which is the mimic. While many species cited above have chemical or physical protection mechanisms, the only hispines studied are *Hispa* species which have mechanical protection by sharp cuticular expansions on the elytra and pronotum (DeRoe & Pasteels, 1982). This mechanical protection occurs primarily in the Old World tribe Hispini. Someone may be interested in investigating the possibility of other chemical or physical protections in the non-spiny hispines.

Acknowledgments

H.A. Hespenheide, UCLA, commented on an earlier draft of this manuscript. S.L. Staines provided editorial assistance.

References

Arnett, R.H. 1973. The beetles of the United States. American Entomological Institute, Ann Arbor. 1112 pp.

Auerbach, M.J. and Strong, D.R. 1981. Nutritional ecology of *Heliconia* herbivores: Experiments with plant fertilization and alternative hosts. Ecol. Monogr. 51: 63-83.

Balsbaugh, E.U. 1967. Possible mimicry between certain Carabidae and Chrysomelidae. Coleopts. Bull. 21: 139-140.

Balsbaugh, E.U. 1988. Mimicry and the Chrysomelidae. *in* Jolivet, P., Petitpierre, E. and Hsiao, T. H., eds., Biology of Chrysomelidae. Kluwer Academic Press, Dordrecht, The Netherlands. pp. 261-284.

Balsbaugh, E.U. 1989. The genus *Phaedon* in middle America (Coleoptera: Chrysomelidae). Entomography 6: 465-485.

Bates, W.H. 1862. Contributions to an insect fauna of the Amazon Valley. Trans. Linn. Soc. Lond. 23: 495-566.

Cott, H.B. 1940. Adaptive coloration in animals. Methuen, London. 508 pp.

Daloze, D. and Pasteels, J.M.. 1979. Production of cardiac glycosides by chrysomelid beetles and larvae. Jour. Chem. Ecol. 5: 63-77.

DeRoe, C. and Pasteels, J.M. 1982. Distribution of adult defense glands in chrysomelids (Coleoptera: Chrysomelidae) and its significance in the evolution of defense mechanisms within the family. Jour. Chem. Ecol. 8: 67-82.

Ekis, G. 1977. Classification, phylogeny, and zoogeography of the genus *Perilypus* (Coleoptera: Cleridae). Smithson. Contr. Zool. 227: 1-138.

Flowers, R.W. and Janzen, D.H. 1997. Feeding records of Costa Rican beetles (Coleoptera: Chrysomelidae). Fla. Ent. 80: 334-366.

Gahan, C.J. 1891. Mimetic resemblances between species of the coleopterous genera *Lema* and *Diabrotica*. Trans. R. ent. Soc. Lond. 1891: 367-374.

Gahan, C.J. 1913. Mimicry in Coleoptera. Proc. S. Lond. ent. nat. Hist. Soc. 1912-13: 28-38.

Hespenheide, H.A. 1995. Mimicry in the Zygopinae (Coleoptera: Curculionidae). Mem. ent. Soc. Wash. 14: 147-156.

Hespenheide, H.A. 1996. Chrysomelidae of the subfamily Clytrinae as models for mimicry complexes. *in* Jolivet, P. H. A. and Cox, M. L., eds., Chrysomelidae Biology, Volume 2: Ecological Studies. SPB Academic Publishing, Amsterdam, The Netherlands. pp. 227-239.

Huheey, J.E. 1964. Studies of warning coloration and mimicry. IV. A mathematical model of model-mimic frequencies. Ecology, Brooklyn 45: 185-188.

Huheey, J.E. 1976. Studies in warning coloration and mimicry. VII. Evolutionary consequences of a Batesian-Müllerian spectrum: A model for Müllerian mimicry. Evolution 30: 86-93.
Huheey, J.E. 1984. Warning coloration and mimicry. *in* Bellad, W. J. and Cardé, R. T. eds., Chemical ecology of insects Sinauer Assoc. Inc. Sunderland, Mass., USA. pp. 257-297
Jolivet, P. 1989. Sélection trophique chez les Hispinae (Coleoptera Chrysomelidae Cryptostoma). Bull. men. Soc. linn. Lyon 58: 297-317.
Lane, F. 1951. Cerambycoidea Neotropica nova II (Coleoptera). Dusenia 2: 1-10.
Lindroth, C.H. 1971. Disappearance as a protective factor. A supposed case of Bates'ian mimicry among beetles [Coleoptera: Carabidae and Chrysomelidae]. Entomologica scand. 2: 41-48.
Linsley, E.G. 1959. Mimetic form and coloration in the Cerambycidae (Coleoptera). Ann. ent. Soc. Am. 52: 125-131.
Linsley, E.G. 1961. Lycidlike Cerambycidae (Coleoptera). Ann. ent. Soc. Am. 54: 628-635.
Massee A.M. 1959. *Campyloneura virgula* (H.-S) (Hem., Miridae) found in association with the beetle, *Malthinus flaveolus* (Payk.) (Co., Cantharidae). Entomologist's mon. Mag. 95: 240.
Maulik, S. 1919. The fauna of British India including Ceylon and Burma. Coleoptera. Chrysomelidae (Hispinae and Cassidinae). Taylor and Francis, London, UK. 439 pp.
Mawdsley, J. 1992. A large coleopteran mimicry ring from eastern North America. Y.E.S. Quart. 9(3): 25-38.
Mawdsley, J. R. 1994. Mimicry in Cleridae (Coleoptera). Coleopts. Bull. 48: 115-125.
Meinwald, J., Meinwald, Y.C., Chalmers, A.M., and Eisner, T. 1968. Dihydromatricaria acid: Acetylenic acid secreted by soldier beetle. Science, Washington 160: 890-892.
Menier, J.J. 1985. Quelques aspects du "mimétisme" chez les Cleridae [Col.]. Bull. Soc. ent. Fr. 90: 1071-1083.
Morrison, G. and Strong, D.R. 1981. Spatial variations in egg density and the intensity of parasitism in a neotropical chrysomelid (*Cephaloleia consanguinea*). Ecol. Ent. 6: 55-61.
Pasteels, J.M., Rowell-Rahier, M. and Raupp, M.J. 1988. Plant-derived defense in chrysomelid beetles. *in* Barbosa,P. and Letourneau, D.K., eds. Novel aspects of insect-plant interactions. John Wiley and Sons, New York, USA. pp. 235-272
Pasteels, J.M., Rowell-Rahier, M., Braekman, J-C. and Daloze, D. 1994. Chemical defense of adult leaf beetles updated. *in* Jolivet, P., Cox, M. L. and Petitpierre, E., eds. Novel aspects of the biology of Chrysomelidae. Kluwer Academic Publishers, Dordrecht, The Netherlands. pp. 289-301
Phillips, W.M. 1977a. Observations on the biology and ecology of the chrysomelid genus *Haltica* Geoff. in Britian. Ecol. Ent. 2: 205-216.
Phillips, W.M. 1977b. Some aspects of the host plant relations of the chrysomelid genus *Haltica* with special reference to *Haltica lythri*. Entomologia exp. appl. 21: 261-274.
Rettermeyer, C. W. 1970. Insect mimicry. A. Rev. Ent. 15: 43-74.
Schaefer, C.W. 1996. Bright bugs and bright beetles: Asopine pentatomids (Hemiptera: Heteroptera) and their prey. *in* Alomar, O. and Wiedenmann, R. W., eds. Zoophytophagous Heteroptera: Implications for life history and integrated pest management. Thomas Say Publ., Ent. Soc. Am. pp. 18-56
Scherer, G. 1969. Die Alticinae des Indischen Subkontinentes. Pacif. Insects Monogr. 22: 1-251.
Seifert, R.P. 1982. Neotropical *Heliconia* insect communities. Q. Rev. Biol. 57: 1-28.
Sexton, O.J. 1960. Experimental studies of artificial Batesian mimics. Behaviour 15: 244-252.
Shelford, R.W.C. 1902. Observations on some mimetic insects and spiders from Borneo and Singapore. Proc. Zool. Soc. Lond. 1902: 230-284.
Shelford, R.W.C. 1916. A naturalist in Borneo. London, Unwin Ltd. 331 pp.
Staines, C.L. 1996. The genus *Cephaloleia* (Coleoptera: Chrysomelidae) in Central America and the West Indies. Special Pub. No. 3 Rev. Biol. Trop. 87 pp.
Strong, D.R. 1977a. Rolled-leaf hispine beetles (Chrysomelidae) and their Zingiberales host plants in Middle America. Biotropica 9: 156-169.
Strong, D.R. 1977b. Insect species richness: Hispine beetles of *Heliconia latispatha*. Ecology, Brooklyn 58: 573-582.
Strong, D.R. 1981. The possibility of insect communities without competition: Hispine beetles on *Heliconia*. *in* Denno, R.F. and Dingle, H., eds. Insect life history patterns habitat and geographical variation. Springer-Verlag, New York, USA. pp. 183-194.
Strong, D.R. 1982a. Potential interspecific competition and host specificity: Hispine beetles on *Heliconia*. Ecol. Ent. 7: 217-220.
Strong, D.R. 1982b. Harmonious coexistence of hispine beetles on *Heliconia* in experimental and natural communities. Ecology, Brooklyn 63: 1039-1049.
Turner, J.R.G. 1984. Mimicry: The palatability spectrum and its consequences. *in* Vane-Wright, R.I. and Ackery, P.R., eds. The Biology of Butterflies. Symposium of the Royal Entomological Society. pp. 141-161.
Uhmann, E. 1930. Hispinen aus Costa Rica aus der Ausbeute des Herrn. Ferd. Nevermann. 20. Beitrag zur Kenntnis der Hispinen (Col. Chrys.). Folia zool. hydrobiol. 1: 209-256.

Uhmann, E. 1957. Coleopterorum Catalogus, Supplementa, Chrysomelidae: Hispinae, Hispinae Americanae. W. Junk, Gravenhage, The Netherlands. Pars 35(1): 1-153.
Waldbauer, G.P. and Sternburg, J.G. 1976. Saturniid moths as mimics: An alternative interpretation of attempts to demonstrate mimetic advantage in nature. Evolution 29: 650-658.
Waldbauer, G.P. and Sternburg, J.G. 1983. A pitfall in using painted insects in studies of protective coloration. Evolution 37: 1085-1086.

ARBOREAL CHRYSOMELID COMMUNITY STRUCTURE AND FAUNAL OVERLAP BETWEEN DIFFERENT TYPES OF FORESTS IN CENTRAL AFRICA

THOMAS WAGNER
Zoologisches Forschungsinstitut und Museum Alexander Koenig, Adenauerallee 160, D-53113 Bonn, Germany

Introduction

Tropical rainforest canopies bear the most species-rich animal communities known on earth. Since the early eighties, when the method of insecticidal tree fogging was introduced, a number of intensive studies on canopy arthropod assemblages were carried out in tropical forests in South East Asia (Stork, 1987; Morse *et al.*, 1988; Floren, 1995; Allison *et al.*, 1997; Hammond *et al.*, 1997), South and Central America (Erwin & Scott, 1980; Erwin, 1983, Davis *et al.*, 1997), Australia (for example Kitching *et al.*, 1997), and also more recently Africa (Wagner, 1996a, 1997). Beetles belong to the most abundant and, if specimens were determined, also to the most species-rich arthropod groups. Among beetles, next to Curculionidae and Staphylinidae, the Chrysomelidae were usually the most abundant and most species-rich taxon in tropical forest canopies. While these correspondences in general may be interpreted as a principal pattern of arboreal arthropod communities in the tropics, the rules which govern the assembly are up to now not well understood. Phytophagous insects were the focus of some studies, including two studies on arboreal chrysomelids on the western edge of the Amazon basin (Farrell & Erwin, 1988) and in Papua New Guinea (Basset, 1996; Basset & Samuelson, 1996).

One principal aspect of the equilibrium theory implies that interspecific competition between species leads to narrow niches and higher specialisation (cf. Schoener, 1983; Denno *et al.*, 1995). This is obviously not decisive for phytophagous, especially not for phyllophagous insects (Lawton & Strong, 1981; Jermy, 1985). I am aware of only one study on interspecific competition in leaf beetles (Neotropical Hispinae), (Seifert & Seifert, 1976), but these interactions were doubted in later studies (Strong, 1982) when no interspecific competition between the same species could be found, even under artificially high population densities. Furthermore, under natural conditions, phyllophagous insects use only a low percentage of the food resource (for example Southwood *et al.*, 1982; Aide 1992) and most of the species especially in tropical forests are extremly rare, and have populations with strong and irregular temporal fluctuations (Wolda, 1992). These features are incompatible with the «equilibrium theory» and interspecific competition (cf. Grant, 1986; Chesson, 1978), but much better concordant with predictions of «predation models» (for example Caswell, 1978; Bernays, 1989). Next to biotic interactions with competitors

Advances in Chysomelidae Biology 1, pp. 247–270
Edited by M.L. Cox

and enemies, the geographical history with its influence on speciation (for example Lawton & Schröder, 1977; Farrell *et al.*, 1992) and effects of disturbance were named as abiotic determinants of local species richness of phytophagous insects.

This paper summarizes data on arboreal Chrysomelidae collected within a project on canopy dwelling arthropods during the last five years in Kivu, Rwanda and predominantly in Uganda. At six sites, insecticidal canopy fogging was carried out in different types of forest comprising rain forest at 670 m altitude to upper montane forest at 2950 m. The bulk of material was collected in Budongo Forest, a semidecidous rain forest in Western Uganda, where the influence of forest structure and forest management on the whole beetle fauna (Wagner, in prep.), and the specificity of chrysomelids to forest types and tree species identity (Wagner, in press) were of special interest. In Budongo, data were collected during the rainy season, as well as the dry season. The taxonomic composition and species richness of the chrysomelid fauna in these forests, effects of altitude and seasonality, leaf density and predation pressure by ants as determinants of chrysomelid density in the tree crowns, and the beta-diversity of leaf beetles on a local and regional scale were analysed.

Materials and methods

Study sites and selected tree species

In the first phase in October/November 1993 one abundant and widespread tree species per forest type was chosen in Rwanda and Kivu (Wagner, 1997). Canopy arthropods were collected in East Rwanda in the valley of the Akagera River on *Teclea nobilis* (Rutaceae) in a gallery forest and on *Lannea fulva* (Anacardiaceae) in a nearby dry forest. When fogging was carried out there was an exceptionally severe dry season in that area. Furthermore, trees of *Carapa grandiflora* (Meliaceae) were fogged in montane rain forests in Western Rwanda (Forêt de Cyamudongo and Nyungwe) and conspecific trees in an upper lowland rain-forest in Kivu (Fig. 1, Table 1, 2).

In the second phase, research was carried out during the rainy season in June/July 1995 in Budongo Forest, a semidecidous rain forest in Western Uganda. This forest consists of logged compartments next to old primary forests and swamp forests (Eggeling, 1947; Plumptre & Reynolds, 1994) which makes it an excellent place to study the effects of forest structure and forest management on the arthropod fauna (for details see Wagner, in press). Tree species selected for fogging in Budongo Forest were abundant in all compartments. Tree species were not closely related taxonomically and differed in shape of leaves, growth, and size (Table 2). *Rinorea beniensis* (Violaceae) has a dense canopy and many small leaves, *Trichilia rubescens* (Meliaceae) has an open canopy with relatively few large leaves, *Cynometra alexandri* (Caesalpiniaceae) and *Teclea nobilis* (Rutaceae) are between these extremes.

In January 1997, during the severe dry season in Budongo Forest, *Rinorea beniensis* trees were fogged again in the same secondary forest compartment as during the rainy season. One month later, research was continued in Semliki Forest, a lowland rainforest situated in the Central African Rift. Data there were collected also under the influence of severe drought. Trees of *Cynometra alexandri, Ficus capensis* (Moraceae) and the oil-palm *Elaeis guineensis* (Arecaceae) were fogged. Finally trees of *Afrocrania volckensii*

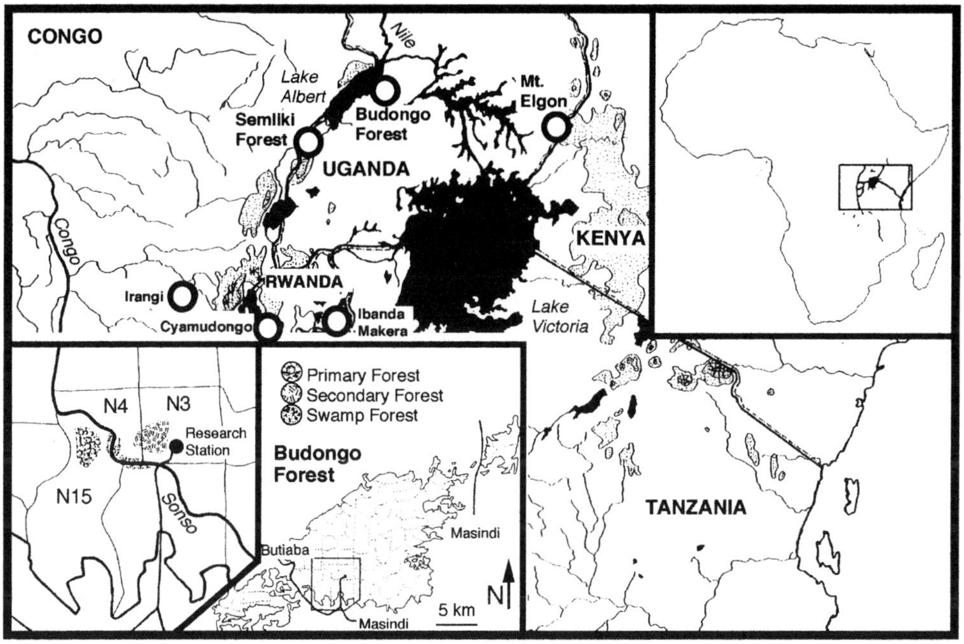

Fig. 1. Study sites. Budongo Forest with location of primary, secondary and swamp forest plots.

(Cornaceae) were investigated in an upper montane forest (*Hagenia abyssinica – Hypericum revolutum* Forest) at Mt. Elgon in September 1997, also during the dry season.

Arthropod collecting

Because fogging was performed from the ground, selected trees were of small stature. Trees fogged in the forest of Rwanda and Kivu were 6-9 m in height, from other sites 7-16 m, with the exception of eight trees of *Cynometra alexandri* in Budongo Forest which were 23-35 m in height. For each forest type usually eight trees were fogged (cf. Table 1, 2).

Table 1. Forests and tree species investigated, density of chrysomelids fogged.

site	date	forest type	altitude (m)	tree species	No.*1	tree height	Chry./m² *2	Chry./m³ *3	abbreviation
Ibanda	X.93	dry	1350	*Lannea fulva*	4	6.0 - 8.0	0.94	-	**L.f.DF**
Makera	X.93	gallery	1300	*Teclea nobilis*	8	6.0 - 9.0	36.4	-	**T.n.GF**
Cyamudongo	X.93	montane rain	1950	*Carapa grandiflora*	9	6.0 - 9.0	3.93	-	**C.g.MrF**
Irangi	X.93	rain	950	*Carapa grandiflora*	5	6.0 - 8.0	0.83	-	**C.g.RF**
Semliki	II.97	rain	670	*Cynometra alexandri*	8	7.0 - 14.5	1.09	0.16	**C.a.RF**
Forest	II.97	rain	670	*Elaeis guineensis*	8	10.0 - 13.0	0.85	0.29	**E.g.RF**
	II.97	rain	670	*Ficus capensis*	8	8.0 - 12.5	0.49	0.15	**F.c.RF**
Budongo F.	I.97	semidecidous	1250	*Rinorea beniensis*	8	7.0 - 12.5	8.40	1.74	**R.b.SF/D**
Mt. Elgon	IX.97	upper montane	2950	*Afrocrania volckensii*	8	8.5 - 11.0	5.16	0.80	**A.v.MF**

*1 No. of fogged trees
*2 No. of chrysomelids per m² collecting sheets
*3 No. of chrysomelids per m³ exploited canopy volume

Table 2. Tree species investigated in Budongo Forest during the rainy season, density of chrysomelids fogged.

Budongo Forest; semidecidous rain forest, 1250m,VI./VII.95, rainy season						
tree species	forest type	No.*1	tree height	Chry./ m² *2	Chry./ m³ *3	**abbre-viation**
Cynometra alexandri	primary	8	11.0 - 35.0	6.14	0.46	**C.a.PF**
	secondary	8	10.0 - 28.0	4.07	0.38	**C.a.SF**
Trichilia rubescens	primary	8	10.0 - 13.0	3.21	0.61	**T.r.PF**
	secondary	8	8.0 - 12.0	3.06	0.57	**T.r.SF**
Rinorea beniensis	primary	8	9.0 - 14.5	4.93	1.13	**R.b.PF**
	secondary	8	7.0 - 12.0	6.91	1.15	**R.b.SF**
	swamp	8	7.0 - 11.0	5.44	1.22	**R.b.SwF**
Teclea nobilis	swamp	8	12.0 - 16.0	6.55	0.95	**T.n.SwF**

*1 – *3 same as Table 1.

Such a group of conspecific trees per forest type is designed as a «collecting unit». Fogged trees were neither flowering nor fruiting. Their canopies were isolated from that of other trees in the same canopy stratum as far as possible. Before fogging, for sites in Rwanda and Kivu eight, for other sites 16 funnel-shaped nylon sheets, one square metre each, were fixed with ropes at waist height under the tree. The square sheets were closely packed around the trunk corresponding to the shape of the canopy. Arthropods on branches in larger tree canopies outside an area of eight or 16 square metres, respectively, were not collected. The total height of fogged trees was measured, for sites in Uganda also the height of the lowest branch. The difference between both figures multiplied by 16 (area of collecting sheets) gives a measurement of the exploited canopy volume. Each tree was fogged for approximately four minutes with a Swingfog SN-50 and a solution of natural pyrethrum with one percent active ingredient in diesel oil. The dropping arthropods were collected for 90 minutes after fogging, and preserved in a mixture of alcohol, acetic acid and water.

Sample analysis and species sorting

The material was sorted into groups, counted, and the beetles assigned to morphospecies by the author at the Museum Koenig. Identification to morphospecies was justified by the high correspondence between the number of morphospecies and the number of «taxonomic» species designated by specialists with beetles in Rwanda and Kivu. Material from 11 taxa with 4305 specimens was first sorted to 180 morphospecies and was later determined to 188 species by various taxonomists (Wagner, 1996b). Various taxonomists recently worked on the chrysomelids collected within this study (see acknowledgements). Determination to genus level of some Alticinae and all Eumolpinae was carried out using identification keys by Scherer (1961) and Selman (1965). With the exception of Diacanthini, Galerucinae were determined by the author (cf. Appendix 1). Bruchinae were included in this study. There is no doubt that seed-beetles are a phylogenetic part of the Chrysomelidae. Furthermore, some species found in remarkably high numbers in the tree crowns.

Species richness, alpha diversity and beta diversity

Fogging allows quantitative enumeration of many groups of tree-dwelling insects, especially for epi-phyllophagous beetles as adult Chrysomelidae are (Basset *et al.*, 1997). Keeping sample size and number of the investigated trees constant per collecting unit, reduces chance effects. Under such conditions the number of species is just as good as other measures of species diversity, and species numbers can be compared directly. Consequently, species diversity is expressed by numbers of species, species accumulation curves, and by rarefaction.

Similarity of the chrysomelid fauna between tree species and forest types were expressed by Morisita-Horn-Indices (see Krebs, 1989, p. 385; Wolda, 1981). These values were calculated between collectings units, that is that all individuals collected within one collecting unit were lumped together, and faunal overlap was calculated between these summarized data (cf. Wagner, in press, where means of faunal overlap between single trees were calculated). The Morisita-Horn-Index ranges between 0, with no species occurring in both samples compared, and 1, when there is total faunal overlap. This is expressed not only by the presence and absence of species, but also takes into account the abundance of species. A rarefaction method for discrete sampling was proposed by Shinozaki (Achtziger *et al.*, 1992), according to rarefaction for continous samplings. The discrete values were numbers of beetle species per tree in this case, and Shinozaki-curves as «rarefied species accumulation-curves» were calculated for some collecting units.

Results and Discussion

Taxonomic composition of chrysomelid communities

On all trees investigated 9909 Chrysomelidae were found which could be assigned to 319 species. The most species-rich and abundant groups were Alticinae, Galerucinae and Eumolpinae, but also Bruchinae reached high numbers at some sites. With the exception of *Afrocrania* at Mt. Elgon, where no Bruchinae were found, seed beetles occurred with low species numbers on all other trees (Fig. 2). During the rainy season seed beetles were the most abundant group of chrysomelids on *Rinorea* in Budongo Forest (Fig. 3), while they were very rare at the same site during the dry season. Bruchinae, with 97.5 % belonging to *Bruchidius latitarsis*, Decelle, reached an average of 7.2 % of all beetle individuals on *Rinorea*, while they were much rarer, comprising 1.4 % on *Cynometra*, a Caesalpiniaceae and therefore a potential food plant. These trees were neither flowering nor fruiting when fogged, and probably not attractive to Bruchinae at that time. No reason can be given why these beetles reach such high percentages on *Rinorea* trees. Bruchinae, especially *Bruchidius ituriensis* (Decelle), and *Callosobruchus rwandanus* Decelle, were also exceptionally abundant on *Teclea*, gallery forest, with eight individuals per square metre collecting sheet.

Also some species of Alticinae reached very high numbers on *Teclea*, gallery forest. One of these species (*Longitarsus sp. 3*) is also the most abundant flea beetle on all trees in Budongo Forest. Other species of *Longitarsus* found on *Teclea* in the gallery forest occurred in Budongo Forest only during the dry season. The chrysomelid fauna on *Afrocrania* at Mt. Elgon is also strongly dominated by only two species of Alticinae (*Hespera* sp. 2,

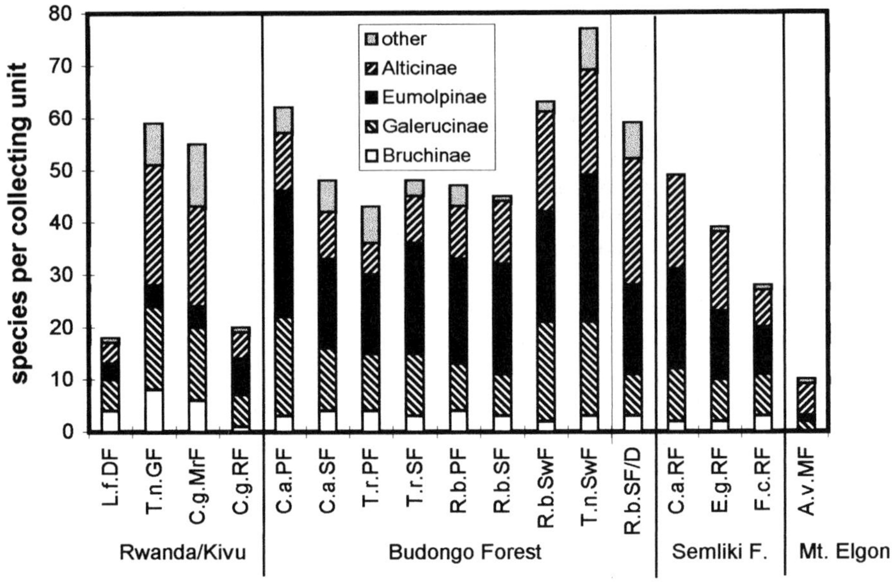

Fig. 2. Numbers of chrysomelid species per collecting unit.

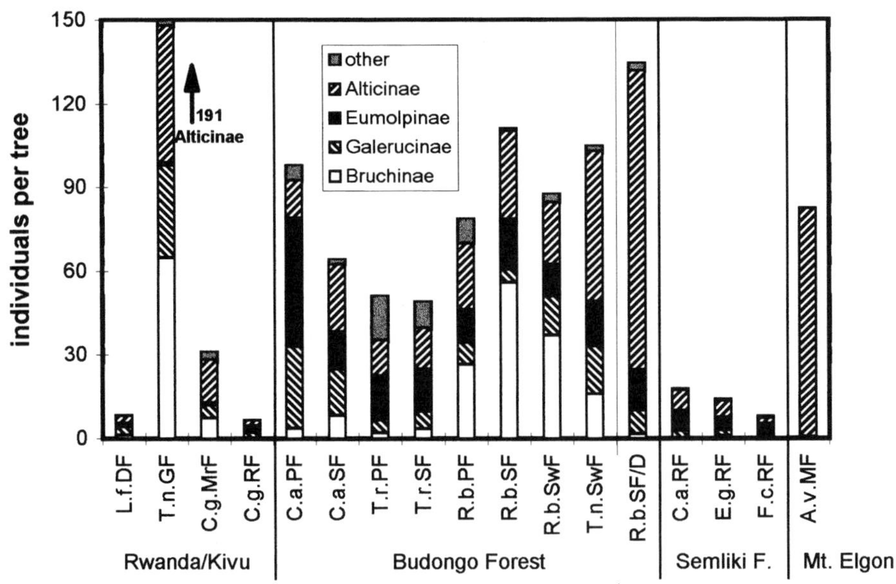

Fig. 3. Mean numbers of chrysomelid individuals per tree.

and *Nisotra* sp. 2), which could frequently be found on all eight trees investigated. Together with *Phygasia* sp. 9., of which 23 individuals were collected on a single tree, 98% of all chrysomelids on *Afrocrania* belonged to these three species.

Eumolpinae, especially many *Paraivongius* sp., and in the swamp forest additionally *Rhembastes* sp., were abundant and species-rich in Budongo and Semliki Forest, but rarer and less abundant at other sites. Within the Galerucinae, *Monolepta vincta* Gerstaecker is a remarkably abundant species in *Teclea*, gallery forest, and in all collecting units from Budongo Forest, where some other *Monolepta* and related species, as well as *Eugaleruca tibialis* Laboissière, were frequently found. Also in Budongo Forest *Linaeidea nubila*, Weise (Chrysomelinae) could be collected predominantly from different tree species in the primary forest, and finally some *Coenobius sp.* (Cryptocephalinae), were very abundant on *Trichilia* trees.

Diversity of chrysomelid communities

In Budongo Forest, species numbers of Chysomelidae were relatively constant on primary and secondary forest trees and ranged between 41 to 49 species per collecting unit, with the exception of *Cynometra*, primary forest (60 species). Despite this exceptional case, the slopes of species accumulation curves on primary and secondary forest trees were similar (Fig. 4). The differences on *Cynometra*, primary forest, is caused by the high number of chrysomelids found on a single tree. With 35 m total height, and 400 m^3 exploited canopy volume, this was by far the tallest fogged tree of the survey. While species richness and

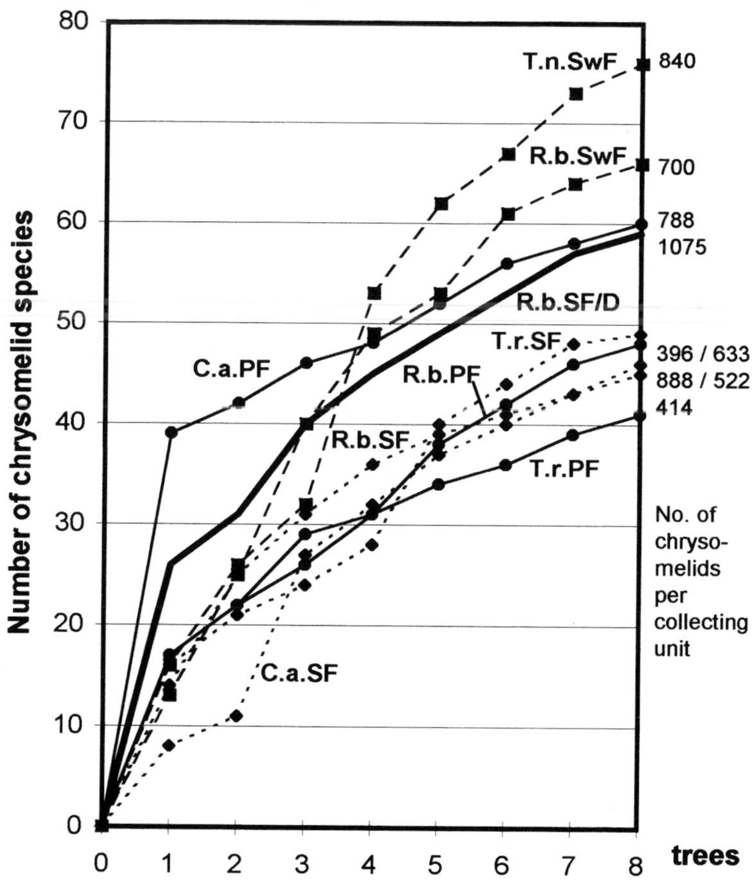

Fig. 4. Species accumulation curves of chrysomelids of different collecting units in Budongo Forest.

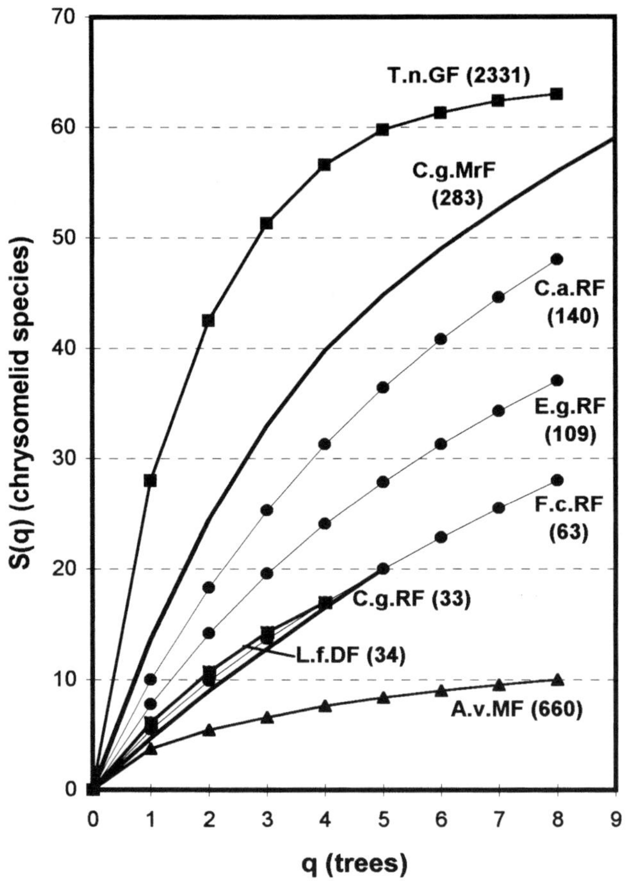

Fig. 5. Shinozaki-curves of chrysomelid species (without Budongo Forest). No. of chrysomelids on all trees per collecting unit in brackets.

distribution patterns of chrysomelids in primary and secondary forest compartments are similar on all tree species investigated, there is a much higher species richness on the trees in the swamp forest with 66 and 76 species per collecting units. During the dry season, the diversity on *Rinorea*, secondary forest, is much lower. The number of specimens collected is more than double, but the number of species is only 25 % higher than during the rainy season at the same forest plot, indicating a much higher dominance of some species during the dry season.

For other collecting units, Shinozaki curves as rarefied species accumulation curves were calculated (Fig. 5). In *Teclea*, gallery forest, the species saturation can be clearly derived from the asymptotic slope, that is the total chrysomelid fauna is approximately registered after fogging eight trees. In comparison to *Teclea*, on *Carapa*, montane rain forest, the chrysomelid fauna reached 93 % of species number with only 12 % of individuals. A much higher number of singletons and fewer abundant species occur on these trees, and thus the curvature is weak and the slope higher. This type of high diversity and incomplete registration of the chrysomelid fauna is also found on lowland rain forest trees, and on *Lannea*, dry forest. On *Afrocrania* at Mt. Elgon the chrysomelid fauna is extremely species poor. The high dominance of three species of Alticinae caused the flat curve. With two trees only, the expected number of species is half the total number found on all eight trees.

Distribution patterns of the canopy arthropod fauna – effect of ant dominance on chrysomelid abundance

The canopy arthropod fauna in general is dominated by ants and beetles. In the dry forest, where the arthropod abundance is lowest of all trees investigated, both ants and beetles reached around 40 % of all arthropods. Chrysomelids were very rare, while the beetle fauna was dominated by two species of ant-mimicking Anthicidae (*Formicomus schimperi,* Pic and *Formicomus spatulus*, van Hille) with an average of 75 % of all beetles. In the nearby gallery forest, the proportion of beetles is similar, but around 50 % of them and 20 % of the entire arthropod fauna belong to the Chrysomelidae. Especially five species of Alticinae (cf. Wagner, 1995) were highly abundant in the gallery forest, 30 % of all beetles collected belong to these species. Ants were comparatively less abundant in the gallery forest, but show constantly higher proportions on the rain forest trees. At medium altitude in Budongo Forest, ants reached on average 29 %, all beetles 22 %, and chrysomelids only 4.5 % of the total arthropod fauna. Both lowland rain forests investigated were very similar with regard to general distribution patterns, with an average of 55 % ants and a very low percentage of beetles. Chrysomelidae are numerically unimportant in comparison to the entire arthropod fauna with on average 12 individuals per tree (14 % of all beetles and 0.62 % of all arthropods) in the lowland rain forests at Irangi and in the Semliki Valley. Ants were much less abundant on *Carapa* in the montane rain forest and absent from montane forest trees at Mt. Elgon. This decrease of ant abundance with altitude is well known from other tropical regions (Brown, 1973; Stork & Brendell, 1990; Samson *et al.*, 1997), and coincides with a significant increase in beetle abundance with forest altitude.

This general influence of ant abundance and their presumably high predacious effect on other arthropods (Wagner, unpublished data) can be also derived from regression analysis between percentages of individuals of chrysomelids and ants. In Budongo Forest and especially in the lowland forests at Irangi and the Semliki Valley, high percentages of ants coincide significantly with low percentages of chrysomelids (Table 3). Also values from other sites show negative, but not significant correlations, probably caused by small sample size, but also low absolute numbers reached by ants on these trees. The decrease of the significance (F- and p-values) with altitude, underlines the strong effect of ant predation, especially in tropical lowland forests (cf. Memmott et al. (1993), Basset & Samuelson (1996) also found no significant correlation between chrysomelid and ant abundance in

Table 3. Regression analysis (of percentage terms) of chrysomelids to several predacious and parasitoid arthropod groups.

		Formicidae			Carabidae/Staphylinidae			Araneae			Parasitoids		
	n	r	F	p	r	F	p	r	F	p	r	F	p
Semliki Forest	24	-0.61	13.2	**<0.001**	-0.04	0.04	0.83	0.49	7.14	**<0.05**	-0.15	0.51	0.48
Carapa, rain forest	5	-0.87	9.47	**<0.05**	-0.84	7.94	0.07	0.27	0.09	0.65	0.32	0.35	0.59
Budongo Forest	64	-0.26	4.19	**<0.05**	-0.18	2.21	0.14	-0.11	0.70	0.39	-0.05	0.16	0.69
Carapa, montane rain f.	9	-0.27	0.58	0.47	-0.22	0.36	0.49	0.12	0.11	0.76	-0.14	0.13	0.72
Teclea, gallery forest	8	-0.62	3.76	0.09	0.81	11.5	**<0.05**	0.53	2.41	0.17	-0.12	0.09	0.77
Lannea, dry forest	4	-0.28	0.14	0.75	0.35	0.28	0.65	-0.61	1.13	0.41	0.35	0.27	0.65

New Guinean forest around 1300 m altitude; but compare Basset (1996) with significantly positive correlations between ants and chrysomelids in branch-clipping samples). High ant dominance in correlation with «non-equlibibrium» features such as low abundance, by-chance-distribution and high beta-diversity (see below) of chrysomelids and other arthropods emphasize the high effect of predation on canopy-dwelling arthropod communities in the tropics.

Correlations of the percentage of chrysomelid individuals with other arthropod predators are indifferent (Table 3). The values of Carabidae/Staphylinidae, webspiders and parasitoid Hymenoptera (predominantly Chalcidoidea) to chrysomelids are slightly negative, but sometimes also significantly positive. Carabidae and predacious Staphylinidae usually feed on soft bodied arthropods, Araneae especially on flying insects and thus both are not typical predators of chrysomelids. Chalcidoidea are egg parasitoids, often with lower specialization to their hosts. If there were strong interactions between a higher number of parasitoids and chrysomelids in these samples, positive correlations would be expected. The slightly negative coefficients, at least at sites with large sample size, underline low interactions between parasitoids and Chrysomelidae (cf. Basset (1996) who found no significant correlation between chrysomelids and parasitoids).

Density of chrysomelids in tree crowns

Next to biotic interactions with predators, the abundance of chrysomelids on trees is influenced by spatial factors such as the density of the foliage and the size of the crowns. Abundance is significantly correlated with the exploited canopy volume (cf. Wagner in press). In Budongo Forest between 0.38 to 1.22, during the dry season even 1.74 leaf beetles could be collected per m^3 exploited canopy volume (Table 1). Low values were found on high trees (*Cynometra*) or on such trees with sparse leaf cover (*Cynometra, Trichilia*), while on *Rinorea* with its very dense foliage of small leaves, the highest densities of chrysomelids could be found. These values are roughly similar to chrysomelid densities in the tropical forests of New Guinea where 0.5 individuals per m^2 leaf foliage were estimated (Basset & Samuelson, 1996). The low number of chrysomelids in Semliki Forest (0.16 to 0.29 per m^3) may be an effect of low abundance in general caused by high ant predation pressure. Densities per square metre collecting sheet, which is an equivalent to chrysomelids captured in a volume of several m^3 canopy, range between values of around 1 in the lowland rain forests and dry forest, to 5 on average in Budongo Forest, with exceptional values of 36.4 in *Teclea*, gallery forest (Table 2).

Influence of drought

On trees of *Teclea*, gallery forest, exceptionally high abundances of Chrysomelidae and other phytophagous beetles such as Apionidae were found. In this forest, *Teclea* has a very dense leaf cover and thus may provide a large spatial resource (cf. Lawton, 1983; Fowler, 1985), but the Rutaceae *Teclea* has effective repellents against phytophagous insects. Actually, the leaves show very little damage by leaf feeders and it is highly questionable if the bulk of Chrysomelidae and Apionidae are feeding on this tree species. Dissected specimens (n=45) of the abundant species of *Longitarsus, Kenialtica, Phyllotreta* and several species of Apionidae show empty mid guts and rarely small particles of faeces in the hind guts. The same could be found by examining specimens from these taxa, collected in Budongo Forest during both the rainy and dry season. Furthermore, fat bodies of these beetles were reduced. Compared to these Alticinae, *Monolepta vinosa* Gerstaecker and *Monolepta vincta* Gerstaecker, abundant Galerucinae on *Teclea* in the gallery forest, had filled mid- and hind-guts. *Monolepta* in principal are very agile beetles and perhaps only occur by chance on the trees when fogged, but some closely related species are known as pests on *Citrus* (Rutaceae) in Southern Africa (for example *Monolepta mashuana* Jacoby) (Jolivet, 1979), or collected frequently feeding on *Citrus* (*Monolepta bioculata* (Fabricius), Wagner unpublished data). As indicated by the gut dissections the bulk of Alticinae were not feeding on *Teclea*. Life cyles of these species are unknown, but *Longitarsus* and *Phyllotreta* from temperate regions predominantly feed on herbaceous Scrophulariaceae, Boraginaceae, Lamiaceae or Brassicaceae, respectively, while Alticinae feeding on Rutaceae are not known in these taxa (Jolivet 1979). Also the Apionidae collected (mainly belonging to *Harpapion, Squamapion* and *Conapion*, M. Wanat, pers. comm.), were up to now only known from Malvales and Fabales and obviously not feeding on *Teclea*.

The high numbers, especially of small beetles in the canopy of dense trees can be explained as an effect of the dry season. Herbaceous food plants are often not available during the dry season. In Semliki Forest, as well as Budongo Forest (dry season) and the gallery forest, trees were fogged at the end of a pronounced dry season, when the whole herbaceous ground vegetation was withered. Under such conditions phytophagous beetles can be found on other plants, sometimes also little feeding on them, a xenophagy especially known from some Alticinae (Jolivet & Hawkeswood, 1995), and even from some *Longitarsus* species on Rutaceae (Jolivet, pers. comm.). During the dry season, the canopies of trees with dense foliage are often the most humid habitats, and beetles accumulate presumably along a gradient of humidity. This aggregation, sometimes combined with dormancy in humid places during the dry season is known from the Coccinellidae (Wolda & Denlinger, 1984), Galerucinae (Masaki, 1980), Cassidinae (Flowers, 1991), and many other insect groups (Denlinger, 1986). In this study, the bulk of the Alticinae do not feed on the tree species investigated. The highest abundances were found in trees with dense leaf cover such as *Teclea*, gallery forest, and *Rinorea*, Budongo Forest. Amazingly one species (*Longitarsus sp. 3*, cf. Appendix 1) was also very abundant during the rainy season on all trees investigated, while species occurring frequently in the dry season in the same forest plot on *Rinorea* were absent there in the rainy season. This aggregation effect during the dry season led to higher similarities of the chrysomelid fauna on different tree species on a regional scale than at the same location between seasons.

Specificity to tree species and forest types

In Budongo Forest, Uganda an average of 19.7 % of the chrysomelid species were found on one tree species only (cf. Wagner, in press; note that Bruchinae were excluded in that study). Nothing is known about the host plants of these leaf-beetles, but at least for the 48 species found as singletons it may be assumed that they are not specialized feeders of the distinct tree species. Without these singletons, the number of species restricted to one tree species are on average only 2.98 %. Compared with these values, the percentage of chrysomelid species restricted to one forest type is about one third higher (31.2 %) for all leaf beetle species, and even about four times higher (12.8 %) without singletons compared. The percentage of species found in the swamp forest only is twice as high as in the secondary forest and primary forest compartment. This is mainly caused by a number of species occurring exclusively in the swamp forest. It indicates higher influences on the distribution of Chrysomelidae probably by factors such as humidity, habitat structure (cover of mosses and lichens), and the availability of minerals which are comparatively higher in the swamp forest than in the adjacent plots of primary and secondary forest (cf. Wagner, in press).

The extent of specificity to tree species is much higher in Semliki Forest, Uganda than in Budongo Forest. An average of 41 % of chrysomelid species were exclusively found on one tree species, without singletons 12.7 %. On one hand, the tree species investigated were extremely different in growth (that is habitat structure), plant chemistry and palatability of leaves, which may cause a higher degree of specificity of the associated chrysomelid fauna. On the other hand, chrysomelids were very rare on these trees, and, as Shinozaki curves reveal (Fig. 5), only a small part of the entire chrysomelid fauna was registered. Since rare species often occurred by chance in a sample, this kind of community «structure» is more influenced by chaos (Vandermeer, 1982) than by any predictability, and further interpretations concerning specificity to food-plant are impossible. In an Amazonian rain forest in Eastern Peru, Farrell & Erwin (1988) found a high «insularity» of chrysomelid species in five types of forests which had stronger differences in floral composition. This distribution pattern was interpreted by a high degree of specialization of leaf beetles to their host plants. In that survey, no single trees, but larger plots of forests were fogged, and thus, differences in the community composition could be explained also as high effect of forest structure on the chrysomelid fauna.

Beta-diversity and influence of the local and regional species pool

The overlap of the chrysomelid fauna between the different forests could be expressed by the percentage of species captured exclusively at one site. Additionally, percentages of individuals belonging to these restricted species were calculated (Table 4). The percentage of species collected exclusively at one site range between 39 % (dry forest) to 65 % (montane rain forest) with an exceptional low value of 15 % in Budongo Forest during the dry season. Most species were also collected during the rainy season in Budongo Forest, indicating a restricted local species pool, but low faunal overlap (Morisita-Horn-Index, see below) between both seasons indicating stronger changes in dominance structure. Only

Table 4. Species numbers and percentages of chrysomelid species and individuals exclusively collected at different sites.

	No. of trees	No. of chryso-melid species	species exclusively found	individuals of species exclusively found
Lannea fulva, dry forest	4	18	39%	32%
Teclea nobilis, gallery forest	8	59	56%	7%
Carapa grandiflora, montane rain forest	9	55	65%	43%
Carapa grandiflora, rain forest	5	20	55%	45%
Budongo Forest (rainy season)	64	159	44%	10%
Budongo Forest (dry season)	8	59	15%	2%
Semliki Forest	24	75	44%	32%
Afrocrania volckensi, upper montane forest	8	10	60%	99%

2 % of individuals collected during the dry season belong to species exclusively found at that time, that is these species are rare, mainly singletons, which were more accidentally collected than being a permanent part of the canopy fauna.

Under this assumption, the relatively high number of chrysomelid species restricted to *Teclea*, gallery forest (56 %), were predominantly rare species, while in *Carapa*, montane rain forest as well as rain forest, also more abundant species were collected exclusively. The upper montane forest at Mt. Elgon (the strongest isolated forest in terms of geographical range, altitude and forest structure), had also the most restricted chrysomelid fauna with a few highly abundant species only occurring there. These basic data reveal a relatively high effect on species distribution patterns on a regional scale, predominantly influenced by biogeographical and historical processes such as speciation rates in isolated areas, effects of extinction and dispersal barriers. The strong climatic changes in Central and East Africa (Hamilton, 1992) with forest fragmentation and shifting of vegetation zones along an altitudinal gradient, may be crucial for the high number of restricted chrysomelid species, especially in isolated areas.

These changes in regional faunal composition can be also derived from beta-diversity measured by Morisita-Horn-Indices (Table 5), a method which combines species identity and species abundance. Taking the results of Table 4 into account, it is not surprising that the highest faunal overlap was found between collecting units in Budongo Forest during the rainy season. Overall faunal similarity ranges from 0.89 between *Rinorea* primary and secondary forest to 0.22 between *Cynometra*, primary forest and *Rinorea*, secondary forest. Many chrysomelids, and even some very abundant species, occur in the entire forest, probably without higher specialization to the tree species from which they were collected. And again, forests which were farthest apart, and also show stronger differences in forest structure and floral composition, as the dry forest in Eastern Rwanda, lowland rain forest in Kivu or the upper montane forest at Mt. Elgon, show the lowest faunal overlap compared with arboreal chrysomelids in other forest.

Differences in species composition on a regional scale may be an effect of the outstanding quarternary changes in climate and vegetation in Central and East Africa, but faunal composition of arboreal chrysomelids is also influenced by local effects (sensu Cornell &

Table 5. Morisita-Horn-Indices of chrysomelid faunal overlap between collecting units.

	L.f. DF	T.n. GF	C.g. MrF	C.g. RF	C.a. PF	C.a. SF	T.r. PF	T.r. SF	R.b. PF	R.b. SF	R.b. SwF	T.n. SwF	R.b. SF/D	C.a. RF	F.c. RF	E.g. RF
T.n.GF	**0.32**															
C.g.MrF	0	**0.31**														
C.g.RF	0	0.04	0.13													
C.a.PF	0.02	0.07	0.11	0.02												
C.a.SF	0.04	0.19	0.13	0.01	**0.56**											
T.r.PF	0.02	0.19	0.13	< 0.01	**0.45**	**0.65**										
T.r.SF	0.03	0.17	0.16	< 0.01	**0.43**	**0.82**	**0.78**									
R.b.PF	< 0.01	0.14	0.12	< 0.01	**0.33**	**0.66**	**0.62**	**0.57**								
R.b.SF	< 0.01	0.08	0.09	< 0.01	0.22	**0.53**	**0.33**	**0.41**	**0.89**							
R.b.SwF	0.04	0.08	0.09	0.02	0.24	**0.58**	**0.34**	**0.44**	**0.86**	**0.85**						
T.n.SwF	0.04	0.19	0.14	< 0.01	**0.36**	**0.83**	**0.62**	**0.76**	**0.79**	**0.68**	**0.71**					
R.b.SF/D	0.15	**0.34**	0.07	0.11	0.21	0.33	0.28	0.30	0.20	0.13	0.14	0.29				
C.a.RF	0.06	0.12	0.04	0.06	0.19	0.27	0.15	0.27	0.09	0.05	0.07	0.25	**0.41**			
F.c.RF	0.02	< 0.01	0	0	0.05	0.02	0.04	0.06	0.03	< 0.01	0.02	0.04	0.01	0.25		
E.g.RF	0.11	**0.34**	0.12	0	0.12	0.19	0.19	0.23	0.12	0.07	0.09	0.23	**0.46**	0.26	0.21	
A.v.MF	0	< 0.01	< 0.01	< 0.01	< 0.01	< 0.01	< 0.01	< 0.01	< 0.01	< 0.01	< 0.01	< 0.01	< 0.01	< 0.01	< 0.01	< 0.01

Legend (shading): < 15; < 30; < 45; > 45

Lawton, 1992). Usually, local species richness depends on biogeographical processes in combination with abiotic fluctuations and biotic interactions. Some effect of (ant-) predation, and tentatively of parasitoids on the chrysomelid fauna are discussed above. Interspecific as well as intraspecific competition in phytophagous insects is in general very low (for example Lawton & Strong, 1981; Jermy, 1985), especially in tropical systems where most communities are «non-interactive» (cf. Cornell & Lawton, 1992), and characterized by the rarity of the bulk of species. However, Morisita-Horn-Indices also indicate that abiotic seasonal fluctuations have a remarkably strong effect on the local composition of the arboreal chrysomelid fauna. Faunal similarity in Budongo Forest between the rainy and dry seasons is much lower than between the collecting units during the rainy season. It is even extremely low (0.13) between conspecific trees in the forest plot investigated during both seasons. Amazingly, the faunal overlap of chrysomelids collected at nearly the same time in Semliki Forest is about twice as high, at least on trees of *Cynometra* and *Elaeis*. These data underline the high seasonal effects on the arboreal chrysomelid fauna. Since the Morisita-Horn-Index is relatively tentative to highly abundant species, the high overlap between chrysomelid communities investigated during the dry season is predominantly caused by the high abundance of Alticinae at that season. In this respect also the high overlap between *Teclea*, gallery forest, to *Carapa*, montane rain forest, *Rinorea*, dry season, and *Elaeis*, Semliki Forest is explicable. The extreme and permanent influence of drought on *Lannea* may be crucial to the high specificity of the chrysomelid fauna, with broad overlap only to *Teclea* trees in the adjacent gallery forest. But also for *Lannea*, the faunal overlap to *Rinorea* in Budongo, a forest of very different habitat structure and plant composition, is much higher during the dry season (0.15) than during the rainy season at the same site (<0.01 – 0.04; for all collecting units).

Summary and conclusions

9909 chrysomelids (including Bruchinae), that could be assigned to 319 species were collected from all 130 trees fogged. Alticinae (96 species), Galerucinae (80 species), and Eumolpinae (66 species) are the most species-rich groups, the latter especially in rain forests at lower altitudes. Next to trees fogged during the dry season, Alticinae were very abundant in the upper montane forest at Mt. Elgon. On *Rinorea* and *Teclea* trees (gallery forest, dry season; seasonal rain forest, rainy season) few species of Bruchinae reached high dominance. Species composition indicates stronger differences of the local species pools between different sites, and a relatively high regional species richness caused by the high geographical fragmentation and strong palaeo-climatical changes in Central and East Africa.

Chrysomelid communities in rain forests are characterized by a high number of rare species. Most of these species have probably no food plant relationship to the trees and occur accidentally in the tree canopy. Comparisons between tree species and local forest types in Budongo Forest (primary, secondary, swamp) reveal a strong influence of forest structure but a low specificity of chrysomelids to tree species.

The individual number of chrysomelids in the crowns of tropical trees is influenced by:
– the spatial resource, given by the density of leaf cover. Chrysomelid numbers range between 0.15 individuals per m^3 exploited canopy volume on *Ficus capensis*, the tree spe-

cies with the most sparse foliation investigated in this survey, to 1.74 individuals on *Rinorea beniensis* which had the most dense cover of small leaves.

– predation pressure of foraging ants. This factors depends on forest type and altitude, because abundance of canopy dwelling ants in moist forests strongly decreases with altitude. Abundance of chrysomelids as well as most other arthropod groups are significantly negatively correlated with ant abundance in the lowland rain forests (670 m and 950 m above sea-level, 50-60 % of all arthropods were ants) and the seasonal rain forest (1250 m, 30 % ants). No such correlation was found in the montane rain forest (1950 m, 6 % ants). The montane forest (2950 m) lacks canopy dwelling ants.

– and depends on season. Especially many Alticinae were only found during the dry season in the tree canopy, when their herbaceous food plants were probably not available and the beetles aggregate along a humidity gradient in the tree crowns. This effect leads to higher faunal overlap expressed by Morisita-Horn-Indices between the chrysomelid fauna in different types of forest, separated by hundreds of kilometres, but higher differences at the same site between both seasons.

Acknowledgements

For cooperation, invitation and manifold logistical help I am very grateful to H. Hinkel, E. Fischer and M. Schmidt (Rwanda, Kivu), A. Plumptre, C. Bakuneeta and V. Reynolds (Budongo Forest Project), and to R. Kidiya and R. Posthouwer (IUCN Kibale-Semliki Forest Conservation and Development Project). Also to the Office Rwandais du Tourisme et des Parcs Nationaux, the Uganda National Council for Science and Technology, the Uganda Forest Department and the Uganda Wildlife Authority for acceptance and permission for the field studies. Many thanks also to M. Cousin, L. Dahlbeck, U. Göllner, S. Hatari, D. Jakisa, C. Kneipp, A. Kurtscheid, A. Melzer, K. Rüsch, B. Schartmann, G. E. Schmitz, A. Silvester, A. Specht, B. Striffler, C. Stuhllemmer, A. Volkwein, A., M., and R. Wagner, B. Wallau and V. Zepp for their valuable, and sometimes enthusiatic help collecting data in Africa and sorting arthropod material in Bonn. I am indebted to taxonomists working on these beetles: Alticinae (M. Biondi, M. Cox, and M. Döberl), Bruchinae (J. Decelle (late)), Cassidinae (L. Borowiec), Chrysomelinae (M. Daccordi), Clytrinae (L. Medvedev), Cryptocephalinae (M. Schöller), Diacanthini (R. Beenen). Without their work, the bulk of specimens would never have been identified. M. Kraemer and F. Krapp made valuable comments on earlier drafts of the manuscript. This research was supported by Deutscher Akademischer Austauschdienst, Stifterverband für die Deutsche Wissenschaft and Deutsche Forschungsgemeinschaft (SCHM-1137/2-1).

References

Achtziger, R., Nigmann, U. and Zwölfer, H., 1992. Rarefaction-Methoden und ihre Einsatzmöglichkeiten bei der zooökologischen Zustandsanalyse und Bewertung von Biotopen. Zeitschrift für Ökologie und Naturschutz 1: 89-105.

Aide, T. M., 1992. Patterns of leaf development and herbivory in a tropical understory community. Ecology 74: 455-463.

Allison, A., Samuelson, G. A. and Miller, S. E., 1997. Patterns of beetle species diversity in *Castanopsis acuminatissima* (Fagaceae) trees studied with canopy fogging in mid-montane New Guinea rainforest. In Stork,

N. E., Adis, J., and Didham, R. K. (eds.). Canopy arthropods. Chapman & Hall, New York, pp. 224-236.
Basset, Y., 1996. Local communities of arboreal herbivores in Papua New Guinea: Predictors of insect variables. Ecology 77: 1906-1919.
Basset, Y. and Samuelson, G. A., 1996. Ecological characteristics of an arboreal community of chrysomelids in Papua New Guinea. In Jolivet, P. H. A. and Cox, M. L. (eds.). Chrysomelidae Biology, vol. 2: Ecological Studies, SPB Academic Publishing, Amsterdam, The Netherlands, pp. 243-262.
Basset, Y., Springate, N. D., Aberlenc, H. P. and Delvare, G., 1997. A review of methods for sampling arthropods in tree canopies. In Stork, N. E., Adis, J. and Didham, R. K. (eds.). Canopy arthropods. Chapman & Hall, New York, pp. 27-67.
Bernays, E. A., 1989. Host range in phytophagous insects: the potential role of generalist predators. Evolutionary Ecology 3: 299-311.
Brown, M. V., 1973. A comparison of the Hylean and Congo-West African rain forest ant faunas. In Meggers, B. J., Ayensu, E. S. and Duckworth, W. D. (eds.). Tropical forest ecosystems in Africa and South America: A comparative review, Smithsonian Institution Press, Washington D.C., pp. 161-185.
Caswell, H., 1978. Predator mediated coexistence: a non -equilibrium model. Am. Nat. 112: 127-154.
Chesson, P. L., 1978. Predator-prey theory and variability. A. Rev. Ecol. Syst. 9: 323-347.
Cornell, H. V. and Lawton, J. H. 1992. Species interactions, local and regional processes, and limits to the richness of ecological communities: a theoretical perspective. J. anim. Ecol. 61: 1-12.
Davies, J. G., Stork, N. E., Brendell, M. J. D. and Hine, S. J., 1997. Beetle species diversity and faunal similarity in Venezuelan rainforest tree canopies. In Stork, N. E., Adis, J. and Didham, R. K. (eds.). Canopy arthropods. Chapman & Hall, New York, pp. 85-103.
Denlinger, D. L., 1986. Dormancy in tropical insects. A. Rev. Ent. 31: 239-264.
Denno, R. F., McClure, M. S. and Ott, J. R., 1995. Interspecific interactions in phytophagous insects: Competition reexamined and resurrected. A. Rev. Ent. 40: 297-331.
Eggeling, W. J., 1947. Observations on the ecology of the Budongo rain forest, Uganda. J. Ecol. 34: 20-87.
Erwin, T. L., 1983. Beetles and other insects of tropical forest canopies at Manaus, Brazil, sampled by insecticidal fogging. In Sutton, S. L., Whitmore, T. C. and Chadwick, A. C. (eds.) Tropical rain forest: ecology and management, Blackwell Scientic Publications, Oxford, pp. 59-75.
Erwin, T. L. and Scott, J. C., 1980. Seasonal and size patterns, trophic structure, and richness of Coleoptera in the tropical arboreal ecosystem: The fauna of the tree *Luehea seemannii* Triana & Planch in the Canal Zone of Panama. Coleopts. Bull. 34: 305-322.
Farrell, B. D. and Erwin, T. L., 1988. Leaf-Beetle community structure in an amazonian rainforest canopy. In Jolivet, P., Petitpierre, E. and Hsiao, T. H. (eds.) Biology of Chrysomelidae, Kluwer Academic, Dordrecht, pp. 73-90.
Farrell, B. D., Mitter, C. and Futuyma, D. J., 1992. Diversification at the insect-plant inferfaces. Insights from phylogenetics. Bio Science 42: 34-42.
Floren, A., 1995. Diversität und Wiederbesiedlungsdynamik arborikoler Arthropodengemeinschaften in einem Tieflandregenwald auf Borneo, Sabah, Malaysia. PhD thesis, University of Würzburg, Germany.
Flowers, R. W., 1991. Aggregations of Cassidinae (Chrysomelidae) in Santa Rosa and Guanacaste National Parks, Costa Rica. Biotropica 23: 308-310.
Fowler, S. V., 1985. Differences in insect species richness and faunal composition of birch seedlings, saplings and trees: The importance of plant architecture. Ecol. Ent. 10, 159-169.
Grant, P. R., 1986. Interspecific competition in fluctuating environments. In Diamond, J. and Case, T. J. (eds.). Community ecology, Harper & Row, New York, pp. 173-191.
Hamilton, A. C., 1992. History of forests and climate. In Sayer, J. A., Harcourt, C. S. and Collins, N. M. (eds.). The conservation atlas of tropical Africa, Macmillan Publishers, Hants, pp. 17-25.
Hammond, P.M., Stork, N. E. and Brendel, M. J. D., 1997. Tree-crown beetles in context: a comparison of canopy and other ecotone assemblages in a lowland tropical forest in Sulawesi. In Stork, N. E., Adis, J. and Didham, R. K. (eds.). Canopy arthropods. Chapman & Hall, New York, pp. 184-223.
Jermy, T., 1985. Is there competition between phytophagous insects? – Z. zool. Syst. Evol.-Forsch. 23: 275-285.
Jolivet, P., 1979. Les Chrysomelidae (Coleoptera) des *Citrus* et apparentés (Rutaceae) en zone tempereé et tropical. Bull. mens. Soc. linn. Lyon 48: 197-200, and 249-256.
Jolivet, P. and Hawkeswood, T. J., 1995. Host-plants of Chrysomelidae of the world. Backhuys Publishers, Leiden.
Kitching, R.L., Mitchell, H., Morse, G. and Thebaud, C., 1997. Determinants of species richness in assemblages of canopy arthropods in rainforests. In Stork, N. E., Adis, J. and Didham, R. K. (eds.). Canopy arthropods. Chapman & Hall, New York, pp. 131-150.
Krebs, C. J., 1989. Ecological Methodology. Harper & Row, New York.
Lawton, J. H., 1983. Plant architecture and the diversity of phytophagous insects. A. Rev. Ent. 28: 23-39.
Lawton, J. H. and Schröder, D., 1977. Effects of host plant type, size of geographic range and taxonomic isolation on number of insect species associated with British plants. Nature (London) 265: 137-140.
Lawton, J. H. and Strong, D. R., 1981. Community patterns and competition in folivorous insects. Am. Nat. 118: 317-337.

Masaki, S., 1980. Summer diapause. A. Rev. Ent. 25: 1-25.
Memmott, J., Godfray, H. C. J. and Bolton, B., 1993. Predation and parasitism in a tropical herbivore community. Ecol. Ent. 18: 348-352.
Morse, D.R., Stork, N. E. and Lawton, H. J., 1988. Species number, species abundance and body length relationship of arboreal beetles in Bornean lowland rain forest trees. Ecol. Ent. 13: 25-37.
Plumptre, A. J., and Reynolds, V., 1994. The effect of selective logging on the primate populations in the Budongo Forest Reserve, Uganda. J. appl. Ecol. 31: 631-641.
Samson, D. A., Richart, E. A. and Gonzales, P. C., 1997. Ant diversity and abundance along an elevational gradient in the Philippines. Biotropica 29: 349-363.
Scherer, G., 1961. Bestimmungsschlüssel der Alticinen-Genera Afrikas (Col. Phytoph.). Ent. Arb. Mus. Georg Frey 12: 251-289.
Schoener, T. W., 1983. Field experiments on interspecific competition. Am. Nat. 122: 240-285.
Seifert, R. P. and Seifert, F. H., 1976. A community matrix of *Heliconia* insect communities. Am. Nat. 110: 461-483.
Selman, B. J., 1965. A revision of the Nodini and a key to the genera of Eumolpidae of Africa (Coleoptera: Eumolpidae). Bull. Br. Mus. nat. Hist. 16: 143-174.
Southwood, T. R. E., Moran, V. C. and Kennedy, C. E., 1982. The richness, abundance and biomass of the arthropod communities on trees. J. anim. Ecol. 51: 635-649.
Stork, N. E., 1987. Arthropod faunal similarity of Bornean rain forest trees. Ecol. Ent. 12: 219-226.
Stork, N. E. and Brendell, M. J. D., 1990. Variation in the insect fauna of Sulawesi trees with season, altitude and forest type. In Knight, W. J. and Holloway, J. D. (eds.). Insects and the rain forest of South East Asian (Wallacea), Roy. ent. Soc. London.
Strong, D. R., 1982. Harmonious coexistence of hispine beetles on *Heliconia* in experimental and natural communities. Ecology 63: 1039-1049.
Vandermeer, J., 1982. To be rare is to be chaotic. Ecology 63: 1167-1168.
Wagner, T., 1995. Verteilungsmuster und Artenvielfalt kronenbewohnender Käfer auf verschiedenen Baumarten in Zentralafrika. Verh. Westdt. Ent. Tag 1994: 79-87.
Wagner, T., 1996a. Artenmannigfaltigkeit baumkronenbewohnender Arthropoden in zentralafrikanischen Wäldern unter besonderer Berücksichtigung der Käfer. PhD thesis, Museum Koenig, University of Bonn, Germany.
Wagner, T., 1996b. Zusammensetzung der baumbewohnenden Arthropodenfauna in Wäldern Zentralafrikas; mit Anmerkungen zur Nebelmethode und zum Morphotypen-Verfahren. Mit. int. ent. Ver. 21: 25-42.
Wagner, T., 1997. The beetle fauna of different tree species in forests of Rwanda and East-Zaire. In Stork, N. E., Adis, J. and Didham, R. K. (eds.). Canopy arthropods. Chapman & Hall, New York, pp. 169-183.
Wagner, T., Influence of tree species and forest type on the chrysomelid community in the canopy of an Ugandan tropical forest. Atti Mus. reg. Sci. nat. Torino, in press.
Wolda, H., 1981. Similarity indices, sample size and diversity. Oecologia 50: 296-302.
Wolda, H., 1992. Trends in abundance of tropical forest insects. Oecologia 89: 47-52.
Wolda, H. and Denlinger, D. L., 1984. Diapause in a large aggregation of a tropical beetle. Ecol. Ent. 9: 217-230.

Appendix 1 (a). List of singletons.

Alticinae:
Altica spec. 1 (A.v.MF); Amphimela spec. 6, spec 7, spec. 8 (all R.b.SwF); Aphthona spec. 5 (L.f.DF); Aphthona spec. 6, spec. 7 (C.a.SF); Aphthona spec. 8 (T.n.SwF); Cercynonia spec. 1 (R.b.SwF); Gabonia rosae (Bryant) (C.g.MrF); Gabonia spec. 7 (T.r.SF); Gabonia spec. 8 (R.b.PF); Gabonia spec. 9 (E.g.RF); Longitarsus spec. 17 (C.g.RF); Nisotra melanopa (C.g.MrF); Phygasia spec. 10 (T.n.SwF); Phygasia spec. 11 (C.a.RF); Phygasia spec. 12 (E.g.RF); Sphaeroderma spec. 7 (R.b.SwF); Sphaeroderma spec. 8 (T.n.SwF); Torodera spec. 1 (C.g.MrF); Alticinae spec. 1, spec. 2 (R.b.SwF); spec. 3 (T.n.SwF); spec. 4 (R.b.SF); spec. 5 (C.a.SF); spec. 6 (F.c.RF).

Bruchinae:
Bruchidius aurivilii (Blanc) (C.g.RF); Bruchidius schoutendeni (Pic) (C.g.MrF); Bruchidius spec. 10 (L.f.DF); Callosobruchus spec. 3 (C.a.RF); Caryedon conformis (Fohr.); Ectacanthus ealensis (Decelle) (T.n.GF).

Cassidinae:
Aspidimorpha mutata Boheman (R.b.SF/D); Cassida humerosa Spaeth (T.n.GF); Cassida imitatrix Spaeth (C.g.MrF); Cassida sublesta Weise (T.n.GF); Chiridopsis aubei Boheman (T.r.SF); Chiridopsis vernicata Fairmaire (C.g.MrF); Chiridopsis spec. 3 (F.c.RF).

Chrysomelinae:
Linaeidea dohrni Jacoby (C.a.PF); Sphaeratrix latifrons Gistel (T.n.GF).

Clytrinae:
Afrophthalma atripennis (T.n.GF); Afrophthalma basalis Jacoby (T.r.SF); Afrophthalma spec. 5 (E.g.RF); Smaragdina spec. 2 (C.a.SF).

Criocerinae:
Lema spec. 2 (T.n.GF); Lema spec. 3 (R.b.SF/D); Lema spec. 4 (Eg.RF).

Cryptocephalinae:
Coenobius spec. 9, spec. 10, spec. 11 (C.g.MrF); Coenobius spec. 12, spec. 13 (R.b.PF); Cryptocephalus spec. 3 (C.g.MrF); Cryptocephalus spec. 4 (F.c.RF); Melixanthus spec.1, spec.2 (T.n.GF).

Eumolpinae:
Afroeurydemus spec. 1, spec. 2 (E.g.RF); Chiridella spec. 3 (C.a.RF); Colposcelis spec. 4 (R.b.SF/D); Corynodes spec. 1 (T.n.SwF); Lefevrea spec. 9 (Cg.RF); Lefevrea spec. 10 (T.n.SwF); Mandollia spec. 1 (C.g.MrF); Melindea spec. 6 (C.g.RF); Melindea spec. 7 (C.a.RF); Paraivongius spec. 17 (T.n.GF); Paraivongius spec. 18 (E.g.RF); Phascus spec. 2 (T.n.GF); Rhembastus spec. 8 (C.a.RF); Scelodonta spec. 2 (F.c.RF).

Galerucinae:
Apophylia cyaneolimbata (Laboissière) (C.a.PF); Bonesioides spec. 2 (F.c.RF); Barombiella

pallida Laboissière (C.a.SF); Chapuisia spec. 3 (T.n.GF); Diacantha clavereaui Weise (L.f.DF); Erythrobapta spec. 1 (R.b.PF); Eugaleruca testacea Laboissière (R.b.SwF); Exosoma dilatatum Jacoby (T.n.GF); Exosoma spec. 3 (T.n.GF); Exosoma spec. 4 (C.g.RF); Luperodes spec. 5 (T.r.PF); Luperodes spec. 6 (R.b.SF); Megaleruca maculiceps Jacoby (T.n.GF); Monolepta elegans Allard (C.g.MrF); Monolepta montana Bryant (A.v.MF); Monolepta pallipes Bryant (R.b.SwF); Monolepta togoensis Laboissière (T.n.SwF); Monolepta spec. 20 (C.a.RF); Monolepta spec. 21 (R.b.SF/D); Phyllobroticella spec. 3 (T.n.GF); Platyxantha spec. 8 (T.r.SF); Stenoplatys spec. 2 (R.b.SwF).

Hispinae:
Callispa spec. 1 (E.g.RF); Dactylispa spec. 3, spec. 4 (R.b.SwF); Dactylispa spec. 5 (C.g.MrF).

Zeugophorinae:
Zeugophora spec. 4 (T.n.GF); Zeugophora spec. 5, spec. 6 (R.b.SF); Zeugophora spec. 7 (R.b.SwF); Zeugophora spec. 8 (T.n.SwF).

Appendix 1 (b). List of chrysomelid species.

	Rwanda/Kivu				Budongo Forest, rainy season different local forest types								dry season	Semliki Forest			Mt. Elgon
Alticinae	L.f.DF	T.n.GF	C.g.MrF	C.g.RF	C.a.PF	C.a.SF	T.r.PF	T.r.SF	R.b.PF	R.b.SF	R.b.SwF	T.n.SwF	R.b.SF/D	C.a.RF	E.g.RF	F.c.RF	A.v.MF
Amphimela nigrocincta Weise	-	6	-	-	-	-	-	-	-	-	-	-	-	-	-	-	-
Amphimela spec. 2	-	10	-	-	-	-	-	-	-	-	-	-	-	-	-	-	-
Amphimela spec. 3	-	7	-	-	-	-	-	-	-	1	-	-	-	-	-	-	-
Amphimela spec. 4	-	-	2	-	-	-	-	-	-	-	-	4	-	-	-	-	-
Amphimela spec. 5	-	-	-	-	-	-	-	-	-	-	3	10	-	-	-	-	-
Aphthona spec. 1	-	-	19	-	-	-	-	-	-	-	-	-	-	-	-	-	-
Aphthona spec. 2	-	-	2	-	-	-	-	-	-	-	-	-	-	-	-	-	-
Aphthona spec. 3	-	-	1	-	-	2	-	2	1	5	1	2	44	-	-	-	-
Aphthona spec. 4	-	-	-	-	-	-	-	-	-	-	-	-	-	-	3	1	-
Afrorestia spec. 1	-	-	33	-	-	-	-	-	-	-	-	-	-	-	-	-	-
Afrorestia spec. 2	-	-	2	-	-	-	-	-	-	-	-	-	-	-	-	-	-
Afrorestia spec. 3	-	-	-	-	1	-	1	-	-	-	-	-	-	-	-	-	-
Chaetocnema spec. 1	-	-	-	-	1	-	-	1	-	-	-	-	-	-	-	-	-
Chaetocnema spec. 2	-	-	-	-	-	-	-	-	-	-	-	-	-	-	10	10	-
Gabonia spec. 1	-	5	-	-	-	-	-	-	-	-	-	-	-	-	-	-	-
Gabonia spec. 2	-	-	-	-	2	6	1	2	6	12	-	-	-	-	-	-	-
Gabonia spec. 3	-	-	-	-	5	4	4	5	3	5	17	12	2	-	-	-	-
Gabonia spec. 4	-	-	-	-	4	-	-	2	1	4	3	63	1	1	1	-	-
Gabonia spec. 5	-	-	-	-	-	-	-	-	-	1	1	8	-	1	-	-	-
Hespera spec. 1	-	-	-	-	3	-	-	-	-	-	-	11	-	10	-	1	-
Hespera spec. 2	-	-	-	-	-	-	-	-	-	-	-	-	-	-	-	-	379
Kenialtica mubalensis (Bechyné)	-	2	-	-	-	-	-	-	-	-	-	-	-	-	-	-	-
Kenialtica muhavura (Bechyné)	-	237	17	-	-	-	-	-	-	-	-	-	4	1	7	-	-
Kenialtica spec. 3	-	-	-	-	-	-	-	-	-	-	-	-	3	-	1	-	-
Kenialtica spec. 4	-	-	-	-	-	-	-	-	-	-	-	-	-	2	1	1	-
Longitarsus spec. 1	1	408	-	-	-	-	-	-	-	-	-	-	-	-	-	-	-
Longitarsus spec. 2	2	257	-	-	-	-	-	-	-	-	-	6	157	6	8	1	-
Longitarsus spec. 3	-	163	11	-	45	107	89	67	165	198	112	179	75	4	5	-	1
Longitarsus spec. 4	-	87	2	-	-	-	-	-	-	-	-	-	14	1	1	-	-
Longitarsus spec. 5	-	45	4	-	-	-	-	-	-	-	-	-	-	-	-	-	-
Longitarsus spec. 6	-	18	2	-	-	-	-	-	-	-	-	-	-	-	-	-	-
Longitarsus spec. 7	-	34	-	-	-	-	-	-	-	-	-	-	97	6	1	-	-
Longitarsus spec. 8	-	7	-	-	-	-	-	-	-	-	-	-	-	-	-	-	-
Longitarsus spec. 9	-	5	-	-	-	-	-	-	-	-	-	-	29	1	-	-	-
Longitarsus spec. 10	-	12	-	-	-	-	-	-	-	-	-	-	-	-	-	-	-
Longitarsus spec. 11	-	-	2	3	-	-	-	-	-	-	-	-	75	-	-	-	-
Longitarsus spec. 12	-	-	2	-	-	-	-	-	-	-	-	-	-	-	-	-	-
Longitarsus spec. 13	-	-	-	-	14	3	-	-	4	7	-	-	124	9	5	-	2
Longitarsus spec. 14	-	-	-	-	-	-	-	-	-	-	7	6	32	1	-	-	-
Longitarsus spec. 15	-	-	-	-	-	-	-	-	-	-	-	-	12	-	-	-	-
Longitarsus spec. 16	-	-	-	-	-	-	-	-	-	-	-	-	7	-	-	-	-
Manobia spec. ?	-	-	-	-	-	-	-	-	-	-	-	-	-	-	2	-	-
Nisotra dalmani (Jacoby)	-	-	-	-	-	-	-	-	-	-	3	-	-	-	-	-	-
Nisotra spec. 2	-	-	-	-	-	-	-	-	-	-	-	-	-	-	-	-	251
Phygasia spec. 1	-	26	-	-	-	-	-	-	-	-	-	-	-	-	-	-	-
Phygasia spec. 2	-	12	-	-	-	-	-	-	-	-	-	-	-	-	-	-	-
Phygasia spec. 3	-	-	-	2	-	-	-	-	-	-	-	-	-	-	-	-	-
Phygasia spec. 4	-	-	-	2	-	-	-	-	-	-	-	-	-	-	-	-	-
Phygasia spec. 5	-	-	-	-	26	67	2	31	3	15	15	46	33	6	-	-	-
Phygasia spec. 6	-	-	-	-	-	-	-	7	-	1		-	-	-	-	-	-
Phygasia spec. 7	-	-	-	-	-	-	-	1	-	-	-	-	1	-	-	-	-
Phygasia spec. 8	-	-	-	-	-	-	-	-	-	-	2	1	1	1	-	-	-
Phygasia spec. 9	-	-	-	-	-	-	-	-	-	-	-	-	-	-	-	-	23

Appendix 1 (b). Cont.

Alticinae	L.f.DF	T.n.GF	C.g.MrF	C.g.RF	C.a.PF	C.a.SF	T.r.PF	T.r.SF	R.b.PF	R.b.SF	R.b.SwF	T.n.SwF	R.b.SF/D	C.a.RF	E.g.RF	F.c.RF	A.v.MF
Phyllotreta spec. 1	7	147	-	-	-	-	-	-	-	-	-	-	-	-	-	-	-
Phyllotreta spec. 2	-	7	3	-	-	-	-	-	-	-	-	-	-	-	-	-	-
Physonychis spec. 1	-	8	-	-	-	-	-	-	-	-	-	-	-	-	-	-	-
Podagrica vulgaris Weise	-	-	6	-	-	-	-	-	-	-	-	-	-	-	-	-	-
Podagrica spec. 2	-	-	-	-	-	-	-	-	-	-	-	40	-	3	-	-	
Sebaethe spec. 1	-	-	-	-	-	-	-	-	-	-	-	-	-	3	1	-	-
Sebaethe spec. 2	-	-	-	-	-	-	-	-	-	-	-	-	-	4	-	-	-
Sphaeroderma exiguum Weise	-	16	-	-	3	-	-	-	-	-	-	-	-	-	-	-	-
Sphaeroderma ugandae Weise	-	-	26	2	-	-	-	-	-	-	-	-	-	-	-	-	-
Sphaeroderma spec. 3	-	4	-	-	-	2	-	-	2	-	1	-	1	-	-	-	-
Sphaeroderma spec. 4	-	2	-	-	-	-	-	-	-	-	-	-	-	-	-	-	-
Sphaeroderma spec. 5	-	-	-	-	-	-	-	-	-	2	1	1	2	-	-	1	-
Sphaeroderma spec. 6	-	-	-	-	-	-	-	-	-	-	-	-	1	-	1	-	-
Alticinae spec. 1	-	-	-	-	4	-	5	-	3	-	-	13	1	-	-	-	-
Alticinae spec. 2	-	-	-	-	-	-	-	-	-	-	6	12	2	-	-	-	-
Alticinae spec. 3	-	-	-	-	-	-	-	-	-	-	-	9	1	-	-	-	-
species	**4**	**24**	**19**	**5**	**11**	**9**	**6**	**9**	**10**	**12**	**19**	**20**	**24**	**18**	**15**	**7**	**6**
individuals	**11**	**1525**	**139**	**10**	**108**	**193**	**102**	**118**	**189**	**252**	**178**	**430**	**853**	**61**	**49**	**16**	**657**

Bruchinae																	
Bruchidius ituriensis (Decelle)	-	217	34	-	-	-	-	-	-	-	-	-	-	-	-	-	-
Bruchidius latitarsis Decelle	-	-	4	-	28	57	10	25	207	446	294	127	6	-	-	-	-
Bruchidius lusingerensis (Decelle)	-	-	1	-	-	-	-	2	-	-	-	1	2	-	-	-	-
Bruchidius mulunguensis (Dec.)	-	8	24	-	-	-	-	-	-	-	-	-	-	-	-	-	-
Bruchidius nigrofasciatus (Dec.)	-	9	-	-	1	2	1	-	1	1	3	2	3	-	3	3	-
Bruchidius spec. (rubicundus gr.)	1	20	-	-	-	-	-	-	-	-	-	-	-	-	-	-	-
Bruchidius spec. 7	-	-	-	-	-	-	-	-	-	-	-	-	-	1	-	2	-
Callosobruchus rwandanus Dec.	-	223	3	-	-	6	4	1	4	-	-	-	-	-	-	-	-
Callosobruchus spec. 2	-	-	-	-	-	-	-	-	-	-	-	-	-	-	2	-	-
Caryedon lunatus (Prevelt)	2	1	-	-	-	-	-	-	-	-	-	-	-	-	-	-	-
Specularius erythraeus (Pic)	-	42	-	-	1	1	1	-	1	2	-	-	-	-	2	-	-
species	**4**	**8**	**6**	**1**	**3**	**4**	**4**	**3**	**4**	**3**	**2**	**3**	**3**	**2**	**3**	**2**	**0**
individuals	**5**	**521**	**67**	**1**	**30**	**66**	**16**	**28**	**213**	**449**	**297**	**130**	**13**	**2**	**7**	**5**	**0**

Eumolpinae																	
Afroeurydemus spec. 1	-	3	-	-	2	-	-	-	-	-	2	2	-	-	-	-	-
Chiridella spec. 1	-	-	-	-	18	4	-	1	-	1	-	5	2	-	-	-	-
Chiridella spec. 2	-	-	-	-	11	4	2	3	5	3	2	6	4	1	-	-	-
Colposcelis spec. 1	-	-	-	3	8	4	2	1	2	1	4	2	-	4	-	-	-
Colposcelis spec. 2	-	-	-	-	-	-	-	-	-	-	-	-	-	-	8	1	-
Colposcelis spec. 3	-	-	-	-	-	-	-	-	-	-	3	-	-	-	4	-	-
Gaberella spec. 1	-	-	-	-	25	9	19	13	13	5	-	20	-	5		4	-
Gaberella spec. 2	-	-	-	-	-	-	-	-	-	-	-	-	-	-	3	3	-
Lefevrea spec. 1	-	2	-	1	-	-	-	-	-	-	-	-	-	-	-	-	-
Lefevrea spec. 3	-	-	-	-	12	-	-	4	-	1	-	1	2	-	-	-	-
Lefevrea spec. 4	-	-	-	-	7	1	1	6	1	1	-	5	5	14	-	-	-
Lefevrea spec. 5	-	-	-	-	2	1	1	2	4	2	1	4	2	8	1	11	-
Lefevrea spec. 6	-	-	-	-	5	1	-	7	-	2	-	-	1	1	-	-	-
Lefevrea spec. 7	-	-	-	-	3	4	1	8	2	1	-	1	3	2	-	1	-
Lefevrea spec. 8	-	-	-	-	1	-	-	-	1	2	-	1	-	-	-	-	-
Lefevrea spec. 9	-	-	-	-	-	-	-	-	-	-	2	-	-	5	-	-	-
Melindea spec. 1	-	-	-	2	-	-	-	-	-	-	-	21	-	1	-	-	-
Melindea spec. 2	-	-	-	-	8	-	2	-	-	-	1	-	-	-	-	-	-
Melindea spec. 3	-	-	-	-	1	-	-	2	1	-	-	-	-	-	-	-	-

Appendix 1 (b). Cont.

Eumolpinae	L.f.DF	T.n.GF	C.g.MrF	C.g.RF	C.a.PF	C.a.SF	T.r.PF	T.r.SF	R.b.PF	R.b.SF	R.b.SwF	T.n.SwF	R.b.SF/D	C.a.RF	E.g.RF	F.c.RF	A.v.MF
Melindea spec. 4	-	-	-	-	-	-	-	-	-	-	-	-	-	3	-	-	-
Melindea spec. 5	-	-	-	-	-	-	-	-	-	-	-	-	-	4	-	-	-
Microeurydemus spec. 1	-	-	-	-	1	-	-	-	-	2	-	-	-	-	-	-	-
Microeurydemus spec. 2	-	-	-	-	7	-	-	-	-	1	-	1	-	-	-	-	-
Paraivongius spec. 1	2	-	-	-	-	-	-	-	-	-	-	-	-	-	-	-	-
Paraivongius spec. 2	-	-	-	2	-	-	-	-	-	-	-	-	-	-	-	-	-
Paraivongius spec. 3	-	-	3	-	-	-	-	-	-	-	-	-	-	-	-	-	-
Paraivongius spec. 4	-	-	-	-	-	-	-	-	-	-	-	4	-	-	-	-	-
Paraivongius spec. 5	-	-	-	-	-	-	-	-	-	-	12	1	-	3	2	2	-
Paraivongius spec. 6	-	-	-	-	-	-	-	-	1	-	-	3	-	-	-	-	-
Paraivongius spec. 7	-	-	-	-	13	-	3	10	1	6	2	11	5	-	9	3	-
Paraivongius spec. 8	-	-	-	-	8	1	6	8	1	-	-	5	2	-	-	-	-
Paraivongius spec. 9	-	-	-	-	32	17	27	18	10	24	28	12	14	-	-	-	-
Paraivongius spec. 10	-	-	3	-	11	22	19	11	14	32	-	-	31	-	-	-	-
Paraivongius spec. 11	-	-	-	-	-	-	16	9	4	-	-	-	8	-	1	-	-
Paraivongius spec. 12	-	-	-	-	61	11	7	8	9	14	-	-	15	-	-	-	-
Paraivongius spec. 13	-	-	-	-	19	6	18	8	12	22	-	-	18	-	-	-	-
Paraivongius spec. 14	-	-	-	-	-	1	-	-	-	-	-	-	1	-	1	-	-
Paraivongius spec. 15	-	-	-	-	-	-	-	-	-	-	-	-	3	-	-	-	-
Paraivongius spec. 16	-	-	-	-	-	-	-	-	-	-	-	-	-	1	2	2	-
Phascus spec. 1	-	-	2	-	-	-	-	-	-	-	-	-	-	-	-	-	-
Pseudivongius spec. 1	-	-	-	-	114	20	2	-	13	-	-	2	-	-	-	-	-
Rhembastus spec. 1	1	1	-	-	-	-	-	-	-	-	-	-	-	-	-	-	-
Rhembastus spec. 2	-	-	-	-	-	1	-	2	1	-	10	3	-	1	-	-	-
Rhembastus spec. 3	-	-	-	-	-	-	-	-	-	-	12	8	-	-	-	-	-
Rhembastus spec. 4	-	-	-	-	-	-	1	1	-	1	2	1	-	-	-	-	-
Rhembastus spec. 5	-	-	-	-	-	-	-	-	-	-	1	3	-	-	-	-	-
Rhembastus spec. 6	-	-	-	-	-	-	-	-	-	-	7	5	-	-	-	-	-
Rhembastus spec. 7	-	-	-	-	-	-	-	-	-	-	-	-	-	1	2	-	-
Scelodonta spec. 1	-	-	-	-	-	-	-	-	-	-	2	-	1	-	-	-	-
Syagrus spec. 1	3	-	-	-	-	-	-	-	-	-	-	-	-	-	-	-	-
Timentes spec. 1	-	-	-	-	-	-	-	-	-	-	-	-	-	-	-	-	2
species	**3**	**5**	**4**	**6**	**22**	**16**	**16**	**19**	**18**	**18**	**17**	**26**	**17**	**19**	**13**	**9**	**1**
individuals	**6**	**8**	**9**	**10**	**369**	**107**	**127**	**122**	**95**	**144**	**92**	**129**	**118**	**57**	**36**	**28**	**2**

Galerucinae																	
Apophylia chloroptera Thomson	-	-	-	-	-	-	1	-	-	-	1	-	-	-	-	-	-
Asbecesta spec. 1	-	-	-	-	-	-	-	-	-	-	-	-	-	-	3	-	-
Asbecesta spec. 2	-	-	-	-	-	-	-	-	-	-	1	-	-	-	2	-	-
Bonesioides laevicollis (Lab.)	-	-	-	-	3	-	-	3	-	-	-	3	-	1	-	-	-
Barombiella mutabilis Lab.	-	3	-	-	-	-	-	-	-	-	-	-		1	-	1	-
Barombiella vicina Laboissière	-	12	-	-	-	-	-	-	-	-	-	-	-	-	-	-	-
Barombiella spec. 3	-	-	4	-	-	-	-	-	-	-	-	-	-	-	-	-	-
Barombiella spec. 4	-	-	-	-	-	2	-	1	-	-	1	-	-	-	-	-	-
Barombiella spec. 5	-	-	-	-	-	-	-	-	-	-	-	-	-	4	-	-	-
Chapuisia usambarica Weise	-	2	-	-	-	-	-	-	-	-	-	-	-	-	-	-	-
Chapuisia spec. 2	-	-	2	-	-	-	-	-	-	-	-	-	-	-	-	-	-
Chapuisia spec. 3	-	-	-	-	1	-	1	-	2	-	-	-	-	-	-	-	-
Diacantha apicata Weise	-	3	-	-	-	-	-	-	-	-	-	-	-	-	-	-	-
Diacantha pauli Weise	-	-	-	-	-	-	-	-	-	-	-	-	-	1	1	-	-
Diacantha spec. 3	-	-	-	-	-	-	-	-	-	-	-	-	-	-	1	1	-
Eugaleruca tibialis Laboissière	-	-	-	-	22	5	11	5	38	5	17	1	4	-	-	-	-
Exosoma ornata Laboissière	-	-	3	-	-	-	-	-	-	-	-	-	-	-	-	-	-
Galerudolphia epipleuralis Jacoby	-	94	-	-	-	-	-	-	-	-	-	-	-	1	-	-	-

Appendix 1 (b). Cont.

Galerucinae	L.f.DF	T.n.GF	C.g.MrF	C.g.RF	C.a.PF	C.a.SF	T.r.PF	T.r.SF	R.b.PF	R.b.SF	R.b.SwF	T.n.SwF	R.b.SF/D	C.a.RF	E.g.RF	F.c.RF	A.v.MF
Hyphaenia africana Laboissière	-	4	-	-	-	-	-	-	-	-	-	-	-	-	-	-	-
Leptaulaca spec. 1	-	-	-	-	-	-	-	-	-	-	11	-	-	-	-	-	-
Leptaulaca spec. 2	-	-	-	-	-	-	-	-	-	-	3	3	-	-	-	-	-
Leptaulaca spec. 3	-	-	-	-	-	-	-	-	-	-	-	-	-	-	-	1	1
Luperodes spec. 1	2	-	-	-	-	-	-	-	-	-	-	-	-	-	-	-	-
Luperodes spec. 2	-	-	7	-	75	52	4	4	9	-	2	1	1	1	-	-	-
Luperodes spec. 3	-	-	-	2	2	-	-	-	-	-	17	16	-	-	-	-	-
Luperodes spec. 4	-	-	-	-	-	-	-	-	-	-	-	-	-	1	1	1	-
Medythia quaterna Fairmaire	-	-	-	-	1	-	-	-	-	-	-	-	1	-	-	-	-
Megaleruca spec.	-	-	-	-	-	-	-	-	-	-	-	-	11	-	-	-	-
Monolepta apicalis Sahlberg	-	-	2	2	-	1	-	1	-	-	-	5	-	-	-	-	-
Monolepta bifossulata Laboissière	-	2	-	-	-	-	-	-	-	-	-	-	-	-	-	-	-
Monolepta burgeoni Laboissière	-	-	-	-	-	-	-	-	-	1	-	-	1	-	-	-	-
Monolepta circumdata Laboissière	-	-	-	-	1	3	-	1	-	3	-	4	-	-	-	-	-
Monolepta conradi Jacoby	-	-	-	-	-	-	-	-	-	-	-	-	-	1	-	1	-
Monolepta dichroa Allard	-	-	-	-	5	-	-	-	-	-	-	-	-	-	-	-	-
Monolepta femoralis Laboissière	3	-	-	-	-	-	-	2	-	-	-	-	-	-	1	1	-
Monolepta fulvosignata Jacoby	-	-	-	-	1	3	1	-	-	-	2	6	-	-	-	-	-
Monolepta testacea Weise	1	12	-	-	-	-	-	-	-	-	-	-	-	-	-	-	-
Monolepta vincta Gerstaecker	2	58	-	-	18	32	9	12	3	17	23	37	43	8	3	-	-
Monolepta vinosa Gerstaecker	3	47	-	-	-	-	-	-	-	-	-	-	-	-	-	-	-
Monolepta spec. 16	-	-	-	-	54	6	1	-	4	1	21	5	6	-	-	-	-
Monolepta spec. 17	-	-	-	-	-	-	-	1	1	-	-	-	-	-	-	-	-
Monolepta spec. 18	-	-	-	-	27	22	6	17	4	5	-	38	-	-	-	-	-
Monolepta spec. 19	-	-	2	-	-	-	-	-	-	-	2	-	-	-	-	-	-
Monoleptoides duplicata (Sahlbg.)	-	-	-	-	1	-	-	-	-	-	-	3	-	-	-	-	-
Parasbecesta ruwensorica Weise	-	-	4	-	-	-	-	-	-	-	-	-	-	-	-	-	-
Phyllobroticella nigripennis Lab.	-	-	3	-	-	-	-	-	-	-	-	-	-	-	-	-	-
Phyllobroticella spec. 2	-	-	-	-	1	-	1	-	-	-	-	2	-	-	-	-	-
Platyxantha haemorrhoidalis Lab.	-	16	-	-	-	-	-	-	-	-	-	-	-	-	-	-	-
Platyxantha spec. 2	-	-	-	2	-	-	-	-	-	-	-	-	-	-	-	-	-
Platyxantha spec. 3	-	-	-	2	-	-	-	-	-	-	-	-	-	-	-	-	-
Platyxantha spec. 4	-	-	-	-	5	-	1	-	-	-	1	1	-	-	-	-	-
Platyxantha spec. 5	-	-	-	-	4	5	-	2	1	4	-	6	-	-	-	-	-
Platyxantha spec. 6	-	-	1	-	11	2	-	-	-	-	1	2	-	-	-	-	-
Platyxantha spec. 7	-	-	-	-	-	-	-	-	-	-	2	-	-	-	2	3	-
Rohania femoralis Laboissière	-	-	4	-	-	-	-	-	-	-	-	-	-	-	-	-	-
Rohaniella megalophthalma Lab.	-	7	-	-	1	-	-	-	-	-	-	-	-	-	-	-	-
Stenoplatys spec. 1	-	-	-	-	-	-	-	-	-	-	3	3	-	-	-	-	-
Xenarthracella sylvatica Lab.	-	-	6	-	-	-	-	-	-	-	-	-	-	-	-	-	-
species	**6**	**16**	**14**	**6**	**19**	**12**	**11**	**12**	**9**	**8**	**19**	**18**	**8**	**10**	**8**	**8**	**2**
individuals	**12**	**264**	**40**	**10**	**233**	**133**	**36**	**49**	**62**	**36**	**108**	**136**	**68**	**20**	**17**	**10**	**2**

smaller groups

Cassidinae																	
Cassida dorsovittata Boheman	-	6	-	-	-	-	-	-	-	-	-	-	-	-	-	-	-
Cassida oberlandi Spaeth	-	-	-	-	-	-	-	-	-	-	1	-	7	-	-	-	-
Notosacantha spec.	-	-	4	-	-	-	-	-	-	-	-	1	-	-	-	-	-
Chrysomelinae																	
Linaeidea decellei Daccordi	-	-	8	1	-	-	-	-	-	-	-	-	-	-	-	-	-
Linaeidea nubila Weise	-	-	-	-	34	-	51	-	63	-	13	5	7	-	-	-	-
Chrysomelinae spec.	-	-	-	-	-	-	-	-	-	-	-	-	2	-	-	-	-
Criocerinae																	
Lema spec. 1	-	-	-	-	-	-	-	-	-	-	1	-	7	-	-	-	-

Appendix 1 (b). Cont.

Clytrinae	L.f.DF	T.n.GF	C.g.MrF	C.g.RF	C.a.PF	C.a.SF	T.r.PF	T.r.SF	R.b.PF	R.b.SF	R.b.SwF	T.n.SwF	R.b.SF/D	C.a.RF	E.g.RF	F.c.RF	A.v.MF
Afrophthalma elongata Jacoby	-	-	-	-	-	1	-	-	1	-	-	1	-	-	-	-	-
Afrophthalma spec. 2	-	-	-	-	1	-	-	-	-	1	-	-	-	-	-	-	-
Smaragdina schoutedeni Burgeon	1	1	-	-	-	-	-	-	-	-	-	1	-	-	-	-	-
Cryptocephalinae																	
Coenobius bicoloriceps Pic	-	-	-	-	8	11	70	67	2	-	-	1	-	-	-	-	-
Coenobius spec. 1	-	2	-	-	-	-	-	-	-	-	-	-	-	-	-	-	-
Coenobius spec. 2	-	-	2	-	-	-	-	-	-	-	-	-	-	-	-	-	-
Coenobius spec. 3	-	-	2	-	-	-	-	-	-	-	-	-	-	-	-	-	-
Coenobius spec. 4	-	-	2	-	-	-	-	-	-	-	-	-	-	-	-	-	-
Coenobius spec. 5	-	-	2	-	-	-	-	-	-	-	-	-	-	-	-	-	-
Coenobius spec. 6	-	-	2	-	-	-	-	-	-	-	-	2	-	-	-	-	-
Coenobius spec. 7	-	-	-	-	-	-	3	4	2	-	-	1	-	-	-	-	-
Coenobius spec. 8	-	-	-	-	-	-	1	2	-	-	-	-	-	-	-	-	-
Cryptocephalus spec. 1	-	2	-	-	-	-	-	-	-	-	-	-	-	-	-	-	-
Cryptocephalus spec. 2	-	-	2	-	-	-	-	-	-	-	-	-	-	-	-	-	-
Hispinae																	
Dactylispa spec. 1	-	-	-	1	-	-	-	-	-	1	2	-	-	-	-	-	1
Dactylispa spec. 2	-	-	-	-	-	-	-	-	-	-	1	-	-	-	-	3	-
Zeugophorinae																	
Zeugophora spec. 1	-	-	-	-	1	-	-	1	-	-	2	-	-	-	-	-	-
Zeugophora spec. 2	-	-	-	-	-	5	-	-	3	-	1	1	-	-	-	-	-
Zeugophora spec. 3	-	-	-	-	-	-	-	-	-	2	-	-	-	-	-	-	-
species	18	59	55	20	60	45	41	49	48	46	66	76	59	49	39	28	10
individuals	34	2331	283	33	788	522	414	396	633	888	700	840	1075	140	109	63	660

FLUCTUATION IN THE ABUNDANCE OF IMMATURE FORMS AND IMAGOES OF EUMOLPINAE (COLEOPTERA: CHRYSOMELIDAE) IN *THEOBROMA CACAO* (STERCULIACEAE) IN THE SOUTH OF BAHIA, BRAZIL

EDNA MARIA DE OLIVEIRA FERRONATTO

Cacao Research Center (CEPLAC/CEPEC-Ministério da Agricultura) Ilhéus – Bahia – Brazil; Current address: Laboratório de Referência Animal – Ministério da Agricultura, Estrada da Ponta Grossa, nº 3036, CEP 91785-340 Belém Novo – Porto Alegre – RS, Brazil

Abstract

This study estimated the population of immature forms of Eumolpinae in the cacao plantations soil. Random samples were taken on a weekly basis, using 15 cm^3 soil cubes as sampling units, from which the insects were manually extracted. Graphs showing the population peaks and fluctuations (of the immature forms) over a 2.5-year period were constructed.

The relative abundance and seasonality of the Eumolpinae imagoes on the crowns of cacao trees were also investigated. The estimates were based on weekly samples from the lower part of the crown for a 1-year period, using an entomological net. The sampled insects were submitted, under laboratory conditions, to a feeding diet of on young cacao leaves. The insects that fed on the leaves were recorded. After death, insects were labelied and sent to a specialist for identification. It was found that the Eumolpinae complex was composed of 18 species, under the experimental conditions. Three of these species were considered to be the dominant species in that crop: *Taimbezinhia theobromae*, *Percolapsis ornata* and *Plaummanita* sp.

1. Introduction

The first significant plantation of *Theobroma cacao* in Bahia (Brazil) was established in 1826, at Almada, a district of Ilhéus. Besides the cultivated ones, cacao trees also grew naturally in the forest, due to cacao seeds spread by a mammal species, *Patos flavus*, the kinkajou, regionally known as "jupará" (Silva, 1944).

Eumolpinae are very common in the cacao trees of the south of Bahia, some species such as *Taimbezinhia theobromae* (Bryant) and *Percolaspis ornata* (Germar) being considered pests (Abreu, 1972; Benton, 1984). However, in addition many others occur, occasionally in abundance, some of them not yet scientifically described, regionally known as "vaquinha".

The adults of this subfamily are polyphagous, showing a probable preference for Asclepiadaceae, Apocynaceae and Convolvulaceae (Jolivet, 1988); the larvae, on the other hand, feed on roots and are completely polyphagous (Chen, 1940; Crowson, 1981; Jolivet, 1988). Eumolpinae eggs, larvae and pupae can be found in the cacao plantation soil. The

Advances in Chysomelidae Biology 1, pp. 271–297
Edited by M.L. Cox

larvae are polyphagous and use the roots of *Theobroma cacao* (Sterculiaceae) as a food substrate (Ferronatto, 1986; 1988).

According to Strong (1974), cacao tree pests are strictly autochthonous. Less than 15% of the total number of 1905 species already recorded in cacao trees are known in more than one cacao-producing region, and less than 1.5% are widely distributed. This shows that the insects belonging to the regional fauna of the place where the cacao tree was introduced rapidly transfer to the cacao trees (Strong, 1974). In the case of the cacao-growing region in the south of Bahia, the Eumolpinae complex appears to originate in the Atlantic forest, since one of the methods used to encourage this crop was to thin out part of the forest, leaving only enough trees to provide shade for the young cacao plants. The Atlantic forest has been described as an endemic center for plant and animal species (Brown, 1975; Prance, 1982; Murphy, 1997; Myers, 1997).

The great speed at which the Atlantic forest was cut down in the south of Bahia, and the replacement of the cacao plantations, still containing remnants of native species, by other crops, has led to a drastic reduction in the fauna associated with this forest. Records showing the situation of the Eumolpinae complex in the cacao plantation agro-ecosystem contribute not only to contemporary research but also as a parameter for future studies regarding this group.

This study shows the fluctuation of the population of immature forms in the subfamily Eumolpinae, in the rhizosphere of the cacao trees, and analyzes the relative abundance, population fluctuation, seasonality and diversity of the numerically most significant species associated with the crown of this plant in Ilhéus, Bahia, Brazil.

2. Characteristics of the cacao plantation studied

The cacao tree plantations at the Cacao Research Center (CEPEC/CEPLAC/ Ministry of Agriculture) are the subject of the present study. They are located in the county of Ilhéus, Bahia, Brazil, latitude 14°45'20"S and longitude 039°15'53"W. The climate in the region is hot and humid, without a defined dry season. It is type AF, according to the classification of Köppen (1948).

Research was performed in a 43-hectare cacao plantation with over 40,000 cacao trees. The area topography is irregular, with a slight, undulating slope. It consists of hydromorphic soil, normally located in the low-lying areas, and red-yellow Podzol located on the slopes. Drainage of hydromorphic soil is slow because of its high clay content. The red-yellow Podzol has an A horizon with a medium texture, a B horizon that is clayey, and the C horizon is friable, providing good drainage. In this area known as plot G', an agroecological reserve of about 10 hectares was established, where it was forbidden to apply chemicals, from the beginning of the CEPEC research activities. These cacao trees were originally planted under a forest where the smaller trees and herbaceous plants were removed and replaced by cacao seedlings. Currently they are over 50 years old, and the old shade trees still remaining from the original forest are numerically insignificant. They are disposed in rows with approximately 3 meter intervals between the trees. The cacao tree crowns touch, and do not allow the penetration of direct light, preventing the growth of bushes or any other type of undergrowth. The site is rather dark, thus preventing the growth of plants requiring light. Therefore, besides the cacao trees and the few remaining shade trees, a herbaceous and bush layer is rare, even in clearings resulting from the death of old shade trees. The soil is covered by a 3 to 6 cm layer of leaf litter. This renders the environment inappropriate for the development of other plants which could be used as a food substrate for the immature forms, rhizophages, even those adapted to little light such as *Paspalum conjugatum* (Poaceae), already found to be a host to Eumolpinae (Ferronatto, 1988). Thus, the only food resource for insects in the experimental area can be considered the cacao tree itself.

Few studies exist on the phenology of cacao tree roots in the region studied. Kummerow *et al.* (1981) stated that the finer roots of the cacao tree tend to concentrate right below the litter, and 90% of them, with a mean diameter of 0.22 mm, concentrate in the upper 10 cm close to the soil surface. Therefore, in cacao plantations the food resource for rhizophagic larvae is located in the first 15 cm of soil, where the highest concentration of roots of that plant occurs. The environment looks like a thick carpet of interlaced roots, of different diameters and lengths.

In the cacao trees the alternation of periods of active growth, with rapid elongation of the shoot and the development of new leaves, alternating with dormant periods. This process of growth is referred to as "flushing" and the young shoot and leaves as the "flush" (Greenwood *et al.*, 1950).

3. Immature forms

3.1. Sampling

The absolute density and population fluctuation of the immature forms of the Eumolpinae were estimated by sampling over a two-and-a-half year period. The subarea selected for this study is located in plot G' (described in item 2). It has a environment with similar characteristics for the Eumolpinae from the standpoint of feeding and shelter. The soil is of the red-yellow Podzol type. In order to perform sampling a subarea of 500 plants was delimited, measuring approximately 4500 m^2. Plants were identified with a numbered metal label. The sampling was randomly performed, at weekly intervals, for a thirty-month period from August 1985 to December 1987. A programe with random numbers chose 50 points at the time of each sampling, each point corresponding to one tree among the 500 numbered ones.

The soil sample was taken from a point equidistant between the randomly selected tree and the next closest one. The sampling unit consisted of a 15 cm^3 cube, with a 3375 cm^3 volume of earth and roots (0.003375 m^3), corresponding to a surface area of 0.0225 m^2. Every time sampling was performed, adding up the volume of the 50 samples, a total of 0.17 m^3 of soil was removed. This corresponds to a surface area of 1.125 m^2. A shovel, 15 cm wide and 15 cm high was used to remove the sampling units.

After the samples were removed from the soil, they were placed in sturdy plastic containers, numbered according to the number of the respective tree that had been randomly chosen, and taken to the laboratory. The samples were examined on the same day, or, when the soil was too wet, the examination was put off for another 24 or 48 hours, keeping the containers with the samples protected under black plastic in a sheltered place.

The manual extraction method was used. Initially the soil is taken apart with the hands and immediately after this put, consecutively, through five and two mm mesh sieves. The Eumolpinae specimens recovered were submitted to rearing methods that will be described in item 3.1 .

Fisher and Bergman (1986) discussed the different known sampling and extraction techniques used to determine the procedures which produce the least variation and highest efficiency in performing population estimates of insects in the soil. According to these authors, the manual extraction method is excessively slow, and the results are highly variable due to the variable efficiency of extraction, since the latter is affected by many environmental factors, including humidity and type of soil. Despite these problems, this method was chosen for the current study because, besides quantifying the larval population, it was necessary to know the species which constituted the Eumolpinae complex. For the latter purpose, the immature forms had to be captured alive, in order to provide the remaining larval and/or pupal development under laboratory conditions. Furthermore, in the manual search no special equipment is required, and all stages, mobile and sedentary, can be recovered.

Parallel to and concurrently with the sampling period for the immature forms, several monthly collections were performed, adding to a preliminary collection of the different species which appeared at each new collection, throughout the period of sampling of immature forms. In this case the same entomological net was used as described in item 4.1, but a variable number of cacao plants was visited for an indeterminate length of time.

3.1. Obtaining the imago under laboratory conditions

Each Eumolpinae larva extracted from the field samples was placed next to the roots of a cacao tree plantule, approximately 2 months old. These plantules, previously sown in plastic bags containing 2 kg of sieved, sterile soil, are maintained individually in screened cages, 1 m high and 50 cm wide. After introducing the larvae, the plantules are kept in a greenhouse, shaded from direct sunlight, under the temperature and relative humidity conditions typical of the region. Once a month the plastic bags were slit open, and the earth around the plantule roots was carefully removed with the help of a paintbrush. If the larva was still alive it was transferred to another plantule, and so on, successively, until its death or until it reached the pupal stage.

The pupae extracted from the samples and obtained by rearing the larvae were transferred to transparent plastic pots, 5 cm high and 7 cm in diameter with lids, and lined with sterilized filter paper and cotton wool, slightly dampened with distilled water and maintained under controlled laboratory conditions in a dark place.

The newly-emerged Eumolpinae imagoes, captured in the soil with their elytra still unpigmented and soft, were transferred to a pot such as the one described for the pupae, but covered with nylon netting in place of the lid. This pot was also lined with filter paper and sterilized cotton wool, slightly dampened with distilled water. In this case young cacao tree leaves were supplied, and the insect was maintained under the environmental conditions of the laboratory. After hardening and definitive pigmentation of the exoskeleton, the insects were killed and fixed in 70% alcohol for later identification.

In the case of the larvae, the pupae and the adults, only a single individual was kept in each "rearing cage ".

After the full development of these immature forms, it became possible to identify the species from their adult form.

3.2. Abundance and fluctuation of immature forms of Eumolpinae in the cacao tree rhizosphere

According to Ferronatto (1986), the numerically most significant contribution in the first 15 cm of soil in the cacao plantations is made by the Chrysomelidae (Eumolpinae), followed by Scarabaeidae and Elateridae. The Neotropical Eumolpinae larvae are not described. Factors inherent to their biology render them difficult to rear them in the laboratory (Ferronatto, 1988). Thus, it is impossible to study the individual population fluctuation of the immature forms of each species.

The method used allowed the location of the following immature forms in the cacao tree roots: *Cayetunya* sp., *Hermesilla* sp., *Percolaspis ornata, Plaumannita* sp., *Podoxenus tornatus* Bechyné, *P. coxalis* Lefèvre, *P. troglodytes* Lefèvre, *Rhabdopterus brasiliensis* Bechyné, and *Taimbezinhia theobromae.* The other species which constitute the complex in the cacao plantation soil, or at least some of them, can only be indirectly deduced from the imagoes that use the cacao tree leaves as a food resource at the same site.

Fig. 1, A, B and C illustrates the absolute density of the last larval and pupal stages of Eumolpinae in the cacao plantation soil during 1985 (August-December), 1986 and 1987.

The population curve trends indicate that, both in 1986 and 1987, the period of greatest abundance of these immature forms occurs between mid-March and the end of August. Beginning in August the numbers diminish, and remain low until February. The highest population peaks of the larvae occurred both in 1986 and in 1987, close to the end of July until the beginning of August. Since the Eumolpinae larvae in a cacao plantation feed from the cacao tree roots, at least part of the contribution to the abundance of adults in the cacao tree crowns is made by the immature stages which inhabit the roots of this plant. Taking into account the occurrence of adults in the cacao tree crown, probably the greatest contribution to the larval peak in the soil of this study was due to the conclusion of the immature phase of *Taimbezinhia theobromae.* The rise in the number of adults of this species in the aerial part of the plant begins in the period following the peak of immature forms in soil, considering the time that the larvae still take to the end of the pupal phase. As will be seen later (item 4.4.2), the peak number of *T. theobromae* adults occurs between October and November, and their numbers, as compared with the other species, are extraordinarily high.

During the population peaks of the larvae, densities of 63 and 57 larvae/0.17 m^3 were found (approximately 350 finals stages larvae per m^3), respectively on July 30, 1986 and August 6, 1987. The first stage larvae are about 0.25 mm long, and due to the difficulty in seeing them, only the last larval stages were sampled. In February the population curve begins to rise, and this means that many of the larvae have already reached the last stages.

The determination of population peaks for the pupae was impaired by the very small number of individuals in the total sampled. A comparatively smaller number of pupae is found as compared to the number of larvae (Fig.1-A, B and C). This fact could indicate two situations: either the larvae have a high mortality rate before they pupate for reasons which have not been examined in this study, or the site where the highest number of these insects remain in the pupal phase lies deeper than the first 15 cm investigated. The second hypothesis appears most feasible, since the sites where there is the greatest abundance of food and

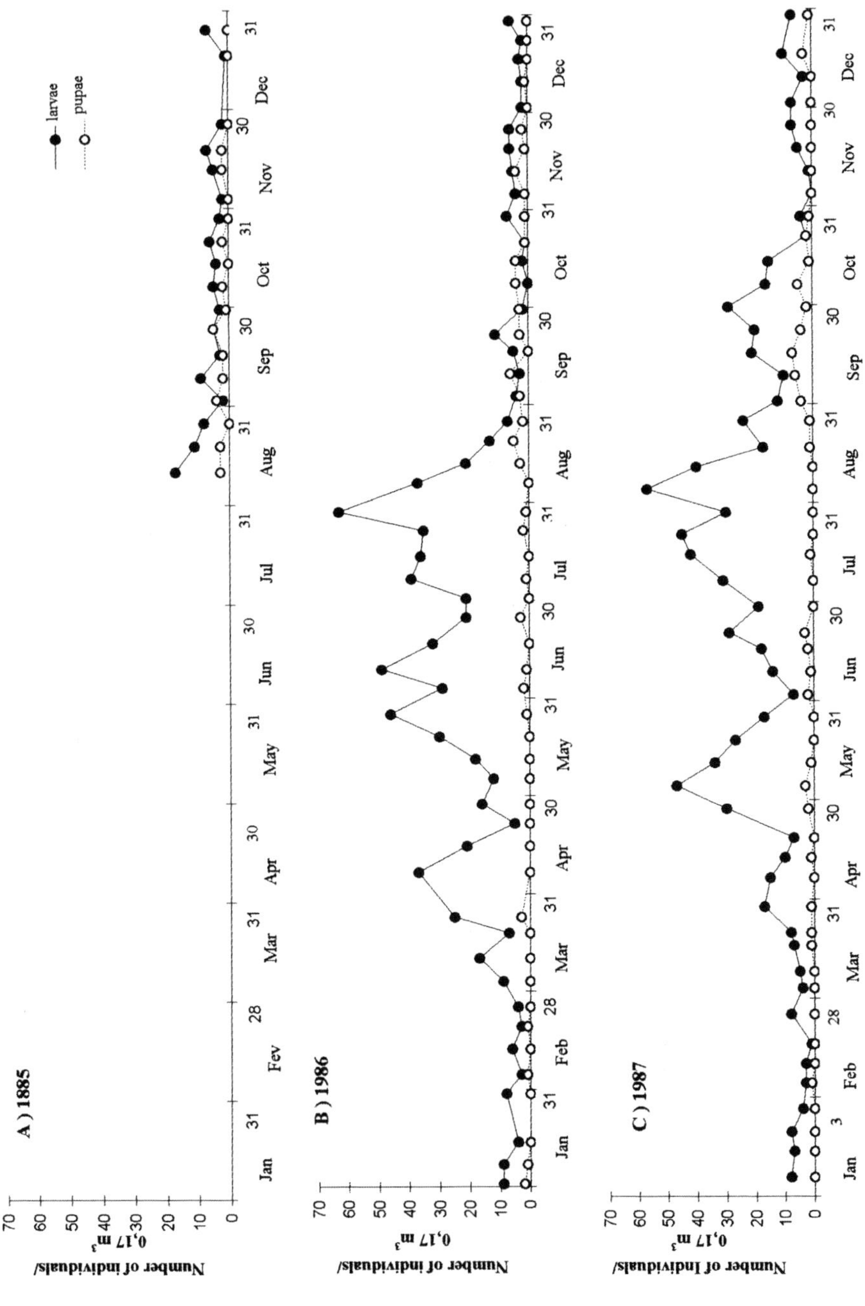

Fig. 1. Abundance of larvae in the finals instars, and Eumolpinae pupae in 0.17 m^3 of the first 15 cm of soil associated with the rhizosphere of cacao trees during the period from A) August to December 1985; B) January to December 1986; C) January to December 1987 (Plot G', CEPEC/CEPLAC, Ilhéus, Bahia, Brazil, 1985/87).

consequently, the greatest abundance of rhizophagous insects, could be less favorable as a shelter for the pupae due to the presence of predators. The highest peak of pupae probably occurs after the first fifteen days in August, since the period immediately before is the highest peak of the final larval stages. Furthermore, for the species whose adults make the best use of the cacao trees as a source of food, the highest pupal population should occur in the period immediately before September-October, when there is the greatest availability of the preferential food for adults in the region, that is, young leaves. This fact is also indicated in the population curve of the Eumolpinae imagoes, and will be discussed in item 4.2.2. The relative population estimates of the imagoes reveal that the highest peak for adults occurred in September-October 1987.

The cacao tree has a peculiar type of very fine roots, the radicles. These, according to Medeiros (1983), present seasonal variations due to the influence of external factors, and the peak of activity in Ilhéus (Bahia) occurs in April-May. In the cacao trees studied, approximately 50 years old, except for the radicles, the other roots are excessively woody for the mandibles of the neonate larvae. For this reason, it is plausible to assume that larval development is related to location and the radicle growth rate.

At no time did we find Eumolpinae larvae inside the cacao tree roots. Gowdey (1926) stated that the larvae of *Colaspis hypochlora* Lefèvre, besides scraping the epidermal tissue from the older banana tree roots, excavate tunnels inside the young roots. Salt (1928), referring to this report, stated that, despite many efforts, he was unable to find even a single larva of *C. hypochlora* inside the roots of banana trees in Colombian plantations, and that both the larvae and the pupae were found at the same soil depth, between 5 and 8 cm. Salt (1928) also mentioned that every time the larvae were extracted from field samples, this invariably occurred close to, or in the roots of *Paspalum conjugatum* (Poaceae). Under laboratory conditions, the same author observed the conclusion of the larval stage with a diet of *P. conjugatum* roots. He stated that *C. hypochlora* larvae feed on the roots of this Poaceae, whether the banana tree roots can be used as food or not.

Ferronatto (1986) tested *P. conjugatum* plantules as diets for the larvae of Eumolpinae collected in the cacao tree rhizosphere. Only a single species, *Colaspis aenea* was able to complete its development in the roots of *P. conjugatum.*

The results of cumulative weekly rainfall (mm) and mean weekly air temperature, dur ing the period in which the samples of the immature forms were taken, illustrated in Figure 2 (A, B and C) are as is expected for the region. Temperature variation is very small and the dry season is not defined. Factors such as high interception of sunbeams by the dense cacao tree crowns, a dark environment with a deep layer of leaves covering the soil, contribute even further to reduce the amplitude of climatic variations and to stabilize the environment. The annual precipitations in 1986 and 1987 were 1214.4 mm and 1400.6 mm respectively. The annual means obtained during the period of the present experiment were lower than those found in the previous periods for the region. According to Miranda (1987), the mean annual precipitation for the period from 1973 to 1986, in the region of Ilhéus, was 1812.6 mm.

The site where the immature forms were sampled has an irregular topography. In the lower regions are areas with hydromorphic , poorly drained soil, with low permeability. In the higher regions there is Podzolic, well-drained soil, with moderate permeability (Departament of Soil, CEPLAC, pers. comm.). Salt (1928) stated that wet soils in a shaded environment present the best conditions for the development of *C. hypochlora* (Eumolpinae)

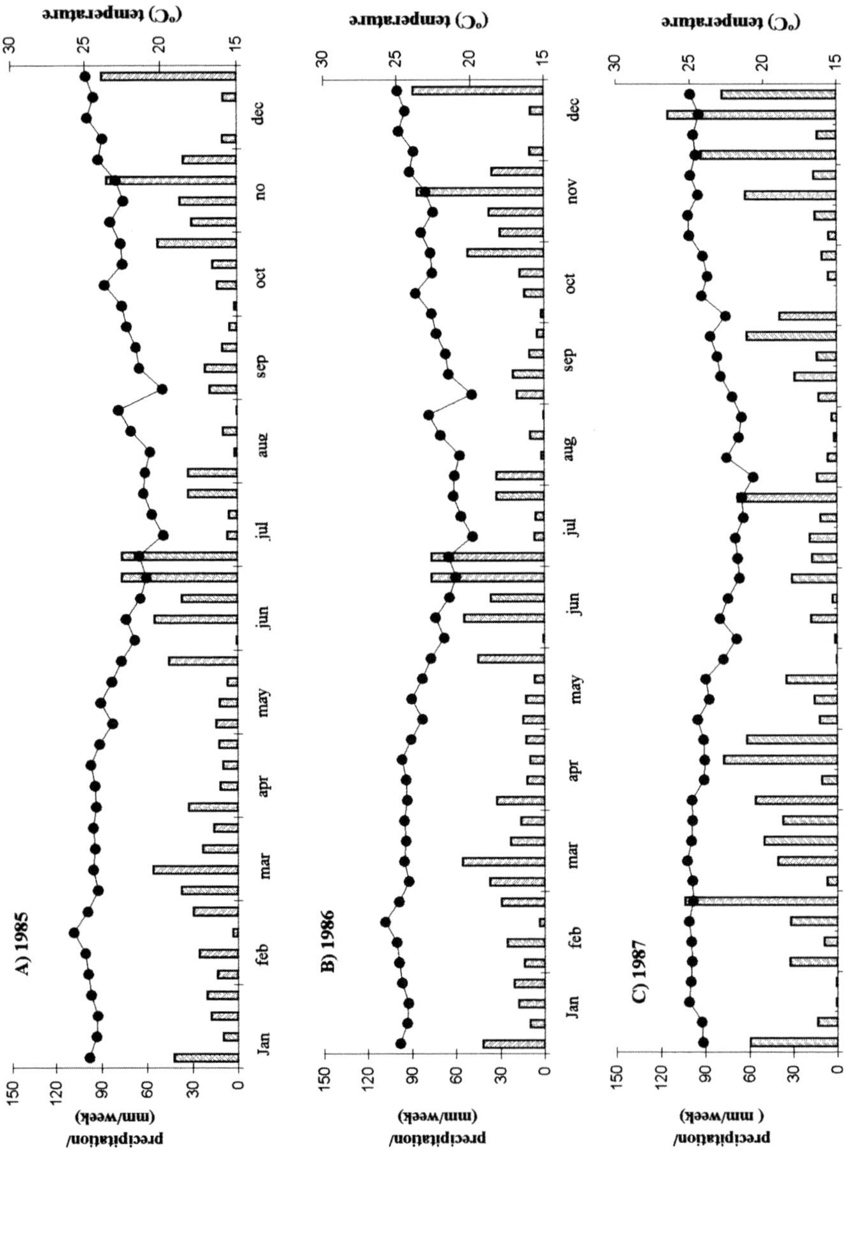

Fig. 2. Mean weekly temperatures (°C) ( ) and sum of weekly precipitation (mm) ( ) from January to December 1986 in the region of Ilheus-Bahia: A) January to December 1985, B) January to December 1986, C) January to December 1987. (Data from the Meteorological Station of CEPEC/CEPLAC, installed 700 m from plot G').

in banana plantations. The author considered soil humidity as a limiting factor in the initial larval stages, and no *C. hypochlora* larvae occur in sandy, well-drained soils.

Taking into account the conclusions of Salt (1928), in plot G', for the Eumolpinae, the most appropriate site for the development of larvae would be the areas with the wettest soil, in regions with moderate slopes. However, due to the problems involved with the extraction of immature forms in very wet soil, because of the manual method adopted in the current investigation, it was decided to perform the study of immature forms in the area with Podzol soil and the best drainage.

4. Imagoes

4.1. Sampling

From June 1987 to May 1988, the relative density of adult Eumolpinae was also assessed by sampling in the lower part of the cacao tree crowns. In this case the sampling area was considered the space taken up by the total number of cacao trees contained in the 43 acres of the cacao plantation.

During the period from June to December 1987, a parallel estimate was made, using visual methods, of the indexes of flushing of new leaves in the cacao trees of plot G'. In order to perform these estimates, the same person with great experience in field work on cacao trees, established visually, every week, the percentages of new leaves in the area, attributing values from 0 to 100 for the foliar flushing observed.

The flushing of new leaves at the site of the experiment, mainly outside the peak periods, besides being different in time were also different in space, forming a mosaic where areas with leaf flushes alternated with areas without any new flushes.

The rate of new leaf flushes in the cacao trees has been studied by many authors (Greenwood and Posnete, 1950; Alvim, 1967, 1977, 1994; Alvim and Alvim, 1975). Alvim (1994) stated that the results on the rhythm of emission of leaf flushes and new flowers in the cacao tree decisively support the "theory of hydroperiodicity" (Alvim 1960, Alvim et al., 1969, 1972; Alvim and Alvim, 1978), considering the changes in available humidity (transition from a dry period to a wet one), the main environmental stimulus to induce flushing. The soil slope at CEPEC in general, and at the experimental site (plot G') in particular, is considered gently rolling, presenting irregular depressions. In plot G' there are two types of soils, red-yellow Podzol, on the slopes, with good permeability, and hydromorphic, in the low-lying areas, with little permeability. Taking into account the theory of Hydroperiodicity, these differences cause distinct phenological responses in the cacao tree, favoring to a heterogeneous occurrence of new leaf flushes in the area. Although the total coverage of each type of soil has not been mapped, it is believed that discontinuity between the "patches" of trees with young leaves, like a mosaic, is related to the characteristics of each soil. In an entirely random sampling, the type of spatial distribution of cacao trees with young leaves could lead to going through very large areas in search of trees that have been drawn randomly, and in many case make it necessary to draw lots again. Therefore it was decided to establish the sampling time beginning from a central point of the cacao plantation area, where the trees already presented leaf flushing ("patch"). These "patches" were irregular-sized, some over 100 m at their longest point. The samples were taken in one of these "patches" in the cacao plantation, chosen randomly, where a cross-section is walked for 30 minutes, a reasonable time for one person to be able to perform the sampling without interruptions and without compromising the precision of climate and relief conditions at the site.

Due to the temporal irregularity in the leaf flushing process, the opportunity for sampling is not the same in the different months. The samples were only taken when the estimate of the leaf flushing indexes was approximately 5%. Based on this percentage, according to previous experience, there are greater chances of collecting Eumolpinae.

On the occasion of each sampling, beginning with a central point of the "patch", chosen randomly, one of the four cardinal points was chosen randomly. Each tree located on an imaginary line formed from this central point, in the direction chosen, was sampled without interruption for 30 minutes. The samplings were performed always by the same person, between 9 and 11 a.m. The entomological net consisted of a circular metal ring, 70 cm in diameter, attached to a short wooden handle and a conical collection bag, in sturdy white cotton, one metre long. The samples were taken only from the branches of the lower level of the crown, at a height accessible to the sampler. The extremity of each branch was shaken three times into the entomological net. For the size of the entomological net used, approximately 33 leaves fit inside it. The branches of the same tree are sampled alternately, leaving aside the immediately subsequent ones. This procedure was adopted to avoid errors resulting from the probable disturbance caused to the insects by the movement of the branches which have already been sampled. After the net was removed, the insects captured were transferred into a screened cage, 40 cm high by 30 cm deep.

In the experimental area, the number of branches checked during a 30-minute period was 82 on average. The number of trees checked depends on the number of branches they have, which in turn is very variable.

The sampling for insects in the aerial part of the cacao trees usually employs the knock-down by pyrethroid insecticides technique, in which a white sheet is placed on the ground, under each tree, to collect the total entomological fauna from all levels of the crown of the cacao trees sampled (Abreu, 1972; Garcia & Silveira Neto, 1980; Garcia & Mendes, 1982; Benton, 1984). When we began sampling we did not know the true number of species which constituted the complex studied. Only later were they sent to a specialist in the Eumolpinae taxonomy. Furthermore, it was forbidden to use insecticides in the experimental area of the current investigation. Our purpose was, besides the quantitative evaluation of Eumolpinae present in the area, to confirm their food association with the cacao tree. Thus, capturing the insect in the entomological net was the best alternative, allowing the verification of the association of the insects with this plant, through a diet of its young leaves, under laboratory conditions.

4.2. Species identification

In March 1987, the adult specimens from the samplings were taken, personally, to the Department of Agricultural Zoology of Universidad Central de Venezuela in Maracay, to be identified or have their identification confirmed by the taxonomist specialising in the Eumolpinae, Mrs. Bohumila Springlova de Bechyné. At the time, specific training was received to recognize the species which constitute the Eumolpinae complex found in the cacao trees in the south of Bahia. A reference collection was made, constituted by the insects identified by Mrs. Bechyné, and these identifications were adopted in the present study.

The specimens studied are presently deposited in the "Gregório Bondar Collection", at the Cacao Research Center (CEPEC/CEPLAC, Ministry of Agriculture) in Ilhéus-Bahia, Brazil. It contains specimens of the most representative insects of the regional fauna of south Bahia. Silva *et al.* (1985) presented a preliminary list of the insects which are part of this collection.

4.3. Estimate of abundance, dominance and diversity of species associated with cacao tree crowns

The abundance of imagoes was estimated by counting the total number of individuals captured during the 30 minutes of sampling. The results concerning the number of individuals are expressed as fifteen-day averages. These fifteen-day averages are transformed into percentages in order to show the relative abundance of each species and in the total sampled.

Dominance is used as a synonym of relative abundance, and the dominance scale of Krogerus (1932) is used for its classification. This scale has been used to classify the species according to their abundance. Species which constitute more than 5% of the captures are considered dominant; those which constitute 2 to 5% are subdominant; those with less than 2% are accessory.

In order to evaluate the diversity of species in the area of study, the index and formula specified below were used.

Species diversity was evaluated by the equation proposed by Margalef (1951):

$$\frac{S-1}{\text{Log } N}$$

where $S =$ number of species
$N =$ number of individuals

4.4. Complex of Eumolpinae associated with cacao trees in the south of Bahia

The taxocenosis of Eumolpinae under study is constituted by the following species: *Antitypona* sp. 1; *Antitypona* sp. 2 ; *Antitypona epiphania* Bechyné; *Coytiera* (*Campylochira*) *pertusa* Lefèvre; *Hylax bahiensis* Bechyné; *Hermesilla* sp.; *Leprocolaspis adducta* Clavareau; *Leprocolaspis sulcata* Lefèvre; *Colaspis aenea, Colapis calcalifera* (Bechyné); *Nycterodina* (*Nycterodinella*) *immetallica* Bechyné & Bechyné; *Percolaspis ornata*; *Podoxenus* sp.; *Taimbezinhia theobromae*. Two species were identified by Bohumila Springlova de Bechyné as new to science. In this study they are designated as *Plaumannita* sp. and *Cayetunya* sp..

The Eumolpinae genera collected in the south of Bahia, belonging to the "Gregório Bondar Collection", are mostly represented by a single species. In the current study three species at most were found for a single genus.

4.5. The taxocenosis of adult Eumolpinae associated with cacao tree crowns

4.5.1. Species Composition

Table 1 lists the species of Eumolpinae already identified, found in the south of Bahia and captured by different methods and on different parts of the cacao plant.

In column A are listed the identified species deposited in the Gregório Bondar Entomological Collection, or which have already been part of it, and that are currently found only in its record books. These records refer to dates before 1985. Most of the insects were collected by knockdown by chemical insecticides, and there are still doubts regarding the connection between non-dominant species and the cacao tree. Thus, the number of Eumolpinae species that really maintain feeding relations with the cacao tree, except for those studied in the current work, is unknown in the region.

In column B are listed the species obtained in the samplings of the rhizosphere of the cacao trees in 1985-1987, identified from the imago. Most of the species from the soil, reared to adulthood in the laboratory , were also found in the crown of the cacao trees. This is not the case for *Rhabdopterus brasiliensis*, *Podoxenus coxalis*, P. *tornatus* and P. *troglodytes*, either because the immature forms of these species occur in strata different from those focused on in this study, or, more likely, due to chance, because of the very small number of individuals in the period studied. On the other hand, some of the species that occur in the aerial part of the cacao tree are not found on the soil. In this case, the reasons are probably related to the techniques used in rearing the immature forms.

The species listed in column C are the result of the preliminary survey performed on different occasions in 1985 and 1986. This survey revealed the presence of 23 species, not all of them coinciding with those recorded in the Gregório Bondar Collection. *Brevicolaspis villosa* Bryant, *Chrysodina* sp. and *Chrysodina corusca* Lefèvre, *Colaspis aeraria* Lefèvre, *C. auripennis* (Germar), *C. obscura* F., *C. olivieri* (Lefèvre), *Eumolpus surinamensis* F., *Habrophora* sp., *Metaxyonycha auripennis* (Germar), *Spintherophyta aurichalcea* (Germar), *S. corusca* (Lefèvre), *Lepronota tessulata* (Germar) and *Typophorus versutus* Lefèvre, which are part of the Gregório Bondar Collection are not represented in these samples. Among the possible explanations for this fact could be the following: a) the species do not occur in the cacao trees of the Ilhéus region; b) the species occur in the cacao trees of the Ilhéus, but in a different height from those focused on in the sampling; c) the species are polyphagous, but have a preference for another type of plant, and are rare and occasional in the cacao tree; d) these species occur in very small numbers during the survey period, and were not captured. On the other hand, species such as *Hermesilla* sp., *Cayetunya* sp., *Plaumannita* sp., currently abundant, do not appear as species identified in the Gregório Bondar Collection until 1985. In the above mentioned preliminary survey, only one or two individuals of *Colaspis aeruginosa* Germar, *C. pertubata phylis* (Bechyné, *Freudeita plaumanni* Bechyné, *Nycterodina* (*N.*) *tincta* Bechyné & Bechyné, *Sphaeropis aeruginosa* Lefèvre, *Podoxenus coxalis, P. tornatus, P. troglodytes* and *Steronoda aerifera* Bechyné, were captured. Besides these, there are also specimens of other species which occur in extremely small numbers, and have not yet been identified because only male specimens were obtained that are not listed in Table 1.

Column D of Table 1 lists the species captured from May 1987 to June 1988 in the samples of adults present in the cacao tree crowns.

Table 1. Eumolpinae species captured in different forms, sites and times in cacao trees in the South Bahia.

Species	A	B	C	D
Antitypona dorsata Baly, 1878	X		X	X
Antitypona epiphania Bechyné, 1955			X	X
Antitypona ventralis Bechyné, 1953	X		X	X
Brevicolaspis villosa Bryant, 1923	X			
Coytiera (C.) pertusa Lefèvre, 1884.	X		X	X
Cayetunya sp.		X	X	X
Chrysodina sp.	X			
Chrysodina corusca Lefèvre, 1877	X			
Colaspis sp.	X			
Colapis aenea (Fabricius, 1801)		X	X	X
Colaspis aeraria Lefèvre, 1885	X			
Colapis aeruginosa Germar, 1824			X	
Colaspis auripennis (Germar, 1814)	X			
Colaspis calcalifera (Bechyné, 1824)			X	X
Colaspis obscura Fabricius, 1801	X			
Colaspis olivieri Lefèvre, 1875	X			
Colaspis perturbata phylis (Bechyné,1953)			X	
Eumolpus surinamensis Fabricius, 1775	X			
Freudeita plaumanni Bechyné, 1951			X	
Habrophora sp.	X			
Hermesilla sp.		X	X	X
Hylax bahiensis Bechyné, 1950	X		X	X
Leprocolaspis adducta Lefèvre, 1884	X		X	X
Leprocolaspis sulcata Lefèvre, 1877	X			X
Metaxyonycha auripennis (Germar, 1824)				
Nycterodina (N.) immetalica B. & B., 1961			X	X
Nycterodina (N.) tincta B & B., 1961			X	
Percolaspis ornata (Germar, 1824)	X	X	X	X
Plaumannita sp.		X	X	X
Pygocolaspis olivieri (Lefèvre, 1885)				
Podoxenus sp.				X
Podoxenus coxalis Lefèvre, 1877		X	X	
Podoxenus tornatus Bechyné, 1950		X	X	
Podoxenus troglodytes Lefèvre, 1877			X	
Rhabdopterus sp.	X			
Rhabdopterus brasiliensis Bechyné, 1950	X	X		
Shaeropis aeruginosa Lefèvre, 1876			X	
Spintherophyta aurichalcea (Germar, 1824)				
Spintherophyta corusca (Lefèvre, 1877)				
Steronoda aerifera Bechyné, 1951			X	
Taimbezinhia theobromae (Bryant, 1824)	X	X	X	X
Lepronota tessulata (Germar, 1824)	X			
Typophorus versutus Lefèvre, 1877	X			
Total	**23**	9	**23**	16

A = Gregório Bondar Entomological Collection; B = Rhizosphere of the cacao tree (1985/87); C = Cacao tree crown (sampling / indeterminate time, 1985/1986); D = Cacao tree crown (sampling / 30 minute period, Jun/87-Maio/88)

The names *Podoxenus* sp. and *Antitypona* sp. could include more species. The species of these exemplars, however, were not identified, because both in the preliminary survey and in the samplings they were very rare. Besides these, it was impossible to identify even the genus of other exemplars, acknowledged as different, since they also occur very rarely. For the reasons mentioned, they have not been included in Table 1.

4.5.2. Relative abundance, seasonality, dominance and diversity of Eumolpinae

Fig. 3 shows the fluctuation in the abundance of imagoes of the subfamily Eumolpinae in the lower stratum of the cacao tree crowns, during the period from June to December 1987, and the visual estimates of the weekly variation of leaf flushing in the cacao trees, expressed as a percentage, during the period from June to December 1987. The mean number of individuals, per fifteen-day period, was plotted in the middle of those fifteen-days. Two clear population peaks are observed in the last fifteen-day period of September and in the last fifteen-day period of October 1987, and another, smaller one in the first fifteen-day period of December.

In the cacao trees the alternation of periods of leaf flush with periods of "rest periods" is found and the plant is most susceptible to insect attack and diseases when flushing (Greenwood *et al*., 1950; Alvim, 1964). Abreu (1972) found a positive, highly significant correlation, between the population occurrence of P. *ornata*, and of foliar flushing in cacao trees at Linhares, Espirito Santo, Brazil. The present results agree with those of Abreu (1972). It was found also that the highest population peaks of Eumolpinae occur during the period in which the highest percentages of flushing of young leaves were found.

The clearest environmental influence on the fluctuation of abundance of this taxocenosis in cacao trees is related to their preferential food, that is, young leaves. According to Alvim (1994), the cacao tree presents typically intermittent growth in Bahia, and there are usually two main peaks during the year: one between March and April, and the other between September and October. The author also stated that hydroperiodicity is the main environmental stimulus to induce foliar flushing in cacao trees. The data show that a period with little available humidity (the period is considered dry when weekly rainfall is less than 25 mm) is a requirement to make the buds "ready to grow "as soon as the plants are rehydrated.

The harvest forecast work at CEPLAC used the foliar flushing indexes for its estimates. These estimates, however, took into account only three situations: trees with foliar flushing less than 50%, above 50% and no flushing (Technical reports, CEPLAC/CEPEC). Although these data do not allow one to distinguish flushing fluctuations, they make it quite clear that for adult Eumolpinae in the south of Bahia , there is almost always the resource "new leaf". What changes, however, is the amount of this resource.

The region characteristically has little temperature variation and lacks defined dry seasons. The values found in the period of the study agree with normal variations for the region. The most rainy periods were the months of March and December 1987 (Fig. 2-C).

Table 2 shows the number of individuals, grouped in fifteen-day means for each of the species, for the period studied. In Table 3 these data have been transformed into percentages as related to the total number of individuals collected per species, sampled every fifteen days to illustrate the relative abundance (%) of each of the species. Figures 4, 5, 6 ,

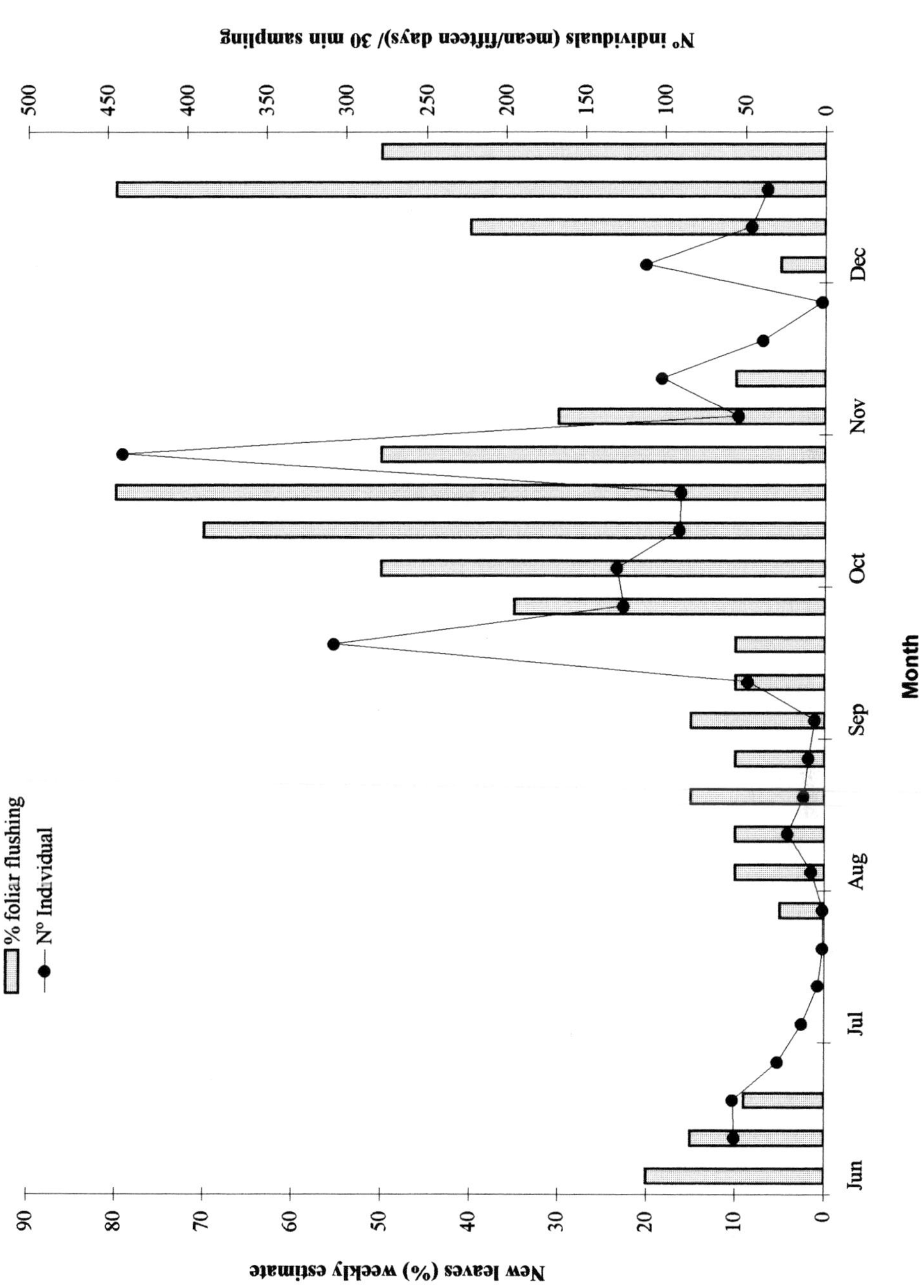

Fig. 3. Percentage of leaf flushing and mean/Fifteen day number of Eumolpinae sampled in the lower level of the cacao trees with an entomological net during 30 minutes (Ilheus-Bahia, from June to December 1987).

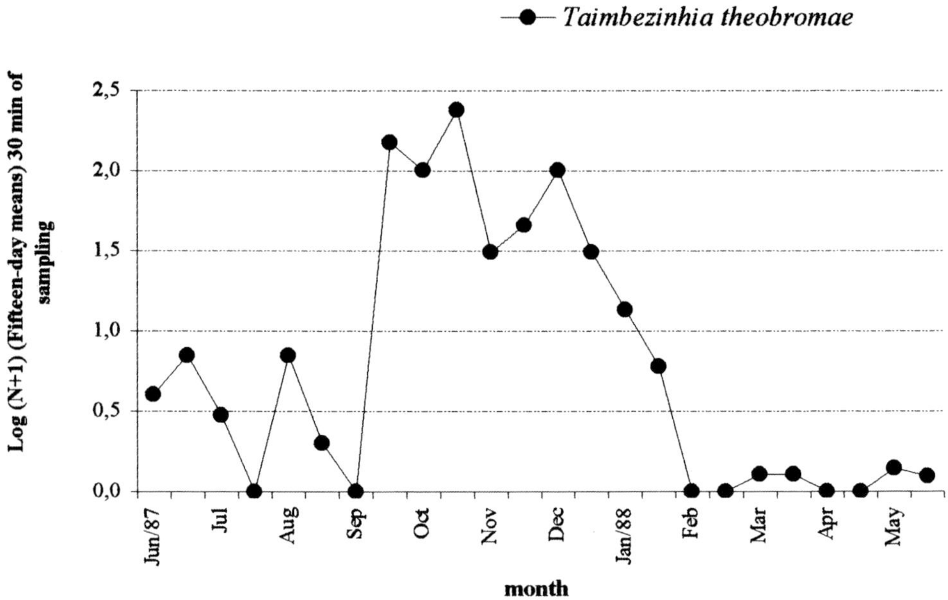

Fig. 4. Fluctuation of the abundance of *Taimbezinhia theobromae* sampled in the lower stratum of the crown of *Theobroma cacao,* during 30 minutes of sampling in the period from June 1987 to May 1988. Fifteen-day means transformed into Log (N+l).

7 (A and B), 8 (A, B and C) and 9 (A and B) (semilogarithmic plot) illustrate the fluctuation of relative abundance (%), based on the mean number of individuals of each of the species sampled every fifteen-days during the period studied. The most outstanding species are *P. ornata* and *T. theobromae*, confirming the conclusions of previous studies (Benton, 1984) on the population abundance of Eumolpinae in the south of Bahia.

The population peak of *T. theobromae* (Fig. 4) coincides with the peak of foliar flushing of the October period. Its occurrence is practically limited to the period between mid-September and the beginning of January. *T. theobromae* represents 43% of the total number of Eumolpinae captured. Its maximum relative abundance occurs in October and in the last fifteen-day period of November, and its contribution from mid-September to the end of December is significant. These results fully agree with those obtained by Benton (1984). This author, using results of samplings performed employing knockdown by chemical insecticides, makes a schematic representation of the seasonal occurrence of the Coleoptera most commonly associated with cacao trees in the south of Bahia and Espirito Santo, and found that the presence of this species is verified in September and October. Benton (1984) also stated that *T. theobromae* has the most outstanding seasonal abundance among the species detected, with marked peaks in September-October.

Percolaspis ornata (Fig. 5), found every month of the year, represents, on average, 28.5% of the total population. Its maximum relative abundance occurs in the second fifteen-day period of August (80.5%), followed by the month of June (73%). Although it is variable, it occurs at considerably high percentages as compared with the other species, except for the months from September to December, when *T. theobromae* is absolutely dominant. According to Benton (1984), *P. ornata* does not present a defined seasonality, and there is an indication that at Mucuri, Bahia (Brazil) it is more common in January. In

Table 2. Fifteen-day means of the number of individuals per species, obtained in samples from the lower level of the crown of *Theobroma cacao*, using an entomological net, during a 30-minute collection (Ilhéus, Bahia, Brazil), from June 1987 to May 1988.

	JUN/87 15-day period		JUL 15-day period		AUG 15-day period		SEP 15-day period		OCT 15-day period		NOV 15-day period		DEC/87 15-day period		Jan/88 15-day period		FEB 15-day period		MAR 15-day period		APR 15-day period		MAY/88 15-day period	
	1st	2nd	1st	2nd	1st	2nd	1st	2nd	1st	2nd	1st	2nd	1st	2nd	1st	2nd	1st	2nd	1st	2nd	1st	2nd	1st	2nd
T. theobromae	3,0	6,0	2,0	0	6,0	1,0	0	150,0	100,0	240,0	30,0	44,7	100,0	30,0	12,5	5,0	0	0	0,3	0,3	0	0	0,4	0,3
P. ornata	41,0	31,5	3,5	0	0	12,3	4,0	11,5	1,5	2,0	3,0	1,3	4,0	1,0	6,0	7,5	0	14,0	34,7	16,9	4,0	16,8	22,6	13,8
Plaumannita sp.	5,0	0,5	0,5	0	1,0	1,3	2,0	7,0	1,0	5,5	3,0	0	2,0	5,0	1,5	0,5	0	2,0	8,1	10,6	2,0	1,3	23,8	7,5
Hermesilla sp.	1,0	0	0,5	0	1,0	0,3	0	5,5	8,0	15,5	10,0	1,7	6,0	10,0	4,0	5,0	0	0	0,7	0	0	0	0,2	0
H.. bahiensis	5,0	5,0	2,5	0	0	0,3	0	3,5	0	0	0	0	0	0	1,5	0	0	0	0	0	0	0	0	0
Cayetunya sp.	0	0	0	0	0	0	0	0	0	0	0	0	0	0	4,5	0	0	3,3	2,0	1,6	1,8	1,3	0,4	0,3
C. pertusa	0	0	0	0	0	0	0	0	0	0	0	0	0	0	0	0	0	2,3	8,4	1,4	1,6	0	0	0
N. immetallica	0	0	0	0	0	0	0	0	0	0	0	0	0	0	0	0	0	0	0	2,7	0	6,0	0	4,5
C. calcalifera	0	0	0	0	0	0	0	0	0	2,0	8,0	0	0	0	0	0	0	0	0	0	0	0	0	0
sp. 1*	0	0	0	0	0	0	0	0	0	0	0	0	0	0	0	0	0	0	0	0	0	0	4,6	2,0
Antitypona spp.	0	0	0	0	0	0	0	0	0	0	0	0	0	0	0	0	0	0	0	0,6	0,6	0,3	4,2	0
L. adducta	0	0	0	0	0	0	0	0	0	0	0	0	0	0	0	0	0	0	0,4	0,4	0,6	1,5	0	1,5
L. sulcata	0	0	0	0	0	0	0	0	0	0	0	0	0	0	0	0	0	0	0	0,1	0,6	0,5	2,2	0
sp. 2*	0	0	0	0	0	0	0	0	0	0	0	0	0	0	0	0	0	0	2,4	0,4	0	0	0	0
C. aenea	0	0	0	0	0	0	0	0	0	0	0	0	0	0	0	0	0	0	0	0,9	0	0,3	1,2	0
Podoxenus sp.	0	0	0	0	0	0	0	0	0	0	0	0	0	0	0	0	0	0	0	0,4	0	0,8	0,4	0,8
sp.3 *	0	0	0	0	0	0	0	0	0	0	0	0	0	0	0	0	0	0	0	0	0	0	0	2,5
sp.4*	0	0	0	0	0	0	0	0	0	0	0	0	0	0	0	0	0	0	0	0	0	0,3	0	0

* GENUS UNIDENTIFIED

Table 3. Relative abundance (%) (dominance), of the Eumolpinae species sampled in the lower level of the crown of *Theobroma cacao*, using an entomological net, during a 30-minute collection (Ilhéus, Bahia, Brazil), expressed as the mean of the data grouped for 15-day periods during the period from June 1987 to May 1988.

	JUN 15-day period		JUL 15-day period		AUG 15-day period		SEP 15-day period		OCT 15-day period		NOV 15-day period		DEC/87 15-day period		JAN/88 15-day period		FEB 15-day period		MAR 15-day period		APR 15-day period		MAY/88 15-day period	
	1st	2nd	1st	2nd	1st	2nd	1st	2nd	1st	2nd	1st	2nd	1st	2nd	1st	2nd	1st	2nd	1st	2nd	1st	2nd	1st	2nd
T. theobromae	5,4	14	22,2	0	75	6,5	0	91	91	90,6	55,6	94	89	65,2	41,7	27,8	0	0	0,5	0,8	0	0	0,7	0,8
P. ornata	73,2	73,3	38,9	0	0	80,5	66,7	1,4	1,4	0,8	5,6	2,8	3,6	2,2	20,0	41,7	0	64,6	60,8	46,5	35,6	58,3	37,7	41,7
Plaumannita sp.	8,9	1,2	5,6	0	12,5	8,7	33,3	0,9	0,9	2,1	5,6	0	1,8	10,9	5,0	2,8	0	9,2	14,3	29,1	17,8	4,3	39,7	22,7
Hermesilla sp.	1,8	0	5,6	0	12,5	2,2	0	7,2	7,2	5,8	18,5	3,5	5,4	21,7	13,3	27,8	0	0	1,3	0	0	0	0,3	0
Hylax. bahiensis	8,9	11,6	27,8	0	0	2,2	0	0	0	0	0	0	0	0	5,0	0	0	0	0	0	0	0	0	0
Cayetunya sp.	0	0	0	0	0	0	0	0	0	0	0	0	0	0	15,0	0	0	15,4	3,5	4,3	15,6	4,3	0,7	0,8
C. pertusa	0	0	0	0	0	0	0	0	0	0	0	0	0	0	0	0	0	10,8	14,8	3,9	14,4	0	0	0
N. immetallica	0	0	0	0	0	0	0	0	0	0	0	0	0	0	0	0	0	0	0	7,5	0	20,9	0	13,6
C. calcalifera	1,8	0	0	0	0	0	0	0	0	0,8	14,8	0	0	0	0	0	0	0	0	0	0	0	0	0
sp. 1 *	0	0	0	0	0	0	0	0	0	0	0	0	0	0	0	0	0	0	0	0	0	0	7,7	6,1
Antitypona spp.	0	0	0	0	0	0	0	0	0	0	0	0	0	0	0	0	0	0	0	1,6	5,6	0,9	7,0	0,0
L. adducta	0	0	0	0	0	0	0	0	0	0	0	0	0	0	0	0	0	0	0,8	1,2	5,6	5,2	0	4,5
L. sulcata	0	0	0	0	0	0	0	0	0	0	0	0	0	0	0	0	0	0	0	0,4	5,6	1,7	3,7	0
sp2*	0	0	0	0	0	0	0	0	0	0	0	0	0	0	0	0	0	0	4,3	1,2	0	0	0	0
C. aenea	0	0	0	0	0	0	0	0	0	0	0	0	0	0	0	0	0	0	0	2,4	0	0,9	2,0	0
Podoxenus sp.	0	0	0	0	0	0	0	0	0	0	0	0	0	0	0	0	0	0	0	1,2	0	2,6	0,7	2,3
sp.3 *	0	0	0	0	0	0	0	0	0	0	0	0	0	0	0	0	0	0	0	0	0	0	0	7,6
sp.4 *	0	0	0	0	0	0	0	0	0	0	0	0	0	0	0	0	0	0	0	0	0	0,9	0	0
(%)	100	100	100	0	100	100	100	100	100	100	100	100	100	100	100	100	0	100	100	100	100	100	100	100

* GENUS UNIDENTIFIED

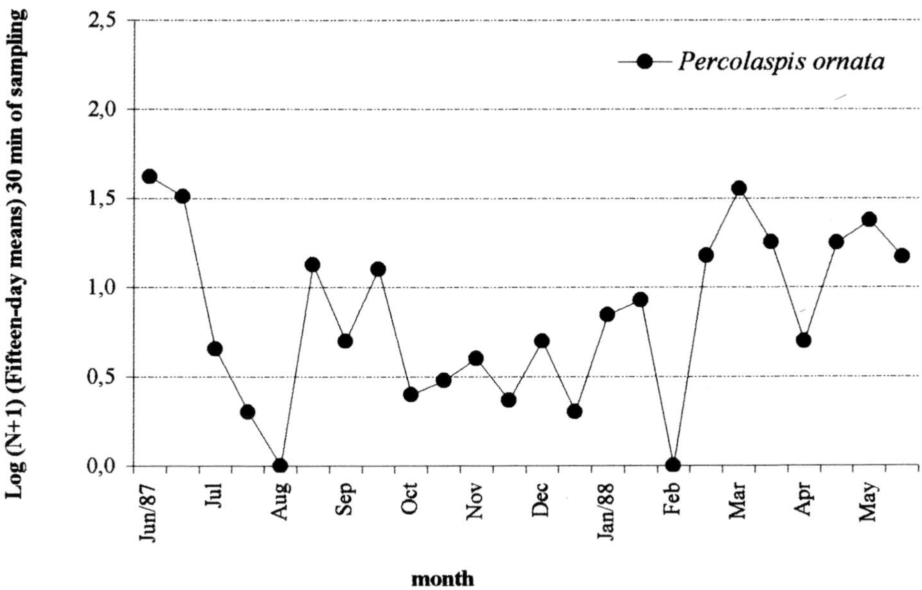

Fig. 5. Fluctuation of the abundance of *Percolaspis ornata* sampled in the lower stratum of the crown of *Theobroma cacao,* during 30 minutes of sampling in the period from June 1987 to May 1988. Fifteen-day means transformed into Log (N+l).

the region of Linhares, Espirito Santo (Brazil), according to the results of Abreu (1972), *P. ornata* is most abundant in December. Benton (1984) stated that local conditions, such as the presence of young leaves on the cacao tree and other uninvestigated biological factors, for instance the existence of parasites and predators, should determine the seasonality of this species. The results of the present study also agree with the conclusions of Benton (1984) as regards the distribution of this species throughout the year. In Figure 5, the semi-logarithmic plot shows that the population of *P. ornata* presents a certain uniformity of distribution, and no very marked population peaks can be perceived.

Plaumannita sp. (Fig. 6) represented 11.5% (Table 3) of the population. The highest relative abundance occurs in May (39.7%). This species was recorded for the first time in the south of Bahia by Ferronatto (1988); its behavior, in terms of distribution , is similar to that of *P. ornata* , although its numbers are much lower than those of the latter species. It was present every month of the year.

Hermesilla sp. and *Coytiera (C.) pertusa* (Fig. 7-A), *Cayetunya* sp. and *Antitypona* spp. (Fig. 7-B) were also recorded for the first time in the south of Bahia, in 1986 (Ferronatto, 1988). These species were not conspicuous in previous times, and, if they had already occurred in the cacao trees, probably were included together with other unidentified species under the name "other Chrysomelidae". Previously, besides *T. theobromae* and *P. ornata,* three species were notable due to their abundance: *Rhabdopterus* sp. near *hypochalceus* Harold, *Colaspis* sp. near *aeraria* (Lefèvre) and *Colaspis* sp. near *metallica* (Lefèvre). (Benton, 1984). In the current investigation, only *Rhadopterus* sp. near *hypochalceus* was found, and even so only in the soil, captured as pupae. The other species, if present on the Ilhéus cacao trees, were not captured in the samples.

The number of individuals sampled of *Nycterodina (N.) immetallica, Hylax bahiensis*

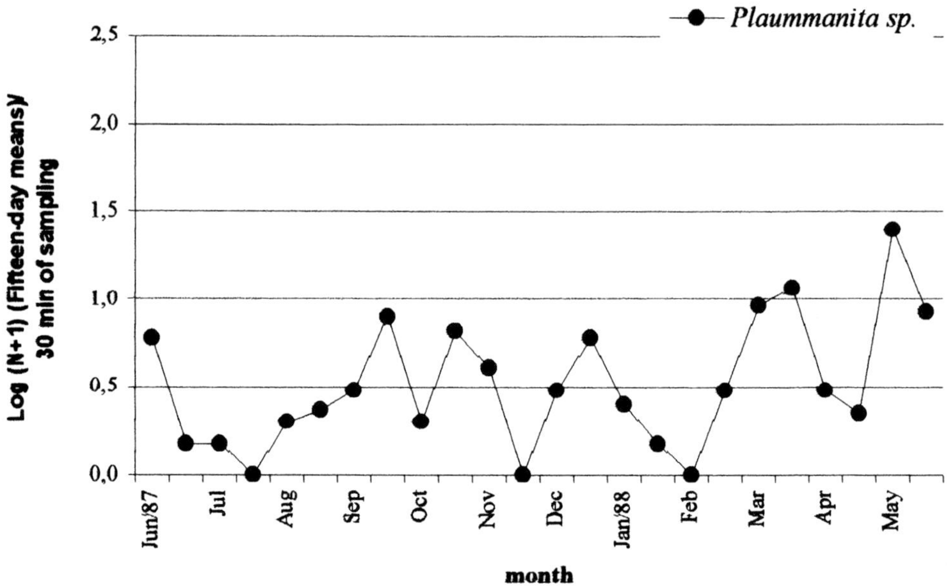

Fig. 6. Fluctuationof the abundance of *Plaumanita* sp. sampled in the lower stratum of the crown of *Theobroma cacao,* during 30 minutes of sampling in the period from June 1987 to May 1988. Fifteen day means transformed into Log (N+1).

and *Colaspis calcalifera* (Fig. 8-A); *Leprocolaspis adducta* and *L. sulcata*; (Fig. 8-B) *Colapis aenea* and *Podoxenus* sp. (Fig. 8-C) and unidentified genera sp. 1, sp. 2 (Fig. 9-A) and sp. 3 and sp. 4 (Fig. 9-B) was very small throughout the sampling period.

Applying the scale of Krogerus, 1932, as related to the total number of species sampled during the period of study, three dominant species were found, *T. theobromae* (42.9%), *P. ornata* (28.5%) and *Plaumannita* sp. (11.5%); three subdominant species, *Hermesilla* sp. (3.7%), *Coytiera* (*C.*) *pertusa* (2.84%), *Cayetunya* sp. (2.1%) and 12 other accessory species. There is a clear dominance of *T. theobromae* over all other species. For *Podoxenus, Antitypona* and *Nycterodina,* the occurrence of three morphospecies was recorded in this same place and period, for the first and second genera marked, and two for the third genus. Thus, the total number of accessory species must be at least 17.

Figure 10 compares the occurrence of the species considered dominant and subdominant according to the scale of Krogerus (1932). The data are expressed in fifteen-day means. It is observed that the times of the main peaks of the dominant species do not coincide: *T. theobromae* in October, *P. ornata* in February, March and June, and *Plaumannita* sp. in May. *Hermesilla* sp., on the other hand, a subdominant species, has its peak of occurrence in the same period as *T. theobromae. Coytiera* (*C.*) *pertusa* has its peaks in the first fifteen-day period in March and second fifteen-day period in April, and *Cayetunya* sp. in the first fifteen-day period in April.

The difference between the first and second dominant species is large over most of the fifteen day periods (Table 3), except for the first fifteen-day period in May,1988, when there is practically no difference between the two main dominant species, *Plaumannita* sp. and *Percolaspis ornata.* Beginning in the second fifteen-day period in September 1987 until the first fifteen-day period in January 1988, *T. theobromae* is the dominant species. In

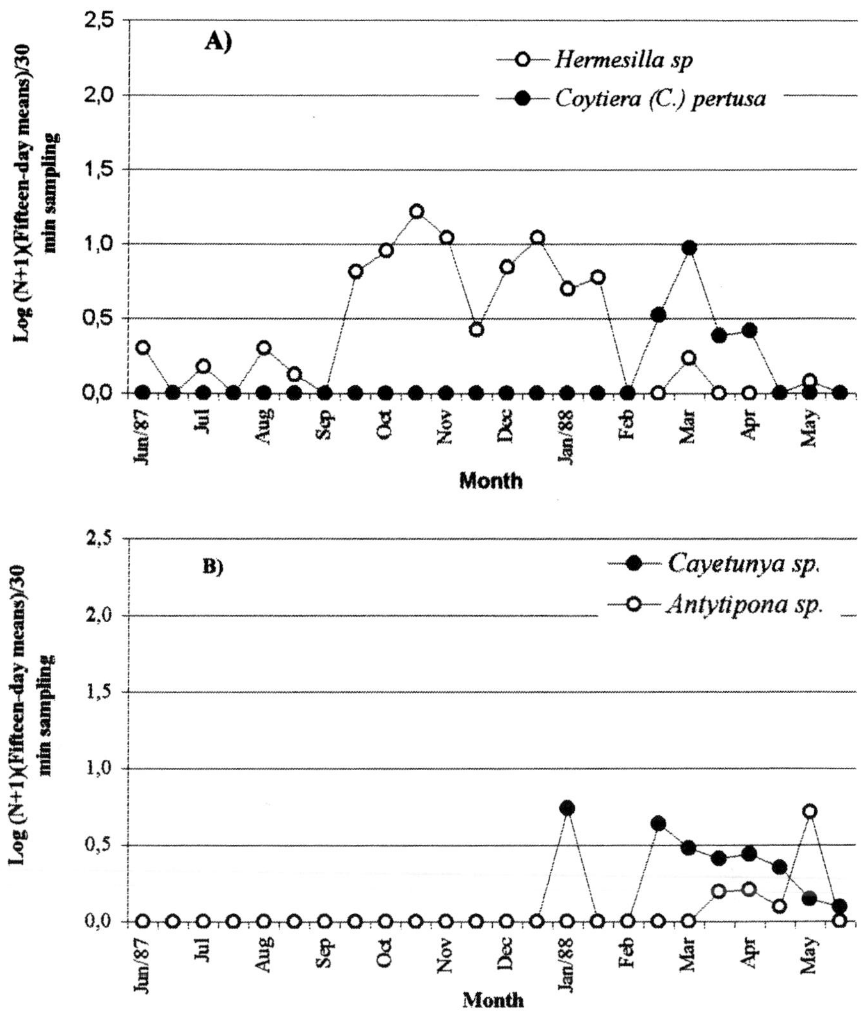

Fig. 7. Fluctuation of the abundance of Eumolpinae sampled in the lower stratum of the crown of *Theobroma cacao*, during 30 minutes of sampling in the period from June 1987 to May 1988. Fifteen-day means transformed into Log (N+1): A) *Hermesilla* sp., *Coytiera (C.) pertusa*, B) *Cayetunya* sp. and *Antytipona* sp.

the second fifteen-day period in January, dominance, also according to the classification of Krogerus (1932), is distributed among three species, and *P. ornata* presents the highest index (42%), whilst *T. theobromae* and *Hermesilla* sp. present the same index (28%). The largest group of dominant species occurred in the first fifteen-day period in April, with six species: *P. ornata*, *Plaumannita* sp., *Coytiera* (*C.*) *pertusa*, *Cayetunya* sp., *Antitypona* sp. and *L. sulcata*.

In order to investigate the entomofauna of the cacao trees in the region of Altamira, Pará, Bicelli (1983) performed random samplings, in fifteen-day periods, for three years (1979 to 1981) . He used knockdown by chemical insecticides as a sampling method and found more than 16 species of Eumolpinae: *Antitypona* spp., *Colaspis* sp., *Ephyraea* sp.,

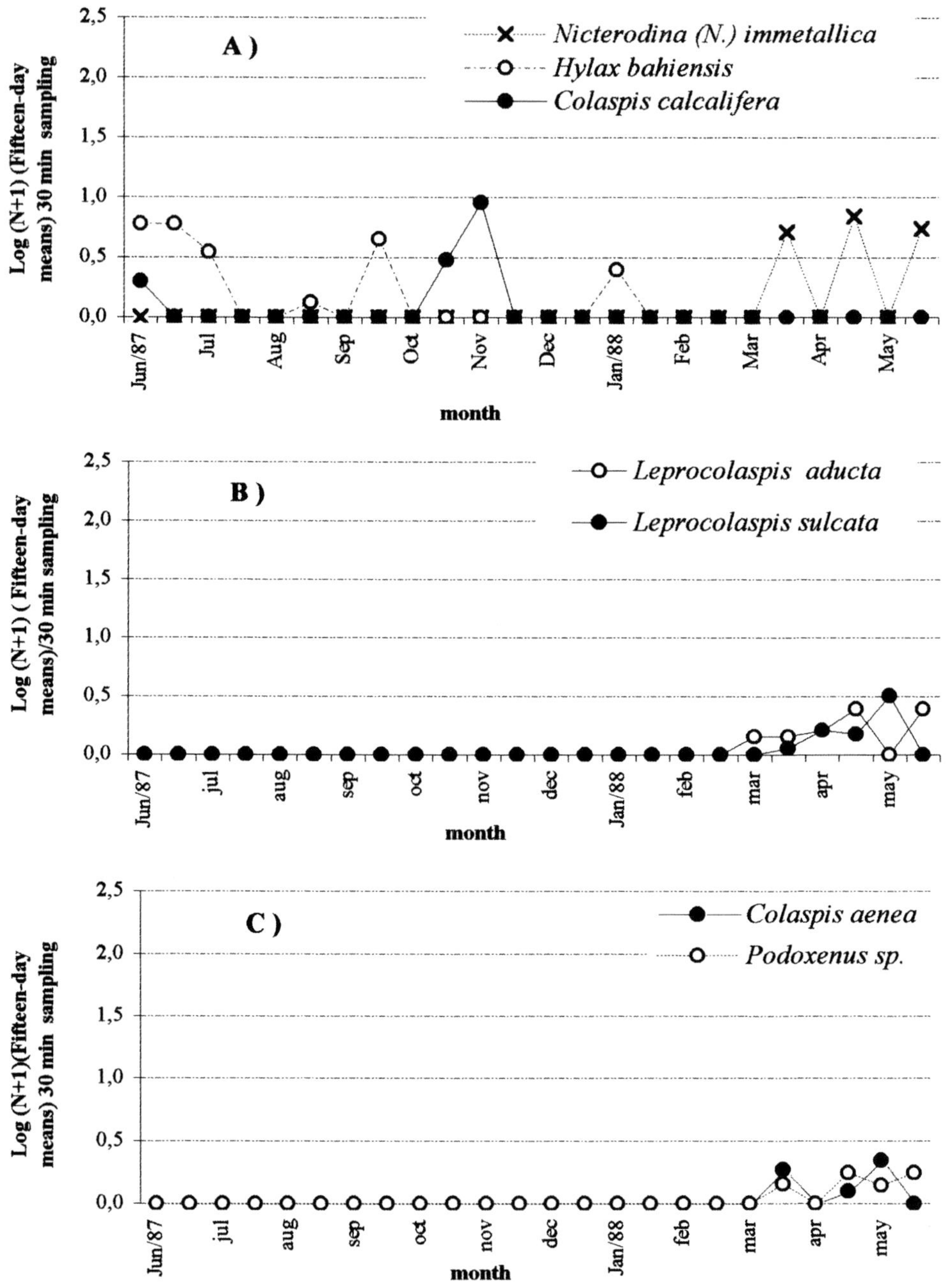

Fig. 8. Fluctuation in the abundance of Eumolpinae sampled in the lower stratum of the crown of *Theobroma cacao*, during 30 minutes of sampling in the period from June 1987 to May 1988. Fifteen-day means transformed into Log (N+1): A) *Nicterodina (N.) immetallica*, *Hylax bahiensis* and *Colapis calcalifera*; B) *Leprocolaspis adducta* and *L. sulcata*, C) *Podoxenus* sp. and *Colaspis aenea*.

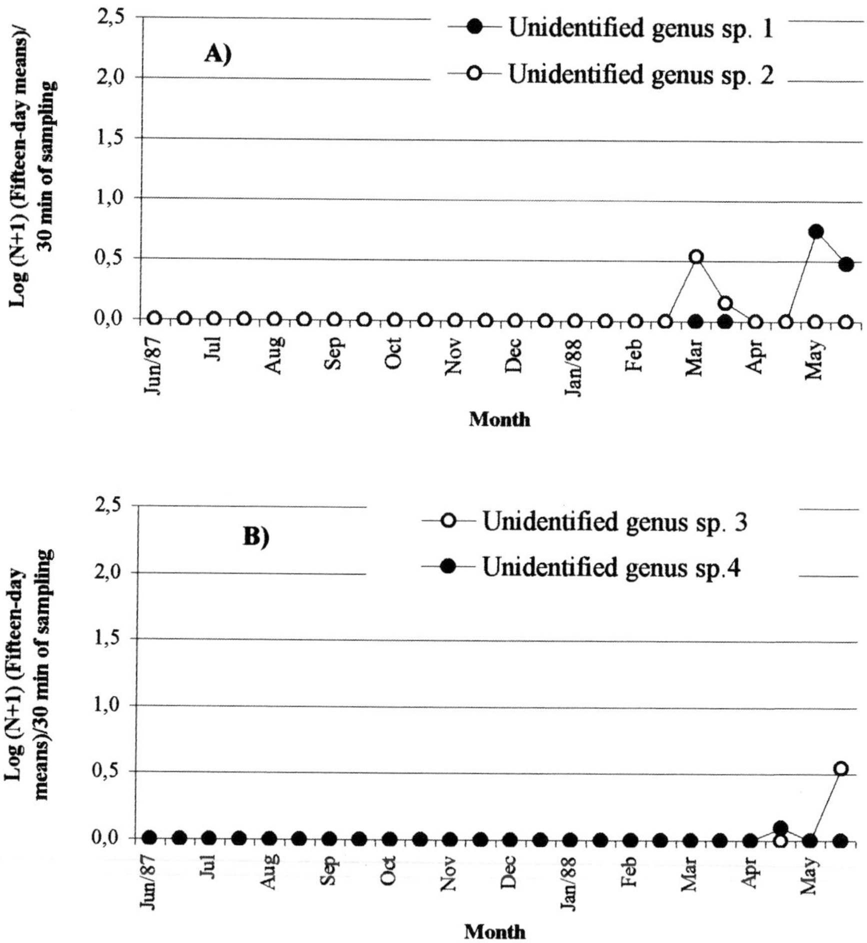

Fig. 9. Fluctuation in the abundance of Eumolpinae sampled in the lower stratum of the crown of *Theobroma cacao*, during 30 minutes of sampling in the period from June 1987 to May 1988. Fifteen-day means transformed into Log (N+1): A) Unidentified genus sp. 1 and sp. 2, B) Unidentified genus sp. 3 and sp. 4.

Hylax sp., *Ischyrolamprina* sp., *Maecolaspis* sp., *Maecolaspis occidentalis* Lefèvre, *Maecolaspis rugosa* Germar, *Metachroma* sp., *Noriaia* sp., *Paria* sp., *Percolaspis* sp., *Percolaspis bicolor* Oliv., *Percolaspis ornata* (Germar) *Prionodera bicolor* Oliv., *Rhabdopterus* sp. and *Typophorus signatus* Lefèvre. Bicelli (1983) stated that the Chrysomelidae are in remarkable abundance. Among them are the Eumolpinae, *Colaspis* sp., *Ischyrolamprina* sp., *Noriaia* sp. and *Rhabdopterus* sp. Bicelli (1983) suggested that the large populations of these species show their adaptability to the cacao tree culture.

Among the species identified of the Eumolpinae complex recorded at Altamira by Bicelli (1983), the only one in common with those in Ilhéus, found in the current study, was *Percolaspis ornata*. *Taimbezinhia theobromae*, the most abundant species in the cacao plantations of the south of Bahia, was not recorded at Altamira, Para. These data agree with

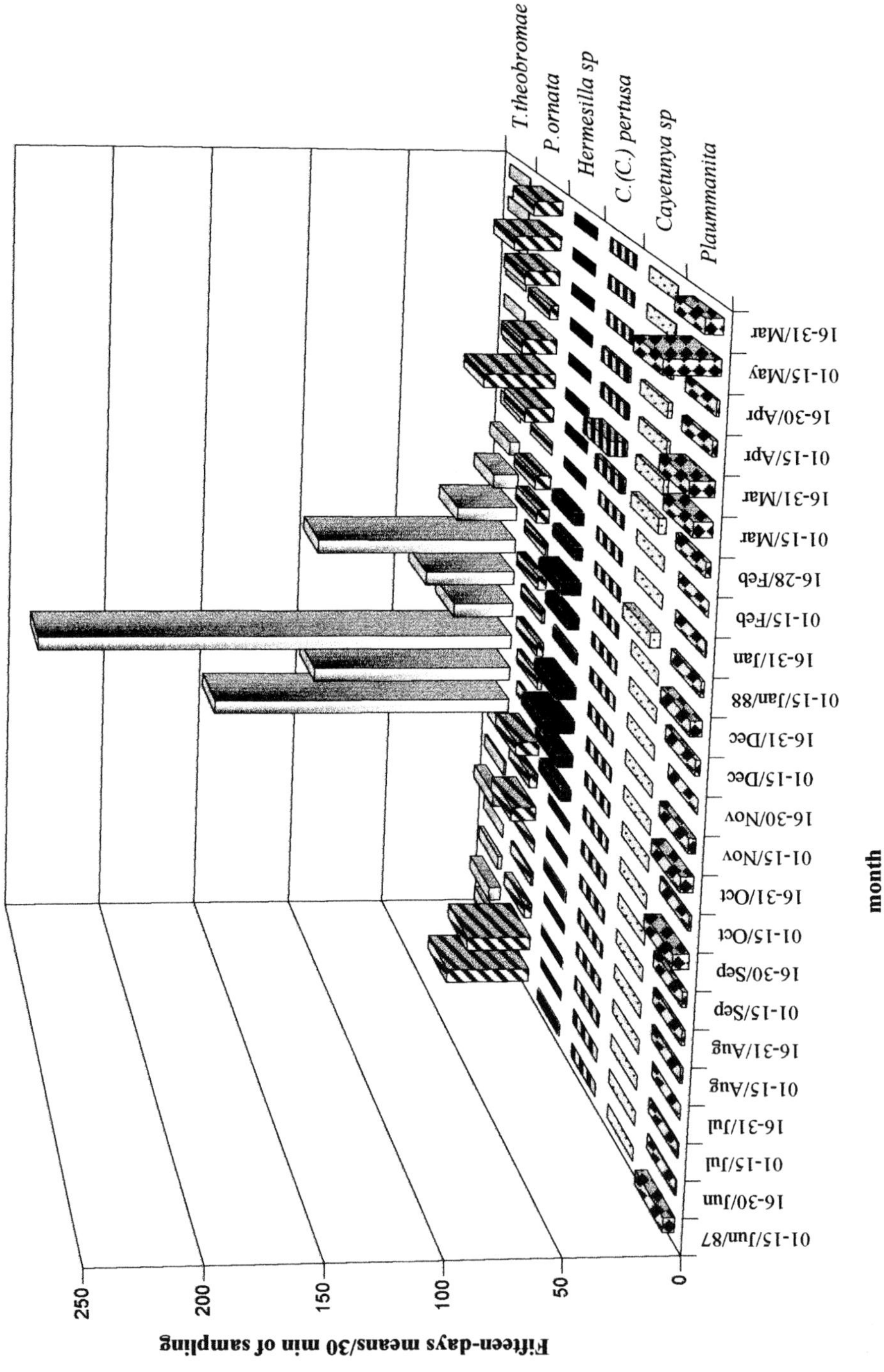

Fig. 10. Comparison between the occurrence of the dominant and subdominant species of Eumolpinae sampled in the lower level of tree crowns. Data expressed as fifteen-day means (Ilheus, Bahia. Brazil. June 1987/May 1988).

the theory of Strong (1974), when he stated that most cacao tree pest insects are autochthonous to the local fauna, and few species are widely distributed.

On the two occasions, March–April and September-October, when the highest peaks of foliar flushing usually occur in the cacao trees in the south of Bahia, two important aspects of the taxocenosis of Eumolpinae became clear: during the first peak of young leaves, the largest number of species occur, and during the second foliar peak, a clear concentration of the dominance of *T. theobromae.* The Margalef (1951) index was calculated in order to study this aspect at different times of the year. These indexes can be compared in Table 4. The month of March presented the greatest number of species (S=13) and the highest diversity index of species (3.99). The lowest diversity indexes were in October 1987 and December 1988, when *Taimbezinhia theobromae* predominated over the others, presenting the highest dominance. It should, however, be stressed that , according to Whittaker (1972) and Hill (1973), this measure depends on the size of the sample, and is very sensitive to sample errors and to differences in susceptibility when sampling rare species.

The Eumolpinae community in the cacao trees is composed of many genera and few species for each genus. Piellou (1975) stated that, in intuitive terms, communities in which the species are distributed into many genera will have greater diversity than communities in which most of the species belong to a single genus.

Table 4. Margalef index (1951) (α) for the taxocenosis of Eumolpinae in the cacao trees of Ilhéus, Bahia, Brazil, in each of the months of the sampling period and number of species (S) observed each month (June 1987 to May 1988).

Month	α	S
Jun/87	1,01	6
Jul	1,33	5
Aug	1,00	5
Sep	0,68	5
Oct	0,6	5
Nob	0,76	5
Dec	0,59	4
Jan/88	1,09	6
Feb	0,72	4
Mar	3,99	13
Apr	1,85	11
May	1,88	12

5. Conclusion

The period of greatest abundance of immature forms of Eumolpinae in the cacao plantation soil is from mid-March until the end of August, and the highest population peak occurs at the end of July and beginning of August.

The highest percentages of foliar flushing of *Theobroma cacao* occurs in the period when the highest population peaks and the highest concentration of Eumolpinae imago dominance are found.

Acknowledgements

To the Directors of the Cacao Research Center (CEPLAC/CEPEC-Ministério da Agricultura) for allowing this study to be performed. To Dr. Miriam Becker, Professor at the Universidade Federal do Rio Grande do sul, Porto Alegre, RS, for supervising the PhD thesis "A taxocenose de Eumolpinae (Coleoptera: Chrysomelidae) em *Theobroma Cacao* (Sterculiaceae) no sul da Bahia" of which this paper is part. To Bohumila Springlova de Bechyné (Universidad Central de Venezuela) for identifying the species. To Dr. Paulo Santos Terra and Dr. Max de Menezes, researchers at CEPLAC/CEPEC, Ilhéus, Bahia for their collaboration and important suggestions made to this study. To Mr. Arlindo Nepomuceno, employees at CEPLAC/CEPEC, for the visual estimate of foliar flushing of the cacao trees. To the Board of Directors of Empresa Brasileira de Aviação VARIG, especially Mr. Geraldo Souza Pinto, for their kindness and courtesy in transporting this collection of Eumolpinae from Caracas to Brazil. To Dr. Ari Bernardes da Silva, at the Faculdade de Veterinária at the Universidade Federal do Rio Grande do sul for his comments on the manuscript.

References

Abreu J. M. , 1972. Flutuações de populações de Coleópteros nocivos a cacaueiro no Espírito Santo, Brasil. Revista Theobroma, Ilhéus, 2: 45-55.

Alvim P. T., 1960. Moisture stress as a requirement for flowering of coffee. Science, 132 (3423): 354.

Alvim, P.T., 1964. Tree growth periodicity in tropical climates. In: Zimmermann, M.H. (Ed.) The formation of wood in forest trees. (Academic Press) New York, pp. 479 –495.

Alvim, P. T. 1967. Ecophysiology of the cocoa tree. *In*: Conference Internationale sur les recherches agronomiques cacaoyères, Abidjan, Côte d'Ivoire,1965. Paris, Institut du Café et du Cacao pp.23 – 35,

Alvim, P. T. 1977. Cacao. *In*: Alvim, P. T.; Kozlowski, T.T., (Eds.).Ecophysiology of tropical crops. Academic Press, New York:, pp. 279-313.

Alvim, P.T. 1994. Hydroperiodicity of flowering and flushing of cacao. *In:* Conference Internationale sur La Rechercher cacaoyère, 11, Yamoussoukro, 1993. Actes. London, J. de Lafforest Transla-Inter., pp. 625-634.

Alvim, P. T., Alvim R. 1978.. Relation of climate to growth periodicity in tropic trees. *In*: Tomlinson, P.B., Zimmermann, M. H. (eds.): Tropical trees as living systems. Cambridge, Univ. Press. pp.445-464.

Alvim R.; Alvim, P.T. 1975. Hidroperiodicidade dos fluxos foliares do cacaueiro. Ilhéus, CEPLAC/CEPEC,. pp. 42-48. (Informe técnico).

Benton, F. 1984. Abundância estacional dos coleópteros fitófagos do cacaueiro do sul da Bahia e Espirito Santo. Revista Theobroma, Ilhéus. 14: 85-102.

Bicelli, C. R. L. 1983 Levantamento e análise faunística de insetos coletados em cultura de cacau na região de Altamira – Pará. 1983. Piracicaba: Escola Superior de Agricultura "Luis de Queiroz". 126p. Dissertação (Mestrado).

Brown, K. S. Jr. 1975; Geographical patterns of evolution in Neotropical Lepidoptera. Systematics and derivation of known and new Heliconiini (Nymphalidae: Nymphalinae). J. Ent. (B) 44 (3):201-242.

Chen, S.H. 1940. Attempt at a new classification of the leaf beetles. Sinensia, Nanking. 11(5/6): 451- 481.

Crowson, R.A. 1981. The biology of Coleoptera. London, Academic Press,. 802p.

Ferronatto, E. M. de O. 1986. Observações preliminares sobre as plantas hospedeiras das formas imaturas dos principais crisomelídeos (Coleoptera: Chrysomelidae) que ocorrem em cacauais. Revista Theobroma, Ilhéus. 16(2):107-110.

Ferronatto, E. M. de O. 1988.. Eumolpinae associated with cacao trees (*Theobroma cacao* L.) in Southeast Bahia. In: Jolivet, P., Petitpierre, E., Hsiao, T. H. (eds.). Biology of Chrysomelidae. Series Entomologica, volume 42, Kluwer Academic Publishers, Dordrecht, pp.553-558.

Fisher, J.R., Bergman, M.K. 1986. Field sampling of larvae and pupae. In: Krysan, J.; Miller T. A. (eds.), Methods for the Study of Pest Diabrotica. (Springer-Verlag) New York. pp.101-121.

Garcia J.J.S., Silveira Neto, 1980.. Estudo faunístico de coleópteros e hemípteros associados ao cacaueiro do Estado do Pará. Revista Theobroma, Ilhéus, Bahia, Brazil. 10:15-23.

Garcia, J.J.S., Mendes, A.C.B. 1982. Insetos nocivos ao cacaueiro na Amazônia. *In:* Conferência Internacional de Pesquisas em Cacao 8ª. 8, 1981, Cartagena, Acta. Lagos, (Nigeria); Cocoa producer's alliance, p.18-23.

Gowdey, C. C. 1926.. The banana fruit-scarring beetle (*Colaspis hyperchlora*). Bull. ent. Res. 17:137.
Greenwood, M.; Posnete, A.F. 1950. The growth flushes of cacao. J. hort. Sci., 25:164-174.
Hill, M.O. 1973. Diversity and evenness: a unifying notation and its consequences. Ecology, 54: 427-432.
Jolivet, P. 1988. Food habits and food selection of Chrysomelidae. Bionomic and evolutionary perspectives. *In:* Jolivet P., Petitpierre E., Hsiao, T.H. (eds.). Biology of Chrysomelidae. Series Entomologica, volume 42, Kluwer Academic Publishers, Dordrecht, pp.1-24.
Köppen, W. 1948. Climatologia.. Buenos Aires, F.C.E., 478 p.
Krogerus, R. 1932. Über die Ökologie und Verbreitung der Arthropoden der Triebsandgebiete an den Küsten Finnlands, Acta Zool. Fenn. 12: 1 –308
Kummerow, J., Kummerow, A., Alvim, P. De T. 1981. Root biomass in a mature cacao (*Theobroma cacao*) plantation. Revista Theobroma, Ilhéus, 11: 77 -86.
Margalef, R., 1951. Diversidade de espécies en las comunidades naturales. Publicaciones Inst. Biol. apl.. Barcelona. 6: 59-72.
Medeiros, Z. M. D. 1983. Variações sazonais do crescimento de raízes finas de cacaueiros adultos. Salvador: UFBA. 51 p. Dissertação (Mestrado em Biologia).
Miranda, R. A. C. de. 1987. Interceptação da chuva por cacaueiros no sudeste da Bahia. Revista Theobroma , Ilhéus, 17:251-259.
Murphy, D.D. 1997. Desafios à diversidade biológica em áreas urbanas. *In*: Wilson E. O. (ed.). Biodiversidade, (Editora Nova Fronteira, S.A.) Rio de Janeiro. pp. 89-97.
Myers, N. 1997. Florestas tropicais e suas espécies Sumindo, sumindo...? *In*: Wilson, E. O. (ed.). Biodiversidade, (Editora Nova Fronteira S.A), Rio de Janeiro. pp.36-45.
Prance, G.T. (ed.). 1982. Biological Diversification in the Tropics. *In*: Ata do Quinto Simpósio Internacional da Association for Tropical Biology [Associação de Biologia Tropical]. 1979, Caracas. Atas. (Columbia Univ. Press) New York, 714 pp.
Piellou, E.C. 1975. Ecological diversity. John Wiley , New York, , 165 p.
Salt, G. 1928. A study of *Colapis hypochlora* Lefèvre. Bul. ent. Res. 19: 295-308.
Silva, P. 1944. Insect pests of cacao in the state of Bahia, Brazil. Trop. Agric., Trin. 21(1): 8-14.
Silva, P.; Soria, S. de J., Santos, G. V. dos, Benton, F. P., Abreu, J. M. de, Menezes, M., Figueroa, G. E. S., Ferraz, E. A., Cruz, P. F. N. da. 1985. A Entomoteca Gregório Bondar do CEPEC, lista preliminar de insetos. Ilhéus: CEPLAC/DIBID, 63 p. (Boletim técnico 125).
Strong, D.R.J.R., 1974. Rapid asymptotic species accumulation in phytophagous insect communities: The pests of cacao. Science, 185:1064-1066.
Whittaker, R. H. 1972. Evolution and measurement of species diversity. Taxon, 21: 213-251.

BEHAVIOR AND POPULATION FLUCTUATION OF *PLAGIOMETRIONA FLAVESCENS* (BOHEMAN) (CHRYSOMELIDAE: CASSIDINAE)

FLÁVIA NOGUEIRA DE SÁ[1,2] and MARGARETE VALVERDE DE MACÊDO[1]
[1]*Laboratório de Ecologia de Insetos – Universidade Federal do Rio de Janeiro – IB – Depto. Ecologia – Rio de Janeiro – RJ – Brazil – 21941-590;* [2]*Present address: Universidade Estadual de Campinas – Inst. Biologia – Depto. Zoologia – Campinas – SP – Brazil 13083-970. E-mail: fnsa@obelix.unicamp.br*

Abstract

The population fluctuation of *Plagiometriona flavescens* on *Aureliana fasciculata* was studied from February 1995 to March 1996 at Poço das Antas Biological Reserve, Rio de Janeiro State, Brazil. During this period, observations of the behavior of this beetle were also conducted. *P. flavescens* is active as immature and adults stages throughout the year, and maintains reproductive activities. On the food plant, each stage of the beetle presents a pattern of localisation on the leaves, which possibly has some relationship with their defensive strategies. The life cycle was also described under laboratory conditions.

1. Introduction

The Cassidinae is considered to be one of the most specialised subfamilies of the Chrysomelidae (Jolivet, 1988) and species are closely associated, almost their entire lives, with their host plants. Generally, all life stages are found on the leaves of the same host plant. These facts represent great advantages for studying the ecology of Cassidinae species. Despite this, there are few studies on tropical Cassidinae species. Most of the studies deals with species from temperate regions and especially those of economic interest such as the Colorado potato beetle, *Leptinotarsa decemlineata* (say for example Groden & Casagrande, 1986; Cappaert *et al.*, 1991).

Information on Chrysomelidae population dynamics is urgently needed, as it would help in understanding the structure and functioning of tropical ecosystem communities. Some of the few studies on the dynamics of tropical chrysomelid populations are those by Nakamura *et al.* (1989), Medeiros & Vasconcellos-Neto (1994) and Becker & Freire (1996).

In the tropics, some factors are suggested to influence insects' population growth. Among the abiotic factors, rainfall and humidity can be cited. The latter, for instance, has a great variation in the tropics and was observed by Hsiao (1986) to induce insect diapause. Host plant characteristics, such as chemistry, temporal and spatial distribution, phenology and nutritional quality can also influence tropical herbivorous insect dynamics.

Solanaceae are known to have many mechanisms to avoid herbivory such as spines, trichomes and highly toxic compounds (they are rich in alkaloids) and this may be the reason why there are some evidences confirming the reduced number of herbivores which

Advances in Chysomelidae Biology 1, pp. 299–305
Edited by M.L. Cox

use this family as hosts (Drummond, 1986). These plants are sometimes hosts for the Cassidinae beetles, although these insects do not appear to have any special adaptations to avoid the many defences of these plants (Hsiao, 1986).

Parasitoids and predators can be frequently found interfering with insects' population numbers. In the majority of the cases Chrysomelidae natural enemies are insect species in the orders Hymenoptera, Diptera and Heteroptera or passeriform birds (Olmstead, 1996). Cassidinae beetles employ a variety of defences against natural enemies including chemical, physical or behavioural attributes. Behavioural defences include larval aggregation, paternal care and movement between feeding and resting sites. Morphological attributes, such as cryptic or aposematic coloration, can also play a role in defence when associated with behaviour.

Plagiometriona flavescens (Boheman, 1855) feeds on two Solanaceae species, *Aureliana fasciculata* and *Acnistus arborescens*, at the studied area (Sá & Macêdo, in press). In this work we describe some aspects of *P. flavescens* natural history, its variation in numbers throughout the year and discuss the factors that affect it.

2. Behaviour of *Plagiometriona flavescens*

Plagiometriona flavescens was studied at Poço das Antas Biological Reserve (20° 30'S and 42° 15'W), Silva Jardim county, Rio de Janeiro State, Brazil. The climate of Poço das Antas Reserve region is tropical. In the study period the mean temperature was 26.4°C and the precipitation was 2464.9 mm. The dry period of that year was from May to September (Fig.1).

Adults, *P. flavescens* have a bright green coloration, which is quite conspicuous. The larvae are pale yellow and carry masses of faecal material attached to the posterior end of the abdomen. The eggs are isolated and sometimes covered by faecal material.

Eggs obtained from females reared in the laboratory, fed with *A. fasciculata*, and maintained at room temperature take about 36.46 days to reach the adult stage. Incubation required 6.52 ± 4.62 days (N = 112); larval and pupal development takes 24.50 ± 2.21 (N = 5) and 5.44 ± 1.60 days (N = 25), respectively. The duration of the immature stages in *P. flavescens*, is longer than for the cassidines *Laccoptera tredecimpunctata* (32.40 days) and *Cassida catenata* (26.4 days) (Nakamura *et al.*, 1994). This difference is clearer when body sizes are considered: *P. flavescens*, the species with the longest developmental duration of immature stage is the smallest species (mean body lenght 5 mm), in comparison to *L. tredecimpunctata* and *C. catenata*, which are 7-8 mm and 5-6 mm, respectively (Nakamura *et al.*, 1994).

At the Biological Reserve, *P. flavescens* was found consuming leaves of *Aureliana fasciculata* and *Acnistus arborescens*, as both larvae and adults. Although the use of these two species by *P. flavescens* was never mentioned in the literature, Buzzi (1994) in a compilation of Cassidinae host plants from the Neotropics showed that *Plagiometriona* is quite specialised in feeding on Solanaceae. Four out of five species of *Plagiometriona* mentioned in that compilation use host plants in that family, and three out of those four species use *Aureliana* or *Acnistus* species. We also observed another *Plagiometriona* species, *P. vigens* Boheman consuming a Solanaceae, *Solanum caeruleum*, in Itatiaia National Park (Rio de Janeiro State), another remnant of Atlantic forest.

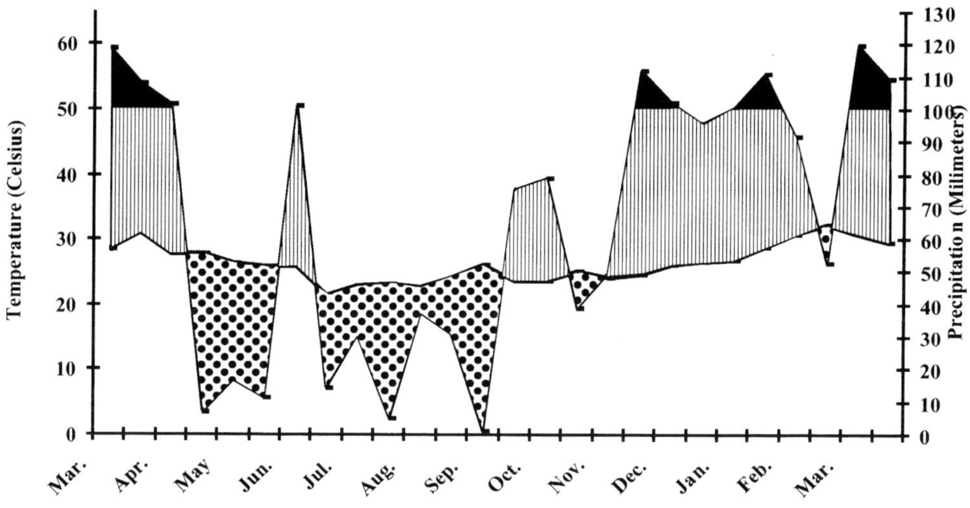

Fig. 1. Climatic diagram for the study period at Poço das Antas Biological Reserve (data from Atlantic Forest Program) from March, 1995 to March 1996.

P. flavescens on *A. fasciculata*, which was the most abundant host plant throughout the Reserve, showed a clear preference for healthy and new leaves. During the day, *P. flavescens* can be frequently found on the lower side of the leaves of *A. fasciculata* (Table 1). This pattern is more frequent for pupae, adults and younger larvae suggesting a defensive behaviour against potential visually oriented predators since conspicuous adults, immobile pupae and small and vulnerable larvae could get some protection by hiding themselves on that side of the leaves. Mature larvae are less frequently found on the lower side of the leaves than the other mentioned stages (Table 1), probably because of a higher effectiveness of the larger faecal shields they carry. These shields provide an effective protection against predators, playing a role not only as camouflage or as mechanical barriers, but they can also serve as chemical defensive agents for the toxic and repellent chemicals they may contain (Olmstead, 1996).

Table 1. Localisation of *Plagiometriona flavescens* on the surface *of Aureliana fasciculata* leaves.

	Under suface	Upper surface
Pupae	100% (N=30)	0%
Adult	96% (N=624)	4% (N=26)
Small larvae	99.60% (N=488)	0.40% (N=2)
Medium larvae	94.87% (N=351)	5.13% (N=19)
Mature larvae	79.13% (N=182)	20.87% (N=48)

3. Population fluctuation of *P. flavescens*

Twenty three individuals of *A. fasciculata* along the border of the roads of the Reserve were marked and checked for *P. flavescens* larval and adult numbers (through direct counts) fortnightly from February, 1995 until March, 1996.

P. flavescens was abundant as larvae and adults throughout the studied period, increasing its population during the winter and spring (Fig.2). Nakamura *et al.* (1989) also observed that the Cassidinae species *Aspidomorpha miliaris* (Fabricius) and *A. sanctaecrucis* (Fabricius) were active throughout the year in Sumatra. Studies on Brazilian Chrysomelidae dynamics shows that they frequently undergo a winter season diapause, such as *P. vigens* in Itatiaia National Park, which is completely inactive during the winter months when its host plant looses all its leaves (Macêdo unpublished data). Most studies, however, were conducted in Brazilian subtropical regions or in areas close by (see Becker & Freire, 1996 and Medeiros & Vasconcellos-Neto, 1994). Our results and those of Nakamura *et al.* (1989) suggest that several tropical Chrysomelidae species commonly do not diapause, occurring throughout the year.

Daily data on mean, maximum and minimum temperatures and precipitation in the Reserve were obtained for the entire study period. The average temperature (mean of maximum and minimum) for each period of time between the checks was obtained by an arithmetic mean of the temperatures from the day following the last check until the present check day. Precipitation was calculated by the sum of every data in the same period. The number of rainy days was also considered for each of those intervals.

The numbers of larvae and adults observed in a given check were not significantly related with climatic factors, measured at the 95% level of confidence (Table 2). We have correlated the number of larvae and adults with the maximum, minimum and the mean temperatures, precipitation (in mm) and the number of rainy days influencing larvae and adults numbers in the 15, 30 (lag 1) and 45 (lag 2) days previous to given check. We have

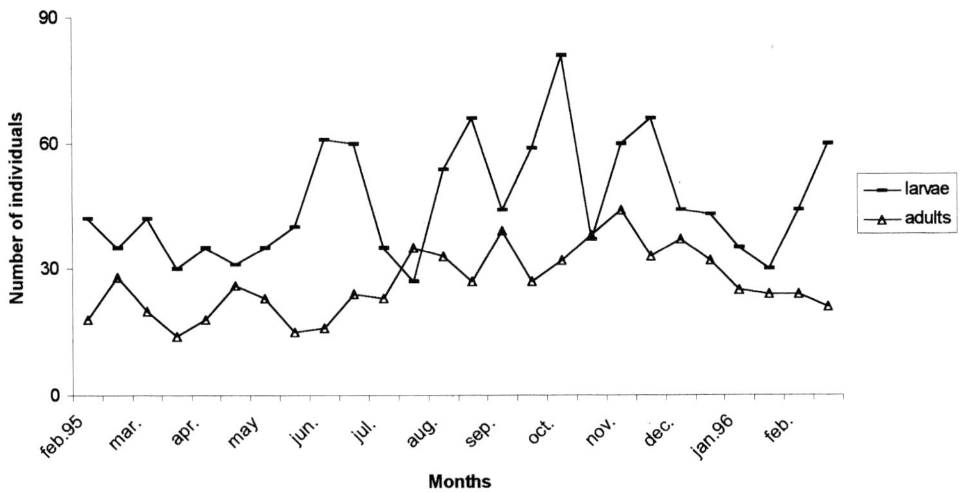

Fig. 2. Fluctuation of the population of *Plagiometriona flavescens* in Poço das Antas Biological Reserve.

Table 2. Pearson correlation matrix testing for influence of climate variables on the numbers of *Plagiometriona flavescens* larvae and adults during the study period at Poço das antas Biological Reserve (N=24 observations). Any correlation was significant after Bonferroni correction at the 95% significance level.

	Maximum temperature	Minimum temperature	Mean temperature	Precipitation temperature	Number of rainy days
Larvae	-0.44	-0.22	-0.37	-0.09	-0.04
Adults	-0.08	-0.06	-0.02	-0.24	-0.35
Lag 1 larvae	-0.30	-0.31	-0.33	-0.12	-0.18
Lag 1 adults	-0.36	-0.14	-0.28	-0.16	0.23
Lag 2 larvae	-0.31	-0.34	-0.35	-0.02	0.02
Lag 2 adults	-0.50	-0.30	-0.44	-0.26	-0.12

used these lag periods to check for influence of weather on reproduction and on the number of individuals on the following developmental stage. The effects of climate on herbivorous insect populations are not common; Rodriguez-Del-Bosque & Magallanes-Estala (1994), also could not explain variability in seasonal abundance of Diabroticina beetles by climate parameters (daily temperature, relative humidity or rainfall). Those abundances were partially attributed only to wind velocity.

P. flavescens also exhibited mating throughout the year. Beetles in copula were observed throughout the entire study period, except during August 1995. In general, this activity seemed to be less frequent in the winter months (Fig. 3). The prolonged reproductive schedule is suggested to have an adaptive value for living in highly disturbed tropical environments, where rainfall is ample but unpredictable (Nakamura *et al.*, 1989). However, it is clear that reproducing all year long is possible only for species which use constantly available food resources.

So, host plant availability must play an important role on *P. flavescens* time distribution, inasmuch as *A. fasciculata* provides the beetle with its preferred age-class foliage. *A. fasciculata* is a shrub that can reach up to 4 metres in height that produces new leaves all year round and is widely distributed along the roads of the Reserve. Moreover, the presence of another food plant species in the area also contributes to the maintenance of the *P. flavescens* population in the field over time.

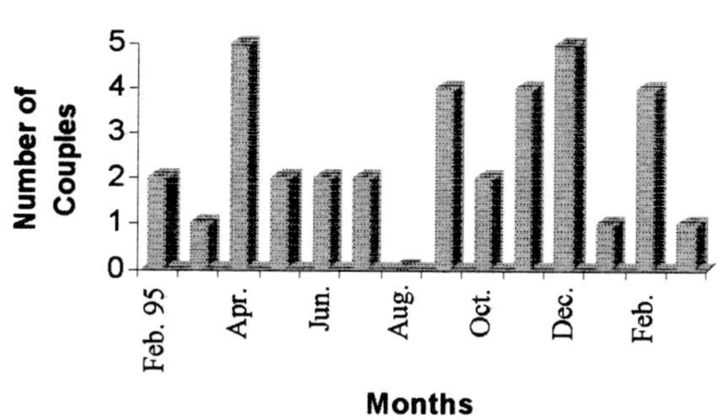

Fig. 3. Numbers of *Plagiometriona flavescens* mating couples in reproductive activities during the study period.

Natural enemies are other factors that can influence *P. flavescens* population numbers. In a previous work three parasitoid species were detected: a Eulophidae (Hymenoptera) as a parasitoid of the eggs, a Tachinidae (Diptera) as a larval parasitoid and a Chalcididae (Hymenoptera) as a pupal parasitoid; and an heteropteran predator species (Sá & Macêdo in press). A more detailed study is necessary to understand the actual influence of these insects on the fluctuation of *P. flavescens*.

4. Concluding remarks

As climate parameters did not seem to influence *P. flavescens* population fluctuation, it is possible that the host plant and natural enemies play important roles in the variation of this beetle's population number. The occurrence of *P. flavescens* throughout the year was possible because of the continuous presence of new leaves. Natural enemies may also influence not only population numbers but also the beetle behaviour concerning its localisation on the host plant leaves.

This type of study is very useful for comparison with other similar ones in different regions of the world, or with different taxonomic groups. Long term studies are also important in understanding the structure and functioning of tropical and temperate insect communities.

5. Acknowledgements

The authors are greatly indebted to Fernando F. F. Ferraz for laboratory and field assistance, Atlantic Forest Program (Rio de Janeiro Botanical Gardens) for providing Poço das Antas Reserve climate data and for the staff of that Reserve. We are grateful to IBAMA for providing the license to work at Poço das Antas Biological Reserve. We also thank Dr. Lúcia d'Ávila Freire de Carvalho and Dr. José Zundir Buzzi who identified the host plant and Cassidinae respectively. Research was supported by O Boticário Fundação de Proteção à Natureza. CNPq/CEPG provided a scholarship to FNS.

References

Becker, M. & Freire, A.J.P. 1996. Population ecology of *Gratiana spadicea* (Klug), a monophagous cassidine on an early successional Solanaceae in southern Brazil. *In* Jolivet, P.H.A. and Cox, M.L. (eds.), Chrysomelid Biology, volume 2: Ecological studies. SPB Academic Publishing bv, Amsterdam, The Netherlands. pp. 271-287.

Buzzi, J.Z. 1994. Host plants of Neotropical Cassidinae. *In* Jolivet, P.H.; Cox, M.L. & Petitpierre, E. (eds.), Novel Aspects of the Biology of Chrysomelidae. Kluwer Academic Publishers, Dordrecht, the Netherlands. pp. 205-212.

Cappaert, D.L.; Drummond, F.A. & Logan, P.A. 1991. Population dynamics of the Colorado potato beetle (Coleoptera: Chrysomelidae) on native host in Mexico. Envir. Ent. 20(6): 1549-1555.

Drummond, B.A. 1986. Herbivory and defense. *In*: D'Arcy, W.G. (ed.), Biology and systematics of Solanaceae. Columbia University Press, New York, USA. pp. 303-305.

Groden, E. & Casagrande, R.A. 1986. Population dynamics of the potato beetle, *Leptinotarsa decemlineata* (Coleoptera: Chrysomelidae), on *Solanum berthaultii*. J. econ. Ent. 79: 91-97.

Hsiao, T.H. 1986. Specificity of certain chrysomelid beetles for Solanaceae. *In*: D'Arcy, W.G. (ed.), Biology and systematics of Solanaceae. Columbia University Press, New York, USA. pp. 345-361.

Jolivet, P. 1988. Food habits and food selection of Chrysomelidae. Bionomic and evolutionary perspectives. *In* Jolivet, P; Petitpierre, E. and Hsiao, T. (eds.), Biology of Chrysomelid. Kluwer academic Publishers, Dodrecht, the Netherlands. pp. 1-24.

Medeiros, L. & Vasconcelos-Neto, J. 1994. Host plants and seasonal abundance patterns of some Brazilian Chrysomelidae. *In* Jolivet, P.H; Cox, M.L. & Petitpierre, E. (eds.), Novel Aspects of the Biology of Chrysomelidae. Kluwer Academic Publishers, Dordrecht, the Netherlands. pp. 185-189.

Nakamura, K.; Abbas, I. & Hasyim, A. 1989. Survivorship and fertility schedules of two Sumatran tortoise beetles, *Aspidomorpha miliaris* and *A. sanctaecrucis* (Coleoptera: Chrysomelidae) under laboratory conditions. Researches on Population Ecology 31: 25-34.

Nakamura, K.; Noerdjito, W.A. & Aswari, P. 1994. Life table parameters of Indonesian tortoise beetles under laboratory conditions. *In* Jolivet, P.H.; Cox, M.L. & Petitipierre, E. (eds.), Novel Aspects of the Biology of Chrysomelidae. Kluwer Academic Publishers, Dordrecht, the Netherlands. pp. 573-578.

Ohmart, C.P. 1996. Population dynamics of chrysomelid beetles feeding on *Eucalyptus*. *In* Jolivet, P.H.A. and Cox, M.L. (eds.), Chrysomelid Biology, volume 2: Ecological studies. SPB Academic Publishing bv, Amsterdam, the Netherlands. pp. 263-269.

Olmstead, K. 1996. Cassidinae defenses and natural enemies. *In* Jolivet, P.H.A. and Cox, M.L. (eds.), Chrysomelid Biology, volume 2: Ecological studies. SPB Academic Publishing bv, Amsterdam, The Netherlands. pp. 3-21.

Rodriguez-Del-Bosque, L.A. & Magallanes-Estala, A. 1994. Seasonal abundance of *Diabrotica balteata* and other Diabroticina beetles (Coleoptera: Chrysomelidae) in Northeastern Mexico. Environ. Entomol. 23: 1409-1415.

Sá, F.N. & Macêdo, M.V.. Host plant preference of *Plagiometriona flavescens* (Coleoptera: Chrysomelidae) for two Solanaceous species. *In* Proceedings of the Third International Symposium on the Chrysomelidae. Mus. Reg. Sci. Nat. Torino. In press.

POSSIBILITIES FOR CONSERVATION AND REHABILITATION OF POPULATIONS OF CHRYSOMELIDAE IN A CULTIVATED ENVIRONMENT

RON BEENEN
Martinus Nijhoffhove 51, NL 3437 ZP Nieuwegein, The Netherlands

Introduction

Recently a great quantity of information concerning the biology of Chrysomelidae has been published in a comprehensive way (for example Jolivet & Petitpierre, 1981; Jolivet *et al*, 1988; Koch, 1992; Jolivet *et al*, 1994; Jolivet & Hawkeswood, 1995; Jolivet & Cox, 1996). These publications give an indication of the vast amount of information available in numerous publications. The availability of so much information should give us a better insight into the habitats and needs of the different chrysomelid species. This insight could be used to conserve or rehabilitate populations of Chrysomelidae.

Still some chrysomelid species seem to be in danger. Signals concerning possible threats to chrysomelid species are being published (for example Jolivet, 1952 [*Macroplea mutica*], Jolivet, 1989 [*Timarcha spp*]; Beenen, 1990 [*Galeruca pomonae*]; Buse, 1993 [*Chrysolina cerealis*]). It is worthwile to combine the available information on the biology of Chrysomelidae to conserve or rehabilitate populations of Chrysomelidae. This chapter is a preliminary effort to present such possibilities. At first a survey will be given of our knowledge concerning the threats to different species. Subsequently a system will be presented that predicts the effects of different types of management on Chrysomelidae (or even insect) populations.

Chrysomelidae under threat

Arguments for insect conservation are often based on conspicuousness (Van Hook, 1997). Although many Chrysomelidae are conspicuous due to their bright colours, they generally are ignored in conservation strategies. Another, rather new, basis for insect conservation could be their contribution to biodiversity. Worldwide there is a growing interest in measures to maintain biodiversity, mainly due to the signing of the Convention on Biological Diversity by a large number of states at the Earth Summit held in Rio de Janeiro in 1992. Leaf beetles are an essential part of the biodiversity of most ecosystems. They are numerous in species (worldwide over 35,000 species described) and occur in almost every ecosystem where vascular plants occur.

Insects face the same anthropogenic threats as vertebrates (Van Hook, 1997) except perhaps, in some groups such as the Chrysomelidae, the threats that are correlated to eco-

Advances in Chysomelidae Biology 1, pp. 307–319
Edited by M.L. Cox

nomic values. Chrysomelidae are generally little valued economically. Trade in specimens of Chrysomelidae is probably not great. Most probably only specimens from the Sagrinae have commercial value.

One of the most apparent anthropogenic threats is habitat deterioration. The loss of habitat is most apparent. In western Europe many habitats have been given up for urban expansion. The remaining habitats, even when they are reserves, will not always be suitable for all of the characteristic species. Suitability is highly dependent on the aims of the management. Management goals are mostly formulated in terms of desired plant or vertebrate species.

Information about population development is available only for very few European chrysomelid species. For some countries Red Data Lists of endangered or extinct species have been published. Red Data Lists give information about population status of the treated species.

In the list of European species of globally threatened plants and animals (Anonymous, 1991) only one chrysomelid species is listed: *Cecchiniola platyscelidina* (Jacobson) from the Ukrainian SSR (Crimea). It is indicated as "indeterminate" which means that this taxon is known to be "endangered", "vulnerable" or "rare" but there is not enough information to indicate which of the three categories is appropriate.

For the European countries a global survey has been made of Red Data Lists of threatened Chrysomelidae. The results are presented in table 1 (for practical reasons administrative areas smaller than countries have not been included).

The authors listed in Table 1 may have used different criteria for their categories of threat. The aim of the lists is to indicate which species are vulnerable or threatened. For this aim the lists usually play an important role in enhancing the public and political awareness. For plants and vertebrates information on Red Data Lists often provide the basis for species conservation plans.

Apart from Red Data Lists, an indication of vulnerability of certain species, can be obtained from legislation. From a compilation of legislation to conserve insects in Europe (Collins, 1987) the legislation in which chrysomelids have been included is extracted: only three countries included some Chrysomelidae in their lists of protected species: Belgium (all species of *Donacia* and *Plateumaris*); Greece (*Chrysolina gypsophilae*) and the United Kingdom (*Chrysolina cerealis*). It is unknown if there have been chrysomelid species included in legislation since.

Both from Red Data Lists and from legislation it is not possible to give a complete picture of threat or vulnerability of European Chrysomelidae. Red Data Lists are often

Table 1. Red Data Lists of European countries

country	author	analysed	listed	percentage
Sweden	Lundberg, 1995	361	81	22 %
Britain	Hyman & Parsons, 1992	255	42	16 %
Belgium	Leclercq et al, 1980	12	4	33 %
Germany W	Blab et al, 1984	463	183	40 %
Austria	Geiser, 1994	521	184	35 %

used to select species to make conservation plans for. However conservation plans for single leaf beetle species may be in conflict with conservation plans for other animal or plant species. Because of the numerous invertebrate species it seems more effective to make plans for groups of species occurring in the same habitat. Moreover, the main threats for chrysomelids are believed to lie in inferior habitat quality. Therefore it might be more efficient to focus upon management of habitats.

Management

There might be some discussion whether habitats should be "left alone" or managed with certain aims. Collins & Luxmoore (1993) suggested that "*habitat conservation may not succeed in conserving all species. However, it provides a pragmatic and achievable goal which does not require endless research to quantify the number of species present before conservation plans can be put into action. Furthermore, habitats contain complex assemblages of species which have evolved together.*". This seems a very appropriate view when natural habitats in a natural environment are concerned. In western Europe, however, we meet almost exclusively natural, semi-natural or even agricultural habitats in a cultivated environment. In these habitats, management (like wood cutting, mowing, grazing or even burning) plays a crucial part in the maintainance of the habitats. Different species often respond in a different way to this management. It is therefore unavoidable to make management choices and explicitly define the aims.

Maes & Decleer (1997) studied the effects of mowing in reedmarshes on carabid beetles and discussed their results in relation to data on other organisms (for example higher plants, birds, spiders and grasshoppers). Not surprisingly, they concluded that overall appropriate management recommendations are not easy to give if the different effects of the cutting management on the different species are taken into account. They proposed a rotational management that must result in reeds of different ages and reedbelts in different successional stages.

Rotational cutting management is often recommended (Decleer, 1989; Kirby, 1992). However this is not applicable in every situation. For example, if the smallest compartment of a rotational management scheme is too small to retain a viable population of a targeted plant or animal species than these species could locally disappear. In such a case it would be wiser to manage larger compartments, even when it means that the number of compartments in a management scheme will be reduced to one or two. If a management plan is developed it is usefull to know the effects on the species occurring in the area. For that reason we need species lists (actual or potential lists) and all available information on life histories of these species.

However, the number of species for which usefull information is available is unknown. In this paper the actual information available on Chrysomelidae of particular habitats, will be elucidated. The concept of "life history types" will be introduced to make use of such information in habitat management.

The concept of "life history types"

Astonishly simple, but very illustrative is the idea proposed by Huston (1994) to describe biodiversity. The total diversity of a community can be described by the number of functional types and the number of functionally analogous species within each functional type. The authors idea behind the use of the functional type approach is that the mechanisms that influence the number of functional types in a community are almost always different from the mechanisms that influence the number of functionally analogous species within a functional type. The functional type can be applied at different ecosystem levels; often it is used to describe "guilds" of species that use the same resources.

Comparable approaches have been made by other ecologists (Bink & Siepel, 1986; Siepel, 1994), but the restriction to functional types or guilds makes these approaches less useful for our purpose. For the purpose of conserving invertebrate species the approach has to be extended to all life history aspects of the species.

The life history types are defined and the analogous species are listed in Fig. 1. If the effect of a management scheme on a life history type can be predicted, it will be possible to give information on the number of species that react in the same way (the analogous species) and it may be possible to give more accurate information on these species. This depends on the available information on life histories.

Schaufuss (1916) added to Calwers Käferbuch an interesting and very useful appendix in which he presented a fact sheet of life history aspects ("*Bionomischer Fragebogen*"). The idea was that collectors should fill in these fact sheets for all species. Unfortunately it did not result in a natural habit of collectors to document their findings on the natural history of beetles in such a systematic way. But progress has been made and in many publications facts on the life history of beetle species can be found.

Life history types in chalk grassland in the south of The Netherlands

In the limestone region, in the south of The Netherlands, arid grasslands occur on the slopes of hills. These grasslands are of national interest because of their deviating flora and fauna. Outside The Netherlands they occur in adjacent parts of Belgium and Germany. In The Netherlands these grasslands used to be managed with mainly floristic aims. In the 1980's there was a growing interest in the invertebrates of these grasslands. It resulted in different management advice for the different Invertebrate Orders.

In this paragraph the life histories of chrysomelids of these grasslands will be analysed. Subsequently, consequences of different management types on the Chrysomelidae diversity will be presented.

In the well known series "Die Käfer Mitteleuropas" some volumes have been published on the ecology of beetles and beetle habitats. In Table 2 the chrysomelids mentioned in this series (Koch, 1995) from arid grasslands on slopes are partly listed. Only those species also recorded from The Netherlands (Beenen & Winkelman, 1993) are listed. In this way Table 2 presents a potential leaf beetle species list for arid grasslands on slopes in The Netherlands. We will use this list to present effects of different management regimes on the Chrysomelidae.

For the definition of the different life history types the following aspects are used: eggs (site, period), larvae and adults (site, period, food [herb, shrub or tree]), pupae (site, pe-

Life history type 1

winter:	stage,	activities,	sites, ...
spring:	stage,	activities,	sites, ...
summer:	stage,	activities,	sites, ...
autumn:	stage,	activities,	sites, ...
	species 1,1		
	species 1,2		
	species 1,.		
	....		
	species 1,n		

Life history type 2

winter:	stage,	activities,	sites, ...
spring:	stage,	activities,	sites, ...
summer:	stage,	activities,	sites, ...
autumn:	stage,	activities,	sites, ...
	species 2,1		
	species 2,2		
	species 2,.		
	....		
	species 2,n		

Life history type 3

winter:	stage,	activities,	sites, ...
spring:	stage,	activities,	sites, ...
summer:	stage,	activities,	sites, ...
autumn:	stage,	activities,	sites, ...
	species 3,1		
	species 3,2		
	species 3,.		
	....		
	species 3,n		

.
.
.
.

Life history type N

winter:	stage,	activities,	sites, ...
spring:	stage,	activities,	sites, ...
summer:	stage,	activities,	sites, ...
autumn:	stage,	activities,	sites, ...
	species N,1		
	species N,2		
	species N,.		
	....		
	species N,n		

Fig. 1. Schematic reflection of the life history concept. For explanation see text.

Table 2. Chrysomelids occurring in arid grasslands on slopes (Koch, 1995) that have been also recorded in The Netherlands (Beenen & Winkelman, 1993).

Oulema gallaeciana (Heyden)	Aphthona euphorbiae (Schrank)
Oulema melanopus (Linnaeus)	Aphthona pygmaea (Kutschera)
Clytra laeviuscula Ratzeburg	Aphthona venustula (Kutschera)
Clytra quadripunctata (Linnaeus)	Longitarsus anchusae (Paykull)
Lachnaia sexpunctata (Scopoli)	Longitarsus exoletus (Linnaeus)
Cryptocephalus aureolus Suffrian	Longitarsus ganglbaueri Heikertinger
Cryptocephalus biguttatus (Scopoli)	Longitarsus kutscherae (Rye)
Cryptocephalus flavipes Fabricius	Longitarsus luridus (Scopoli)
Cryptocephalus hypochaeridis (LInnaeus)	Longitarsus melanocephalus (De Geer)
Cryptocephalus moraei (Linnaeus)	Longitarsus membranaceus (Foudras)
Cryptocephalus nitidus (Linnaeus)	Longitarsus nasturtii (Fabricius)
Cryptocephalus pygmaeus Fabricius	Longitarsus nigrofasciatus (Goeze)
Cryptocephalus sericeus (Linnaeus)	Longitarsus obliteratus (Rosenhauer)
Timarcha goettingensis (Linnaeus)	Longitarsus pellucidus (Foudras)
Timarcha tenebricosa (Fabricius)	Longitarsus rubiginosus (Foudras)
Leptinotarsa decemlineata (Say)	Longitarsus succineus (Foudras)
Chrysolina cerealis (Linnaeus)	Longitarsus tabidus (Fabricius)
Chrysolina fuliginosa (Olivier)	Altica brevicollis Foudras
Chrysolina geminata (Paykull)	Altica oleracea (Linnaeus)
Chrysolina kuesteri (Helliesen)	Podagrica fuscicornis (Linnaeus)
Chrysolina limbata (Fabricius)	Crepidodera nitidula (Linnaeus)
Chrysolina polita (Linnaeus)	Asiorestia transversa (Marsham)
Chrysolina sanguinolenta (Linnaeus)	Mantura chrysanthemi (Koch)
Chrysolina staphylaea (Linnaeus)	Mantura rustica (Linnaeus)
Chrysolina varians (Schaller)	Sphaeroderma testaceum (Fabricius)
Gonioctena olivacea (Forster)	Batophila rubi (Paykull)
Galeruca pomonae (Scopoli)	Chaetocnema hortensis (Fourcroy)
Galeruca tanaceti (Linnaeus)	Psylliodes attenuata (Koch)
Lochmaea crataegi (Forster)	Psylliodes dulcamarae (Koch)
Pyrrhalta viburni (Paykull)	Psylliodes isatides Heikertinger
Sermylassa halensis (Linnaeus)	Psylliodes napi (Fabricius)
Calomicrus circumfusus (Marsham)	Hispa atra Linnaeus
Calomicrus pinicola (Duftschmid)	Cassida margaritacea Schaller
Phyllotreta astrachanica Lopatin	Cassida murraea Linnaeus
Phyllotreta cruciferae (Goeze)	Cassida nebulosa Linnaeus
Phyllotreta nodicornis (Marsham)	Cassida prasina Illiger
Aphthona cyparissiae (Koch)	Cassida rubiginosa Müller
	Cassida stigmatica Suffrian

riod). The information on activities are grouped in seasons. The seasons are defined as follows: winter (December – February); spring (March – May); summer (June -August); autumn (September – November).
Table 2 lists 75 species. Of these, 43 have been included in our list of analogous species for the different life history types. That means that of almost 60 % of the species of arid grasslands on slopes in The Netherlands enough information is available to describe the effects of certain management regimes. In addition, for each species, information on dispersal capacity (D) and ecological type (ET: st=stenotopic, eu=eurytopic) is listed in Table 3 (when available).

Using the information from Table 3 we can advise managers of this habitat type. To demonstrate some possibilities we will discuss the effects of two different kinds of man-

Table 3. Life history types

1

winter:	eggs in the soil; adults in soil/litter.
spring:	larvae feeding on herbs; adults feeding on herbs, eggs in soil
summer:	larvae feeding on herbs; pupae in the soil; adults feeding on herbs; eggs in the soil
autumn:	adults feeding on herbs; eggs in the soil
	Timarcha goettingensis (D:low; ET:st)
	Timarcha tenebricosa (D:low; ET:eu)

2

winter:	eggs in soil/litter
spring:	larvae feeding on herbs
summer:	larvae feeding on herbs; pupae in the soil
autumn:	adults feeding on herbs; eggs in soil/litter
	Sermylassa halensis (D:?; ET:st)

3

winter:	eggs on herbs
spring:	larvae feeding on herbs
summer:	adults feeding on herbs; adults in the soil/litter
autumn:	adults feeding on herbs; eggs on herbs
	Galeruca pomonae (D:low; ET:st)
	Galeruca tanaceti (D:low; ET:eu)

4

winter:	eggs in shrubs
spring:	larvae on shrubs
summer:	pupae in the soil/litter; adults on shrubs
autumn:	adults on shrubs; eggs on shrubs
	Pyrrhalta viburni (D:high; ET:eu)

5

winter:	larvae in the soil
spring:	larvae in the soil; pupae in the soil
summer:	adults feeding on herbs; eggs on herbs, larvae feeding on/in herbs.
autumn:	larvae feeding on/in herbs
	Chrysolina cerealis (D: ?; ET: st)
	Sphaeroderma testaceum (D: ?; ET: eu)

5'

All Cryptocephalinae feeding on herbs; although they do not eat fresh plant material they depend on fallen leaves. It is assumed that Cryptocephalinae larvae react the same as species from group 5

Cryptocephalus aureolus (D: ?; ET: eu)
Cryptocephalus hypochaeridis (D: ?; ET: eu)
Cryptocephalus moraei (D: ?; ET: eu)
Cryptocephalus pygmaeus (D: ?; ET: st)
Cryptocephalus sericeus (D: ?; ET: eu)

6

winter:	larvae in the soil
spring:	larvae in the soil; pupae in the soil
summer:	adults feeding on deciduous trees; eggs in litter, larvae feeding on litter.
autumn:	larvae feeding on litter.
	Cryptocephalus biguttatus (D: ?; ET: eu)
	Cryptocephalus flavipes (D: ?; ET: eu)
	Cryptocephalus nitidus (D: ?; ET: eu)

Table 3. Cont.

7	
winter:	larvae feeding on roots/hibernate
spring:	larvae feeding on roots; pupae in the soil
summer:	adults feeding on herbs
autumn:	adults feeding on herbs; eggs in soil; larvae feeding on roots
	Aphthona cyparissiae (D:?; ET:st)
	Longitarsus rubiginosus (D:?; ET:eu)
8	
winter:	larvae feeding on roots/hibernate
spring:	larvae feeding on roots; pupae in the soil
summer:	adults feeding on herbs; eggs in soil
autumn:	adults feeding on herbs; eggs in soil; larvae feeding on roots
	Longitarsus succineus (D: ?; ET:eu)
9	
winter:	larvae feeding on roots/hibernate
spring:	pupae in the soil
summer:	adults feeding on trees; eggs in the soil
autumn:	larvae feeding on roots
	Calomicrus pinicola (D:?; ET: st)
10	
winter:	larvae inside ants nest
spring:	larvae inside ants nest; adults feeding on leaves of deciduous trees; eggs dropped in the vicinity of ants nests.
summer:	larvae inside ants nest; adults feeding on leaves of deciduous trees; eggs dropped in the vicinity of ants nests.
autumn:	larvae inside ants nest
	Clytra laeviuscula (D:high; ET:st)
	Clytra qaudripunctata (D:high; ET:st)
11	
winter:	larvae in the soil/litter; adults in the soil/litter
spring:	pupae in the soil; adults feeding on herbs; larvae feeding on herbs; eggs on herbs.
summer:	adults in soil/litter; larvae feeding on herbs
autumn:	adults feeding on herbs; larvae feeding on herbs
	Chrysolina sanguinolenta (D: low; ET: eu)
12	
winter:	adults in the soil/litter
spring:	adults feeding on herbs; eggs on herbs
summer:	larvae feeding on herbs; eggs on herbs; adults feeding on herbs; pupae in the soil
autumn:	adults feeding on herbs
	Leptinotarsa decemlineata (D:high; ET:eu)
	Chrysolina staphylaea (D:intermediate; ET:eu)
	Chrysolina varians (D:low; ET:eu)
	Chaetocnema hortensis (D: ?; ET:eu)
	Psylliodes dulcamarae (D: ?; ET:eu)
13	
winter:	adults in soil/litter
spring:	adults feeding on herbs; eggs on herbs; larvae feeding on/in herbs; pupae in soil
summer:	adults feeding on herbs
autumn:	adults feeding on herbs
	Mantura chrysanthemi (D: ?; ET: st)
	Mantura rustica (D: ?; ET: eu)
	Oulema melanopus (D: high; ET: eu)
	Oulema gallaeciana (D: high; ET: eu)

Table 3. Cont.

14	
winter:	adults in soil/litter
spring:	adults feeding on shrubs
summer:	eggs on shrubs; larvae feeding on/in shrubs, adults feeding on shrubs, pupae in the soil
autumn:	adults in soil/litter
	Gonioctena olivacea (D: high; ET: eu)
	Lochmaea crataegi (D: ?; ET st)
	Altica brevicollis (D:?; ET: eu)
15	
winter:	adults in soil/litter
spring:	adults on herbs; eggs in soil
summer:	larvae feeding on roots; pupae in soil; adults on herbs
autumn:	adults on herbs
	Phyllotreta cruciferae (D:?; ET:eu)
	Aphthona euphorbiae (D:?; ET:eu)
	Longitarsus ganglbaueri (D:?; ET st)
	Longitarsus kutscherae (D: ?; ET: eu)
	Longitarsus melanocephalus (D:?; ET:eu)
	Longitarsus nasturtii (D:?; ET:eu)
	Longitarsus nigrofasciatus (D: ?; ET:st)
	Psylliodes attenuata (D:?; ET: eu)

agement. For this type of grassland, consistent management seems to be more important than in any other habitat type (Kirby, 1992). For our purpose, it will be interesting to see what the effects might be if the management is changed. In the past some recommendations have been made to change the management of (some of) these grasslands in The Netherlands. An attempt is made to discuss the effects of these changes in management on Chrysomelidae, using the concept of life history types.

I Removal of all shrubs (in some "unmanaged" chalk grasslands shrubs have greatly expanded; to "restore" the chalk grassland vegetation, shrubs have to be removed). This type of restoration management gave rise to serious discussion when a population of the butterfly *Thecla betulae* L. completely disappeared from a chalk grassland after removing shrubs of *Prunus spinosa* (Felix, 1986).

II Cutting at the begin of August to prevent dominance of the grass *Brachypodium pinnatum* is recommended by Bobbink (1988). For decades it has been the practice to cut the vegetation in the autumn.

Removing all shrubs from the grassland (I) will affect all species that rely on parts of the shrubs. These are the species from life history types 4, 6, 9, 10 and 14. This group of life history types contains 10 species. Of these, 4 species are stenotopic. For two of these species only, the dispersal capacity is known. These two species have high dispersal capacity. If similar grasslands with shrubs remain in the vicinity then it is not unlikely that, after some shrubs have regrown, these beetle species will recolonize the area. For the other two stenotopic species this is unknown.

Mowing in August (II) will have a negative effect on life history types 1, 2, 3, 5, 5', 7, 8, 11, 12, 13 and 15. This group of life history types contains 33 species. It is likely that these species will disappear when mowing in August is practiced. Of these, 9 species are stenotopic. Of these stenotopic species the dispersal capacity is poorly known. Two species have low dispersal capacity. It is unlikely that recolonization of these species will happen, unless a similar untreated habitat adjoins the treated one.

However, the presentation here gives the effects of mowing in August compared with an ideal situation (potential species list). In the situation where mowing in August is recommended, the general management has been mowing in autumn. It would be best to use an actual species list of the area where the management will be altered. If this is not available we have to incorporate the effects of the management that has been practiced for years. It is possible to do so by only describing the effect of mowing in August that is attendant. In the situation of mowing in August instead of mowing in autumn it will give an extra negative effect on life history types 1 and 8. For these types mowing in August will prevent egg deposition completely whereas mowing in autumn has no influence on eggs deposited in the soil in late summer. Life history types 1 and 8 contain 3 species of which one is stenotopic. This species has poor dispersal capacity. It will be lost unless it can colonize from an adjacent similar habitat.

Discussion

Morris (1968) demonstrated the extreme complexity of the relationship between invertebrate populations and habitat in English chalk grassland. He concluded that no single method of management is likely to maintain the habitats of all or even many of the characteristic invertebrate species. Choices must be made and because of this it seems necessary to gain insight into the kind of species that will vanish or that will benefit from the measures. This insight can provide a profound basis to the choices to be made. The concept of life history types that is presented here can provide a basis for such management choices.

Managers of nature reserves can be advised on the basis of the life history concept without troubling them with scientific names of species that are too cryptic for them to be recognised. Advice will be in terms of biodiversity (numbers of species) and in the long term effect (possibility of recolonizing). By restricting to the stenotopic species the advice can be very specific. In this way, management schemes can be evaluated also on their effects on Chrysomelidae. It will not guarantee that all Chrysomelidae species will be maintained; it is inherent to any management type that some species will be lost and others be conserved or rehabilitated. Because the concept stresses the effects on stenotopic species with poor dispersal capacities it provides a good basis for the conservation and rehabilitation of populations of these, mostly, vulnerable species (Jolivet & Hawkeswood, 1995). At least managers can't say afterwards that they "were not aware".

The concept of life history types can be easily expanded to include other species (other taxonomic groups). There is no limit to the number of life history types although a larger number will make it more complex. If the life history information presented here could be applied in a computer programme it would be more applicable. It allows rapid comparisons at many levels and using many different combinations of categories as has been demonstrated by Speight (1997) for hover flies (Syrphidae). He made comparisons between observed and predicted hover fly faunas and related it to management and restoration.

If the concept of life history types could be expanded to include other groups then it would be best to formalise it. The national organisations of the European Invertebrate Survey, are appropriate institutions to maintain databases of life history types (for their experience with large databases concerning invertebrates and for their contacts with invertebrate specialists).

From the example of arid grasslands in the south of The Netherlands presented here we may conclude that for a substantial part of the chrysomelid species (almost 60 %) enough information is available for advising on management. Because the family Chrysomelidae is very diverse (both in species as in life histories) they can be of great value in such advice. However, although much information is available this is still not enough. For at least 40 % of the species the dataset is incomplete. Furthermore, information on dispersal capacity of Chrysomelidae is very limited. There is a lot of information on wing morphology and wing venation of Chrysomelidae (Jolivet, 1957, 1959) but factual information on flight ability is hardly available. The results of studies such as those on the dispersal capacity of Carabidae (see for example, den Boer, 1977) and the results of studies using window traps (van Huizen, 1980) are essential for a better understanding of Chrysomelidae biology.

Information on the viability of Chrysomelidae populations is not available. In rotational management the assumption is made that if in some parts of a reserve the vegetation is maintained, or if in a nature reserve all successional stages are available, populations of insects must live on. Actually this greatly depends on the viability of a population. This is partly dependent on the number of individuals in a population.

Although we actually do not have any information on minimum viable populations of Chrysomelidae there are some general indications based on theoretical evidence and on empirical information of other organisms. Thomas (1990) suggested that a population of the order of 5000 individuals is needed to survive disturbances ranging from short-term environmental fluctuations to long term variation and catastrophes (see also Ryan & Siegfried, 1994). If this number is applicable to the Chrysomelidae it is at present not possible to indicate what area is needed to maintain a population of 5000 individuals. This seems even more difficult if fluctuations are to be taken into account. Fluctuations, as demonstrated for the field-cricket (*Gryllus campestris* L.), where a population grew within a year from 200 specimens to 25.000 specimens (Remmert, 1979), might also occur within the Chrysomelidae (for example for thermophilous species that reach their range boundaries in Central Europe).

Acknowledgements

Fruitful discussions with Jaap Winkelman (Amsterdam) on the subject of protection of chrysomelids have certainly contributed to this chapter. Jaap Winkelman (Amsterdam) and Roy Kleukers (European Invertebrate Survey – The Netherlands, Leiden) made useful comments on the draft of this article. The author thanks Michael Cox (London) for the invitation to contribute to this book and for his critical evaluation of the manuscript.

References

Anonymous, 1991. European red list of globally threatened animals and plants. Economic commission for Europe, Geneva / United Nations, New York, pp. i-v, 1-153.

Beenen, R., 1988. Het genus *Timarcha* in Nederland. Ent. Ber., Amst 48: 153-158.

Beenen, R., 1990. *Galeruca pomonae* (Scopoli), een in Nederland zeer bedreigde diersoort. Natuurh. Maandbl. 79: 276-278.

Beenen, R. & Winkelman, J., 1993. Naamlijst van de Nederlandse bladkevers. Nederlandse Faunistische Mededelingen 5: 9-18.

Bink, F.A. & Siepel, H., 1986. Life history tactics and strategies in butterflies. Proceedings of the 3rd European Congress of Entomology 3: 409-412.

Biondi, M., 1996. Proposal for an ecological and zoogeographical categorization of the Mediterranean species of the flea beetle genus *Longitarsus* Berthold. In: Jolivet, P. & Cox, M.L. (eds.). Chrysomelidae Biology Volume 3. SPB Academic Publishing, Amsterdam, The Netherlands, pp. 13-35.

Blab, J., 1993. Grundlagen des Biotopschutzes für Tiere. Schriftenreihe für Landschaftspflege und Naturschutz 24: 1-479.

Blab, J., Nowak, E., Trautmann, W. & Sukopp, H. (eds), 1984. Rote Liste der gefährdeten Tiere und Pflanzen in der Bundesrepublik Deutschland. Kilda Verlag, Greven. pp. 1-270.

Boer, P.J.den, 1977. Dispersal power and survival. Carabids in a cultivated countryside. Misc. Pap. LandbHoogesch. Wageningen 14: 1-190.

Bobbink, R., 1988. De toename van Gevinde kortsteeel in zuidlimburgse kalkgraslanden. Publties natuurh. Genoot. Limburg 37(2): 1-72.

Buse, A., 1993. Life-cycle and behaviour of the British population of *Chrysolina cerealis*: the implications for survival. Entomologist 112: 105-117.

Chevin, H., 1985. Contribution à la biologie des *Timarcha* II – *Timarcha tenebricosa* F. Cah. Liaison O.P.I.E. 19: 7-14.

Chevin, H., 1987. Contribution à la biologie des *Timarcha* IV – *Timarcha goettingensis* L. Cah. Liaison O.P.I.E. 21: 21-25.

Collins, N.M., 1987. Legislation to conserve insects in Europe. The Amateur Entomologist's Society Pamphlet 13: 1-77.

Collins, N.M. & Luxmoore, R., 1993. Saving habitats first ?. Naturopa 73:8-9.

Cox, M.L., 1994. Diapause in Chrysomelidae. In: Jolivet, P.H., Cox, M.L. & Petitpierre, E. (eds.) Novel aspects of the biology of Chrysomelidae. Kluwer Academic Publishers, Dordrecht, The Netherlands, pp. 469-502.

Decleer, K., 1989. Kansen voor ongewervelden in het huidige natuurbeheer. Verhandelingen van het symposium "Invertebraten van België" 1989: 447-453.

Doguet, S., 1994. Coléoptères Chrysomelidae. Volume 2 Alticinae. Faune de France 80: I-IX, 1-694.

Felix, C., 1986. Dagvlinders van de St.-Pietersberg. Natuurhistorisch Maandblad 75: 146-151.

Geiser, E., 1994. Coleoptera: Chrysomelidae. In: Gepp, H. (ed.) Rote Listen gefährdeter Tiere Österreichs. Grüne Reihe des Bundesministriums für Umwelt, Jugend und Familie (Wien) 2: 176-185.

Hiller, E., 1975. Zur Kenntnis des Schwarzbraunen Kiefernblattkäfers, *Luperus pinicola* (Dftsch.). Z. angew. Ent. 79: 398-421.

Huizen, T.H.P. van, 1980. Why not use window traps for collecting Coleoptera and other flying insects. Ent. Ber., Amst. 40: 131-132.

Huston, M.A., 1994. Biological diversity. The coexistence of species on changing landscapes. Cambridge University Press, Cambridge, USA. i-xix, 1-681.

Hyman, P.S. & M.S.Parsons, 1992. A review of the scarce and threatened Coleoptera of Great Britain. Part 1. UK Nature Conservation 3: 1-484.

Jolivet, P., 1952. Les Chrysomèles rares de la Belgique III. Les *Haemonia*. Les Naturalistes Belges Bull. 33: 92-97.

Jolivet, P., 1957. Recherches sur l'aile des Chrysomeloidea 1. Mém. Inst. r. Sci. nat. Belg., deuxième série, fasc. 51: 1-180.

Jolivet, P., 1959. Recherches sur l'aile des Chrysomeloidea 2. Mém. Inst. r. Sci. nat. Belg., deuxième série, fasc. 58: 1-152.

Jolivet, P., 1989. Un genre a danger de mort: *Timarcha*. Entomologiste 45: 301-310.

Jolivet, P. & Cox, M.L. (eds.), 1996. Chrysomelidae Biology Volumes 1, 2, 3. SPB Academic Publishing, Amsterdam, The Netherlands.

Jolivet, P, Cox, M.L. & Petitpierre, E. (eds.), 1994. Novel aspects of the biology of Chrysomelidae. Kluwer Academic Publishers, Dordrecht, The Netherlands.

Jolivet, P. & Hawkeswood, T.J., 1995. Host-plants of Chrysomelidae of the world. Backhuys Publishers, Leiden. 1-281.

Jolivet, P. & Petitpierre, E., 1981. Biology of Chrysomelidae. Butll. Inst. catal. Hist. nat. 47: 105-138.

Jolivet, P., Petitpierre, E. & Hsiao, T.H. (eds.), 1988. Biology of Chrysomelidae. Kluwer Academic Publishers, The Hague, The Netherlands.

Kirby, P., 1992. Habitat management for invertebrates: a practical handbook. Royal Society for the Protection of Birds.

Kloet, G.S. & Hincks, W.D., 1977. A check list of British insects. Second edition (completely revised) part 3. Royal ent. Soc., London.

Koch, K., 1992. Die Käfer Mitteleuropas, Ökologie. E3: 1-389. Goecke & Evers, Krefeld.

Koch, K., 1995. Die Käfer Mitteleuropas, Ökologie. Artenassoziationen in Makrohabitaten E7: 1-334. Goecke & Evers, Krefeld.

Leclercq, J., Gaspar, C., Marchal, J.L., Verstraeten, C. & Wonville, C., 1980. Analyse des 1600 premieres cartes de l'atlas provisoire des insectes de Belgique, et premiere liste rouge d'insectes menaces dans la fauna Belge. Notes fauniques de Gembloux 4: 1-104.

Lundberg, S., 1995. Catalogus Coleopterorum Sueciae. Naturhistoriska Riksmuseet / Entomologiska Föreningen, Stockholm.

Maes, D. & Decleer, K., 1997. The effects of nature management on reedmarsh carabid beetles in the Blankaart Nature Reserve. Colloquy on conservation management and restoration of habitats for invertebrates: enhancing biological diversity. Environmental Encounters (Council of Europe Publishing) 33: 64-73.

Mohr, K.-H., 1966. Familie Chrysomelidae. Die Käfer Mitteleuropas 9: 95-299.

Morris, M.G., 1968. Populations of invertebrate animals and the management of chalk grassland in Britain. Biol. Conserv. 1: 225-231.

Perner, J., 1996. Life cycle and population dynamics of selected flea beetles. In: Jolivet, P. & Cox, M.L. (eds.). Chrysomelidae Biology Volume 3. SPB Academic Publishing, Amsterdam, The Netherlands, pp. 271-283.

Remmert, H. 1979. Grillen, oder wie gross müssen Naturschutzgebieten sein ?. Nationalpark 22: 6-9.

Ryan, P.G. & Siegfried, W.R., 1994. The viability of small populations of birds: an empirical investigation of vulnerability. In: Remmert, H (ed.). Minimum animal populations. Ecological studies 106. Springer Verlag.

Schaufuss, C., 1916. Calwer's Käferbuch. 6th edition. E.Schweizerbart'sche Verlagsbuchhandlung, Stuttgart, pp. 1-1390.

Siepel, H., 1994. Structure and function of soil microarthropod communities. PhD Thesis. pp. 1-136.

Speight, M.C.D., 1997, Invertebrate species lists as management tools: an example using databased information about Syrphidae. Colloquy on conservation, management and restoration of habitats for invertebrates: enhancing biological diversity. Environmental encounters (Council of Europe Publishing) 33: 74-83.

Steinhausen, W.R., 1996. Biological remarks on rearing and collecting of Middle European Leaf beetle larvae. In: Jolivet, P.H.A. & Cox, M.L. (eds) Chrysomelidae Biology 3. SPB Academic Publishing, Amsterdam, The Netherlands, pp. 93-105.

Thomas, C.D., 1990. What do real population dynamics tell us about minimum viable population sizes. Conserv. Biol. 4: 324-327.

Van Hook, T., 1997. Insect coloration and implication for conservation. Fla Ent. 80: 193-210.

Verdyck, P. 1996. Differentiation and evolution in two groups of closely related *Phyllotreta* species. PhD Thesis. XXII +217 pp.

Verstraeten, C., Boosten, G. & Caspar, C., 1973. Cartes 749-752 in: Leclercq, J. (ed.) Cartographie des invertebres europeens. Atlas provisoire des insectes de Belgique.

Verstraeten, C. & Boosten, G., 1979. Cartes 1249-1262 in: Boosten, G., Cammaerts, R., de Prins, W. & Verstraeten, C. Atlas provisoire des insectes de Belgique (et des regions limitrophes).

Winkelman, J., 1993. *Chrysolina limbata* in Nederland. Ent. Ber., Amst. 53: 44-48.

EVOLUTION OF HOST-PLANT AFFILIATION AND CHEMICAL DEFENSE IN *CHRYSOLINA-OREINA* LEAF BEETLES AS REVEALED BY mtDNA PHYLOGENIES

T.H. HSIAO[1] AND J.M. PASTEELS[2]
[1]*Professor emeritus, Department of Biology, Uthah State University, U.S.A;* [2]*Laboratoire de Biologie Animale et Cellulaire, CP 160/12, ULB, 50 Av. F.D. Roosevelt, 1050 Bruxelles, Belgium*

Introduction

Chrysolina Motschulsky *sensu lato* is a very large genus including more than four hundred species and about two hundred subspecies, most of them living in the Palaearctic Region (Daccordi, Bourdonné, pers. com.). It has been divided into numerous subgenera (73 in Daccordi, 1994), but, not surprisingly, without a general consensus. For example, *Oreina* Chevrolat or *Fastuolina* Warchalowski (= *Dlochrysa* Motschulsky) classified as subgenera by Daccordi (1994) were often considered by other authorities as distinct genera, for example both taxa by Burakowski *et al.* (1990), and *Oreina sensu lato* (seven subgenera in Daccordi, 1994) by Kühnelt (1984) and Kippenberg (1994) to quote only some recent references.

Chrysolina leaf beetles are highly specialized herbivores. Although food-plant affiliation often seems rather conservative at the subgeneric level, there are numerous exceptions and members of no less than eight different plant families are listed as host plants of *Chrysolina* spp. as a whole (Jolivet and Petitpierre, 1976; Bourdonné and Doguet, 1991).

These beetles are often brightly coloured and such aposematic coloration is linked to chemical defense. The toxins secreted by pronotal and elytral glands are either autogenous (synthesized by the beetles) or host-derived (Pasteels *et al.*, 1988, 1994). Plant-derived chemical defense is restricted to some *Oreina* species which sequester pyrrolizidine alkaloids (PAs) from Asteraceae (Pasteels *et al.*, 1995; Hartmann *et al.*, 1997). This restricted distribution of host-derived defense suggests that sequestration is secondary to *de novo* synthesis, and evolved in species which specialized on *Senecio* and *Adenostyles* food plants (Pasteels & Rowell-Rahier, 1991; Pasteels, 1993). However, the present confusing state of the classification of *Chrysolina* spp. prevents firm conclusions on the evolution of host-plant affiliation and chemical defense in this taxon. Phylogeny of *Oreina* spp., based on allozymes, suggests that PA sequestration is secondary to the *de novo* synthesis of cardenolides, but allozyme data failed to suggest a single unambiguous evolutionary scenario (Dobler *et al.*, 1996, Pasteels *et al.*, 1996).

The main purpose of the present study is to further explore the possible evolution of host-plant specialization and chemical defense in *Chrysolina*, and especially *Oreina*, by comparing mtDNA sequences. Also, current classifications of *Chrysolina sensu lato* will

Advances in Chysomelidae Biology 1, pp. 321–342
Edited by M.L. Cox

be assessed in the light of these analyses. In a first step, 12S and 16S mtDNA sequences are compared in 30 *Chrysolina-Oreina* species (21 subgenera listed by Daccordi, 1994). *Ambrostoma quadriimpressum* Motsch., which belongs to the same subtribe, Chrysolinina, was used as an outgroup. Phylogenetic trees constructed from these sequences allow taxonomic inferences and a first approach to the evolution of host-plant relationship and chemical defense.

In a second step, we focus on *Oreina sensu lato* (14 species, all 7 subgenera of Kühnelt, 1984). CO1 sequences were added in these analyses and selected *Chrysolina* spp. and *A. quadriimpressum* were included as outgroups.

Material and methods

Leaf beetles and their host plants

Leaf beetles and their collection data are listed in Table 1. Current classification of the beetles and the main toxins produced are given in Table 2. For the clarity of further discussion, *Oreina sensu lato* is considered as a distinct genus.

In Table 3, the reported host plants of each species are listed. Since errors are frequent in the literature, only well-established records were used, that is personal observations or plant association certified by colleagues with personal experiences or from recent publications specifically dealing with leaf beetle host plants. Genera and families are specified. For large plant families (Apiaceae, Lamiaceae and Asteraceae), subfamilies, tribes or subtribes are also indicated.

Insect source and DNA extraction

Beetles were field collected, starved, and preserved either in 95% ethanol or stored at –70 °C until use. The extraction procedure used in this study is described in Hsiao (1994a). Briefly, the standard protocol involved homogenization of a single adult insect in dry ice, followed by incubation of homogenate in proteinase K, and phenol extraction. Total DNA of some insect samples were extracted with benzyl chloride, a protocol modified from Zhu *et al.*, (1993). Individual beetles were homogenized in a microfuge tube with 300μl of extraction buffer (100mM TRIS-HCl, 40mM EDTA, pH 9.0). 60 μl of 10% SDS (pH 7.5) and 180 μl of benzyl chloride were added to the homogenate and the mixture was incubated at 50°C for 30 min under stirring. 180 μl of 3M sodium acetate was then added and the mixture was kept on ice for 15 min and then centrifuged at 6,000 rpm for 15 min. The supernatant was precipitated with 1 ml isopropanol, centrifuged at 12,000 rpm for 10 min, and washed twice with 70% ethanol. The DNA pellet was dissolved in water for PCR. This protocol produced adequate yield of genomic DNA without enzyme digestion and phenol extraction.

DNA primers and PCR amplification

The primers were named using the convention of Simon *et al.* (1994). Two sets of mtDNA primers were used. The 12S RNA gene primers were designed by Simon *et al.* (1991) and

Table 1. Beetle species included in this study and collecting information.

Species	Locality	Collector - Date	Host
Ambrostoma quadriimpressum Motsch.	Great Wall, Beijing, China	T. Hsiao, 07/05/92	*Alnus sp.*
Chrysolina americana (L.)	Toulouse, France	J.M. Pasteels, 06/07/92,	*Rosmarinus officinalis*
	Son Serra, Mallorca, Spain	E. Petitpierre, 05/01/92	unidentified
C. aurichalcea (Mannerheim)	Beijing, China	T. Hsiao, 06/05/92	*Artemisia sp.*
C. banksi (Fabricius)	Val Verde (Evora), Portugal	J.M. Pasteels, 06/01/92	*Mentha suavolens*
C. brunsvicensis (Gravenhorst)	Le Champ du Feu (Alsace), France	J.M. Pasteels, 6/04/92	*Hypericum perforatum*
C. coerulans (Scriba)	Lessines (Brabant), Belgium	J.M. Pasteels, 08/03/91	*Mentha suavolens*
C. fastuosa (Scopoli)	Hohegeiss (Harz), Germany; Subotica, Serbia	J.M. Pasteels, 06/14/92,	*Galeopsis tetrahit*
	Subotica, Serbia	P. Verdyck, 09/18/94	unidentified
C. geminata (Psykull)	Hohegeiss (Harz), Germany	J.M. Pasteels, 06/14/92	*Hypericum perforatum*
C. graminis (L.)	Ax (Ariege), France	T. Hsiao, 08/30/93	*Mentha sp.*
C. grossa (Fabricius)	Val Verde (Evora), Portugal	J.M. Pasteels, 06/01/93	*Mentha suavolens*
C. haemoptera (L.)	Côte de Locmariaquer (Golfe du Morbihan,Bretagne), France	J.M. Pasteels, 05/01/94	*Plantago lanceolota*
C. herbacea (Duftschmid)	Añon (Zaragoza), Spain	E. Petitpierre, 05/27/92	*Mentha suavolens*
C. hyperici (Forster)	Treignes, Belgium	J.M. Pasteels, 06/10/92	*Hypericum perforatum*
C. lepida (Olivier)	Vauvert (Gard), France	J.-C. Bourdonné, 05/24/95	*Mantisalca salmantica*
C. peregrina (Herrich-Schäffer)	Val Verde (Evora), Portugal	J.M. Pasteels, 06/02/93	*Foeniculum vulgare*
C. polita (L.)	Buizingen (Brussels), Belgium	J.M. Pasteels, 06/15/92	*Lycopus europeus*
C. varians (Schaller)	Le Champ du Feu (Alsace), France	J.M. Pasteels, 06/04/92	*Hypericum perforatum*
Oreina bidentata Bontemps	Bosco Gurin (Tessin), Switzerland	B. Haegele, 07/93	*Leuzea (=SD Centaurea) rhapontica*
O. bifrons (Fabricius)	Le Hohwald (Alsace), France	J.M. Pasteels, M.Rahier, 06/04/92	*Chaerophyllum hirsutum*
O. cacaliae (Schrank)	Lieserwasser (Alsace), France	J.M. Pasteels, 05/25/91	*Adenostyles alliariae*
O. coerulea (Olivier)	Calmbach (Black Forest), Germany	S. Dobler, 07/93	*Centaurea nemoralis*
O. elongata (Suffrian)	Parc National du Mercantour, France	J.M. Pasteels, 07/25/91	*Adenostyles leucophylla, Cirsium spinosissimum*
O. frigida (Weise)	Hörnlihütte (Aroza, Graubünden), Switzerland	J.M. Pasteels, M. Rahier, 07/14/94	under stones
O. gloriosa (Fabricius)	Saas-Grund (Valais), Switzerland	J.M. Pasteels, 07/88	*Peucedanum ostruthium*
O. intricata (Germar)	Guarda (Unter-Engadin), Switzerland	S. Knoll, 06/93	*Senecio nemorensis, Adenostyles alliariae*
O. melanocephala (Duftschmid)	Hörnlihütte (Aroza, Graubünden), Switzerland	J.M. Pasteels, M. Rahier 07/14/94	under stones
O. speciosa (L.)	Lötschental (Wallis), Switzerland	S. Dobler, 05/93	*Chaerophyllum hirsutum*
O. speciosissima (Scopoli)	Les Moulins (Vaud), Switzerland	J.M. Pasteels, M. Rahier 06/05/92	*Petasites albus*
O. variabilis (Weise)	Zastler (Black Forest), Germany	J.M. Pasteels, M. Rahier 06/07/92	*Chaerophyllum hirsutum*
O. virgulata (Germar)	Hörnlihütte (Aroza, Graubünden), Switzerland	J.M. Pasteels, M. Rahier 07/14/94	*Cirsium spinosissimum*
O. viridis (Duftschmid)	Hörnlihütte (Aroza, Graubunden), Switzerland	J.M. Pasteels, M. Rahier 07/14/94	*Meum athamanticum*

Table 2. Current classification of the taxa and type of chemical defense.

Genera, subgenera and species	Chemical defense
Ambrostoma Motsch.	
A. quadriimpressum	cardenolides
Chrysolina Motsch.	
Allochrysolina Bechyné	
C. lepida	unknown
Anopachis Motsch	
C. aurichalcea	unknown
Chrysolina s.str.	
C. banksi	cardenolides
Colaphodes Motsch.	
C. haemoptera	cardenolides
Erythrochrysa Bechyné	
C. polita	cardenolides
Euchrysolina Bechyné	
C. graminis	cardenolides
Fastuolina Warchalowski	
C. fastuosa	cardenolides
Hypericia Bedel	
C. brunsvicensis	polyoxygenated steroids
C. geminata	polyoxygenated steroids
C. hyperici	polyoxygenated steroids
Melasomoptera Bechyné	
C. grossa	cardenolides
Menthastriella Bechyné	
C. herbacea	cardenolides
Minckia Strand	
C. peregrina	unknown
Sphaeromela Bedel	
C. varians	polyoxygenated steroids
Synerga Weise.	
C. coerulans	cardenolides
Taeniochrysea Bechyné	
C. americana	cardenolides
Oreina Chevrolat	
Allorina Weise.	
O. bidentata	cardenolides
O. coerulea	cardenolides
Chrysochloa Hop	
O. cacaliae	PAs
O. elongata	PAs, cardenolides
O. speciosissima	PAs, cardenolides
Frigidorina Kühnelt	
O. frigida	PAs, cardenolides
Intricatorina Kühnelt	
O. intricata	PAs, cardenolides
Oreina s. str.	
O. bifrons	cardenolides
O. gloriosa	cardenolides
O. speciosa	cardenolides
O. variabilis	cardenolides
O. viridis	cardenolides
Protorina Weise.	
O. melanocephala	unknown
Virgulatorina Kühnelt	
O. virgulata	cardenolides

Nomenclature and classification combined from Bourdonné and Doguet (1991) and Daccordi (1994). Defensive chemistry from Pasteels *et al.* 1988, 1994, 1996 and unpublished results. PAs: pyrrolizidine alkaloids, sequestered from host plant; unknown: unidentified compounds, but neither cardenolides, nor PAs.

Table 3. Host-plant affiliations.

	Genera	Families, subfamilies or tribes, subtribes
Genus Chrysolina		
C. americana	Lavandula L.	Lamiaceae, Lavanduloideae
	Rosmarinus L.	Lamiaceae, Rosmarinoideae
C. aurichalcea	Artemisia L.	Asteraceae, Arthemideae
C. banksi	Ballota L.	Lamiaceae, Stachyeae, Lamiinae
	Calamintha Mill.	Lamiaceae
	Marrubium L.	Lamiaceae, Marrubieae
	Mentha L.	Lamiaceae, Saturejeae, Menthinae
	Plantago L.	Plantaginaceae
	Ranunculus L.	Ranunculaceae
C. brunsvicensis	Hypericum L.	Hypericaceae
C. coerulans	Mentha L.	Lamiaceae, Saturejeae, Menthinae
C. fastuosa	Galeopsis L., Lamium L.	Lamiaceae, Saturejeae, Menthinae
C. geminata	Hypericum L.	Hypericaceae
C. graminis	Lycopus L., Mentha L.,	Lamiaceae, Saturejeae, Menthinae
	Stachys L	Lamiaceae, Stachyeae, Lamiinae
	Tanacetum L.	Asteraceae, Anthemideae
C. grossa	Mentha L.	Lamiaceae, Saturejeae, Menthinae
	Origanum L.	Lamiaceae, Saturejeae, Thyminae
	Satureja L.	Lamiaceae, Saturejeae, Saturejinae
C. haemoptera	Plantago L.	Plantaginaceae
C. herbacea	Mentha L.	Lamiaceae, Saturejeae, Menthinae
C. hyperici	Hypericum L.	Hypericaceae
C. lepida	Centaurea L., Mantisalca Cass.	Asteraceae, Cardueae
C. peregrina	Foeniculum Mill.	Apiaceae, Apieae
C. polita	Lycopus L., Mentha L.	Lamiaceae, Saturejeae, Menthinae
	Glechoma L.	Lamiaceae, Nepeteae
	Origanum L.	Lamiaceae, Saturejeae, Thyminae
C. varians	Hypericum L.	Hypericaceae
Genus Oreina		
O. bidentata	Centaurea L., Leuzea DC.	Asteraceae, Cardueae
O. bifrons	Anthriscus Pers., Chaerophyllum L.	Apiaceae, Scandiceae
	Peucedanum L.	Apiaceae, Peucedaneae
O. cacaliae	Adenostyles Cass., Petasites Mill., Senecio L.	Asteraceae, Senecioneae
O. coerulea	Centaurea L.	Asteraceae, Cardueae
O. elongata	Adenostyles Cass., Senecio L.	Asteraceae, Senecioneae
O. coerulea	Centaurea L.	Asteraceae, Cardueae
O. elongata	Adenostyles Cass., Senecio L.	Asteraceae, Senecioneae
	Cirsium Mill.	Asteraceae, Cardueae
O. frigida	Achillea L.	Asteraceae, Anthemideae
	Adenostyles Cass., Homogyne Cass.	Asteraceae, Senecioneae
	Meum Mill.	Apiaceae, Apieae
O. gloriosa	Peucedanum L.	Apiaceae, Peucedaneae
O. intricata	Adenostyles Cass., Senecio L.	Asteraceae, Senecioneae
O. melanocephala	Doronicum L.	Asteraceae, Senecioneae
O. speciosa	Aegopodium L., Meum Mill.	Apiaceae, Apieae
	Angelica L., Peucedanum L.	Apiaceae, Peucedaneae
	Anthriscus Pers., Chaerophyllum L.	Apiaceae, Scandiceae
	Laserpitium L.	Apiaceae, Laserpitieae

Table 3. Cont.

	Genera	Families, subfamilies or tribes, subtribes
Genus Oreina (Cont.)		
O. speciosissima	Achillea L.	Asteraceae, Anthemideae
	Adenostyles Cass., Doronicum L., Senecio L., Petasites Mill., Tussilago, L.	Asteraceae, Senecioneae
	Cirsium Mill.	Asteraceae, Cardueae
O. variabilis	Angelica L.	Apiaceae, Peucedaneae
	Anthriscus Pers., Chaerophyllum L.	Apiaceae, Scandiceae
O. virgulata	Adenostyles Cass. (occasional)	Asteraceae, Senecioneae
	Carduus L., Cirsium Mill.	Asteraceae, Cardueae
O. viridis	Meum Mill.	Apiaceae, Apieae

Compiled from Bourdonné and Doguet (1991), Bourdonné pers.com., Cox pers. com. Kippenberg (1994), Dobler *et al.* (1996), and personal observations. Plant nomenclature according to Brummitt (1992) (Genera) and Cronquist (1981) (Families). Infrafamilly classification according to Engler (1964). All Apiaceae genera are classified in the subfamily Apioideae and all Asteraceae genera, in the Asteroideae. Except for *Lavandula* and *Rosmarinus*, all Lamiaceae genera are classified in the subfamily Stachyoideae.

amplified a portion of the 12S region. The primer sequences were: SR-N-14759 (12SH), 5'-GACAAAATTCGTGCCAGCAGT-3' and SR-J-14233 (12SB), 5'-AAGAGCGACGGGCGATGTGT-3'. The 16S RNA gene primers were modified from those used by Xiong and Kocher (1991) for the black flies. The primer sequences were: LR-J-12883 (16SL), 5'-CACCGGTTTGAACTCAGATC-3'and LR-N-13398 (16SH), 5'-CACCTGTTTATCAAAAACAT-3'. Cytochrome oxidase 1 primers were C1-J-1751, (CO1r, Ron) 5'-GGATCACCTGATATAGCATTCCC-3' and C1-N-2191 (CO1n, Nancy) 5'-CCCGGTAAAATTAAAATATAAACTTC-3' (Simon *et al.*,1994). The primers were dissolved in distilled water at 20 μM concentration for PCR amplification and 100 μM for sequencing.

A typical PCR amplification protocol for a 100 μl reaction required 5-20 ng of DNA template, 2 μl of a 20 μM solution of each opposing primer, 8 μl of 10 μM dNTPs, 10 μl of 10X taq polymerase PCR buffer, and 3 μl of 0.1M $MgCl_2$. The mixture was placed in a 0.6 ml microfuge tube, overlaid with mineral oil, preheated in an Ericomp thermocycler at 95°C for 3 min, and 2 units of a Taq polymerase (Epicentre Tfl DNA polymerase or Promega Taq polymerase) were added. The thermocycling programs consisted in the following cycles, repeated 35 times: one denaturation step at 93°C for 35 sec, one annealing step at 55°C for 35 sec, and one extension step at 72°C for 2 min for 12S RNA and CO1 primers; one denaturation step at 93°C for 35 sec, one annealing step at 49°C for 35 sec, and one extension step at 72°C for 2 min for 16S RNA primers. A final extension at 72°C for 5 min was used in all programs. The PCR DNA fragments were purified by agarose gel electrophoresis and recovered by DEAE cellulose stripe or GeneClean (Bio101 Inc.), and precipitated with ethanol. The amount of DNA amplified during one PCR reaction was sufficient for 2 to 4 sequencing reactions. The procedure of direct double-stranded DNA sequencing of PCR products was identical to that described by Hsiao (1994a). Dideoxy chain termination reaction was carried out with Sequenase version 2.0 kit (USB-Amersham Corp.). Both strands were sequenced using PCR primers and/or new internal primers. Two to three individuals from each species were used to confirm the accuracy of the sequence data.

Sequence data analysis

DNA sequences were aligned initially using Clustal W (Higgins *et al.*, 1992) and the GCG pileup programs (Wisconsin Sequence Analysis Package, 1992). All alignments were inspected for obvious misaligments and corrected manually. Aligned sequences are available upon request

Phylogenetic reconstruction of sequence data was conducted by the neighbour-joining method using Kimura's 2-parameter distances (MEGA program, Kumar *et al.*, 1993), and maximum parsimony method (PAUP v. 3.1.1, Swofford, 1993) (heursistic search, 100 random addition sequences, TBR swapping). Gaps were coded as a fifth state (see results). To test for a possible saturation of one substitution type relative to the other, we plotted the number of transitions (Ti) and transversions (Tv) against the % divergence of each pair of taxa. Using MacClade 3.04 (Maddison and Maddison, 1992), we also generated a chart of the distribution of the number of changes per character.

A maximum-likelihood analysis of the three genes sequences of the *Oreina sensu lato* set of species, using the HKY model (Hasegawa *et al.*, 1985), was performed using PAUP* version 4 d 61 (Swofford, in press). During the search, the following parameters of the model were estimated from the data: nucleotide frequencies, proportion of invariable sites, and Ti/Tv ratio.

Reconstruction of ancestral host-plant association and defensive chemistry

The ancestral host-plant association and defensive chemistry were reconstructed using the criterion of maximum parsimony with MacClade 3.04. Each insect taxon was assigned a "host-plant" or "defensive" state corresponding to the family of its host plant(s) or class(es) of toxins secreted. If a species was associated with more than one plant family or class of toxins, a state was defined containing all plant families or toxins associated with the species. A step matrix was used to assign a cost to each transformation from one state to another by adding one step for each gain or loss of a plant family or class of toxins.

Results and discussion

Sequence divergence between taxa

Nucleotide sequences of the small subunit ribosomal RNA (12S) and the large subunit ribosomal RNA (16S) were generated from 31 species including 16 *Chrysolina*, 14 *Oreina*, and 1 *Ambrostoma* (outgroup). The sequence length of 12S varied between 499 and 505 base pairs (bp) and the 16S between 494 and 502 bp. The aligned sequences consisted of 513 and 506 bp for the 12S and 16S segments respectively. There were 20 gaps in the 12S and 8 gaps in the 16S sequences. Most gaps (73%) had a length of a single nucleotide. Only two gaps of two nucleotides and one of three were present in more than one taxon. The base composition of the 12S (A: 39.9%, T: 43.1%, C: 5.2%, G: 11.8%) and the 16S (A: 37.1, T: 39.7, C: 8.5, G: 14.7) segments showed similar high proportions of A and T and low proportions of C and G, which is typical for insect mtDNA. Cytochrome oxidases 1 (CO1) sequences of 411 nucleotides (no gap included) were obtained from 19 species (14

Oreina, 4 *Chrysolina* and 1 *Ambrostoma*). The base composition of CO1 (A:28.6, T: 37.0, C: 19.1, G: 15.3) showed higher C and G content relative to the 12S and 16S segments. All sequences are deposited in Genbank under accession numbers AF097028-AF097108.

Sequence divergences were compared between *Oreina* and *Chrysolina* species for all three segments. In the 12S sequences, uncorrected sequence divergences between the *Oreina* species varied from 0.2 to 8.4% with a mean of 4.0 ± 1.6% (S.D), and from 0.8 to 12.0% between *Chrysolina* species with a mean of 8.6 ± 1.9%. In the 16S segments, a lower sequence divergence was observed: from 0.1 to 5.0% with a mean of 3.0 ± 0.9% between the *Oreina* species, from 0.4 to 10.4% with a mean of 7.0 ± 1.6% between the *Chrysolina*. These values of sequence divergence were similar to values recorded from several other chrysomelid genera (Hsiao, 1994a). The CO1 sequences were also translated into AA sequences (137 amino acids). The AA sequence divergence between the 18 *Oreina* and *Chrysolina* taxa varied from 0.7 to 10.2% with a mean of 4.0 ± 2.0%. The nucleotide differences varied from 3.0 to 20.0% with a mean of 12.3 ± 5.3%. As shown by the separate analysis of the three codon position sites, only the third codon position was phylogenetically informative.

The 12S and 16S aligned sequences were combined for phylogenetic analyses of all 31 species. In both sequences, no evidence of saturation of nucleotide substitution types was observed on the Ti-Tv graphs (not shown). Also the majority (64%) of the variable characters display one or two changes. Accordingly Ti and Tv were equally weighted in the maximum parsimony analyses.

The neighbour-joining tree is given in Fig.1. The maximum parsimony analysis generated five equally parsimonious trees, the strict consensus tree is given in Fig. 2. Both trees of Figs 1 and 2 are very similar. Deep nodes, unresolved in Fig. 2, are not supported by high bootstrap values in the neighbour-joining tree. The position the *O. intricata-O. melanocephala* clade somewhat differs in both trees, but in neither case is its position supported by the bootstrap values.

To examine the relationship between *Oreina sensu lato* species, the CO1 segments were combined with the 12S and 16S for the analysis of 19 species. Identical trees were generated by neighbour-joining (Fig. 3) and maximum-likelihood (not shown) methods. The maximum parsimony analysis generated two equally parsimonious trees. One tree is identical to the one in Fig. 3. In the second, *O. intricata* (Germar) is placed at the base of the other *Oreina*, followed by *O. melanocephala* (Duftschmid). The strict consensus tree is illustrated in Fig.4.

Generic and subgeneric classification

In all trees the *Oreina sensu lato* (including *C. fastuosa* (Scopoli), see below) form a clade supported by bootstrap values ranging from 87 to 100%. The neighbor-joining tree of Fig.1 suggests that the remaining *Chrysolina* taxon could be paraphyletic, but this is only supported by low bootstrap values. Since the branching pattern between the *Chrysolina* subgenera remains poorly resolved (see also Fig. 2), it is not possible to decide if *Chrysolina* spp. form a single sister genus of the genus *Oreina sensu lato* (as in Kippenberg, 1994), or if it should be further subdivided into several genera.

A totally unexpected result is that in all trees, *C. fastuosa* is included in the *Oreina sensu lato* clade. Both 12S and 16S combined sequences and CO1 sequences suggest this place-

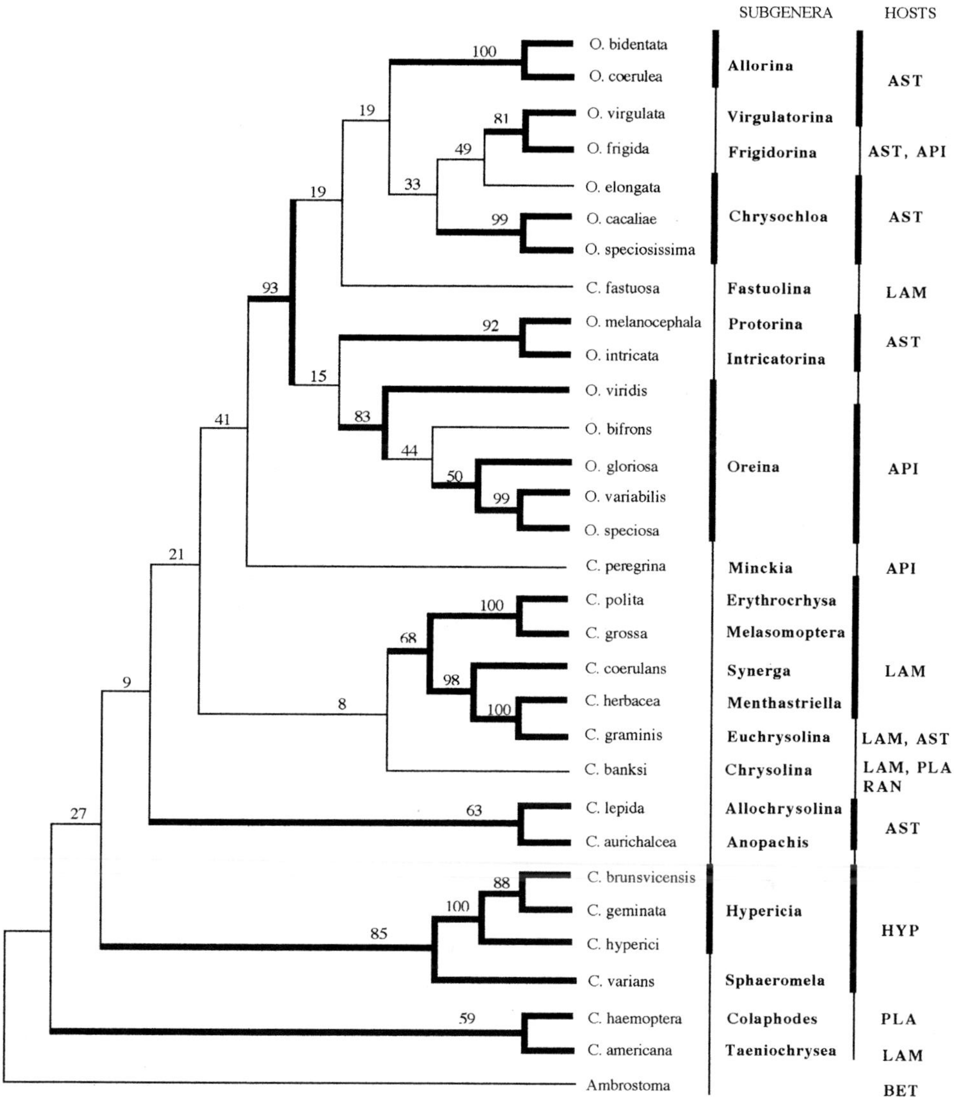

Fig. 1. Neighbour-joining tree (12S and 16S sequences combined). Bootstrap proportions are indicated on branches (500 replicates). Thick branches are supported by bootstrap value ≥ 50 %. API: Apiaceae, AST: Asteraceae, BET: Betulaceae, HYP: Hypericaceae, LAM: Lamiaceae, PLA: Plantaginaceae, RAN: Ranunculaceae.

ment. This is surprising since it contradicts all previous classifications, as well as previous phylogenenetic hypotheses based on allozymes (Dobler *et al.*, 1996). Moreover, *C. fastuosa* is the only species feeding on Lamiaceae in a clade of species feeding either on Apiaceae or Asteraceae. *C. fastuosa* was sometimes classified as a monospecific genus, or more often subgenus, but never suggested to be an *Oreina*. Bourdonné and Doguet (1991) placed it near most other *Chrysolina* feeding on Lamiaceae (their "groupe 2"), which include all those considered in the present study. These authors reached their conclusion on the basis

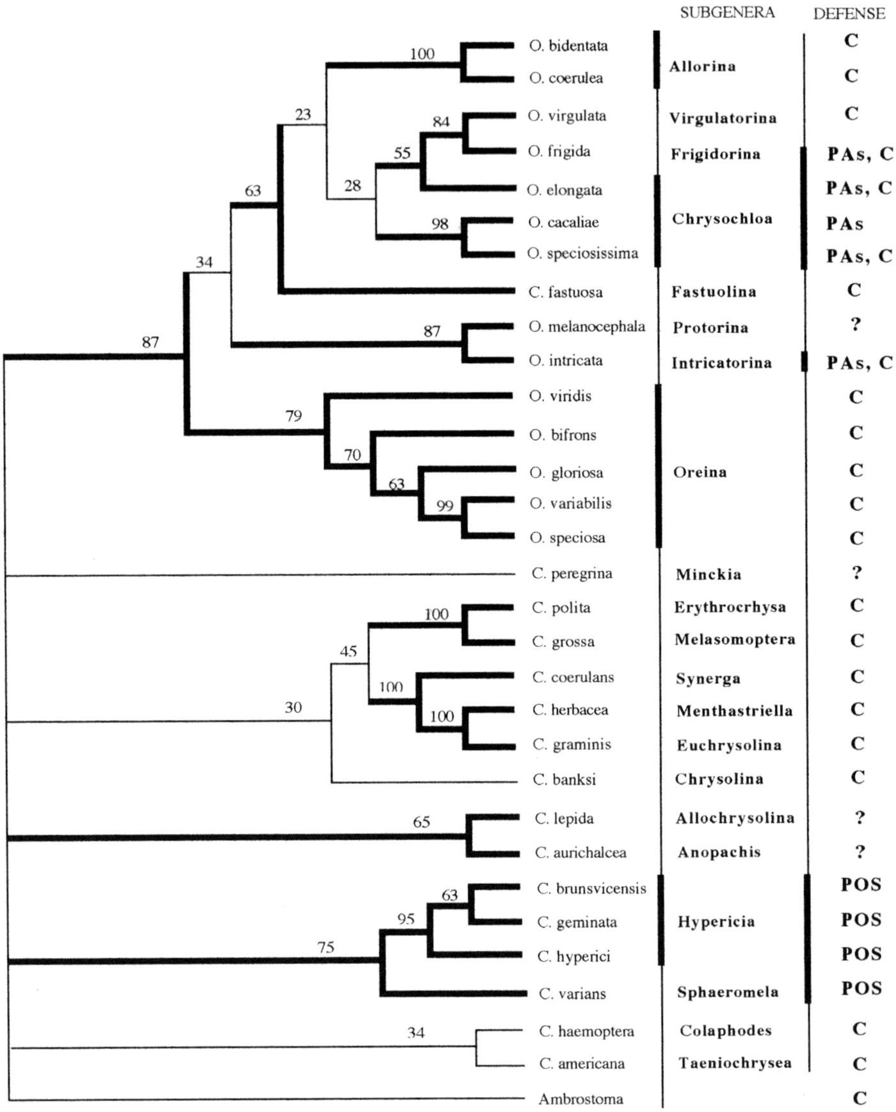

Fig. 2. Strict consensus of five most-parsimonious trees (12S and 16S sequences combined). Bootstrap proportions are indicated on branches (400 replicates). Thick branches are supported by a bootstrap value ≥ 50 %. C: cardenolides, PAs: pyrrolizidine alkaloids, POS: polyoxygenated steroids, ?: unknown, but neither C, nor PAs.

of similarities in male genitalia and spermatheca.

The unambiguous placement of *C. fastuosa* within the *Oreina sensu lato* clade in this study is not an artefact due to an unfortunate mixing of species or misidentification. The analyses were duplicated on beetles collected in two distant localities (in Germany and Slovenia), by two different collectors who independently identified this easily recognizable species (Table 1).

The precise position of *C. fastuosa* within the *Oreina sensu lato* clade remains uncertain. All trees suggest that *C. fastuosa* belongs to the clade *Allorina-Chrysochloa*, and is

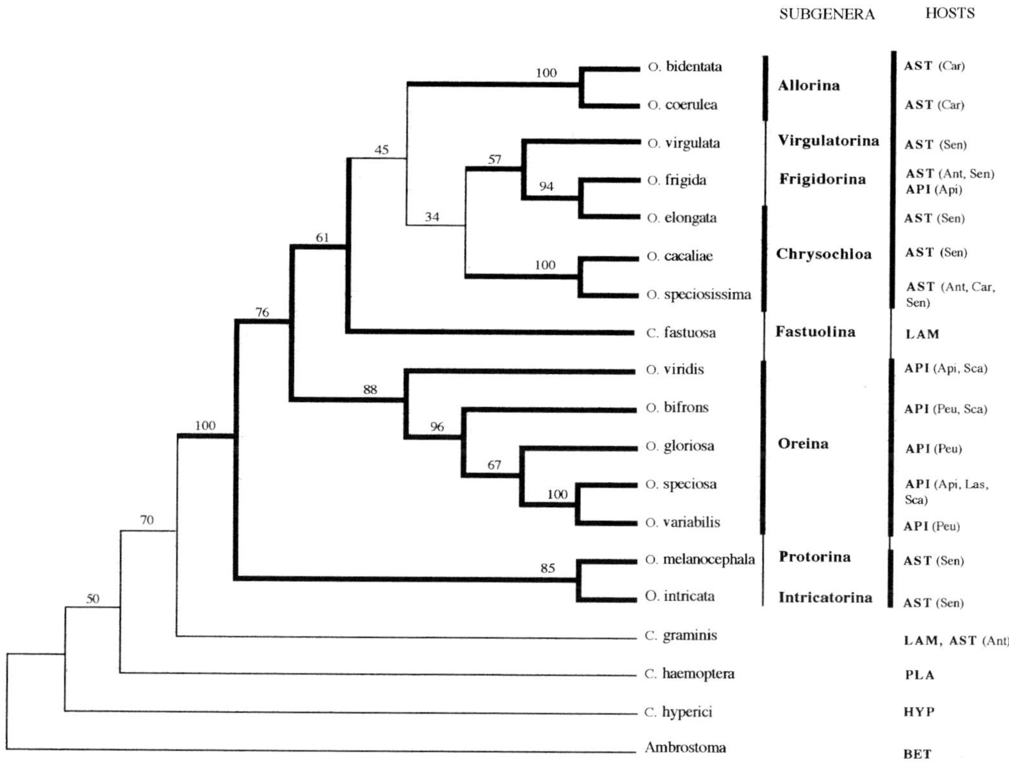

Fig. 3. Neighbour-joining tree (12S, 16S and CO1 sequences combined). Same legend as in Fig.1. Api: Apieae, Ant: Anthemideae, Car: Cardueae, Las: Laserpitieae, Peu: Peucedaneae, Sca: Scandiceae, Sen: Senecioneae.

less related to the *Oreina sensu stricto* and *Intricatorina-Protorina* clades, but the bootstrap values supporting this are low (from 19 to 63%). Clearly the position of *C. fastuosa* deserves additional investigations.

Within the *Oreina sensu lato* the subgenus *Oreina sensu stricto*, feeding on Apiaceae, is a well-defined clade, the branching pattern of which seems well- established (that is, *viridis* (Duftschmidt), *bifrons* (Fabricius), *gloriosa* (Fabricius), *variabilis* (Weise), *speciosa* (L.)), with bootstrap values most of the time above 70%, especially in the maximum parsimony trees (Figs 2 and 4). As expected, the two very similar *Allorina* species are grouped together, but the status of *Chrysochloa* is more confusing. *O. cacaliae* (Schrank) and *O. speciosissima* (Scopoli) are undoubtedly closely related species, but *O. elongata* (Suffrian) appears closer to *O. frigida* (Weise) and *O. virgulata* (Germar), both placed in other monospecific subgenera. The 12S and 16S sequences support the grouping of *O. frigida* with *O. virgulata*, whereas CO1 sequences strongly favour the grouping of *O. frigida* with *O. elongata*. *O. intricata* (Germar) and *O. melanocephala* (Duftschmid) are grouped together with high bootstrap values (from 85 to 92%) in Figs 1 to 3, but the maximum parsimony analysis of Fig. 4 failed to give an unambiguous support to this grouping.

The branching pattern of the *Oreina sensu lato* clade differs in several respects to that obtained with allozymes (Dobler *et al.*, 1996). In the allozyme tree, the subgenus *Oreina sensu stricto* is a clade as in the present study. However, the subgenus *Chrysochloa* forms

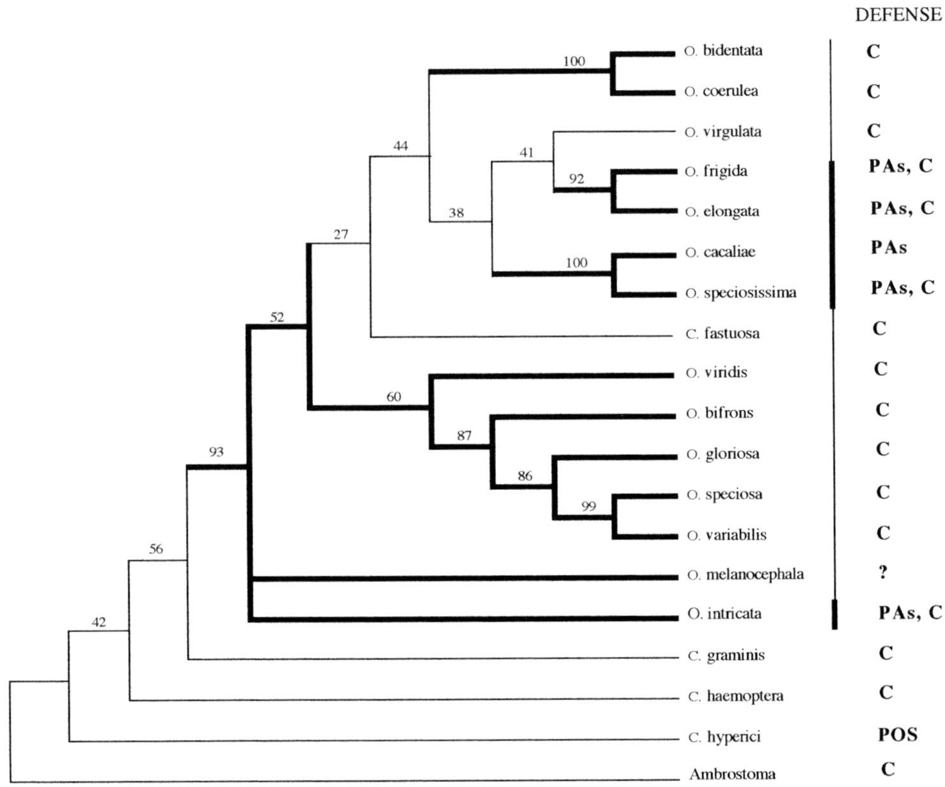

Fig. 4. Strict consensus of two most-parsimonious trees (12S, 16S and CO1 sequences combined). Same legend as in Fig.2.

a distinct clade in the allozyme tree, but is paraphyletic in this study. Contrary to our results, *O. melanocephala* and *O. intricata* were not grouped together in the allozyme tree, *O. melanocephala* appearing related to *Oreina sensu stricto* and *O. intricata* to *Chrysochloa*, although this pattern was supported by very low bootstrap values. Also, the grouping of *O. virgulata* with the *Oreina sensu stricto* suggested by allozyme data, is contradicted by the DNA trees, which are supported by higher bootstrap values.

Considering the *Chrysolina* subgenera, most are monospecific in our study, either because they are monospecific in the present classification (for example, *Colaphodes*, *Taeniochrysea*, *Erythrochrysa*, *Synerga* and *Sphaeromela*) or because of our limited sampling. Moreover as stated before, the branching pattern between subgenera is poorly resolved in many instances. Despite these limitations, some trends emerge from our analyses.

The three *Hypericia*-feeding species and the monospecific *Sphaeromela* form a well supported clade (Figs 1 and 2). Besides, all these species feed on *Hypericum* and are characterized by a distinct chemical defense (see below). Bourdonné and Doguet (1991) already mentioned that all *Chrysolina* species feeding on *Hypericum* are closely related on the basis of their male genitalia morphology and by the absence of a spermatheca. They suggested that these species could be classified in a genus distinct from the other *Chrysolina*.

C.coerulans (Scriba), *C. herbacea* (Duftschmid) and *C. graminis* (L.) also form a well-defined clade, as well as *C. polita* and *C. grossa*. Moreover, both groups possibly belong to the same lineage, although the bootstrap value supporting this is rather low (45%). All these species feed on Lamiaceae, frequently on *Mentha* spp (although in England, *C. graminis* is reported to feed also on *Tanacetum vulgare* L., Cox, pers. com.). Presently, the five species are each classified in a different subgenus, two of which being monospecific (*Erythrochrysa* and *Synerga*). Obviously, a lesser splitting attitude seems possible in the subgeneric classification of *Chrysolina*. In an independent study, Bourdonné (pers. com.) considers that *C. coerulans, C. herbacea* and *C. graminis* should be classified in the same subgenus, and *C. polita* and *C. grossa* in another.

Host-plant affiliation

With the exception of *C. banksi* (Fabricius), *C. graminis* and *O. frigida*, all beetle spp. feed on plants within a single family (table 3). The level of dietary specialization, however, varies from broad oligophagy to monophagy sensu Bernays and Chapman (1994), that is, species feeding on plants within a single genus: for example, compare the host ranges of *O. speciosa* and *O. gloriosa* among the Apiaceae feeders.

C. banksi , *C. graminis* and *O. frigida* feed on plants in three or two families. However, they feed on plants in a restricted number of genera and are thus oligophagous. Quite intriguingly, most host plants of *C. banksi and O. frigida* are also listed as hosts of other, more specialized, *Chrysolina* or *Oreina* spp. It is tempting to hypothesize that these species exemplify intermediate steps in the evolution of host affiliation. If confirmed, these beetles could prove to be key species for the understanding of the mechanisms underlying host switches and of the genetic basis of dietary specialization.

As observed in other leaf beetle genera (*Leptinotarsa*, Hsiao, 1994a; *Gonioctena*, Mardulyn *et al.*, 1997; and *Phratora*, Köpf *et al.*,1998), host-plant affiliation tends to be conservative during leaf beetle speciation, but abrupt switches repeatedly occurred.

Conservatism in host-plant selection is considered to be the rule during leaf beetles speciation (for example, *Phyllobrotica*, Farrell and Mitter, 1990; *Ophraella*, Funk *et al.*, 1995; *Blepharida*, Becerra, 1997). If parallel cladogenesis was observed in the evolution of the *Phyllobrotica*-Lamiales association (Farrel and Mitter, 1990), chemical similarity between plants, more than plant cladogenesis, appears to have played a significant role in the evolution of the *Blepharida-Bursea* association (Becerra, 1997). Conservatism is demonstrated in the present study by the observation that well-defined clades are consistently associated with plants within a single family, that is, *Oreina sensu stricto* with Apiaceae spp., the clade *C. polita-C. graminis* with Lamiaceae (with the exception of *C. graminis* in England, see above), and the clade *C. brunsvicensis-C.varians* with Hypericaceae (only with *Hypericum*). This is in agreement with previous studies, based on the morphological classification of the beetles (Bourdonné and Doguet, 1991), as discussed above.

If we admit that the *Chrysolina-Oreina* spp are monophyletic, as suggested by current classifications and supported by previous mt DNA phylogenies (Hsiao, 1994a, b), the repeated occurrence of abrupt host switches during their evolution is demonstrated by the number of food- plants families of these beetles. To the families listed in table 3, two must be added: the Scrophulariaceae and the Asclepiadaceae, in which are classified the food plants of *Chrysolina* spp. not available for this study. Food plants of at least 10 species of the subgenus *Stichoptera* Motschulsky are within the Scrophulariaceae, another example of conservatism (Bourdonné and Doguet, 1991).*C. (Anopachis) asclepiadis* (Villa) feeds on the asclepiad *Vincetoxicum hirundinaria* (Bourdonné and Doguet, 1991; Pasteels, pers.obs.), although *Anopachis* spp. mainly feed on Asteraceae (including *C. aurichalcea* with which it was long confused), a clear example of an abrupt shift within this subgenus. In total, eight different families are reported among the food plants of the *Chrysolina-Oreina* spp. These eight families are classified in four out of the six different subclasses of dicots.

Our results are too preliminary for suggesting an evolutionary scenario of host shifts during speciation within the whole *Chrysolina-Oreina* complex, but some trends are apparent in the genus *Oreina sensu lato* for which more comprehensive data are available. If we neglect *C. fastuosa* feeding on plants within the Lamiaceae, all species in the *Oreina sensu lato* feed on Asteraceae and/or Apiaceae spp., *O. frigida* being the only species feeding on plants in both families. Different trees suggest conflicting hypotheses on the evolution of host-plant association. The strict concensus tree of Fig. 2 (12S+16S sequences), but not the neighbour-joining tree of Fig.1, suggests that the ancestral hosts of the *Oreina sensu lato* were Apiaceae. A shift to Asteraceae would have occurred at the origin of the *Allorina-Intricatorina* clade, and a secondary (but partial) return on Apiaceae in *O. elongata*. However in this phylogeny estimate, the position of the *Protorina-Intricatorina* clade remains uncertain, being supported by a low bootstrap value (34%), and its position is critical for supporting this evolutionary scenario.

Trees of Figs 3 and 4 (based on combined 12S, 16S and CO1 sequences) suggest that the original host plants of *Oreina sensu lato* were within the Asteraceae. We tend to favor this scenario supported by phylogeny estimates deduced from longer mtDNA sequences, showing identical topologies obtained by neighbour-joining and maximum-likelihood methods, and supported by higher bootstrap values, at least in the neighbour-joining tree. A switch on Apiaceae would have occurred at the origin of the subgenus *Oreina sensu stricto*. Ancestral feeding on Apiaceae would have required at least three evolutionary steps: a shift to Asteraceae in *O. intricata* and *O. melanocephala,* and a similar shift once or several times in the other Asteraceae-feeders, depending upon they represent a single lineage or not. Also, a secondary recolonization of Apiaceae plants by *O. frigida* should be postulated. The most-parsimonious hypothesis on the evolution of host-plant associations in the *Oreina sensu lato* clade is given in Fig.5. The ambiguity subsisting in the concensus tree of Fig. 4 does not make the scenario itself ambiguous.

The phylogeny estimates of Figs 3 and 4 suggest that feeding on Senecioneae spp. is ancestral in the *Oreina sensu lato* clade. Plants within the Senecioneae are hosts of all the Asteraceae-feeding *Oreina* investigated here, with the exception of the *Allorina* spp. only affiliated with Cardueae. The broader oligophagy of *O. frigida* and *O. speciosissima* is secondary. Ancestral feeding on Senecioneae is important to stress since among the food plants of the *Oreina*, only Senecioneae contain PAs available for sequestration.

Speciation sequence in the clade *Oreina sensu stricto*, feeding on Apiaceae, seems well resolved, and the broad host range of *O. speciosa* appears secondary. It would be hazardous to speculate further on the evolution of host shifts, for example, from plants in one subtribe to plants in another, or in narrowing or expanding host ranges, for three main reasons. First, the branching pattern of the species feeding on Asteraceae remains poorly resolved. Second, if our study covers the majority of the *Oreina* species living in the Alps,

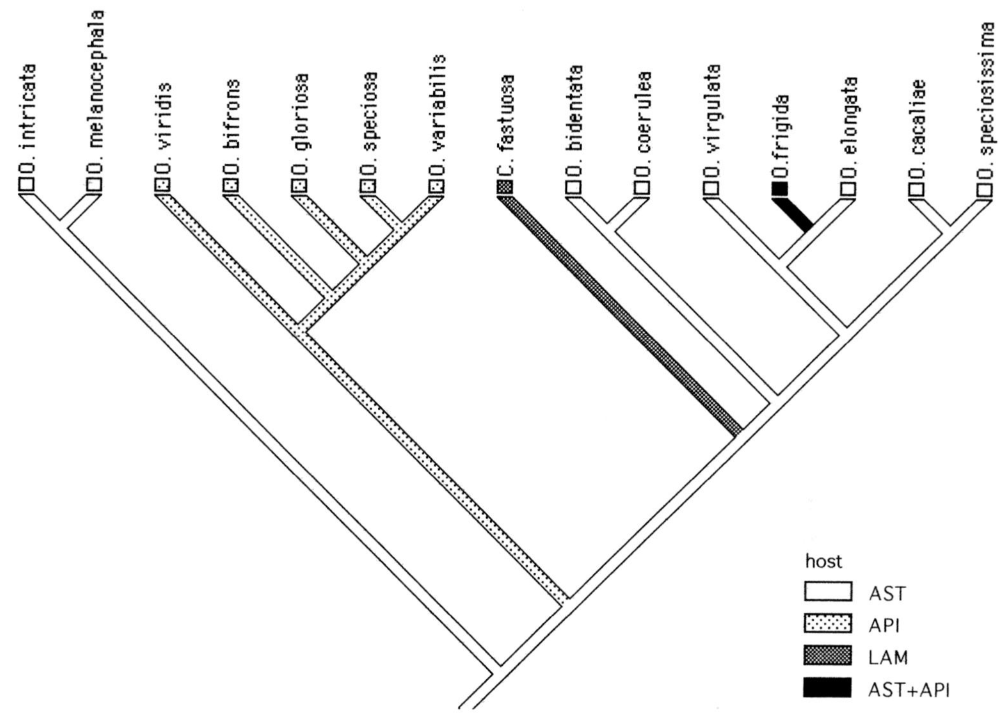

Fig. 5. Most-parsimonious reconstruction of ancestral host-plant associations based on the phylogenetic relationships shown in Fig.3. AST: Asteraceae, API: Apiaceae, LAM: Lamiaceae.

it includes only about half of the species in the genus (14 out of 26, Bourdonné, pers. com.). Finally, these mountain species are subdivided into numerous local populations, some recognized as subspecies (up to ten in *O. cacaliae,* Bourdonné, pers.com.), and our knowledge of the food plants is at best approximate for many of these local populations.

Chemical defense

The secretion of autogenous cardenolides appears to be ancestral in the *Chrysolina-Oreina* complex (Fig. 2). Not only does this chemical defense characterize most clades, but it is present in the outgroup, *Ambrostoma*. Moreover, *de novo* cardenolides are secreted by still other leaf beetles classified in the Chrysolinina sensu Daccordi (1994) (including the Doryphorina), but belonging to other lineages: for example,*Calligrapha*, *Zygogramma* and *Stilodes* species (Timmermans *et al.*, 1992; Pasteels and Daloze, unpublished results).

Cardenolides were lost in several clades demonstrating that defensive chemistry was rather labile in the course of evolution. Cardenolides are neither produced by beetles in the clade *C. varians* – *C. brunsvicensis* in which autogenous polyoxygenated steroids are secreted (Daloze et al., 1985; Randoux *et al.*, 1990), nor by members of the clade *C. lepida* – *C. aurichalcea*, nor by *C. peregrina*. These beetles secrete other unidentified compounds (Daloze and Pasteels, unpublished results). Since the phylogenetic relationships between these clades remains unresolved, it is impossible to decide how many times cardenolides

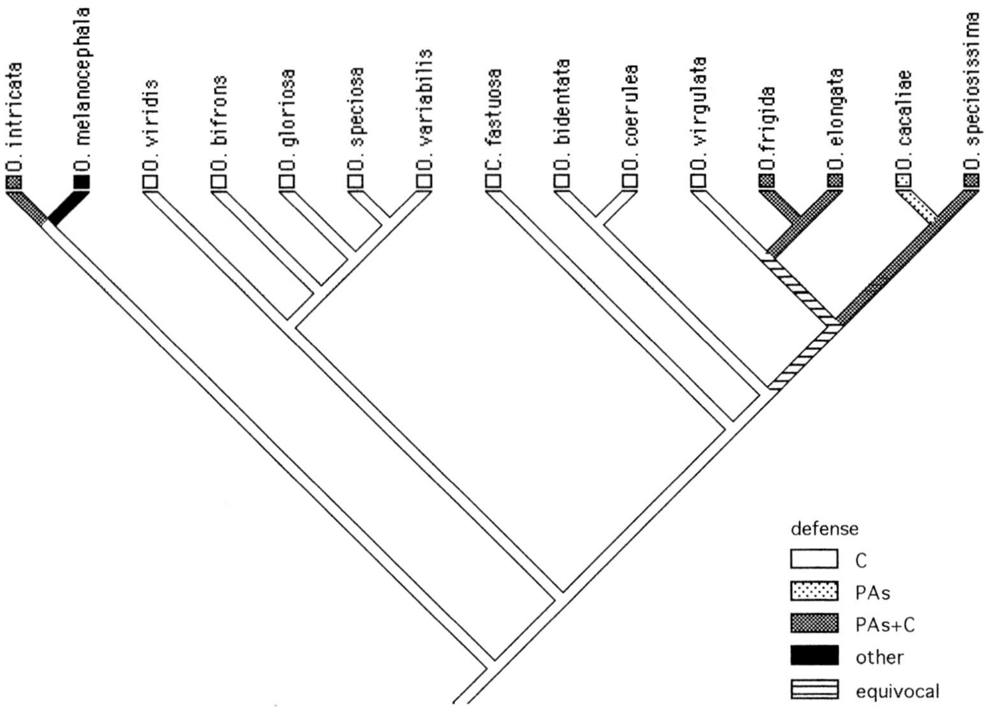

Fig. 6. Most-parsimonious reconstruction of ancestral chemical defense based on the phylogenetic relationships shown in Fig.3. C: cardenolides, PAs: pyrrolizidine alkaloids.

were lost, and even more to speculate on the causes of such changes in defensive chemistry.

All phylogenetic estimates suggest that the sequestration of pyrrolizidine alkaloids (PAs) by some *Oreina* species was a secondary event and first appeared as an addition to the production of autogenous cardenolides. Cardenolide secretion was lost in *O. cacaliae*, which so far is the only species known to only sequester PAs, but this was a more recent evolutionary event. These conclusions agree with those deduced from allozyme electrophoresis (Dobler *et al.*, 1996). However, since the branching pattern deduced from allozymes differs from that obtained here with mtDNA sequences, there is no agreement between the two studies on the number of times sequestration of PAs evolved and if this ability could have been lost secondarily.

In the present study, the most-parsimonious hypothesis is suggested by the tree based on the combined sequences of 12S, 16S and CO1 genes (Fig. 3) and illustrated in Fig.6. Sequestration would have evolved at least twice independently: once in *O. intricata* and at least once in the clade *O. virgulata – O. speciosissima*. This most-parsimonious reconstuction of the ancestral states holds even if *C. fastuosa* is excluded from the analysis. Only if *O. intricata* branches at the basis of the *Oreina sensu lato* clade without being grouped with *O. melanocephala* (a possibility in Fig.4), the ancestral defense of*Oreina* (cardenolides alone or PAs in admixture with cardenolides) remains equivocal. In any case, however, the loss of the ability to sequester PAs and one (or two) secondary evolutions of this ability must be postulated. Independent evolutions of PAs sequestration is thus supported by all acceptable phylogenetic hypotheses.

PAs were not present in the secretions of our sample of *O. virgulata*, which was found feeding on *Cirsium*. Sequestration of PAs could have been secondarily lost in this species after it shifted on Cardueae or on Senecioneae devoid of PAs (see table 3), but its position in the tree needs to be more firmly established. Moreover, it is reported that individuals of this species were occasionally observed feeding on *Adenostyles*, which contains PAs (Kippenberg, 1994). Secretions of *O. virgulata* beetles feeding on *Adenostyles* should be checked before concluding that this species lost the ability to sequester PAs. Also, a better resolution of the sister clade of the *Oreina sensu stricto* clade is needed before deciding if sequestration ability evolved more than twice.

Conclusions and perspectives

Host-plant association

Conservatism in host-plant association during the evolution of the *Chrysolina-Oreina* complex is exemplified by many clades being consistently associated with a single plant family and sometimes even with a single genus (for example, the *Hypericia-Sphaerolema* clade with *Hypericum*). This conservatism was already noticed by colleagues who systematically compared beetles' taxonomy and host-plant associations in this group of species (for example, Jolivet & Petitpierre, 1976; Bourdonné & Doguet, 1991). Host-plant affiliation is so constant in some taxa that it offers an additional character to the more traditional morphological ones in the diagnosis of these taxa. Evolution being gradual, conservatism in host-plant association is not surprising. More surprising and more interesting in our opinion, is that many exceptions are observed to this trend, as exemplified by the broad range of plant taxa used as food by the whole *Oreina-Chrysolina* complex of species (8 plant families classified in 4 out of the 6 subclasses of dicots). All beetles, however, are specialized in their feeding, although to variable degrees. The challenge is to trace the evolutionary history of these host shifts, to understand their ultimate causes and to disclose the underlying mechanisms.

Our analyses are too preliminary to offer any scenario for the evolution of host shifts in the whole *Oreina-Chrysolina* complex, but one evolutionary hypothesis is favoured for the *Oreina sensu lato* clade. The original host plant(s) would have been Asteraceae belonging to the Senecioneae, shifts to other Asteraceae subfamilies occurred either as a new specialization (for example, *Allorina* species feeding exclusively on Cardueae) or as a widening of host range (for example, in *O. speciosissima* and *O. elongata*, see Table 3). Shift to the Apiaceae occured as a new specialization in *Oreina sensu stricto* and as a widening of its range of food plants, including both Senecioneae and Apiaceae, in *O. frigida* .

It is of course difficult to answer the how and why questions just on the basis of phylogenetic estimates, but some trends may point to possible directions that deserve further investigation. First, beetles do not shift to any plant available in their environment. Moreover, independent shifts to the same plant taxon occurred, as already observed in the *Gonioctena* leaf beetles (Mardulyn *et al.*, 1997). Some shifts are more likely than others and constraints are present. As expected, moves to plants within the same family are more frequent than moves to plants in a different family, indicating that some « similarity » in plants (most probably chemical) facilitates shifts. We cannot rule out that "cryptic" chemical similarities facilitated what appears to be more abrupt shifts. If plant chemistry is an obvious constraint stressed many times (see Futuyma & Keese, 1992, for a review), other constraints are present. For example, no *Chrysolina-Oreina* species, to our knowledge, were observed feeding on shrubs or trees. Plant stratum is not necessarily a barrier for host shifts, since other beetle taxa feed on herbs, shrubs or trees (but usually first on plants within a single family, such as *Phratora* or *Chrysomela* species on Salicaceae, some species in both genera moving later to Betulaceae; Brown, 1956; Köpf *et al.*, 1998). It is tempting to speculate that plant chemistry, as well as plant availability or plant habitus, constrain host shifts and that one predominant similarity in a local environment could offer an opportunity for invading a new host, although quite dissimilar in other respects to the original food plant. This new host would then offer a "bridge" for further radiation on a different set of plants.

Second, our phylogeny estimates suggest that a wider host range often evolved in recent branches, from narrow, ancestral, specialization. This is exemplified by beetles which broaden their diet within the same plant family, although on plant taxa in different subfamilies (for example, in *O. speciosa* among the Apiaceae-feeders or in *O. speciosissima* and *O. elongata* among the Asteraceae-feeders) or even in different plant families (for example, Asteraceae and Apiaceae in *O. frigida*). Other examples are found in the *Chrysolina sensu stricto* (such as *C. banksi* and *C. graminis*), but this needs more detailed phylogenetic analysis including additional species. We would like to speculate that this broadening of host range, mainly observed in recent branches of phylogenetic trees, illustrates a transient step in shifts from one specialization to another. Pressure due to local intraspecific competition (any leaf beetle collector has observed this in nature) could be sufficient to force the beetles to move onto non-hosts and to enlarge their diet. New narrow specialization could then evolve in beetles isolated in habitats with more restricted food conditions (allopatric evolution). Alternatively, feeding on one specific plant could limit gene flow between beetles developing and mating on different hosts (sympatric evolution), with all possible intermediates between the two possibilities. It must be stressed, however, that the host plants listed in Table 3 lump together hosts observed in different beetle populations, sometimes considered as subspecies. A better knowledge of local host-availablity and of the whole spectrum of plants accepted as food, even if not available, is needed for beetles collected in different localities to get more insight into the evolution of host shifts. If natural enemies are omitted in this speculation, it is not because their role is considered as irrelevant, the less so that defense can be derived from host plants. It is simply because natural enemies and the possible pressure they could exert remain practically unknown.

Chemical defense

Sequestration of PAs is secondary to the original production of *de novo* cardenolides. It occured in beetles that moved on Senecioneae containing PAs and started by adding PAs to the autogenous cardenolides in the defensive secretions. This sequestration ability evolved (or was reacquired) at least twice independently in *O. intricata* and in *Chrysochloa* species.

Several questions remain unresolved. First, did the ancestral species of the *Oreina sensu lato* already sequester PAs or did sequestration first evolve in *O. intricata*, which colonized Senecioneae containing PAs from Senecioneae devoid of them? A better resolution of the basal branches of the *Oreina sensu lato* clade by incorporating additional *Protorina* species in the analysis could help to answer this question. To our knowledge all *Protorina* feed on *Doronicum* (Senecioneae). Although some *Doronicum* contain PAs (Hartmann & Witte, 1995), these were not listed as food-plant of *Protorina*. Also an analysis of the mechanism by which PAs are sequestered and how this mechanism could have evolved, would help to gain insight into this question.

Second, how many times did this sequestration ability evolve? A better resolution of the branching pattern of the Asteraceae-feeding *Oreina* could answer this question, by including additional *Allorina* and above all *Chrysochloa* species in the analysis.

Third, did some *Oreina* lose the ability to sequester PAs? Obviously, beetles which feed on plants without PAs cannot sequester, but this does not mean that they cannot do it, if the opportunity is present. *O. bifrons* , an Apiaceae-feeder, is unable to sequester PAs even if PAs are ingested by the beetle or injected in it. In similar experiments, *O. coerulea* feeding on Cardueae, proved unable to sequester injected PAs in its secretion, but PA excretion appeared much slower than when injected in *O. bifrons* (Ehmke *et al.*, 1991 and unpublished results). Sequestration potentiality should be investigated in other Asteraceae-feeders, especially in *O. virgulata*. Moreover, answers to the first two questions are needed to fully answer the third one.

The secondary loss of cardenolide production in *O. cacaliae* which exclusively sequesters PAs is firmly established. This species feeds temporarily on *Petasites* in early spring (Rowell-Rahier *et al.* 1991). Although *Petasites* is devoid of PAs in its leaves, chemical defense is still present thanks to the long term accumulation of PAs in the hemolymph, allowing the beetles to still secrete PAs when an external source of PAs is temporarily lacking (Pasteels *et al.*, 1992).

Chrysolina-Oreina systematics

It is of course premature to draw firm conclusions from phylogenic estimates, in which only a limited number of possible in-group species was included. However, even if preliminary, molecular phylogenetic estimates are helpful in trying to resolve conflicting attitudes in traditional taxonomy.

Oreina sensu lato forms a well-established clade and must be considered as a natural taxon. In the absence of any objective definition of the "genus" category, it is a matter of pragmatism to classify it as a genus or as a subcategory of a larger genus. Taking into account the number of species involved, we feel more convenient and in any case more informative to consider *Oreina senso lato* as a genus. In this, we agree with Kippenberg (1994), for example, but not with Daccordi (1994). Since it seems to be contradicted by strong mophological evidence (Bourdonné, pers. com.), the inclusion of *C. fastuosa* within the genus *Oreina*, strongly suggested by our results, must await further study.

Within the genus *Oreina*, the subgenera *Oreina sensu stricto*, *Allorina* and most probably *Protorina* (for which a single species was available) seem well defined. If *O. intricata* is often grouped with *O. melanocephala* in our trees, it does not mean that both species are close enough to be classified together. Indeed *O. intricata* is morphologically quite different from *Protorina*. However, *O. intricata* and the other Asteraceae-feeders are not grouped together and the subgenus *Intricatorina* suggested by Kühnelt (1984) seems justified. Other subgeneric divisions look more unsatisfactory. The subgenus *Chrysochloa* appears paraphyletic. The relationship between the *Chrysochloa*, *Frigidorina* and *Virgulatorina* species remains unresolved, and the subgeneric classification of these species needs additional studies.

Our results suggest that some *Chrysolina* subgenera can be lumped together, still forming natural taxa, without deciding at this stage if these taxa should be considered as genera or subgenera. More specifically, the subgenera *Hypericia* and *Sphaeromela* can be united in a single taxon. Similarly, *Synerga, Menthastriella* and *Euchrysolina* could be united, as well as *Erythrochrysa* and *Melasomoptera*. These last five subgenera could even been grouped together, but this is less firmly established. Also, there are some indications that *Allochrysolina* and *Anopachis* could be grouped together, as well as *Colaphodes* and *Taeniochrysea*, but this definitively needs to be substantiated by studying additional species.

Acknowledgments

This work was supported by grants of the Belgian Fund for Joint Basic Research and the Communauté Française de Belgique (ARC 93-3318) to JMP.The CO1 primers were given to us by Dr Francesco Frati of the University of Siena, Italy. Several colleagues sent specimens to us or helped us to collect them: J.-C. Bourdonné, B. Haegele, S. Knoll, E. Petitpierre, M. Rahier and P. Verdijck. Valuable information on the systematics of *Oreina-Chrysolina* leaf beetles were provided by J.-C. Bourdonné and M. Daccordi, and on host plants, by J.-C. Bourdonné and M. Cox. M. Cox, T. Hartmann, P. Mardulyn, M. Rahier and A. Termonia made useful comments on earlier drafts of the manuscript.

References

Becerra, J.X., 1997. Insects on plants: macroevolutionary chemical trends in host use. Science 276: 253-256.

Bernays, E.A. and Chapman, R.F. 1994. Host-plant selection by phytophagous insects. Chapman & Hall, New York. 312 pp.

Bourdonné, J.-C. and Doguet, S. 1991. Données sur la biosystématique des *Chrysolina* L.S. (Coleoptera: Chrysomelidae: Chrysomelinae). Annls Soc. ent. Fr. (N.S.) 27: 29-64.

Brown, W. J. 1956. The New World species of *Chrysomela* L. (Coleoptera: Chrysomelidae). Can. Ent. 88: 225-229.

Brummit, R.K., 1992. Vascular Plant Families and Genera. Royal Botanic Gardens, Kew. 804 pp.

Burakowski, B.M., Mroczkowski, M. and Stefanska, J. 1990. Coleoptera-Chrysomelidae 1. Katalog Faunae Polski.Institute of Zoology of the Polish Academy of Science, Warsaw. 23 (16), 279 pp.

Cronquist, A. 1981. An Integrated System of Classification of Flowering Plants. Columbia University Press, New York. 1962 pp.

Daccordi, M., 1994. Notes for phylogenetic study of Chrysomelinae, with descriptions of new taxa and a list of all the known genera (Coleoptera: Chrysomelidae, Chrysomelinae). In: Furth, D. G. (ed.) Proc. 3th Symp. Chrysomelidae, Beijing, 1992. Backhuys Publishers, Leiden, pp.60-84.

Daloze, D., Braekman, J.-C. and Pasteels, J.M. 1985. New polyoxygenated steroïdal glucosides from *Chrysolina hyperici* (Coleoptera: Chrysomelidae). Tet. Lett. 26: 2311-2314.

Dobler, S., Mardulyn, P., Pasteels, J.M. and Rowell-Rahier, M. 1996. Host-plant switches and the evolution of chemical defense and life history in the leaf beetle genus *Oreina*. Evolution 50: 2373-2386.

Ehmke, A., Rowell-Rahier, M., Pasteels, J.M. and Hartmann, T. 1991. Sequestration of ingested [^{14}C]senecionine

N-oxide in the exocrine defensive secretions of chrysomelid beetles. J. Chem. Ecol.17: 2367-2379.
Engler, A. 1964. Syllabus der Planzenfamilien. Gebrüder Borntraeger, Berlin. 666 pp.
Farrell, B.D. and Mitter, C. 1990. Phylogenesis of insect/plant interactions: have *Phyllobrotica* leaf beetles (Chrysomelidae) and the Lamiales diversified in parallel ? Evolution 44: 1389-1403.
Funk, D.J., Futuyma, D.J., Orti, G. and Meyer, A. 1995. A history of host associations and evolutionary diversification for *Ophraella* (Coleoptera: Chrysomelidae). Evolution 44: 1885-1905.
Futuyama, D.J. and Keese, M.C. 1992.Evolution and coevolution of plants and phytophagous arthropods. In: Rosenthal, G.A. and Berenbaum, M.R.(eds). Herbivores. Their Interactions with Secondary Pant Metabolites. Academic Press, San Diego, pp.439-475.
Hartmann, T. and Witte, L. 1995. Chemistry, biology and chemoecology of the pyrrolizidine alkaloids. In: Pelletier, S. W. (ed). Alkaloids: chemical and biological perspectives. Vol. 9. Pergamon Press, Trowbridge, pp. 155-233.
Hartmann, T., Witte, L., Ehmke, A., Theuring, C., Rowell-Rahier, M. and Pasteels, J.M. 1997. Selective sequestration and metabolism of plant derived pyrrolizidine alkaloids by chrysomelid leaf beetles. Phytochemistry 45: 489-497.
Hasegawa, M., Kishino, H. and Yano, T. 1985. Dating of the human-ape splitting by molecular clock of mitochondrial DNA. J. Mol. Evol. 21: 160-174.
Higgins, D.G. , Bleasby, A.J. and Fuchs, R. 1992. CLUSTALV: improved software for multiple sequence alignment. CABIOS 8: 189-191.
Hsiao, T.H. 1994a. Molecular techniques for studying systematics and phylogeny of Chrysomelidae. In: Jolivet, P.H., Cox, M.L. and Petitpierre (eds). Novel Aspects of the Biology of Chrysomelidae, Kluwer Academic Publishers, Dordrecht, pp.237-248.
Hsiao, T.H. 1994b. Molecular phylogeny of chrysomelid beetles inferred from mitochondrial DNA sequence data (Coleoptera: Chrysomelidae). In: Furth, D.G. (ed.) Proc. 3th Int. Symp. Chrysomelidae, Beijing, 1992. Backhuys Publishers, Leiden, pp.9-17.
Jolivet, P. and Petitpierre, E. 1976. Les plantes hôtes connues des *Chrysolina* (Col. Chrysomelidae). Essais sur les types de sélections trophiques. Annls Soc. ent. Fr. (N.S.) 12: 123-149.
Kippenberg, H. 1994. Familie: Chrysomelidae. In: Lohse, G.A. and Lucht, W.H. (eds.) Die Käfer Mitteleuropas. 3. Supplementband. Goecke & Evers, Krefeld. pp.17-92.
Köpf, A., Rank, N.E., Roininen, H., Julkunen-Tütto, R., Pasteels, J.M. and Tahvanainen, J. 1998. The evolution of host-plant use and sequestration in the leaf beetle genus *Phratora* (Coleoptera: Chrysomelidae). Evolution 52: 517-528.
Kühnelt, W. 1984. Monographie der Blattkäfergattung *Chrysochloa* (Coleoptera, Chrysomelidae). Anz.öst. Akad. Wiss. Mathematische- Naturwissenschaftliche. Kl.1984: 171-287.
Kumar, S., K. Tamura and Nei, M. 1993. MEGA: Molecular Evolutionary Genetic Analysis. Version 1.0. Pennsylvania State University, University Park, PA.
Maddison, W. P. and Maddison, D. R. 1992. MacClade: Analysis of phylogeny and character evolution, version 3.0. Sinauer, Sunderland, Massachusetts.
Mardulyn, P., Milinkovitch, M. and Pasteels, J.M. 1997. Phylogenetic analyses of DNA and allozyme data suggest that *Gonioctena* leaf beetles (Coleoptera; Chrysomelidae) experienced convergent evolution in their history of host-plant family shifts. Syst. Biol. 46: 722-747.
Pasteels, J.M 1993 The value of defensive compounds as taxonomic characters in the classification of leaf beetles. Biochem. Syst. Ecol. 21: 135-142.
Pasteels, J.M. and Rowell-Rahier, M. 1991. Proximate and ultimate causes for host plant influence on chemical defense of leaf beetles (Coleoptera: Chrysomelidae). Entomologia gen. 15: 227-237.
Pasteels, J.M., Braekman, J.-C. and Daloze, D. 1988. Chemical defense in the Chrysomelidae. In: Jolivet, P., Petitpierre, E. and Hsiao, T.H. (eds). The Biology of Chrysomelidae. KLuwer Academic Publisher, Dordrecht, The Netherlands, pp.233-252.
Pasteels, J.M., Rowell-Rahier, M., Braekman, J.-C. and Daloze, D. 1994. Chemical defense of adult leaf beeltes updated. In: Jolivet, P.H., Cox, M.L., Petitpierre, E. (eds). Novel aspects of the biology of Chrysomelidae. Kluwer Academic Publisher, Dordrecht, The Netherlands, pp. 289-301.
Pasteels, J.M., Rowell-Rahier, M., Ehmke, A. and Hartmann, T. 1996. Host-derived pyrrolizidine alkaloids in *Oreina* leaf beetles: physiological, ecological and evolutionary aspects. In: Jolivet, P.H.A. and Cox, M.L. (eds). Chrysomelidae Biology, vol. 2: Ecological Studies. SPB Academic Publishing bv, Amsterdam, pp.213-225.
Pasteels, J.M., Eggenberger, F., Rowell-Rahier, M., Ehmke, A. and Hartmann, T. 1992. Chemical defense in chrysomelid leaf beetles. Storage of host-derived pyrrolizidine alkaloids versus de novo synthesized cardenolides. Naturwissenschaften 79: 521-523.
Pasteels, J.M., Dobler, S., Rowell-Rahier, M., Ehmke, A. and Hartmann, T. 1995. Distribution of autogenous and host-derived chemical defenses in *Oreina* leaf beetles (Coleoptera, Chrysomelidae). J. Chem. Ecol. 21: 1163-1179.
Randoux, T., Braekman, J.-C., Daloze, D., Pasteels, J.M. and Riccio, R. 1990. New polyoxygenated steroid glycosides from the defense glands of several species of Chrysolinina beetles (Coleoptera: Chrysomelidae). Tetrahedron 46: 3871-3888.

Rowell-Rahier, M., Witte, L., Ehmke, A., Hartmann, T. and Pasteels, J.M. 1991. Sequestration of plant pyrrolizidine alkaloids by chrysomelid beetles and selective transfer into the defensive secretions. Chemoecology 2: 41-48.
Simon, C., Francke, A. and Martin, A. 1991. The polymerase chain reaction: DNA extraction and amplification. In: Hewitt, G.M., Johnston, A.W.B. and Young, J.P.W. (eds). Molecular Techniques in Taxonomy. NATO Advanced Studies Institutes. Springer Verlag, Berlin. H57, pp. 329-355.
Simon, C., Frati, F., Flook, P., Beckenbach, A., Crespi, B. and Liu, H. 1994. Assessing the phylogenetic usefulness of specific mitochondrial genes and a compilation of conserved PCR primers. Ann. ent. Soc. Am. 87:1-57.
Swofford, D.L. 1993. PAUP: Phylogenetic analysis using parsimony, version 3.1.1. Illinois Natural History Survey, Champaign.
Swofford, D.L. in press. PAUP* (version d 6 1). Washington D.C. Laboratory of Molecular Systematcis, Smithsonian Institution, Sinauer Associates, Sunderland, MA.
Timmermans, M., Randoux, Th., Daloze, D.,Braekman, J.-C., Pasteels, J. M. and LeSage, L. 1992. The chemical defense of Doryphorina beetles (Coleoptera: Chrysomelidae). Bioch. Syst. Ecol. 20: 343-349.
Xiong, B. and Kocher, T.D. (1991) Comparison of mitochondrial DNA sequences of seven morphospecies of black flies (Diptera: Simuliidae). Genome 34, 306-311.
Zhu, H., Feng, Q. and Zhu, L-H. 1993. Isolation of genomic DNAs from plants, fungi and bacteria using benzyl chloride. Nucleic Acids Research 22 :5279-5280.

HOST USE BY CHRYSOMELID BEETLES FEEDING ON MORACEAE AND EUPHORBIACEAE IN NEW GUINEA

VOJTECH NOVOTNY[1], YVES BASSET[2], G.A. SAMUELSON[3] and SCOTT E. MILLER[4]
[1]*Institute of Entomology, Czech Academy of Sciences and Biological Faculty, University of South Bohemia, Branisovska 31, 370 05 Ceske Budejovice, Czech Republic, novotny@entu.cas.cz;* [2]*Smithsonian Tropical Research Institute, Apartado 2072, Balboa, Ancon, Panama, bassety@tivoli.si.edu;* [3]*Bishop Museum, Dept of Natural Sciences, 1525 Bernice Street, Honolulu 96817-0916, U.S.A., alsam@bishop.bishop.hawaii.org;* [4]*Biodiversity and Conservation Programme, International Centre of Insect Physiology and Ecology, Box 30772, Nairobi, Kenya, scottm@bishop.bishop.hawaii.org*

Abstract

The composition and host specificity of chrysomelid communities, feeding on 15 species of Moraceae and 15 of Euphorbiaceae in New Guinea is detailed. All 30 plant species were trees or shrubs, growing in rainforest and coastal habitats. Chrysomelids were collected alive and tested in the laboratory for feeding on the plant species from which they were obtained. Only specimens which fed were considered in the analyses. Sampling spanned almost over three years and 9,186 individuals from 93 species, belonging to 5 subfamilies were collected. Eumolpinae and Galerucinae dominated in terms of abundance and species richness. The foliage-feeding community consisted almost exclusively of adults, as eumolpine and most of the galerucine larvae were probably root-feeders. Species richness of chrysomelid communities varied from 1 to 27 species per host plant species, and abundance from 8 to 1,462 individuals. Trees without diverse chrysomelid communities were fed upon mostly by generalists, whilst high density chrysomelid communities were dominated by specialists. Usually, Galerucinae were more host specific than Eumolpinae, but the overall host specificity of chrysomelids was rather low. In particular, there was a large overlap in the chrysomelid communities among congeneric plants. Most of the chrysomelid species fed on several congeneric hosts and were not strict monophages. Within *Ficus*, where such evaluation was possible, the similarity of chrysomelid communities did not reflect phylogenetic relationships among the host plants. Considered across plant genera, specialisation to a single genus was the most frequent case. Plant palatability, estimated from feeding experiments using a generalist species of weevil, was negatively correlated with the average host specificity in the chrysomelid community supported by this plant species; polyphagous chrysomelid species tended to be concentrated on palatable hosts. In conclusion, the patterns of host use by Eumolpinae and Galerucinae appear to be shifted towards oligophagy, in comparison with most of other subfamilies, feeding on Temperate herbs. Low host specificity of adults feeding on the foliage may be caused by the predominance of species with root-feeding larvae, and may explain the importance of plant palatability for the composition of chrysomelid communities.

1. Introduction

Many species of Chrysomelidae are distributed in the tropics and represent conspicuous elements of the fauna of tropical rain forests. Undoubtedly, one of the most significant issues when discussing the host specificity of insect herbivores such as the Chrysomelidae is the extent of specialisation of the numerous species that thrive in tropical rain forest and, particularly, that of species feeding in the canopy (Farrell and Erwin, 1988; Jolivet and Hawkeswood, 1995).

Advances in Chysomelidae Biology 1, pp. 343–360
Edited by M.L. Cox

Perhaps with the notable exceptions of *Eucalyptus*- and *Salix*-feeding chrysomelids (for example, Ohmart, 1996; Rank *et al.*, 1996), the host ranges of species feeding on trees appear to be less well known than those feeding on herbs, particularly in the tropics. In part, this may have resulted from the relative ease of collection and observation of chrysomelids feeding on herbs, in contrast with the same of live specimens from trees. In the tropics, few studies of host specificity of tree-feeding chrysomelids have been conducted, particularly at the community level. For example, Basset and Samuelson (1996) presented some data on the host specificity of species found on ten distantly-related tree species in Papua New Guinea. Flowers and Janzen (1997) listed feeding records for Costa Rican chrysomelids, with many species feeding on trees.

Earlier studies of insect-plant interactions in the Temperate zone showed that, in general, trees support more species of insect herbivores than herbs, due to their higher architectural complexity (for instance, Lawton and Schröder, 1977). Although this observation has not been tested for Chrysomelidae in particular, it is probable that tropical trees may often harbour more species of chrysomelids than tropical herbs.

The host specificity of tropical species that belong to subfamilies such as the Hispinae, Cassidinae, Chrysomelinae, or Sagrinae is reasonably well-known and often appears to be rather narrow (for example, Gressitt, 1959; Jolivet and Petitpierre, 1976; Strong, 1977; Jolivet, 1989; Jolivet and Hawkeswood, 1995). For instance, Hispinae appear to be often associated with Monocotyledons (for instance, palms, *Heliconia*, etc.), Cassidinae with Convolvulaceae, while Chrysomelinae feed collectively on many plant families, but each species is rather specialised (Jolivet and Hawkeswood, 1995). However, the dominant chrysomelid groups in the canopy of tropical rain forests are often the Eumolpinae, Galerucinae and Alticinae (Farrell and Erwin, 1988; Basset and Samuelson, 1996). The host specificity of these groups in particular is poorly known; the host ranges of especially the first two subfamilies appear also to be more varied, involving both specialists and generalist of various degree (Jolivet and Hawkeswood, 1995). New data on the host specificity of these groups, and their communities feeding on tropical trees in particular, is necessary for analysis of the host specificity of Chrysomelidae.

Any study of host specificity greatly depends on the scale of the analysis, particularly on the identity and phylogenetic relatedness of the host plants being considered. The study of chrysomelids from numerous, closely related, co-occurring species of trees in rain forests could be informative, particularly with reference to problems such as the patterns of species packing of insect herbivores, and the maintenance of their diversity on the local scale. In Temperate ecosystems, studies on many related, for instance congeneric, plants are usually possible only using herbaceous species. Yet, in the tropics, they have been rarely attempted (but see Marquis, 1991, for such data on weevils). For instance, the above cited studies on chrysomelids in Papua New Guinea (Basset and Samuelson 1996) and Costa Rica (Flowers and Janzen 1997) included a selection of phylogenetically rather distant hosts. Tropical studies on closely related plants are indispensable since we want to know whether trees in the tropics support a rather specialised chrysomelid fauna, perhaps depending on the chemical characteristics of the host, as is known for the chrysomelids feeding on certain herbs in the Temperate zone (for instance, Tahvanainen, 1983).

Given the number of both chrysomelid and plant species in the tropical rainforest, dissecting pair-wise interactions one by one is not a feasible proposition. Instead, broader-based studies, searching for general patterns in herbivore communities on a wide array of host plants may represent a more promising approach to the study of tropical communities. Numerous plant traits, potentially important for chrysomelid host choice, exist. The challenge is to identify those which are important for a wide range of chrysomelid species on a variety of plants, thus having a pervasive influence on the structure of the whole chrysomelid community on the diverse tropical vegetation. It is argued here that plant palatability for generalist herbivores might be one such generally-important trait. Plant palatability is a composite result of the plant's qualities and its herbivores' preferences, so it may differ from one plant-herbivore interaction to another. However, there is an interesting possibility that while specialists are, by definition, idiosyncratic, generalists may respond similarly to certain plant qualities, making some plants generally palatable or impalatable to them (Cates and Orians, 1975; Grime, 1979; Price, 1991; Herms and Mattson, 1992). If this were found to be so, the palatability for generalists might become a good predictor of at least the generalist part of herbivorous communities.

The present contribution details information about the composition and host specificity of chrysomelid communities, feeding on 30 tree and shrub species from the Moraceae and Euphorbiaceae in the Madang area of Papua New Guinea. It concentrates on the host specificity of chrysomelids, and the importance of plant palatability and their phylogeny (or taxonomic classification) on the structure of chrysomelid communities. This study of a fraction of plants potentially available to chrysomelids in a single rain forest ecosystem is a small step towards understanding the host specificity of chrysomelids feeding on tropical trees and it is hoped that it will stimulate similar studies at other tropical locations, and in different types of rain forests.

2. Methods

2.1. Study area and insect sampling

The study area was situated in the Madang province of Papua New Guinea and was delimited as part of the lowlands (0-400m asl), stretching from the coast to the slopes of the Adelbert Mts, which lie between the Gogol and Sempi Rivers. This approximate rectangular area of 17 x 31 km encompasses about 434 km^2 of primary and secondary forests, 21 km^2 of coastal habitats (including seashore and ruderal vegetation in coconut plantations) and the town of Madang. Field work was concentrated in primary and secondary lowland forests near Baitabag, Ohu and Mis Villages, and in a coastal area near Riwo village (145°41-8' E, 5°08-14' S). The average annual rainfall in the Madang area is 3,558 mm, with a moderate dry season from July to September, mean air temperature is 26.5 °C which varies little throughout the year (McAlpine *et al.*, 1983).

Chrysomelid beetles were collected from the foliage of trees individually, by hand. Taller trees were often climbed into, or only lower, accessible branches were examined. Insects were sampled by five collecting teams usually including three people per team. Insect collecting continued from July 1994 to March 1996 on 15 species of *Ficus* and from August 1996 to August 1997 on 15 species of Euphorbiaceae.

Sampling effort was recorded as the time spent searching the foliage of each of the plant species for insects, which was approximately proportional to the leaf area examined. The number of tree inspections, that is, a particular tree sampled at a particular time, was also recorded. Overall, the sampling programme, which surveyed individual trees in a variety of age classes and growing in various habitats, was optimised towards the estimation of the total number of chrysomelid species feeding on each of the tree species studied. For species of *Ficus*, collecting effort varied from 22.9 to 25.6 hours (average 24.7 hours per species), while for all Euphorbiaceae it amounted to exactly 24.7 hours per species. The total number of tree inspections was 25,152. This sampling effort represented approximately 950 days of field work.

In the laboratory, each chrysomelid was provided with fresh, as far as possible young foliage of the plant species it was collected from and kept on it until it fed or died. Only the specimens which fed were considered in the analyses, to exclude transient species from the samples. All insects were assigned to morphospecies and identified as far as possible. Voucher specimens are deposited in Bishop Museum, Honolulu.

2.2. Host plants

Thirty locally abundant species of trees and shrubs, involving 15 species of *Ficus* (Moraceae), 6 species of *Macaranga* and 9 species representing 9 other genera of Euphorbiaceae, were selected for the study of their chrysomelid fauna (Table 1). Moraceae and Euphorbiaceae, with 3,000 and 5,000 species worldwide respectively (Heywood, 1993), were selected for the study since they represent important components of tropical floras (for instance Gentry, 1990). They are also abundant and species-rich in the Madang area, as well as in other lowland rain forests in New Guinea (for instance, Oatham and Beehler, 1998).

Ficus is an exceptionally large, pantropical genus, with over 700 species (Berg, 1989). New Guinea is one of the main centres of its diversity with 135 described species (Corner, 1965). In the lowlands around Madang, there is at least 48 species of *Ficus* (G. Weiblen, pers. comm.), from which 15 were selected for the present study. This selection included one monoecious strangling fig (*F. microcarpa*) and 14 gynodioecious free standing trees and shrubs. A cladistic analysis, using molecular and morphological characters (G. Weiblen, pers. comm.), did not confirm the monophyly of all Corner's (1965) species series. The results of this analysis were used to classify the present species to monophyletic groups (Table 1A).

Both in PNG (Höft, 1992) and in the Madang area (pers. obs.), Moraceae other than *Ficus* are of lesser importance in terms of species richness and biomass (with the exception of *Artocarpus*). Thus, the following data on insects feeding on *Ficus* may be used to study the local distribution of leaf-chewing insects on the family Moraceae.

There are 461 species of Euphorbiaceae reported from New Guinea, 73% of them being endemic (van Welzen, 1997). *Macaranga*, with over 280 species distributed from West Africa to the islands of the western Pacific, is by far the largest genus of early successional (pioneer) trees in the world (Whitmore, 1979). New Guinea is the main centre of diversity for this genus, with 82 species described (van Welzen, 1997) and some more awaiting description. [It should be noted that the discovery of 2 new species (Whitmore, pers. comm.) resulted from the present study alone]. The six species of *Macaranga* studied belong to the three largest species groups, of the nine present in New Guinea (Whitmore, 1980).

The other nine species of Euphorbiaceae, one per genus, included representatives of

Table 1. Host plants studied.

A) *Ficus* spp.

Code	Species	Phy.	Pal.	Habitat
BE	*Ficus bernaysii* King	2.2	0.10	primary forest
BO	*Ficus botryocarpa* Miq.	2.2	0.09	secondary forest
CO	*Ficus conocephalifolia* Ridley	2.1	0.10	primary forest
CP	*Ficus copiosa* Steud.	2.1	0.24	secondary forest
DA	*Ficus dammaropsis* Diels	2.2	0.52	secondary forest
HI	*Ficus hispidioides* S Moore	2.2	0.05	secondary forest
MI	*Ficus microcarpa* L. f.	1	0.00	sea coast
NO	*Ficus nodosa* Tejsm. & Binn.	2.2	0.24	secondary forest
PH	*Ficus phaeosyce* Lauterb. & K. Schum.	2.1	0.19	primary forest
PU	*Ficus pungens* Reinw. ex Bl.	2.2	0.13	secondary forest
SE	*Ficus septica* Burm. f.	2.2	0.00	secondary forest
TI	*Ficus tinctoria* G. Forst.	2.1	0.09	sea coast
TR	*Ficus trachypison* K. Schum.	2.1	0.13	secondary forest
VA	*Ficus variegata* Bl.	2.2	0.44	secondary forest
WA	*Ficus wassa* Roxb.	2.1	0.25	secondary forest

B) Euphorbiaceae spp.

Code	Species	Tax.	Pal.	Habitat
MA	*Macaranga aleuritoides* F. Muell.	1.1.a	0.69	secondary forest
MB	*Macaranga brachytricha* Airy Shaw	1.1.b	0.63	secondary forest
MD	*Macaranga densiflora* Warb.	1.1.c	0.70	secondary forest
MQ	*Macaranga quadriglandulosa* Warb.	1.1.c	0.62	secondary forest
M2	*Macaranga* sp. nov. 2 (Whitmore, in litt.)	1.1.c	0.88	primary forest
M3	*Macaranga* sp. nov. 3 (Whitmore, in litt.)	1.1.c	0.70	secondary forest
BR	*Breynia cernua* (Poir.) Muell. Arg.	4	0.00	secondary forest
CD	*Codiaeum ludovicianum* Airy Shaw	2	0.11	rocky river beds
EN	*Endospermum labios* Schodde	2	0.58	secondary forest
EX	*Excoecaria agallocha* L.	3.1	0.08	sea coast
HO	*Homalanthus novoguineensis* (Warb.) K. Schum	3.1	0.26	secondary forest
ML	*Mallotus mollissimus* (Geiseler) Airy Shaw	1.1	0.62	secondary forest
ME	*Melanolepis multiglandulosa* (Reinw.ex Bl) Rchb.&Zoll	1.2	0.16	secondary forest
PH	*Phyllanthus lamprophyllus* Muell. Arg.	4	0.62	rocky river beds
PI	*Pimelodendron amboinicum* Hassk.	3.2	0.11	primary forest

Phy. – monopyletic groups, based on a cladistic analysis of morphological and molecular characters (Weiblen, pers. com.); Tax. – taxonomic classification; numbers refer to subfamily.tribe: 1 – Acalyphoideae (1 - Acalypheae, 2 – Chrozophoreae), 2 – Crotonoideae 3 – Euphorbioideae (1 – Hippomaneae, 2 - Stomatocalyceae), 4 – Phyllanthoideae; letters refer to species groups in *Macaranga*: a – longistipulata, b – tanarius, c – dioica; Pal. – palatability (see Methods).

four, of five, subfamilies of Euphorbiaceae recognized by Webster (1975). No cladistic analysis is available for all the genera in our study. Therefore, we have used the recent suprageneric classification of Euphorbiaceae (Webster, 1975) and the classification of species groups in *Macaranga* (Whitmore, 1980) as surrogates for phylogeny (Table 1B). The placement of three genera related to Hippomaneae in a more recent phylogenetic analysis (Esser and van Welzen, 1997) agrees with that in the Webster's classification.

2.3. Host specificity

The host specificity of each species was quantified by Lloyd's index, which characterises the variance of a species' distribution among the 30 plant species studied (Lloyd, 1967; see also Lepš, 1993). Its value is minimum for an equitable distribution (that is, indiscriminate polyphagy) and increases with increasing host specificity. Many species were too rare to derive any estimates of their host specificity with confidence. Potentially spurious correlation of host specificity, expressed by Lloyd's index, on species abundance (that is, sample size) disappeared once species collected as less than 15 individuals were excluded from the analysis. This threshold was used to divide species into those common and rare; all analyses concerning host specificity were based on the former only.

Lloyd's index may be influenced by imbalanced choice of host plants in the present study, which include many species of *Ficus* and *Macaranga,* but only a single species from any other genus. Species indiscriminately polyphagous with respect to congeneric plants, but restricted to a single plant genus would thus attain very different values of Lloyd index, dependent on the genus on which they were specialised. Thus, host specificity was characterised also by the number of host plant genera used by a chrysomelid species. All common species were further classified into specialists (at least 90% of individuals concentrated on a single host plant genus) and generalists (remaining species).

Overall host specificity in a chrysomelid community feeding on a particular tree species was quantified as the average Lloyd's index and the average number of host genera for an individual chrysomelid feeding on a given tree species. These characteristics are averages of the Lloyd's index and the number of host genera, respectively, weighted by species abundance. Again, only common species of chrysomelids were used for the calculations.

The average number of host plants per chrysomelid species appeared to be the most straightforward characteristic of the host specificity in the entire chrysomelid community. However, the number of host plants recorded for a chrysomelid species is determined by its host specificity as well as its abundance. Obviously, the number of hosts may be underestimated severely for rare species. In order to separate host specificity from abundance, the number of hosts was calculated only for the species collected as at least 15 individuals from the set of plant species considered.

Host plant is defined either as a plant suitable for the completion of the herbivore's life cycle, or a plant fed upon by at least one of the herbivore's developmental stages. The latter meaning was used throughout this study as virtually all host plant records concerned feeding by adults, while feeding preferences by their larvae remained unknown.

2.4. Plant palatability

Highly generalist weevils (*Oribius* sp., species CURC012) were starved for 24h and then placed on potted saplings of the tree species studied. Each weevil was introduced inside a plastic vial which contained one leaf *in situ*. Feeding damage was scored visually after 24h on a logarithmic scale, as follows: 0 – no feeding; 1 – attempting to feed; 10 – moderate feeding; 100 – extensive feeding. During these experiments, two to five saplings were used for each plant species. On average, the feeding damage of 25 different weevils was measured for each plant species. An index of palatability was derived by recording the number of cases in which a score of 100 was attributed, divided by the number of cases tested. This procedure emphasised regular feeding as compared to food-probing.

2.5. Statistical analyses

The similarity in the composition of chrysomelid communities was analysed by the agglomerative cluster analysis (Ward's method) and by direct (Canonical Correspondence Analysis, CCA) and indirect (Correspondence Analysis, CA) ordinations, using the Euclidean distance on $\log_{10}(N+1)$ transformed data. The CCA relates the composition of samples to environmental variables. Palatability was used as the sole environmental variable in order that the first (constrained) ordination axis was constructed to maximise fit between tree palatability and the composition of its chrysomelid communities. The CCA score on this axis, obtained for each chrysomelid species, represents a species' optimum with respect to the host plant palatability. The remaining (unconstrained) axes capture residual variation in the species data after extracting the constrained axis. The CA orders samples only on the basis of their species composition; environmental variables, such as palatability, are only correlated with the ordination axes *a posteriori*, reflecting the sample composition. The Canoco 4.0 and CanoDraw 3.1 software was used for the ordinations.

The species and singleton accumulation curves were calculated as the average from 100 replicates, each accumulating samples in a random sequence.

3. Results

3.1. Taxonomic composition

The sampling produced 9,186 individuals of chrysomelids with confirmed ability to feed on the plant species from which they were collected. Almost all of these individuals were adults; larvae of only *Atysa* and *Sastra* spp. were found freely feeding on the foliage. In total, 93 species from five subfamilies were collected (Appendix 1). The only two important subfamilies were Eumolpinae (45 spp., 6,977 individuals) and Galerucinae (37 spp., 2,023 individuals), together accounting for 88% of species and 98% of individuals. Eumolpinae species fed both on *Ficus* and Euphorbiaceae (171 and 164 host plant records, respectively), while Galerucinae were more concentrated on *Ficus* hosts (83 host plant records, vs. 36 on Euphorbiaceae; the difference between subfamilies was significant, $P<0.01$, Fisher's exact test).

Eumolpinae were less host specific than Galerucinae. For common species, the median number of host plants was 10 in eumolpines and 5 in galerucines, the number of host genera 4 and 1, and Lloyd's index 5.9 and 18.4, respectively (all differences significant, $P<0.05$, Mann-Whitney tests, $n=38$ species).

3.2. Host specificity

Many species were found as only a few individuals (20 species as singletons), so that their host range could not be estimated. The profound effect of insect abundance on their appar-

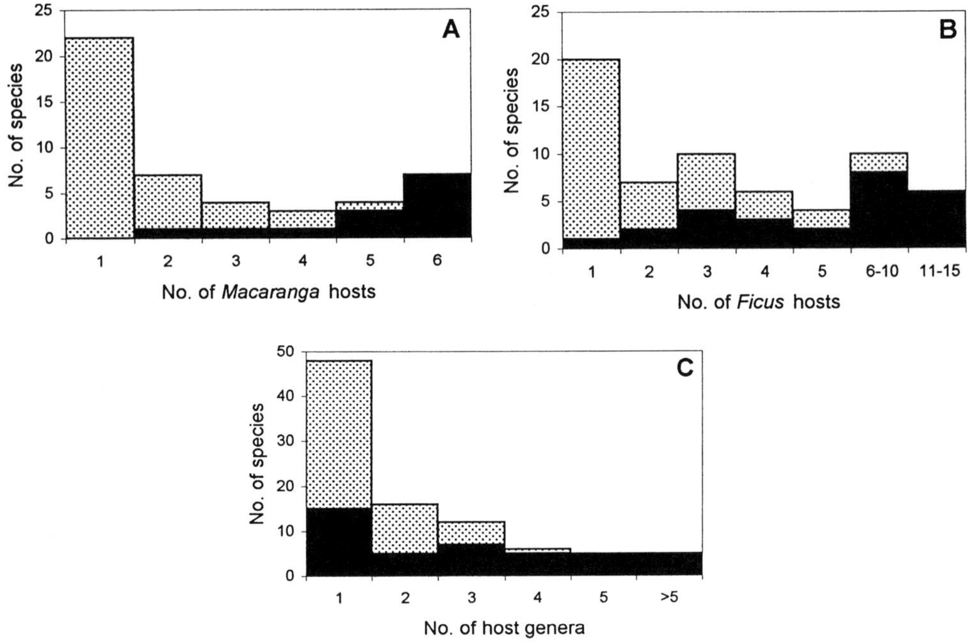

Fig. 1. Host specificity of chrysomelids on *Macaranga* spp. (A), *Ficus* spp. (B) and the whole set of studied tree species (C).
Black – common species, stippled – rare species.

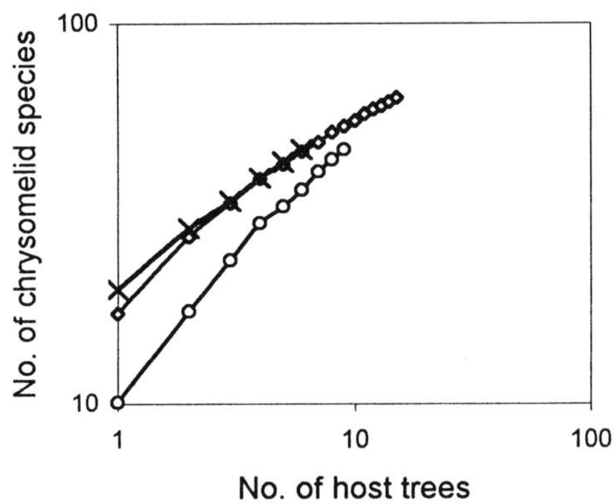

Fig. 2. Accumulation of chrysomelid species with increasing number of plant species studied.
Diamonds – *Ficus* spp., crosses and thick line – *Macaranga* spp., circles – Euphorbiaceae other than *Macaranga*.

ent host specificity is obvious from the number of hosts recorded for rare and common species (Fig. 1). Clearly, the prevalence of apparently monophagous species is caused by the rare species. Data for common species indicate a large overlap in their chrysomelid fauna among congeneric plants. This is particularly true for *Macaranga*, where the modal host specificity was polyphagy on all species studied and the average was 5.1 host species (Fig. 1a). Similar trends can be seen in *Ficus*, where the average was 7.0 hosts per chrysomelid species (Fig. 1b). In contrast, the specialisation to a single genus appeared to be

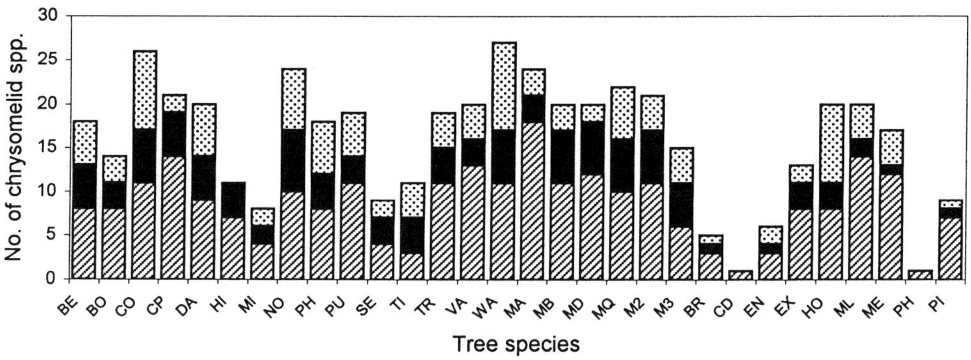

Fig. 3. Species richness of chrysomelids on various species of *Ficus* and Euphorbiaceae. Hatched - generalists, black - specialists, stippled - rare species; see Table 1 for full names of host plants.

most common in the whole data set, and the average was only 3.0 host genera (Fig 2c). Despite such specialisation, the overlap between the chrysomelid fauna on Moraceae and Euphorbiaceae was not negligible, as 23 out of the 42 common species used hosts from both families.

The species accumulation curves (Fig. 2) indicate that the average number of chrysomelid species found on a tree from either of the two exceptionally large genera, *Ficus* and *Macaranga*, is higher than the average number for the remaining 9 genera. The slope of log (number of species) on log(number of hosts) regression with a 95% confidence interval was 0.46 (0.42 – 0.50) for *Ficus*, 0.37 (0.28 – 0.47) for *Macaranga*, and 0.73 (0.65 – 0.81) for Euphorbiaceae other than *Macaranga* (Fig. 2). This indicates that the rate of species accumulation with increasing number of tree species was slower for congeneric trees than for those from different genera, but all three accumulation rates were very high.

3.3. Abundance of generalists and specialists

The species richness of chrysomelid communities varied from one species (*Codiaeum* and *Phyllanthus*) to 27 species (*F. wassa*). Their composition in terms of specialists, generalists, and rare species is depicted in Fig. 3. There was a positive correlation between the number of specialists and generalists ($r=0.473$, $P<0.01$, $n=30$ trees) so that both groups contributed to an exceptionally high (or low) species richness on certain trees.

The abundance of chrysomelids on different tree species varied from 8 (*Codiaeum*) to 1,462 (*Excoecaria*) individuals. There was a well-defined group of plant species with low abundance of chrysomelids (less than 100 individuals), most of which (>90%) were generalists (*Codiaeum, Endospermum, Phyllanthus, Pimelodendron, Melanolepis*). For the remaining trees, the proportion of specialists was increasing with the abundance of their chrysomelid community (Fig. 4). *Mallotus*, with a large chrysomelid community, composed mostly (99.5%) by generalists, is a notable exception to this trend. This is because its chrysomelid fauna overlaps with that of closely related *Macaranga* (see below). From the viewpoint of chrysomelids, *Macaranga* and *Mallotus* probably constitutes a single host genus. If this is accepted and the *Macaranga* and *Mallotus* amalgamated to a single data point, a very close relationship between the proportion of specialists and the abundance of

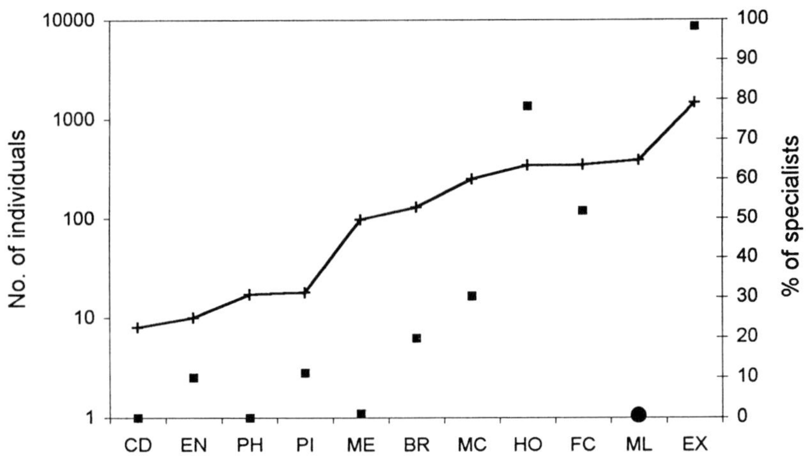

Fig. 4. Abundance of chrysomelids (line, left y axis) and the proportion of specialized individuals (squares, right y axis) on various plant genera.
FC – *Ficus*, MC – *Macaranga*; see Table 1 for other generic names.

chrysomelids across host plant genera ensues (r= 0.852, P<0.01, n=10).

Although the 22 specialist and 20 generalist species did not differ from each other in abundance (geometric mean 88 and 96 individuals per species, respectively; P>0.6, t-test), the 4,157 generalist chrysomelids were spread among 257 chrysomelid-host plant combinations, while the 4,844 specialists among only 108 such combinations. That suggests the largest chrysomelid populations feeding on a particular host belonged mostly to specialist species. Indeed, 8 from the 10 most abundant chrysomelid-plant combinations involved specialists.

The most abundant species, found as more than 500 individuals, were *Rhyparidella sobrina* (Bryant) gr. (a specialist on *Ficus*), *Rhyparida coriacea* Jacoby (a generalist), *Stethotes lateralis* Baly (a generalist), an unidentified galerucine sp. and *Rhyparida* sp. (both specialists on *Excoecaria*), and *Rhyparida huona* Gressitt (a generalist). The two specialists on *Excoecaria* also represent the two most abundant species-plant combinations.

3.4. Similarity of communities from different hosts

Cluster analysis separated the 13 forest species of *Ficus* and the rest of the data, which was, in turn, divided into a cluster involving all coastal species (*F. tinctoria, F. microcarpa* and *Excoecaria*), and a cluster of the forest Euphorbiaceae. The coastal species were further split into *Ficus* spp. and *Excoecaria*. The cluster of Euphorbiaceae was split into the *Macaranga* & *Mallotus* cluster and another, containing the remaining 8 species (Fig. 5). Further division of neither the forest species of *Ficus*, the *Macaranga* and *Mallotus* group, nor the other genera reflected their phylogenetic (or taxonomic) relationships (cf. Table 1). However, the only two plants from the riverine habitat (*Codiaeum* and *Phyllanthus*) formed a separate cluster.

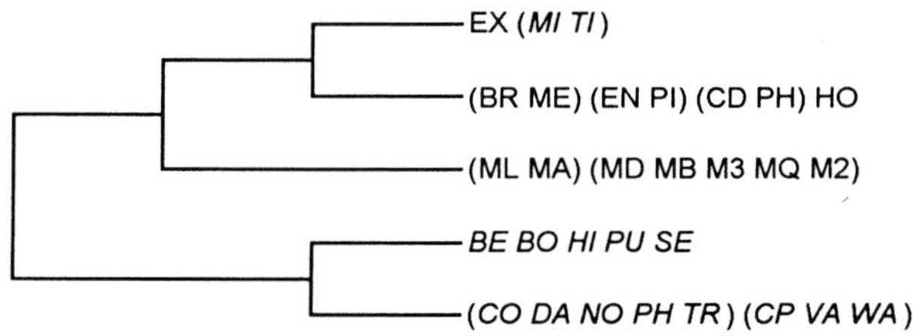

Fig. 5. Cluster analysis of chrysomelid communities from various host plants.
Ficus hosts in italics; see Table 1 for full names of host plants.

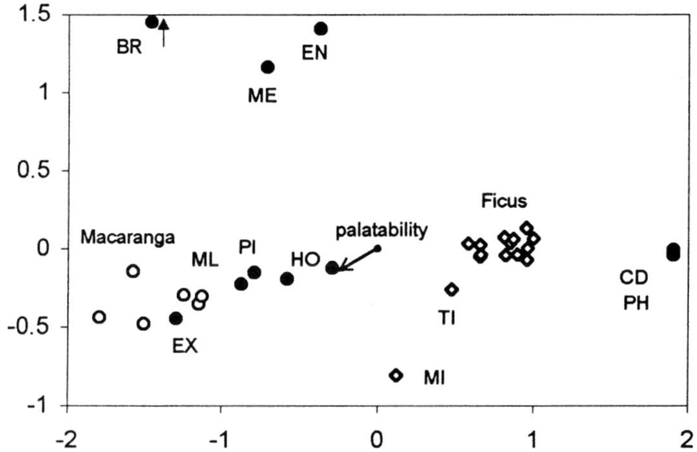

Fig. 6. Correspondence analysis of chrysomelid communities from various host plants.
Each point represents a chrysomelid community from one plant species. Palatability depicted as a passive environmental variable. Diamonds – *Ficus* spp., empty circles – *Macaranga* spp., solid circles – other euphorb species. See Table 1 for full names of host plants.

The CA analysis revealed a compact cluster of the 13 forest species of *Ficus*, with the 2 coastal fig species nearby (Fig. 6). *Macaranga* spp. were close to each other, but the spread of the other Euphorbiaceae was wide. The two riverine species were again close to each other. These results suggest that two factors, *viz.* plant taxonomy, especially at suprageneric levels, and plant habitat, are important in shaping the composition of chrysomelid communities.

3.5. Plant palatability and chrysomelid host specificity

Plant palatability was negatively correlated with the average host specificity of chrysomelids feeding on them. This relationships was found both for Lloyd's index ($r=-0.514$, $P<0.01$, Mann-Whitney test, $n=30$ trees) and the average number of host genera ($r=0.551$, $P<0.01$). This result suggests that palatable plants are preferably fed upon by generalists, and *vice versa.* Consistent with this pattern, there was a positive correlation between plant palatability and the number of generalist species of chrysomelids ($r= 0.420$, $P<0.05$, $n=30$ tree species), but analogous correlation with specialists was not significant ($P>0.1$). Hosts with

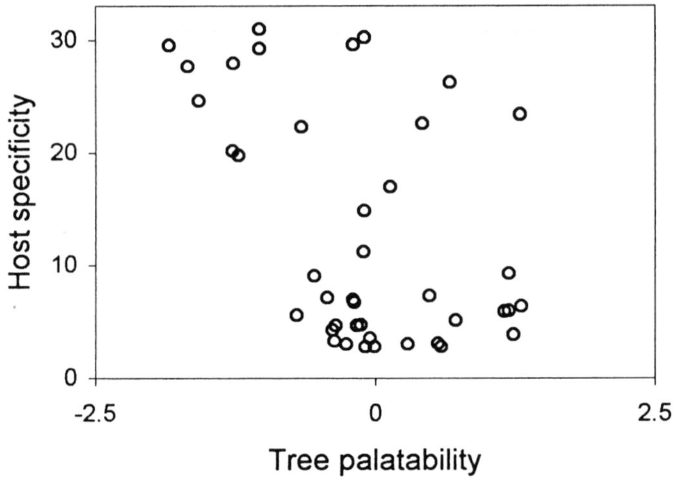

Fig. 7. Relationships between host specificity of chrysomelid species and their distribution optimum with respect to plant palatability.
Each circle represents one chrysomelid species. Tree palatability – species' score on the CCA constrained axis with palatability as the environmental variable (higher scores mean higher palatability); Host specificity - Lloyd's index (higher values mean higher host specialisation).

especially specialised communities (average Lloyd's index > 15) included *Excoecaria, F. microcarpa, Breynia,* and *Homalanthus*; the former three species have low palatability as well.

Plant palatability proved to be a significant explanatory variable for the species composition of chrysomelid communities (the significance of canonical axis $P<0.001$, Monte Carlo permutation test, 499 permutations, correlation of palatability with this axis $r=0.872$), but explained only 10% variability in the chrysomelid communities. The scores of chrysomelids on the canonical ("palatability") axis were correlated with their Lloyd's index of host specificity (Spearman $r=-0.417$, $P<0.05$; $n=42$ common species; Fig. 7). In other words, polyphagous species tended to be concentrated especially on palatable hosts.

4. Discussion

The prevalence of Eumolpinae and Galerucinae appeared to be a common feature of arboreal communities in the tropics. For example, chrysomelid communities on 9 of 10 tree species, each from a different family, studied in New Guinea by Basset and Samuelson (1996), were dominated by eumolpines and, similarly as in the present study, galerucines ranked second in terms of species richness. Galerucinae and Eumolpinae were also the most prominent chrysomelid subfamilies in samples from 5 rainforest trees in Borneo (Stork, 1987), while Amazonian rainforest samples were dominated by eumolpines and alticines (Erwin, 1983; Farrell and Erwin, 1988). This taxonomic composition has important ecological consequences, namely the scarcity of larvae in the foliage-feeding chrysomelid communities in the tropics as all species of Eumolpinae and many Galerucinae have subterranean, root-feeding larvae (Jolivet and Hawkeswood, 1995). Predominance of adults in the foliage-feeding chrysomelid communities may also contribute to their apparently low host specificity, as adults are often more likely to feed on marginal hosts (Jolivet and Hawkeswood, 1995).

Although taxonomy is a very unsatisfactory surrogate of phylogeny, and taxonomic ranks such as genera or families are somewhat arbitrary, there was still a clear difference in patterns of host specificity of chrysomelids when considered within and among plant genera. Within both *Ficus* and *Macaranga*, chrysomelids showed a clear tendency towards oligophagy, feeding often on numerous congeneric hosts. As this pattern was similar in both, mutually independent, cases, as well as in a similar study on weevils feeding on *Piper* spp. (Marquis, 1991), it may be more common. This would have important consequences for the large scale patterns of host specificity in the tropics since species-rich genera are an important part of the tropical vegetation. For instance, three genera (*Macaranga, Glochidion* and *Phyllanthus*) account for 40% of all Euphorbiaceae in New Guinea (van Welzen, 1997), whereas the position of *Ficus* within Moraceae is even more prominent.

The present host specificity estimates are particularly unsatisfactory with regard to the host plant use by widely generalist species. The present study includes only two plant families, while there are a conservative 64 families of woody plants present in one of our study plots (that at Baitabag; L. Balun, pers. comm.). Plant diversity in rain forests is extremely high; for instance, censuses restricted to one hectare and to large trees (DBH>10cm) yielded 97-182 species for a lowland forest in New Guinea (Oatham & Beehler, 1998; Reich, 1998), around 200 species for mid-elevation New Guinean forests (Paijmans, 1970; Wright et al., 1997; Weiblen in press), and a cosmopolitan estimate of between 60 to 150-300 species in the rain forests world-wide (Gentry, 1990; Richards, 1996), belonging often to as many as 50 different families. In comparison with these numbers, the limitations of the present study are obvious.

The similarity of chrysomelid communities does not closely follow the phylogeny (*Ficus*) or taxonomic classification (Euphorbiaceae) of their hosts, although the smaller differences among congeneric hosts and large differences between the families are consistent with the plant taxonomy. It is possible that for congeneric hosts, chrysomelids respond to ambient ecological conditions and quality of various plants, rather than being constrained in their host choice by evolutionarily fixed preferences.

The similarity among chrysomelid communities from different host plant genera was generally low, as they were dominated numerically by specialists and as a consequence of idiosyncratic composition. This is illustrated by a low proportion of variance in their composition, captured by the first two CA axes, and makes the patterns of their similarity difficult to interpret. There were only two notable exceptions to this pattern, where plant genera contained similar chrysomelid communities – the pair *Mallotus* and *Macaranga*, possibly reflecting a close phylogenetic relationship between the two genera, and the pair *Codiaeum* and *Phyllanthus*, which are species from an extreme habitat of rocky river beds. The vegetation in this habitat is often, at least several times a year, submerged after a sudden increase in the water level, following torrential rains. This unpredictable disturbance may explain the low numbers of chrysomelids, as well as a complete absence of specialists, feeding on these plants.

Despite differences between chrysomelid communities on Moraceae and Euphorbiaceae, the number of chrysomelid species able to feed on plants from both families, found in this study, was relatively large. In contrast, Basset and Samuelson (1996) found 70% of the arboreal chrysomelids able to feed on a single plant family only.

An interesting question is how many rainforest trees are not colonised by chrysomelids. Although all species of *Ficus* had appreciable chrysomelid communities, there were four Euphorbiaceae, each from a different genus, which were practically devoid of chrysomelids. These species represent a heterogeneous group with respect to their habitat, growing in riverine, secondary and primary forest habitats, taxonomy, representing three subfamilies, size, including the smallest and largest Euphorbiaceae studied, as well as palatability. The proportion of unexploited hosts in rainforest vegetation is largely unknown, despite it being an important determinant of the overall species richness and abundance of chrysomelids in the rainforest ecosystems. For example, as reported by Jolivet and Hawkeswood (1995) and confirmed by a current study in the Madang area (Novotny, pers. obs.), Rubiaceae, one of the most species-rich families, has almost no chrysomelids. Basset and Samuelson (1996) found only scarce chrysomelid communities on rainforest trees from several families, with *Ficus* and *Pipturus* (Urticaceae) being the only two hosts, from the ten studied, harbouring species-rich communities.

Farrell and Erwin (1988) suggested that many chrysomelid species in the Amazonian canopy could be host-specific, in comparison with species living in the canopies of temperate forests. As possible causal effect, they cited the higher proportion of tree genera in the tropics that have herbaceous representatives in the temperate zone and the likelihood that these tropical hosts bear some chemical similarity with the chemistry of temperate herbs, often dominated by qualitative defences which could select for host-specific herbivores. This appears to be an open question as their own data do not seem to support the notion of narrow host specificity. They sampled five distinct types of rain forest and found that, on average, chrysomelid species were present in 2.2 (or 44%) from these 5 habitats. This seems to indicate a rather low host specificity, given a probably very low overlap in the species composition of plants sampled in these different forest types.

Polyphagous chrysomelids prefer palatable hosts and palatable hosts are colonised mostly by polyphagous species (note that these are two separate, partly independent patterns). The correlation between plant palatability and host specificity of chrysomelids, reported here, is of interest because the palatability index was derived using a weevil, rather than a chrysomelid, species. Although more extensive screening by various generalist species is necessary in order to establish their validity, such indices might be useful predictors of the community structure of herbivores. There are several theories, suggesting that certain integrated sets of life history traits of plants (life history syndromes) will make them palatable to a wide range of herbivores [for instance, R-strategists (Grime, 1979), growth-dominated plants (Herms and Mattson, 1992), rapidly-growing plants (Price, 1991) and early successional plants (Cates and Orians, 1975)]. Such hypotheses are rarely tested using tropical trees (but see Coley, 1983). The present study indicates that further study of relationships between plant palatability, other life history traits of plants, and the structure of their herbivorous communities may be a promising direction towards better understanding of rainforest insect communities.

Acknowledgements

We are pleased to acknowledge the help of Chris Dal, Martin Kasbal, Keneth Molem, William Boen, John Auga and Markus Manumbor, who assisted with most of the technical aspects of the project. The important contribution of insect collectors, too numerous to be listed here, will be acknowledged elsewhere. The landowners Kiatik Batet, Hais Wasel and Sam Guru kindly allowed us to collect in Baitabag, Ohu and Mis, respectively. Larry Orsak, Anne Borey-Kadam and the staff of the Christensen Research Institute provided invaluable assistance with the logistics of the project. Plants were identified by C. Berg, Peter van Welzen, T. C. Whitmore, George Weiblen and Paul Katik. Pavel Drozd wrote several computer programmes used in the analysis. The manuscript benefited from discussions with and comments by Neil Springate, Olivier Missa, Petr Svacha, David Furth and Lawrie Springate. The project was funded by the National Science Foundation (DEB-94-07297, DEB-96-28840 and DEB-97-07928), the Christensen Fund (Palo Alto, California), the National Geographic Society (grant no. 5398-94), the Czech Academy of Sciences (grants no. C6007501 and A6007705), and the Czech Ministry of Education (grant ES 041).

References

Basset, Y. and Samuelson, G.A. 1996 Ecological characteristics of an arboreal community of Chrysomelidae in Papua New Guinea. In: Jolivet, P. H. A. and Cox, M. L. (eds), Chrysomelidae Biology. Volume 2: Ecological Studies, SPB Academic Publishing, Amsterdam, pp. 243-262.

Berg, C. C. 1989 Classification and distribution of *Ficus*. Experientia 45: 605-611.

Cates, R. G. and Orians, G. H. 1975 Successional status and the palatability of plants to generalized herbivores. Oikos 65: 97-106.

Coley, P. D. 1983 Herbivory and defensive characteristics of tree species in a lowland tropical forest. Ecol. Monogr. 53: 209-233.

Corner, E. J. H. 1965 Check-list of *Ficus* in Asia and Australasia with keys to identification. Gdns' Bull. Straits Settl. 21: 1-186.

Erwin, T. L. 1983 Beetles and other insects of tropical forest canopies at Manaus, Brazil, sampled by insecticidal fogging. In: Sutton, S. L., Whitmore, T. C. and Chadwick, A. C. (eds), Tropical Rain Forest: Ecology and Management, Blackwell Sci. Publ., Oxford, pp. 59-75.

Esser, H.-J., and van Welzen, P. 1997 A phylogenetic classification of the Malesian Hippomaneae (Euphorbiaceae). Syst. Bot. 22: 617-628.

Farrell, B. D. and Erwin, T. L. 1988 Leaf-beetle community structure in an amazonian rainforest canopy. In: Jolivet, P., Petitpierre, E. and Hsiao, T. H. (eds), Biology of Chrysomelidae, Kluwer Academic Publishers, Dordrecht, pp. 73-90.

Flowers, R. W. and Janzen, D. H. 1997 Feeding records of Costa Rican leaf beetles (Coleoptera: Chrysomelidae). Fla. Ent. 80: 334-366.

Gentry, A.H. (ed) 1990 *Four Neotropical Forests*. Yale University Press, New Haven.

Gressitt, J. L. 1959 Host relations and distribution of New Guinea Hispine beetles. Proc. Hawaii. ent. Soc. 17: 70-75.

Grime, J. P. 1979 Plant strategies and vegetation processes. Wiley, Chichester.

Herms, D. A. and Mattson, W. J. 1992 The dilemma of plants: to grow or defend. Q. Rev. Biol. 6: 283-335.

Heywood, V. H. (ed) 1993 Flowering Plants of the World. B. T. Batsford Ltd, London.

Höft, R. 1992. Plants of New Guinea and the Solomon Islands. Dictionary of the Genera and Families of Flowering Plants and Ferns. Wau Ecology Institute (Handbook No. 13), Wau.

Jolivet, P. and Petitpierre, E. 1976 Les plantes-hôtes connues des *Chrysolina* (Col. Chrysomelidae) : essai sur les types de selection trophique. Annls Soc. ent. Fr. 12: 123-149.

Jolivet, P. 1989 Sélection trophique chez les Hispinae (Coleoptera Chrysomelidae Cryptosoma). Bull. mens. Soc. linn. Lyon 58: 297-317.

Jolivet, P. and Hawkeswood, T. J. 1995 Host-plants of Chrysomelidae of the World. An Essay about the Relationships between the Leaf-beetles and their Food-plants. Backhuys Publishers, Leiden.

Lawton, J. H. and Schröder, D. 1977 Effects of plant type, size of geographical range and taxonomic isolation on number of insect species associated with British plants. Nature 265: 137-140.

Lepš, J. 1993 Taylor's power law and measuring variation in the size of populations in space and time. Oikos 68: 349-356.

Lloyd, M. 1967. Mean crowding. J. anim. Ecol. 36: 1-30.

Marquis, R. J. 1991. Herbivore fauna of *Piper* (Piperaceae) in a Costa Rican wet forest: diversity, specificity and impact. In: Price, P. W., Lewinsohn, T. M., Fernandes, G. W., Benson, W. W. (eds.) Plant - Animal Interactions: Evolutionary Ecology in Tropical and Temperate Regions. London, John Wiley & Sons, pp 179-208.

McAlpine, J. R., Keig, G., Falls, R. 1983 Climate of Papua New Guinea. CSIRO and Australian National University Press, Canberra

Oatham, M. and Beehler, B. M. 1998 Richness, taxonomic composition and species patchiness in three lowland forest tree plots in Papua New Guinea. In Dallmeier, F. and Comiskey, J. A. (eds.) Forest biodiversity research, monitoring and modelling: Conceptual background and Old World case studies. UNESCO and Parthenon Publishing Group, Paris and New York, pp 613-631.

Ohmart, C. P. 1996 Population dynamics of chrysomelid beetles feeding on *Eucalyptus*. In: Jolivet, P. H. A. and Cox, M. L. (eds), Chrysomelidae Biology. Volume 2: Ecological Studies, SPB Academic Publishing, Amsterdam, pp 263-269.

Paijmans, K. 1970. Analysis of four tropical rain forests in Papua New Guinea. J. trop. Ecol. 58: 77-101

Price, P. W. 1991 The plant vigor hypothesis and herbivore attack. Oikos 62: 244-251.

Rank, N., Smiley, J. and Koepf, A. 1996 Natural enemies and host plant relationships for Chrysomeline leaf beetles feeding on Salicaceae. In: Jolivet, P. H. A. and Cox, M. L. (eds), Chrysomelidae Biology. Volume 2: Ecological Studies, SPB Academic Publishing, Amsterdam, pp 147-171.

Reich, J. A. 1998. Vegetation. Part 1: A comparison of 2 two-hectare tree plots in the Lakekamu basin. In: Mack, A. (ed.), A Biological Assessment of the Lakekamu Basin, PNG. Conservation International, Washington, pp 25-35.

Richards, P. W. 1996 The Tropical Rain Forest. Cambridge University Press, Cambridge.

Stork, N. E. 1987 Arthropod faunal similarity of Bornean rain forest trees. Ecol. Ent. 12:219-226.

Strong, D. R. J. 1977 Rolled-leaf hispine beetles (Chrysomelidae) and their Zingiberales host plants in middle America. Biotropica 9: 156-169.

Tahvanainen, J. 1983 The relationship between flea beetles and their cruciferous host plants: the role of plant and habitat characteristics. Oikos 40: 433-437.

van Welzen, P. C. 1997 A preliminary checklist of Euphorbiaceae of Thailand and the Malesian area. Malesian Euphorbiaceae Newsletter 6: 1-31.

Webster, G. L. 1975 Conspectus of a new classification of the Euphorbiaceae. Taxon 24: 593-601.

Weiblen, G. in press. Forest composition and structure of a one hectare plot in the Crater Mountain Wildlife Management Area, Papua New Guinea. Science in New Guinea.

Whitmore, T. C. 1979 Studies in *Macaranga*, X. – potentially commercial species in new Guinea. Commonw. For. Rev. 58: 271-272.

Whitmore, T. C. 1980 *Macaranga*. In: Airy Shaw, H. K., The Euphorbiaceae of New Guinea, Kew Bulletin Additional Series 8, Royal Botanic Gardens, Kew, pp. 123-162.

Wright, D. D., Jessen, J. H., Burke, P. and Garza, H.G.D.S. 1997 Tree and liana enumeration and diversity on a one-hectare plot in Papua New Guinea. Biotropica 29, 250-260.

Appendix 1. Chrysomelids feeding on the studied trees.

Morphospecies code is followed by name (if known), host specificity category (G – generalist, S – specialist, r – too rare to decide), the number of individuals, the number of hosts, and, for specialists, the principal host plant genus.

Alticinae:
CHRY070 (r, 1, 1); CHRY068 – *Nisotra* sp. (G, 84, 5); CHRY069 – *Sutrea* sp. (r, 1, 1); CHRY058 – *Xenidea* sp. (r, 9, 2); CHRY096 (r, 2, 1).

Criocerinae:
CHRY053 – *Lema connectens* Baly (r, 1, 1); CHRY093 – *Lema staudingeri* Jacoby (r, 14, 4)

Cryptocephalinae:
CHRY085 – *Cadmus multinodosus* (Gressitt) (r, 5, 2); CHRY043 – *Coenobius* sp. (S, 21, 1, *Ficus*); CHRY099 – *Melatia ?glochidionis* (Gressitt) (S, 39, 1, *Homalanthus*); CHRY090 – *Melatia* sp. (r, 3, 2).

Eumolpinae:
CHRY072 – *Cleorina* sp. (S, 228, 7, *Macaranga*); CHRY061 – *Deretrichia* sp. (r, 7, 5); CHRY076 – *Deretrichia* sp. (S, 17, 5, *Macaranga*); CHRY054 – *Deretrichia* sp. (S, 261, 12, *Homalanthus*); CHRY023 – *Deretrichia* sp. (r, 14, 9); CHRY108 – *Deretrichia* sp. (G, 18, 6); CHRY124 – *Deretrichia* sp. (r, 7, 3); CHRY027 – *Rhyparida* sp. (G, 106, 8); CHRY025 – *Rhyparida basalis* Baly (G, 148, 5); CHRY064 – *Rhyparida* nr. *basalis* Gressitt (r, 1, 1); CHRY031 – *Rhyparida cacaovora* Gressitt (G, 180, 24); CHRY019 – *Rhyparida calami* Gressitt (G, 52, 14); CHRY001 – *Rhyparida coriacea* Jacoby (G, 1041, 27); CHRY042 – *Rhyparida dejecta* Gressitt (r, 2, 2); CHRY024 – *Rhyparida duni* Gressitt (S, 106, 3, *Ficus*); CHRY052 – *Rhyparida fasciata* Baly (G, 41, 10); CHRY021 – *Rhyparida fruhstorferi* Jacoby (G, 50, 13); CHRY063 – *Rhyparida fulvicornis* Jacoby (r, 3, 3); CHRY046 – *Rhyparida huona* Gressitt (G, 634, 17); CHRY026 – *Rhyparida* nr. *impressipennis* Bryant (G, 25, 11); CHRY079 – *Rhyparida lineolata* Gressitt (S, 26, 6, *Macaranga*); CHRY018 – *Rhyparida ?normalis* (r, 5, 4); CHRY014 – *Rhyparida picticollis* Gressitt (G, 123, 19); CHRY084 – *Rhyparida pruinicollis* Gressitt (G, 68, 10); CHRY037 – *Rhyparida sinuata* Gressitt (G, 133, 16); CHRY078 – *Rhyparida* sp. (S, 673, 1, *Excoecaria*); CHRY113 – *Rhyparida* sp. (S, 19, 3, *Macaranga*); CHRY071 – *Rhyparidella casuarinae* Gressitt (r, 4, 3); CHRY082 – *Rhyparidella sewana* Gressitt group (G, 70, 7); CHRY002 – *Rhyparidella sobrina* (Bryant) group (S, 1803, 20, *Ficus*); CHRY032 – *Rhyparidella* sp. (r, 10, 3); CHRY012 – *Rhyparidella* sp. (r, 6, 5); CHRY022 – *Rhyparidella* sp. (r, 4, 3); CHRY044 – *Rhyparidella* sp. (r, 4, 2); CHRY006 – *Rhyparidella* sp. (r, 1, 1); CHRY028 – *Rhyparidella* sp. (r, 1, 1); CHRY013 – *Stethotes* nr *integra* Jacoby (r, 1, 1); CHRY030 – *Stethotes lateralis* Baly (G, 879, 19); CHRY017 – *Stethotes* nr *nigritula* Jacoby (r, 4, 3); CHRY056 – *Thyrasia?* sp. (G, 82, 14); CHRY050 (r, 4, 4); CHRY029 (S, 118, 5, *Macaranga*); CHRY098 (r, 1, 1); CHRY104 (r, 2, 2); CHRY119 (r, 1, 1).

Galerucinae:
CHRY004 – *Atysa* sp. (S, 125, 6, *Ficus*); CHRY051 – *Atysa* sp. (S, 31, 2, *Ficus*); CHRY059 – *Atysa* sp. (r, 3, 3); CHRY040 – *Aulacophora* sp. (r, 8, 5); CHRY066 – *Aulacophora* sp. (r, 2, 1); CHRY067 – *Aulacophora* sp. (r, 2, 1); CHRY110 – *Aulacophora* sp. (r, 2, 1); CHRY073 – *Aulacophora* sp. (r, 1, 1); CHRY127 – *Aulacophora* sp. (r, 1, 1); CHRY007 – *Aulacophora* sp. (G, 17, 4); CHRY041 – *Aulacophora* sp. (G, 106, 7); CHRY055 – *Aulacophora* sp. (G, 33, 6); CHRY091 – *Cassena* sp. (r, 7, 4); CHRY097 – *Cassena* sp. (r, 1, 1); CHRY016 – *Lomirana* sp. (G, 138, 12); CHRY047 – ?*Monolepta* sp. (r, 1, 1); CHRY074 – ?*Monolepta* sp. (r, 1, 1); CHRY003 – *Sastra* sp. (S, 316, 4, *Ficus*); CHRY039 – *Sastra* sp. (S, 21, 7, *Ficus*); CHRY033 – *Sastra* sp. (r, 13, 7); CHRY038 – *Sastra* sp. (r, 7, 5); CHRY057 – *Sastra* sp. (r, 4, 2) CHRY081 (S, 748, 1, *Excoecaria*); CHRY009 (S, 163, 12, *Ficus*); CHRY020 (S, 85, 9, *Ficus*); CHRY060 (r, 1, 1); CHRY075 (r, 1, 1); CHRY083 (r, 1, 1); CHRY087 (S, 17, 1, *Excoecaria*); CHRY088 (S, 27, 2, *Breynia*); CHRY089 (G, 129, 3); CHRY105 (r, 1, 1); CHRY111 (r, 2, 1); CHRY117 (r, 1, 1); CHRY120 (r, 5, 2); CHRY122 (r, 1, 1); CHRY123 (r, 1, 1).

HOST PLANT BREADTH AND IMPORTANCE OF CUCURBITACINS FOR THE LARVAE OF DIABROTICITES (GALERUCINAE: LUPERINI)

ASTRID EBEN
Instituto de Ecología, A.C., Km. 2.5 antigua carretera a Coatepec, 91000 Xalapa, Veracruz, Mexico

Abstract

The extant literature on biology and ecology of Diabroticite beetles has mainly been concerned with the adult insects. This review compiles information on Diabroticite larval biology, especially host plant location and diet breadth. The main focus is on different aspects of larval interactions with Cucurbitaceae. Earlier hypotheses on the evolutionary history of this insect-plant interaction and recent data on second and third trophic level effects of secondary plant compounds are discussed.

Introduction

Host plant breadth within groups of closely related insects can be highly variable (Scriber, 1983; Scriber & Hainze, 1987). Some species may be specialists on a certain plant species, genus or family, while others in the same genus feed and develop on a wide variety of unrelated plant families (Singer, 1983; Jermy, 1984). An example of feeding specialization on different host plants in closely related herbivore species are Galerucinae in the subtribe Diabroticites.

Diabroticites are of Neotropical origin, the adults feed mainly on the pollen of a number of different plant families, larvae are root feeders (Krysan & Branson, 1983). The genera *Acalymma* and *Diabrotica* are native to Mexico and Central America (Webster, 1895). The 338 described *Diabrotica* species are mainly distributed in South and Central America (Wilcox, 1972). About 338 species belong to the Neotropical *fucata* group, whereas 24 of the remaining species belong to the *virgifera* group, found in more temperate regions and 11 species belong to the univoltine *signifera* group which is restricted to South America (Wilcox, 1972; Krysan & Smith, 1987). For *Diabrotica*, published data indicate, that species in the *virgifera* group are univoltine and the larvae are oligophagous on Poaceae, whereas adults feed on pollen of a variety of plant families (Krysan & Branson, 1983; Krysan & Smith, 1987). Species in the *fucata* group are, as far as data are available, multivoltine and polyphagous on more than 54 plant families. Their hypothesized original host plants are found in the Cucurbitaceae. The genus *Acalymma* is multivoltine (Sell, 1915). Nineteen species are described (Munroe & Smith, 1979) for North America and mainly distributed in tropical regions of Mexico. *Acalymma* spp. are considered oligophagous on Cucurbitaceae as adults and larvae. Host plant relationships are known with certainty only

Advances in Chysomelidae Biology 1, pp. 361–374
Edited by M.L. Cox

for five species found in the US (Smith, 1966). Studies of host plant suitability and preference have mainly compared different cucurbit cultivars as larval hosts of *Acalymma* spp. (Bach, 1977; Reed *et al.*, 1984).

The greatest difficulty in determining the evolutionary history and patterns of host use within the Luperini is the scarce knowledge regarding larval host plants. The Luperini is a tribe with close to 4,000 species (Wilcox, 1972). All Luperini are root feeders as immatures, but larval host associations are known only for a small fraction of these species (Krysan & Branson, 1983; Yaro & Krysan, 1986). Much is known about the feeding habits of the adults of a number of species, especially those of economic importance in the US and Mexico (Krysan & Branson, 1983; Anaya-Rosales *et al.* 1992). Very little information is available about the host plants of larval Diabroticites. To date, larval host associations have mainly been inferred from adult feeding behaviour. Those data are not reliable because adults are pollen feeders and pharmacophagous on bitter cucurbits. Consequently, many of the host plant records for adults are not related to potential larval hosts or oviposition hosts.

Oviposition and larval attraction to host roots

Few data are available on oviposition behaviour in the field, as well as on larval development on different host plants for most of the Diabroticite species distributed in the US, Mexico and Central America (Meinke, 1984; Yaro & Krysan, 1986; Branson & Krysan, 1987; Golden & Meinke, 1991). Again, observations are almost exclusively restricted to species of economic importance. Meinke (1984) described for the southern corn rootworm, *Diabrotica undecimpunctata howardi* Barber, oviposition on corn, peanuts and soybeans under field and laboratory conditions. He found no significant preferences for any of the crops in the laboratory. However, in the field females laid significantly more eggs on peanuts than on the other two crops. The same preference for peanuts was reported by Eben & Barbercheck (1992) for a system containing corn, bitter (cucurbitacin containing) squash and a non bitter squash (without cucurbitacins) in field cages.

Selection of oviposition hosts by the northern corn rootworm, *D. barberi* Smith & Lawrence, was compared under field conditions between corn, soybean, oat stubble, foxtail, lambsquarters, pigweed and smartweed. Significantly higher numbers of eggs were recovered from corn than any of the other hosts (Boetel *et al.*, 1992). Siegfried & Mullin (1990) described that female *D. barberi* oviposited repeatedly on sunflower (*Helianthus annuus*), golden rod (*Solidago canadensis*) and cultivated squash (*Cucurbita maxima*), which are less suitable hosts than corn in terms of adult longevity, but not in terms of larval development. Adults of western and northern corn rootworm were found in equal numbers in corn and squash flowers, but adult emergence was highest in corn. This might indicate that larvae developed principally on corn, whereas corn leaves were not a preferred food for the adults of this species.

Despite its Neotropical origin, the genus *Diabrotica* contains in the *virgifera* group six univoltine species *D. longicornis barberi* (Say), *D. l. longicornis* (Say), *D. v. virgifera* LeConte, *D. v. zeae* Krysan & Smith, *D. cristata* (Harris) and *D. lemniscata* LeConte. The larval hosts of these species are probably restricted to certain Gramineae (Smith, 1966; Branson & Ortman, 1970). Golden & Meinke (1991) verified corn as a potential host for

D. l. longicornis. Branson & Krysan (1981) proposed that the ancestor of the *virgifera* group specialized as larvae on the roots of grasses, and subsequently became adapted to the annual life cycle of those grasses. This hypothesis is supported by the fact that species in the *virgifera* group are univoltine even in the humid tropical regions in Central America. One of the few reports on both, larval and adult hosts, is the observation of Branson & Krysan (1981) who found *D. virgifera* larvae and adults respectively on roots and pollen of *Setaria* sp. (Gramineae). *Diabrotica cristata*, a species distributed in North American prairies, was reared in the laboratory from corn roots, and could be captured after emergence from three species of prairie grasses: *Bromus inermis, Panicum virgatum* and *Andropogon gerardi* (Yaro & Krysan, 1986). This establishes grasses as probable larval hosts, but adults were generally found on flowers of Asteraceae, which again demonstrates the habit of pollen feeding on plants unrelated to the larval hosts, that can be observed for most adult Diabroticites. *Diabrotica tibialis* Jacoby, a species common in Mexico and southern Texas (Krysan & Smith, 1987), could be reared on the roots of several different plants in the laboratory. Larvae survived on corn (*Zea mays*), a cultivated cucurbit (*Cucurbita* sp.) and soybean (*Glycine max*), but not on beans (*Phaseolus vulgaris*). Corn was the best host in terms of larval development time and survival to adulthood (Branson & Krysan, 1987). Under field conditions in eastern Mexico, adults were rarely found on corn, but very commonly on cucurbit flowers (Eben & Barbercheck, 1996).

In studies of host plant recognition and acceptance by larval Diabroticites, Branson & Ortman (1970) found that *D. virgifera* larvae were not able to distinguish between the roots of host and non-host plants, but rather were attracted to the roots of a number of plants from different families. The authors concluded that host plant recognition is made entirely by taste and not by long range olfactory cues. For example, carbon dioxide is known to have behavioural importance to all *Diabrotica* spp. larvae observed, but since it is a volatile released by the roots of most plant species it is not a specific host finding cue for rootworm larvae and therefore does not provide a basis to distinguish between host and non-host plants. In order to determine if larvae are attracted to specific root volatiles, Hibbard & Bjostad (1988) collected volatile compounds of corn seedlings and observed significantly stronger attraction of *D. virgifera virgifera* larvae to an extract of seedling volatiles compared to carbon dioxide alone. In a subsequent experiment, Hibbard & Bjostad (1990) isolated corn root volatiles and used behavioural choice-tests to test the biological activity of whole extracts and certain fractions. They found two fractions of the extracts attractive and another two fractions repellent for the corn rootworms. The semiochemicals in the attractive fractions were effective even without the presence of carbon dioxide. In conclusion, the authors proposed that western corn rootworm larvae may use carbon dioxide for longer distance orientation and more specific cues for short range orientation. Recently, Bjostad & Hibbard (1992) reported 6-methoxy-2-benzoxazolinone as a semiochemical for corn rootworms. MBOA is produced by corn seedlings and has been proven to be the most important root semiochemical for rootworms. MBOA occurs almost exclusively in corn and other grasses, and might therefore be the most informative compound for host searching neonate rootworms in the soil. Toxicity and anti-feeding properties of MBOA have been reported for a number of herbivorous insects, but not for western corn rootworms (Abou-Fakhr *et al.*, 1994). Infochemicals other than carbon dioxide are therefore equally important for host plant detection by rootworm larvae in the soil.

In fact, host specificity of Diabroticite larvae and the chemical and behavioural basis for the potential specificity has not yet been determined. Branson & Ortman (1970) tested 44 grass species and found that 18 supported rootworm larval development for at least 10 days, whereas no larvae survived on any of the Angiosperm species tested. More larvae survived on corn roots, but females reared on alternative hosts produced an equal number of viable eggs to females reared on corn (Branson & Ortman, 1970). Branson (1982) observed in olfactometer choice tests that larvae clearly distinguished between an odour source and the blank arm of the olfactometer. He also found that significantly more larvae were attracted to volatiles from corn than from squash, soybeans, and sunflowers, which are non-hosts of this species, but larvae were equally attracted to corn than to several other host and non-host grasses. In spite of the pest status of the larvae of several species in the *Diabrotica virgifera* group (Metcalf, 1986), the subterranean search behaviour of larval rootworms is poorly understood. Strnad & Dunn (1990) studied the host search behaviour of neonate western corn rootworms and reported an obvious switch from ranging to localized search when larvae had come into contact with corn roots. By contrast, when presented with a non-host (oats) reduced search area and velocity were observed.

Association of Diabroticites with Cucurbitaceae

Numerous studies have described the association of Diabroticites with plants of the family Cucurbitaceae (Chambliss & Jones, 1966; Howe *et al.*, 1976; recent review Metcalf, 1994). Both partners of this plant-insect interaction evolved in southern Mexico and Central America (Webster, 1895; Metcalf, 1986). 118 genera with more than 900 species in the family Cucurbitaceae produce cucurbitacins which are among the most bitter and toxic (LD_{50} intraperitoneal mice 1.0-6.8 mg/kg body wt., David & Vallance, 1955) plant secondary compounds known (Rehm *et al.*, 1957) and effectively deter non adapted herbivores (Tallamy *et al.*, 1998). Host preferences of Diabroticites are strongly influenced by the presence of cucurbitacins in cucurbit hosts. These non volatile compounds act as arrestants and feeding stimulants for adult Diabroticites.Their response has been interpreted as evidence for their evolutionary relationship with Cucurbitaceae and it was further hypothesized that ancient Diabroticites were cucurbit specialists in their native habitat (Metcalf, 1986). Species of the genus *Diabrotica* introduced into the US about 100 years ago, may have more recently broadened their host range or specialized on grasses, as reported for species in the *fucata* and *virgifera* group, respectively. While in the native habitat in Mexico and Central America, where cucurbits have been grown together with corn, beans or peanuts since pre-Columbian times, these species were cucurbit specialists (Branson & Krysan, 1981; Risch, 1981). *Cerotoma arcuata* Olivier and *D. speciosa* Germar, two South American Diabroticites, were found to sequester cucurbitacins in the adult stage, but no data on larval development and host plants have so far been published (Nishida *et al.*, 1992). *Cerotoma ruficornis* Olivier and *C. facialis* Erickson are common pests on beans as adults and larvae (Cardona *et al.*, 1982; Anaya-Rosales *et al.*, 1992). *Cerotoma ruficornis* has been repeatedly found on bitter wild *Cucurbita martinezii* in eastern Mexico (Eben, unpubl. data).

Aulacophorites, a subtribe within the Luperini, distributed in the Old World tropics, and New World Diabroticites have similar life histories and feeding habits (Maulik, 1936). The genus *Aulacophora* contains 535 species (Wilcox, 1972). Adults show a remarkably similar, oligophagous association with Cucurbitaceae as New World Diabroticites, but no data on larval development and preference are available (Sinha & Krishna, 1969, 1970; Nishida *et al.*, 1992; Lewis & Metcalf, 1996).

Cucurbitacins as pharmacophagous agents for diabroticite larvae

Diabroticites have been described as pharmacophagous insects (Nishida & Fukami, 1990), because the adults seek cucurbitacins for reasons other than primary metabolism or food recognition (Boppré, 1984). In laboratory feeding choice tests adults of the polyphagous *Diabrotica* spp. in the *fucata* group and grass specialists in the *virgifera* group left suitable hosts when presented with bitter cucurbits (Eben *et al.*, 1997*a*). In the field, adult Diabroticites seem to seek the acquisition of the chemical compounds as defense against natural enemies. To date, little evidence exists that Diabroticites obtain fitness benefits from this feeding behaviour. The bitter compounds are sequestered and incorporated into the hemolymph and elytra of the adult insects, and further transferred through the ovarioles to the eggs (Ferguson *et al.,* 1985). Thus, adult feeding on bitter cucurbits provides potential chemical defense not only for the beetles themselves, but also for the eggs since they acquire distastefulness. Moreover, under field conditions, the majority of the compounds are possibly ingested by the root-feeding larvae since the roots are generally the most bitter part of a number of cucurbit species (Metcalf *et al.*, 1982), followed by fruits, cotyledons, leaves and seeds, but very limited data are available of larval development on bitter cucurbit roots.

More recently, a few studies have investigated the physiological effects of a cucurbitacin containing diet on Diabroticite development and fitness. Tallamy & Gorski (1997) reared *Acalymma vittatum* (F.), the striped cucumber beetle, a common pest on cucurbits in North America, on roots with high concentrations of cucurbitacin B (*Citrullus colocynthis*) and on roots of non bitter *C. pepo* var. Crookneck. The authors tested the effect of a continuous cucurbitacin rich diet, an interrupted cucurbitacin ingestion, and a cucurbitacin free diet on larval growth rate, survival, adult longevity, and age at first reproduction. Larvae developed more quickly on a cucurbitacin containing diet and survived equally well on all diets. No significant differences on female fitness were found, but males that fed only in intervals on cucurbitacin containing diet lived significantly longer than males that ate no cucurbitacins or males that ate cucurbitacins throughout their adult lives. Their results indicate that the physiological costs of a prolonged cucurbitacin rich diet are negligible. In an earlier experiment Ferguson *et al.* (1985) reported that *A. vittatum* males had significantly shorter life spans after feeding on *Cucurbita* fruit (*C. andreana x C. maxima)* containing cucurbitacins, whereas females showed no effects, *D. balteata* LeConte and *D. virgifera* also had a significantly shorter life span after ingesting bitter cucurbits. No data on larval development on cucurbits are available for these two species.

Eben *et al.* (1997*a*, 1997*b*) assayed adults of 11 Mexican Diabroticite species and found a significant preference for bitter *C. pepo* (var. Ambassador) cotyledons over non bitter *C. pepo* (var. Crookneck) cotyledons, corn and bean leaves for all adult insects tested. However, in an assay using neonates *of D. undecimpunctata howardi, D. balteata* and *A. blomorum* Munroe & Smith a significant preference for corn roots over bitter *C. pepo* var. Ambassador and bitter over non bitter *C. pepo* var. Crookneck roots was observed (Figs. 1

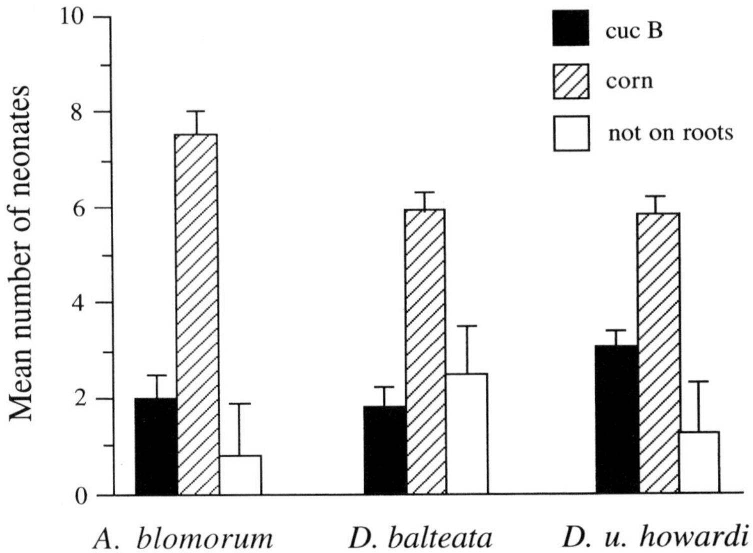

Fig. 1. Mean ± SEM number of neonate rootworms (n=10) observed on roots of bitter ('cuc B') *C. pepo* var. Ambassador and corn throughout the 36 h tests. Means were separated using a PROC GLM procedure in an analysis of variance ($P < 0.05$, F-test) (*A. blomorum*: N=200, *D. balteata*: N=300, *D. undecimpunctata howardi*: N=300). (from Eben *et al.*, 1997*a*).

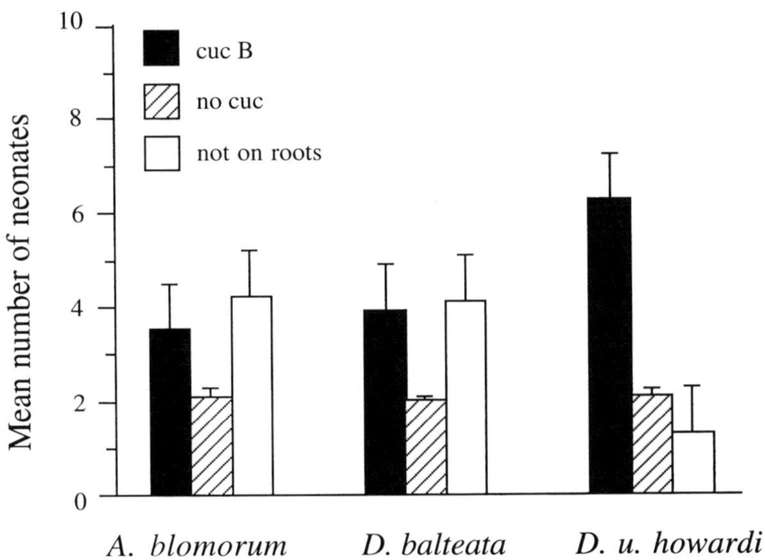

Fig. 2. Mean ± SEM number of neonate rootworms (n=10) observed on roots of bitter ('cuc B') *C. pepo* var. Ambassador and non-bitter ('no cuc') *C. pepo* var. Crookneck throughout the 36 h tests. Means were separated using a PROC GLM procedure in an analysis of variance ($P < 0.05$, F-test)(*A. blomorum*: N=140, *D. balteata*: N=480, *D. undecimpunctata howardi*: N=300). (from Eben *et al.*, 1997*a*).

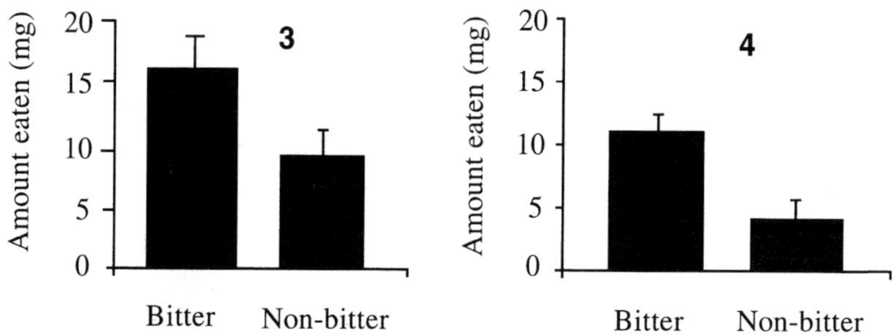

Figs. 3 and 4. Comparisons of *D. undecimpunctata howardi* larval consumption of (Fig. 3) bitter *C. andreana* versus slightly bitter *C. palmeri* roots (n=22) and (Fig. 4) nonbitter *C. sativus* 'Marketmore 80' roots exogenously coated with an ether solution of 1.0 mg/ml cucurbitacin B versus roots of the same cultivar coated only with ether (n=32). Statistical intervals are SEs. (from Deheer & Tallamy, 1991).

& 2). In order to determine larval attraction to bitter cucurbit roots, Deheer & Tallamy (1991) conducted bioassays with southern corn rootworm larvae (*D. undecimpunctata howardi*). The neonates tested showed significant preference for bitter *C. andreana* over non bitter *C. palmeri* roots, and for non bitter roots which were coated with cucurbitacin solution (Figs. 3 & 4).

Studies comparing suitability of several host plants for the development of southern corn rootworm (*D. undecimpunctata howardi)* larvae indicated that corn followed by peanuts are the best host plants, followed by non bitter *C. pepo* var. Crookneck and bitter *C. pepo* (var. Ambassador)(Hirsh & Barbercheck, 1996). In this study larval growth was significantly affected by host plant diet (Fig. 5). The results suggest that cucurbitacins had no obvious detrimental effect on larval growth, but that they did on survival to adulthood, since after four weeks adult insects emerged only from corn and peanuts. To interpret these data it should be noted that cucurbits produce large quantities of nitrogen and non protein amino acids, which may not contribute to the insects nutrition, whereas corn and peanuts are richer in carbohydrates. The same authors found after conducting experiments with an artificial cucurbitacin-containing diet, that an equal amount of cucurbitacin D was sequestered in the larvae fed on the low and high concentration diet. They concluded that there might be a threshold for the accumulation of cucurbitacins beyond which the insect must excrete the compound to avoid self-intoxication. This possibly explains the slower growth of the larvae reared on a cucurbitacin-containing diet and roots.

With adult beetles the defense hypothesis was tested, but obtained contradictory results. In feeding assays bitter beetles were offered to Chinese mantids (Ferguson & Metcalf, 1985) and sparrows (Nishida & Fukami, 1990). Both predators were repelled by the bitter tasting, cucurbitacin-containing beetles. In contrast, a study by Gould & Massey (1984) found that other vertebrate predators (toads and passerine birds) were not deterred from predating bitter adult *D. undecimpunctata howardi*. The hypothesis that the acquisition of cucurbitacins serves the purpose of chemical defense was further tested with eggs containing cucurbitacins. In laboratory, greenhouse and field experiments eggs *from D. undecimpunctata howardi* females, that had fed on a cucurbitacin C containing cucumber (*Cucumis sativus*) variety, were exposed to water stress and arthropod predators (Brust & Barbercheck, 1992) to test for potential benefits acquired through the sequestration of cucurbitacins. Common soil arthropods, such as carabid larvae (*Pterostichus chalcites* Say),

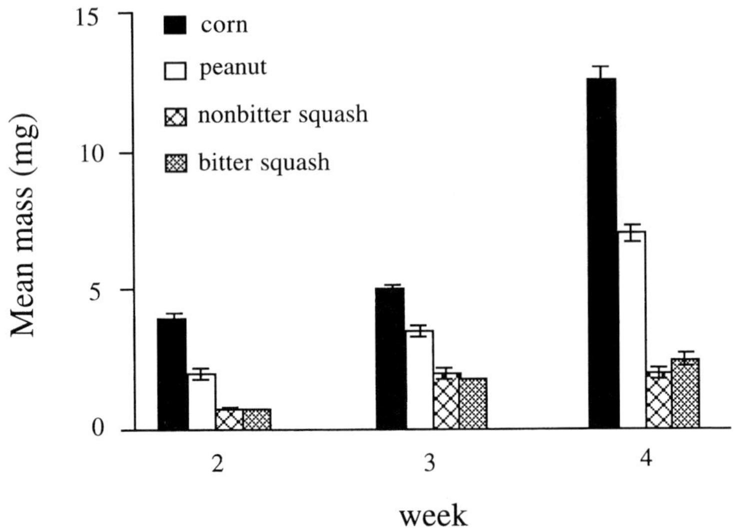

Fig. 5. Means mass of larval *D. undecimpunctata howardi* growing on corn, peanuts, or squash containing or free from bitter cucurbitacins. (from Hirsh & Barbercheck, 1996).

mites (*Proctolaelaps* sp., *Tyrophagus putrescentiae* Schrank) and centipedes, were used in these assays. No difference in predation preference on bitter or non bitter eggs was observed for any of the predators tested. In the water stress experiments bitter eggs hatched significantly earlier than non bitter eggs. The earlier hatch might be beneficial for the eggs, since the egg stage is the most vulnerable stage to soil arthropod predation in the life cycle of Diabroticites. It may also be of advantage for the larvae to begin earlier feeding on host plant roots, especially during the summer months in the so called «corn belt» of the United States when heavy showers flood the corn fields, followed by prolonged dry periods. Eggs which begin to develop earlier may take advantage of favourable environmental conditions and avoid exposure to fluctuating moisture conditions in the soil. This study showed that the transfer of cucurbitacins to eggs did not confer chemical defense from predation by several generalist soil predators, but the chemicals provided increased egg survival based on a different response to environmental conditions in the soil.

The defense hypothesis was further tested against entomopathogenic nematodes (Barbercheck *et al.*, 1995; Eben & Barbercheck, 1997) and entomopathogenic fungi (Tallamy *et al.*, 1998). Eben & Barbercheck (1997) conducted bioassays to determine the effects of host plant and substrate type on the susceptibility of *D. undecimpunctata howardi* larvae to entomopathogenic nematodes. Larvae were reared on corn, peanuts, and two Cucurbit species. One cucurbit contained cucurbitacin E (*Citrullus colocynthis*), the other lacked cucurbitacins (*Cucurbita pepo* var. Crookneck). The nematodes *Heterorhabditis bacteriophora* (Heterorhabditidae) and *Steinernema carpocapsae* (Steinernematidae) were applied at two concentrations to soil and sand, where rootworms were transferred from the different host plants. Host plant and substrate type significantly affected the susceptibility of rootworm larvae to both nematode species. The hypothesis that secondary compounds in cucurbits serve as defense compound against natural enemies of rootworm larvae was not supported by the results obtained in this study. In contrast, *D. undecimpunctata howardi*

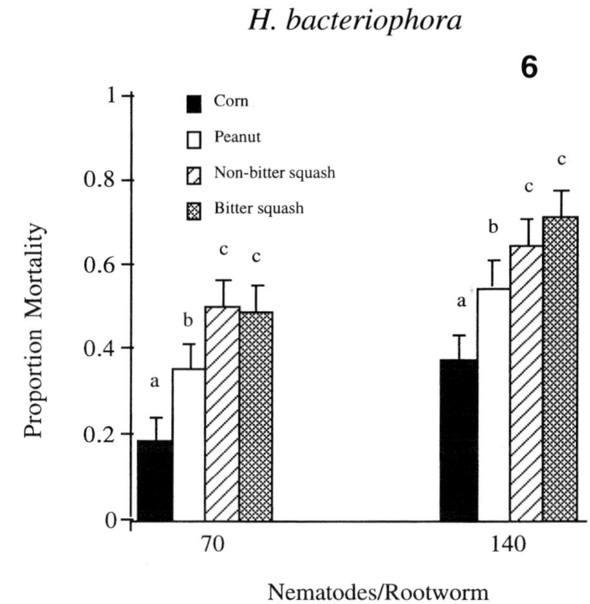

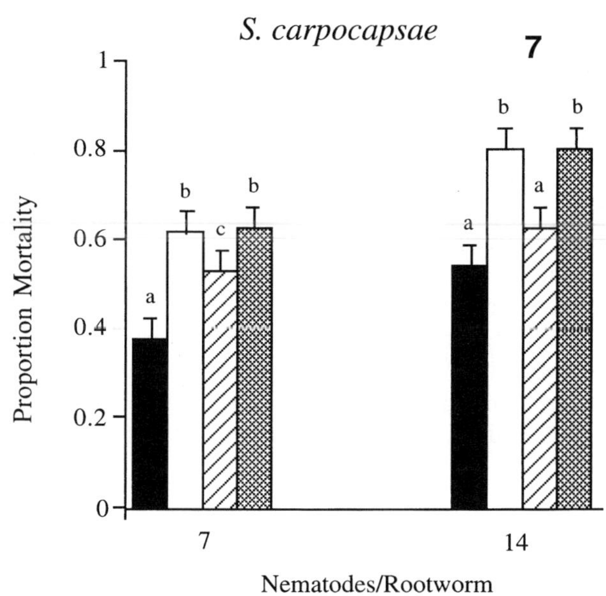

Figs. 6 and 7. Proportion mortality of southern corn rootworm larvae in soil after exposure to 70 and 140 *Heterorhabditis bacteriophora*/rootworm (Fig. 6.) or to 7 and 14 *Steinernema carpocapsae*/rootworm (Fig. 7) for 72 h. Rootworms fed for 7 days on bitter *Citrullus colocynthis*, nonbitter *Cucurbita pepo*, corn, or peanuts. Because rootworm weight affected rootworm mortality from *H. bacteriophora*, data shown are least square means of mortality adjusted for rootworm weight. Means in each concentration superceded by the same letter are not different according to comparison of least square means of arcsine transformed mortality data. Rootworm weight did not affect rootworm mortality from *S. carpocapsae* and unadjusted means are shown. Means in each concentration superceded by the same letter are not different ($P > 0.05$) according to Fisher's Protected LSD. (from Eben & Barbercheck, 1997).

larvae were significantly more susceptible to infection by *H. bacteriophora* after feeding on cucurbit hosts. Rootworms that had fed on bitter *C. colocynthis* were smaller that rootworms reared on corn and suffered higher mortality compared to larvae reared on corn, when exposed to both species of entomopathogenic nematodes tested (Figs. 6 & 7). Rootworm weight significantly influenced mortality by both nematode species, which indicated that higher mortality rates of rootworms reared on cucurbitacin-producing cucurbits cannot be interpreted as an effect of secondary compounds alone. Rootworms may have been nutritionally stressed after having been reared on cucurbit roots which could have contributed to higher mortality rates. In general, larval mortality rates were significantly higher in soil. In this study ingestion and sequestration of cucurbitacins resulted in costs for the insects in terms of sublethal effects. In a similar study by Barbercheck *et al.* (1995) feeding on *C. pepo* var. Ambassador, which contained cucurbitacin B, also rendered rootworm larvae more susceptible to the same species of entomopathogenic nematodes. Furthermore, the plants on which the rootworms had fed, significantly affected nematode progeny production. Progeny production of nematodes from rootworms which had fed on the bitter cucurbit was lower than from rootworms fed on the non bitter cucurbit. Progeny production was highest from rootworms that had fed on corn, lower for peanuts and lowest for rootworms that had fed on cucurbits. The entomopathogenic nematodes tested have symbiotic bacteria, *Xenorhabdus* spp. and *Photorhabdus* spp., which are important for the development and reproduction of nematodes within the insect host (Woodring & Kaya, 1988). Barbercheck & Wang (1996) found retarded growth of *Xenorhabdus* spp. in in vitro experiments with cucurbitacin D. Inhibition of the bacterial symbiont may be a factor that contributed to the observed effects on nematode progeny production in rootworms that had fed on bitter cucurbits.

Tallamy *et al.* (1998) examined the effect of sequestered cucurbitacins in eggs and larvae of *D. undecimpunctata howardi* to reduce pathogenicity of the entomopathogenic fungus *Metarhizium anisopliae*. Bitter and non bitter eggs and larvae were exposed to various concentrations of *M. anisopliae* conidia and a significantly greater percentage of eggs and larvae containing cucurbitacins survived infestation by the fungus (Fig. 8). Their data suggest that sequestration of cucurbits conferes an antibiotic benefit to Diabroticites.

All the studies reviewed support the hypothesis that pharmacophagous feeding on bitter cucurbits observed for adult Diabroticites, followed by the sequestration of the chemicals and the transfer to eggs and to the females via sperm from mating with «bitter» males might have considerable adaptive value for certain life stages of Diabroticites.

Conclusions and suggestions for further research

In the experimental studies published to date, important effects of cucurbitacin sequestration on the second and third trophic level were found. Since no data are available on oviposition and larval development under field conditions, and the distribution of adult beetles in cucurbit flowers, as well as their preference for bitter cucurbits are the most remarkable observations commonly reported for all species so far studied, the pharmacophagy of Diabroticites could be a relict of an ancestral host plant relationship with Cucurbitaceae, which is maintained through natural selection based on the advantage gained by the sequestration of the bitter compounds. Many species that have no obligate association with

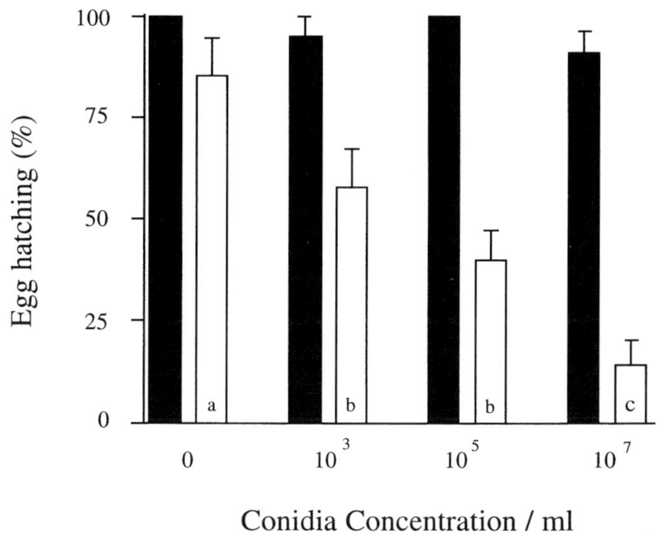

Fig. 8. Effect of various doses of *M. anisopliae* conidia on the survivorship of spotted cucumber beetle eggs in which cucurbitacins had (solid bars) and had not (clear bars) been sequestered. Means with letters different from the control concentration (0 conidia per milliliter) denote significant pathogenicity (Ryan Q multiple comparison test at P = 0.05). Statistical intervals = standard errors. (from Tallamy *et al.*, 1998).

Cucurbitacae in either life stage were generally found on bitter, wild cucurbits in the field in eastern Mexico (Eben & Barbercheck, 1996).

These observations and previously reported data from laboratory experiments with field collected insects, where adult beetles left suitable host leaves untouched and only ate bitter cucurbit cotyledons indicate, that the pharmacophagy might be restricted to adult insects, whereas eggs are deposited on host plants more suitable for larval development. Further studies on oviposition and larval development are necessary to understand the patterns of the Diabroticite-cucurbit associations observed under field conditions.

Tallamy & Pesek (1996) reported how carbon isotopic signatures of Diabroticites elytra can be used to determine larval diet in rootworms. They used *D. undecimpunctata howardi* and *Acalymma vittatum* as model organisms and fed them a diet of either C_3 plants throughout their life, or C_3 plants as larvae and C_4 plants as adults, or C_4 plants exclusively. The ratio between ^{12}C and ^{13}C reflected accurately whether an insect consumed C_3 or C_4 plants in either life stage. This assay could be used to determine the host plant breadth for mono- and dicotyledonous hosts of Diabroticites collected in the field from different plant families and in distinct habitats.

Above all, tropical species of Diabroticites need to be more thoroughly investigated with regard to the above mentioned questions, since these species are still living in close association with both wild and cultivated species of Cucurbitaceae.

Acknowledgement

I wish to express my sincerest gratitude to Dr. Pierre Jolivet for encouraging me to write this review.

References

Abou-Fakhr, E.M., Hibbard, B.E. & Bjostad, L.B. 1994. Tolerance of western corn rootworm larvae (Coleoptera: Chrysomelidae) to 6-methoxy-2-benzoxazolinone, a corn semiochemical for larval host location. J. econ. Ent. 87: 645-652.

Anaya-Rosales, S., Martinez, N.B. & Ruiz, B.D. 1992. Manejo fitosanitario de las hortalizas en Mexico. Colegio de Postgraduados Chapingo, Mexico.

Bach, C.E. 1977. Distribution of *Acalymma vittatum* and *Diabrotica virgifera* (Coleoptera: Chrysomelidae) on cucurbits. Grt. Lakes Ent. 10: 123-125.

Barbercheck, M.E., Wang, J. & Hirsh, I.S. 1995. Host plant effects on entomopathogenic nematodes. J. Invert. Path. 66: 169-177.

Barbercheck, M.E. & Wang, J. 1996. Effect of cucurbitacin D on in vitro growth of *Xenorhabdus* and *Photorhabdus* spp., symbiotic bacteria of entomopathogenic nematodes. J. Invert. Path. 66: 169-177.

Bjostad, L.B. & Hibbard, B.E. 1992. 6-methoxy-2-benzoxazolinone: a semiochemical for host location by western corn rootworm larvae. J. Chem. Ecol. 18: 931-944.

Boetel, M.A., Walgenbach, D.D., Hein, G.L., Fuller, B.W. & Gray, M.E. 1992. Oviposition site selection of the northern corn rootworm (Coleoptera: Chrysomelidae). J. econ. Ent. 85: 246-249.

Boppré, M. 1984. Redefining «pharmacophagy». J. Chem. Ecol. 10: 1151-1154.

Branson, T.F. 1982. Olfactory response of larvae of *Diabrotica virgifera virgifera* to plant roots. Entomologia exp. appl. 31: 303-307.

Branson, T.F. & Krysan, J.L. 1981. Feeding and oviposition behavior and life cycle strategies of *Diabrotica*: an evolutionary view with implications for pest management. Envir. Ent. 10: 826-831.

Branson, T.F. & Krysan, J.L. 1987. Laboratory biology of *Diabrotica tibialis* and a closely related *Diabrotica* species (Coleoptera: Chrysomelidae). J. Kans. ent.. Soc. 60: 446-450.

Branson, T.F. & Ortman, E.E. 1967. Host range of larvae of western corn rootworm. J. econ. Ent. 60: 201-203.

Branson, T.F. & Ortman, E.E. 1970. The host range of larvae of the western corn rootworm: further studies. J. econ. Ent. 63: 800-803.

Brust, G.E. & Barbercheck, M.E. 1992. Effect of dietary cucurbitacin C on southern corn rootworm (Coleoptera: Chrysomelidae) egg survival. Envir. Ent. 21: 1466-1471.

Cardona, C., Gonzalez, R. & Schoonhoven, A.V. 1982. Evaluation of damage to common beans by larvae and adults of *Diabrotica balteata* and *Ceratoma facialis*. J. econ. Ent. 75: 324-327.

Chambliss, O.L. & Jones, C.M. 1966. Cucurbitacins: specific insect attractants in Cucurbitaceae. Science 153: 1392-1393.

David, A. & Vallance, D.L. 1955. Bitter principles of Cucurbitaceae. J. Pharm. Pharmacol. 7: 295-296.

Deheer, C.J & Tallamy, D.W. 1991. Affinity of spotted cucumber beetle (Coleoptera: Chrysomelidae) larvae to cucurbitacins. Envir. Ent. 20: 1173-1175.

Eben, A. & Barbercheck, M.E. 1992. Host plant mediation of *Diabrotica undecimpunctata howardi* ecology. Paper presented at the Annual Meeting of the Entomological Society of America, Baltimore, MD.

Eben, A. & Barbercheck, M.E. 1996. Field observations on host plant associations and natural enemies of diabroticite beetles (Chrysomelidae: Luperini) in Veracruz, Mexico. Acta zool. mex. 67: 47-65.

Eben, A. & Barbercheck, M.E. 1997. Host plant and substrate effects on mortality of southern corn rootworm from entomopathogenic nematodes. Biol. Control 8: 89-96.

Eben, A., Barbercheck, M.E. & Aluja, M. 1997*a*. Mexican diabroticite beetles: I. Laboratory test on host breadth of *Acalymma* and *Diabrotica* spp. Entomologia exp. appl. 82: 53-62.

Eben, A., Barbercheck, M.E. & Aluja, M. 1997*b*. Mexican diabroticite beetles: II. Test for preference of cucurbit hosts by *Acalymma* and *Diabrotica* spp. Entomologia exp. appl. 82: 63-72.

Ferguson, J.E. & Metcalf, R.L. 1985. Cucurbitacins: plant derived defense compound for diabroticites (Coleoptera: Chrysomelidae). J. Chem. Ecol. 11: 311-318.

Ferguson, J.E., Metcalf, R.L. & Fisher, D.C. 1985. Distribution and fate of cucurbitacin B in five species of diabroticites. J. Chem Ecol. 11: 1307-1321.

Fukami, H. & Nishida, R. 1990. Sequestration of distasteful compounds by some pharmacophagous insects. J. Chem. Ecol. 16: 151-164.

Golden, K.L. & Meinke, L.J. 1991. Immature development, fecundity, longevity, and egg diapause of *Diabrotica longicornis* (Coleoptera: Chrysomelidae). J. Kans. ent. Soc. 64: 251-256.
Gould, F. & Massey, A. 1984. Cucurbitacins and predation of the spotted cucumber beetle, *Diabrotica undecimpunctata howardi*. Entomologia exp. appl. 36: 273-278.
Hibbard, B.E. & Bjostad, L.B. 1988. Behavioral responses of western corn rootworm larvae to volatile semiochemicals from corn seedlings. J. Chem. Ecol. 14: 1523-1539.
Hibbard, B.E. & Bjostad, L.B. 1990. Isolation of corn semiochemicals attractive and repellent to western corn rootworm larvae. J. Chem. Ecol. 16: 3425-3439.
Hirsh, I.S. & Barbercheck, M.E. 1996. Effects of host plant and cucurbitacins on larval *Diabrotica undecimpunctata howardi* (Coleoptera: Chrysomelidae). Entomologia exp. appl. 81: 47-51.
Howe, W.L., Sanborn, J.R. & Rhodes, A.M. 1976. Western corn rootworm adults and spotted cucumber beetle association with *Cucurbita* and cucurbitacins. Envir. Ent. 5: 1043-1048.
Jermy, T. 1984. Evolution of insect/host plant relationships. Am. Nat. 124: 609-630.
Krysan, J.L. & Branson T.F. 1983. Biology, ecology, and distribution of *Diabrotica*. In Gordon, D.T. , Knoke, J.K., Nault, LR. & Ritter, R.M. (eds.). Proceedings International Maize Virus Disease Colloquium and Workshop, 2-6 August 1982. The Ohio State University, Ohio Agric. Res. and Development Center, Wooster.
Krysan, J.L. & Smith, R.F. 1987. Systematics of the *virgifera* species group of *Diabrotica* (Coleoptera: Chrysomelidae: Luperini). Entomography 5: 375-484.
Lewis, P.A. & Metcalf, R.L. 1996. Behavior and ecology of Old World Luperini beetles of the genus *Aulacophora* (Coleoptera: Chrysomelidae). Chemoecology 7: 150-155.
Maulik, S. 1936. The fauna of British India. Including Ceylon and Burma. Coleoptera: Chrysomelidae: Galerucinae. Taylor & Francis, London.
Meinke, L.J. 1984. The phenology and ovipositional ecology of southern corn rootworm, *Diabrotica undecimpunctata howardi* Barber (Coleoptera: Chrysomelidae) in eastern North Carolina. Ph.D. dissertation, North Carolina State Univ., Raleigh.
Metcalf, R.L. 1986. Coevolutionary adaptations of rootworm beetles (Coleoptera: Chrysomelidae) feeding upon *Cucurbita* spp. J. Chem. Ecol. 12: 1109-1124.
Metcalf, R.L. 1994. Chemical ecology of diabroticites. In: Jolivet, P.H., Cox, M.L. & Petitpierre, E. (eds.) Novel aspects of the biology of Chrysomelidae. Kluwer Academic Publ. Dordrecht, pp. 153-169.
Metcalf, R.L., Rhodes, A.M., Metcalf, R.A., Ferguson, J.E., Metcalf, R.E., & Lu, P.Y. 1982. Cucurbitacin contents and diabroticite (Coleoptera: Chrysomelidae) feeding upon *Cucurbita* spp. Envir. Ent. 11: 931-937
Munroe, D.D. & Smith, R.F. 1979. A revision of the systematics of *Acalymma sensu stricto* Barber (Coleoptera: Chrysomeliae) from North America including Mexico. Mem. ent. Soc. Can. 112.
Nishida, R. & Fukami, H. 1990. Sequestration of distateful compounds by some pharmacophagous insects. J. Chem. Ecol. 16: 151-164.
Nishida, R., Yokoyama, M. & Fukami, H. 1992. Sequestration of cucurbitacin analogs by New and Old World chrysomelid leaf beetles in the tribe Luperini. Chemoecology 3: 19-24.
Reed, G.L., Myers, H.S. & Powell, J.D. 1984. Comparison of cucurbit cultivars as hosts and description of a technique for rearing the striped cucumber beetle (Coleoptera: Chrysomelidae). J. econ. Ent. 77: 337-338.
Rehm, S., Enslin, P.R., Meeuse, A.D.J. & Wessels, J.H. 1957. Bitter principles of the Cucurbitaceae. VII. The distribution of bitter principles in this plant family. J. Sci. Food Agric. 8: 679-686.
Risch, S.J. 1981. Insect herbivore abundance in tropical monocultures and polycultures: an experimental test of two hypotheses. Ecology 62: 1325-1340.
Scriber, J.M. 1983. The evolution of feeding specialization, physiological efficiency, and host races in selected Papilionidae and Saturniidae. In Denno, R.F. & McClure, M.S. (eds.). Variable plants and herbivores in natural and managed systems. Academic Press, N.Y. pp. 373-412.
Scriber, J.M. & Hainze, J.H. 1987. Geographic invasion and abundance facilitated by differential host plant utilization abilities. In Barbosa, P. & Schultz, J.C. (eds.). Insect outbreaks. Academic Press, N.Y. pp. 433-468.
Sell, R.A. 1915. Some notes on the western twelve-spotted and the western striped cucumber beetle. J. econ. Ent. 8: 515-520.
Siegfried, B.D. & Mullin, C.A. 1990. Effects of alternative host plants on longevity, oviposition, and emergence of Western and Northern Corn Rootworms (Coleoptera: Chrysomelidae). Envir. Ent. 19: 474-480.
Singer, M.C., 1983. Determinants of multiple host use by a phytophagous insect population. Evolution 37: 389-403.
Sinha, A.K. & Krishna, S.S. 1969. Feeding of *Aulacophora foveicollis* on cucurbitacin. J. econ. Ent. 62: 512-513.
Sinha, A.K. & Krishna, S.S. 1970. Further studies on the feeding behaviour of *Aulacophora foveicollis* on cucurbitacin. J. econ. Ent. 63: 333-334.
Smith, R.F. 1966. Distributional patterns of selected western North American insects: the distribution of diabroticites in western North America. Bull. ent. Soc. Am. 27: 108- 110.
Spencer, K.C. (ed.). 1988. Chemical mediation of coevolution. Academic Press, New York.
Strnad, S.P. & Dunn, P.E. 1990. Host search behaviour of neonate western corn rootworm (*Diabrotica virgifera virgifera*). J. Insect Physiol. 36: 201-205.

Tallamy, D.W. & Gorski, P.M. 1997. Long- and short-term effect of cucurbitacin consumption on *Acalymma vittatum* (Coleoptera: Chrysomelidae) fitness. Envir. Ent. 26:672-677.

Tallamy, D.W. & Krischik, V.A. 1989. Variation and function of cucurbitacins in *Cucurbita*: an examination of current hypotheses. Am. Nat. 133: 766-786.

Tallamy, D.W. & Pesek, J.D. 1996. Carbon isotopic signatures of elytra reflect larval diet in Luperine rootworms (Coleoptera: Chrysomelidae). Envir. Ent. 1167-1172.

Tallamy, D.W., Whittington, D.P., Defurio, F., Fontaine, D.A., Gorski, P.M. & Gothro, P.W. 1998. Sequestered cucurbitacins and pathogenicity of *Metarhizium anisopliae* (Moniliales: Moniliaceae) on spotted cucumber beetle eggs and larvae (Coleoptera: Chrysomelidae). Envir. Ent. 27:366-372.

Tallamy, D.W., Stull, J. Ehresman, N.P., Gorski, P.M., & Mason, C.E. 1998. Cucurbitacins as feeding and oviposition deterrents to insects. Envir. Ent. 26: 678-683.

Webster, F.M. 1895. On the probable origin, development and diffusion of North American species of the genus *Diabrotica*. J.N.Y. ent. Soc. 3: 158-166.

Whitaker, T.W. & Bemis, W.P. 1965. Evolution in the genus *Cucurbita*. Evolution 18: 553-559.

Wilcox, J.A. 1972. Chrysomelidae: Galerucinae, Luperini: Diabroticina and Aulacophorina. Pars. 78. Fasc. 2. In Wilcox, J.A. (ed.). Coleopterum Catalogus. Supplementa. 2nd ed., pp. 296-343. Uitgerij Dr. W. Junk Gravenhage, Netherlands.

Woodring, J.L. & Kaya, H.K. 1988. Steinernematid and heterorhabditid nematodes: a handbook of biology and techniques. South. Coop. Ser. Bull. 331.

Yaro, N. & Krysan, J.L. 1986. Host relationships of *Diabrotica cristata* (Coleoptera: Chrysomelidae). Ent. News 97: 11-16.

BIOLOGY AND ECOLOGY OF LEAF-MINING HISPINAE (COLEOPTERA, CHRYSOMELIDAE) OF THE LA SELVA BIOLOGICAL STATION, COSTA RICA

HENRY A. HESPENHEIDE and VAN DANG
Department of Organismic Biology, Ecology and Evolution, University of California Los Angeles, CA 90095-1606, USA

Abstract

Leaf-mining species of Hispinae have been collected at the La Selva Biological Station in Costa Rica for 20 years as parts of projects studying leaf-mining beetles and inventorying the arthropod fauna. In all, 135 hispine species have been collected, 77 of which are leaf-miners, of which in turn 35 have been reared. Hosts and brief descriptions of the mines are given for 31 species, as well as data on extent of damage and parasitoids. Overall, leaf-mining hispines were found to use 5 monocot and 17 dicot families as hosts, of which 9 are reported here for the first time. Most species are monophagous based on limited data, but host families are quite diverse within genera. Although there is a diversity of mine types, larvae most commonly form solitary blotch mines from eggs laid on the upper surface of the leaf. Mines tend to be infrequently encountered and rarely cause significant damage to a host plant. Limited rearing has produced fewer pupal parasitoids than for other beetle leaf-miners, but inspection of egg masses of communal species suggests levels of egg parasitism may be higher.

Introduction

The New World members of the chrysomelid subfamily Hispinae include two basic groups of species defined by larval ecology, external feeding species and leaf-miners. Externally feeding species may be exposed or hidden in rolled leaves or bracts and are associated primarily with monocots – grasses, palms, and families in the Zingiberales (Maulik, 1937; Strong, 1977) – and in the New World tropics are dominated by the large genus *Cephaloleia* (Staines, 1996). Leaf-miners are somewhat more diverse overall (see below) and have the advantage of being easier to study by rearing (Hespenheide, 1991). For the last several years the arthropods of the La Selva Biological Station in Costa Rica have been the object of extensive ecological sampling by the Arthropods of the La Selva (ALAS) project (see below), and the Hispinae have been one of the focal taxa for analysis (Staines, 1997). To date, 135 hispine species have been collected, of which 58 are external feeding species and 77 are leaf-miners. Despite these efforts, slightly fewer than half of the leaf-mining species have been reared (36) and additional species continue to be found, both by field collecting and rearing, so that the present summary must be considered provisional.

There have been several reviews of the host plants of hispines (Maulik, 1937; Jolivet, 1988; Santiago Blay MS), and other papers give rearing records (Bondar, 1938; Uhmann, 1944). Maulik (1937) summarized data on the food plants of 148 (out of a described 2570) species of hispines, which were found to use 27 plant familes. Of the 71 New World spe-

Advances in Chysomelidae Biology 1, pp. 375–389
Edited by M.L. Cox

cies whose hosts were reported, 57 were leaf-miners which used 17 families of host plants. Costa Rican species of Hispinae were well represented in that group because of the publications of Uhmann (especially 1934,1937; see also Staines & Staines 1989) on the collections and rearings of Schmidt, Nevermann, and others. Working in Brasil, Bondar (1938) increased to 23 the number of recorded plant families mined by hispines.

The present paper gives the results of collections and rearings at La Selva by the senior author and his graduate students as part of a more general survey of leaf-mining beetles (Hespenheide, 1991, 1994). In addition to host plant data, which add nine families to the list of those mined by the Hispinae, characteristics of mines are described briefly for most reared species, and ecological notes on extent of damage and parasitism are included. General patterns of host use, mine structure and ecology are given in the concluding section.

La Selva

The La Selva Biological Station is located in Heredia Province, Costa Rica, about 3 km S of the town of Puerto Viejo de Sarapiqui, between the confluence of the Rio Sarapiqui and the Rio Puerto Viejo. The station is run by the Organization for Tropical Studies, a consortium of universities and research organizations in the United States and Costa Rica, and is located at the foot of Volcan Barva at one extension of the Braulio Carillo National Park. Ecologically the station consists of an older forested section and more recently acquired disturbed tracts in various stages of succession or used for experimental studies. The natural vegetation is lowland tropical wet forest with an average annual rainfall of about 3.5 m. Rainfall peaks in July and again in October, and there is a relatively dry season from mid-December to May, but average monthly rainfall is not less than 0.2 m for any month. There has been a recent, extensive description of the ecology of the station and a review of research there (McDade *et al.*, 1994). This paper principally reports on collections from La Selva, but also includes a few rearings made within 2 km of the station. Within the station, most collections have been made at forest edges, at clearings and along trails, and in the successional plots on the northeast corner of the oldest part of the property.

Methods

The collections and rearings upon which this paper is based have been made more or less opportunistically over the last 20 years. The rearing protocol is simple (Hespenheide, 1991; Hespenheide & Kim, 1992): Mines with pupae or late instar larvae are collected in the field; pupae are isolated in small vials to allow them to eclose; and mined leaves are pressed and dried for identification and later analyses. Leaves with late instar larvae are placed in resealable plastic bags until larvae complete feeding and pupate. Leaves with mines are then pressed and dried for later examination and measurement. In a few cases old mines were collected for study. Measurements for leaf size are for the blade only, exclusive of the petiole.

Because the Hispinae have not been the primary taxon of interest among leaf-mining beetles, there has been some bias in the collection of mines. The most important family of plant hosts for the Hispinae are the grasses (Poaceae; Maulik, 1937), but collections of mines at La Selva have focussed principally on dicots or monocot families other than grasses. For this reason, it is important to note that in this paper there is a disproportionate scarcity of rearings of grass-feeding hispines, especially those of the genus *Chalepus*. The ALAS project has used four major ecological sampling methods: canopy fogging, malaise trapping, light trapping and litter extraction. Except for the presence of a few species on the overall list of La Selva Hispinae, results of this sampling will be reported elsewhere.

Results – Species accounts:

Baliosus productus (Baly)

Host: undetermined Bignoniaceae – liana; leaves compound, mined leaflet 4.8 cm wide, 13.2 cm long, with an area of 40.7 cm^2.

Mine characteristics (n = 1): full-depth blotch mine at apex of leaf on both sides of rachis, about 13.4 cm^2 in area; egg placed on ventral surface, 6.4 cm from leaf base, 1 cm from rachis.

Phenology: active late July; pupal period at least 8 days.

Baliosus sp. 1

Host: *Urera bogotaense* (Urticaceae) – liana, moderately common along forest edges; leaves simple, mined leaf 6.8 cm wide, 11.9 cm long, with an area of 55.7 cm^2.

Mine characteristics (n = 1): full-depth blotch mine at apex of leaf on one side of rachis, about 7.4 cm^2 in area; egg not recognized.

Phenology: active late July; pupal period at least 8 days.

Baliosus sp. 3

Host: *Odontonema tubaeforme* (Acanthaceae) – herbs in secondary growth, leaves narrowly obovate, acuminate, mined leaves 6.8-10.2 cm wide, 19.6-25.6 cm long with areas of 70.3-188 cm^2.

Mine characteristics (n = 3+): full-depth serpentine mines on any part of leaf, 7.5-9.7 cm long, 5.2-6.4 cm^2 in area, initially separate but frequently joining when, as is usual, several larvae are in a leaf; eggs placed on dorsal surface; mines lack frass, and there is no special larval resting or pupation site.

Phenology: mines collected July.

Carinispa nevermanni Uhmann

Host: reported as *Malphigia* sp. (Malphigiaceae; Uhmann, 1937); collected but not yet reared at La Selva.

Chalepus consanguineus Baly

Host: undetermined Poaceae – mined leaves 9.5-12 mm wide, 13.3-16.7 cm long, with an area of 8-13.2 cm^2.

Mine characteristics (n = 2): full-depth mine occupying entire width of blade in apical portion of leaf to within 12-28 mm of tip, about 6.1 cm^2 in area; egg on ventral surface at basal 1/3 of mine within 2 mm of blade margin.

Phenology: pupae collected late July, adults emerged early August.

Chalepus digressus Baly
Host: an early rearing of this species at La Selva gave the host as "Tiliaceae;" Memmott *et al.* (1993) recorded hosts as members of the genus *Lasiacis* in the Poaceae. The mine was not available for study.

Glyphuroplata nigella (Weise)
Host: undetermined Poaceae – blade scabrous, 0.6 cm wide, 10.2 cm long.
Mine characteristics (n = 1): mine occupies all but apical 1.0 cm of blade, mostly superficial, but small sections full-depth; egg not recognized.
Phenology: active August; pupal period 5-8 days.
Literature: recorded mining *Valota* (Poaceae; Riley, 1985).

Heptispa limbata (Baly)
Hosts: *Cassia fruticosa* (Caesalpiniaceae) – shrub or small tree, common in successional plots and along forest edges; leaves pinnately compound with four leaflets, mined leaflets (n = 4) 6.7-10.5 cm wide, 10.2-21.0 cm long, and with an area of 50-148 cm^2.
– *Inga edulis* (Mimosaceae) – tree to 20m, commonly seeding into Successional Plots and along forest edges; leaves pinnately compound with 3-6 pairs of leaflets, mined leaflets (n = 5) 9.7-15.9 cm wide, 23.5-33.2 cm long, and with an area of 142-188 cm^2.
– *Mimosa* sp. (Mimosaceae) – vine or liana along forest edges; leaves bipinnately compound with 6 pairs of leaflets, mined leaflet (n = 1) 2.5 cm wide, 4.8 cm long, and with an area of 8.1 cm^2.
Mine characteristics: superficial irregular blotch mines, usually in apical half of leaf (*Inga*) or not (*Cassia*); eggs placed on dorsal surface, at the edge of the mine; small amounts of frass at edges of mine, and there is no special larval resting or pupation site. Mines on *Cassia* about 1.8 cm^2 in area and conspicuously darkened; on *Inga*, about 3.6 cm^2 in area, superficial and transparent; on *Mimosa* about 8.1 cm^2 in area and darkened.
Phenology: mines collected March, May, July, August.
Literature: Uhmann (1934, 1937) reported *Inga, Cassia grandis,* and *Machaerium* as hosts. He also described and illustrated (Fig. 2, 1937) a mine in a host identified as *Serjanea* sp. (Sapindaceae), but the leaf appears to be misidentified and to be a legume instead.

Heterispa vinula (Erichson)
Host: *Apeiba membranacea* (Tiliaceae) – secondary growth tree to 20 m; leaves obovate, entire; mined leaves 9.7-13.8 cm wide, 17.8-21.7 cm long, and with areas of 116-201 cm^2.
Mine characteristics (n = 2): superficial to full-depth communal mines, starting as serpentine but becoming an irregular blotch, in all parts of leaf but more frequently in apical half, 34.2-47.0 cm^2 in area; clutches of eggs placed on ventral surface, along major lateral veins; frass more concentrated and conspicuous in serpentine portion of mine, and there is no special larval resting or pupation site.
Phenology: abundant on a fallen tree in late July with many fresh egg masses.
Parasitoids: eggs were frequently parasitized; an early-instar larva was found parasitized by a small hymenopteran parasitoid which made a cocoon; numbers of a braconid species were reared from late instar larvae, and a eulophid was found in an old braconid cocoon, probably as a hyperparasitoid.

Literature (as *Heterispa westwoodi* Baly): Uhmann (1934, 1937) recorded hosts as *Triumfetta* (Tiliaceae) and *Sida* (Malvaceae).

Octhispa bimaculata Uhmann

Host: *Stigmatophyllum lindenianum* (Malphigiaceae) – vine, common in second growth and at forest margins; leaves ovate, three-lobulate or entire; mined leaf lobulate, 12.4 cm wide, 12.1 cm long, and with an area of about 62.3 cm^2.

Mine characteristics (n = 1): irregular blotch mine at apex of leaf about 3.8 cm^2 in area; egg not found; small amounts of frass at edges of mine, and there is no special larval resting or pupation site.

Phenology: mine collected May.

Octhispa decepta (Baly)

Host: *Stigmatophyllum lindenianum* (Malphigiaceae) – see *Octhispa bimaculata*; mined leaf entire, 6.2 cm wide, 10.2 cm long, and with an area of about 42 cm^2.

Mine characteristics (n = 1): full-depth communal blotch mine with 3 larvae, occupying more than half the leaf area; egg mass placed on ventral surface, along lateral vein near base of leaf about 1.5 cm along vein; some frass placed on dorsal surface of mine, and there is no special larval resting or pupation site.

Phenology: mine collected May.

Octhispa elegantula (Baly)

Host: *Serjanea* sp. (Sapindaceae) – liana, leaves palmately compound; mined leaflet 6.4 cm wide, 12.1 cm long, with an area of about 38.6 cm^2.

Mine characteristics (n = 1): superficial blotch mine, occupying most of one side of leaf, area about 10.4 cm^2; clutch of three eggs on dorsal surface 2/3 distance from rachis to margin; thin lines of frass placed on dorsal surface of mine, and there is no special larval resting or pupation site; only one adult produced from the clutch of three eggs.

Phenology: mine collected July.

Parasitoids: one of three eggs appeared to be parasitized.

Literature: Uhmann (1937) reported the host as *Paullinia* (also Sapindaceae).

Octhispa elevata (Baly)

Host: probably *Paullinia* sp. (Sapindaceae) – liana, leaves palmately compound; mined leaflet 4.6 cm wide, 13.7 cm long, with an area of about 33.3 cm^2.

Mine characteristics (n = 1): full-depth blotch mine, occupying most of one side of leaf, area about 11.3 cm^2; egg not found; some frass placed on dorsal surface of mine, and there is no special larval resting or pupation site.

Phenology: mine collected July.

Literature: Uhmann (1934) reported hosts as *Paullinia costaricensis* (Sapindaceae) and *Pithoctenium* (Bignoniaceae; Uhmann, 1937).

Octhispa haematopyga (Baly)

Host: *Colubrina spinosa* (Rhamnaceae) – shrubs in secondary growth; leaves obovate, acuminate, mined leaves 5.0-9.2 cm wide, 12.3-19.0 cm long with areas of 41.6-118.3 cm^2.

Mine characteristics (n = 4): more or less full-depth lobulate blotch mines near middle of leaf, single mines about 7.4 cm^2 in area; eggs on dorsal surface; larvae construct a special round or elliptical resting area beneath the egg (about 1.7 cm^2 in area), cover the dorsal surface with frass, and feed in tunnels which radiate outward from the center of the mine; this resting area is also the pupation site

Phenology: mines collected July and August; pupal period at least 10 days.

Parasitoids: a rearing from this host produced eulophids as pupal parasitoids.

Octhispa nevermanni Uhmann

Host: *Ochroma lagopus* (Bombacaceae) – trees to 30 m tall in secondary growth. Leaves large, simple, palmately lobed.

Mine characteristics (n = 1): full-depth blotch mine on dorsal surface, about 3.5 cm^2 in area; egg placed next to rachis very near a lateral vein; mine has holes at the margins for voiding frass, and there is no special larval resting or pupation site.

Phenology: mine collected July; pupal period about 4 days.

Oxychalepus alienus (Baly)

Host: *Cassia fruticosa* (Caesalpiniaceae) – see description under *Heptispa limbata*; mined leaflet 7.6 cm wide, 13.3 cm long, and with an area of 68.3 cm^2.

Mine characteristics (n = 1): full-depth blotch mine occupying apical 1/3 of leaf, about 20 cm^2 in area; egg on ventral surface, at the tip of the leaflet; mine lacks frass and special larval resting or pupation site.

Phenology: mine collected April.

Oxychalepus posticatus (Baly)

Host: *Cassia fruticosa* (Caesalpinaceae) – see description under *Heptispa limbata*; mined leaflets 5-10.3 cm wide, 9.2-18.3 cm long, and with an area of 42-113 cm^2.

Mine characteristics (n = 13): full-depth blotch mines, usually at apex of leaf, about 27 cm^2 in area; eggs placed on ventral surface, usually at the edge and near the tip of the leaflet; mines lack frass, and there is no special larval resting or pupation site. In one case there were two mines on one leaflet.

Phenology: mines collected March, May, July; pupal period 5-8 days

Oxyroplata nr. *bellicosa* (Baly)

Host: undetermined – mined leaf obovate, acuminate and 5.8 cm wide, 18 cm long with an area of about 67 cm^2.

Mine characteristics (n = 1): full-depth communal blotch mine with 3 larvae beginning as a serpentine mine, about 30.3 cm^2 in area; eggs placed in group on dorsal surface midway between rachis and margin beyond middle of leaf; some fine frass placed on dorsal surface of mine, and there is no special larval resting or pupation site.

Phenology: mine collected July, August.

Literature: Uhmann (1937) recorded the host of *Uroplata bellicosa* Baly as *Banisteria argentea* (Bignoniaceae). The mine he described seems somewhat different from the La Selva rearing.

Pentispa explanata (Chapuis)

Host: Uhmann (1934) recorded *Pithoctenium* sp. (Bignoniaceae) as the host; it has been collected but not yet reared at La Selva.

Pentispa fairmairei (Chapuis)

Host: reported hosts include *Calea axillaris, C. urticifolia, Clibadium* sp., *Elephantopus* sp., *Eupatorium populifolium, Verbesina* sp., and *Vernonia* sp. (all Asteraceae; Uhmann, 1934, 1937); collected but not yet reared at La Selva.

Platocthispa championi (Baly)

Hosts: *Piper* sp. (Piperaceae) – shrubs in understory or at forest margins. Leaves simple, in mined species hastate and about 13 cm wide, 26 cm long with an area of about 220 cm^2.

– *Calathea* sp. (Marantaceae) – small to large understory herbs. Leaves simple, ovate, acuminate, large, mined leaves about 10-16 cm wide, 21-27.5 cm long with areas of about 137-318 cm^2.

Mine characteristics (*Piper*; n = 2): superficial blotch mines on dorsal surface, about 9.6 cm^2; eggs not found; mines lack frass and have holes at the margins, and there is no special larval resting or pupation site. Two separate mines were found in the same leaf.

Mine characteristics (*Calathea*; n = 4): superficial irregular/lobulate blotch mines on dorsal surface near margins usually at or beyond the middle, about 5.4-6.5 cm^2; one egg on dorsal surface at edge of mine; mines have sparse frass and holes at the margins, and there is no special larval resting or pupation site. Two leaves had 3 or 4 separate mines.

Phenology: mines found in April, August; pupation takes 8-10 days.

Platocthispa sp.1

Host: *Ochroma lagopus* (Bombacaceae) – see *Octhispa nevermanni*; mine not available for description.

Phenology: mine collected July.

Probaenia armigera (Baly)

Host: *Piptocarpha chontalensis* (Asteraceae) – lianas; leaves ovate, entire, mined leaves 10.8-17.5 cm wide, 21.4-29.3 cm long, with an area of about 230 cm^2.

Mine characteristics (n = 6): more or less full-depth blotch mines, usually somewhat confined between stronger lateral veins on one side of leaf, near or just beyond the middle, 21.5-29 cm^2 in area; eggs placed on dorsal surface, usually adjacent to rachis; mines with little frass and no special larval resting or pupation site.

Phenology: mines collected May, July, October; pupal period of 10-11 days.

Probaenia pici Uhmann

Host: *Mikania guaco* (Asteraceae) – vine; leaves ovate, entire, mined leaf 5.9 cm wide, 13.8 cm long with an area of 52.6 cm^2.

Mine characteristics (n = 1): full-depth blotch mine, beyond the middle, 11.6 cm^2 in area; egg not found; mine with little frass and no special larval resting or pupation site.

Phenology: mine collected July.

Probaenia sp.1

Host: *Arrabidaea chica* (Bignoniaceae) – vine; leaves compound with two leaflets and median tendril, leaflets ovate, mined leaflets 3.5-7.1 cm wide, 8.0-9.8 cm long with areas of 19.0-39.5 cm^2.

Mine characteristics (n = 4, 2 reared): full-depth lobulate blotch mines near middle on one side of leaf, single mature mines 6.1-8.5 cm^2 in area; eggs on dorsal surface within 1 mm of a secondary vein; larvae construct a special round or elliptical resting area beneath the egg (0.7-1.2 cm^2 in area), cover the dorsal surface with frass, and feed in broad tunnels which radiate outward from the center of the mine; this resting area is also the pupation site.

Phenology: mines collected July.

Sceloenopla erudita (Baly)

Host: *Cupania* sp. (Sapindaceae) – understory tree, leaves serrate, entire, mined leaf 7.3 cm wide, 16.8 cm long; mine not available for study.

Phenology: mine in May; pupal period 8 days.

Literature: Uhmann (1944) reported *Anthurium* (Araceae) as a host.

Sceloenopla godmani (Baly)

Host: *Clusia flava* (Clusiaceae) – epiphytic shrubs, leaves very succulent, obovate, mined leaves 4.2-5.1 cm wide, 9.3-11.1 cm long.

Mine characteristics (n = 2): Full-depth blotch mines occupying most of apex of leaf; larvae apparently need several leaves to complete development and move between leaves.

Phenology: mines in July.

Parasitoid: an undetermined Tachinidae was reared from one pupa.

Sceloenopla longula (Baly)

Host: undetermined Araceae – epiphytic vine, mined leaf elliptical, acute, 5.2 cm wide, 17.7 cm long.

Mine characteristics (n = 1): more or less superficial lobulate blotch mine at apex on both sides of rachis, 8.5 cm^2 in area; egg placed on dorsal surface, midway between margin and rachis; mines with little frass and no special larval resting or pupation site.

Phenology: mine in August; pupal period 9 days.

Sceloenopla obscurovitatta (Baly)

Host: *Philodendron radiatum* var. *radiatum* (Araceae) – epiphytic vine, leaves very large, elliptical, pinnately dissected.

Mine characteristics (n = 5): elongate full-depth communal blotch mines. Egg mass usually placed on ventral surface, several observed with exit holes of egg parasitoids; larva leaves mine and constructs separate cell in leaf a few cm from mine in which to pupate. Adults make linear feeding lesions on leaves.

Phenology: mines collected July.

Sceloenopla scherzeri (Baly)

Host: *Davilla nitida* (Dilleniaceae) – liana; leaves somewhat scabrous, mined leaves 6-9.5 cm wide, 13.8-30 cm long.

Mine characteristics (n = 2): full-depth serpentine mines around margin of leaf, 1-2 cm wide; larvae apparently need more than one leaf to complete development and move between leaves; eggs placed on ventral surface at edge of leaf; mines with slits at margin where frass is voided; a special larval resting and pupation site along rachis at tip of leaf, or between two lateral veins and reaching rachis.

Phenology: mines in July; pupal period at least 14 days.

Sceloenopla sp.3

Host: *Sterculia recordiana* var.*papyracea* (Sterculiaceae) – small to large understory trees; leaves palmately lobed, large, one mined leaf 29.3 cm wide, 26.4 cm long, with area of about 506 cm^2.

Mine characteristics (n = 5, one reared): full-depth elongate blotch mines along margins of major veins or the edge of leaf, single mines about 18.6 cm^2 in area; of five eggs found, all were on dorsal surface at edges of mine, three along major veins, one along a secondary vein and one at the tip of a leaf lobe; mines have holes at the margins from which frass is voided, and there is no special larval resting or pupation site. Although eggs are laid singly, leaves appear frequently to have more than one mine – two leaves had 2-5 separate mines – and adjacent mines appear to join.

Phenology: active mine found in August.

Sumitrosis amica (Baly)

Host: *Heliconia* spp. (Musaceae) – large herbs to 4 m; leaves large, mined leaves 25-38 cm wide.

Mine characteristics (n = 6): superficial to more or less full-depth transversely elongate blotch mines, usually confined between stronger lateral veins, 8-19 cm^2 in area; eggs placed on dorsal surface, usually at the edge of a lateral vein about 41% of the distance from rachis to margin; mines lack frass, and there is no special larval resting or pupation site.

Phenology: mines collected March, June, July; pupal period at least 6 days.

Parasitoid: *Conura* sp. (Chalcididae) from pupae.

Sumitrosis pallescens (Baly)

Host: "pea," probably Fabaceae, mine not available for study.

Literature: *Cassia* (Caesalpiniaceae) has been reported as a host in the United States (Cavey, 1994).

Sumitrosis terminata (Baly)

Host: undetermined Fabaceae – herbaceous vine at forest edge; trifoliate, mined leaflet 3.3 cm wide, 5.2 cm long with an area of 8.8 cm^2.

Mine characteristics (n = 1): full-depth blotch mine occupying tip and most of one side of leaflet, about 2.7 cm^2 in area; egg on dorsal surface, 3 mm from tip of the leaflet; mine has little frass and no special larval resting or pupation site.

Phenology: mine collected July, adult emerged August.

Uroplata fusca Chapuis

Host: Uhmann (1934) recorded *Pithoctenium echinatum* (Bignoniaceae) and an unidentified plant as hosts. Although not yet reared at La Selva, an adult was collected on a plant in the Bignoniaceae.

Uroplata sculptilis Chapuis

Hosts: *Inga edulis* (Mimosaceae) – see *Heptispa limbata* above; mined leaflets 12-22 cm wide, 21.4-41.8 cm long with areas of 285-591 cm^2.

– *Gouania adenophora* (Rhamnaceae) – liana, ovate, toothed leaves; mine not available for study.

Mine characteristics (*Inga*; n = 7, 5 reared): small full-depth lobulate blotch mines in almost any part of leaf, more frequently in apical 1/2, single mines about 2.58 cm^2 in area; eggs on dorsal surface along a secondary vein; larvae construct a special round or elliptical resting area beneath the egg, cover the dorsal surface with frass, and feed in broad tunnels which radiate outward from the center of the mine; this resting area is also the pupation site. Leaves frequently appear to have more than one mine – two leaves had 2-4 separate mines.

Phenology: pupae collected late July, adults emerged early August; pupal period at least 8 days.

Parasitoid: a *Conura* sp. (Chalcididae) was reared from a pupa in a second mine on *Gouania* (mined leaf 6.4 cm wide, 9.0 cm long) that was probably of this species.

Literature: Uhmann (1934, 1937) recorded *Clibadium asperum* (Asteraceae) and "*Sidedrella*" (perhaps *Sida*, Malvaceae) as hosts. The mine in "*Sidedrella*" Uhmann (1937) pictured is quite different from the mines described here.

Xenochalepus amplipennis (Baly)

Host: undetermined Fabaceae – vine, trifoliate, mined lateral leaflet broadly, obliquely ovate, acute, 7.1 cm wide, 7.9 cm long, 40.5 cm^2 in area.

Mine characteristics (n = 1): full-depth communal blotch mine; all but about 5 cm of area mined; clutch of four eggs laid along rachis on underside of leaf, of which three hatched.

Phenology: pupae collected late August.

Xenochalepus erythroderus (Chapuis)

Hosts: *Coussapoa nymphaeifolia, C. villosa, Cecropia insignis, Pourouma bicolor* (all Cecropiaceae); *Coussapoa* – hemiepiphytic tree to 15 m tall; leaves ovate, simple, large

– *Cecropia* – tree, usually in forest gaps to 30 m; leaves large, peltate, 7-9-lobed; mine not available for study.

– *Pourouma* – tree, usually in secondary growth to 30 m; leaves large, palmately, 3-5-lobed; mine typical of this species observed (L. La Pierre, pers. comm.) but not available for study.

Mine characteristics (*Coussapoa*; n = 6): superficial, somewhat elongate communal blotch mines, usually confined between stronger lateral veins, 38-40 cm^2 in area; eggs placed in oblique stack on dorsal surface, usually at the edge of a lateral vein about 5.3 cm from rachis to margin; mines lack frass, and there is no special larval resting or pupation site.

Phenology: mines collected January, July.

Parasitoids: an adult eulophid was observed on an egg mass (L. La Pierre, pers. comm.); pupal parasitoids include undetermined Tachinidae, *Conura* sp. (Chalcididae), and a pimpline ichneumonid.

Xenochalepus sp.

Host: *Anthurium* sp. (Araceae) – epiphytic vine.

Mine characteristics (n = 1): superficial serpentine mine; eggs placed on dorsal surface, about 1.2 cm from rachis to margin; mine lacks frass, and there is no special larval resting or pupation site.

Phenology: mine collected July.

Unsuccessful rearings from other hosts:
Clibadium pittieri (Asteraceae) – herb in secondary growth. *Clibadium* sp. was reported as a host of *Pentispa fairmairei* Chapuis (see above).

Discussion

Hosts
As indicated in the introduction, Maulik (1937) reported that 57 New World species of leaf-miners used 17 families of host plants and Bondar (1938) increased the number of host plant families to 23. The present study of just 34 hispine species whose hosts could be determined at least to the family level (22 families – 5 monocots, 17 dicots; Table 1) increases this number by nine to 32. The added families are the monocots Marantaceae, Musaceae, and Zingiberaceae, and the dicots Acanthaceae, Bombacaceae, Clusiaceae, Dilleniaceae, Piperaceae, and Urticaceae, although the monocot families are known to be hosts of externally-feeding hispines. Donald M. Windsor (unpublished) and others in Panamá have studied 37 species of leaf-mining Hispinae that use 19 families and that include at least five additional families.

The table shows that there are very few patterns in the use of hosts. There are nine (of 14 total) hispine genera for which hosts are known for more than one species. Of these nine, five genera (including the one species, *Platochispa championi*) use both monocots and dicots as hosts and all of the nine except *Oxychalepus* (two species) use more than one family. Most species have been reared from only a single host, but the small total number of rearings makes it difficult to asses the true extent of monophagy. Two species have been reared from more than one host genus in a single family, and three have been reared from more than one family.

The Hispinae are only one of several groups of beetle leaf-miners at La Selva, which include members of the Buprestidae, Curculionidae, and the subfamily Alticinae of the Chrysomelidae (Hespenheide, 1991). In terms of species richness, they are second only to the Buprestidae (117 species). Although hispines use some hosts that are not known to be used by other beetles (Araceae, Asteraceae, and Poaceae), in a number of cases they share hosts with one or more other beetle taxa, including hosts in the plant genera *Arrabidaea, Cecropia, Cupania, Coussapoa, Gouania, Inga, Mimosa, Ochroma, Sterculia, Stigmatophyllum,* and *Urera*. Although hispines at La Selva rarely reach ecologically important densities (see below), resource competition and within-leaf specialization (Nielson, 1978) may be important in some cases and are currently being studied in *Cecropia, Coussapoa* and *Pourouma* (Cecropiaceae; L.M. La Pierre, unpublished).

Life histories
There is considerable diversity of types of mines in this small sample of hispines. Eggs are placed on the dorsal (adaxial) surface of the leaf in 16 of the 25 species (64%) for which eggs could be recognized and 9 (36%) on the ventral (abaxial) surface. This is nearly opposite the trend found by Reavey and Gaston (1991) for leaf-mining Lepidoptera in Great Britain, where 73% of 227 species lay only on the lower surface, 21% on the upper

Table 1. Larval host plants of leaf-mining Hispinae reared at La Selva Biological Station, Costa Rica.

Hispine species	Larval host genus (family)
Baliosus productus (Baly)	undetermined (Bignoniaceae)
Baliosus sp. 1	*Urera* (Urticaceae)**
Baliosus sp. 3	*Odontonema* (Acanthaceae)**
Chalepus consanguineus Baly	undetermined (Poaceae)
Chalepus digressus Baly	undetermined (Tiliaceae)
Glyphuroplata nigella (Weise)	undetermined (Poaceae)
Heptispa limbata (Baly)	*Inga, Mimosa* (Mimosaceae)
	Cassia (Caesalpiniaceae)
Heterispa vinula (Erichson)	*Apeiba* (Tiliaceae)
Octhispa bimaculata Uhmann	*Stigmatophyllum* (Malphigiaceae)
Octhispa decepta (Baly)	*Stigmatophyllum* (Malphigiaceae)
Octhispa elegantula (Baly)	*Paullinia, Serjanea* (Sapindaceae)
Octhispa elevata (Baly)	undetermined
Octhispa haematopyga (Baly)	*Colubrina* (Rhamnaceae)
Octhispa nevermanni Uhmann	*Ochroma* (Bombacaceae)**
Oxychalepus alienus (Baly)	*Cassia* (Caesalpinaceae)
Oxychalepus posticatus (Baly)	*Cassia* (Caesalpinaceae)
Oxyroplata nr. *bellicosa* (Baly)	undetermined
Platocthispa championi (Baly)	*Piper* (Piperaceae)**
	Costus (Zingiberaceae)*
	Calathea (Marantaceae)*
Platocthispa sp.1	*Ochroma* (Bombacaceae)
Probaenia armigea (Baly)	*Piptocarpha* (Asteraceae)
Probaenia pici Uhmann	*Mikania* (Asteraceae)
Probaenia sp. 1	*Arrabidaea* (Bignoniaceae)
Sceloenopla erudita (Baly)	*Cupania* (Sapindaceae)
Sceloenopla godmani (Baly)	*Clusia* (Clusiaceae)**
Sceloenopla longula (Baly)	undetermined (Araceae)
Sceloenopla obscurovitatta (Baly)	*Philodendron* (Araceae)
Sceloenopla scherzeri (Baly)	*Davilla* (Dilleniaceae)**
Sceloenopla sp .3	*Sterculia* (Sterculiaceae)
Sumitrosis amica (Baly)	*Heliconia* (Musaceae)*
Sumitrosis pallescens (Baly)	undetermined (Fabaceae?)
Sumitrosis terminata (Baly)	undetermined (Fabaceae)
Uroplata sculptilis Chapuis	*Inga edulis* (Mimosaceae)
	Gouania (Rhamnaceae)
Xenochalepus amplipennis (Baly)	undetermined (Fabaceae?)
Xenochalepus erythroderus Chapuis	*Cecropia, Coussapoa, Pourouma* (Cecropiaceae)
Xenochalepus sp.	*Anthurium* (Araceae)

**New host plant family for Hispinae; *new host plant family for leaf-mining Hispinae.

surface, and 6% on both. They expected most species to lay on the lower surfaces for reasons of microclimate or leaf structure and suggested that those which lay on the upper surface do so because "oviposition is time-limited and laying on the upper surface is quicker and simpler." Laying several eggs in a clutch takes longer, and of the seven hispine species that lay multiple eggs, three lay on the upper, four on the lower surface, somewhat different from the 18 solitary layers (13 on the upper, 5 on the lower).

Hispines produce several different types of mines between the extremes of serpentine and blotch mines. Of the 32 species for which mine structure could be studied, only 3 are narrowly serpentine in form. Communal miners in some cases begin as serpentine mines but eventually become a blotch mine as most of the leaf area is consumed (*Heterispa*, *Oxyroplata*). Some species have elongate blotch mines, in some cases constrained by leaf shape (*Sceloenopla obscurovitatta*; species using grasses – *Chalepus consanguineus*, *Glyphuroplata*) or venation (*Sumitrosis amica*), and in other cases not obviously constrained (*Sceloenopla* sp.3, *Xenochalepus erythroderus*). The most complex mines can be called lobulate blotch mines. In these mines larvae construct a round central chamber beneath the egg placement and coat the dorsal part of the chamber with fecal material; from the chamber the larva feeds in tunnels that radiate outwards, and pupation takes place in the chamber. In one species the larva leaves the mine and makes a small chamber separate from the mine in which it pupates (*Sceloenopla obscurovitatta*). Another species of *Sceloenopla* makes a separate chamber for resting and pupation at the edge of the mine (*S. scherzeri*).

Ecology

The ecological effect of hispines on their hosts is probably usually minimal at La Selva (but see Hespenheide, 1991, for literature on economically important species elswhere). In a few cases populations on a single plant have been observed to be high enough to cause significant damage to a number of leaves (*Baliosus* sp.3 on *Odontonema*, *Probaenia* sp.1 on *Arrabidaea*, *Sceloenopla obscurovittata* on *Philodendron*, *Sceloenopla* sp.3 on *Sterculia*), but mines are more often few and difficult to find. Most mines have a single larva and consume only a small part of a leaf or leaflet (0.5-37% for 16 species measured; mean = 15%), except for the two species that use grasses, which consume most of the leaf blade (61-98%). Larvae of two species use more than one leaf (*Sceloenopla godmani*, *S. scherzeri*), and some species have been observed to have several mines on the same leaf (damaging species above, and *Platochispa championi* and *Uroplata sculptilis*). Communal miners also have higher rates of damage to single leaves or leaflets (16-87% for 5 species measured; mean = 48%). Because intensive collections have been made in only a few months of the year, no conclusions can be drawn about the importance of seasonal phenology to the Hispinae. Seasonality is important at other tropical sites, such as Barro Colorado Island in Panamá (Hespenheide, 1994), but rainfall is more evenly distributed at La Selva as noted above.

Parasitoids

The parasitoids of hispines have been recently reviewed (Mariau, 1988; Cox, 1994). Ecologically, rearings at La Selva have yielded few parasitoids, although because only later instar larvae or pupae are collected, parasitism at earlier stages would be undetected. Although egg parasistism is difficult to detect in solitary laying species, inspection of clutches of eggs of communal miners (*Heterispa*, *Sceloenopla obscurovitatta*, *Xenochalepus erythroderus*) shows evidence of moderate levels of parasitism, although the parasitoids have not been reared. Dissection of larger samples of old mines also indicates parasitism of early-instar larvae (*Heterispa*). Because of the rearing protocol, pupal parasitoids are most easily detected and include Tachinidae, undetermined eulophids, braconids, *Conura* spp. (Chalcididae), and a pimpline ichneumonid. Overall, levels of pupal parasitism seem relatively low for hispines compared to leaf-mining buprestid beetles (Hespenheide, unpublished). Memmott *et al.* (1993) reared 21 species of parasitoids from 3 *Chalepus* species in an extensive ecological study in Guanacaste, Costa Rica, and also observed high levels (29-61%) of predation from crawling predators. Memmott *et al.* (1994) also found that the number of parasitoids per leaf-miner host increased with sample size. These data and the more intensive study of single species at La Selva (*Xenochalepus erythroderus*; L.M. La Pierre unpublished) suggest our knowledge of hispine parasitoids is very incomplete.

Acknowledgements

The authors are greatly indebted to Charles L. Staines, Jr., for identifications of the Hispinae and comments on the manuscript. Graduate students of HAH – Louis LaPierre, Grant Gentry, and Thomas W. Sherry – have each contributed several rearings of species he had missed. Plants have been identified by a number of individuals over the years, especially by Orlando Vargas R., Naturalist at the La Selva Biological Station. Donald M. Windsor supplied unpublished rearing records from Panamá. HAH has been supported throughout the period by the Research Committee of the UCLA Academic Senate, and more recently by grants in support of the ALAS Project (National Science Foundation grants BSR 9025024, DEB 9401069 and DEB 9706976), as well as from personal funds.

References

Bondar, G. 1938. Notas entomologicas da Bahia. II. Revta Ent., Rio de J. 8:1-24.

Cavey, J.F. 1994. Annotated new distributional records for North American Chrysomelidae (Coleoptera). Coleopts. Bull. 48:1-9.

Cox, M.L. 1994. The Hymenoptera and Diptera parasitoids of Chrysomelidae. In: Jolivet, P., Cox, M.L. and Petitpierre, E. (eds.); Novel Aspects of the Biology of Chrysomelidae, Kluwer Academic Publishers, pp. 419-467.

Hespenheide, H.A. 1991. Bionomics of leaf-mining insects. A. Rev. Ent. 36: 535-560.

Hespenheide, H.A. 1994. The fauna of La Selva. In, *La Selva: Ecology and History of a Neotropical Rainforest*, L. McDade, K.S. Bawa, H.A. Hespenheide and G.S. Hartshorn, eds. pp. 238-242. Chicago: University of Chicago Press.

Hespenheide, H.A., and C. Kim. 1992. Clutch size, survivorship, and larval biology of *Pachyschelus psychotriae* Fisher (Coleoptera, Buprestidae). Ann. ent. Soc. Am. 85: 48-52.

Jolivet, P. 1988. Food habits and food selection of Chrysomelidae. Bionomic and evolutionary perspectives. In: Jolivet, P., Petitpierre, E, and Hsiao, T.H. (eds.); Biology of Chrysomelidae, Kluwer Academic Publishers, pp. 1-24.

Mariau, D. 1988. The parasitoids of Hispinae. In: Jolivet, P., Petitpierre, E, and Hsiao, T.H. (eds.); Biology of Chrysomelidae, Kluwer Academic Publishers, pp. 449-461.

Maulik, S. 1937. Distributional correlation between hispine beetles and their host-plants. Proc. zool. Soc. Lond., Ser. A, 107: 129-159.

McDade, L., K.S. Bawa, H.A. Hespenheide and G.S. Hartshorn, eds. 1994. *La Selva: Ecology and History of a Neotropical Rainforest*. x + 478 pp. Chicago: University of Chicago Press.

Memmott, J., H.C.J. Godfray, and B. Bolton. 1993. Predation and parasitism in a tropical herbivore community. Ecol. Ent. 18: 348-352.

Memmott, J., H.C.J. Godfray, and I.D. Gauld. 1994. The structure of a tropical host- parasitoid community. J. Anim. Ecol. 63: 521-540.

Nielson, B.O. 1978. Food resource partition in the beech leaf-feeding guild. Ecol. Ent. 3: 193-201.

Reavey, D., and K.J. Gaston. 1991. The importance of leaf structure in oviposition by leaf-mining microlepidoptera. Oikos 61: 19-28.

Riley, E.G. 1985. Review of the North American species of *Glyphuroplata* Uhmann, 1940 (Coleoptera: Chrysomelidae). J. Kans ent. Soc. 58: 428-436.

Santiago-Blay, J.A. MS Leaf mining in the Chrysomelidae.

Staines, C.L. 1996. The genus *Cephaloleia* (Coleoptera: Chrysomelidae) in Central America and the West Indies. Revta Biol. trop., Publ. Esp. 3: 3-87.

Staines, C.L. 1997. The Hispinae (Coleoptera: Chrysomelidae) known from the La Selva Biological Station, Costa Rica. Website: www.evergreen.edu/user/serv_res/research/arthropod/hispinae/hispinae.html (temporary).

Staines, C.L., Jr., and S.L. Staines. 1989. A bibliography of New World Hispinae (Coleoptera: Chrysomelidae). Maryland Entomologist 3: 83-122.

Strong, D.R., Jr. 1977. Rolled-leaf hispine beetles (Chrysomelidae) and their Zingiberales host plants in Middle America. Biotropica 9: 156-169.

Uhmann, E. 1930. Hispinen aus Costa Rica aus der Ausbeute des Herrn Ferd. Nevermann. 20. Beitrag zur Kenntnis der Hispinen (Col. Chrys.). Folia zool. hydrobiol. 1: 209-256.

Uhmann, E. 1934. Hispinen-Minen aus Costa Rica. I Teil. 48. Beitrag zur Kenntnis der Hispinen. Arb. angew. Ent.Berlin-Dahlem 1: 272-277.

Uhmann, E. 1937. Hispinen-Minen aus Costa Rica. II Teil. 62. Beitrag zur Kenntnis der Hispinen. Arb. morph. taxon. Ent. Berl. 4: 61-67.

Uhmann, E. 1944. Hispinen-Minen aus Costa Rica. (Coleoptera: Chrysomelidae) 104. Beitrag zur Kenntnis der Hispinen. Arb. morph. taxon. Ent. Berl. 11: 59-61.

SEXUAL BEHAVIOUR AMONG CHRYSOMELIDAE

PIERRE JOLIVET
Florida State Collection of Arthropods, POBox 147100,Gainesville, FL 32614-7100, USA

Abstract

Mating among leaf-beetles is sometimes preceded by a more or less lengthy courtship of the male with a responding female. Sometimes the courtship happens only during amplexus. Competition among males is the rule. A female can reject the male if not ready or on an empty stomach and mating can last a few minutes, hours or even days and can be repeated. Cryptic female choice is often the case, and intromission does not always result in insemination, nor does insemination always result in fertilization of eggs. Mating between different species or different genera occurs when the specimens are abundant, feed on the same plant, have related colours, but so far, in nature, very few cases of hybrid species have been reported among the leaf-beetles. Pheromones, plant odour, size and genitalic compatibility have more impact than colour on interspecific crossings. Multiple mating among males or females is rather frequent. Homosexuality happens in confinement but seems absent in nature.

Introduction

Several chrysomelids are parthenogenetic (Isono, 1988; Furth, 1994; Cox, 1996), but separate sexes are the rule. Only recently, serious observations on the courtship and mating behaviour have been made. Before, pioneer reports were written, dealing exclusively with the mating process and not with the preliminaries before copulation (Gadeau de Kerville, 1900). The last paper concerns various beetle families and leaf-beetles are only briefly reviewed. Written in the pompous style of that period, the paper had however the great merit to review the topic for the first time among the Coleoptera.

Courtship sequence is common among animals. It has been especially developed among birds, but, many insects sometimes show before mating rather complicated procedures, including lek behaviour, that is a specific site where males display and compete for the attention of females. That is known for insects, mostly among Apidae (Hymenoptera), Odonata, Drosophilidae (Diptera) and Lampyridae (Coleoptera). but probably it occurs among some Alticinae, such as *Gabonia*.

Chrysomelidae have a peculiar mating behaviour, often specific, not especially complicated, and followed sometimes, among a few species, by maternal care (Goidanich,1956; Jolivet, 1988; Windsor & Choe, 1994; Kudo & al., 1995-1996). To date, additional paternal care has not been demonstrated, but could eventually be shown to exist among some Cassidinae. Maternal care seems restricted to some South American leaf-beetles (several Cassidinae, few Doryphorina) and the Holarctic *Gonioctena*. However, it is very probable that parental care is more common than generally supposed in various parts of the world with other genera. Maternal care is often, but not always, associated with cycloalexy or

Advances in Chysomelidae Biology 1, pp. 391–409
Edited by M.L. Cox

gregarious grouping of the larvae (Vasconcellos-Neto & Jolivet, 1994; Jolivet, et al.,1990). When the species are viviparous (*Platyphora*), cycloalexic species deposit their larvae together and those larvae group themselves in an approximate ring very rapidly.

Egg laying by the female can be also very complicated, as in the Megalopodinae (Monros, 1954; Yu & Xingke, 1994; Schulze,1996) and the Sagrinae. It probably derives from the Cerambycidae. Mating habits in the Megalopodinae have been recently studied by Eberhard & Marin (1996), but it seems that preliminaries in Sagrinae have never been seriously observed. However, we are dealing here only with the special behaviour of the male before and during mating, somewhat similar between the subfamilies, sometimes very simple, sometimes less. Roughly, it is related to cerambycid general behaviour and to all Phytophaga.

Generally the males are smaller than the females in the Coleoptera, but there are cases where the females are smaller (scarabaeids sensu lato). Among the chrysomelids, sexual dimorphism is also generally in favour of the female (Sacchi & Busardo, 1935), but there are cases (Clytrinae, Megalopodinae) where males are bigger due to their big sexual dimorphism.

Cassidinae females are often 10 to 20% longer than males. Sexual dimorphism, generally small in some genera, like *Chrysolina* and related genera, can be important in some subgenera of the above genus, like *Caudatochrysa* Bechyné. *Caudatochrysa* females possess a kind of ovipositor, such as in *C. angusticollis* (Motschulsky) (Bechyné, 1950). Among the genus *Timarcha* Latreille, male tarsi are greatly enlarged to facilitate grasping of the female during the amplexus. Others, like some male Camptosoma can have elongated front legs, allowing a vertical position of the male above the female during mating. Many Camptosomata males have bigger head, eyes, mandibles, and legs, a modified pronotum, enlargement of the anterior tarsi, etc. Apterous beetles, like *Timarcha*, *Elytrosphaera*, *Iscadida*, etc. have a reduced metasternum, but it is common to males and females. Some leaf-beetles have thorny appendices in the males, but modified sternites and abdominal depressions among the females. Physogastry is frequent among females of Chrysomelinae (*Gastrophysa, Oreomela*, *Xenomela*, etc.) or Galerucinae (*Aulacophora, Metacycla, Galeruca, Coelomera*, and many others). Apterism or brachelytrism (*Metacycla,Galeruca*) helps in the inflation of the gravid abdomen.

Certain genera with species with normal genitalia have species with enormous ones, such as *Monoxia puncticollis* Say (Jolivet, 1957-59). Females of such species must be deeply modified to harbour such an aedeagus. The spermatheca and bursa copulatrix, must be all adapted in consequence. The antennae also sometimes differ strongly among the species of the same genus (*Agetocera*, for instance) and among the sexes.

In certain cases, alticine eggs can be enormous (some *Longitarsus*, like *L. pratensis* (Panzer), *Clavicornaltica takimotoi* LeSage) (LeSage, 1997). It is certain that, as for cavernicolous carabids, the size of the eggs must have some relationships with the biology of the beetle concerned and the reproductive habits. Nothing really is known about the biology of the litter or canopy frequenting *Clavicornaltica* beetles. It is also evident that the ovaries, spermatheca and abdominal plates are modified to adapt to the large size of the egg. A study of the spermatophora of the males must be also interesting. Spermatozoa can be big but not necessarily, but they can be enormous among many leaf-beetles. Strategy differs enormously about reproduction from the thousands of eggs laid by the Colorado potato beetle to the hundreds of *Diabrotica virgifera* LeConte, to the dozens of *Timarcha* spp , to the few eggs of *Longitarsus* spp.,and to only one enormous egg of a *Clavicornaltica*, extreme K-strategist. This egg must be very well protected either by its situation in the soil or by its toxicity. Probably life metabolism in the soil or in the litter or in the suspended soils must be very slow and duration very long. In all cases cited above, mating apparently speeds up ovarian development. Egg protection is important to leaf-beetles and involves secretive, faecal and maternal protective devices (Damman & Cappuccino, 1991).

Yang (1992) described a new feature on the eighth tergite of male leaf-beetles, namely in the genera *Paridea* and *Pseudocophora* (Galerucinae). It must be related to the type of mating, acting as harpagones. Medvedev (1962) insisted on the functional importance of secondary sexual characters in Camptosoma as did Eberhard (1981) when referring to the sternal horn of *Doryphora* sp. (Chrysomelinae) and to the horned beetles in general (Eberhard, 1980). So, it is surprising that Monros (1953) does not see any functional role to the above characters, either for copulation preliminaries or for inter-male fighting.

Daccordi (1994,1996) mentioned special enigmatic abdominal cavities among several genera of South African and Australian Chrysomelinae. Those pouches could carry in certain cases mites but the observations were made on dead specimens only. The physiological meaning of the cavities, perhaps provided with glands or secreting hairs, remains uncertain and could be linked with hygrometry, sex pheromones or infra-red detectors. Crowson (in litt.) sees them as possible organs of pollen transfer. As long as the biology of those cryptic beetles is not known all hypotheses are possible. Fresh specimens must be studied on the spot and slide preparations made of the abdomens. *Gasterantodes*, like *Iscadida*, could feed on small bushes or more probably lead, like *Cyrtonus*, a hidden life under stones or at the foot of low grassy plants. In the case of *Gasterantodes* Daccordi, from South Africa, the pouches exist in both sexes, but in the case of the Australian *Callidemum* Blanchard and *Strumatophyma* Baly, the structures are special to males and represent a secondary sexual character, surely with an unknown physiological (sexual) function. .

1. Courtship and mating behaviour among Coleoptera

Among most of the beetles, when there is a certain number of males, mating with a female is preceded by a serious fight between males themselves or an intruder tries to dislodge a male already copulating or trying to do so. Coleoptera with big mandibles or cephalic or sternal horns are well equiped for fighting between themselves. Sexual conflict among insects has been studied in detail by Brown et al. (1997). The authors see even the possibility of female-female conflict, competing for opportunities to mate with males bearing good genes. Choe (1994,1995) has mostly studied the mating of Zoraptera but also the behaviour of Formicidae (Choe et al., 1997 a, b).

Fights among male scarabaeids, lucanids, dynastids, cerambycids, passalids, and many others are common when females are accessible in the vicinity (Palmer,1978; Eberhard, 1980-1982). Pheromones, now increasingly identified, are attracting males towards females and used as a trap, in the case of agricultural pests,such as *Diabrotica* spp., among the chrysomelids. Their use is more important in the case of lepidopterous pests. Sometimes, the pheromones are also emitted by the males which in that case attract the females.

As already mentioned by Brown & Siegfried (1983), the males of the tenebrionid beetle *Bolitotherus cornutus* Panzer vary greatly in both horn and body size. Relative horn size is an important determinant of male access to females among males of equivalent body sizes.

Horns serve for fighting and larger males are stronger and thus better fighters, says Eberhard (1982). He studied the dimorphic horned dynastine beetle, *Podischnus agenor* Olivier and suggested that minor forms are a facultative adaptation to reduce direct competition with major males. Whatever is the reason, males fight for the possession of the females. Curiously, Zelazny (1975) studying the behaviour of *Oryctes rhinoceros* (L.), also a dynastine, did not mention the common male fight known in this species. Among *Onthophagus acuminatus* Harold, a scarabaeine, male dimorphism induces alternative reproductive tactics, but both types of males have direct or indirect access to the females (Emlen, 1997). In the cerambycid *Acrocinus longimanus* (L.), sexual selection favours large size in males which are always strong fighters (Zeh & al., 1992).

Among the bees, males of *Amegilla dawsoni* (Rayment) occur in two size classes, large majors and small minors (Alcock, 1996, 1997 a & b). Major males compete aggressively for females. Majors, perhaps because they fight with rivals on the ground, are slightly shorter-lived than minor males. The two forms of males survive because of differential mortality. For the Meloidae, Bologna & Marangoni (1986) classified the sexual behaviour as follows : preliminary activity, dorsal phase, genital phase. Meloids show a quite complex courtship pattern according to the species and homosexual behaviour exists, but is only shown in overcrowded conditions, mostly in the laboratory. A dorsal position is followed by a linear position in this family. Cockchafers (Scarabaeidae, Melolonthinae) are well known for the inverted position of the males after mating. Among most of the beetles, however, females normally carry the males on their back. Passalids and scolytids, copulate sternal side against sternal side, which is possibly an adaptation to the subcortical life. Male and female passalids cooperate in the rearing of the offspring. In *Lytta nuttali* Say, a meloid, Gerber & Church (1973) showed that the male assumes the active role : antennal manipulation, abdominal vibration, and attempted genital insertion. Courtship lasts from 8 to 10 hours, the male and the female being aligned end to end.

As already mentioned by H. Gadeau de Kerville in 1900, the common mating position in the Coleoptera is the male above the female, but there are cases where the male remains on its back in the last phase (*Melolontha*), where the male and female are in line, opposed against each other (Bostrichidae, Scydmaenidae, Cantharidae). We will see that among the chrysomelids in general the normal position is the rule but male clytrines with their long anterior legs remain almost vertical above the female.

For the aquatic beetle, *Dytiscus alaskanus* J. Balfour-Browne, Aiken (1991) discerned eight patterns : mounting, lateral shake, swimming, copulation, plug visible, probing, female breathing and female swim. Copulation lasts 8 hours and the male seems to guard the female during the post-copulatory period. Mating can be very long also among the chrysomelids, but for terrestrial insects the patterns seem less complicated. Mating among Donaciinae is done outside water only.

2. Behaviour of the Chrysomelidae during mating

There are several steps before and after the amplexus and chrysomelids, if they behave grossly like curculionids, cerambycids, bruchids, or coccinellids, can present different courtship and mating systems.

2.1. Competition between males

Not only does the male of Dynastinae (Eberhard, 1977) and Scarabaeinae (Sato & Hiramatsu, 1993) fight for the possession of a female, but the chrysomelids can be also strong fighters. For instance, Eberhard (1981) mentioned that the chrysomeline *Doryphora* sp. nr. *punctatissima* (Olivier) from Colombia uses its sternal horn as a weapon in aggressive encounters on the food-plant, an Apocynaceae. Aggressive behaviour of the males seems also to be related to territory defence. The beetles display little apparent courtship behaviour and the males detect the female from 5-10mm away. Copulation lasts 85 minutes on average and females mate more than once.

Windsor (1987) studied the tortoise beetle *Acromis sparsa* Boheman in Panama and showed that courtship and mating occurs on the apical foliage of the vine, *Merremia umbellata*, a Convolvulaceae. Severe fighting occurs between males, attempting to chase, immobilize or dislodge competing males. The extension of the humeral angle of the elytra is used for fighting and often causes oblong holes in the elytra.

Practically, the males of all chrysomelids seem to fight for females or territory, but some groups such as the Megascelinae, the Aulacoscelinae, the Palophaginae, the Orsodacninae and the Zeugophorinae, have not yet been observed fighting, but they probably do it like the others. *Aulacoscelis* seems to mount the female without apparent preliminaries and I did not see the males fighting. Since the species are gregarious it must happen in the wild on *Zamia*, their host-plant. Sagrinae and Donaciinae, like the cerambycids, must be strong fighters, but surprisingly enough the literature is very poor on the topic. In some species,such as *Labidomera clivicollis* Kirby, Dickinson (1996) stated that males locate mates via scramble competition, but there is little conflict among males in the field, females being generally in sufficient numbers. In cages, when many males are gathered together, that can be different. Shimizu and Fujiyama (1986) described aggressive encounters between males of *Chrysolina aurichalcea* (Mannerheim) and sperm competition is certainly involved. An unpaired male is occasionally seen attacking a mating male on the host-plant, an *Artemisia*. Intruders resulted in take-over (10.4%), loss (36.5%), draw (39.0%) or non-interference (14.1%). The size of the males can be related to success and frontal attack has, according to Shimizu & Fujiyama (1986), a 5 times higher probability of take-over than a rear one. It is certain that Sagrinae males fight for females, but this fighting has been rarely observed, perhaps because hatching of the imagos outside the stem galls is very rapid and generally follows the first rains. Mating follows almost immediately. The giant hind legs of the male of *Megalopus armatus* Lacordaire (Megalopodinae) function as weapons in battles between males over sites where feeding and mating occur. Those places (Solanaceae or Asteraceae) are generally not used to court females, but males court females actively during copulation (Eberhard & Marin, 1996). Eberhard (1981) found similar habits with *Doryphora* sp. in Colombia: adults also display little apparent courtship behaviour. This can vary greatly among chrysomelids, but a courtship ritual is always dispensed with amplexus. Fighting among males occurs first for the possession of females, but also for defending territories on stalks or leaves (Mitchell, 1980). Males of *Chelymorpha alternans* Boheman in Costa Rica court females both before and during copulation (Rodriguez, 1995). Rodriguez (1993,1994a) also studied courtship procedures among several Central America Cassidinae : *Charidotella* sp.nr. *sexpunctata* (F.), *Omaspides convexicollis* Spaeth and *O. bistriata* Boheman.

The cerambycid *Acrocinus longimanus* (L.) shows intense male competion to monopolize suitable sites for egg deposition (Zeh & al., 1992) and that behaviour resembles the fighting of male *Doryphora* sp. on their host-plants for territory defence. According to Medvedev & Pavlov (1988), there is fierce competition between the males in Clytrinae, such as *Labidostomis longimana* L. and *Coptocephala* sp. A single isolated male will try to dislodge the male from a copulating pair. For that, the intruder will use its forelegs and mandibles. As we see, the behaviour can vary intergenerically but also intraspecifically.

2.2. *Preliminaries and courtship behaviour*

Courtship can be different according to the genus or the species. In certain cases, except for the pheromone emitted by the female, there are no apparent preliminaries at all. Generally, however there is a ritual dance around and on the female to make her receptive.

In *Longitarsus jacobaeae* (Waterhouse), the females emit a sex pheromone that is attractive to male beetles on the tansy ragwort leaves, *Senecio jacobaea*. According to Brown (1993), a size-assortive mating exists in the galerucine *Trirhabda canadensis* Kirby. That can be due to the greater ease with which size-matched pairs can achieve intromission. Anyhow, the difference in size between different males or different females is not very important in most of the Chrysomelidae, except for the Clytrinae and related Camptosomata, where sexual dimorphism is more developed.

The male specific antennal glands of *Psylliodes chrysocephala* L. may secrete a sex pheromone and the antennae are very active at the beginning and at the end of copulation (Bartlet et al., 1994). Such preliminaries linked with a pheromone must be common among many leaf-beetles, mostly Galerucinae and Alticinae. Many Galerucinae possess very complex structures on the antennae, such as the Asian *Agetocera* (Laboissière, 1927; Maulik, 1936). The physiology of those structures has never been studied. The antennae differ generally between males and females. Medvedev (1962) clearly demonstrated that secondary sexual characters of Camptosomata, for instance, have a sexual function.

Courtship and mating of *Labidomera clivicollis* Kirby has been extensively studied by Dickinson (1986-1997). Males antennate and climb rapidly on female backs. Females are often refractory to mounting and drop to the ground. Even virgin females kick at male genitalia with their hind legs, pull their abdomen and walk between branches to avoid intromission. Once mounted, a refractory female can even inhibit insemination. *Labidomera clivicollis*, like *Chrysolina herbacea* (Duftschmidt), exhibits prolonged mating (up to 2.5 days). A complicated courtship maintains the male with the female and it takes between 24

→

Figs. 1–6. Mating habits of various chrysomelids.
Fig. 1. Mating habits of the Colorado Potato Beetle, *Leptinotarsa decemlineata* (Say). (Chrysomelinae). A starving female retracting her abdomen under the elytra refusing the male penis.
Fig. 2. The same. Mating of the male with a well-fed female, accepting the male.
Fig. 3. Two males of *Acromis sparsa* Boheman (Cassidinae). One male during the fighting has been lifted from the surface.
Fig. 4. Starting of the mating process in *Gastrophysa polygoni* (L.) (Chrysomelinae).
Fig. 5. Maximum activity of *Gastrophysa polygoni* during mating.
Fig. 6. Two males of *Doryphora* sp., battling on the stem of *Prestonia isthmica* Woodson (Apocynaceae) in Colombia. They use the mesosternal horn during the fight.
1. & 2., after Thibout, 1982; 3., after Windsor,1987. 4. & 5., after Medvedev & Pavlov, 1988; 6., after Eberhard, 1981.

1

2

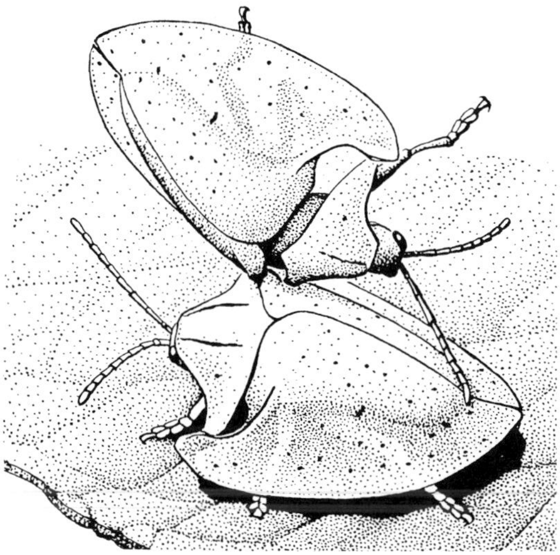

3

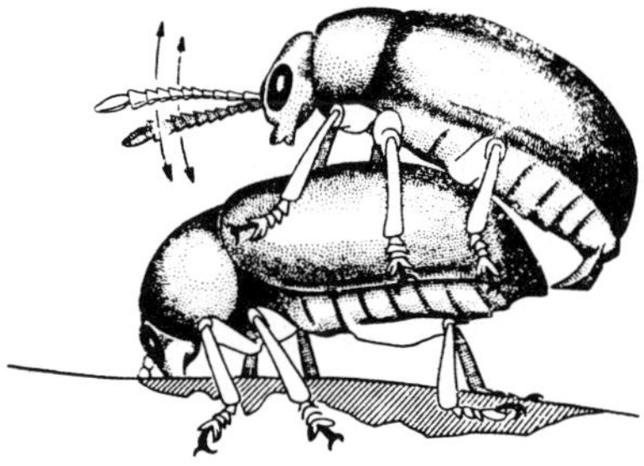

4

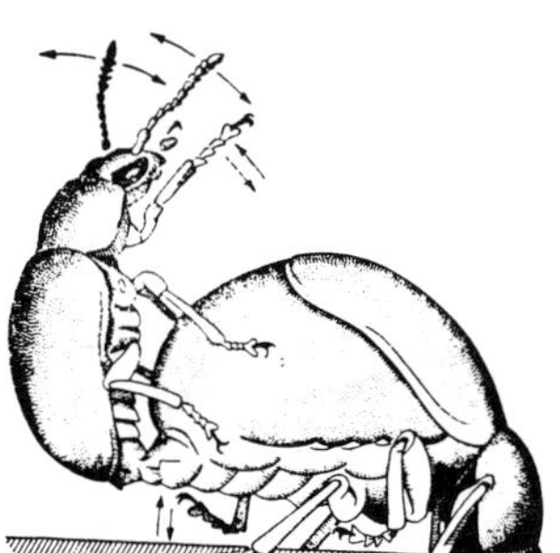

5

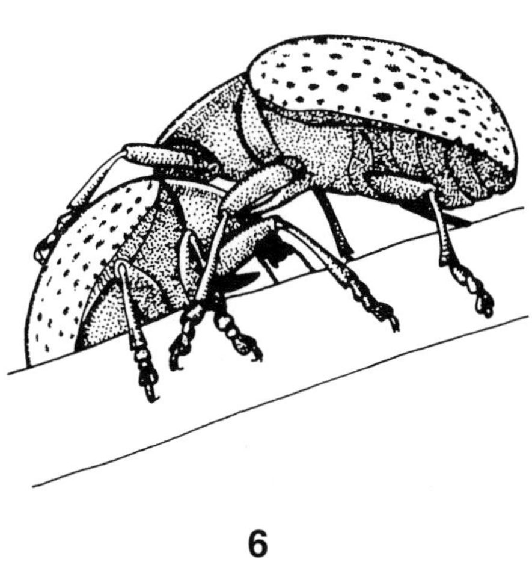

6

to 45 hours to inseminate the female to capacity. As for many leaf-beetles, males and females of *L. clivicollis* mate with multiple partners and form prolonged associations.

2.3. Pairing, amplexus and copulation

Pairing seems to be classical among the leaf beetles (male above the female and the tarsi against the female elytra, sometimes the last pair along the abdomen), but, as mentioned already, in the case of the Clytrinae and relatives, the male remains vertical behind the female, the huge forelegs on the substrate. The hairy tarsal brushes, mostly on the enlarged anterior tarsi of the male (*Timarcha* etc.) help the male to hold the female elytra. The hind tarsi can be also against the dilated female abdomen (*Gastrophysa*, *Galeruca*, and many others). There is really a slight variation from genus to genus, as for *Chrysochus asclepiadeus* Pallas, a eumolpine, where the hind legs of the male are on the substrate (Medvedev & Pavlov, 1988).

After the courtship, exchange of various stimuli between the male and the female seems to be the rule. If the female is receptive and does not reject the male or prevent the extension of the internal sac, the male is able to control the female from the start of the amplexus to the end of the copula. Among the long-horned beetles, Michelsen (1963, 1966 a, b) described several tactile stimulatory actions such as licking, tapping and abdominal movements. Similar actions are known among leaf-beetles but those can differ from group to group.

Thibout (1982) described the sexual behaviour of the Colorado potato beetle (*Leptinotarsa decemlineata* (Say) and its control by juvenile hormone. He noted that a young virgin, female, not yet fed, is not receptive and rejects the male. An old, well fed female, has an inflated abdomen and copulates immediately after the male antennae vibrate. Otherwise, the undisturbed female does not stop feeding during the mating.

The mating behaviour in *Diabrotica virgifera* (LeConte), a galerucine, starts by preliminary courtship and mounting by the male, physical stimulation of the female, and copulation. Antennal palpitation, stroking activities are performed by the male during the mating process (Lew & Ball, 1979). Mating frequency in *Diabrotica virgifera* is 8.2 times during 41.6 days on average and females mate more than once (Branson et al., 1977). Mating in this species lasts one to two hours. It stimulates an increase in weight, egg laying and ovarian development (Sherwood & Levine, 1993). They are shorter matings but it requires some time before sperm is deposited into the spermatheca. Coupling can last much longer as in some *Chrysolina*, up to, or more than 24 hours. It can be interrupted or, if the female is not receptive, the male, we have observed, can be kicked out. Rodriguez (1994 a, b,1995 a, b) studied especially one Cassidinae, *Chelymorpha alternans* Boheman, concerning cryptic choice of the female. Females of this tortoise beetle seem to favour males with longer flagella storing much more sperm in their spermatheca. Cutting the spermathecal muscle reduces egg fertility, probably by stopping partially ejection of sperm. Villavaso (1975 a) showed that, for the boll weevil, *Anthonomus grandis* Boheman, spermathecal filling was not affected by severing the spermathecal muscle, but, as for Rodriguez, the physical displacement of the sperm was largely dependent on functional spermathecal muscle. In *Labidomera clivicollis* Kirby, females control the ability of males to fully evert their genitalia and inseminate (Dickinson, 1997). They respond to copulation attempts by positioning their abdomens out of reach of the male's genitalia or by kicking males with their legs.

Spermathecal muscles play a role in sperm retention and fertilization of eggs. That way, the females control emission and retention of the male's sperm.

In some species of tortoise beetles,such as *Omaspides bistriata* Boheman and *O. convexicollis* Spaeth, males court females during, rather than before copulation (Rodriguez, 1994 a). Courting behaviour differs slightly between the two species. Among *Charidotella* spp., males do not perform any precopulatory courtship (Rodriguez, 1993). Copulation for those species can last up to 6 hours. Mating is done the classical way : the hind legs of the male on the substrate, the forelegs on the female's elytra near the scutellum and the middle legs on the margins of the elytra.

In *Dicladispa armigera* (Olivier), an hispine, mating can last more than 2 hours and the male initiates the copulation without much preliminaries (Deka & Hazarika,1996).

Phillips (1979) described in detail the reproductive behaviour of *Altica lythri* Aubé. When a male encounters a female, it taps the anterior region of her elytra with its own antennae and then, rapidly, mounts, tapping the elytra and thorax with the maxillary palpi. The aedeagus is quickly extended. The male grips the dorsal surface of the female's elytra with fore and middle tarsi, the hind tarsi gripping the lateral margin of the abdomen. During copulation, the antennae and palpi of the male are active on the head and antennae of the female. Unreceptive females violently extend the ovipositor preventing the entry of the aedeagus or they close the valve. Among other species, the unreceptive female prevents the expansion of the internal sac inside her vagina.

The alticine, *Macrohaltica jamaicensis* (Fabricius) feeds on various plants (*Ludwigia octovalvis*, Onagraceae; *Cuphea* sp., Lythraceae; *Gunnera insignis*, Gunneraceae) in Panama and Costa Rica. The type is metallic blue and there is a metallic purple variation or a distinct species (?). Both varieties copulate with each other. On the highlands of Panama, on the « continental divide », the insects can be very abundant. Females apparently need to mate repeatedly to sustain egg production (Eberhard et al., 1993) and females very probably mate with different males in the field. Females discard the spermatophore in which sperm is transferred a few hours after copulation. In the case of *Chrysomela crotchi* Brown, a univoltine species of Ontario, Canada, mating occurs throughout the day, but usually stops before 22.00 hours (Smereka, 1965). It lasts from a few minutes to over one hour. Males and females mate more than once.

Multiple mating is common among chrysomelids, and McCauley and O'Donnell (1984) showed that wild caught females of *Plagiodera versicolora* (Laicharting) produce egg clutches expressing multiple sires (see also Wade, 1994).

Medvedev & Pavlov (1988) described the selection of the position of the male towards the female (ritual dance) and the male climbing the female elytra from the side of the apex, rarely from the front. Continuous stimulation of the female by the male during the amplexus is the rule.The female rapidly starts feeding during the process. They describe also among Galerucinae and several Alticinae special glands, which they name odorators, and the male tarsi serve to impregnate the male antennae and to excite the female during mating.

2.4. Termination of mating

In *Donacia aquatica* (L.), the female is restless after the end of the copula and performs pendulum movements to get rid of the male (Michelsen, 1966 a). This behaviour is common to most of the leaf-beetles. In *Diabrotica virgifera*, when the male dismounts, the female walks away immediately (Lew & Ball, 1979). Not only do unresponsive females of *Sphondylia tomentosa* (Lacordaire) (Megalopodinae) kick the male on its head (Schulze, 1996), but, after mating, many female leaf-beetles do the same.

Large sperm cells of certain Alticinae (Virkki & Bruck, 1994) are probably also used as mating plugs by the male (Ladle & Foster, 1992), so the female cannot mate again. Mating can last from 12 minutes to one day according to the species involved . Medvedev & Pavlov (1988) quoted 12-25 minutes for *Donacia marginata* Hoppe, 50-60 minutes for *Chrysomela populi* L., 60-70 minutes for *Cryptocephalus cordiger* (L.), etc. It is much longer for many species, like *Chrysolina* and relatives (often 24 hours or more). A prolonged copulatory period of 3-4 hours seems necessary for maximal insemination of the female of *Diabrotica virgifera* LeConte (Ball, 1957) during a single copulation. It seems a minimum for most of the species. Spermatozoa are present and remain viable within the spermatheca up to 40 days following mating (Lew & Ball, 1980). Age and ovarian development are correlated (Short & Hill, 1972) and only a receptive female will respond to the male. In *Agathomerus sellatus* (Germar), Santos (1981) found that copulation duration was on average 86 minutes, a minimum for Megalopodinae. In all cases, termination of copulation remains similar and the female, restless after the amplexus, walks away as soon as possible.

It is well known that species react differently when disturbed during mating. Some remain together, even when dropped in killing jars, others, the majority, separate, Medvedev & Pavlov (1988) stated that several *Donacia* fly away separated, some *Cryptocephalus* separate and fall from the plant, *Clytra*, *Labidostomis* and *Cassida* spp. fall to the ground independently or in pairs. A couple of *Donacia dentata* Hoppe, when disturbed, remained attached, but tried to escape by dropping into the water. Even in the case of the fully aquatic *Neohaemonia* and *Macroplea*, mating is always outside the water and both sexes fly away for copulation. Separation of certain beetles is also dependant on the time of mating, and often at the end they remain temporarily firmly attached to each other.

2.5. Conditions for mating

In the Tropics, beetles mate when they are mature, but in temperate climates, mating occurs when the air temperature is high, usually above 20-23°C and under moderate sunlight, often in the morning or towards the end of the afternoon and under moderate wind (Medvedev & Pavlov,1988). Mating happens openly on the leaves or on the stems of the plants. Copulation also occurs under a cloudy sky and, if very windy, the pair can find shelter amongst vegetation.

Under cold climates, such as Western Siberia or Wrangel Island, mating and activity of *Chrysolina cavigera* (J. Sahlberg) occurs in early « summer », when the temperature is not above 6°C (Khruleva,1996). The average duration of one mating is 24 hours, as in many other *Chrysolina*. To be successful, matings of *Chrysolina cavigera* must concern fully developed partners with a hard exoskeleton.

2.6. Interspecific and intergeneric matings

I have already mentioned (Jolivet, 1942,1949,1997) interspecific matings between different species of *Chrysolina* : *C. herbacea* (Duftschmidt) female x *C. polita* (L.) male; *C. banksi* (F.) female x *C. staphylaea* (L.) male; *C. coerulans* (Scriba) male x *C. herbacea* female; *C. graminis* (L.) male x *C. herbacea* female; etc. Many other aberrant copulations are mentioned in the literature: *Chrysomela populi* L. x *C. haemorrhoidalis* L.; *Cryptocephalus labiatus* L. x *C. nitidus* L., etc.

Except perhaps for the cross between the green *C. herbacea* and the blue *C. coerulans* (Jolivet, 1952), no hybrids have been found in nature. Most of the unnatural matings happen when the two species are feeding on the same plant, for instance, in Belgium. The orange *C. polita* males often mate with the green *C. herbacea* and colour vision, in that case, does not seem to be involved. Both species feed on *Mentha* spp. and are different in size, but never a big male of *C. herbacea* has been seen coupled with a small female of *C. polita*, and the contrary is common. Size relationships are respected. Medvedev & Pavlov (1988) mentioned several cases of pairing attempts between the alticine *Crepidodera aurata* (Marsham) male with the female of *Plagiodera versicolora* Laicharting, which can be due to the similarity of the host-plant, a *Salix*, and the similarity of colour. Surprisingly enough, a interspecific cross has never been observed between two sympatric species of *Timarcha*, for instance between *T. goettingensis* and *T. tenebricosa*, even though both species are blue-black and feed on various species of *Galium*. However, in the laboratory, in cages, very often matings have been observed between different species, different genera, different families, different Phytophaga (for instance between a chrysomelid and a curculionid), and even between adults and larvae or pupae. Abnormal conditions, such as confinement in a small place and overcrowding may be responsible for this aberrant behaviour.

Wüest (1996) quoted a mating between *Linaeidea aenea* (L.) male and *Chrysomela lapponica* L. female in France. The former one is blue-green and the latter red and black.Once more, the colour does not seem to play any role, but *L. aenea* feeds usually on *Alnus* and *C. lapponica* on *Alnus*, *Salix*, etc. and both species have common food-plants. In that case, as for *Chrysolina*, the host-plant seems to have priority over color.

Franz (1938) has seen *Chrysolina* hybrids everywhere (*C. limitata* Küster, *C. gypsophilae* Küster), but there is no experimental proof in favour of any hybridization between these species of *Chrysolina,* at least in the wild. The flightless *Chrysolina angusticollis* Motschulsky complex in Sapporo (Japan) shows a parapatric distribution of the species and probably hybrid unfitness. There could be, however, natural hybrids at the boundaries of their distribution but it remains to be proved and the concept is purely academic (Saitoh & Katakura, 1996). Genetic divergence in populations of leaf-beetles, *Leptinotarsa decemlineata* (Say), for instance, is small (Jacobson & Hsiao, 1983). So, crossing is free between all the races, and even subspecies when they exist, as in *Diabrotica*. It is only limited by spatial distribution.

Species hybrids exist however. They have been found among gryllids and *Carabus*, for instance. They do not seem to be common among chrysomelids, often differing in many ways : genitalic structure, morphology, chromosome formula, size, food-plant, etc. Eggs from interspecific laboratory crosses between *Diabrotica virgifera* LeConte and *D. longicornis barberi* Smith & Lawrence had a much lower percentage of hatch than eggs from conspecific pairings (Krysan & Guss, 1978). Interspecific insemination occurred when beetles were confined together for periods of 11 to 30 days. Two barriers to hybridization were detected : one involving mating behaviour, the other a post- insemination barrier which reduced egg viability. Once, I have tried to hybridize the European *Gastrophysa viridula* (De Geer) with the American *G. cyanea* Melsheimer, obtaining mating on *Rumex crispus*, but no viable offspring, except perhaps for dysplasic specimens (modified elytra).

2.7. Homosexual mating

In the laboratory, abnormal matings between males and females occur frequently. Since, several chrysomelids are parthenogenetic, namely several Alticinae, pairing between females occur eventually. Lesbian oriented strains of vinegar or fruit flies have been isolated and homosexuality is known among the males of many beetles, including cockchafers. Homosexual matings are frequent among males of *Meloe* spp. (Bologna & Marangoni, 1986) and females of *Otiorhynchus* (Pardi, 1987). That can happen exceptionally in nature and must be extremely rare there, courtship and mating being dominated by female and male pheromones. In captivity, pseudocopulation of parthenogenetic females of *Otiorhynchus pupillatus* Gyllenhal has been often observed and the females search and mount other females in the way characteristic of the males. Homosexuality remains possible among leaf-beetles, but in confinement. Thibout (1982) noted that, among 80 males of the Colorado potato beetle in cages, 67.5% mounted each other.

Discussion

A summary of the mating habits among leaf-beetles has been made previously (Jolivet, 1997). We refer to this book for further details.

Among all living things, there seems to be a choice between mating partners. Even, the primitive Rotifers seem to select their mates as a function of age (Gomez & Serra, 1996). Courtship seems to be very common everywhere. Assortive mating in *Trirhabda canadensis* Kirby, a galerucine, probably is due to the greater ease with which size-matched pairs can achieve intromission (Brown, 1993).

In general, among animals, females often prefer to mate with the most flamboyant males (Dugatkin & Godin, 1998). Finding Mr. Right is clear among the birds and fishes. However, it will take time and long observations to make sure that insects proceed the same way, but it is not impossible. According to Brooks (1998), even females of certain species copy the behaviour of other females with their partners, but that seems impossible to detect among insects and very improbable at this scale.

According to Eberhard (1990), rapid evolutionary divergence of male genitalia may be explained by the ability of females to choose the paternity of their offspring. Certain structures of the internal sac were once interpreted as designed to empty the spermatheca of the sperm of the previous males.That interpretation seems very questionable. However, removal of rival sperm during copulation in a beetle, *Tenebrio molitor* L., seems possible and the ability of the males of this beetle to manipulate ejaculate size adaptatively according to the risk of sperm competition has been demonstrated (Gage & Baker, 1991; Gage, 1992). Sperm removal by males has been found in the longicorn beetle, *Psacothea hilaris* Pascoe, in Japan (Yokoi, 1989, 1990). Sperm utilization strategies are well known among all insects (Walker, 1980). Sperm transfer and sperm competition were studied in a bruchid, *Callosobruchus maculatus* (F.) and the system is not at all isolated among Chrysomeloidea (Eady, 1994).

What Thornhill (1983) has termed « cryptic female choice » is however sometimes evident. According to this author, female choice during and after mating leads to a stronger sexual selection on males than premating choice. This cryptic female choice is widespread among beetles (Eberhard,1991 et seq.), among crickets (Sakaluk, 1977), and many other insects.. Now it is fully confirmed by experiments and direct observation. For Eberhard (1985 et seq.), the diversity of male genitalia morphology is a direct result of female choice for stimulation and during copulation. Also, copulation does not always result in insemination, nor does insemination always result in fertilization of eggs. Eberhard research in chrysomelids covers *Megalopus armatus* Lacordaire, *Macrohaltica jamaicensis* (F.), *Doryphora* sp.and others (Eberhard, 1981, 1996, 1997). Many other beetles were also studied. Variation in fertilization success is a fundamental component of sexual selection for Hosken & Stockley (1998). Sperm competition among males exists and increases ejaculate expenditures for copulation with high risk females.

Generally, courtship and mating are very simple among chrysomelids, but complex internal sac and complex female tracts can mean more complex behaviour among Cassidinae, for instance, and other subfamilies.

Loosing the spermatheca is frequent among viviparous species (Bontems, 1988) and this must have some impact on the use and selection of the male sperm in case of polyandrous females. For a successful mating, a female must remain still, or at least not terminate copulation prematurely. Sperm must be moved to storage and fertilization sites and this transport depends to a large extent on female behaviour (Eberhard, 1991). Some male alticines have enormous spermatozoa and probably use them partly as nourishment of the eggs or as a plug to prevent other males from fertilizing the same female (Ladle & Foster,1992). To Virkki & Bruck (1994), the big spermatozoa carry proteins, and to others (Anonymous, 1998, Karr & Pitnick, 1996) they play some vital role in embryo development. However, the basic phenomenon of giant sperm is still a mystery. Bacetti & Daccordi (1988) studied in detail the sperm structure of the Chrysomelidae in general. In both species studied, *Asiorestia transversa* (Marsham) and *Podagrica malvae* Illiger, the sperm are extremely long and very distinct from other chrysomelid subfamilies.

Intromission of the spermatophora in *Acanthoscelides obtectus* Say (Huignard, 1969), a bruchid, and, in various leaf-beetles, can influence oogenesis by chemical action and increases fecundity. Accessory gland nutrients are common in many insects (Bissoondath & Wiklund, 1997), including chrysomelids and function as paternal investment in the male mating efforts. Large males may have a selective advantage over small males in sperm competition because they produce more sperm. MacCauley & Lawson (1986) believed that mating reduces predation on milkweed beetles, as they name a cerambycid, *Tetraopes tetraophthalmus* Forster. I am rather sceptical over those conclusions.

A last point : apterous, brachypterous and brachelytrous species are on the verge of extinction everywhere due to the abuse of urbanization and pesticides. Remaining isolated patches of populations lead to inbreeding, and inbreeding leads to extinction. That is true for the Holarctic *Timarcha* or *Chrysolina* and the Tropical *Elytrosphaera, Iscadida,* and many others in America, South Africa and Australia. Loss of flight means the loss of recolonization and loss of new blood. Conservationists must add those beetles to the list of endangered species.

Acknowledgements

I want to thank here Dr. Michael L. Cox from the Natural History Museum, London, for linguistic help and Mr. Jean-Claude Bourdonné who kindly provided colour pictures of French chrysomelids. Unfortunately, they could not be used in this paper as well as my own colour pictures.

References

Aiken, R. B. 1992. The mating behaviour of a boreal water beetle, *Dytiscus alaskanus* (Col. Dytis.). Ethology , Ecology & Evolution 4 : 245-254.

Alcock, J. 1996. Male size and survival : the effects of male combat and bird predation in Dawson's burrowing bees, *Amegilla dawsoni*. Ecol. Entomol. 21 : 309-316.

Alcock, J. 1997a. Competition from Large Males and the Alternative Mating Tactics of Small Males of Dawson's Burrowing Bee (*Amegilla dawsoni*) (Apidae, Apinae, Anthophorini). Journal Insect Behavior 10 (1) : 99-113.

Alcock, J. 1997b. Small males emerge earlier than large males in Dawson's burrowing bee (*Amegilla dawsoni*). (Hym. Anthophorini). J. Zool., Lond. 242 : 453-462.

Alexander,R. D., Marshall, D.C. and Cooley,J.R. 1997. Evolutionary perspectives on insect mating. In Choe J.C. & Crespi, B.J. (eds.). The Evolution of Mating Systems in Insects and Arachnids. Cambridge University Press,Cambridge : 4-31.

Anonymous. 1998. Monster Sperm. New Scientist, April : 40-43.

Bacetti, B. & Daccordi,M. 1988. Sperm structure and phylogeny of the Chrysomelidae. In Jolivet,P., Petitpierre,E. & Hsiao, T. H. (eds.). Biology of Chrysomelidae. Kluwer Academic Publsh., Dordrecht, The Netherlands: 357-378.

Ball, H.J. 1957. On the biology and egg-laying habits of the Western Corn Rootworm. J. econ. Ent. 50(2) : 126-128.

Bao,S.N. 1996. Spermiogenesis in *Coelomera lanio* (Col. Chrys.) : Ultrastructural and cytochemical studies. In Jolivet,P. & Cox,M.L. (eds.) Chrysomelidae Biology. III. SPB Academic publs., Amsterdam, The Netherlands : 119-132.

Bartlet, E., Isodoro, N. & Williams, I.H. 1994. Antennal glands in *Psylliodes chrysocephala*, and their possible role in reproductive behavior. Phys. Entomol. 19 (4) : 241-250.

Bechyné, J. 1950. 7° contribution à la connaissance du genre *Chrysolina* Motsch. Ent. Arb. Mus.Georg. Frey 1 : 47-185.

Bissoondath, C. J. & Wiklund, C. 1997. Effect of male body size on sperm precedence in the polyandrous butterfly, *Pieris rapae*. Behav. Ecol. 8 (5) : 518-523.

Blum, M.S. & Blum, N.A. 1979 (eds.). Sexual selection and reproductive competition in Insects. Academic Press, N.Y., USA, 463 pp.

Bologna, M.A. & Marangoni, C. 1986. Sexual behaviour in some Palaearctic species of *Meloe* (Col. Mel.). Bol. Soc. ent. Ital., Genova 118 (4-7) : 65-82.

Bontems, C. 1988. Localization of spermatozoa inside viviparous and oviparous females of Chrysomelinae. In Jolivet,P., Petitpierre,E. & Hsiao, T.H. (eds.) Biology of Chrysomelidae. Kluwer Acad. Publsh., Dordrecht, The Netherlands : 299-316.

Boppre, M. & Scherer, G. 1981. A new species of flea-beetle (Alticinae) showing male-biased feeding on withered *Heliotropium* plants. Syst. Entom. 6 : 347-354.

Branson, T.F., Guss,P.L. & Jackson, J.J. 1977. Mating frequency of the Western Corn Rootworm. Ann. ent. Soc. Am. 70(4) : 506-508.

Branson, T.F. & Johnson, R.D. 1973. Adult Western Corn Rootworms : Oviposition, fecundity, and longevity in the laboratory. J. econ. Ent. 66(2) : 417-418.

Brooks, R. 1998. The importance of mate copying and cultural inheritance of mating preferences. TREE 13 (2) : 45-46.

Brown, L. & Siegfried, B.D. 1983. Effects of male horn size on courtship. Activity in the Forked Fungus Beetle, *Bolitotherus cornutus* (Col. Ten.). Ann. ent. Soc. Am. 76 : 253-255.

Brown, W.D. 1993. The cause of size-assortative mating in the leaf-beetle, *Trirhabda canadensis* (Col. Chrys.).Behav. Ecol. Sociobiol. 33 (3) : 151-157.

Brown, W.D., Crespi, B.J. & Choe, J.C. 1997. Sexual conflict and the evolution of mating systems. In Choe J.C. & Crespi, B.J. (eds.). The Evolution of Mating Systems in Insects and Arachnids. Cambridge University Press : 352-377.

Choe, J.C. 1994a. Sexual selection and mating system in *Zorotypus gurneyi* Choe (Ins. Zoroptera) : Dominance, hierarchy and mating success. Behav. Ecol. Sociobiol. 34 : 87-93.
Choe, J.C. 1994b. Sexual selection and mating system in *Zorotypus gurneyi* Choe (Ins. Zoroptera) II. Determinants and dynamics of dominance. Behav. Ecol. Sociobiol. 34 : 233-237.
Choe, J.C. 1995. Courthip feeding and repeated mating in *Zorotypus barberi* (Ins. Zoroptera). Anim. Behav. 49 : 1511- 1520.
Choe, J.C. 1997. The evolution of mating systems in the Zoroptera : mating variations and sexual conflicts. In Choe, J.C. & Crespi, B.J. (eds.). The Evolution of Mating Systems in Insects and Arachnids. 7. Cambridge University Press, Cambridge, UK : 130-145.
Choe, J.C. & Crespi, B.J. 1997.The Evolution of Mating Systems in Insects and Arachnids. Cambridge University Press, Cambridge,UK : 387 pp.
Choe, J. C. & Crespi, B.J.1997. The Evolution of Social Behavior in Insects and Arachnids. Cambridge University Press, Cambridge.
Choe, J.C. & Perlman, D.L. 1997. Social conflict and cooperation among founding queens in ants (Hym. Formicidae). In Choe, J.C. & Crespi, B.J. (eds.). The Evolution of Social Behavior in Insects and Arachnids. Cambridge University Press, Cambridge,UK : 392-406.
Cinereski,J.E. & Chiang, H.C. 1968. The pattern of movements of adults of the Northern Corn Rootworm inside and outside of corn fields. J. econ. Ent. 61 (6) : 1531-1536.
Cox, M.L. 1996. Parthenogenesis in the Chrysomelidae. In Jolivet, P. & Cox, M.L. (eds.). Chrysomelidae Biology. III. SPB Acad. Publsh., Amsterdam, The Netherlands: 133-151.
Crowson, R.A. 1981. The Biology of the Coleoptera. Academic Press, London : 802 pp.
Daccordi, M. 1994. Notes for phylogenetic study of Chrysomelinae with descriptions of new taxa and a list of all the known genera. In Furth, D.G. (ed.) Proc. 3th Int. Symp. Chrysomelidae, Beijing,1992. Backhuys Publ., Leiden, The Netherlands : 60-84.
Daccordi, M. 1996. Notes on the distribution of the Chrysomelinae and their possible origin. In Jolivet,P. & Cox, M.L. (eds.). Chrysomelidae Biology. I. SPB Acad. Publsh., Amsterdam, The Netherlands : 399-412.
Damman, H. & Cappucino, N. 1991. Two forms of egg defence in a chrysomelid beetle : egg clumping and excrement cover. Ecol. Entomol. 16 : 163-167.
Deka, M. & Hazarika, L.K. 1996. Mating behavior of *Dicladispa armigera* (Col. Chrys.). Ann. ent. Soc. Am. 89 (1) : 137-141.
Dickinson, J.L. 1986. Prolonged mating in the milkweed leaf beetle (*Labidomera clivicollis*) : a test of the « sperm loading » hypothesis. Behav. Ecol. Sociobiol. 18 : 331-338.
Dickinson, J.L. 1988. Determinants of paternity in the milkweed leaf-beetle. Behav. Ecol. Sociobiol. 23 : 9-19.
Dickinson, J.L. 1992. Scramble competition, polygyny in the milkweed leaf beetle : combat, mobility, and the importance of being there. Behav. Ecol. 3 : 32-41.
Dickinson, J.L. 1995. Trade-offs between postcopulatory riding and mate location in the blue milkweed beetle. Behav. Ecol. 6(3) : 280-286.
Dickinson, J.L. 1996. The behavior and ecology of *Labidomera* Chevrolat (Chrys.) In Jolivet,P. & Cox, M.L. (eds.) Chrysomelidae Biology. II. SPB Acad. Publsh., Amsterdam, The Netherlands : 323-335.
Dickinson, J.L. 1997. Multiple mating, sperm competition, and cryptic female choice in the leaf beetles (Col. Chrys.). In Choe, J.C. & Crespi, B.J. (eds.). The Evolution of Mating Systems in Insects and Arachnids. 9. Cambridge University Press, Cambridge, The Netherlands : 164-183.
Dugatkin, L.A. & Godin, J.G.. 1998. How females choose their mates. Scient. Amer. 278 (4) : 56-61.
Eady,P. 1994. Sperm transfer and storage in relation to sperm competition in *Callosobruchus maculatus*. Behav. Ecol. Sociobiol. 35 : 123-129.
Eberhard, W.G. 1977. Fighting behavior of male *Golofa porteri* beetles (Scarab. Dynast.).
Psyche 83 : 292-298.
Eberhard, W.G. 1979. The function of horns in *Podischnus agenor* and other beetles. In Blum,M. & Blum,N. (eds.). Sexual Selection and Reproductive Competition in Insects. Academic Press, New York, USA : 231-258.
Eberhard, W.G. 1980. Horned beetles. Scient. Amer. 242 :166-182.
Eberhard, W.G. 1981. The natural history of *Doryphora* sp. (Col. Chrys.) and the function of its sternal horn. Ann. ent. Soc. Am. 79 : 445-448.
Eberhard, W.G. 1982. Beetle horn dimorphism : making the best of a bad lot. Amer. Nat. 119 (3) : 420-426.
Eberhard, W.G. 1985. Sexual selection and animal genitalia. Harvard University Press, Cambridge.
Eberhard, W.G. 1990. Animal genitalia and female choice. Amer. Scientist. 78 (2) : 134-141.
Eberhard, W.G. 1991. Copulatory courtship aand cryptic female choice in insects. Biol. Rev. 66 : 1-31.
Eberhard, W.G. 1994. Evidence for widespread courtship during copulation in 131 species of insects and spiders, and implications for cryptic female choice. Evolution 48 : 711-733.
Eberhard, W.G. 1996. Female control : sexual selection by cryptic female choice. Princeton University Press, Princeton, USA : 501 pp.
Eberhard, W.G. 1997. Sexual selection by cryptic female choice in insects and arachnids. In Choe, J.C. & Crespi,

B.J. (eds.). The Evolution of Mating Systems in Insects and Arachnids. Cambridge University Press, Cambridge, UK : 32-57.

Eberhard, W.G., Achoy, R. Marin, M.C. & Ugalde, J. 1993. Natural history and behavior of two species of *Macrohaltica* (Col. Chrys.). Psyche 100 : 93-119.

Eberhard, W.G. & Cordero, C. 1995. Sexual selection by cryptic female choice on male seminal products- a new bridge between sexual selection and reproductive physiology. TREE 10 (12) : 493-496.

Eberhard, W.G. & Karibo, S.J. 1996. Copulation behavior inside and outside the beetle *Macrohaltica jamaicensis* (Col. Chrys.). J. Ethol. 14 : 59-72.

Eberhard, W.G. & Marin, M.C. 1996. Sexual behavior and the enlarged hind legs of male *Megalopus armatus* (Col. Chrys. Megal.). J. Kans. ent. Soc. 69 (1) : 1-8.

Emlen, D.J. 1997. Alternative reproductive tactics and male-dimorphism in the horned beetle, *Onthophagus acuminatus* (Col. Scar.). Behav. Ecol. Sociobiol. 41 : 335-341.

Evans, G. 1975. The Life of Beetles. George Allen & Unwin Ltd., London : 232 pp.

Evers, A.M.J. 1956. Uber die Funktion der Excitatoren beim Liebenspiel der Malachiidae. Ent. Bl. 52 (3) : 165-169.

Feytaud, J.1930. Recherches sur le Doryphore (*Leptinotarsa decemlineata* Say). Ann. Epiphyt. 16 : 303-390.

Frankham, R. & Rolls, K. 1998. Inbreeding leads to extinction. Nature 392 : 441-444. .

Franz, H. 1938. Revision der Verwandtschaftsgruppe der *Chrysomela gypsophilae* Küst. Ent. Blätt. 34 :190-210 & 249-273.

Furth, D.G. 1994. A new case of parthenogenesis in beetles *: Longitarsus melanurus* (Melsheimer) (Col. Chrys.). Jl. N. Y. ent. Soc. 102 (3) : 310-317.

Gadeau de Kerville, H. 1900. L'accouplement des Coléoptères. Bull. Soc.ent. Fr. : 101- 107.

Gage, M.J.G. 1992. Removal of rival sperm during copulation in a beetle, *Tenebrio molitor*. Anim. Behav. 44 : 587-589.

Gage, M. J. G. & Baker, R. 1991. Ejaculate size varies with socio-sexual situation in an insect. Ecol. Entomol. 16: 331-337.

Gerber, G.H. & Church, N. S. 1973. Courtship and copulation in *Lytta nuttalli* (Col. Meloidae). Can. Ent. 105 : 719-724.

Goidanich, A. 1956. Gregarismi ed individualismi larvali e cure materne nei Crisomelidi. Mem. Soc. ent. ital. 35 : 151-183.

Gomez, A. & Serra, M. 1996. Mate choice in male *Brachionus plicatilis* rotifers. Functional Ecology 10 : 681-687.

Gwynne, D.T. 1998. Genitally does it. Nature 393 : 734-735.

Hill, R.E. 1975. Mating, oviposition patterns, fecundity and longevity of the Western Corn Rootworm. J. econ. Ent. 68 (3) : 311-315.

Hinton, H.E. 1981. Biology of Insect Eggs. 3 vols. Pergamon Press, N.Y. : 1125 pp.

Hosken, D.J.& Stockley, P. 1998. Sperm counts. TREE 13 (3) : 91-92.

Huignard, J. 1969. Action stimulatrice du spermatophore sur l'ovogénèse chez *Acanthoscelides obtectus* Say (Ins. Col.). C. R. Acad. Sc., Paris 268 (D) : 2938-2940.

Isono, M. 1988. Differentiation in life history pattern and oviposition behavior, and thelytoky in *Demotina* and *Hyperaxis* beetles (Col. Chrys.) in Western Japan. Kontyû. 56 : 402-409.

Jacobson, J.W. & Hsiao, T.H. 1983. Isozyme variation between geographic populations of the Colorado potato beetle, *Leptinotarsa decemlineata* (Col. Chrys.). Ann. Entomol. Soc. Amer. 76 : 162-166.

Jolivet, P. 1942. Hybridation probable de deux Chrysomèles : *C. polita* et *C. menthastri*. Bull. Soc. ent. Fr. 47 (9) : 141.

Jolivet, P. 1949. Les hybrides d'espèces chez les Chrysomélides.Entomologiste 5 (3-4) : 77-80.

Jolivet, P. 1952. Notes biométriques sur quelques Chrysomélides. Inst. R. Sc. Nat. Belg. Bull. 28 (44) : 1-8.

Jolivet, P. 1957-1959. Recherches sur l'aile des Chrysomeloidea (Col.). Inst. R. Sc. Nat. Belg. Mem. 2 (51) : 1-180 & 2 (58) : 1- 152.

Jolivet, P. 1988. Les soins parentaux chez les Chrysomélides.Entomologiste 44 : 93-101.

Jolivet, P. 1994. Physiological colour changes in Tortoise Beetles. In Jolivet, P.,Cox, M.L. & Petitpierre, E. (eds.). Novel Aspects of the Biology of Chrsyomelidae. Kluwer Acad. Publsh., Dordrecht, The Netherlands : 331-335.

Jolivet, P. 1997. Biologie des Coléoptères Chrysomélides. Boubée & Co. publsh., Paris, France : 275 pp.

Jolivet, P. & Cox, M. L. (eds.). 1996. Chrysomelidae Biology. I- II- III. SPB Acad. Publsh., Amsterdam, The Netherlands: 444 pp. + 465 pp. +365 pp.

Jolivet,P., Cox, M.L. & Petitpierre, E. (eds.) 1994. Novel Aspects of the Biology of Chrysomelidae. Kluwer Acad. Publsh., Dordrecht, The Netherlands : 582 pp.

Jolivet, P. & Petitpierre, E. 1981. Biology of Chrysomelidae (Col.). Butll. Inst. Cat. Hist. Nat. 47 (Sec. Zool. 4) : 105-138.

Jolivet, P., Petitpierre, E. & Hsiao, T.H. (eds.). 1988. Biology of Chrysomelidae. Kluwer Acad. Publsh., Dordrecht, The Netherlands : 615 pp.

Jolivet, P., Vasconcellos-Neto, J. & Weinstein, P. 1990. Cycloalexy : a new concept in the larval defense of insects. Insecta Mundi. 4 (1-4) : 133-142.
Karr, J. & Pitnick, S. 1998. The ins and outs of fertilization. Nature 379 : 405-406.
Khruleva, O.A. 1996. Biology of arctic leaf-beetle *Chrysolina cavigera* on Wrangel island. In Jolivet, P. & Cox, M. L. (eds.). Chrysomelidae Biology. III : 259-270.
Krysan, J.L. & Guss, P.L. 1978. Barriers to hybridization between *Diabrotica virgifera* and *D. longicornis barberi* (Col. Chrys.). Ann. ent. Soc. Am. 71 : 931-934.
Kudo, S.I. & Ishibashi, E. 1995. Notes on maternal care in the ovoviparous beetle, *Gonioctena japonica* (Col. Chrys.). Can. Ent. 127 : 275-276.
Kudo, S.I. & Ishibashi, E. 1996. Maternal defence of a leaf-beetle is not effective against parasitoids but is against pedestrian predators. Ethology 102 : 560-567.
Kudo, S.I., Ishibashi, E. & Makino, S.J. 1995. Reproductive and subsocial behaviour in the ovoviviparous leaf-beetle, *Gonioctena sibirica* (Col. Chrys.). Ecol. Entomol. 20 : 367-377.
Laboissière, V. 1927. Contribution à l'étude des Galerucini de l'Indo-Chine et du Yunnan. Ann. Soc. ent. Fr. 96 : 37-62.
Ladle, R.J. & Foster, E. 1992. Are giant sperm copulatory plugs ? Acta Oecologica 13 (5) : 635-638.
Leopold, R. A. 1976. The role of male accessory glands in insect reproduction. Ann. Rev. Entomol. 21 : 199-221.
LeSage, L. 1997. Description d'une nouvelle espèce de *Clavicornaltica* en provenance de Taiwan (Col. Chrys.). Nouv. Rev. Ent. (N.S.). 14 (3) : 239-247.
Lew, A.C. & Ball, H.J. 1979. The mating behavior of the Western Corn Rootworm, *Diabrotica virgifera* (Col. Chrys.). Ann. Entomol. Soc. Amer. 72 (3) : 391-393.
Lew, A.C. & Ball, H.J. 1980. Effect of copulation time on spermatozoan transfer of *Diabrotica virgifera* (Col. Chrys.). Ann. ent. Soc. Am. 73 : 360-361.
Linsley, E.G. 1959. Ecology of Cerambycidae. Ann. Rev. Entomol. 4 : 99- 138.
MacCauley, D.E. 1983. An estimate of the relative opportunities for natural and sexual selection in a population of milkweed beetles. Evolution 37 (4) : 701-707.
MacCauley, D.E. & Lawson, E.C. 1986. Mating reduces predation on male milkweed beetles. Am. Nat. 127 (1) : 112-117.
MacCauley, D.E. & O'Donnell, R. 1984. The effect of multiple mating on generic relatedness in larval aggregations of the imported willow leaf-beetle (*Plagiodera versicolora*). Behav. Ecol. Sociobiol. 15 : 287-291.
MacCauley, D.E., Wade, M.J., Breden, F.J. & Wohltman, M. 1988. Spatial and temporal variation in group relatedness : evidence from the imported willow leaf beetle. Evolution 42 (1) : 184-192.
Magis, N. 1991. Les parades nuptiales chez les Coléoptères. Verviers : 10 pp.
Matthes, D. 1959. Das Paarungsverhalten (Paarungssspiel und Kopulation des Malachiiden *Troglops albicans* L.) Zool. Anz. 163 (5-6) : 53-160.
Maulik, S. 1936. Coleoptera Chrysomelidae Galerucinae. In Fauna of British India, Taylor & Francis, London : 648 pp.
Medvedev, L.N. 1962. On the functional importance of secondary sexual characters in chrysomelid beetles (Col. Chrys.). Zool. Zh. 41 (1) : 77-84.
Medvedev, L.N. & Pavlov, S.J. 1988. Mating behavior of the Chrysomelidae. Ent. Review 67 (3) : 100-108.
Mendoza, C.E. & Peters, D.C. 1968. Morphology and histology of the reproductive systems of adult Southern Corn Rootworm. Ann. ent. Soc. Am. 61 (5) : 1279-1284.
Michelsen, A. 1963. Observations on the sexual behaviour of some longicorn beetles, subfamily Lepturinae (Col. Ceram.). Behaviour 22 : 152-166.
Michelsen, A. 1966a. On the evolution of tactile stimulatory actions in longhorned beetles (Ceram. Col.). Z. Tierpsychol. 23 : 257-266.
Michelsen, A. 1966b. The sexual behaviour of some longhorned beetles (Col. Ceram.). Entomol. Meddel. 34 (4) :329-355.
Mitchell, P.L. 1980. Combat and territorial defense of *Acanthocephala femorata* (Hem. Coreidae). Ann. ent. Soc. Am. 73 : 404-408.
Monros, F. 1953. Revision sistematica de las especies de Clytrinae de la Argentina, Paraguay, Uruguay and Chile. Acta zool. lilloana 14 : 5-324.
Monros, F. 1954. *Megalopus jacobyi*, nueva plaga de Solanaceas en el noroeste argentino, con notas sobre biologia y taxonomia de Megalopodinae. Revta. Agron. NE Argent. 1 (2) : 167-177.
Naranjo, S.E. & Sawyer, A.J. 1987. Reproductive biology and survival of *Diabrotica barberi* (Col. Chrys.) : Effect of temperature, food and seasonal time of emergence. Ann. ent. Soc. Am. 80 (6) : 841-848.
Oddie,K. 1998. Sex discrimination before birth. TREE 13 (4) : 130-131.
Palmer, T.J. 1978. A horned beetle which fights. Nature 274 : 583- 584.
Pardi, L. 1987. La « pseudocopula » delle femmine di *Otiorhynchus pupillatus cyclophthalmus* (Sol.) (Col. Curc.). Boll. Ist. ent. Guido Grandi, Univ., Bologna 41 : 355- 363.
Paulian, R. 1943. Les Coléoptères. Formes-Mœurs-Role. Payot Publsh.,Paris : 396 pp.
Paulian, R. 1988. Biologie des Coléoptères. Lechevalier Publsh., Paris : 719 pp.

Phillips, W.M. 1979. A contribution to the study of species relations within the chrysomelid genus *Altica* Müller in Britain. Zool. J. Linn. Soc. 66 : 289-308.
Rodriguez, V. 1993. Mating behavior in *Charidotella* sp. nr. *sexpunctata* (Col. Cass.). Coleopts. Bull. 41 (7) : 37-38.
Rodriguez, V. 1994. Sexual behavior in *Omaspides convexicolllis* Spaeth and *O. bistriata* Boheman with notes on maternal care of eggs and young. Coleopts. Bull. 48 (2) : 140-144.
Rodriguez, V. 1994. Function of the spermathecal muscle in *Chelymorpha alternans* Boheman (Col. Chrys. Cass.). Physiol. Entomol. 19 (3) : 198-202.
Rodriguez, V. 1995. Relation of flagellum length to reproductive success in male *Chelymorpha alternans* Boheman (Col. Cass.). Coleopts. Bull. 49 (3) : 201-205.
Rodriguez, V. 1995. Copulatory courtship in *Chelymorpha alternans* Boheman (Col. Chrys. Cass.). Coleopts.Bull. 49 (4) : 327-331.
Sacchi, R. & Busardo, G. 1935. Variabilita e correlazione in species molto affini. Noti Appunti sper. Ent. Agraria 3 : 1-24.
Saitoh,S. & Katakura, H. 1996.Strictly parapatric distribution of flightless leaf beetles of the *Chrysolina angusticollis* species complex (Col. Chrys.) in the vicinity of Sapporo, Northern Japan. Biol. J. Linnean Soc. 57 (4) : 371-384.
Sakaluk, S.K. 1977. Cryptic female choice predicated on wing dimorphism in decorated crickets. Behav. Ecol. 8 (3) : 326-331.
Santos, R. dos, 1981. Biologia de *Agathomerus sellatus* (Germar, 1824) (Col. Chrys. Megal.), broca de tomateiro. Revta. bras. Ent. 25 (2) : 165-170.
Sato,H. & Hiramatsu, K. 1993. Mating behaviour aand sexual selection in the African ball-rolling scarab, *Khepher platynotus* (Bates) (Col. Scarab.). J. nat. Hist. 57 : 657-658.
Schulze,L. 1996. Life-history and description of early stages of *Sphondylia tomentosa* (Luc.) (Col. Chrys. Megal.). In Jolivet,P. & Cox, M.L. (eds.). Chrysomelidae Biology. III. SPB Acad. Publsh.,Amsterdam, The Netherlands : 187-199.
Sherwood, D.R. & Levin, E. 1993. Copulation and its duration affects female weight, oviposition, hatching patterns, and ovarian development in the Western Corn Rootworm (Col. Chrys.). J. econ. Ent. 86 (6) : 1664-1671.
Shimizu,N. & Fujiyama, S. 1986. Aggressive encounters between males in *Chrysolina aurichalcea*. J. Ethol. 4 (1) : 11-15.
Short, D.E. & Hill, R.F. 1972. Adult emergence, ovarian development and oviposition sequence in the Western Corn Rootworm in Nebraska. J. econ. Ent. 65 (3) : 685-689.
Smereka, E.P. 1965. The life-history and habits of *Chrysomela crotchi* Brown (Col. Chrys.) in Northern Ontario. Can. Ent. 97 (5) : 541-549.
Thibout, E. 1982. Le comportement sexuel du Doryphore, *Leptinotarsa decemlineata* (Say) et son possible contrôle par l'hormone juvénile et les corps allates. Behaviour 80 (1-2) : 199-217.
Thornhill, R. 1983. Cryptic female choice and its implication in the scorpionfly, *Harpobittacus nigriceps*. Am. Nat. 122 : 765-788.
Thornhill, R. & Alcock, J. 1983. The evolution of Insect mating systems. Harvard University Press, Cambridge,USA : 547 pp.
Vasconcellos-Neto, J. & Jolivet, P. 1994. Cycloalexy among chrysomelid larvae. In . Jolivet,P., Cox , M.L. & Petitpierre, E. (eds.). Novel Aspects of the Biology of Chrysomelidae. Kluwer Acad. Publsh., Dordrecht, The Netherlands : 303-309.
Villavaso, E.J. 1975a. Functions of the spermathecal muscle of the boll weevil, *Anthonomus grandis*. J. Insect Physiol. 21 : 1275-1278.
Villavaso, E.J. 1975b. The role of the spermathecal gland of the boll weevil, *Anthonomus grandis*. J. Insect Physiol. 21 : 1457-1462.
Virkki, N. & Bruck, T. 1994. Unusually large sperm cells in Alticinae : their formation and transportation in the male genital system, and their evolution. In Jolivet,P.,Cox, M.L. & Petitpierre, E. (eds.). Novel Aspects of the Biology of Chrysomelidae. Kluwer Acad. Publsh., Dordrecht, The Netherlands : 371-381.
Wade, M.J. 1994. The biology of the Imported Willow Leaf Beetle, *Plagiodera versicolora* (Laicharting). In Jolivet, P., Cox, M.L. & Petitpierre (eds.). Novel Aspects of the Biology of Chrysomelidae. Kluwer Acad. Publsh., Dordrecht, The Netherlands: 541-547.
Walker, W.F. 1980. Sperm utilization strategies in nonsocial insects. Am. Nat. 115 (6) : 780-799.
Waloff, N. & Richards, O.W. 1958. The biology of the chrysomelid beetle, *Phytodecta olivacea* (Forster). Trans. R. ent. Soc. Lond, 110 (5) : 99-116.
Windsor, D.M. 1987. Natural history of a subsocial tortoise beetle, *Acromis sparsa* Boheman (Chrys. Cass.) in Panama. Psyche 94 (1-2) : 127-150.
Windsor, D.M. & Choe, J.C. 1994. Origins of parental care in chrysomelid beetles. In Jolivet, P., Cox, M.L. & Petitpierre, E. (eds.). Novel Aspects of the Biology of Chrysomelidae. Kluwer Acad. Publsh., Dordrecht, The Netherlands : 111-117.
Windsor, D.M.,Riley, E.G. & Stockwell, H.P. 1992. An introduction to the biology and systematics of Panamanian

tortoise beetles. In Quintero,D. & Aiello, A. (eds.). Insects of Panama and Mesoamerica. Selected Studies. Oxford University Press, Oxford : 372-391.

Wüest, J. 1996. Accouplement interspécifique chez des Chrysomèles. Bull.Soc. Romande Ent. 14 : 160.

Yang, X. 1992. A new feature of the eighth tergite of male in Coleoptera and its function. Sinozoologia 9 : 191-196.

Yokoi, N. 1989. Observation on the mating behavior of the yellow spotted longicorn beetle, *Psacothea hilaris* Pascoe (Col. Cerambycidae). Jpn. J. Appl. Ent. Zool. 33 : 175-179.

Yokoi, N. 1990. The sperm removal behavior of the yellow spotted longicorn beetle, *Psacothea hilaris* (Col. Ceramb.). Jpn. J. Appl. Ent. Zool. 25 : 383-388.

Yong, H.S. 1993. Biochemical genetic differentiation between two *Aulacophora* leaf beetles (Col. Chrys.) from peninsular Malaysia. Comp. Biochem. Physiol. 106 (2) : 317-319.

Yu, Peiyu & Yang, Xingke. 1994. Biological studies on *Temnaspis nankinea* (Pic) (Chrys. Megal.). in Jolivet,P., Cox, M.L. & Petitpierre, E . (eds.) Novel Aspects of the Biology of Chrysomelidae. Kluwer Acad. Publsh., Dordrecht, The Netherlands : 527-531.

Zeh, D.W., Zeh, J.A. & Tavakilian,G. 1992. Sexual selection and sexual dimorphism in the Harlequin beetle, *Acrocinus longimanus*. Biotropica 24 (1) : 86-96.

Zelazny,B. 1975. Behavior of young rhinoceros beetles, *Oryctes rhinoceros*. Entom. Exper. & Appl. 18 : 135-140.

Zhang, Z. & Mc Evoy, P.B. 1994. Attraction of *Longitarsus jacobaeae* males to ones associated with conspecific females. Environ. Entom. 23(3) : 732-737.

Zunino, M. 1988. La coevolucion de los aparatos copuladores. Elytron 1 : 105-107.

MATING BEHAVIOUR IN DONACIINAE (COLEOPTERA, CHRYSOMELIDAE)

A.O. BIEŃKOWSKI
Zelenograd 1121-107, 103460 Moscow K-460, Russia

Abstract

The mating behaviour in 14 Palaearctic species of Donaciinae was studied in captivity. A preliminary stage is separated within the mating in 12 species. During this stage the male occupies a position on the female's elytra, but does not attempt to copulate for a long time. Precopulatory courtship is observed during the preliminary stage in nine species: male rubs female's antennal or vertex calli with his fore-tarsi. During the copulation the males of 10 species perform a courtship: they rub the female's antennal or vertex calli, or behind the eyes, or pronotum with their fore tarsi. A courtship by the rubbing of head is reported for the first time in the Coleoptera. Well developed convex vertex calli are present only in some of the species, in which the courtship by rubbing of them occurs. The position and behaviour of the male during the mating is described for each species in detail. Females of several species can feed or move here and there during the mating. At the end of mating a male usually leaves a female without the prior attempts by the female to drive him away. A female rarely pulls off a male with her hind tarsi.

Introduction

The reproductive behaviour in insects is of special interest and is intensively studied presently. A knowledge of reproductive peculiarities provides additional insights into the phylogeny as demonstrated, for example, in Orthoptera (Alexander, 1964) and scarabaeine beetles (Halffter & Edmonds, 1982). Secondary sexual characters are of a great value in a systematics. The mating behaviour of leaf beetles is characterized by a great diversity (Eberhard, 1991, 1994, Medvedev & Pavlov, 1985, 1987, Michelsen, 1963, 1966). As regards Donaciinae, this question has been studied extremely poorly. Olsoufieff (1903) noted that two or three males of *Donacia fennica* (Paykull) position themselves alongside the mating pair and patiently wait their turn. Wesenberg-Lund (1943) mentioned that males of *D. crassipes* Fabricius and *D. cinerea* Herbst actively pursue females on leaves of their host plants. Michelsen (1963, 1966) in articles devoted to the mating of long-horned beetles (Cerambycidae) dealt briefly with the separate phases of copulation behaviour in *D. aquatica* (Linnaeus). Medvedev & Pavlov (1985) described a search behaviour of the males in *D. crassipes* and *D. marginata* Hoppe. Medvedev & Pavlov (1987) observed a copulation act in *D. marginata*.

Material and methods

The observation of the mating behaviour has been performed in the captivity. Beetles were captured in the Moscow region at Glubokoe lake (23 km west of Zwenigorod) and at small

Advances in Chysomelidae Biology 1, pp. 411–420
Edited by M.L. Cox

neighbouring waterbodies (ponds, swamps and ditches) in 1995-1998. I investigated the reproductive behaviour in 14 species of *Donacia* and *Plateumaris*. I usually placed from one to four pairs of one species in a petri dish provided with leaves, stems and flowers of the host plant. The petri dishes were protected from sunlight. I studied the behaviour of beetles under the stereoscopic microscope at 8-16 x magnification through the transparent cover of the dish. The results obtained in laboratory were supplemented by field observations. The terms "antennal calli" and "vertex calli" of the head and the "anterolateral calli" of the pronotum are used after Askevold (1991).

Results

Behaviour prior to mating

During the flight period the adults of a number of Donaciinae species occur in abundance on their host plants. Therefore, a meeting of sexes is easy. On account of this I failed to observe a search behaviour in nature. Medvedev & Pavlov (1985) noted that males of *D. crassipes* raise themselves on their fore and middle legs above the surface of floating leaves of water lily and usually orientate themselves against the wind. In this position the beetle turns forward, raises and separates its antennae. These authors suggested that the males use olfactory information from the females (pheromones of Donaciinae are still unknown). However, I registered this behaviour in the both sexes. I did not observe an aggressive behaviour of the males either in captivity, or in the field. However, a single male, which meets a mating pair, usually attempts to perch upward and disturbs the copulating animals. As a result the copulation is often interrupted.

Mating

The majority of Donaciinae species usually copulate openly on the leaves and stems of their host plants. Mating individuals of *D. clavipes* Fabricius crawl foremost into the leaf axils of *Phragmites australis*. The length of the mating period was measured as 12-25 minutes for *D. marginata* by Medvedev & Pavlov (1987), and about 15 minutes for *D. cinerea* by me. When a male meets a female, he perches on her elytra from behind. The male always orientates itself immediately rightly. The antennae of the both partners are directed forward and separated. The males of *D. bicolor* Zschach and *D. brevitarsis* Thomson alternately hit the outer sides of the females antennae with their antennae during the initial period of mating. The same action takes place during the copulation act in *D. sparganii* Ahrens and when the female attempts to drive the male away in different species. I recognize a preliminary stage in the mating process in Donaciinae. The males of 12 species do not set about copulation immediately they get into position on the elytra of the female, and the aedeagus is still hidden. In addition, the males of nine of these species perform a pre-copulatory courtship. In all species examined, a courtship implies that the male rubs the antennal or vertex calli of the female with his fore-tarsi alternately. Males of *D. thalassina* Germar, *D. sparganii* and *D. bicolor* execute these motions continuously, others do it now and then. The position of the male on the female elytra differs during the preliminary stage and the copulation act in six species. The importance of the preliminary stage to the successful copulation is confirmed by the following observation. One male of *D. brevitarsis* attempted to copulate without a precopulatory courtship. The female clasped the male elytra with her hind tarsi and tried to pull him off backwards. The male held on to the female strongly, but could not copulate in this position. Alternatively, beetles of *D. cinerea* and D. *vulgaris* Zschach always copulate without a preliminary stage, and females usually do not interfere with the copulation. I observed the both types of mating, with the preliminary stage and without it, in *D. crassipes*. During copulation the male of the majority of species extends to rub female's head or anterolateral calli of the pronotum. The male of *D. bicolor* rubs the female's vertex calli continuously during copulation, while others, including *D. sparganii* and *D. thalassina* perform this courtship now and then. Males of *D. vulgaris* and *D. cinerea* perform a similar copulatory courtship. The rubbing of the head as a variation of the copulatory courtship is found in the Coleoptera for the first time. It was recently pointed out that beetle males from different families rub the female elytra, thorax, femora and pygidium, or hit the antennae with his fore-tarsi (Eberhard, 1994). I have examined vertex relief in detail in the females of all the species observed and found that well developed vertex calli present in some of the species, in which a courtship by rubbing of them occurs. I have never found convex calli in the species, in which the male tarsi do not touch the female head. This suggests that the courtship behaviour was primary and respective morphological structures arose as a result of the sexual selection. It is interesting to note that the shape of vertex and pronotal calli are used as taxonomic characters for the *Donacia* and *Plateumaris* species (Medvedev, 1965, Lopatin, 1986, Lopatin & Kulenova, 1986, Askevold, 1991). However, a function for these calli in the females during a mating was not known before. A function of the male middle and hind legs is to maintain the mating position.

Table 1. Shape of vertex calli in females of some Donaciinae species.

Species	Male rubs calli	shape of calli	number of specimens
Donacia			
aquatica	+	C,W	10
bicolor	+	C	10
brevitarsis	+	W,A	10
cinerea	+	W	8
clavipes	–	W	10
crassipes	+	C,W	10
*dentata**	–	C,W	10
obscura	+	C,W	4
simplex	+	W	10
sparganii	+	C	9
thalassina	+	C	9
versicolorea	–	W	8
vulgaris	–	W	10
Plateumaris discolor	–	A	10

+ = yes; – = no. Shape of vertex tubercles: C = convex; W = weak; A = absent. *male rubs antennal calli.

Position and behaviour of male in different species during the mating

Donacia aquatica (Linnaeus)
Mating was observed in two pairs. The male performs the same behaviour during the preliminary stage and copulation act. He weakly rubs vertex calli of female with fore-tarsi; his middle legs clasp her body laterally, with tarsi resting between the middle and hind coxae, or in front of the middle coxae; his hind legs rest on substratum. Michelsen (1963) described another facet of the courtship in this species: the male strikes the female's body with his antennae.

Donacia bicolor Zschach
The preliminary stage and copulation act were observed in eight and four pairs, respectively. The male occupies the same position during the preliminary stage and copulation act. The male rubs female's vertex calli with fore-tarsi. His middle legs clasp her humeri at the front, with the middle tarsi placed between fore and middle coxae or behind the latter, hind tarsi rest on the substratum behind the female when she is motionless, and clasp her abdomen or drag behind when she crawls.

Donacia brevitarsis Thomson
The preliminary stage and copulation act were observed in 12 and eight pairs, respectively. The position of the male during the preliminary stage and copulation act does not differ. The male puts his fore-tarsal segments 2-4 or 3-4 on the female's head between the eyes and sometimes rubs the vertex calli, segment 1 or 1-2 on her pronotum interior to the anterolateral calli. His middle legs clasp her humerae at the front, with tarsi meeting or crossing on the metasternum, his hind legs clasp her abdomen, with tarsi crossing or turning forward.

Donacia cinerea Herbst
The preliminary stage was not found, although 12 pairs were under my observation. During copulation the male puts his fore-tarsi laterally on the female's pronotum with an inclination in front and down or dorsally interior to the anterolateral calli. In some cases the fore-tarsi are placed asymmetrically: one laterally and the other dorsally on the pronotum or on the head of the female. The male rubs the female's head below the eyes, the vertex calli or anterolateral calli of the pronotum now and then. His middle legs clasp the female's body laterally, with the tarsi placed between the middle and hind coxae, or on the elytral epipleura above the hind coxae. The male hind legs rest on substratum behind the female elytral apex or are brought together over her abdominal sternites.

Donacia clavipes Fabricius
The preliminary stage and copulation act was observed in three and two pairs, respectively. During the preliminary stage the male placed segment 1 of the fore-tarsi on the anterior margin of the female's pronotum, and the remaining segments on her head between the eyes. His middle legs clasp the female's body behind the middle coxae. Male hind legs rest on the substratum behind the female elytral apex. During the copulation act the male places his fore-tarsi laterally on the female's pronotum with an inclination in front and down, or dorsally on the pronotum interior to the anterolateral calli, with the claws catching on the anterior margin of the pronotum. The position of the middle and hind legs of the male is similar to that during the preliminary stage.

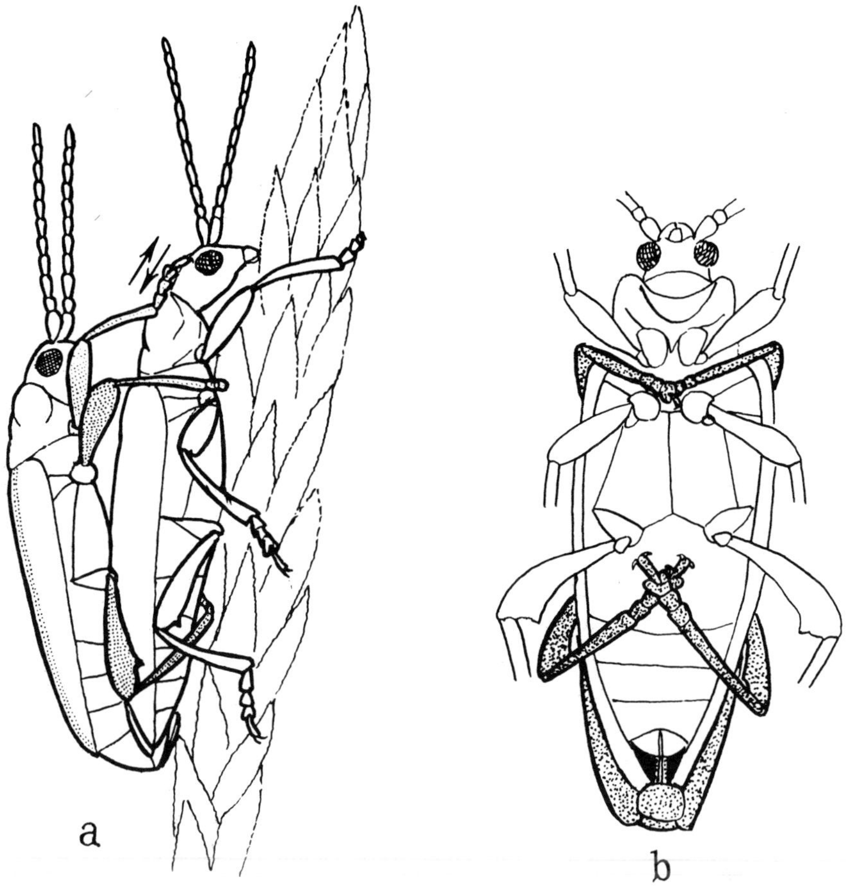

Fig. 1. Donacia brevitarsis: a) copulation on a stamen spike of *Carex vesicaria*, lateral view, b) position of middle and hind legs of male during copulation, ventral view. Arrows show the direction of male fore-tarsi rubbing.

Donacia crassipes Fabricius

In this species mating was observed in four pairs. The male performs the same courtship during the preliminary stage and copulation act. He puts his fore-tarsi on the female's vertex calli or on the pronotum interior to the anterolateral calli, or on the elytral base. The male rubs the female's vertex calli with his fore-tarsi now and then. His middle legs clasp the female body laterally, with his tarsi resting on her elytral epipleuri above the middle coxae, or on metasternum; his hind legs widely separated, with the tarsi placed on the substratum.

Donacia dentata Hoppe

The preliminary stage and copulation act were observed in four and one pair, respectively. During the preliminary stage the male puts his fore-tarsi on the female's head along the longitudinal axis with the segment 1 on the antennal calli, segments 2 and 3 on the antenna or between antennae. Sometimes the male places the segment 1 of one of his fore-tarsi on female's eye. The male rubs the female's antennal calli now and then. His middle legs clasp the female's humerae in front, with the tarsi meeting in front of her middle coxae and directing backwards between them. The male's hind tibiae are brought together over the female's abdominal sternites, and the tarsi are directed backwards and placed on the substratum. During the preliminary stage the female is motionless or crawls. During the copulation act, the male places his fore-tarsi asymmetrically: the segment 1 of one tarsus on the female's antennal callus and that of the other tarsus on her eye. The position of the middle legs of the male is similar to that during the preliminary stage. His hind tarsi rest on the substratum when the female is motionless, and trail behind when she crawls.

Donacia marginata Hoppe
I failed to observe the mating behaviour of this species. Therefore, I cite a report by Medvedev & Pavlov (1987): "The male puts his fore-tarsi on the basal margin of the female's elytra, somewhat outside the pronotum", the middle legs "can settle on the female's hind femora in the central or apical part". The male "separates his antennae upwards and then brings them together in front of himself and touches the outside of the female's antennae".

Donacia obscura Gyllenhal
The preliminary stage and copulation act were observed in one and five pairs, respectively. During the preliminary stage the male places the fore-tarsi on the female head: the left tarsus between the antennal insertions, the other, right tarsus on the outside of the antennae. The male rubs the female head ever so slightly, but usually holds his fore-tarsi motionless. The male middle legs clasp the female humeri at the front, with the tarsi meeting on the mesosternite; his hind legs clasp her body, with the tarsi meeting on the abdominal sternite 2 and orientated almost perpendicularly to the longitudinal body axis. During the copulation act the male places his fore-tarsi in the following way: segments 1-3 on the female pronotum interior to the anterolateral calli, segment 4 on her head behind the vertex calli or eyes, or tarsi entirely on the female pronotum, or segments 1 and 2 on the pronotum and 3 and 4 on the head, and the claws on the vertex calli. The male rubs the female head now and then. His middle legs clasp the female body, with the claws catching in front or behind her middle coxae; the hind tibiae meet each other below her abdomen, and the tarsi are directed backwards and rest on the substratum or are directed forward and settle on abdominal sternites 1-3.

Donacia simplex Fabricius
The preliminary stage and copulation act were observed in three and one pair, respectively. During the preliminary stage the male places his fore tarsi on female's vertex calli or behind the eyes and weakly rubs her head now and then. His middle legs clasp her humeri in front or behind; the hind tarsi cross below her abdomen or catch on the elytral epipleuri. During the copulation act, segments 1 and 2 of the male's fore-tarsi rest on the female's pronotum interior to the anterolateral calli, segments 3 and 4 (or 4 only) are placed on her head. The male sometimes slowly rubs the female's head behind the eyes with the fore-tarsi. His middle legs clasp her body at the level of the metathorax, with the tarsi resting in front of, or immediately behind the hind coxae.

Donacia sparganii Ahrens
Mating was observed in two pairs. During the preliminary stage the male places his fore-tarsi on the female's head and rubs the vertex calli with segment 1. The male's middle legs

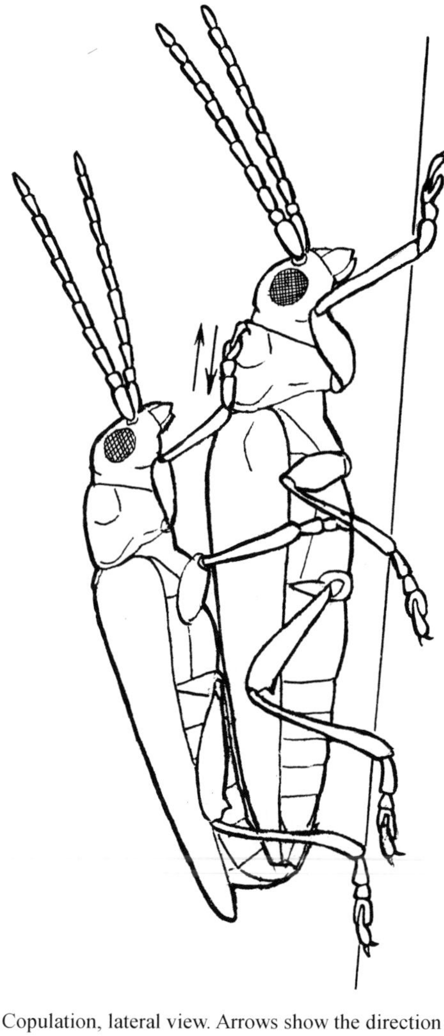

Fig. 2. Donacia thalassina. Copulation, lateral view. Arrows show the direction of male fore tarsi rubbing.

clasp the female's humeri in front or behind; his hind tarsi meet below her abdomen. During the copulation act the male puts segments 1 and 2 of the fore-tarsi along the midline of the female's pronotum, segments 3 and 4 on her head behind the eyes, or one tarsus between the antennae and the other between the antenna and eye. The male rubs the female's head and pronotum with the fore-tarsi and touches her antennae with his antennae in turn, as fast as he moves his fore-tarsi. The position of the male's middle and hind legs is the same as that of the preliminary stage, or the hind tarsi rest on the substratum.

Donacia thalassina Germar

Mating was observed in seven pairs. During the preliminary stage, the male rubs the female's vertex calli with segments 1 and 2 of the fore-tarsi, clasps her body between the fore and middle or between the middle and hind coxae with his middle legs. The male's hind tarsi cross below the female's abdomen. During the copulation act the male places his fore-tarsi on the female's pronotum, or segments 2-4 on her head, behind the eyes and rubs them now and then. The position of the male's middle legs is similar to that during the preliminary period. The male's hind tarsi rest on the substratum when the female is motionless and trail, rarely cross below her abdomen, when she moves.

Donacia versicolorea (Brahm)
The preliminary stage and copulation act was observed in one and two pairs, respectively. The position of the male's fore and middle legs does not differ during the preliminary stage and the copulation act. The male puts his fore-tarsi on the female's pronotum interior to the anterolateral calli, with the claws catching on the anterior margin of the pronotum; his middle legs clasp her body between the middle and hind legs. The male's hind tarsi rest on the substratum during the preliminary stage, and trail freely during the copulation.

Donacia vulgaris Zschach
The preliminary stage did not occur in the three pairs observed. During the copulation act the male puts his fore-tarsi laterally on the female's pronotum with an inclination in front and down, or dorsally on the pronotum, interior to the anterolateral calli, or on the elytral base interior to the humeri. The male rubs the female's anterolateral calli now and then. His middle legs clasp the female's body laterally, with the tarsi positioned between her middle and hind coxae. The male hind tarsi rest on the substratum.

Plateumaris discolor (Panzer)
Mating was observed in one pair only. The male occupies the same position during the preliminary stage and copulation act. He puts his fore-tarsi laterally on the female's pronotum below the anterolateral calli with an inclination in front and down, and with the claws catching on the anterior margin of the proepimerae. The male's middle legs clasp the female's body between the middle and hind legs, the hind tarsi cross below her abdominal sternites.

Feeding during mating

At times, females of some species were observed to feed during the preliminary stage and copulation act. Namely, females of *D. bicolor* fed on leaves of *Sparganium erectum*, *D. versicolorea* – on leaves of *Potamogeton nutans*, *D. brevitarsis* – on spikes of *Carex vesicaria* and P. *discolor* – on spikes of *Eleocharis palustris*. Medvedev & Pavlov (1987) reported on the feeding during the mating, in females of *D. dentata* and *D. crassipes*.

Obviation danger during mating

In nature, the copulating pairs of *D. cinerea*, *D. thalassina*, *D. aquatica*, *D. brevitarsis*, *D. bicolor* and *D. crassipes* get unlinked and try to disappear independently (fall from the plant or fly upwards) as someone approaches. Olsoufieff (1913) and Medvedev & Pavlov (1987) noted that the beetles *D. dentata* and *D. malinovskyi* Ahrens, which copulate on floating leaves, crawl in water on the lower leaf surface and do not interrupt the mating in danger.

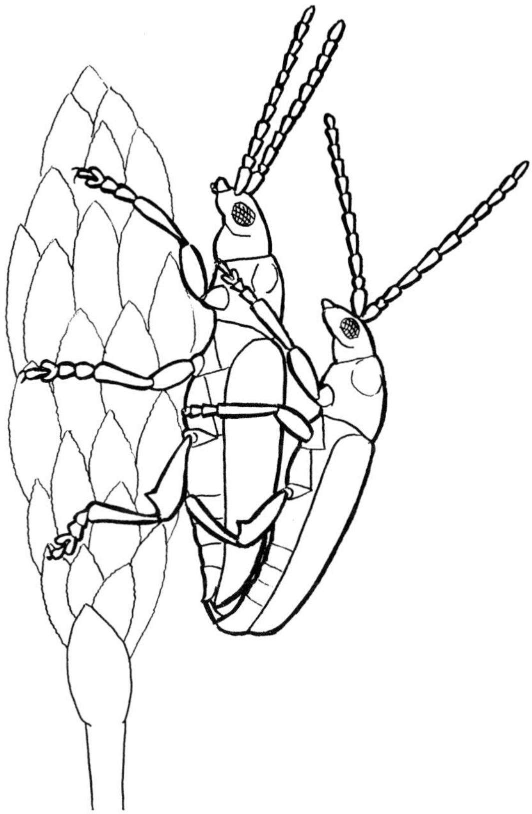

Fig. 3. Plateumaris discolor. Copulation on a spike of *Eleocharis palustris*, lateral view.

Completion of mating

From my observations the male usually leaves the female without the prior attempts by the female to drive him away at the end of mating. In one case, I observed a female of *D. sparganii* to push off a male with her right hind leg. In D. *aquatica* the female sometimes pushes her hind tarsi against the sides of the male's elytra and tries to pull off him backwards. Michelsen (1966) described a different manner by which a female releases herself from the male's embrance: "In *Donacia aquatica* the female is very restless after the end of copula and often performs "pendulum movements", i.e. the female stands upon the front and middle pair of legs, lifts her hind legs, and turns her body rapidly from side to side like a pendulum centred near the middle of the pronotum". Although I studied mating in this species, such interesting female behaviour was not noticed.

Conclusions

1. Mating behaviour was studied in 14 Palaearctic species of Donaciinae. It is subdivided into the following consecutive stages: 1) male takes the mating position, 2) preliminary

stage (+ precopulatory courtship), 3) copulation act (+ copulatory courtship), and 4) end of mating.
2. One characteristic of mating in Donaciinae is the preliminary stage before the copulation act: the male occupies a certain position on the female's elytra and mostly performs a precopulatory courtship.
3. The copulatory courtship was found in 10 species.
4. Precopulatory and copulatory courtship consists of the rubbing of the female's antennal and vertex calli, the head behind the eyes, or the anterolateral calli of pronotum by the male's fore-tarsi. A courtship by the rubbing of the head is reported for the first time in the Coleoptera.
5. The well developed convex vertex calli are apparently present only in some species in which the courtship by rubbing of them occurs. They never occur in species in which the male does not touch the female's head.

Acknowledgements

I am greatly indebted to Dr. M.L. Cox for his invitation to contribute to this book and correcting the manuscript. I thank my wife M.Ja. Orlova-Bieńkowskaja for checking of the manuscript.

References

Alexander, R.D., 1964. The evolution of mating behaviour in Arthropods. In: K.C. Highnam (ed.). Insect reproduction. Symposium No. 2. London, Royal Entomological Society: 78-94.
Askevold, I.S., 1991. Classification, reconstructed phylogeny and geographic history of the New World members of *Plateumaris* Thomson, 1859 (Coleoptera: Chrysomelidae: Donaciinae). Mem. ent. Soc. Can. 157: 5-175.
Eberhard, W.G., 1991. Copulatory courtship and cryptic female choice in insects. Biol. Rev. 66: 1-31.
Eberhard, W.G., 1994. Evidence for widespread courtship during copulation in 131 species of insects and spiders, and implications for cryptic female choice. Evolution 48: 711-733.
Halffter, G. & Edmonds, W.D., 1982. The nesting behaviour of dung beetles (Scarabaeinae): an ecological and evolutive approach. Mexico, Instituto de Ecologia. Publication 10. Part of the Man and the Biosphere Program, UNESCO: 1-176.
Lopatin, I.K., 1986. [Leaf-beetles of Byelorussia and Baltic republics: keys to determination.] Minsk, Vyshejshaja shkola: 1-131. [in Russian].
Lopatin, I.K. & Kulenova, K.Z., 1986. [Leaf-eating beetles (Coleoptera, Chrysomelidae) of Kazakhstan: keys to identification.] Alma-Ata, Nauka Kazakhskoi SSR: 1-200. [in Russian].
Medvedev, L.N., 1965. [Chrysomelidae. In: Keys to insects of European part of URSS 2. Coleoptera and Strepsiptera.] Moscow-Leningrad, Nauka: 1-668. [in Russian].
Medvedev, L.N. & Pavlov, S.I., 1985. [Reproductive behaviour of leaf beetles (Coleoptera, Chrysomelidae).] Zool. Zh. 64 (8): 1168-1178. [in Russian].
Medvedev, L.N. & Pavlov, S.I., 1987. [Mating behaviour of the leaf-beetles (Coleoptera, Chrysomelidae).] Revue Ent. URSS 66 (4): 746-753. [in Russian].
Michelsen, A., 1963. Observations on the sexual behaviour of some longicorn beetles, subfamily Lepturinae. Behaviour 22: 152-166.
Michelsen, A., 1966. The sexual behaviour of some longicorned beetles (Col., Cerambycidae). Ent. Meddr. 34 (4): 329-355.
Olsoufieff, G., 1903. Notes sur quelques *Donacia* de Russie [Col.]. Bull. Soc. ent. Fr.: 302-303.
Olsoufieff, G., 1913. [*Donacia malinowskii* Ahr. (Coleoptera, Chrysomelidae). Its life and habits.] Ent. Rev. Russ. 13 (2): 285-289. [in Russian].
Wesenberg-Lund, C., 1943. Biologie der Süsswasserinsekten. Kopenhagen, Gyldendalske Boghandel. Nordisk Forlag und Berlin – Wien, Verlag J. Springer, 682 pp.

FIELD STUDIES OF CRYPTOCEPHALINAE BIOLOGY

MATTHIAS SCHÖLLER
Biologische Bundesanstalt für Land- und Forstwirtschaft, Institut für Vorratsschutz, Königin-Luise-Str. 19, D-14195 Berlin, Germany

Abstract
Field data on the biology and ecology of a western Palaearctic and a Neotropical species of *Cryptocephalus*, *C. moraei* (L., 1758) and *C. anceps* Suffrian, 1852, are presented. The adults of both species are phytophagous, the larvae detrivorous. In *C. moraei*, the mortality was 50 %. Five species of parasitoids were reared from the immature stages of *C. moraei*: *Gelis acarorum* (L.), *Gelis forticornis* (Foerster), *Theroscopus esenbeckii* (Gravenhorst), *Dimophora evanialis* Gravenhorst and *Psilocera punctifrons* Thoms. The total parasitism was 20.1 %. Considerable differences were found in the mortality between the three years of investigation, ranging from 30 to 74 %. In *C. anceps*, the total mortality was 86 %. At least 15.2 % of the immature stages were parasitised. Larval cases of both *Cryptocephalus*-species were used as hibernating sites by spiders and as a place for colony founding by myrmicine ants. The possibilities and limitations of the Cryptocephalinae as model organisms for the study of population dynamics and ecology are discussed.

Introduction
Usually, both the larvae and adults of leaf beetles, Chrysomelidae, feed on living parts of plants, such as the leaves, stems and roots. On the contrary, the larvae of most of the species belonging to the tribe Cryptocephalini are found in leaf litter, where they feed on decaying, dead plant material (phytosaprophagy). The adults feed on living leaves (phyllophagy) or petals (petalophagy) of the host plant. The female covers her eggs with faecal plates, building an egg case. This is subsequently worn as a protective case by the larva and is enlarged as it grows. In the laboratory, the host range of the larvae is wider compared to that of the adults, as dead leaves originating from plants of various families are accepted (Rosenhauer, 1852; Erber, 1988; Schöller, 1996a,b).

Almost nothing is known about the consequences of this exceptional life-history on the population dynamics, spatial distribution within the habitat, predation and parasitism by natural enemies and other interactions with species occurring in the biocoenosis together with the Cryptocephalini. There are about 1400 species described in the Cryptocephalini. In this study, field data on the biology and ecology of a western Palaearctic and a Neotropical species of *Cryptocephalus*, *C. moraei* (L., 1758) and *C. anceps* Suffrian, 1852, are presented. *C. moraei* is a xerophilous, oligophagous species feeding on *Hypericum* spp.. The geographical distribution ranges from the Iberian Peninsula in the west to the Ural mountains in the east (Warchalowki, 1991). *C. anceps* was described from Venezuela, but no information on the biology is available.

Advances in Chysomelidae Biology 1, pp. 421–436
Edited by M.L. Cox

Materials and Methods

From 1993 to 1995, two populations of *C. moraei* were studied in Berlin-Johannisthal and Berlin-Frohnau. The localisation of the habitats within the city of Berlin was described in Schöller (1997). The disturbed areas are characterised by sparse vegetation on sandy soil, annual and perennial herbs. It is assumed, that the sparse vegetation is partly due to the application of herbicides in the past. Before the German reunification in 1989, the Berlin wall was located in these places, and the frontier was kept free from shrubs and trees. Patches of St. John's Wort, *Hypericum perforatum*, the host plant of *C. moraei*, are present in both habitats. St. John's Wort is a perennial which grows 1-3 feet tall. Larvae of *C. moraei* were sampled in April and May. Additional visits to the habitats took place in June, August and September. At four sampling plots within the habitats, the upper 2 cm of the soil with a surface of 2500 cm^2 was carefully collected. The number of *H. perforatum*-plants on the surface was counted. In the laboratory, the larvae of *C. moraei* were sorted out. Additional larvae were collected by hand, both in Johannisthal and Frohnau. The larval cases were examined under a dissecting microscope and the presence or absence of a living larva within and the condition of the case was documented. The larvae were kept singly in 5 cm petri-dishes in a climatic chamber at 26 °C and 75 % RH and fed with litter of *H. perforatum*. The petri-dishes were checked at least once a week for the eclosion of adult *C. moraei* or parasitoids. Larvae which did not develop into adults were determined with the key by Steinhausen (1994). In the case of the eclosion of parasitoids, these were fed with honey and provided with host larvae in order to study the oviposition behaviour.

A population of *C. anceps* was sampled in Cali, Colombia, March 9-11, 1998 at the International Centre for Tropical Agriculture (CIAT) in Cali. Cali is set in the Cauca Valley, about 1000 m NN. The adults of *C. anceps* (Fig. 1) were beaten from *Acacia* sp. and the larvae were collected in the *Acacia*-litter. To date, no larva of a Neotropical species of *Cryptocephalus* has been described. The head of the larva of *C. anceps* showed a characteristic sculpturation (Fig. 2). At the time of collection, the Cryptocephalinae were predominant on *Acacia*. Together with *C. anceps*, adults of *Euryscopa cingulata* Latreille, *Saxinis* sp. (Clytrini) and *Pachybrachis dissolutus* Suffrian (Cryptocephalini) were present. A total of 79 larval cases was collected by hand.

Results

Observations on the biology of C. moraei *and its parasitoids*

Adults of *C. moraei* were found from mid-May until mid-August. In 1995, the occurrence of adults was delayed with the first occurring in the first week of June. The female *C. moraei* laid their eggs on the host plant, holding them with the hind legs and covering them with faecal plates. Subsequently, the eggs were dropped to the ground.

The larvae were most frequently found in the vicinity of the stem, and some larvae attached their larval cases to the stem prior to pupation. Only once, a larva of *C. moraei* was found up on a *H. perforatum* plant in Johannisthal. A maximum density of 24.4 larvae / 1000 cm^2 was found. The number of larvae of *C. moraei* was significantly correlated with the number of *H. perforatum*-plants on the 2500 cm^2 sampling-plots (multiple regres-

Fig. 1. The adult of *Cryptocephalus anceps*

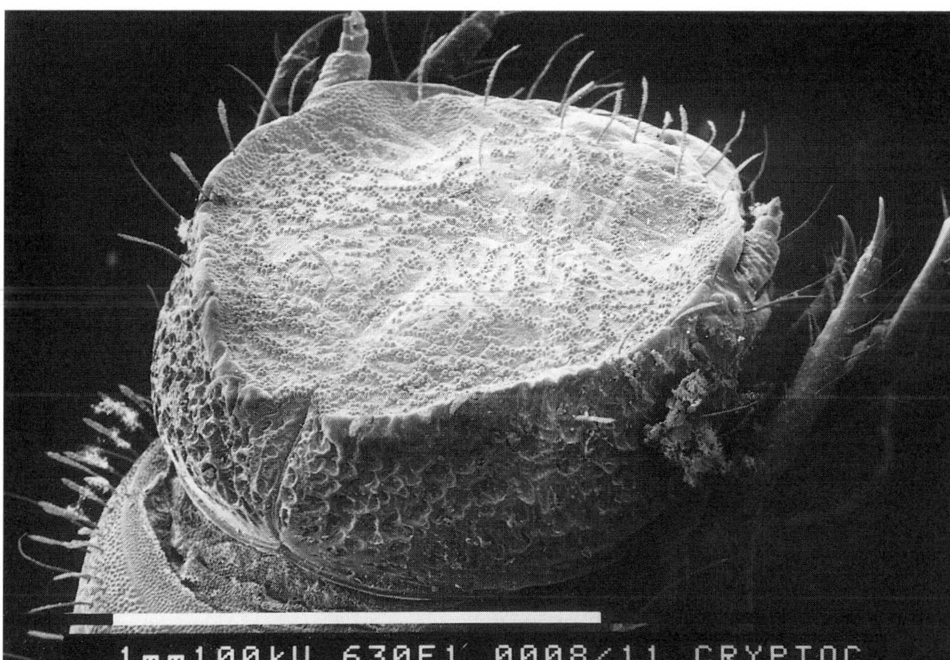

Fig. 2. The head of the larva of *Cryptocephalus anceps*

sion, $r^2 = 0.895$; analysis of variance, $F = 34.15$; Signif $F = 0.0043$) (Fig. 3). In the laboratory, the larvae fed on living and dead leaves of *H. perforatum* and of a *Rubus* sp. The last instar larval cases of male *C. moraei* were smaller, on an average 2.3 x 4.9 mm, than those of the females (2.9 x 6.0 mm). The fate of a larva could partly be determined by the condi-

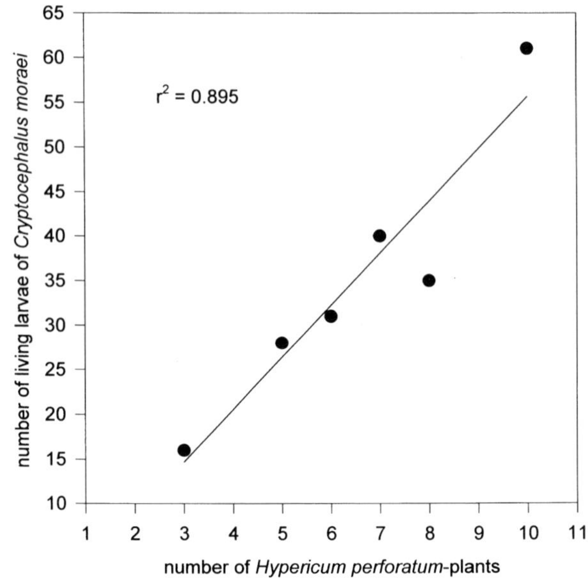

Fig. 3. Number of living larvae of *Cryptocephalus moraei* within the the upper 2 cm of soil with a surface area of 2500 cm² depending on the number of *Hypericum perforatum*-plants

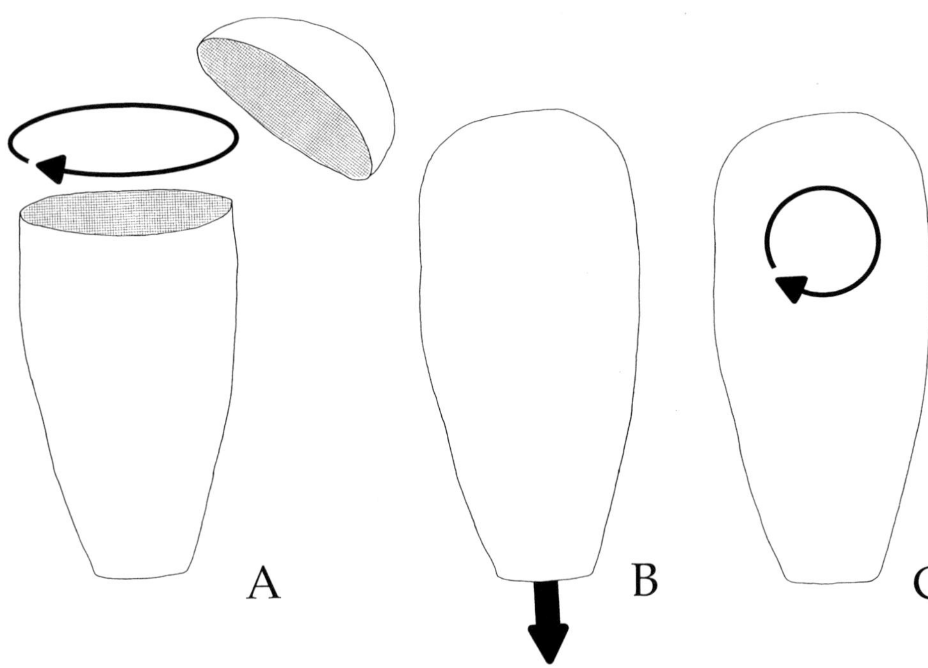

Fig. 4. Methods of emergence from a larval case of *Cryptocephalus moraei* (A) by the adults of *C. moraei* (B) by the adults of *Psilocera punctifrons* and (C) by the adults of *Dimophora evanialis*, *Gelis acarorum*, *Gelis forticornis* and *Theroscopus esenbeckii*

tion of the larval case. After eclosion, the adult *C. moraei* rested inside the larval case until the cuticle was hardened. Then, using its mandibles the beetle gnawed through the larval case in a regular way, leading to the separation of the bottom end of the larval case (Fig. 4A). All other conditions of the larval cases were rated as mortality of the larvae, because they are not capable of constructing a new case once it is lost. Within the group of dead larvae, it was possible to distinguish between parasitism by hymenopterous parasitoids and other causes of mortality. Five species of parasitoids were reared from immature stages of *C. moraei*. The species names are given, followed by the numbers of the voucher specimen which were deposited in the insect collection of the Senckenbergische Naturforschende Gesellschaft, Frankfurt am Main, Germany (in brackets): *Gelis acarorum* (L.) (SMFH 2401), *Gelis forticornis* (Foerster) (SMFH 2400), *Theroscopus esenbeckii* (Gravenhorst) (Ichneumonidae: Phygadeuontinae), *Dimophora evanialis* Gravenhorst (Ichneumonidae) (SMFH 2402) and *Psilocera punctifrons* Thoms. (Pteromalidae) (SMFH 2403). *D. evanialis* was the dominant parasitoid, only one specimen of *T. esenbeckii* was found in Johannisthal and three and four *G. acarorum* and *G. forticornis*, respectively, at both sites.

The examination of the larval cases from which the parasitoids emerged showed that the ichneumonid parasitoids left the larval cases in a different way than the pteromalid. The ichneumonids gnawed circular emergence holes 1-1.3 mm in diameter into the bottom end of the larval case (Fig. 4C). These larval cases were already closed at the top end by the *C. moraei*-larva with secretions and substrate, indicating that the larva was ready for pupation, or the case contained a pupa. Furthermore, the larva of *D. evanialis* spins a cocoon inside the cryptocephaline larval case (Fig 5). The pteromalid *P. punctifrons* left the larval case through an opening at the top end (Fig. 4B). Prior to pupation, the parasitoid larva closed the cryptocephaline larval case with the head capsule of the *C. moraei*-host larva and a secretion.

The larval cases which showed no symptoms of parasitoid attack were either empty or contained dead larvae. There were intact and broken empty cases. Dead larvae in various stages of decay were found, some were subject of fungal attack, others mummified. About 5 % of the larval cases were covered externally with green algae.

In addition, starting from the end of May, egg cases were found in the soil samples, containing first instar larvae, or empty. The empty egg cases were not damaged, but were covered with mycelia of fungi (Fig. 6).

The oviposition behaviour of Psilocera punctifrons

The foraging behaviour of the female *P. punctifrons* was characterised by slow walking and rapid drumming of the substrate with the antennae. When a larva of *C. moraei* was encountered, the larval case was drummed with the antennae (host recognition). Then, the female mounted the larval case, still drumming its surface (host examination) (Fig. 7A). The opening of the larval case was examined intensively. Subsequently, the female walked on the opening of the larval case and was arrested there for some time, while the antennae were vibrating (Fig. 7B), followed by the oviposition. For oviposition, the female inserted the ovipositor into the opening of the larval case (Fig. 7C) and afterwards, the whole abdomen (Fig. 7D). The host larva retracted deeply inside the larval case. It was not observed if *P. punctifrons* oviposited on or into the host larva. However, the observation that host larvae from which afterwards *P. punctifrons* emerged were found active and feeding, sug-

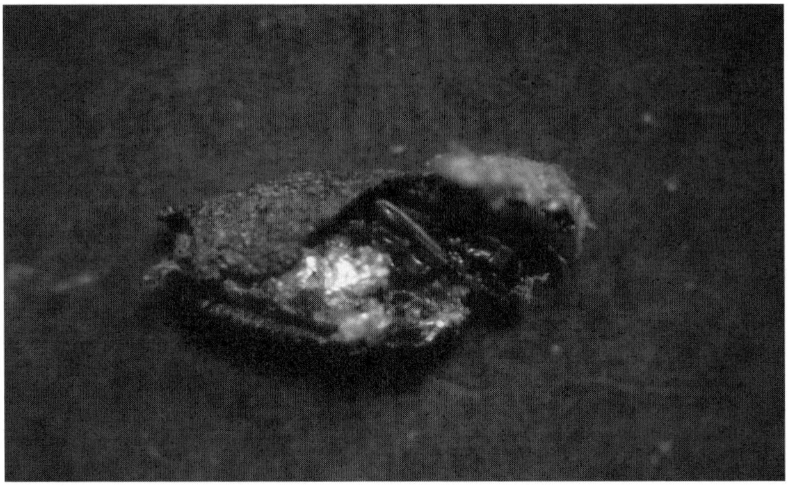

Fig. 5. Larval cocoon of *Dimophora evanialis* inside the larval case of *Cryptocephalus moraei*

Fig. 6. Egg case of *Cryptocephalus moraei* covered with mycelia of fungi.

gests that at least the first larval instars of *P. punctifrons* are endoparasitic. *P. punctifrons* were never found to parasitise final, but early instar larvae. The larva of *P. punctifrons* did not spin a cocoon, the pupa is shown in Fig. 8A and the position of the pupa and the meconium within the *C. moraei*-larval case in Fig. 8B.

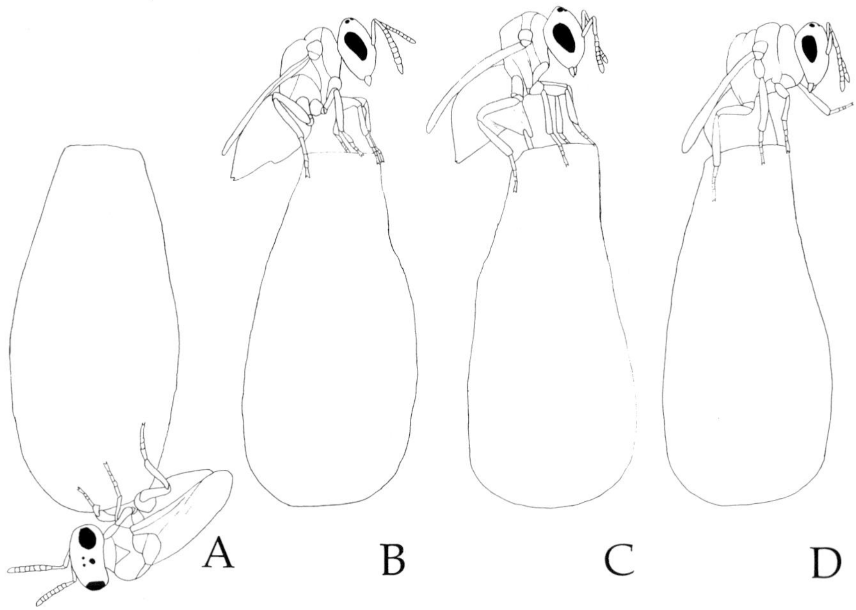

Fig. 7. The parasitisation behaviour of *Psilocera punctifrons* (A) host examination (B) arrestment at the opening of the larval case (C) insertion of the ovipositor into the larval case and (D) insertion of the abdomen into the larval case

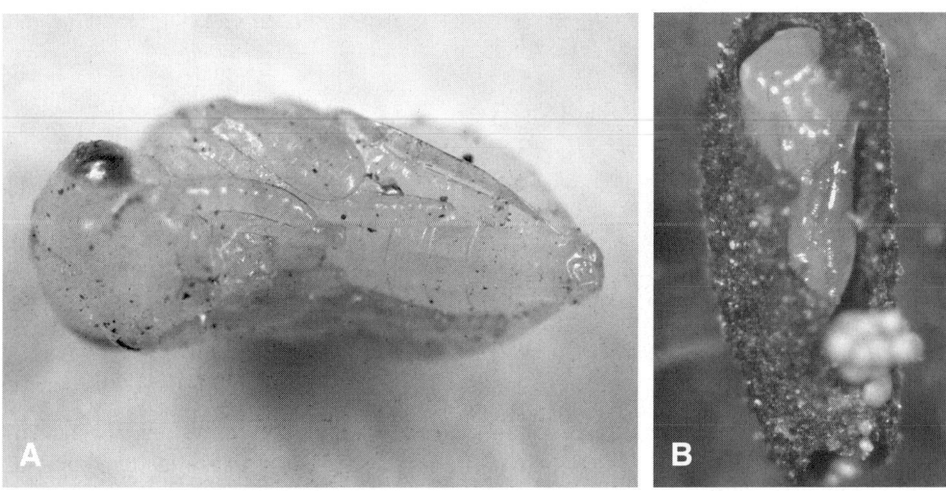

Fig. 8. A. The pupa of *Psilocera punctifrons*; B. the position of the pupa and the meconium inside the larval case of *Cryptocephalus moraei*

Population dynamics of C. moraei

About half of the sampled larvae were dead. The total parasitism for both locations was 20.1 % (n=897) (Fig. 9). For the remaining dead larvae, no causes of mortality were iden-

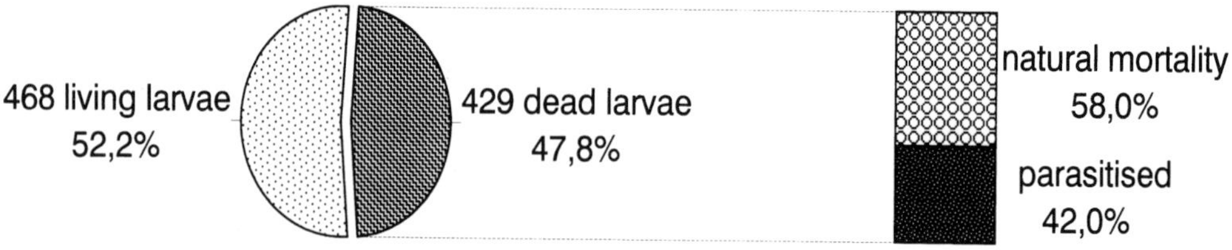

Fig. 9. Total number of living and dead larvae of *Cryptocephalus moraei* collected in 1993-95 at two locations in Berlin, Germany, and percentage parasitism (n=897)

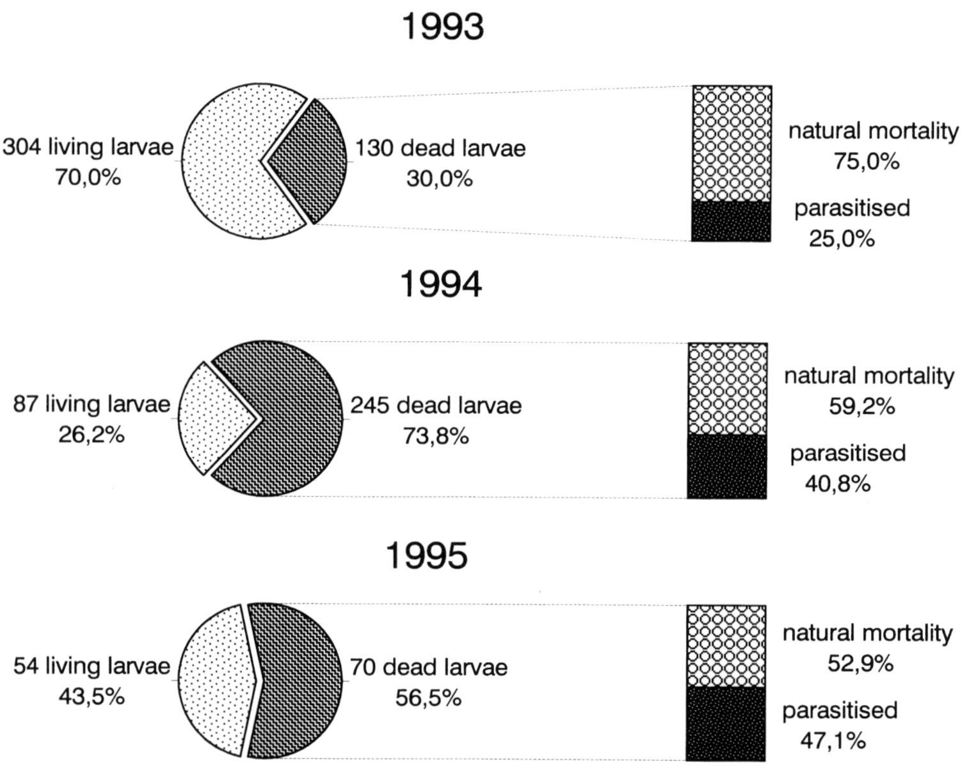

Fig. 10. Mortality of *Cryptocephalus moraei* in 1993-95 in Berlin, Germany, and percentage parasitism

tified. Considerable differences were found in the mortality between the years of investigation, ranging from 30 to 74 % (Fig. 10). The natural mortality of the larvae was relatively stable, ranging from 4.3 % to 7.1 % (Fig. 11). Comparing the mortality at the two locations, Frohnau and Johannisthal (Fig. 12), similar figures of approximately 50 % were found. However, the contribution of parasitism to larval mortality of *C. moraei* was different between the two locations, with 29 % in Frohnau and 53 % in Johannisthal. This was due to a higher percentage of parasitism by the ichneumonid parasitoids (Fig 13). The parasitism by *Psilocera punctifrons* was low, 3.1 % in Frohnau and about half of this, that is 1.4 %, in Johannisthal. In 1995, no *P. punctifrons* were found at the two locations (Fig. 14).

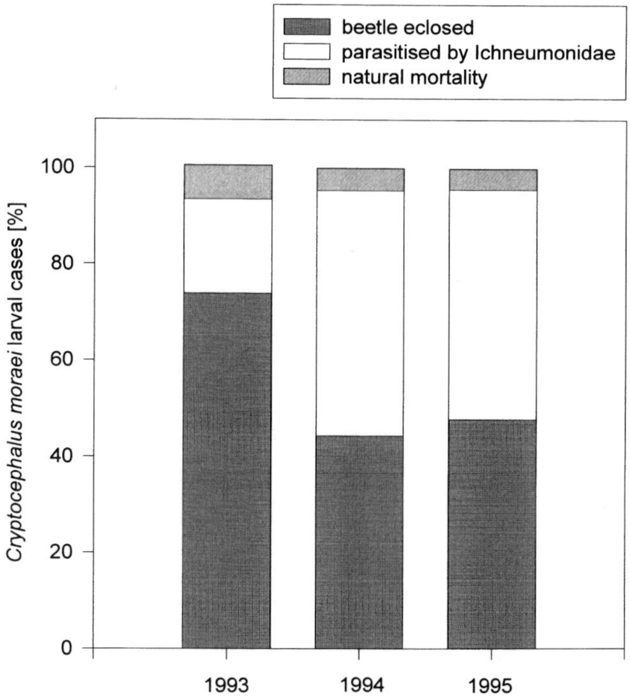

Fig. 11. Causes of mortality of *Cryptocephalus moraei* in 1993-95 at two locations in Berlin, Germany

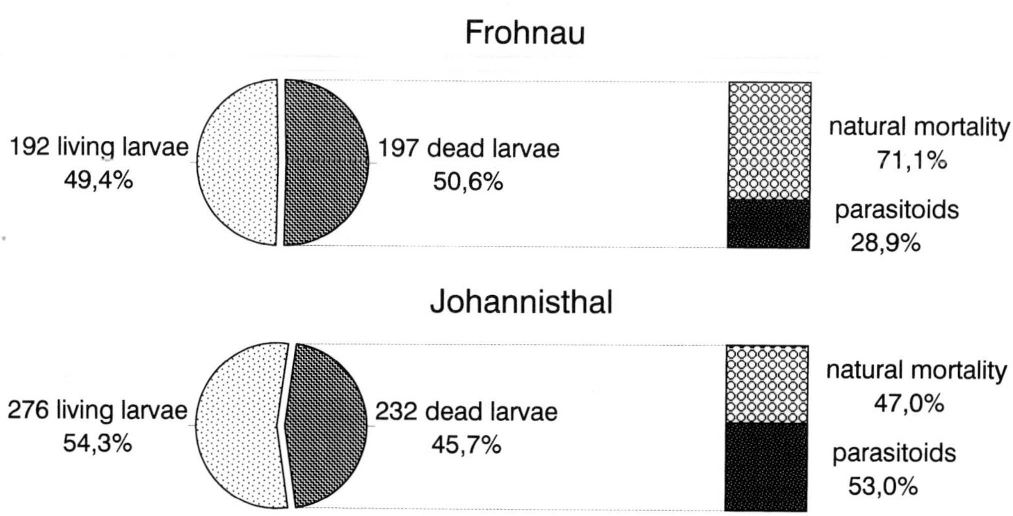

Fig. 12. Comparison of mortality of two populations of *Cryptocephalus moraei* in 1993-95 at Frohnau and Johannisthal, Berlin, Germany, and percentage parasitism

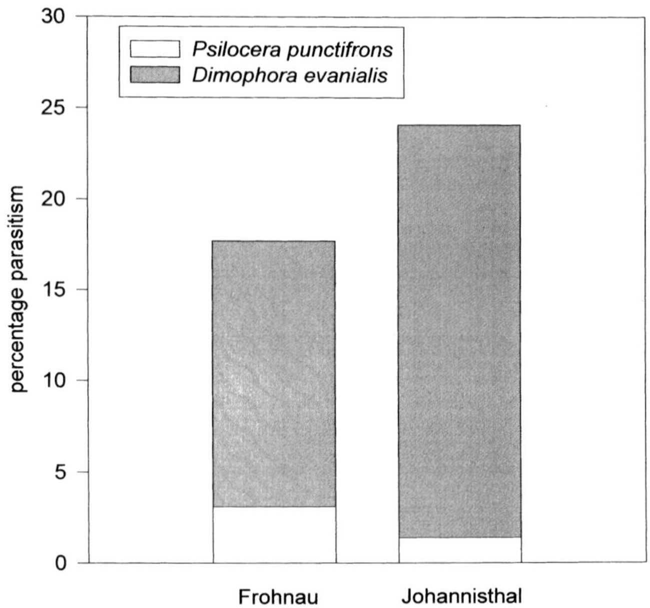

Fig. 13. Comparison of mortality due to parasitoids of the genera *Dimophora* and *Psilocera* at two populations of *Cryptocephalus moraei* in 1993-95 at Frohnau and Johannisthal, Berlin, Germany

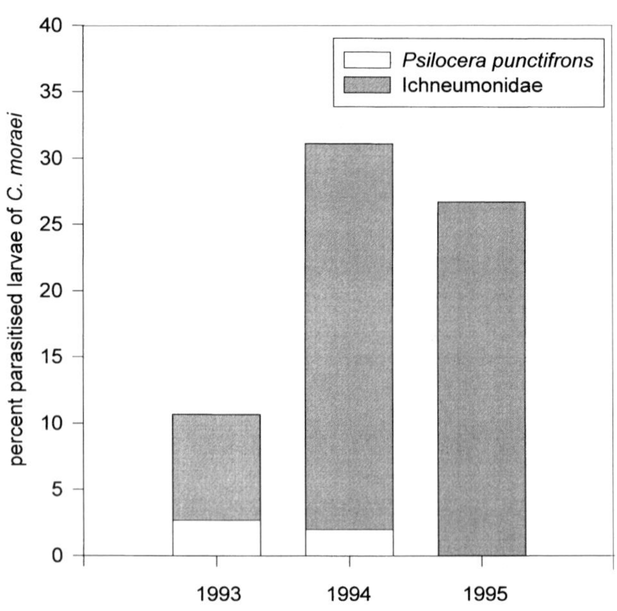

Fig. 14. Percentage parasitism of immature stages of *Cryptocephalus moraei* by *Psilocera punctifrons* and Ichneumonidae in 1993-95 in Berlin

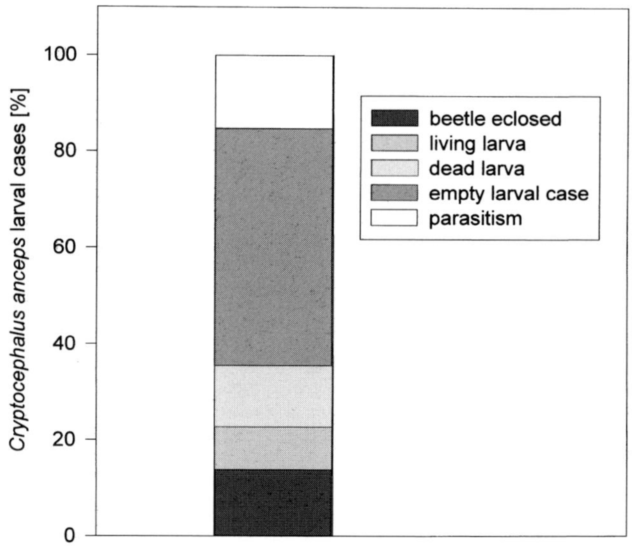

Fig. 15. Mortality of *Cryptocephalus anceps* in 1998 in Cali, Colombia, and percentage parasitism

Population dynamics of C. anceps

13.9 % of the larval cases of *C. anceps* showed the mode of opening which indicated that the adult beetles had emerged (Fig. 15). 21.5 % of the larval cases contained larvae, 41.2 % of which were alive. Many of the larval cases were empty (49.4 %), 69 % of these were intact and 31 % broken. The typical emergence holes of parasitoids were found in 15.2 % of the larval cases.

Interactions of Cryptocephalus moraei *and* C. anceps *with other organisms*

The empty larval cases of *C. moraei* and *C. anceps* were found to be microchorions for other organisms, namely Formicidae and Araneae.

Ants of the subfamily Myrmicinae used *Cryptocephalus* larval cases as a place for colony founding. In a late instar larval case of *C. moraei* a queen of *Myrmica rugulosa* Nylander and in a larval case of *C. anceps* a queen of *Pheidole* sp. was found with about 15 eggs and 5 workers.

Webbings of spiders, and occasionally living spiders were found in the larval cases, both in the empty cases and in cases, from which adult *C. moraei* and *C. anceps* had already emerged. Moreover, both in a larval case of *C. moraei* and of *C. anceps*, unidentified insect eggs were found (Fig. 16), therefore the larval case was used as a site for oviposition. Exuviae of heteropterans were once found inside a larval case, indicating that it was used as a site for moulting.

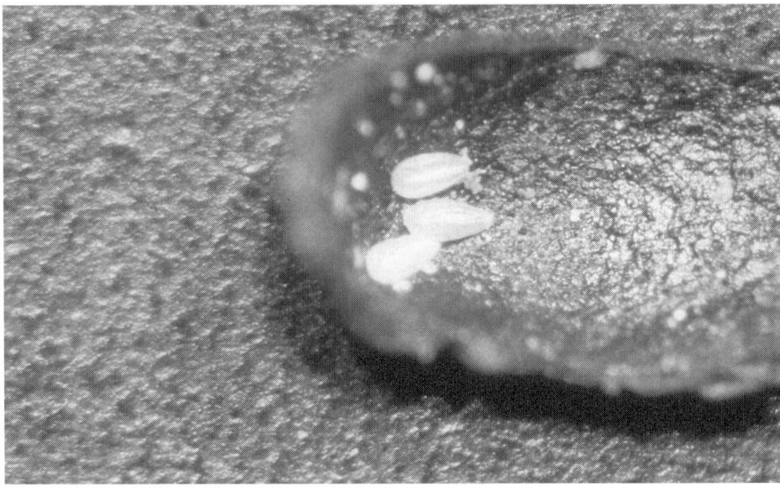

Fig. 16. Unidentified insect eggs inside the the larval case of *Cryptocephalus moraei*

Further observations

In Frohnau and Johannisthal, *C. fulvus* Goeze occurred in addition to *C. moraei*. *C. fulvus* was also a host for *Dimophora evanialis*. From *C. fulvus* larval cases, only male *D. evanialis* emerged, the emergence holes had a diameter of 0.9 mm.

Discussion

Generally, a polyphagous host selection by larvae of species of *Cryptocephalus* species was reported in the literature (Erber, 1988). To date the relation of host selection of the larvae and adults has not been studied systematically in this genus. As reported, female *C. moraei*, standing on the host plant, dropped their eggs to the ground. The aggregation of the larvae at the base of the host-plant of the adults may be due to reduced dispersal from the adult's oviposition site. Within a *H. perforatum*-patch, the *H. perforatum* leaf-litter was distributed all over ground. There may be differences in the microclimate, for example increased humidity at the bottom of the plants was observed. However, this was not studied systematically. Another possible explanation for the aggregation of the larvae is a preference for a certain food plant. No preference tests were conducted, this should be done in future studies. For a chrysomeline leaf beetle, it was supposed that a secondary plant substance of *Hypericum hirsutum*, hypericin, mediated the choice of the feeding site (Rees, 1969).

As a larva of *C. moraei* was only once found to climb up a plant, it is rather unlikely that the larvae feed on green leaves in the field, at least in the daytime. However, they are able to climb up plants, as are the larvae of *Cryptocephalus quadruplex* Newman and *C. venustus* Fabricius (LeSage, 1986).

All the host-parasitoid records in this study are new. In this study, *P. punctifrons* was reared from both *C. moraei* and *C. fulvus*. Another species of the genus, *Psilocera rufipes*

(Ashmead), was recorded as an endoparasitoid of *Chlamisus gibbosus* F., 1777 (Coleoptera, Chrysomelidae) (Sorauer, 1954; Peck, 1963). Traditionally, *Chlamisus* is placed in the subfamily Chlamisinae (Seeno & Wilcox, 1982), but recently this subfamily was ranked as a tribe, Chlamisini, within the subfamily Cryptocephalinae (Reid, 1995). There are no other host records available for the genus *Psilocera* from rearing experiments (Vidal, pers. communication). As the leaf beetle *C. gibbosus* is closely related to the species of *Cryptocephalus*, beetles of this group are possibly the common hosts for *Psilocera* spp.. The number of larvae parasitised by *P. punctifrons* may have been underestimated, because for intact, empty larval cases of early instar larvae, parasitism could have been confused with other causes of mortality.

Parasitoids of the genus *Gelis* are almost all idiobiont ectoparasitoids, the commonest hosts are prepupae or pupae enclosed in cocoons, including those of other ichneumonoids. Hosts in other cocoon-like structures were described to be the subject of attack by *Gelis* (Gauld & Bolton, 1988), which is in accordance with the record of attack of the *Cryptocephalus*-larval cases reported in this study. *Gelis* are either primary parasitoids or pseudo-hyperparasitoids. The pseudo-hyperparasitoids oviposit into cocoons of Braconidae or Ichneumonidae, which are located out of their host, behaving therefore like primary parasitoids. As four species of Ichneumonidae were found to parasitise *C. moraei*, among the species of *Gelis* and *Theroscopus* may have been pseudo-hyperparasitoids and primary parasitoids of *C. moraei*, or both. No host is known from *G. forticornis*, nor from closely related species (Schwarz, per. communication). The Phygadeuontinae is the largest subfamily of the Ichneumonidae with about 100 genera worldwide. Three other records of species belonging to the Phygadeuontinae as parasitoids of Cryptocephalinae were published. *Bathyrix pellucidator* (Gravenhorst) was reared from *Cryptocephalus janthinus* Germar in Berlin (Reineck, 1913). Voucher specimens of larval cases showing the emergence holes of the parasitoids are preserved in the Reineck collection, which is located today in the insect collection of the Forschungsinstitut Senckenberg, Frankfurt am Main, Germany. *Gelis* sp. was reared from *Cryptocephalus octacosmus* Bedel, 1891 (Duhaldeborde, 1994). Jolivet (1950) listed *Hemiteles pedestris* Fabricius as an ectoparasitoid of *Cryptocephalus octomaculatus* Rossi, 1790. According to Cox (1994), there are no other records available. All other records of species of Chrysomelidae as hosts for *Gelis* were from the genus *Oulema*, subfamily Criocerinae (Dysart et al., 1973; Momoi, 1973). *Dimophora evanialis* emerged from host larvae which were active and feeding at the time of collection and therefore at least the first larval instars of *D. evanialis* are endoparasitic. The finding that only male *D. evanialis* emerged from *Cryptocephalus fulvus*-larvae may be due to the smaller size of this host. The fitness of female parasitoids increases with increasing size in most species studied so far, and female eggs are laid preferably in large hosts (Gauld & Bolton, 1988).

Emergence holes of parasitoids in larval cases were illustrated by Rosenhauer (1852). A larger sample size was used only in a few studies of Cryptocephalinae to determine the parasitism. In a field study of the chlamisine *Exema canadensis*, Root & Messina (1983) collected 355 larvae at two sampling sites and recorded 27 % parasitism. About one third of the larvae of Neotropical Cryptocephalinae and Lamprosomatinae which were studied by Fiebrig (1910) failed to develop into adults and the occurrence of parasitoids was reported, but without details. Seven out of 12 larvae of *Cryptocephalus octacosmus* which were collected by Duhaldeborde (1994) were parasitised. The lowest parasitism of about

5 %, was found in a population of the chlamisine *Chlamisus cribripennis* (LeConte) in Canada (Wood, 1966). The parasitism found in the study presented here was therefore within the range reported for these other cryptocephaline leaf beetles. According to a recent review, the overall parasitism found for European chrysomelids belonging to other subfamilies ranged between 5 and 50 % (Schöller, 1996b).

An unsolved problem is the interpretation of the empty larval cases. As the larvae are not capable of constructing a new case once it is lost, these larvae were not able to develop into adults. In addition to natural mortality due to developmental disorders, many causes of mortality such as diseases and predation are possible. Dead larvae shrink, and unlike living larvae, they can fall out of the larval case. Dead larvae may have also been taken out of the larval case by scavengers. No indications of predation were found. General predators found in the habitat were carabid beetles, ants and earwigs. According to the reviews of insect predators (Erber, 1988; Cox, 1996), no predators of the larvae of Cryptocephalinae were identified so far. Root & Messina (1983) observed no predation in the field, but in no-choice-experiments in the laboratory, the pentatomid *Podisus maculiventris* (Say) and the nabid *Nabicula subcoleoptrata* were capable of predating the case-bearing larvae of *Exema canadensis*.

Dead larvae which remained in the larval case were attacked by fungi. Fungal attack may have been the consequence of mortality of the larvae rather than the cause. All organisms living in leaf-litter are potential subjects of fungal attack, and many of these organisms are protected by mycotoxic secretions. The inner side of the *Cryptocephalus*-larval cases are shiny, suggesting that the larvae also secrete substances inside the larval cases. However, this aspect has still to be studied.

The 5 % of larval cases which were covered with green algae may have been larval cases which persisted for two years or more. The method used in this study did not allow the discrimination between larval cases from different generations.

The observation of *Cryptocephalus*-larval cases seem to be the first record of the use of structures of beetle origin as a place for colony founding by ants. In the ant colony life cycle, the founding stage is the most vulnerable. It is therefore hypothesised, that the larval cases generally offer good mechanical protection against adverse influences. This hypothesis is supported by the finding that spiders used the *Cryptocephalus*-larval cases as hibernating sites.

Some common characteristics of the biology of *C. moraei* and *C. anceps* were found. In both species

- the adults are phytophagous
- the larvae are detrivorous
- the larvae were hosts for parasitoids
- the larval cases were used as hibernating sites by spiders
- the larval cases were used as a place for colony founding by ants

Due to the large geographic distance between the two species, it is hypothesised that the common characteristics are general traits in the biology of *Cryptocephalus*.

In this study, it was supposed that all larval cases had the same probability to persist in the habitat, regardless if they had emergence holes from adult *Cryptocephalus* or parasitoids, or were empty or broken. If this assumption is correct, it would be possible to study the percentage and causes of mortality of every population of *Cryptocephalus* or of other genera of Cryptocephalinae with a similar life history to a certain extent. The mortality of field populations could be estimated by taking only one sample, as it was done here for *C. anceps*. Moreover, the larval cases offer the possibility to mark individual larvae. Marking would also allow the discrimination between the different generations. The Camptosomata are in general a group within the Chrysomeloidea which seems to be suitable for mark-release and recapture studies, to study the dispersal and fate of individuals. To date this possibility was only described for the Cassidinae, the larvae of which can be marked on the caudal fork (Buzzi, 1988).

Possible limitations of the Cryptocephalinae as model organisms include the relatively low intrinsic rate of natural increase and the fact that rarely large populations are described in the literature. However, systematic collecting may reveal that large populations of more species additional to the two described in this study can be found and sampled. Therefore, the "dwarfs that live in their hat" (Jenks, 1940) are a very promising group within the Chrysomeloidea for the study of population dynamics and ecology.

Acknowledgements

For the determination of specimens and provision of valuable information, the author would like to thank Jens Dauber, Giessen (Myrmica), Dr. K. Horstmann, Würzburg (*Dimophora*), Dr. Martin Schwarz, Kirchschlag, (*Gelis* and *Theroscopus*) and Prof. Dr. Stefan Vidal, Gießen, (*Psilocera*). Inga Mewis, Berlin, assisted in the preparation of the REM-photographs.

References

Buzzi, Z. J., 1988. Biology of Neotropical Cassidinae. In: Jolivet, P. H., Petitpierre, E. & Hsiao, T. H. (eds.) Biology of Chrysomelidae. Kluwer Academic Publishers, Dordrecht, The Netherlands, pp. 559-580.

Cox, M. L, 1994. The Hymenoptera and Diptera parasitoids of Chrysomelidae. In: Jolivet, P. H., Cox, M. L. & Petitpierre, E., (eds.) Novel aspects of the biology of Chrysomelidae. Kluwer Academic Publishers, Dordrecht, The Netherlands, pp. 419-467.

Cox, M. L, 1996. Insect predators of Chrysomelidae. In: Jolivet, P. H. A. & Cox, M. L., (eds.) Chrysomelidae biology vol. 2: Ecological studies. SPB Academic Publishing, Amsterdam, The Netherlands, pp. 23-91.

Duhaldeborde, F., 1994. *Cryptocephalus octacosmus* Bedel: régime alimentaire et parasitologie (Coleoptera, Chrysomelidae). Bull. Soc. ent. Fr. 99: 447-454.

Dysart, R. J., Maltby, H. L. & Brunson, M. H., 1973. Larval parasites of *Oulema melanopus* in Europe and their colonization in the United States. Entomophaga 16: 445-452.

Erber, D., 1988. Biology of Camptosomata Clytrinae – Cryptocephalinae – Chlamisinae – Lamprosomatinae. In: Jolivet, P., Petitpierre, E. & Hsiao, T. H. (eds.) Biology of Chrysomelidae. Kluwer Academic Publishers, Dordrecht, The Netherlands, pp. 514-552.

Fiebrig, K., 1910. Cassiden und Cryptocephaliden Paraguays. Ihre Entwicklungsstadien und Schutzvorrichtungen. Zool. Jb. Suppl. 12: 161-264.

Gauld, I., & Bolton, B., 1988. The Hymenoptera. Oxford University Press, Oxford, 332 pp.

Horstmann, K. 1993. Revision der brachypteren Weibchen der westpaläarktischen Cryptinae (Hymenoptera, Ichneumonidae). Entomofauna 14: 85-148.

Jenks, G. E., 1940. Dwarfs that live in their hats. Nature Mag. 33: 337-340.

Jolivet, P., 1950. Les parasites, prédateurs et phorétiques des Chrysomeloidae (Coleoptera) de la faune Franco-Belge. Bull. Inst. r. Sci. nat. Belg. 26: 1-39.

LeSage, L., 1986. The eggs and larvae of *Cryptocephalus quadruplex* Newman and *C. venustus* Fabricius, with a key to the known immature stages of the Nearctic genera of Cryptocephalinae (Coleoptera: Chrysomelidae). Can. Ent. 118: 97-111.

Momoi, S., 1973. On ichneumonid parasites of the rice leaf-beetle, *Oulema oryzae*, in Japan and Korea (Hymenoptera: Ichneumonidae). Kontyû 41: 444-445.

Peck, O., 1963. A catalogue of the Nearctic Chalcidoidea (Insecta: Hymenoptera). Can. Ent., Suppl. 30: 1-1092.

Rees, J. C., 1969. Chemoreceptor specificity associated with choice of feeding site by the beetle *Chrysolina brunsvicensis* on its foodplant, *Hypericum hirsutum*. Entomologia exp. appl. 12: 565-583.

Reid, C. A. M., 1995. A cladistic analysis of subfamilial relationships in the Chrysomelidae sensu lato (Chrysomeloidea). In: Pakaluk, J. & Slipinski, S. A. (eds.) Biology, phylogeny, and classification of Coleoptera: Papers celebrating the 80th birthday of Roy A. Crowson. Muzeum i Instytut Zoologii PAN, Warszawa, pp. 559-631.

Reineck, G., 1913. Beitrag zur Lebensweise von *Cryptocephalus janthinus* Germ. Dt. ent. Z. 39: 163-168.

Root, R. B. & Messina, F. J., 1983. Defensive adaptions and natural enemies of a case-bearing beetle, *Exema canadensis* (Coleoptera: Chrysomelidae). Psyche 90: 67-80.

Rosenhauer, W. G., 1852. Über die Entwicklung und Fortpflanzung der Clythren und Cryptocephalen, einer Insektengruppe aus der Ordnung der Coleopteren. PhD-Dissertation, Philosophische Fakultät Universität Erlangen, Germany.

Schöller, M., 1996a. Petalophagie bei *Cryptocephalus nubigena* Franz 1982 aus La Palma (Kanarische Inseln) (Coleoptera: Chrysomelidae: Cryptocephalinae). Mitt. int. ent. Ver. 20: 119-127.

Schöller, M., 1996b. Ökologie mitteleuropäischer Blattkäfer, Samenkäfer und Breitrüssler (Coleoptera: Chrysomelidae einschließlich Bruchinae, Anthribidae). EVCV, Bürs, ISBN 3-9500146-6-7, 65 pp.

Schöller, M., 1997. Räumliche Verteilung der Larven von *Cryptocephalus moraei* (Linnaeus, 1758) (Coleoptera, Chrysomelidae, Cryptocephalinae). Novius 22: 511-514.

Seeno, T. N. & Wilcox, J. A., 1982. Leaf beetle genera (Coleoptera: Chrysomelidae). Entomography Publ., Sacramento, 221 pp.

Sorauer, P., 1954. Handbuch der Pflanzenkrankheiten. Band 5. Tierische Schädlinge an Nutzpflanzen; Teil 2. Zweite Lieferung, 2. Coleoptera. H. Blunck, Berlin, pp. viii+599.

Steinhausen, W. 1994. 17. Überfamilie: Chrysomeloidea, 116. Familie: Chrysomelidae. In: Klausnitzer, B. (ed.) Die Larven der Käfer Mitteleuropas. 2. Band Myxophaga, Polyphaga, Teil 1. Goecke & Evers, Krefeld, pp. 231-314.

Warchalowski, A., 1991. Chrysomelidae, Stonkowate (Insecta: Coleoptera), Czesc II, podrodziny: Clytrinae i Cryptocephalinae. Fauna Polski, Tom 13, Polska Akademia Nauk Instytut Zoologii, Warszawa, 346 pp.

Wood, G. W., 1966. Life history and control of a casebearer, *Chlamisus cribripennis* (Coleoptera: Chrysomelidae), on blueberry. J. econ. Ent. 59: 823-825.

REVISION OF THE GENUS *LABIDOMERA* DEJEAN WITH A DESCRIPTION OF TWO NEW SPECIES (COLEOPTERA: CHRYSOMELIDAE: CHRYSOMELINAE)

MAURO DACCORDI[1] and LAURENT LESAGE[2]
[1]*Museo Regionale di Scienze Naturali, 10123 Torino, Via Giolitti, 36, Italy;* [2]*Agriculture and Agri-Food Canada, Eastern Cereal and Oilseed Research Centre, Ottawa, Ontario, Canada K1A 0C6*

Abstract

The genus *Labidomera* Dejean (1836) is redescribed and discussed. Species included are described, illustrated, and keyed out. *Labidomera trimaculata* Linnaeus (1767) *sensu* Fabricius (1775), *Doryphora rogersii* LeConte (1858), and *Labidomera mimica* Brown (1961) are placed in synonymy with *Labidomera clivicollis* (Kirby, 1837). Lectotypes are designated for *Chrysomela clivicollis* Kirby and *Doryphora rogersii* LeConte. The authorship of *Labidomera suturella* should be attributed to Guérin-Méneville, not to Chevrolat as generally reported by authors. Two new species from Mexico, *Labidomera olmeca* and *L. maya* are described and illustrated. New information on biology and host plants is also reported.

Introduction

Several hundred new species of Nearctic and Neotropical Chrysomelinae were described since Stål (1862-1865) wrote his monograph on this subfamily of leaf beetles. Previous authors tried to split the subfamily into consistent genera, and in some cases, even recognized subgenera. However, it is not always easy to distinguish the different genera using the usual external characters of shape, colour or sculptures because convergent evolution is frequent between genera. For instance, *Euryceraea wagneri* Steinheil is superficially identical to *Platyphora paradoxa* Achard although clear morphological differences can be found under closer examination.

In order to solve such problems, authors have looked at different characters. This is the case of Bechyné (1980) who recognized eight lineages using characters of the intercoxal process. According to these characters, *Labidomera* falls into the lineage of *Cryptostetha*, close to *Cryptostetha* Baly and *Proseicela* Dejean with its mesosternum produced into a short lobe is distinguished from both of these by its unique indented apophysis of the front femora in the male (Figs. 12-17).

The purpose of the present paper is to redescribe and illustrate the genus *Labidomera* Dejean including, for the first time, features of the male and female genitalia, and to describe two new species.

Advances in Chysomelidae Biology 1, pp. 437–461
Edited by M.L. Cox

Materials

Specimens examined in the present study are deposited in the collections listed below. Addresses of institutions from which material was borrowed, and the names and addresses of curators can be found in LeSage (1995). Specimens studied here were obtained from:

ACHC : Achard Collection, Prague
CAS : California Academy of Sciences
CDFA : California Department of Food and Agriculture
CMNC : Canadian Museum of Nature Collection
CNC : Canadian National Collection, Ottawa
FMNH : Field Museum of Natural History, Chicago
FREY : Frey collection, Bale
LEM : Lyman Entomological Museum
LLC : Laurent LeSage private collection
MCZ : Museum of Comparative Zoology, Cambridge
MDC : Mauro Daccordi private collection
USNM : United States National Museum
UD : Unknown depository. This acronym will be used for specimens which have been identified several years ago, but for which the information on the depository has not been recorded.

Labidomera Dejean, 1836

Labidomera Dejean 1836:397; 1837:421; Neave 1939:841.
Labidomera Chevrolat 1843:656, 1846:183; Guérin-Méneville 1838:301; Crotch 1873:47; Chapuis 1874:394; Jacoby 1883:240; Linell 1896:195; Blatchley 1910:1152; Weise 1916:28; Leng 1920:294; Blackwelder 1946:672; Bechyné 1952:42; Wilcox 1954:413; Chagnon & Robert 1962:300; Hatch 1971:188; Balsbaugh & Hays 1972:92; Wilcox 1972:11, 1974:65; Seeno & Wilcox 1982:79; 1991:311; Downie & Arnett 1996:1336.

DIAGNOSIS. Front femora of male deeply indented on inner side into one or two apophyses (Figs. 12-17). Mesosternum forming blunt tubercle between middle coxae. Apex of metasternum not bordered.

DESCRIPTION. Size large, 8-15 mm. Body robust, shape ovoid. Antennae 11-segmented, segments 6-11 hairy and slightly enlarged. Pronotum not sulcate laterally; procoxal cavities open behind. Mesosternum forming blunt tubercle between middle coxae, apex of metasternum not bordered. Elytra glabrous, with punctures arranged in irregular double rows. Tarsal claws simple. Sexual dimorphism important: front femora deeply indented in male (Figs. 12-17), without such indentation in female. Aedeagus with flagellum always exposed (Figs. 18, 19). Spermatheca hook-shaped (Figs. 10, 11).

REMARKS. Although the description of Chevrolat (1846) in d'Orbigny's *Dictionnaire d'histoire naturelle* has been accepted by most authors (see above) to validate the genus *Labidomera*, earlier mentions can be found in the literature. Actually, under the heading

"Chrysomélines", Chevrolat (1843) already mentioned, in the same dictionary, the indented front femora of the male, and designated *L. germari* Chevrolat as the type species of *Labidomera*. The earliest mention of *Labidomera* had appeared in the second and third versions of Dejean's catalogue (Dejean 1836:397; 1837:421) as properly recorded by Neave (1939). Although Dejean did not specifically describe *Labidomera* in his catalogues, he must be credited for its authorship because he fulfills the provisions of Articles 11 and 12b(5) of the Code for names published before 1931 by using "one or more available species-group names in combination with, or clearly included under it" [a generic name]. The fact that Dejean, in his catalogue, erroneously credited Chevrolat for the authorship of *Labidomera* also contributed to the confusion.

The publication year of *Labidomera* is also difficult to establish accurately because the fifth fascicle of the second edition of Dejean's catalogue (where the name *Labidomera* was supposed to appear) was destroyed in a fire on December 12, 1835. Following this disaster, the editor asked Dejean to publish an updated version of the lost fascicle. Except for the pagination, the new fifth fascicle of the second and third versions are essentially similar. According to Sherborn (1922), the fifth fascicle of the second edition was published at the end of 1835. However, Madge (1988), who studied in detail the publication dates of Dejean's catalogues, concluded that this fascicle was published in 1836, although most copies are dated 1834, very likely because the printer used covers of previous fascicles without changing the date. Also, some copies are erroneously dated 1837 since early workers, or book dealers, bound together any available fascicles to complete their set.

Labidomera Dejean presently contains two Mexican species (*maya* and *olmeca*), one primarily Meso-american species (*suturella*), and a largely northern species (*clivicollis*). Species of *Labidomera* feed exclusively on Asclepiadaceae (genera *Asclepias*, and *Cynanchum*) and Solanaceae (genus *Witheringia*). The bright red colouration on greenish blue or black background in live specimens appears aposematic (Jones 1932). It has been demonstrated that several Doryphorina rely upon the production of amino acid derivatives for their chemical defence (Timmermans *et. al.* 1992). Among these, *Labidomera* is particularly noteworthy because it uses its own amino acid-derived allomones instead of the cardenolides of the host- plants (Isman *et. al.* 1977).

Phylogenetically, *Labidomera* is allied to *Proseicela* Dejean and *Cryptostetha* Baly whose species inhabit only the Neotropical region (Bechyné 1980). Chapuis (1874) considered *Labidomera* a subgenus of *Cryptostetha*, whereas Weise (1916) treated *Cryptostetha* as a subgenus of *Labidomera*. On the other hand, Blackwelder (1946) had a different opinion and listed *Cryptostetha* as as a synonym of *Labidomera*.

Because of the peculiar structure of the male anterior femora, we agree with Stål (1862), Linell (1896), and Bechyné (1952, 1980) in considering *Labidomera* a valid genus. Actually, the inner side of the anterior femora of the male possesses a deep cavity delimited proximad by a rather strong apophysis and distad by one or two symmetrical spines (Figs. 12-17). This hypertely probably relates to some mating function. The male of *Leptinotarsa stali* Jacoby also possesses two distinct distal spines on the front femora, but each lacks a central cavity or proximal tooth.

Based on the presence of one or two teeth near the apex of the anterior femora in the male, we can distinguish two groups of species in the genus *Labidomera*. The first group with one tooth includes only *L. clivicollis* (Kirby), whereas the second, with two teeth, consists of three species distributed in the Mexican region and Central America.

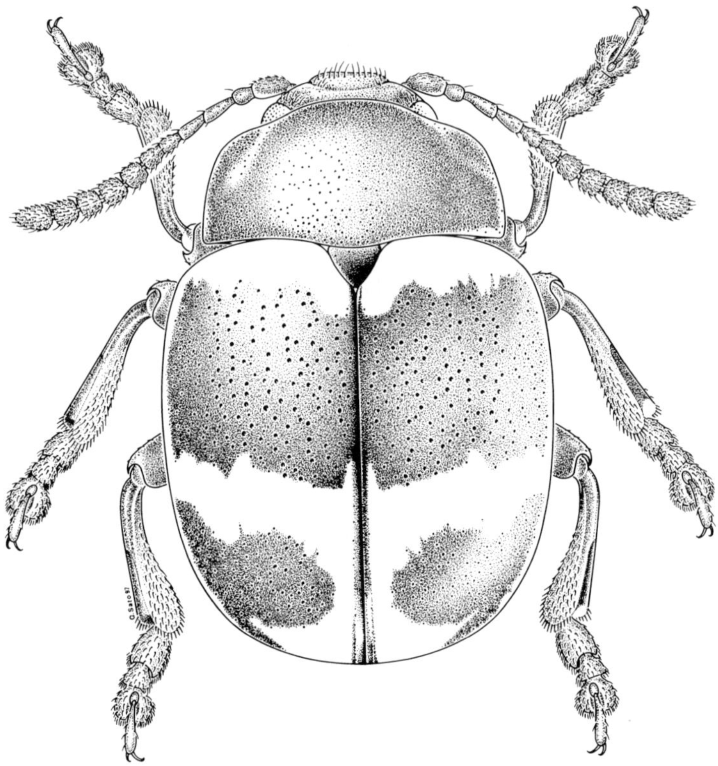

Fig. 1. Labidomera clivicollis, habitus, dorsal view.

Key to the species of *Labidomera* Dejean

1. Inner border of elytral epipleura ciliate along entire length; in male, front femora with two apical apophysis ("spines") (Figs. 14-17) 2
– Inner border of epipleura ciliate only in apical half or third; in male, apex of front femora of male with one apophysis only (Figs. 12, 13) .... ***clivicollis* (Kirby)**
2. Prothorax with yellow margins 3
– Prothorax unicolourous, without pale margins .... ***suturella* Guérin-Méneville**
3. Lateral margins of prothorax straight (Fig. 6); disc of pronotum shiny, with fewer punctures; frons shiny, very finely, sparsely punctate; clypeus concave, with large dense shallow punctures giving rugose aspect to surface; elytral punctation denser (Fig. 8) .... ***olmeca* n. sp.**
– Lateral margins of thorax distinctly arcuate (Fig. 7); disc of pronotum opaque, sparsely punctate; frons opaque with distinct punctation near eyes and vertex; clypeus flat, shiny, with row of large punctures; elytral punctation sparser, and more or less arranged in pseudo-geminate rows (Fig. 9) .... ***maya* n. sp.**

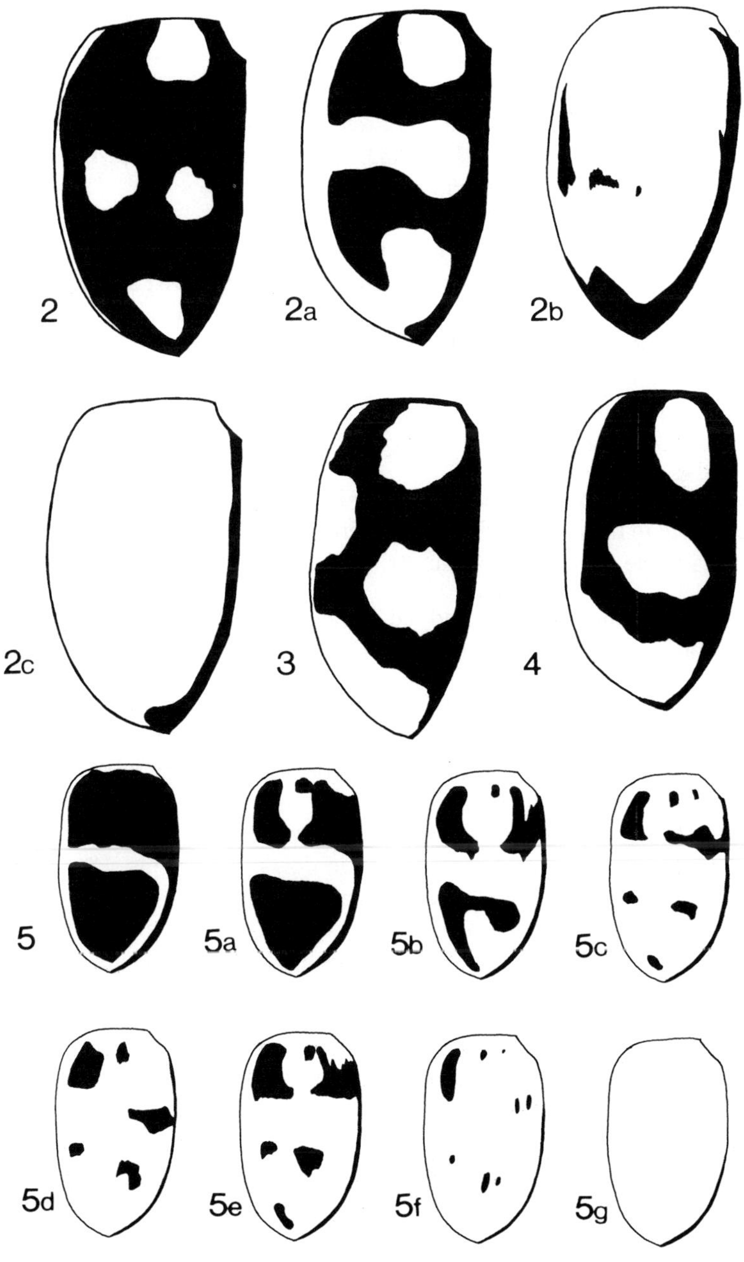

Figs. 2–5. Elytral pattern variations in *Labidomera* species. 2- 2a, *L. suturella germari* auctorum; 2b- 2c, *L. suturella*, typical form; 3, *L. maya*; 4, *L. olmeca* ; 5, *L. clivicollis* from various locations in North America: 5, Mer Bleue (SE Ottawa, Ontario, Canada); 5a, Ottawa (Ontario, Canada); 5b, Forest Hill (Massachusetts, U.S.A.); 5c, Columbia (Missouri, U.S.A.); 5d, Brownsville (Cameron Co. Texas, U.S.A.), Holt Co. (Missouri, U.S.A.); 5e, Columbia (Boone Co. Missouri, U.S.A.); 5f, Holt Co. (Missouri, U.S.A.); 5g, [holotype of *mimica* Brown] (White Rose Canyon, Jeff Davis Co., Texas, U.S.A.).

Labidomera clivicollis (Kirby, 1837)
(Figs. 1, 5-5g, 11-13, 18-18e, 22-24)

Chrysomela clivicollis Kirby 1837:213.
Chrysomela trimaculata Linnaeus 1767:592 *sensu* Fabricius 1775:95; 1781:117; 1787:67; 1792:312; 1801:424; Gmelin 1788:1683; Coquebert 1804:123; Olivier 1807:515; Schönherr 1808:239. Misidentifications.
Doryphora rogersii LeConte 1858:26. **New synonymy**.
Labidomera clivicollis: Stål 1863:144; Crotch 1873:47; Blatchley 1910:1152; Weise 1916:28; Blackwelder 1946:672; Chagnon & Robert 1962:30; Hatch 1971:188; Balsbaugh & Hays 1972:93; Wilcox 1972:11; 1974:65. Generic transfers.
Labidomera clivicollis rogersii: Linell 1896:195; Weise 1916:28; Leng 1920:294; Wilcox 1972:11; 1974:65. Subspecific status.
Labidomera mimica Brown 1961:973. **New synonymy**.
Labidomera trimaculata (Fabricius): LeSage 1991:311. Synonym.

ETYMOLOGY. From the combination of two Latin names *clivus* (= slope) and *collis* (= neck) in reference to the unusual shape of the pronotum as clearly stated by Kirby (1837): "... in the remarkable peak of the prothorax" (see comments on lectotype).

DIAGNOSIS. Medium-large black and red ovoid species, with lateral margins of prothorax abrubtly converging in apical third (Figs. 1, 5-5g), and with only one apophysis in front femora of male (Figs. 12, 13).

DESCRIPTION. Good descriptions of external features are found in Stål (1865), Blatchley (1910), Balsbaugh & Hays (1972), and Wilcox (1954, 1972). The following characters on the genitalia of both sexes can be added.

In the male, the median lobe of the aedeagus is evenly curved in lateral view (Figs 18, 18d) while the apical portion appears convex in dorsal view (Figs. 18a, 18b, 18c, 18e); sometimes, the flagellum almost reaches the tip (Fig. 18c).

In the female, the spermatheca (Fig. 11) is always of small size (‹ 0.5 mm long), hook-shaped, and the pump does not extend beyond the middle of the length .

TYPE DESIGNATION. The type specimen of *Chrysomela clivicollis* Kirby is preserved in the British Museum, in London. It bears three labels: a round type label circled in red "Type H.T.", a second round label with "Canada", and a handwritten identification label "*Chry. clivicollis* Kirby/ Canada 5933. rev. W. Kirby".

The colouration of the type (examined by L.L.) fits well within the known variations (Fig. 22): elytral suture and elytral spots black, median spots large and extending beyond middle but not reaching base of elytra, basal lateral pair about 1/3 length of elytra and narrowly separated from median pair; and finally, apical pair deeply notched on inner side.

The aberrant pronotum is produced into a large volcano-like tubercle grooved with several concentric furrows (Fig. 22). Normally, the pronotum is smooth and moderately convex (Fig. 23).

The following labels have been added to the specimen: a red designation label "LECTOTYPE *Chrysomela clivicollis* Kirby/ Des. L. LeSage '96", and a determination label "*Labidomera clivicollis* (Kirby)/ Det. L. LeSage'96".

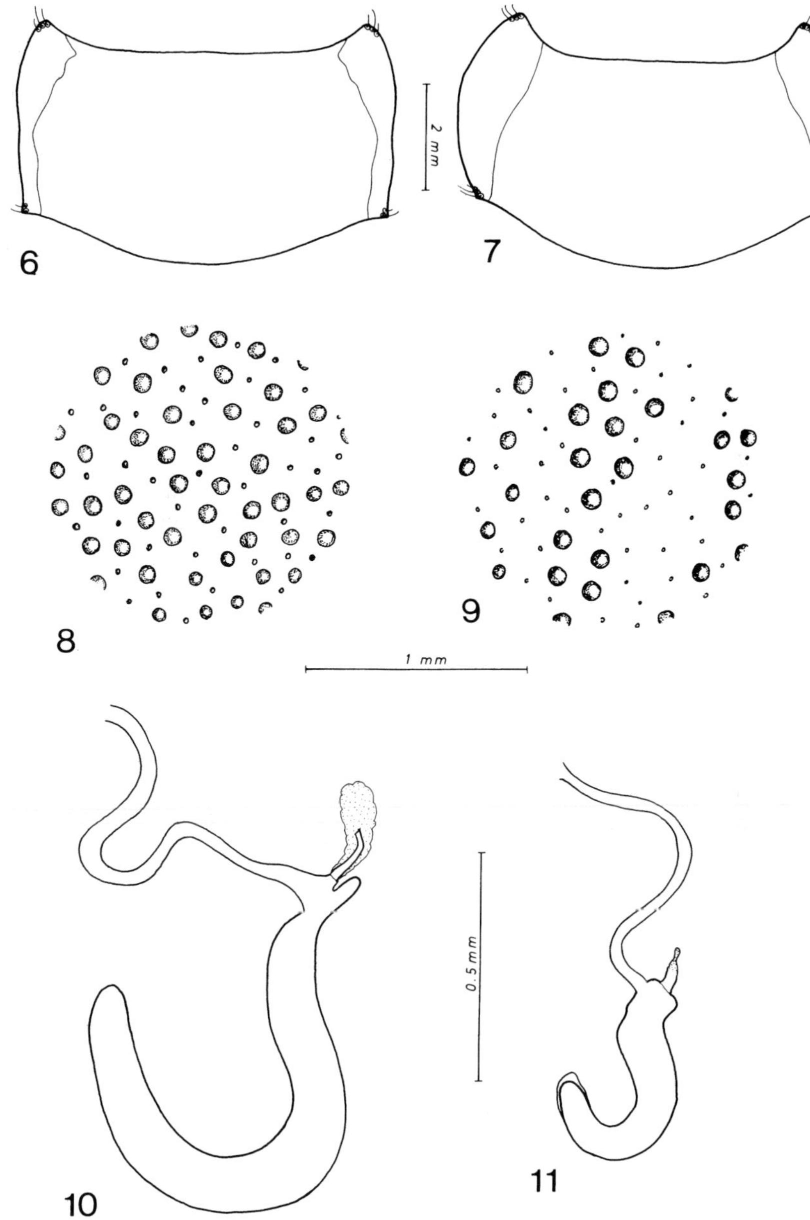

Figs. 6–11. Diagnostic structures in *Labidomera* species: 6, pronotum of *L. olmeca.* n. sp; 7, pronotum of *L. maya* n. sp.; 8, elytral punctation of *L. olmeca* n. sp.; 9, elytral punctation of *L. maya* n. sp.; 10, spermatheca of *L. suturella*.; 11, spermatheca of *L. clivicollis*.

NEW SYNONYMY.

a) *Doryphora rogersii* LeConte

The type specimen of *Doryphora rogersii* LeConte (examined by L.L.) is preserved in the Museum of Comparative Zoology, in Cambridge, Massachusetts (U.S.A.). The specimen is accompanied with three labels: a locality label "N.M." (= New Mexico), a determination label "*D. Rogersii* Pope Lec.", and a red type label "Type 4326".

It is a female which has already been dissected. Unfortunately, the genitalic structures have not been preserved, and the right front tarsus is missing (Fig. 23). The elytra are pale orangish with 7 pairs of black spots, those near the scutellum and at the apex of elytra not being as dark and sharply defined as those on disc of elytra. Otherwise, the elytral pattern fits well within the known colour variations (Figs. 5-5g).

This specimen is designated here as the lectotype of *Doryphora rogersii* LeConte. A designation label "LECTOTYPE *Doryphora rogersii* LeConte/ Des. L. LeSage '96", and a determination label "*Labidomera clivicollis* (Kirby), Det. L. LeSage '96" have been added to it.

The following new synonymy can therefore be established:

Doryphora rogersii LeConte (1858) = *Labidomera clivicollis* Kirby (1837).

b) *Labidomera mimica* Brown

The holotype of *Labidomera mimica* Brown (1961) [from White Rose Canyon, Jeff Davis co., Texas], preserved in the Canadian National Collection, in Ottawa, is a male entirely ochraceous. The paratype is also a male with the same data, but its elytra are ornamented with only a few very small dark spots. In both, the structure of the median lobe is identical to that of *L. clivicollis*. (Figs. 18d, 18e). Consequently, we can establish the following new synonymy:

L. mimica Brown (1961) = *L. clivicollis* Kirby (1837).

REMARKS.

a) Nomenclatorial comments

Fabricius (1775) cited Linnaeus (1767) as the author of "*Chrysomela trimaculata* " but evidently misinterpreted Linnaeus's description since he considers *C. trimaculata* to be a large black spotted species ("*Magna caput et thorax cyanea* ..."), whereas Linnaeus described this species as a small, black, coccinellid-shaped beetle ("*corpus nigrum minutum forma Coccinellae* ..."). This error was overlooked by Fabricius (1781, 1787, 1792, 1801) and repeated by subsequent authors (Gmelin 1788; Coquebert 1804; Olivier 1807; Schönner 1808). In his monumental work on the American Chrysomelinae, Stål (1862-1865) pointed out this mistake and the lack of relationship between *trimaculata* Linnaeus and *trimaculata sensu* Fabricius ("*Chrysomela trimaculata Linnei ad auctoribus plurimis ad Chrysomelam clivicollem falsissime est relata*"). Since it is not possible to clearly establish the identity of *Chrysomela trimaculata* Linnaeus from the original description, and because the type specimens cannot be located (lost ?) the status of this name still remains doubtful for the time being (*nomen dubius*).

b) Morphological polymorphism

Labidomera clivicollis appears to be clearly a polymorphic taxon. Among the abundant material examined, we noted that the darker individuals come mostly from the colder regions of its large range. The black elytral pattern varies greatly in extension (Figs. 5-5f),

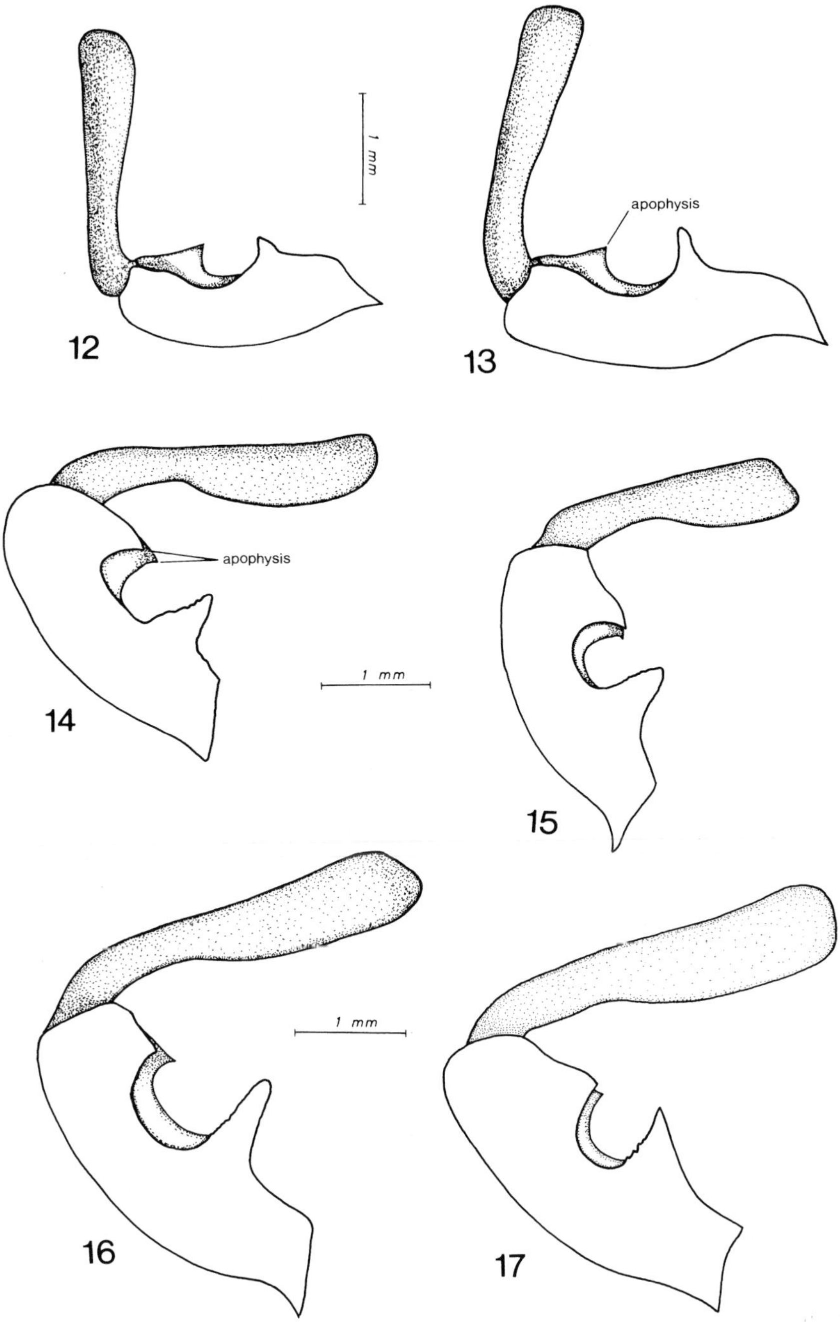

Figs. 12–17. Front tibia and femur of male in *Labidomera* species: 12, *L. clivicollis* ; 13, *L. clivicollis* (type of *L. mimica*); 14, *L. suturella.*; 15, *L. suturella.*; 16, *L. olmeca*; 17, *L. maya.*

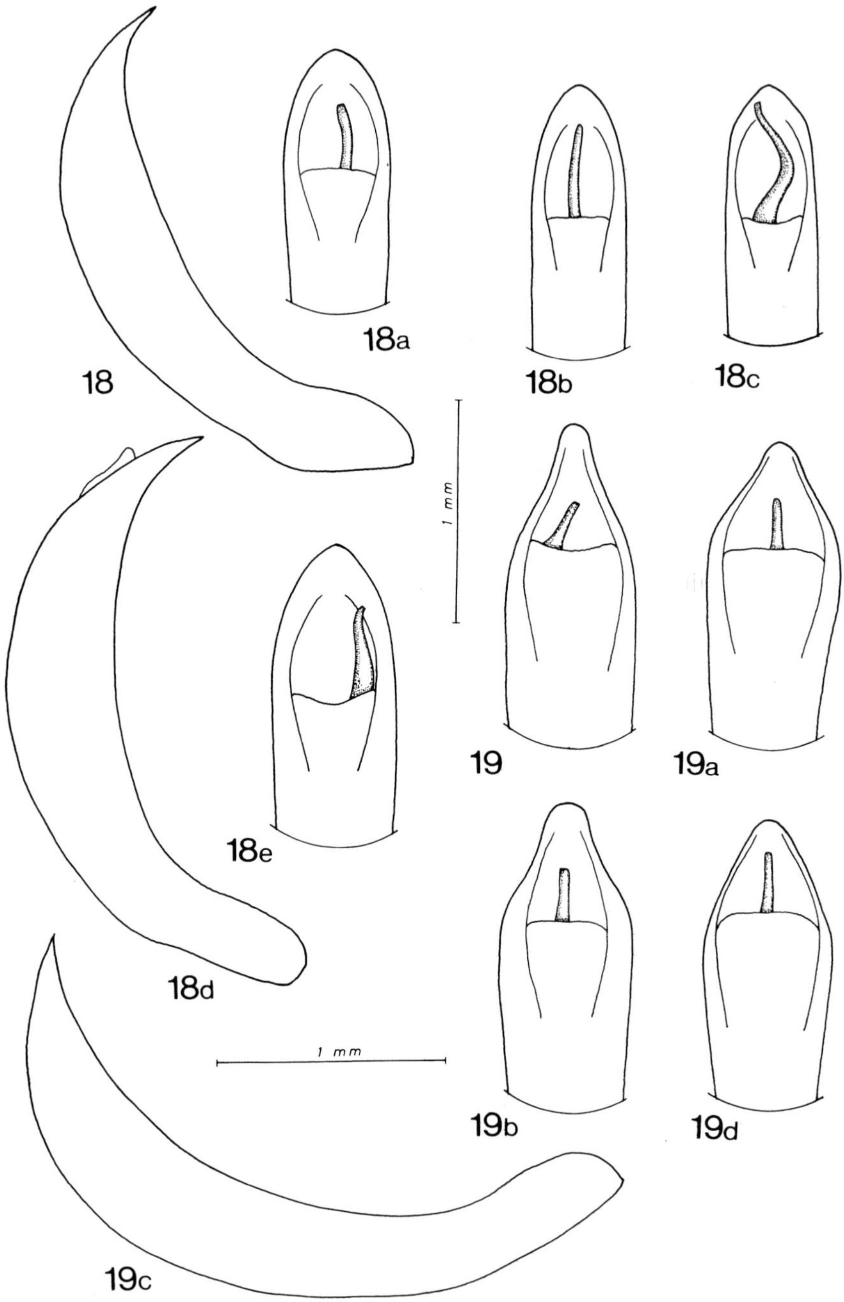

Figs. 18–18e. Variations of the median lobe of aedeagus in *Labidomera* species: 18, *L. clivivollis*, lateral view; 18a, dorsal view (Columbia [Boone Co.], Missouri, U.S.A.); 18b, *L. clivicollis* (Bentson Rio Grande State Park [Hidalgo Co.], Texas, U.S.A.); 18c. *L. clivicollis* (Newark [Wayne Co.], New York, U.S.A.). 18d- 18e, holotype of *L. mimica* [= *clivicollis*] (Rose Canyon, Jeff Davis Co., Texas, U.S.A.).

Figs. 19–19d. Median lobe of aedeagus in *Labidomera suturella*: 19, dorsal view (city of Mexico, Mexico); 19a, dorsal view (Taumalipes, Mexico); 19b, *germari* auctorum, dorsal view; 19c, idem, lateral view; 19d, idem, dorsal view (Taumalipes, Mexico).

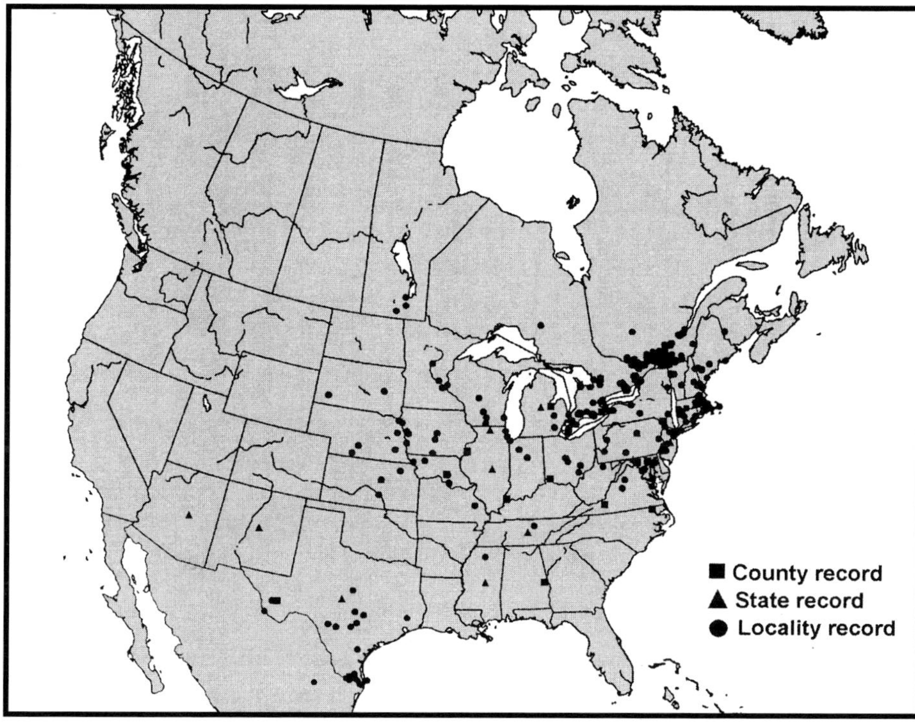

Fig. 24. Distribution of *Labidomera clivicollis* in North America.

New York. *Bronx co.*: Van Cortland Park (CAS); *Erie co.*: Buffalo (CAS), Lancaster (CAS); *New York co.*: New York (MDC); *Richmond co.*: Staten Island (USNM); *Suffolk co.*: Cold Spring Harbour (CAS); *Sullivan co.*: county record only (MDC); *Tompkins co.*; Cornell University (in Ithaca) (CAS), Ithaca (ACHC, CAS, USNM); *Warren co.*: Brant Lake (CAS); *Wayne co.*: Newark (MDC, USNM). Locality not found: "Nealham Lake" (LEM).
Ohio. State record only (CAS). *Athens co.*: Athens (CAS), / Airport (USNM); *Delaware co.*: Delaware (CAS); *Franklin co.:* Columbus (CMNC); *Hamilton co.*: county record only (CAS); *Knox co.*: county record only (FREY); *Morgan co.*: Eagleport (CAS). Location not found: "Clifty Falls" (CAS).
Oregon. *King co.*: Seattle (USNM).
Pennsylvania. State record only (CAS, MDC); *Allegneny co.*: Pittsburg (MDC), Swissvale (FREY); *Bedford co.*: Finleyville (CAS); *Clarion co.:* Easton (CAS); *Clearfield or Northampton co.* [two possibilities since county not recorded, also not on map]: Bethlehem (CAS); *Dauphin or Philadelphia co.* [two possibilities since county not recorded, also not on map]: Harrisburg (CAS); *Delaware co.*: Honeybrook Chester (USNM); *Montgomery co.*: Abington (MDC); *Northampton co.*: Bethlehem (MDC); *Philadelphia co.*: Angora (MCZ, MDC).
Rhode Island. State record only (CAS).
South Dakota. *Beadle co.*: Huron (CNC, MDC); *Pennington co.*: Badlands/ cedar gap (CNC).
Tennessee. State record only (CAS). *Smith co.*: Elmwood (CAS).

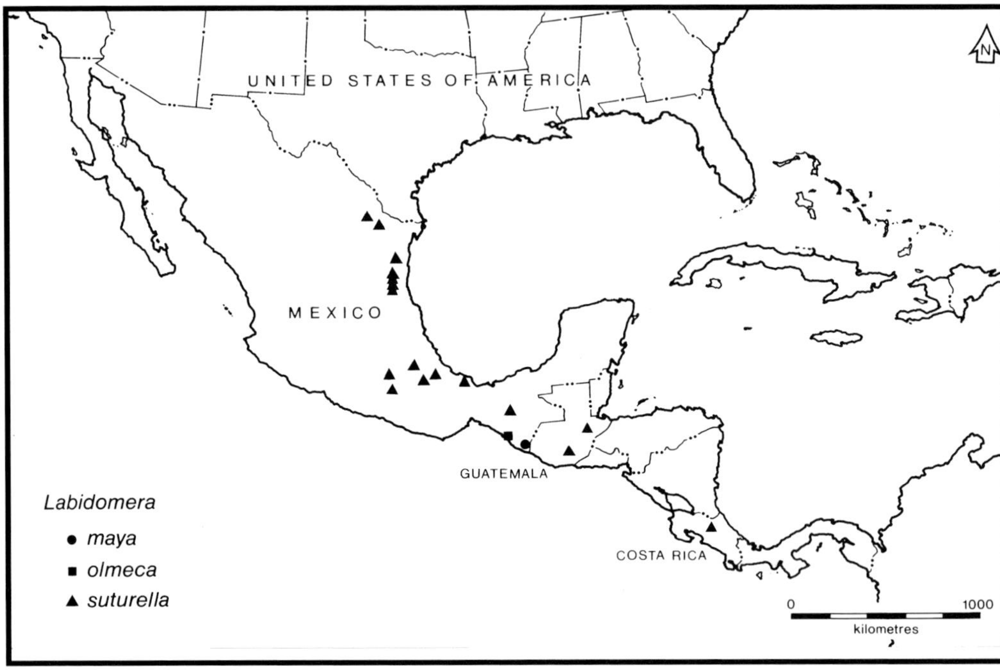

Fig. 25. Distribution of *Labidomera maya*, *L. olmeca* and *L. suturella* in Central America.

Texas. State record only (CAS, MDC, USNM). *Bexar co.*: San Antonio (CAS); *Blanco co.*: Dan's Ranch [NW of county] (MDC), Davis Ranch (CNC); *Brewster co.*: county record only (CAS); *Cameron co.*: Brownsville (CNC, MDC, USNM), Harlingen (USNM), Sabal Palm Grove near Southmost (MDC), San Benito (USNM); *Comal co.*: New Braunfels (CAS, USNM); *Hidalgo co.*: county record only (MDC, USNM), Bentson Rio Grande Valley (MDC), Pharr (USNM), Santa Ana Refuge (CAS); *Jeff Davis co.*: White Rose Canyon (CNC); *Kleberg co.:* 10 km E Riviera (CMNC), Kingsville (USNM); *Liberty co.*: Fort Sam Houston (CAS); *Llano co.*: Llano (USNM); *Presidio co.*: 11 mi. W Marfa (MDC), Presidio (CNC); *San Patricio co.*: Lake Corpus Christi State Park near Mathis (MDC); *Travis co.*: Austin (CNC, MDC); *Ulvalde co.*: Sabinal (USNM), Laguna (CAS). Location not found "Weslace" (USNM).

Virginia. *Falls Church co.*: Falls Church (MDC); *Floyd co.*: W Tarpley (USNM); *Montgomery co.*: county record only (USNM); *Nelson co.*: Blue Ridge Parkway (CMNC); *Page co.*: Shenandoah (CAS). Location not found: "Williamport" (MDC); "Hertford co., Va" (MCZ): this state does not exist in Virginia but there one in North Carolina; not on map).

Vermont. *Chittenden co.*: Burlington (CNC, LEM), 2 mi. S Jonesville/ Gillette road (CAS); *Windsor co.*: Woodstock (MDC).

West Virginia. *Greenbrier co.*: White Sulphur Springs (FREY); *Marion co.*: Fairmont (CAS).

Wisconsin. *Baraboo co.*: Parfrey's Glen (CNC); *Dane co.*: Madison (CAS, MDC, MDC); *Sauk co.*: Parfrey's Glen Portage (CNC, MDC); *Green co.*: Albany (USNM); *Milwaukee co.*: Milwaukee (CAS); *Wood co.*: Cranmoor (USNM).

– *MEXICO.*
Country record only (USNM) (not on map).
Nuevo Léon. Monterrey (MDC).

– *BRAZIL.*
Country record only (USNM). Old specimen very likely mislabeled. Record not included on map (Fig. 24).

HOST PLANTS. On *Asclepias incarnata* L. and *A. syriaca* L. in the northeastern part of its distribution. Eickwort (1977) reported that he could rear this species on *Asclepias amplexicaulis* Sm. with no significant differences in growth. It is commonly found on *Cynanchum unifarium* (Scheele) in Texas (Palmer 1982).

BIOLOGY. *Labidomera clivicollis*, the well known *Milkweed Leaf Beetle* hibernates in the adult stage. Overwintered individuals come out from the ground in early summer when milkweed plants sprout.

Females usually lay clutches of 15-30 eggs in more or less regular double rows. On occasion, the number of eggs deposited can exceed 60. During their life span females lay between 200 and 500 eggs (Eickwort, 1971). Eggs are often cannibalized by newly hatched larvae. Cannibalism is an important mortality factor for this life stage. Dickinson (1992a) recognized three types of egg cannibalism: siblicide, eggs eaten by older larvae from earlier egg clutches, eggs eaten by adult females. Intense cannibalism may favour egg guarding and increase the survival of larvae (Eickwort, 1973, 1981).

According to Eickwort (1977), larvae are subject to high mortality (as much as 54%) by ground predators, mainly Nabidae, Pentatomidae and syrphid larvae (Diptera), and at a lesser rate by scorpion flies (*Panorpa* spp.), harvestmen, and spiders. This author also interprets the apparent higher density of larvae on *Asclepias incarnata* L., a swamp milkweed, by the better protection provided by a swampy habitat against ground predators in general. On the other hand, Palmer (1982) demonstrated that larvae can pupate under short-day photoperiod thus enhancing the moulting of larvae in the fall.

Jones (1932) has shown that adults are systematically avoided by wild birds because of their aposematic colours and repulsive taste. *Chrysomelobia labidomerae* Eickwort, a parasitic mite hiding under the elytra, feeds on the haemolymph with no significant ill effects upon the health of the host (Baker & Eickwort, 1975). Eickwort (1977) reported the emergence of a tachinid fly *Myiopharus aberrans* (Townsend) from the abdomen of a caged adult, killing the host.

Adult size may vary greatly because last instar larvae can metamophose at weights much lower than those normally attained under favourable conditions, given the amounts of food available during the last larval instar (Palmer, 1984). Larger females lay approximately twice as many eggs as do smaller ones, these are heavier and result in higher larval survivorship to pupation (Palmer, 1985a).

The reproductive behaviour of the *Milkweed Leaf Beetle* is unusual among leaf beetles since males fight for mates. Their deeply indented front femora very likely correspond to a morphological adaptation to this behaviour but we still lack accurate information on this subject. Dickinson (1992b) concluded that an earlier date of emergence from overwintering sites is more important for males than their larger body size or their better vagility in order to achieve successful mating. Prolonged matings enable males to remove and replace the sperm of previous mates with their own sperm, and to prevent females from remating prior to oviposition (Dickinson, 1988). Photoperiod is another factor which affects reproduction. According to Palmer (1983), longer photoperiods induce reproduction while shorter ones induce diapause. This may be significant ecologically, since it permits a substantial portion of the autumn generation to initiate a third generation in favourable years.

In most of its range, *Labidomera clivicollis* has only one generation per year, the adult being the overwintering stage, but Palmer (l985b) concluded that it is bivoltine in Texas, with two reproduction periods: one in the spring and one in autumn, a bimodal scheme which corresponds to the annual bimodal rainfall pattern in this state.

The *Milkweed Leaf Beetle* was one of our commonest beetles in the past. More than 30 years ago, it was not rare to observe spectacular aggregations of hundreds of individuals on a single plant. In the seventies, such large colonies were not observed. In the eighties, the species was still relatively common, but never in large numbers. In the early nineties it became one of the rarest species in eastern Canada. And we have not seen any individual since 1995 in the Ottawa region. The causes of such a drastic decline are very likely of environmental origin, but so far, we have been unable to identify one factor in particular.

Labidomera suturella Guérin-Méneville, 1838
(Figs 2-2c, 10, 14, 15, 19-19d, 25)

Labidomera suturella Guérin-Méneville 1838:301; Dejean 1836:397 (*nomen nudum*); 1837:421 (*nomen nudum*); Stål 1863:143; Jacoby 1883:240.
Labidomera germari Guérin-Méneville 1838:301; Dejean 1836:397 (*nomen nudum*); 1837:421 (*nomen nudum*); Chevrolat 1843:656; 1846:184; Stål 1863:144 (synonymy); Jacoby 1883:240; Weise 1916:28; Blackwelder 1946:673; Bechyné 1952:42.

ETYMOLOGY. Derived from the Latin name *sutura* (= suture) in reference to the elytral suture which is distinctly darker than the rest of the elytra in the typical colour pattern (Fig. 2c).

DIAGNOSIS. Large ovoid species, generally with dark elytral suture and pale elytra (Figs 2b-2c); in the male, apex of front femora with two apophyses (Figs. 14-15); lateral margins of pronotum evenly arcuate.

DESCRIPTION. A short description is found in Stål (1865), and a colour illustration showing the usual pattern is given by Jacoby (1883). To these contributions we can add that the colour pattern is much more variable than previously described by authors. In fact, it ranges from the typical pale elytra, ornamented with a narrow dark suture (Fig. 2c), to elytra largely pale provided with a few small spots (Fig. 2b), to darker patterns composed of several spots fused together (Fig. 2a), or to unusual cases where the dark portions become much larger than the pale background which is reduced to small pale areas (Fig. 2).

In the male, the median lobe of the aedeagus is evenly convex in lateral view (Fig. 19c), the tip is rounded, and the lateral margins of the apical portion are usually concave (Figs. 19-19c), although, sometimes, they appear almost straight (Fig. 19d).

In the female, the spermatheca is large (> 5 mm) and strongly bent, with the pump extending beyond the middle of the length (Fig. 10).

REMARKS. This species varies in the extent and number of elytral spots which can be numerous as in *germari* auctorum (Figs.2, 2a) or be totally lacking as in the holotype (as in Fig. 2c). The elytral punctation, the shape of the pronotum, and the structure of the median lobe of the aedeagus (Figs. 18- 18d) may also present slight individual variation.

Labidomera suturella differs from *L. clivicollis* by its larger size, the characters of the male genitalia (Figs. 19-19d), as well as by its much larger arcuate spermatheca (Fig. 10) which possesses a longer and wider spermathecal pump than in *L. clivicollis* (Fig. 11).

One of us (M.D.) had the opportunity to examine the De Breme Collection at the Regional Museum in Turin where some of Dejean's types are preserved, and found a specimen of *Labidomera* identified as *L. troberti* sp. nov. from Mexico. This (invalid) name has remained *in litteris* and the specimen examined undoubtedly belongs to the dark variation of *L. suturella* (= *germari* auctorum).

The early mentions by Dejean (1836, 1837) of *suturella* or *germari* are not since these species were still undescribed when Dejean used these names. Consequently, the authorship of both *suturella* and *germari* should be attributed to Guérin-Méneville (1838) who provided the first valid illustrations and descriptions of these two species. According to Cowan (1971), the text of Guérin-Méneville's work was completed only on September 7, 1844, although the volume is dated 1838 (plates were released from 1829-1837).

DISTRIBUTION. From southeastern Mexico to Costa Rica (Fig. 25).

Specimens examined: 67.

– *MEXICO.*

Country record only (ACHC, FREY, MCZ, USNM).

Chiapas. Bochil, 8 km N (CNC).

Hidalgo. Rancho Viejo, 24 mi. NE (CNC).

Mexico. Tierra Blanca, 34.4 km N Mexico City (LLC).

Nuevo Leon. State record only (MDC); Nocimiento del Rio Sabinas (MDC).

Puebla. Teziutlán, 15 mi. E. (CNC).

San Luis Potosi. Tamazunchate, 25 mi. N (USNM).

Tamaulipas. Bacatame/ 6 mi. S Gomez Farias (MDC), / 7 km SSE Gomez Farias (MDC), Gomez Farias, 1-3 mi. W, along road to Rancho de Cielo (MDC), Huauchinango (MCZ), Ocampo, 1-3 mi. E (MDC), Rio Corono, 30 km NE Ciudad Victoria (MDC).

Veracruz. Catemaco (USNM), Catemaco Lake (CNC), Orizaba (ACHC, CAS, MDC), San Rafael, Jicaltepee (USNM), Veracruz (USNM), Veracruz, E (MDC).

Note: specimen in ACHC is erroneously reported from "Yucatan".

– *COSTA RICA*.

Country record only (USNM).

Cartago. Turrialba (USNM).

Guanacaste. Tilaran (USNM).

Heredia. Finca La Selva (specimens not seen but locality reported by Choe (1989).

– *GUATEMALA*.
Country record only ((MCZ).
Baléu/ San Cristóbal (LLC), Rio Polochique [= Rio Polochic] (MDC).

HOST PLANTS. *Asclepias curasavica* L. (Asclepiadaceae), *Witheringia heteroclita* (Solanaceae) (Choe, 1989). According to F. Hrusa (pers. com.) the latter name is a *nomen nudum* and the proper citation should be *Witheringia heteroclita nom. nudum* = *Lycianthes heteroclita* (Sendtner) Bitter to indicate that this name has been in print but not legitimately published according to the botanical nomenclatorial rules.
BIOLOGY. Larvae of *Labidomera suturella* live grouped in a circle in order to defend themselves against predators, a phenomenon known as *cycloalexy* (Jolivet 1997). Choe (1989) reported that females protect their larvae by gripping firmly on leaves with their bodies held over the larvae.

Labidomera olmeca Daccordi & LeSage, **new species**
(Figs. 4, 6, 8, 16, 20, 20a, 25, 26)

ETYMOLOGY. The species takes its name from the *Olmecs*, an ancient Central American people.

DIAGNOSIS. Large ovoid species with pronotum subrectangular (Figs. 6, 26). Median lobe of aedeagus with apical lateral margins distinctly sinuous (Fig. 20a).

DESCRIPTION. Winged; body elongate, ovoid, with maximum width in posterior third. Colour. Head, most of pronotum, dark elytral spots, ventral portions and legs all dark bluish green with metallic hues. Lateral margins of pronotum, most of proepisterna, epipleura, lateral margins and posterior third of elytra and two large pale spots (one near scutellum and the other past midway point of elytra) all ochre yellow.

Size. Length 12.0 mm; width 7.7 mm.

Head. Mandibles large, protruding at sides; maxillary palps with penultimate segment enlarged and with last segment shorter and narrower than penultimate. Labrum straight with row of dense punctures in middle. Clypeus slightly concave, with numerous large and shallow punctures giving rugose aspect to surface. Frons shiny and sparsely punctate, with two large irregular depressions in middle and with very fine punctation becoming thicker near anterior margin of each eye; antennal sockets (toruli) wide and smooth. Antennae extending well beyond base of pronotum, even though with only nine antennomeres enlarged; segment 3 twice as long as segment 2. Eyes elongate, narrow, slightly elevated.

Thorax. Pronotum clearly transverse, width/length ratio 1.77 (6.64/ 3.76 mm), with lateral margins finely bordered, almost right angled with hind margin; maximum width in anterior third; 3-4 setigerous pores present at anterior angles, always 3 at posterior ones; disc with few scattered idisodiametric punctures, those with largest diameters merging near posterior angles in marginal depression (Fig. 6). Scutellum wide, smooth, ogival. Ventral portion of prothorax (hypomeron), and proepisterna smooth, without furrow and punctation; prosternal process short, flat, concave only at apex; mesosternum with wide pit posteriorly elevated into short lobe; metasternum with process not bordered, rather narrow

Fig. 26. Labidomera olmeca, dorsal view.

and elongate. Mesepisterna with large sparse punctures; mesepimera with silky slightly rugose surface, and with smaller punctures along margins; metepisterna almost smooth, with dense punctation only in posterior portion; metepimera smooth, and rounded.

Legs. Elongate, robust; anterior femora each with one deep cavity (Fig. 16); tibiae widened and enlarged from middle and beyond; claws large, robust, simple, divaricate; onychia as long as last two segments combined.

Elytra. Irregularly punctate, with punctures dense and much more uniformly distributed than those on pronotum (Figs. 8, 26); epipleura wide, flat, smooth and ciliate on entire length; humeral callus wide, slightly elevated.

Abdomen. First abdominal sternites with shallow cavities along anterior margin. Abdominal sternites subequal, clearly punctate; last sternite with depression in central portion. Pygidium without median furrow.

Male genitalia. Median lobe evenly curved in lateral view (Fig. 20); apical lateral margins sinuous (Fig. 20a); flagellum thin, protruding outside ostium (Fig. 20a).

Female genitalia. Unknown (female not available).

TYPE MATERIAL. HOLOTYPE. Male labeled as follows: "[Mexico] Soconusco Distr. [= District], Hotzen S.V." (no date given).

This specimen is preserved in the Daccordi Collection (DMC), Torino, Italy.

DISTRIBUTION. Presently known only from the type locality, Socomusco District, in Mexico (Fig. 25)

HOST PLANTS. Unknown.

BIOLOGY. Unknown.

Fig. 27. labidomera maya, dorsal view.

Labidomera maya Daccordi & LeSage, **new species**
(Figs. 3, 7, 9, 17, 21-21a, 24, 25, 27)

ETYMOLOGY. The species takes its name from the *Mayas*, an ancient Central American people.

DIAGNOSIS. Large species with ellipsoidal body and sides of pronotum broadly and regularly arcuate (Figs. 7, 27). Median lobe of aedeagus with apex evenly convex laterally (Fig. 21a).

DESCRIPTION. Winged; body ellipsoidal, broadest in middle.

Colour . Body bright violet-blue, becoming more or less dark on ventral parts, legs and antennae. Head, disc of pronotum, scutellum and elytral pattern bluish. Lateral margin of elytra, epipleura, two large spots on elytra, sides of pronotum, and ventral portion of pronotum ochre yellow.

Size. Length 12.0 mm; width 8.0 mm.

Head. Mandibles narrow, pyramidal, not protruding at sides, with entire length of external surface hollowed; last segment of maxillary palps shorter and narrower than penultimate one. Antennae narrow, extending well beyond base of pronotum, with segment 3 twice as long as segment 2; segments 7-11 transverse, thick, covered with short hairs. Clypeus shiny, separated from frons by distinct arcuate line, and with row of large punctures along free margin. Frons opaque, surface shagreened, with sparse punctures becoming denser and roughly the same size as those on clypeus only near eyes and on apex; antennal sockets wide, elevated, and smooth.

Thorax. Pronotum transverse, width/length ratio 1.95 (7.04/ 3.60 mm), with broadly arcuate sides (Fig. 7), and scattered punctures smaller than those on frons; anterior and posterior angles with deep setigerous pores. Scutellum wide, ogival, impunctate, shining. Ventral portions of prothorax smooth, not separated from respective proepimera and proepisterna. Prosternal process short, flat, bilobate, and free at apex. Metasternum slightly elevated, short, widened; metasternal appendix not bordered, wide, smooth at sides; metasternum distinctly shagreened with occasional punctures and with fine striations in central portion. Mesepisterna with large dense punctures; mesepimera with slightly rugose surface with superficial irregular punctation along margins; metepisterna with sparse punctation becoming denser posteriorly; metepimera short, smooth, rounded.

Legs. Elongated, robust; anterior femora deeply hollowed (Fig. 17); tibiae strongly widened beyond distal half. Claws thick, simple, widened; onychia as long as last two tarsomeres combined.

Elytra. With irregular punctation, punctures forming short pseudo-geminate rows only in central portion (Figs. 9, 27); punctation finer and quite irregular in apical portion; humeral calluses smooth, raised; epipleura smooth, slightly widened, flat, entire length of inner margin ciliate.

Abdomen. Abdominal segments almost rugose because of striations, with deep punctation. Last sternite with deep pit medially.

Male genitalia. Median lobe slightly enlarged in middle in lateral view (Fig. 21); apex ogival in dorsal view (Fig. 21a); flagellum protruding outside ostium.

Female genitalia. Unknown (female unavailable).

TYPE. HOLOTYPE. Male labelled as follow: "Mexico, Chiapas, Sumidero Cyn [= canyon] nr. Tuxtl. [= Tuxtla], Gutz., 4000', 19 June, D.B. Thomas coll."

The specimen is preserved in the Daccordi Collection (DMC), Torino, Italy.

DISTRIBUTION. Presently, known only from the type locality, Sumidero in Mexico (Fig. 25).

HOST PLANTS. Unknown.

BIOLOGY. Unknown.

Acknowledgements

We want to express our thanks to our friends and colleagues who have made this paper possible by loans of specimens or by their precious comments and suggestions on the manuscript: Yves Bousquet (ECORC, Ottawa), Jean Buckley (ECORC), Henri Goulet (ECORC), M.S. Kelley (Museum of Comparative Zoology, Cambridge), Giachino Pier Mauro (Museo Regionale, Turin, Italy), Ole Martin (Zoologisk Museum, Copenhagen, Denmark), Edward G. Riley (Texas A & M University, College Station, Texas, U.S.A) and Robert Schuster (University of California, Davis, U.S.A). The habitus drawing of Go Sato and the technical support of Jocelyn Denis for processing the locality data and the preparation of maps were also greatly appreciated.

References

Baker, T.C. and Eickwort, G.C. 1975. Development and bionomics of *Chrysomelobia labidomerae* (Acari: Tarsonemina: Podapolipidae), a parasite of the milkweed leaf beetle (Coleoptera: Chrysomelidae). *Can. Ent.* **107**: 627-638.

Balsbaugh, E. U. and Hays, K.L. 1972. The leaf beetles of Alabama (Coleoptera: Chrysomelidae). *Bull. agric. Exp. Stn Auburn Univ.* **441**. 223 pp.

Bechyné, J. 1952. Nachträge zu den Katalogen von Blackwelder und Junk-Schenkling der neotropischen echten Chrysomeliden. *Ent. Arb. Mus. Georg Frey* **3**(1): 1-62.

—, 1980. El Jeannelismo y la evolution conceptode las leyes orgánicas sin excepción. Grafindustrial Aragua S.R.L., Maracay. 181 pp.

Blackwelder, R. E. 1946. Checklist of the coleopterous insects of Mexico, Central America, the West Indies, and South America. Part 4. Smithsonian Institution, *Bull. U.S. natn. Mus.* **185:** 551-763.

Blatchley, W. S. 1910. The Coleoptera or beetles of Indiana. *Bull. Indiana Dept. Geol. nat. Res.*.**1**: 1-1386.

Brown, W. J. 1961. Notes on North American Chrysomelidae (Coleoptera). *Can. Ent.* **93**: 967-977.

Chagnon, G. and Robert, A. 1962. Principaux Coléoptères de la province de Québec. Les Presses de l'Université de Montréal, Montréal. 439 pp.

Chapuis, F. 1874. Famille des Phytophages. *In* Lacordaire, T. and Chapuis, F.(Eds.), *Genera des Coléoptères*. Librairie encyclopédique de Roret, Paris. Volume X. 455 pp.

Chevrolat, L. 1843. Chrysomélines [pp. 654-657]. *In* d'Orbigny, M. C. (Ed.), *Dictionnaire universel d'histoire naturelle*. M. M Renard, Martinet et Cie, Paris. Tome troisième. 744 pp.

—, 1846. *Labidomera* [pp. 183- 184]. *In* d'Orbigny, M. C. (Ed.). *Dictionnaire universel d'histoire naturelle*. MM. Renard, Martinet et Cie, Paris. Tome septième. 808 pp.

Choe, J.C. 1989. Maternal care in *Labidomera suturella* Chevrolat (Coleoptera: Chrysomelidae) from Costa Rica. *Psyche* **96**: 65-67.

Coquebert, A. J. 1804. *Ilustratio Iconographica Insectorum. Tabularum decas prima*. P. Didot, Paris. 142 pp.

Cowan, C. F. 1971. On Guérin's *Iconographie*, particularly the insects. *J. Biblio. nat. Hist.* **6:** 18-29.

Crotch, G. R. 1873. Material for the study of the Phytophaga of the United States. *Proc. Acad. nat. Sci. Philad.* **25**: 19-83.

Daccordi, M. 1994. Notes for phylogenetic study of Chrysomelinae with description of new taxa and a list of all the known genera [pp. 60-84]. In Furth, D. (Ed.) *Proc. 3rd int. Symp. Chrysomelidae, Beijing*, 1992. 150 pp.

Dejean, M. 1833-1836. Catalogue des Coléoptères de la collection de M. le Comte Dejean. Deuxième édition. Méquignon-Marvis Père & Fils, Paris. 443 pp.

—, 1837. Catalogue des Coléoptères de la collection de M. le Comte Dejean. Troisième édition. Méquignon-Marvis Père et Fils, Paris. 503 pp.

Dickinson, J. L. 1988. Determinants of paternity in the milkweed leaf beetle. *Behav. Ecol. Sociobiol.* **23**: 9-19.

—, 1992a. Egg cannibalism by larvae and adults of the milkweed leaf beetle (*Labidomera clivicollis*, Coleoptera: Chrysomelidae). *Ecol. Ent.* **17**: 209-218.

—, 1992b. Scramble competition polygyny in the milkweed leaf beetle: combat, mobility, and the importance of being there. *Behavl Ecol.* **3**: 32-41.

Downie, N. M. and Arnett, R. H.1996. The beetles of Northeastern North America. The Sandhill Crane Press, Gainesville, Florida. 1721 pp. (2 volumes)

Eickwort, G. C. 1971. The ecology of *Labidomera clivicollis*, a relatively rare milkweed beetle closely related to the Colorado potato beetle (Coleoptera: Chrysomelidae, Doryphorini). Ph.D. thesis, Cornell Univ. 123 pp.

—, 1973. Cannibalism and kin selection in *Labidomera clivicollis* (Coleoptera: Chrysomelidae). *Am. Nat.* **107**: 452-453.

—, 1977. Population dynamics of a relatively rare species of milkweed beetle (*Labidomera*). *Ecology* , Brooklyn **58**: 527-538.

—, 1981. Presocial insects. *In* Hermann, H.R. (Ed.). Social insects. Academic Press, New York. Vol. 2, pp.199-280.

Fabricius, J. C. 1775. *Systema entomologiae*. Korte, Flensburgi et Lipsiae. 832 pp.

—, 1781. *Species insectorum*. C. E. Bohn, Hamburgi et Kilonii. Tomus II. 552 pp.

—, 1787. *Mantissa Insectorum*. Proft, Hafniae. Tomus I. 348 pp.

—, 1792. *Entomologia systematica*. Proft, Hafniae. Pars I. 330 pp.

—, 1801. *Systema Eleutheratorum*. Bibliopolii Academic, Kiliae. 687 pp.

Gmelin, J. F. 1788. *Coleoptera* [pp.1526-2040]. *In Caroli a Linnaeus Systema Naturae*. XIII edition. G. F. Beer, Lipsia. Tomus I, Pars IV, pp. 1517-2224.

Guérin-Méneville, F. E. 1844. *Insectes*. *In* Iconographie du règne animal de G. Cuvier. J. B. Baillière, Paris. 576 pp. (Text was published in 1844 but illustrations appeared in 1829-1838).

Hatch, M. H. 1971. The beetles of the Pacific Northwest. Part V. Univ. of Washington Press, Seattle and London. 662 pp.

Isman, M. B., Duffey, S.S. and Scudder, G.G.E. 1977. Cardenolide content of some leaf- and stem-feeding insects on temperate North American milkweeds (*Asclepias* spp.) *Canadian Jl of Zoology* **55**: 1024-1028.
Kirby, W. 1837. *Fauna Boreali-Americana*, or the zoology of the Northern parts of British America. J. Fletcher, Norwich. 325 pp.
Jacoby, M. 1880-1892. *Biologia Centrali-Americana. Insecta Coleoptera*. Volume VI, Part 1. 625 pp.
Jolivet, P. 1997. Biologie des Coléoptères Chrysomélides. Boubée, Paris. 279 pp.
Jones, F. M. 1932. Insect colouration and the relative acceptability of insects to birds. *Trans. R. ent.Soc. Lond.* **80**: 345-386.
LeConte, J. L. 1858. Catalogue of Coleoptera of the regions adjacent to the boundary line between the United States and Mexico. *Jl. Acad. nat. Sci. Philad.* (new series) **4**: 9-42.
Leng, C. W. 1920. Catalogue of the Coleoptera of America, north of Mexico. Mount Vernon, New York. 470 pp.
LeSage, L. 1991. Family Chrysomelidae . *In* Bousquet, Y. (Ed.). Checklist of beetles of Canada and Alaska. Research Branch of Agriculture Canada, Publication 1861/E. pp. 301- 323.
—, 1995. Revision of the costate species of *Altica* Müller of North America North of Mexico (Coleoptera: Chrysomelidae). *Can. Ent.* **127**: 295-411.
Linnaeus, C. 1767. *Systema Naturae*. XII edition. Holmiae. Volume II, pp. 533-1327.
Linell, M. L. 1896. Short review of the Chrysomelas of North America. *Jl NY ent. Soc* **4**: 195-200.
Madge, R. B. 1988. The publication dates of Dejean's catalogues. *Arch. Nat. Hist.* **15**: 317-321.
Neave, S. A. 1939. *Nomenclator Zoologicus*. The Zoological Soc. of London, London. Volume II. 1025 pp.
Olivier, A. G. 1807. No. 91. Chrysomèle. Entomologie, ou histoire naturelle des insectes. Desray Librairie, Paris. Tome cinquième. 612 pp.
Palmer, J. O. 1982. Photoperiodic effect on size-related metamorphosis in the milkweed leaf beetle, *Labidomera clivicollis. Physiol. Ent.* **7**: 37-41.
—, 1983. Photoperiodic control of reproduction in the milkweed leaf beetle *Labidomera clivicollis*. *Physiol. Ent.* **8**: 187-194.
—, 1984. Environmental determinants of seasonal, body size variation in the Milkweed Leaf Beetle, *Labidomera clivicollis* (Kirby) (Coleoptera: Chrysomelidae). *Ann. ent. Soc. Am.* **77**: 188-192.
—, 1985a. Life-history consequences of body-size variation in the Milkweed Leaf Beetle, *Labidomera clivicollis* (Coleoptera: Chrysomelidae). *Ann. ent. Soc. Am.* **78**: 603-608.
—, 1985b. Phenology and dormancy in the Milkweed Leaf Beetle *Labidomera clivicollis* (Kirby). *Am. Midl. Nat.* **114**:13-18.
Schönherr, C. J. 1808. *Synonymia Insectorum*. Zweiter Theil. C. F. Marquard, Stockholm. 424 pp.
Seeno, T. N. and Wilcox, J. A. 1982. Leaf beetle genera (Coleoptera: Chrysomelidae). *Entomography* **1**: 1-222.
Sherborn, C. D. 1922. *Index Animalium* 1801-1850. British Museum, London. Volume I: cxxxix + 128 pp.
Stål, C. 1862-1865. Monographie des Chrysomélides de l'Amérique. C.A. Leffler, Upsala. 365 pp.
Timmermans, M., Randoux, T., Daloze, D., Brekman, J.C., Pasteels, J.M. and LeSage, L. 1992. The chemical defence of Doryphorina beetles (Coleoptera: Chrysomelidae). *Biochemi. Syst. Ecol.* **20**:343-349.
Weise, J. 1916. Pars 68. Chrysomelidae 12: Chrysomelinae. *In Coleopterorum Catalogus*. Junk, W. and Schenling, S. (Eds.). W. Junk, Berlin. 255 pp.
Wilcox, J. A. 1954. The leaf beetles of Ohio (Chrysomelidae: Coleoptera). *Bull. Ohio biol. Surv.* **43**: 353-506.
—, 1972. A review of the North American Chrysomeline leaf beetles (Coleoptera: Chrysomelidae). The Univ. of the State of New York. The State Education Dept. State Museum and Science Service. Albany, New York. Bull. 421. 37 pp.
—, 1974. Checklist of the beetles of Canada, United States, Mexico, Central America, and the West Indies. Volume 1, Part 7, The Leaf Beetles (Red Version). The Biological Research Institute of America, Inc. Latham New York. 166 pp.

BIOLOGY OF THE PAROPSINI (CHRYSOMELIDAE: CHRYSOMELINAE)

TARA L. SIMMUL[1] and DAVID. W. de LITTLE[2]
[1]*Cooperative Research Centre for Sustainable Production Forestry and University of Tasmania, GPO Box 252-12, Hobart, Tasmania, Australia 7001;* [2]*Cooperative Research Centre for Sustainable Production Forestry, PO Box 63, Ridgley, Tasmania, Australia, 7321*

Abstract

There are twelve paropsine genera and over 400 species but detailed biological information is restricted to those species considered to be pests. This paper is a review of the distribution, host plants, behaviour, lifecycle, natural enemies and pest status of these better known species.

1. Introduction

"Unfortunately we are not yet able to trace the influence the Flora exercises on the insect Fauna: to do so we require to have a knowledge of every plant which affords nourishment to each species of insect, both in its first and last stage. Of this knowledge we are now wholly deficient, and yet it is of the greatest importance in our study of the geographical distribution of insects, for by it alone can we determine the connection between Flora and Fauna. For example, were the numerous examples of the genus Paropsis (Blattkafer) to be found only on one family or genus of plants in New Holland, the knowledge of the limits of this family or genus would have a double signification. We must hope for future elucidations of the intimate connection between the Flora and Fauna, . . ."

W.F. Erichson (1842) (translated from the original in German by S.A. Fogg (1859))

It is only in recent years that there has been widespread intensive cultivation of Australian native plants within Australia. This has led to the emergence of some Australian insect groups as pests, with a consequent requirement to understand their biology. Foremost among these insect groups are the "paropsine chrysomelids" which from time to time inflict significant defoliation on native tree and shrub species in the families Myrtaceae and Mimosaceae in Australia and in other places where both paropsines and their hosts are introduced.

As many of the paropsine species are relatively large and brightly coloured in life, they attracted the attention of the early explorer-naturalists of Australia and their European patrons from a taxonomic perspective. In an age where naturalists sought to delineate the "perfection of the Creator" by describing the Creation, only the "perfect" (adult) stage of the insect was considered worthy of description. Thus the extensive taxonomic literature of the nineteenth century contains no information concerning the juvenile stages, hosts, lifecycles or habits of the paropsines.

Advances in Chysomelidae Biology 1, pp. 463–477
Edited by M.L. Cox

It took a priest, the most thorough and committed Australian taxonomist of his day (Canon Thomas Blackburn), to observe in 1899 when dealing with the group of species later to be placed in the genus *Chrysophtharta*, "I am of the opinion that the determination of the species is *impossible*, except by the means of some student with plenty of leisure time at his command devoting himself to a study of their characters and habits in their larval and pupal stages" (Blackburn, 1899). Indeed much of the taxonomic confusion which has reigned within the group might have been avoided if more attention had been paid to their immature stages and biology earlier on. Significantly also, it was a female entomologist (Margaret Cumpston) who 40 years later, in an age when entomology was still an unusual profession for a woman, provided us with the first published information concerning the biology of paropsine species (Cumpston, 1939).

The pest potential of the paropsines first became apparent in New Zealand, where *Paropsis charybdis* Stål, a native of the eastern Australian seaboard first appeared around 1916 and spread unchecked wherever Australian eucalypts of the subgenus *Symphyomyrtus* were planted (Styles, 1970). Recognition of pest status did not occur in Australia until post World War 2 development put increasing pressure on native forests with the consequent emergence of significant pests *Chrysophtharta bimaculata* (Olivier) in Tasmania (Greaves, 1966) and *P. atomaria* Olivier in mainland SE Australia (Carne, 1966a).

2. Distribution

Since it is not our purpose to enter further into taxonomic discussion, we shall confine our attention in this contribution to the following genera:

Paropsis Olivier
Paropsides Motschulsky
Paropsisterna Motschulsky
Dicranosterna Motschulsky
Chrysophtharta Weise
Trochalodes Weise
Faex Weise
Philhydronopa Weise
Acacicola Lea (= *Pyrgoides* Aslam = *Pyrgo* Weise)
Trachymela Weise
Sterromela Weise
Procrisina Aslam (= *Procris* Weise)

The Coleopterorum Catalogus (Weise, 1916) enumerates in excess of 400 species in these genera, however subsequent taxonomic research may substantially increase this number (B.J. Selman, pers. comm.). With the exception of *Paropsides*, the group has an almost entirely Australian natural distribution. Following species localities recorded in the Coleopterorum Catalogus, the greatest diversity of species occurs in mainland SE Australia, where species diversity in the large genera *Paropsis*, *Paropsisterna*, *Chrysophtharta*, and *Trachymela* reaches its peak. This region of Australia possesses the greatest amount of habitat diversity, with climates varying from cool temperate through Mediterranean to warm temperate, and also includes alpine, humid and desert environments. Species of the

Myrtaceous genera *Eucalyptus, Corymbia, Tristania, Leptospermum* and *Melaleuca,* as well as *Acacia* (Mimosaceae) are the hosts of species of these paropsine genera (Jolivet and Hawkeswood, 1995).

The paropsine faunas of Tasmania and SW Australia show a similarity in that both are species depauperate when compared with the mainland SE Australian fauna, and a degree of endemism occurs in each. Many of the endemic species in these regions have allopatric sibling species in the mainland SE, for example *Paropsis deboeri* Selman (Tasmania) and *P. atomaria* (Olivier) (SE mainland) (Selman,1983).

According to the Coleopterorum Catalogus, diversity in the large genus *Acacicola* is approximately equivalent in SW and SE Australia. Species in this genus feed mainly on species of *Acacia* (Mimosaceae), and the relatively large number of species of *Acacicola* in SW Australia may reflect the remarkable diversification of *Acacia* species in that region (Barlow, 1981).

Paropsine diversity is reduced in northern Australia, however the *Acacia* feeding genus *Dicranosterna* is more diverse in this region. Both *Paropsisterna* and *Paropsis* just extend to New Guinea where a couple of species feed on *Eucalyptus* (Jolivet & Petitpierre, 1981). *Paropsides* has a wide distribution throughout eastern Asia, Indonesia, New Guinea and Eastern Australia (Daccordi, 1994) with one species (*Paropsides umbrosa*) (Chapuis) even occuring in Tasmania where it feeds on flowers of *Oxylobium* and *Pultenea* (Fabaceae) (D.W. de Little, pers. comm.). The Asian species feed mainly on rosaceous hosts (Jolivet and Hawkeswood, 1995).

3. Hosts and feeding

3.1. Hosts and Interactions

As mentioned in the previous section, paropsine beetles feed on a diverse range of plants including species of *Acacia, Leptospermum, Melaleuca, Angophora, Tristania, Callistemon, Eucalyptus* and *Corymbia*. These are all Myrtaceous species, except for *Acacia*, which is leguminous. Within the genus *Eucalyptus*, younger trees (1-5m) appear to be favoured as a food source, although mature trees are also attacked (de Little, 1983; Kile, 1974). Details of damage to non-eucalypt species are uncommon in the literature, but Elliott (1978) reported that 34 out of 200 *Acacia dealbata* trees died after severe defoliation by *Acacicola orphana* (Erichson) and Maddox (1995) provided anecdotal evidence on several hectares of *Melaleuca* plantation which were stripped of all new growth over a one month period in 1992 by *Paropsisterna tigrina* (Chapuis).

Many of the paropsines observed feeding on *Eucalyptus* have a wide host range. A single species can feed and develop on a variety of hosts, but may be more frequently observed on some hosts than others. Larval performance has been found to vary between eucalypt species as shown by Carne (1966a) who observed larvae of *P. atomaria* matured faster when feeding on *E. blakelyi* than on *E. maculosa*. S.C. Baker *et al.* (unpublished) found larvae of *C. bimaculata* (Olivier) developed faster and consumed less foliage of the Symphyomyrtus species *E. nitens*, than those on the Monocalyptus species *E. regnans*, even though adults oviposit on expanding leaves of *E. regnans* when given a choice between the two species (Steinbauer *et al.*, 1998). This may be due to the faster growth rate of

E. nitens which results in a smaller canopy with less expanding leaves for oviposition (Steinbauer *et al.*, 1998). De Little and Madden (1975) found *C. bimaculata* and *C. agricola* (Chapuis) consumed both *E. dalrympleana* and *E. delegatensis* foliage, but survival of *C. bimaculata* larvae was significantly less on *E. dalrympleana*. Both species consumed more and matured faster on *E. delegatensis*, but when provided with a choice, *C. agricola* oviposited on *E. dalrympleana*, possibly indicating adaptation to an alternative host to avoid resource competition.

Insect feeding may also cause enhancement of the food resource later in the season, as the tree produces nutritious regrowth following defoliation. Severe defoliation however, directly reduces the photosynthetic capacity of a tree and hence its ability to recover after a defoliation event (Leon, 1989).

Physical characteristics of foliage may also affect insect-host interactions. Some adult beetles for example, *P. charybdis* are unable to cling to glaucous juvenile foliage (Edwards, 1982; Li, 1993). However, that adults of some species can feed and oviposit on glaucous foliage suggests the development of behavioural characteristics such as clinging to leaf margins or tips of leaves, rather than the flat leaf surface (Edwards and Wanjura, 1990).

3.2. Leaf chemistry

Leaf chemistry influences feeding behaviour and development of insects (Feeny, 1970; Li, 1993). In particular, foliar nitrogen content has been related to larval development of paropsines (Fox and Macauley, 1977; Morrow and Fox, 1980; Ohmart *et al.*, 1987). On young foliage, Patterson *et al.* (1996) found leaf toughness and nitrogen levels were relatively unimportant for survival and growth of *C. bimaculata* larvae. When fed on older, tougher leaves which are lower in nitrogen, Ohmart *et al.* (1987) found that *P. atomaria* larval growth slowed below a threshold concentration of 1.7% N. They suggested that early instar larvae may have difficulty feeding on these leaves due to toughness and that while later instar larvae may be able to feed on these leaves, they may need to consume more foliage to meet their nutrient requirements. High foliar 1,8-cineole levels have been found to correlate with reduced paropsine herbivory in *E. camaldulensis* (Stone and Bacon, 1994) and *E. obliqua* (Li, 1993). However, no effects on larval feeding rates have been found for condensed tannins and other phenols (Fox and Macauley, 1977).

3.3. Feeding behaviour

Adult and larval feeding behaviour and patterns are distinctly different (Cumpston, 1939; Greaves, 1966; Kile, 1974). Adults feed alone, chewing from the edge of a leaf towards the midrib which results in a scalloped effect on the leaf margins (Greaves, 1966; de Little, 1983). Large numbers of beetles result in the host tree appearing ragged and defoliation in successive seasons can result in the tree developing a "broom-topped" appearance (Elliott *et al.*, 1993; Leon, 1989). Intense adult feeding early in the season promotes development of "flush" foliage which is the preferred food source for larvae developing later in the season (Selman, 1994a). The most vigorous feeding by adults is just prior to overwintering (Cumpston, 1939; Carne, 1966a; Styles, 1970). Feeding at this time may prevent refoliation of branches in the upper parts of the crown, leaving, in some areas, exposed shoots susceptible to further damage from severe cold in winter (Elliott *et al.*, 1993).

Larvae feed preferentially on new growth, stripping the host tree of young leaves and shoots, both apically and laterally (Cumpston, 1939). *P. atomaria* larvae migrate together to a terminal leaf after hatching, where they commence feeding on the leaf margins, from the apex to the petiole. Branches become progressively defoliated as the larvae grow and move on to older and less succulent leaves (Greaves, 1966; Carne, 1966a). As larvae develop, their rate of consumption rises steeply (Carne, 1966b). For example, Greaves (1966) found the two later instars of *C. bimaculata* caused 90% of defoliation even though they were present for only 55% of the total larval period. Paropsine larvae do not feed only on foliage, with some authors observing larvae of *C. bimaculata* and *A. orphana* consuming green bark of the host trees when all foliage has been consumed (Greaves, 1966; Elliott, 1978). New (1981) also found that day-old pupae of *Acacicola hamadryas* (Stål.) (formerly *Pyrgoides*) fed on *A. baileyana* flowers alone, were heavier than those fed on foliage, suggesting that flowers were an important part of the diet of this insect.

4. General behaviour

Defence behaviour in paropsines has been widely observed (Cumpston, 1939; Greaves, 1966; Carne, 1966a; Moore, 1967; Tribe and Cillie, 1997). Adult paropsines feign death and drop to the ground if the foliage on which they are resting is disturbed (Greaves, 1966; Edwards, 1982). They are able to protect themselves from some smaller predators (for example, predatory ants) by pulling their legs under the body and clinging tightly to the leaf surface (Selman, 1994a).

The main paropsine larval defence mechanism is a pair of glands, present in each larval stage and everted from the base of the 8th tergite (T8) when the larva is threatened (Moore, 1967; Tribe and Cillie, 1997). These glands are particularly well developed in the larvae of *Paropsis* and *Chrysophtharta* (Selman, 1994a). Carne (1966a) elaborated on this defence mechanism, having observed three levels of response to predator disturbance by *P. atomaria* larvae. Slight disturbance causes larvae to stop feeding, raise their abdomens and tap on the leaf. This behaviour continues for several minutes, until the threat retreats. Violent disturbance (for example, a bird disturbing the foliage) causes larvae to regurgitate a drop of yellow fluid. If the predator contacts the larvae they vigorously flex their abdomens so that the everted abdominal glands on T8 come into contact with the attacker. These glands secrete a fluid containing hydrogen cyanide (Moore, 1967) which is toxic to predators (Cumpston, 1939; Greaves, 1966). Ants, for example *Iridomyrmex rufoniger* (Lowne) and *I. detectus* (Smith), were found to die within minutes of being struck (Carne, 1966a). If the predator withdraws, the glands retract. Up to 10 minutes may pass before the larvae return to normal feeding or resting behaviour (Carne, 1966a). Larvae of the genera *Trachymela* and *Sterromela* have less well developed defence glands and rely instead on cryptic colouring to camouflage them from predators (Selman, 1994a).

Defensive grouping behaviours have been given the terms 'cycloalexy' and 'alytoalexy'. Vasconcellos-Neto and Jolivet (1988) used the term cycloalexy to describe the formation of larvae in a circle for defence purposes. This circle may often contain larvae of different developmental stages and species. Selman (1994a) suggested the term alytoalexy for larvae which form into defensive rows, usually in the angle between a lateral shoot and a stem of the host plant.

When not under threat from predators, some paropsine larvae still exhibit grouping behaviour. Initially larvae are gregarious and this may last throughout the four stages as in *P. atomaria* and *P. tasmanica* (Baly) (Carne, 1966a; de Little, 1979). Species not gregarious in the later stages may still reform into groups to rest and migrate. This grouping behaviour may be diagnostic of some species. For example, *Chrysophtharta obovata* (Chapuis) (= *C. variicollis* Chapuis) (B.J. Selman, pers. comm.) are gregarious, arranging themselves on the leaf in a circular fashion with their heads pointing inward toward a central focal point (Cumpston, 1939). *Paropsisterna liturata* (Marsham) (Cumpston, 1939) and *Paropsis rubidipes* (Blackburn) (de Little, 1979) are almost solitary in habit, as are larvae of *Trachymela* and *Sterromela* which feed individually and are nocturnal. *Trachymela* and *Sterromela* larvae rest under bark during daylight (de Little, 1979; Tribe and Cillie, 1997).

Adult flight behaviour can result in groups of insects moving large distances in a single season. *P. charybdis*, *P. atomaria* and *C. bimaculata* adults have been especially observed to fly during warm summer days (Clarke *et al.*, 1997; Greaves, 1966; de Little, 1979; White, 1973; Carne, 1966a). Adults of *C. bimaculata* dwell in an area for approximately 10 days, during which time they feed and lay eggs. Before the eggs hatch they move to another area (Clarke *et al.*, 1997). This behaviour results in movement over several kilometres. In Tasmania, Australia, no genetic differences were observed between populations of *C. bimaculata* greater than 160 km apart, suggesting the insects had dispersed over this distance (Congdon *et al.*, 1997). In South Africa, *Trachymela tincticollis* (Blackburn) has been observed to move average distances of 130km annually, with dispersal over 200km in one area (Tribe and Cillie, 1997). Wind may also aid dispersal over long distances as suggested by White (1973) when discussing movement of *P. charybdis* from one island to the other in New Zealand.

Recently Clarke *et al.* (in press) attempted to quantify an unusual behaviour first noted by Greaves (1966) who made the observation that larvae having completed their development *"fall to the ground"* from the canopy of the host trees. Clarke *et al.* (in press) supported this statement, having found final instar larvae of *C. bimaculata* falling from trees 50-65m in height. It appears that to accumulate enough food reserves for pupation the larvae consume food until they are unable to grip the leaves. At this stage they simply drop from the trees to the ground where they burrow into the soil to pupate (A.R. Clarke, pers. comm.).

5. Life cycle

Of the 12 paropsine genera, detailed life cycle information is known only for the genera that include economic pests; *Paropsis*, *Chrysophtharta* and *Paropsisterna*. Some information on the lifecycle of *Trachymela* and *Acacicola* has been published, but little is known regarding the other 7 genera.

Under field conditions mature *Paropsis* and *Chrysophtharta* adults emerge from overwintering in the spring (Greaves, 1966; Styles, 1970; Kile, 1974; Tanton and Khan, 1978) and feed on "flush" foliage. In sunny weather they fly short distances and feed, but on cool or windy days they drop from the foliage and shelter in leaf litter on the forest floor (Greaves, 1966).

In the majority of species, eggs are laid late in the spring or early in summer on new or one year old foliage (Greaves, 1966; Styles, 1970; Tribe and Cillie, 1997). If new growth is absent, females may refrain from ovipositing for 2-3 weeks until favourable growth has developed (Carne, 1966a). Eggs are laid in a variety of patterns, depending on the species (de Little, 1979). For example, *P. atomaria* lays on the stems of terminal shoots or occasionally around the extremities of sharply pointed leaves. Eggs are laid in successive rings, aligned in rows parallel to the shoot axis, or following a slightly spiral path depending on whether or not the shoot circumference is an exact multiple of basal egg width (Carne, 1966a). The number of eggs laid by this species is negatively correlated with increasing twig diameter, suggesting that branch tips are preferred oviposition sites (Tanton and Khan, 1978). In comparison, *P. charybdis* (Stål) lays eggs on the underside of leaves (Styles, 1970), as does *A. orphana* (Elliott, 1978). Oviposition behaviour of insects in the genus *Trachymela* is different to that of other paropsines, as these insects oviposit into crevices on the trunk of the host *Eucalyptus* (de Little, 1979). *Trochalodes* produces eggs on the tips of long stalks which hold the eggs high above the leaves and away from predators (Selman, 1994b). *Chrysophtharta lignea* (Erichson) is virtually ovoviviparous, with the larvae bursting through a thin membrane within minutes of being deposited on a leaf (de Little, 1979).

Egg batch numbers vary according to the species, but averages range between 9.8 (*T. tincticollis*) (Tribe and Cillie, 1997) and 26.1 eggs per batch (*C. bimaculata*) (Greaves, 1966). Immediately after laying a single batch, females may feed and mate again (Carne, 1966a) resulting in a single female laying over 1000 eggs in a 3-4 month season (de Little, 1983; Styles, 1970). Females need to be inseminated periodically to lay fertile eggs, but may lay infertile eggs if this does not occur (Greaves, 1966).

Eggs of *P. atomaria, C. bimaculata* and *A. orphana* hatch within three weeks in the field (Carne, 1966a; Greaves, 1966; Elliott, 1978). From within the egg, the larva concentrates pressure on the thorax so the hatching spines on each side of the body pierce the chorion. This creates two slits, through one of which it forces its way out (Cumpston, 1939; Greaves, 1966). After approximately 30 minutes, the larva becomes pigmented (Cumpston, 1939). Newly emerged larvae feed on egg-shells (Cumpston, 1939; Carne, 1966a; Greaves, 1966; Tribe and Cillie, 1997) and then migrate as a group to the new "flush" foliage of the tree (Cumpston, 1939; Greaves, 1966; Carne, 1966a; Styles 1970; de Little, 1983). Larvae of *T. tincticollis* hide in crevices or bark curls on the trees, either singly or gregariously depending on the size of the crevice, and emerge to feed on young foliage just before sunrise (Tribe and Cillie, 1997).

Larvae develop through four instars. For *P. atomaria* and *C. bimaculata* each stage takes 5-6 days (Carne, 1966a; Greaves, 1966). *P. charybdis* develops from egg to adult in 7-9 weeks (Styles, 1970). *A. orphana* is a winter insect and develops over a period of 7-8 months (Elliott, 1978), due to the seasonally cooler temperatures.

Some larvae remain gregarious throughout the larval stages (for example, *P. atomaria*) (Carne, 1966a) whilst others such as *Paropsisterna tigrina* (Chapuis) (Maddox, 1995), *C. bimaculata* (Greaves, 1966) and *P. charybdis* (Styles, 1970) may disperse individually or in small groups following the first or second instars.

When larval feeding is completed, larvae of some species drop directly from the leaves or crawl down the tree stems to pupate in the soil below (Carne, 1966a; Styles, 1970; Elliott, 1978; Tribe and Cillie, 1997; Clarke *et al.*, in press). After burrowing into the soil to a depth of 2-8 cm, depending on the soil hardness (Elliott, 1978), larvae construct a pupal case consisting of earth particles (Cumpston, 1939). The pupal period lasts for up to three weeks (Greaves, 1966; Carne, 1966a; Elliott, 1978) with the newly formed adult emerging during summer.

Adults fly upon leaving the pupal cell, making short abrupt flights from one part of the tree to another (Cumpston, 1939). Then, in preparation for overwintering, the adults congregate on the host trees in large numbers, where they intensively feed and accumulate fatbodies (Cumpston, 1939; Greaves, 1966; Carne, 1966a; Styles, 1970).

During the overwintering or reproductive diapause phase, adult insects can be found sheltering behind bark, under leaves and stones on the ground or in splits and cracks in the wood of dead trees (Cumpston, 1939; Greaves, 1966; Carne, 1966a; Styles, 1970; Tanton and Khan, 1978). Recent habitat studies have shown that more *C. bimaculata* adults can be found in clumps of cutting grass, *Gahnia grandis* than *Eucalyptus* leaf litter when overwintering (Clarke *et al.*, 1998), indicating a need for more research on overwintering sites of paropsines. Overwintering insects are not gregarious (Cumpston, 1939; Clarke *et al.*, 1998) and on warm days individuals may emerge and make short flights, or remain motionless for long periods on leaves exposed to sunshine (Carne, 1966a).

Paropsisterna tigrina has a similar lifecycle to *Paropsis* and *Chrysophtharta*, with Maddox (1995) reporting that in northern New South Wales, Australia, significant rainfall events appear to start each period of larval activity. This insect has two peak periods, a small spring population and a much larger more damaging autumn population. *A. orphana* differs to the other genera, as adults of this species mate and lay eggs in the autumn, and the larvae develop through the winter months. Adults emerge from pupation in early summer (Elliott, 1978).

6. Natural enemies

Predators and parasitoids have an important influence on the mortality of paropsine beetles with authors reporting up to 97% mortality from egg to adult stage (Greaves, 1966; Tanton and Khan, 1978; de Little, 1982; Tanton and Epila, 1984; de Little, *et al.*, 1990; Mo and Farrow, 1993). Natural enemies of the insect can be divided into three categories; a) egg parasitoids, b) larval parasitoids and c) predators. Predators may consume eggs, larvae or adults.

All egg parasitoids recorded are hymenopterous species, and cause up to 20% mortality (Tanton and Khan, 1978). Parasitised eggs are dull brown in appearance, instead of glossy (Tanton and Khan, 1978). Table 1 lists known insect parasitoids of the Paropsini. Other egg parasitoids observed include three unknown genera reared from *P. atomaria* (Tanton and Khan, 1978; Tanton and Epila, 1984).

Tachinids account for up to 62% of mortality caused by parasitisation in paropsine larvae (Greaves, 1966; Tanton and Epila, 1984; Tanton and Khan, 1978; de Little *et al.*, 1990). Tachinids lay a small, white, oval egg on the cuticle of the host larva (Greaves, 1966; Elliott, 1978; de Little, 1982) from which the parasitoid larva emerges and burrows into the body of the host (Greaves, 1966). Parasitoid eggs may not always be apparent on the cuticle, as they are lost with the cuticle when the insect moults (de Little, 1982). Usually only one parasitoid larva develops inside the host (Tanton and Epila, 1984; Cox, 1994), although Elliott (1978) noted up to five eggs on some host larvae. The parasitoid does not

Table 1. Paropsine hosts and known parasitoids

Host	Stage	Parasitoid	Reference
P. atomaria	Egg	*Neopolycystus insectifurax* (Girault) (Pteromalidae)	1, 2, 7
	Egg	*Enoggera* sp. (Pteromalidae)	7, 11
	Egg	*Aphaneromella ovi* (Dodd) (Playtgasteridae)	7, 1, 2
	Egg	*Baeoanusia albifunicle* (Girault) (Encyrtidae)	7
	Larval	*Froggattimyia anguliventris* (Mall.) (Tachinidae)	1
	Larval	*Froggattimyia tillyardi* (Mall.) (Tachinidae)	1
	Larval	*Paropsivora* sp. (Tachinidae)	1
	Larval	*Eadya paropsidis* (Braconidae)	2
P. charybdis	Egg	*Neopolycystus insectifurax*	8
	Egg	*Enoggera* sp.	8, 9
C. obovata (=*C. variicollis)*	Egg	*Neopolycystus insectifurax*	7
	Egg	*Enoggera* sp.	7
C. bimaculata	Egg	*Enoggera* sp.	4, 11
	Egg	*Aphaneromella* sp.	4
	Larval	*Anagonia rufifacies* (Macquart) (Tachinidae)	3, 4
	Larval	*Paropsivora* sp.	3, 4
	Larval	*Eadya paropsidis*	3
T. tincticollis	Egg	*Neopolycystus insectifurax*	10
	Egg	*Enoggera reticulata* (Naumann)	10, 11
	Egg	*Enoggera nassaui* (Girault)	11
A. orphana	Larval	*Lixophaga* sp. (Tachinidae)	5
	Larval	*Deltomyza australiensis* (Malloch) (Tachinidae)	5
Trochalodes pedestris	Larval	*Anagonia scutellata* (Malloch) (Tachinidae)	6
Paropsisterna beata	Egg	*Enoggera nassaui*	11

Reference key: [1]Tanton and Khan, 1978; [2]Tanton and Epila, 1984; [3]de Little, 1982; [4]de Little *et al.*, 1990; [5]Elliott, 1978; [6]Cantrell, 1986; [7]Mo and Farrow, 1993; [8]Bain and Kay, 1989; [9]Kay, 1990; [10]Tribe and Cillie, 1997; [11]Naumann, 1991.

kill the host until it nears the prepupal stage, at which time affected larvae become darker in colour. The parasitoid then forms a puparium either inside the skin which may protrude from the dead hosts' body (Greaves, 1966; Tanton and Khan, 1978) or it may bore a hole to get out of the hosts' body prior to pupation. Parasitoids which pupate inside the body of the host may be more susceptible to attack from secondary parasites (Tanton and Khan, 1978).

Larval parasitisation can occur in all stages, but some parasitoids are stage-specific such as the braconid *Eadya paropsidis* (Huddleston and Short) which achieves maximum parasitisation by the second larval instar of *C. bimaculata* (de Little, 1982). De Little (1982) also found that the tachinid, *Anagonia rufifacies* (Macquart) prefers fourth instar hosts while *Paropsivora* sp. exhibits little differentiation between host stages.

Eadya paropsidis is the dominant larval parasitoid of *C. bimaculata* and *P. atomaria* (de Little, 1982), causing up to 93% parasitisation (Tanton and Epila, 1984). Usually six eggs are laid per host, although as many as 13 have been observed (Tanton and Epila, 1984). Tanton and Khan (1978) incorrectly identified this as *Aridelus* sp.

Hyperparasitoids have also been identified from the tachinid parasitoids (Tanton and Khan, 1978; Tanton and Epila, 1984; de Little, 1982). Tanton and Epila (1984) found tachinid larvae within *P. atomaria* containing 1-4 larvae of the ichneumonid *Mesochorus* sp. Also identified was *Perilampus tasmanicus* (Cameron) (Pteromalidae), a very efficient hyperparasitoid due to its high reproductive capacity and ability to parasitise both the braconid and tachinid hosts.

Another natural enemy was observed by Selman (1989) who found nematodes of the family Mermithidae infesting populations of paropsine beetles. These nematodes parasitise the larval stages of the host and emerge either from final instar larvae or adults. If they emerge from larvae, the larvae are killed, but in fast developing, multivoltine species, adult nematodes may emerge from adult beetles without causing death. Initially host beetles appear sick and discoloured, but under laboratory conditions they have been observed to recover and lay eggs.

Insect predators cause substantial loss of paropsine eggs and early instar larvae (Elliott and de Little, 1980; Selman, 1989). De Little *et al.* (1990) recorded up to 77.6% predation of eggs at one site in the north-west of Tasmania, Australia and observed that the main predator was a coccinellid, *Cleobora mellyi* (Mulsant). *C. mellyi* and *Harmonia conformis* (Boisduval) another coccinellid, have been noted to consume eggs of *P. atomaria* (Mo and Farrow, 1993), *P. charybdis* (Bain *et al.*, 1979), *C. variicollis* (Mo and Farrow, 1993) and *C. bimaculata* (Greaves, 1966; de Little *et al.*, 1990; Mensah and Madden, 1994). Greaves (1966) found that they also consumed early instar larvae and are predators in both the larval and adult stage. Other important predators include the cantharid *Chauliognathus lugubris* (F.) which feeds on the eggs, larvae and pupae of *C. bimaculata* (Madden and Mensah, 1994 – wrongly identified as *C. pulchellus*; Shohet and Clarke, 1997), the pentatomid *Oechalia schellenbergii* (Guérin-Méneville) which consumes eggs of *P. atomaria* (Mo and Farrow, 1993) and *P. charybdis* (Bain and Kay, 1989). Minor predators observed feeding on *P. charybdis* are *Cermatulus nasalis* (Westwood) (Pentatomidae), *Vespula germanica* (Fabricius) and birds (Styles, 1970).

7. Pest potential

As mentioned in the introduction, the intensive cultivation of Australian native plants, especially eucalypts and acacias within Australia is a recent development, with the consequence that new pests are still emerging from among the native paropsine fauna. This process is analogous to the chrysomelid species *Leptinotarsa decemlineata* (Colorado potato beetle) achieving pest status only after cultivation of potatoes began in Colorado in the middle of last century (Johnson, 1969).

Those paropsine species so far regarded as significant pests are listed in Table 2, with their main hosts, states of Australia and other countries where they are regarded as pests. The species listed for New Zealand and South Africa are exotic pests in those countries, and it is noteworthy that *Trachymela tincticollis*, which first appeared in the latter country in 1981 is regarded as a rare insect in its native environment in Western Australia (Tribe and Cillie,1997). Other paropsines to become established in exotic locations include *Trachymela sloanei* (Blackburn) in New Zealand (1976) and California (1998), *T. catenata* (Chapuis) in New Zealand (1992) and *Dicranosterna semipunctata* (Chapuis) in New

Table 2. The main paropsine pests of commercial tree crops in the various Australian States, and other countries where they are significant pests.

Commercial Tree Crop	Queensland[1,2]	New South Wales[3,4,5]	Victoria[6]	Tasmania[7]	South Australia[8]	Western Australia[9,10]	New Zealand[11,12]	South Africa[13]
Eucalyptus grandis	*C. cloelia**	*C. cloelia* *C. obovata* *P. atomaria**						*T. tincticollis**
Eucalyptus pellita	*C. cloelia*							
Eucalyptus urophylla	*C. cloelia*							*T. tincticollis*
Eucalyptus botryoides	*D. picea**							
Eucalyptus camaldulensis		*C. obovata* *P. atomaria*						*T. tincticollis*
Eucalyptus globulus				*C. agricola* *P. porosa*	*C. obovata*	*C. amoena*		
Eucalyptus nitens				*C. bimaculata* *C. agricola* *C. obovata* *P. porosa* *P. charybdis* *P. delittlei*			*P. charybdis*	
Eucalyptus dunnii		*C. cloelia* *C. obovata* *P. atomaria*						
Eucalyptus viminalis			*C. agricola* *C. obovata*					
Eucalyptus pilularis	*P. atomaria*	*P. atomaria* *P. variolosa*						
Eucalyptus regnans				*C. bimaculata*				
Eucalyptus obliqua				*C. bimaculata* *P. delittlei*				
Eucalyptus delegatensis				*C. bimaculata*				
Corymbia torelliana	*P. ornata*	*P. variolosa*						
Corymbia maculata	*P. ornata* *D. picea*							
Melaleuca alternifolia	*Pa. tigrina**							
Acacia dealbata				*A. orphana**				
Acacia mearnsii			*A. orphana*					
Acacia auriculiformis	*D. picea*							
Acacia holosericea	*D. picea*							
Acacia mangium	*D. picea*							

*A., Acacicola; C., Chrysophtharta; D., Dicranosterna; P., Paropsis; Pa., Paropsisterna; T., Trachymela.
Reference key: [1]R. Wylie (pers. comm.), [2]S. Lawson (pers. comm.), [3]C. Stone (pers. comm.), [4]R. Cant (pers. comm.), [5]R. Floyd (pers. comm.), [6]N. Collett (pers. comm.), [7]de Little (1979, 1989), [8]C. Phillips (pers. comm.), [9]R. Fremlin (pers. comm.), [10]T. Burbidge (pers. comm.), [11]Styles (1970), [12]M. Kay (pers. comm.)
[13]Tribe and Cillie (1997)

Zealand (1994) (M. Kay, pers. comm.). To date, these species have had only minor impact at their exotic locations (M. Kay, pers. comm.).

The main damage to commercial crops caused by paropsine pests is the loss of pulpwood and timber production resulting from heavy and repeated defoliation of the host trees (Carne 1996a; Bain and Kay, 1989; de Little, 1989; Neumann, 1993). In some cases (for example, *Acacicola orphana* on *Acacia dealbata* in Tasmania) heavy and repeated paropsine attack results in host tree mortality (Elliott, 1978). De Little (1989) estimated that in a severe defoliation of a *Eucalyptus nitens* plantation in Tasmania by *Chrysophtharta bimaculata* the equivalent of two years growth or 22 m^3ha^{-1} was lost. Elliott *et al.* (1993) found that defoliation of one year old *E. regnans* trees resulted in a 55% reduction in height compared to undamaged (protected) trees. After eight years, Elek (1997) observed that the same insect-protected trees were 25% taller than unprotected trees, with almost three times the wood volume.

Eucalypt species are now being widely planted in Australia as well as in New Zealand and South Africa for commercial wood and fibre production (Tibbits *et al.* 1997; Poynton, 1979). The economics of this operation demand high productivity and the loss of one or more year's growth has serious implications on profitability. For this reason, broadacre control operations such as aerial applications of organophosphate and synthetic pyrethroid insecticides have been used in New Zealand and Tasmania to successfully protect eucalypt plantations from paropsine attack (B. Poole, pers.comm., de Little, 1989).

Conclusion

Ever increasing intensive cultivation of the host plants of the paropsines, both within Australia and overseas is leading to a growing number of these species being regarded as pests. This number is however, still small compared with the total number of species and it is unfortunate that knowledge of paropsine biology is still largely restricted to those species accorded pest status.

Selman (1994a) predicted that adult paropsine beetles would increasingly be introduced via growing international traffic into overseas countries where they would become pests of the ever-growing areas of exotic eucalypt plantations. This prediction is being borne out with 3 of the 6 recorded exotic occurrences of paropsines having been reported in the past 6 years. Currently only 2 of these species are regarded as significant pests at exotic locations and the third species (*Dicranosterna semipunctata*) attacks *Acacia melanoxylon* (M. Kay, pers. comm.). It is noteworthy that all recent emigrant species are from genera with more cryptic adults, viz., *Trachymela* and *Dicranosterna.*

It is likely that more paropsine species will become significant pests in Australia as the domestication and cultivation of their host plants proceeds. While biological control, through the introduction of egg parasitoids has been successful in reducing pest paropsine populations in exotic locations (Tribe and Cillie, 1997; M. Kay, pers. comm.), natural biological control in Australia has not been sufficiently successful to obviate the need for pesticide usage. Research aimed at more sophisticated methods of control is continuing.

Acknowledgments

The authors thank Dr. Anthony Clarke, Ms Susan Baker, Dr. Tom Burbidge, Mr Steve Candy, Mr Russell Cant, Mr Nick Collett, Dr. Jane Elek, Mr Ray Fremlin, Dr. Rob Floyd, Dr. Malcolm Kay, Dr. Simon Lawson, Mr Brendan Murphy, Mrs Charlma Phillips, Mr. Barry Poole, Dr. Brian Selman, Dr. Christine Stone and Dr. Ross Wylie for their assistance and unpublished material. Drs. Anthony Clarke, John Madden, Clare MacArthur and Geoff Allen provided useful comments. One of us (T.S.) was in receipt of an Australian Postgraduate Award (Industry) partnered by the Tasmanian Forest Research Council and the Australian Research Council.

References

Bain J.; Faulds, W. and Alma, P. (1979). Biological control for eucalypt defoliator? What's New in Forest Research 75. Forest Research Institute, Rotorua, New Zealand, 4 pages.

Bain, J. and Kay, M.K. (1989). *Paropsis charybdis* Stål, *Eucalyptus* tortoise beetle (Coleoptera: Chrysomelidae). In Cameron, P.J.; Hill, R.L.; Bain, J. and Thomas, W.P. (Eds.) *A review of biological control of pests and weeds in New Zealand 1874-1987*. Technical Communication 10, CAB International and DSIR, Wallingford, U.K. pp. 281-287.

Baker, S.C; Elek J.A. and Candy, S.G.(unpublished). Feeding, development time and mortality of Tasmanian *Eucalyptus* leaf beetle larvae on foliage of *Eucalyptus nitens* and *E. regnans.*

Barlow, B.A. (1981). The Australian Flora: its origin and evolution. In George, A.S. (Ed), *Flora of Australia,* Volume 1. Australian Government Publishing Service. Canberra. pp. 25-75.

Blackburn, T. (1899). Revision of the genus *Paropsis*, Part V. Proc. Linn. Soc. N.S.W. 24: 482-521.

Cantrell, B.K. (1986). An updated host catalogue for the Australian Tachinidae (Diptera). J. Aust. ent. Soc. 25:255-265.

Carne, P.B. (1966a) Ecological characteristics of the eucalypt-defoliating chrysomelid *Paropsis atomaria* Olivier. Aust. J. Zool. 14: 647-672.

Carne, P.B. (1966b). Growth and food consumption during the larval stages of *Paropsis atomaria* (Coleoptera: Chrysomelidae). Entomologia exp. appl. 9:105-112.

Clarke, A.R.; Paterson, S.C. and Steinbauer, M.J. (In press). Larval trapping as a means of sampling paropsina (Coleoptera: Chrysomelidae: Chrysomelinae) from *Eucalyptus* forests. Coleopts Bull. 52 (2).

Clarke, A.R.; Shohet. D; Patel, V.S. and Madden, J.L. (1998) Overwintering sites of *Chrysophtharta bimaculata* (Olivier) (Coleoptera: Chrysomelidae) in commercially managed *Eucalyptus obliqua* forests. J. Aust. ent. Soc. 37: 149-154.

Clarke, A.R.; Zalucki, M.P.; Madden, J.L.; Patel, V.S. and Paterson, S.C. (1997). Local dispersion of the Eucalyptus leaf-beetle *Chrysophtharta bimaculata* (Coleoptera: Chrysomelidae), and implications for forest protection. J. appl. Ecol. 34: 807-816.

Congdon, B.C.; Lange, C.L. and Clarke, A.R. (1997). Geographical variation and gene flow in the eucalyptus defoliating beetle *Chrysophtharta bimaculata* (Coleoptera: Chrysomelidae). J. appl. Ecol. 34:1287-1292.

Cox, M. L. (1994). The Hymenoptera and Diptera parasitoids of Chrysomelidae. In Jolivet, P.H.; Cox, M.L. and Petitipierre, E. (Eds), *Novel aspects of the biology of Chrysomelidae*. Kluwer Academic Publishers, Netherlands. pp. 555-565

Cumpston, D.M. (1939). Observations on the bionomics and morphology of seven species of the tribe *Paropsini* (Chrysomelidae). Proc. Linn. Soc. N.S.W. 64:353-66.

Daccordi, M. (1994). Note for the phylogenetic study of Chrysomelinae, with descriptions of new taxa and a list of all the known genera (Coleoptera: Chrysomelidae, Chrysomelinae). In Furth, D.G. (Ed.) *Proceedings of the 3rd international symposium on Chrysomelidae*, Beijing. 1992. Backhuys Publishers, Leiden. pp. 61-84.

de Little, D.W. (1979). Taxonomic and ecological studies of the *Eucalyptus*-defoliating paropsids (Coleoptera: Chrysomelidae), with particular reference to *Chrysophtharta bimaculata* (Olivier). PhD. thesis, University of Tasmania; Hobart.

de Little, D.W. (1982). Field parasitization of larval populations of the *Eucalyptus*-defoliating leaf-beetle, *Chrysophtharta bimaculata* (Olivier) (Coleoptera: Chrysomelidae). Gen. appl. Ent. 14:3-6.

de Little, D.W. (1983). Lifecycle and aspects of the biology of Tasmanian *Eucalyptus* leaf beetle, *Chrysophtharta bimaculata* (Olivier) (Coleoptera: Chrysomelidae). J. Aust. ent. Soc. 22:15-18.

de Little, D.W. (1989). Paropsine chrysomelid attack on plantations of *Eucalyptus nitens* in Tasmania. N.Z. Jl. For. Sci. 19(2-3): 223-227.

de Little, D. W.; Elliott, H.J.; Madden, J.L and Bashford, R. (1990). Stage-specific mortality in two field populations of immature *Chrysophtharta bimaculata* (Olivier) (Coleoptera:Chrysomelidae). J. Aust. ent. Soc. 29:51-55.
de Little, D.W. and Madden, J.L. (1975) Host preference in the Tasmanian eucalypt defoliating Paropsini (Coleoptera: Chrysomelidae) with particular reference to *Chrysophtharta bimaculata* (Olivier) and *C. agricola* (Chapuis). J. Aust. ent. Soc. 14: 387-394.
Edwards, P.B. (1982). Do waxes on juvenile *Eucalyptus* leaves provide protection from grazing insects? Aust. J. Ecol. 7: 347-352.
Edwards, P.B. and Wanjura, W.J. (1990). Physical attributes of Eucalypt trees and the host range of chrysomelid beetles. Symp. Biol. Hung. 39: 227-236.
Elek, J.A. (1997). Assessing the impact of leaf beetles in eucalypt plantations and exploring options for their management. Tasforests, 9:139-154
Elliott, H.J. (1978). Studies on the fireblight beetle, *Pyrgoides orphana* (Erichson) (Coleoptera:Chrysomelidae) and its effect on the growth of silver wattle in Tasmania. Aust. For. 41 (3):160-166.
Elliott, H.J. and de Little, D.W. (1980). Laboratory studies on predation of *Chrysophtharta bimaculata* (Olivier) (Coleoptera: Chrysomelidae) eggs by the coccinellids *Cleobora mellyi* Mulsant and *Harmonia conformis* (Boisduval). Gen. appl. Ent. 12: 233-36.
Elliott, H.J.; Bashford, R and Greener, A. (1993). Effects of defoliation by the leaf beetle, *Chrysophtharta bimaculata*, on growth of *Eucalyptus regnans* plantations in Tasmania. Aust. For. 56 (1): 22-26.
Erichson, W.F. (1842). Beitrag zur Fauna von Vandiemensland mit besonderer Rucksicht auf die Geographische Verbreitung der Insecten. Arch. Naturgesch. 8:83-287.
Feeny, P. (1970). Seasonal changes in oak leaf tannins and nutrients as a cause of spring feeding by winter moth caterpillars. Ecology 51 (4):565-581
Fogg, S.A. (1859). XVII. – Contribution to the insect fauna of Van Diemen's Land, (now Tasmania), with particular reference to the geographical distribution of insects. By Dr. W.F. Erichson. Pap. Proc. R. Soc. Tasm. 3(2): 298-338.
Fox, L.R. and Macauley, B.J. (1977). Insect grazing on *Eucalyptus* in response to variation in leaf tannins and nitrogen. Oecologia 29:145-162.
Greaves, R. (1966) Insect defoliation in the eucalypt regrowth in the Florentine Valley, Tasmania. Appita 19:119-126.
Johnson, C.G. (1969). Migration and dispersal of insects by flight. Methuen, London.
Jolivet, P. and Hawkeswood, T. J. (1995). Host-plants of Chrysomelidae of the world. Backhuys: Leiden.
Jolivet, P. and Petitpierre, E. (1981). Biology of the Chrysomelidae (Coleoptera). Bull. Inst. Cat. Hist. Nat., 47(Sec. Zool., 4): 105-138.
Kay, M. (1990). Success with biological control of the eucalyptus tortoise beetle, *Paropsis charybdis*. What's New in Forest Research 184. Forest Research Institute, Rotorua, New Zealand. pp. 1-4.
Kile, G.A. (1974) Insect defoliation in the eucalypt regrowth forests of southern Tasmania. Aust. For. Res. 6 (3) : 9-18.
Leon, A. (1989). The Tasmanian *Eucalyptus* leaf beetle, *Chrysophtharta bimaculata*: An overview of the problem and current control methods. Tasforests, September. pp. 33-37.
Li, H. (1993). Phytochemistry of Eucalyptus spp. and its role in insect-host-tree selection. PhD. thesis, University of Tasmania; Hobart.
Maddox, C.D.A. (1995). Aspects of the biology of *Paropsisterna tigrina* (Chapuis) (Coleoptera: Chrysomelidae) the major pest species of *Melaleuca alternifolia* (Cheel) [Myrtaceae]. Masters thesis; University of Queensland; Brisbane.
Mensah, R.K. and Madden, J.L. (1994). Conservation of two predator species for biological control of *Chrysophtharta bimaculata* (Col.: Chrysomelidae) in Tasmanian forests. Entomophaga 39 (1): 71-83.
Mo, J.H. and Farrow, R.A. (1993). Estimation and correction of egg mortality rates from sample data of two coexisting leaf beetles (Coleoptera: Chrysomelidae: Paropsini). J. Aust. ent. Soc. 32:85-92.
Moore, B.P. (1967). Hydrogen cyanide in the defensive secretions of larval Paropsini (Coleoptera: Chrysomelidae). J. Aust. ent. Soc. 6: 36-38.
Morrow, P.A. and Fox, L.R. (1980). Effects of variation in *Eucalyptus* essential oil yield on insect growth and grazing damage. Oecologia 45:209-219.
Naumann, I.D. (1991). Revision of the Australian genus *Enoggera* Girault (Hymenoptera: Pteromalidae: Asaphinae). J. Aust. ent. Soc. 30 (1): 1-17.
Neumann, F.G. (1993). Insect pest problems of eucalypt plantations in Australia. 3. Victoria. Aust. For. 56 (4):370-374.
New, T.R. (1981). Paropsine beetle larvae as possible pollinators of *Acacia baileyana* in Victoria. Aust. ent. Mag. 8(4): 55.
Ohmart, C.P.; Thomas, J.R. and Stewart, L.G. (1987). Nitrogen, leaf toughness and the population dynamics of *Paropsis atomaria* Olivier (Coleoptera: Chrysomelidae) – A hypothesis. J. Aust. ent. Soc. 26: 203-207.
Patterson, K.C.; Clarke, A.C.; Raymond, C.A. and Zalucki, M.P. (1996). Performance of first instar *Chrysophtharta*

bimaculata larvae (Coleoptera:Chrysomelidae) on nine families of *Eucalyptus regnans* (Myrtaceae). Chemoecology 7:94-106.

Poynton, R. J. (1979). Report to the Southern African Regional Commission for the conservation and utilization of the soil on tree planting Southern Africa. Volume 2. The Eucalypts. Dept. of Forestry. Republic of South Africa. pp. 350-378.

Selman, B.J. (1983). The naming of the Tasmanian species of Paropsis Olivier (Coleoptera: Chrysomelidae). J. Aust. ent. Soc. 22:333-339

Selman, B.J. (1989). Some factors including nematodes influencing the population levels of chrysomelid beetles. Entomography 6: 403-406.

Selman, B.J. (1994a). The biology of the paropsine eucalyptus beetles of Australia. In Jolivet, P.H.; Cox, M.L. and Petitipierre, E. (Eds), *Novel aspects of the biology of Chrysomelidae*. Kluwer Academic Publishers, the Netherlands. pp. 555-565.

Selman, B.J. (1994b). Eggs and oviposition in chrysomelid beetles. In Jolivet, P.H.; Cox, M.L. and Petitipierre, E. (Eds), *Novel aspects of the biology of Chrysomelidae*. Kluwer Academic Publishers, the Netherlands. pp. 69-74.

Shohet, D. and Clarke, A.R. (1997). Life history of *Chauliognathus lugubris* (F.) (Coleoptera: Cantharidae) in Tasmanian forests. J. Aust. ent. Soc. 36: 37-44.

Steinbauer, M.J.; Clarke, A.R. and Madden, J.L. (1998). Oviposition preference of a *Eucalyptus* herbivore and the importance of leaf age on interspecific host choice. Ecol. Ent. 23: 201-206.

Stone, C. and Bacon, P.E. (1994). Relationships among moisture stress, insect herbivory, foliar cineole content and the growth of red river gum *Eucalyptus camaldulensis*. J. Appl. Ecol. 31:604-612.

Styles, J.H. (1970). Notes on the biology of *Paropsis charybdis* Stål (Coleoptera: Chrysomelidae). N.Z. Ent. 4(3): 103-11.

Tanton, M.T. and Epila, J.S.O. (1984). Parasitization of larvae of *Paropsis atomaria* Ol. (Coleoptera: Chrysomelidae) in the Australian Capital Territory. Aust. J. Zool. 32:251-259.

Tanton, M.T. and Khan, S. M. (1978). Aspects of the biology of the eucalyptus-defoliating chrysomelid beetle, *Paropsis atomaria* Ol. in the Australian Capital Territory. Aust. J. Zool. 26:113-120.

Tibbits, W.N.; Boomsma, D.B. and Jarvis, S. (1997). Distribution, biology, genetics and improvement programs for *Eucalyptus globulus* and *E. nitens* around the world. In *Proceedings of the 24th Biennial Southern Forest Tree Improvement Conference*, Orlando, Florida, June 9-12 1997. Southern Forest Tree Improvement Committee, sponsored publication No. 46. pp 81-95.

Tribe, G.D. and Cillie, J.J. (1997). Biology of the Australian tortoise beetle *Trachymela tincticollis* (Blackburn) (Chrysomelidae: Chrysomelini: Paropsina), a defoliator of *Eucalyptus* (Myrtaceae), in South Africa. Afr. Ent. 5(1): 109-123.

Vasconcellos-Neto, J. and Jolivet, P. (1988). Une nouvelle strategie de defense: la strategie de defense annulaire (cycoalexie) chez quelques larves de Chrysomelides Bresiliens. Bull. Soc. ent. Fr. 92:291-299.

Weise, J. (1916). Chrysomelidae: 12. Chrysomelinae. Coleopt. Cat. Junk: Berlin 68:1-255.

White, T.C.R. (1973). The establishment, spread and host range of *Paropsis charybdis* Stål (Chrysomelidae) in New Zealand. Pacif. Insects 15(1): 59-66.

SELECTED TOPICS IN THE BIOLOGY OF *DIABROTICA*

JAMES L. KRYSAN
1048 Cherokee Road, Apt. 3., Louisville, KY 40204, USA, E-mail: Jkrysan@aol.com

1. Introduction

This review will emphasize the related events of mating, oviposition, flight, dormancy, and selected basic topics that underpin those issues. Information on natural enemies is summarized. Host plant associations will be treated only in the context of the emphasized subjects. Larval hosts are the subject of a chapter by Eben in the present volume and Metcalf (1994) recently reviewed adult host relationships from the perspective of chemical ecology. An extensive and fascinating literature centers around the strong predilection for *Diabrotica* species to consume cucurbitacins, a chemical characteristic of plants in the family Cucurbitaceae but, like other host- related phenomena, it will only appear in this review in the context of the specified subjects.

The 354 described species of *Diabrotica* (Wilcox, 1972; Bechyné, 1997; Bechyné & Springlova-Bechyné, 1970, 1976; Krysan & Smith, 1987a,b) are overwhelmingly Neotropical. Introduced populations occur in the Easter Islands [*D. viridula* (Fabricius), (Olalquiaga, 1980)] and Europe [*D. virgifera virgifera* LeConte, (Baca, 1994)]. But, most research concerns the eight species (and five subspecies) that occur in North America north of Mexico. For those species, range, voltinism, and larval hosts follow taxonomic lines. Taxa in the *fucata* species group (*Diabrotica balteata* LeConte, *D undecimpunctata howardi* Barber and *D. u. undecimpunctata* Mannerheim) are multivoltine, survive winter as adults where subfreezing temperatures are moderate and infrequent, and are polyphagous as larvae. Nearctic species in the *virgifera* group have no more than one generation per year, overwinter as eggs in the temperate climate, and as larvae are known (*D. barberi* Smith and Lawrence, *D. v. virgifera*, and *D. v. zeae* Krysan and Smith) or inferred [*D. cristata* (Harris), *D. lemniscata* LeConte, *D. longicornis* (Say)] to be either oligophagous or essentially monophagous on maize (Branson & Krysan, 1981; Krysan *et al.*, 1989; Eben, this volume).

2. The egg stage

2.1. Survival of eggs under chill and sub-freezing conditions

The ability of eggs to survive periods when the temperature is either below the rate for development (chill) or well below freezing, reflects adaptive traits. In the north, species that overwinter as eggs routinely face long periods of sub-freezing temperatures. Eggs of *D. barberi* are more cold-hardy than those of *D. v. virgifera* (Gustin, 1983). A series of

Advances in Chysomelidae Biology 1, pp. 479–513
Edited by M.L. Cox

studies by Gustin and coworkers showed that sub-freezing temperatures sometimes occur in soil in the northern corn belt of the U. S. that can kill eggs of *D. v. virgifera*. In one of four years soil temperatures stayed below –7.5 °C for the length of time that caused reduced survival in laboratory studies. Following that winter, *D. v. virgifera* populations over the northern corn belt were drastically lower than normal (Gustin 1981). Subsequent laboratory studies have refined the relationships between the depth and duration of subfreezing temperatures, on the one hand, and survival on the other hand (Gustin & Wilde, 1985; Gustin, 1986; Woodson & Gustin, 1993; Woodson & Ellsbury, 1994). As expected, survival declines as exposure time lengthens and temperatures decline. Among species which overwinter as adults, optimal time/temperature combinations for long term storage of eggs for mass production have been determined for *D. balteata* (Schalk & Jones, 1982) and *D. u. howardi* (Jackson *et al.*, 1995). The data establish the ecologically signficant fact that eggs of these species are not capable of dormancy via chill quiescence. Conversely, *D. v. virgifera* eggs are capable of long chill quiescence as part of the dormancy process but survival was reduced when chill quiescence was imposed after post-dormancy development had begun (Palmer *et al.*, 1977). The ability to survive quiescence imposed by chill is an evolved, stage specific, adaptive trait.

2.2. Moisture relationships of eggs

Egg weight approximately doubles during embryogenesis, the result of water uptake: weight, shortly after oviposition and near the time of hatch, was 38 and 72 micrograms respectively for *D. v. virgifera* (Krysan, 1978), and 48 and 89 micrograms respectively for *D. u. howardi* (Krysan, 1976). Water uptake by *D. v. virgifera* occurs during post-diapause embryogenesis. Most water uptake by eggs of *D. u. howardi* occurs between 24 and 72 h of age; while absorbing water they are extremely susceptible to both water loss and death from desiccation. Eggs can accumulate moisture from air at RH of 100% but not 96%. After three days of age some eggs complete embryogenesis at RH as low as 75%. Eggs can adapt to the moisture level of their environment; eggs acclimated in water vapor only (at 100% RH) were subsequently more resistant to water loss than eggs acclimated in contact with water droplets on moist paper (Krysan, 1976; 1977). These features may explain why eggs of *D. u. howardi* were better able than larvae to adjust to soil that was drying (Brust & House, 1990a). *D. u. howardi* eggs high in cucurbitacin C (which, coincidentally, had greater mass) hatched earlier than those with no detectable cucurbitacin and the delay in hatch was exaggerated by 6 h exposure to an RH of 94-96% (Brust & Barbercheck, 1992).

3. Larval and pupal stages

Larvae occupy roots and the soil around roots; the pre-pupal stage forms a discrete earthen cell for pupation.

3.1. Development time

The time for completion of development, from larval hatch to adult emergence at approximately 25 °C, for the following species, is (days, in brackets); *D. barberi* [male, 35; female, 36] (Woodson & Jackson, 1996); *D. speciosa* [26] (Silva-Werneck *et al.,* 1995); *D. tibialis* Jacoby [29] (Branson & Krysan, 1987); *D. v. virgifera* [26.3 male, 28.9 female] (Jackson & Elliott, 1988). The effect of temperature on larval development has been thoroughly characterized by Jackson and coworkers. The temperature range of 21° to 30 °C yields maximal values for mean head width and survival of *D. v. virgifera* larvae/pupae. The lower threshold of development is near 9 °C. Males complete larval/pupal development faster than do females; the time difference varies from 0.9 to 2.8 days, depending on temperature (Jackson & Elliott, 1988). Jackson and Elliott observed more rapid development than did earlier workers and reasonably ascribed it to improvements in diet. For *D. barberi* the lower developmental threshold is near 10.2 °C for the period from hatch to adult emergence. Extreme temperatures were detrimental; the optimal range for *D. barberi* was 18° to 30 °C. In the range of 21° to 30 °C females took 0.1 to 1.5 days longer that males to develop from hatch to the adult stage (Woodson & Jackson, 1996). Thus, all stages of immature development contribute to the protandry seen in the field.

3.2. Longevity of unfed neonatal larvae

In any one maize field the relationship between adult numbers at oviposition time, and the extent of larval feeding the following season, is feeble at best (Stamm *et al.,* 1985) and is only rarely explainable by egg mortality (Gustin, 1981). MacDonald and Ellis (1990) suggested that mortality of neonatal larvae accounts for most of the difference. Data on neonatal larval survival varies within and between studies. For example, survival time of starved larvae varied among egg "sources" when the sources were separate dishes of eggs collected from the same set of females at slightly differing times (Oloumi-Sadeghi & Levine, 1989b). In another case, longevity was reduced some 24 hours by the act of examining the larvae for survival (MacDonald & Ellis 1990).

For *D. v. virgifera* Strnad and Bergman (1987a) found that nearly all neonate larvae survived starvation for 12, 24 or 36 h and then survival declined until all were dead at 96 h, but the ability to burrow into a root declined significantly after only 12 h. MacDonald and Ellis (1990) obtained maximal survival of 120 to 144 h under conditions similar to those of Strnad and Bergman (1987a). Larvae survived longer as temperature was decreased and relative humidity was increased. Maximal survival time was 13 d at 15 °C with contact moisture. Mean survival times at 15 °C varied from 28 h at 80% RH to 86 h on wet paper. Oloumi-Sadeghi and Levine (1989b) also found that unfed *D. v. virgifera* larvae survive longer at lower temperatures and that, if larvae locate a root within 72 h after hatching, starvation did not reduce survival when the criterion for survival was being alive after two weeks (27 °C). When the criterion for survival was reaching adulthood, *D. v. virgifera* survival was significantly reduced (45%) even when neonate larvae were denied access to food for only 1 d; when starved 3 d, only 4.7% reached adulthood (Branson, 1989). In the latter study, larvae that commenced feeding did not necessarily survive. Neonatal *D. barberi* died within 1 or 2 d if they did not establish on a host (Dominique & Yule, 1984).

3.3. Mobility

Eggs are laid in the soil, not on host tissue, so survival of unfed larvae depends on mobility. In laboratory tests, neonate *D. v. virgifera* larvae readily moved from sandy soil to clay soil, but seldom moved in the opposite direction (Turpin & Peters, 1971). The distance neonate larvae can move in search of a host is affected by the type, moisture, porosity, and bulk density of the soil. In 6 hours *D. v. virgifera* larvae readily moved toward a carbon dioxide source through 30 cm of non-compacted soil, that is, soils with bulk densities of 0.76 and 0.89 g/cm^3 for sandy loam and silt loam, respectively, (Strnad & Bergman, 1987a). Some movement occurred at bulk density of 1.1 g/cm^3 but with bulk densities of 1.3 and 1.5 g/cm^3 movement was limited to <5 cm after 20 hours. Gustin and Schumacher (1989) found that size, geometry and continuity of pores were more important than simple bulk density as determinants of larval mobility in soil. Larvae moved through pores of 0.15 mm despite a head width of ca. 0.2 mm (George & Hintz, 1966), suggesting larvae displace some soil particles to allow movement. But, the distance moved through soil with 0.15 mm pores was signficantly less than that through soil with pores of 0.2 and 0.3 mm; with the latter pore sizes there was no significant interaction between pore size and distance moved. MacDonald and Ellis (1990) used methods of soil preparation designed to retain more soil structure than those of earlier workers and observed that movement was unaffected by bulk densities of 1.2 to 1.8 g/m. However, in field studies Ellsbury *et al.* (1994) found that soils with bulk densities >1.4 g/cm^3, that were artificially compacted, reduced survival, presumably by reducing mobility. Movement of *D. v. virgifera* larvae was significantly restricted in very wet (36% moisture) soils (MacDonald & Ellis, 1990), a finding that may well explain reduced rootworm damage in field plots saturated with water during the period of hatch (Riedell & Sutter, 1995). Clearly, neonate larvae have severely limited longevity and range so they are dependent for survival on oviposition near a place where a suitable host plant is, or will occur.

3.4. Mortalilty factors for post-neonatal immature stages

As the clay content of the soil occupied increased, survival of *D. u. howardi* larvae increased (Turpin & Peters, 1971; Lummus *et al.*, 1983) and when plant-available water ranged from 70 to 100% immatures survived best but survival was significantly poorer below 55% (Lummus *et al.*, 1983). Mortality was greater for eggs and first instar larvae than for older larvae, and irrigated systems generally had lower mortality than non-irrigated systems. Pupal mortality was intermediate between that of first instars, on the one hand, and second and third instars, on the other hand (Brust, 1991). Presumably some of the first instar larval mortality in this study was assignable to neonate larvae. Soil type and moisture availability clearly affect immature survival.

Pupae of *D. v. virgifera* and *D. u. howardi* had morphological deformations when maintained at relative humidity below 100%. Survival declined as relative humidity was lowered from 96 to 32.5 %; no *D. v. virgifera* and a few *D. u. howardi* survived at 32.5% relative humidity. As temperature increased, survival of pupae decreased at relative humidity less than 75% (Fisher, 1986).

3.5. Feeding sites

Larvae feed only on the thin-walled parenchyma cells of the root cortex (Riedell & Kim, 1990). First instar larvae feed on seminal roots and roots of whorls 1 through 7, burrowing into root branches as small as 0.5 mm in diameter (Strnad & Bergman, 1987b). The great majority was found in roots 2.0 mm or less in diameter. Second instars feed on seminal roots and whorls while third instars were only found in whorls. Larvae are usually oriented toward the root tip, the zone of actively growing tissue. When the root tip was excised, the orientation became equal, suggesting that a chemical gradient originating in the tip attracts the feeding larva. Larvae abandon older roots and move through the soil to reach younger roots, so older larvae tend to congregate at the plant base (Strnad & Bergman, 1987b). The damage to maize by rootworm feeding is reviewed thoroughly by Reidell (1993). For other aspects of larval host relationships see the chapter by Eben in this volume.

3.6. Interspecific larval competition

In the middle of the 20th Century two previously allopatric species came to occupy the same range in the corn belt of the USA; some field observations showed displacement of *D. barberi* by *D. v. virgifera.* On potted plants, when survival was determined by recovering old larvae and pupae, the presence of *D. v. virgifera* decreased the survivorship of *D. barberi* and caused spatial displacement of the recovered individuals within the root system. Conversely, the presence of *D. barberi* did not affect *D. v. virgifera* survivorship (Piedrahita *et al.,* 1985). When a different worker used different soil and environmental conditions, and adult emergence was the criterion for survivorship, the superiority of *D. v. virgifera* as a competitor was confirmed (Woodson, 1994) but some differences between studies appeared. At higher innoculation levels of both species, *D. barberi* did reduce survivorship of *D. v. virgifera*, specifically the females (Woodson, 1994). Competitive effects depended on both the total density of the two species and on their relative frequency. At least some of the widespread occurrences of *D. barberi* displacement by *D. v. virgifera* can be explained by larval competition.

4. Mating

4.1. Sex pheromones

Sex pheromones have been isolated and identified from females of *D. balteata*, *D. u. howardi,* and *D. v. virgifera;* attractiveness of the optical isomers and analogs of these compounds has been determined for several taxa [Table 1, after Chuman *et al.* (1987) with additions).

4.2. Pheromone structural response and systematics

Pheromone structure follows systematic arrangement; pheromones of species in the *fucata* group are methyl ketones with a methyl branch on the fourth carbon from the hydrocarbon end, while those of the *virgifera* group species are all esters of 8-methyl-2-decanol.

Within the *virgifera* group the responses to the compounds and isomers relate well to phylogenetic placement and putative roles in reproductive isolation (Krysan *et al.,* 1989). Cross attractancy of the subspecies of *D. virgifera* conforms to their conspecific status. The partially sympatric sibling species *D. barberi* and *D. longicornis* are especially well isolated; each is attracted by a different stereoisomer and that attraction is inhibited by the antipode attractive to the sib (Krysan *et al.,* 1986). In *D. barberi* that inhibition has been

Table 1. Sex pheromones of *Diabrotica* (after Chuman *et al.*, 1987 and Krysan *et al.*, 1989).

Taxon	Compound	Stereoisomer response
fucata species group		
D. balteata[1)]	6,12 dimethylpentadecan-2-one	Unknown
D. undecimpunctata howardi [1)]	10methyltridecan-2-one	10R: (AA)[a]
D. u. undecimpunctata [1)]	10methyltridecan-2-one	10R: (AA)
D. u. duodecimnotata Harold [1)]	10methyltridecan-2-one	Unknown
virgifera species group		
D. cristata[1)]	8methyl-2-decanol acetate	2S, 8R: (AA)
D. lemniscata[1)]	8methyl-2-decanol propanoate	2S, 8R: (AA); 2R, 8R: (A)
D. longicornis[1)]	8methyl-2-decanol propanoate	2S, 8R: (AA); 2R, 8R: (II)
D. barberi[1)]	8methyl-2-decanol propanoate	2R, 8R: (AA); 2S, 8R: (II); 2S, 8S: (I)
D. virgifera virgifera[1)]	8methyl-2-decanol propanoate	2R, 8R: (AA); 2S, 8R: (A)
D. v. zeae[1)]	8methyl-2-decanol propanoate	2R, 8R: (AA); 2S, 8R: (A)
D. porracea Harold [1)]	8methyl-2-decanol propanoate	2S, 8R: (AA)
D. mapiriensis gussi Krysan and Smith[2)]	8methyl-2-decanol propanoate	Unknown

[a] AA = highly attractive; A = weakly attractive; II = strongly inhibits attraction by antipode; I = weakly inhibitory.
[1)] Chuman *et al.* (1987).
[2)] Krysan *et al.* (1989).

shown to be concentration, not ratio, dependent (Dobson & Teal, 1987). *D. cristata*, which is broadly sympatric with *D. barberi*, *D. lemniscata*, *D. longicornis* and *D. v. virgifera* (Krysan & Smith, 1987), is attracted by the acetate of 8-methyl-2-decanol while all others are attracted by the propanoate ester. In stark contrast is the striking cross attractancy between the distantly related pair *D. barberi* and *D. v. virgifera* (Krysan *et al.*, 1989), a characteristic that provides a chemical-ecological dimension to decades of observations of interspecific mountings in maize fields (Chiang, 1973; Sammons *et al.*, 1997). This sexual confusion between the species surely stems from the recent sympatry; they were allopatric until the middle of the 20th century (Krysan & Smith, 1987). Aside from the pheromone response, there are functional pre-and postmating reproductive isolation mechanisms between these species (Krysan & Guss, 1978), and given their genetic distance (Krysan *et al.*, 1989) the prospect that hybridization occurs seems remote. That *D. barberi* and *D. v. virgifera* are now sympatric and share a common sex pheromone, but have other barriers to hybridization provides an opportunity for evolutionary change to occur in activities surrounding the sex pheromone response. Study of putative changes would be facilitated by the presence of widely separated areas where only one species occurs; for example, *D. v. virgifera* in Mexico, the southwest and intermountain west, and *D. barberi* in the southeast.

The attraction of both *D. v. virgifera* and *D. longicornis* to some preparations is a problematic issue that review of extant data resolves. Synthetic (2S, 8R) 8-methyl-2-decanol propanoate was highly attractive to *D. longicornis* (Krysan *et al.*, 1986) and weakly attractive to *D. v. virgifera* (Guss *et al.*, 1984) as bait in field traps. However, in a wind tunnel, a sample from the same (2S,8R) preparation used by Guss *et al.* (1984) was unattractive to *D. v. virgifera* (Dobson & Teal, 1987). Analysis of further data yields a plausible explanation. Natural pheromone isolated from a laboratory colony of *D. longicornis* simultaneously attracted males of both *D. longicornis* and *D. v. virgifera* to field traps while a sample of natural *D. v. virgifera* pheromone collected with the same procedure attracted only *D. v.*

virgifera (Krysan *et al.,* 1986). But, traps baited with virgin females attracted only conspecific males (Krysan *et al.*, 1983). Each inconsistency involves possible concentration effects. Based on the concentration of pheromone in the rubber septa, the threshold of attraction in the wind tunnel (Dobson & Teal, 1987) was 10 to 25X higher than that in the field (Guss *et al.*, 1982). Under wind tunnel conditions, the already weak response of *D. v. virgifera* to (2S, 8R) may have been lost entirely. Both the natural pheromone collected from *D. longicornis* and the synthetic (2S, 8R) apparently made a strong enough plume to attract *D. v. virgifera* males in the field. Assuming that the pheromone collected from *D. longicornis* was (2S, 8R), then it can be concluded that *D. v. virgifera* males can be attracted by (2S, 8R) sources, but that the response is probably of no biological significance, given lack of response to virgin female *D. longicornis*. Most importantly, the response of *D. v. virgifera* to the 2S, 8R isomer may reflect lack of stereospecificity in its pheromonal receptors. However, lack of specificity is unlikely to explain the contrasting responses to two different isomers by *D. barberi* and *D.longicornis*. Those responses, falling in the same concentration range, are most surely biologically based and relate to reproductive isolation. Metcalf (1994) recently presented an interpretation of pheromone response in *Diabrotica* from the perspective of evolutionary chemical ecology.

4.3. Diel Pattern in sex pheromone response

Catch of *D. barberi* in pheromone traps in the field was greatest around midnight and was negligible at other times (Bartelt & Chiang, 1977) while most catch of *D. v. virgifera* and *D. v. zeae* occurred after sunrise and before sunset (Bartelt & Chiang, 1977; Krysan *et al.*, 1980). These general patterns prevailed in later studies when field temperatures were warm, but cooler temperatures heavily modulated the pattern of catch (Dobson & Teal, 1986). Particularly, the lower temperatures that typically occur at night shifted *D. barberi* response to early morning and, sometimes, evening as well. Additionally, the evening trap catch of male *D. v. virgifera* was not reflected in the response to pheromones when measured in a wind tunnel (Dobson & Teal, 1987).

4.4. Sex pheromone-releasing behavior

Females assume a characteristic stationary posture, often soon after emergence from the soil, that is properly deemed a calling, that is, sex pheromone-releasing behavior, based on events associated with it. That is, (1) females displaying the posture released a chemical that attracted males in the field, (2) when placed with males, females that had been calling mated more quickly and in greater numbers than did non-calling individuals, and (3) females ceased calling within 24 hours of insemination (Hammack, 1995). The calling posture exposes integument between abdominal segments 7 and 8 bearing tissue that Lew and Ball (1978) concluded were sex pheromone glands, based on histological examination. The attractiveness of individual females to males varied significantly. The diel periodicity of mating in the field (Quiring & Timmins, 1990) was not completely reflected in calling behavior as observed with laboratory populations (Hammack, 1995).

4.5. Post-pheromone mating activity

The mating process in *D. v. virgifera* has been characterized (Lew & Ball, 1979; Hammack, 1995). Females probably mate only once (Hill, 1975) but can mate again during the pre-ovipositonal period, apparently if spermatozoa are inadequate (Branson *et al*., 1977) Males will mate as often as twice in a 24 h period (Quiring & Timmins, 1990). Females will mate soon after emerging from the soil (Ball, 1957), about half begin calling on the day they emerge (Hammack, 1995), and 96.6% of the females observed in copula in the field were teneral (Quiring & Timmins, 1990). (Teneral females remain pale for ≤ 24 h). From 42 to 71.4% of all teneral females collected in 8 field samples were mated; for non-teneral females, from 68.6 to 98.4% in 5 samples were mated. Males begin mating within 2 days of emergence, mating frequency peaked during the second and third weeks, and declined thereafter. In the laboratory *D. u. howardi* paired during the first 3 days after eclosion, mating frequency peaked after 7 days and essentially ceased after 15 days (Tallamy & Halaweish, 1993). In the presence of attractive females, males antennate rapidly, the antennae eventually contacting the apex of the females abdomen; copulation ensues after a courtship of 10 to 60 min (Lew & Ball, 1979). Copulation continues for 3 to 4 h during which the male forms a spermatophore in the bursa copulatrix, a process that takes no less than 1.5 h (Lew & Ball, 1980). Females rarely had spermatozoa after 1 h of copulation, most had spermatozoa after 2 h and essentially all had been inseminated if copulation ended spontaneously (Sherwood & Levine, 1993). Sperm migrate to the spermatheca during the 3 days following copulation, by day 5 the spermatophore begins to degrade, and it is gone by day 7 (Lew & Ball, 1980). In females, the process of copulation causes increases in weight, oviposition, and ovarian development. If spermatozoa are transferred the increases become permanent and females continue to lay eggs. Without spermatozoa, the process is transitory and the increases reverse in one to two weeks (Sherwood & Levine, 1993).

4.6. Cucurbitacins and reproductive activity

Cucurbitacin affects the reproductive activity of *D. u. howardi*. Males and females that were "reproductively active" became less responsive to cucurbitacin as a feeding stimulant after having cucurbitacin available in their diet for six days. Conversely, decline in response to cucurbitacin took 21 and 31 d in unmated male and female beetles, respectively. Furthermore, mated females never exposed to cucurbitacin became increasingly able to detect lower concentrations of cucurbitacin as they aged, a pattern that did not prevail for mated males. Unmated males with prior access to cucurbitacin could detect lower concentrations of cucurbitacin than could such males that had mated (Tallamy & Halaweish, 1993). The observations on all females and on males that had mated are consistent with the hypothesis that beetles seek cucurbitacins until their bodies contain a certain level; a level not reached by virgin males until they were so old that the prospect for mating was nil. Continued sensitivity to cucurbitacins by ovipositing females is consistent with the presence of cucurbitacins on or in eggs (Ferguson and Metcalf, 1985; Brust & Barbercheck, 1992) and with its role in protecting eggs from an entomopathogen (Tallamy *et al.*, 1998); ovipositing females presumably need to replenish supplies. These observations, together with the fact that males pass cucurbitacins to females during copulation (Tallamy & Halaweish, 1993), may well be the prelude to a new dimension in *Diabrotica* reproductive biology and contribute a theoretical basis for cucurbitacin pharmacophagy.

5. Fecundity

Mean egg production per female, as determined in the laboratory under presumably optimal conditions, was: *D. balteata*, 415 (Cuthbert *et al.*, 1968); *D. barberi*, 274 (Naranjo & Sawyer, 1987) and 312 (Boetel & Fuller, 1997); *D.* sp. near *D. tibialis*, 3673 and *D. tibialis*, 1044 (Branson & Krysan, 1987); *D. v. virgifera*, 1023 (Branson & Johnson, 1973) and 794 (Boetel & Fuller, 1997). When temperatures deviated from the optimum (22.5 °C), eggs per female for *D. barberi* fell to 137 at 15 °C and 118 at 30 °C (Naranjo & Sawyer, 1987). Adults of both *D. barberi* and *D. v. virgifera* that eclosed in the field later in the season, had reduced longevity and fecundity when observed under "optimal" conditions in the laboratory (Boetel & Fuller, 1997). Both *D. barberi* and *D. v. virgifera* lay eggs in clutches. At about 25 °C, *D. barberi* laid a mean of 8.7 clutches of 31.5 eggs every 6.4 days (Naranjo & Sawyer, 1987) while *D. v. virgifera* laid a mean of 13 clutches of 79.7 eggs at intervals of 5 d (Hill, 1975). Elliott *et al.* (1990) provided *D. v. virgifera* beetles with maize tissues representative of those available if the beetles had emerged (1) with tassels present but before silking and pollen shed, (2) with plants silking and shedding pollen, and (3) after pollination was complete and silks had turned brown. Those results were compared to (4) the fecundity of beetles provided with a continuous supply of maize pollen, green silks, and leaves. Mean number of eggs per female was 125, 235, 179, and 441 for conditions 1, 2, 3, and 4, respectively. Females fed leaves and immature tassels did not initiate reproduction and if that was the exclusive diet for the first two weeks, reproductive potential was permanently reduced. Similarly, if initial diet was tissue after silk had turned brown, overall fecundity was reduced, even when the optimal diet was supplied later in life. Hence, stage of maize at adult eclosion could affect overall fecundity. Also, larval population can affect fecundity; beetles collected where populations were unusually high tended to have lower fecundity; when maize rows were artificially infested with differing numbers of eggs, rows with high egg numbers had larvae with smaller head capsules and adults with lower fecundity (Branson & Sutter, 1985).

6. Oviposition

Eggs are laid in the soil and the specific site of placement depends on moisture, soil type and structure, odors, surface debris, and the species of plant growing nearby. These factors vary with weather, location, and agronomic practices. For a thorough review of the early literature on egg distribution in the soil, and egg sampling methods, see Ruesink (1986). The species for which most is known are *D. barberi, D. v. virgifera* and *D. u. howardi.*

6.1. Subsurface access

It has been said that these beetle cannot burrow so crevices near plant stems, earthworm burrows, and drought cracks allow access to suitable sites under the soil surface (Kirk, 1979;1981a). Despite this generalization, it is peculiar that most *D. v. virgifera* eggs were laid in moist soil (among maize plants) that had a smooth and firm surface compared with drier soil (among small grain stubble) that appeared to have access points (Gustin, 1979). Furthermore, Gray *et al.* (1992), and Gray and Tollefson (1987), found no obvious extensive or deep drought cracks to account for oviposition. Further study is needed.

6.2. Depth, soil type, moisture and organic residue

The depths at which eggs have been found in soil are quite variable but certain generalizations can be made. Eggs of *D. v. virgifera* are generally found deeper in the soil than are those of *D. barberi* (Weiss *et al.*, 1983, Gray *et al.*, 1992). Under very dry conditions eggs of *D. v. virgifera* and *D. v. zeae* have been found as deep as 40 cm (Branson *et al.*, 1982, Weiss, *et al.*, 1983). Most eggs of *D. barberi* are usually found in the top 10 cm of soil. When eggs of both species were collected in the same soil samples, 66% of *D. barberi* eggs were in the upper 10 cm core while 60 % of the *D. v. virgifera* eggs were in the core 20 to 30 cm deep. Since deeper soil is generally more moist, these species differences in depth of oviposition conform to the origin of *D. virgifera* in the highlands of Mexico (a dry climate) and *D. barberi* in north central United States, a climate where soil is relatively moist year-around (Krysan & Smith, 1987).

In choice tests, *D. u. howardi* preferred to oviposit in moist dark colored soils, favoring soils with organic matter and clay over sandy soils (Brust & House, 1990b). With *D. u. howardi,* Brust (1991) detected three levels of egg numbers dependent on soil conditions. Dry soil bearing little or no organic residue had few eggs, and density depended on soil moisture. An intermediate number of eggs was present with either organic residue or greater soil moisture, while the highest density level occurred when both high soil moisture and organic residue were present. Beetles avoided oviposition in otherwise moist soil if the top cm or two was very dry. *D. barberi* lays more eggs in soil under a residue of corn ears or rye grass than on bare soil (Dominique & Yule, 1984).

6.3. Odors

A laboratory search for a putative ovipositional pheromone was negative but did establish that excised maize roots, elevated levels of carbon dioxide, and the odors from certain bacterial cultures, all elicit egg laying from *D. v. virgifera* (Lance, 1992). Study of sensory mechanisms related to oviposition should be revealing.

6.4. Vegetation

For the next generation to survive, eggs must be laid where larvae can reach a suitable host. Multivoltine species lay eggs in soil next to the plant that just days later will host the larval offspring, but univoltine species lay eggs in soil where a larval host *might* occur the following year. These differences affect oviposition site selection.

6.4.1. Multivoltine species

D. u. howardi females lay eggs near the bases of plants; 91% of the eggs were oviposited within 3 cm of maize plants and less than 1% were oviposited >10 to 15 cm therefrom (Brust & House, 1990b). In relation to weeds, the ovipositon rate changed in the order broadleaved weeds>wheat straw>grassy weeds>bare soil. The beetles fed heavily on the broadleaved weeds but not on the grassy weeds, suggesting in this case the same plant types (broadleaved weeds) provoked both adult feeding and oviposition. However, Barbercheck *et al.* (1995) suggested that *D. u. howardi* oviposits near peanut, a favored larval host, but adults rarely feed on peanut; in this case different plants were chosen for feeding and oviposition.

In Costa Rica, *D. balteata* caused six times more damage to seedling maize plants in ploughed ground compared with 5 different no-till systems. Ploughed soil was judged to be a primary factor in oviposition site selection (Schenk & Saunders, 1984). In the laboratory *D. balteata* beetles laid large numbers of eggs in dishes containing the foliage of sweet corn (Teng *et al.*, 1984) and in field studies Risch (1981) found that *D. balteata* adults flew from plots of mature maize. Furthermore, *D. balteata* beetles fed only corn foliage did not produce eggs (Teng *et al.*, 1984). Since maize roots are an excellent and widely used larval host for *D. balteata* (Melhus *et al.*, 1954; Schalk, 1986), it follows that *D. balteata* adults make different plant choices for oviposition, on the one hand, and feeding on the other.

6.4.2. Univoltine species

Egg placement with regard to vegetation is less precise for these species. Eggs are placed in the general area or field of choice but not necessarily immediately next to the plants in place (Weiss *et al.*, 1983). Univoltine species such as *D. barberi* and *D. virgifera* s. l., are essentially monophagous on maize, and oviposit in soil during late summer or autumn but the eggs do not hatch until the following spring. Furthermore, maize is an annual whose spatial location from year to year can change drastically due to crop rotation, a control strategy for *Diabrotica* larvae based on these traits. The simplest rotation is annual; a field is planted to maize on alternate years and to a non-host in the intervening years. With this strategy maize and the alternate crop(s) are in nearby fields each year. Larval infestations of maize roots in crops under annual rotation provoked studies on the effect of vegetation on *Diabrotica* oviposition.

6.4.2.1. *D. barberi*

This section requires a brief aside on feeding behavior of adult *D. barberi*. They eat succulent reproductive tissues of maize, but the ovipositional period exceeds the time when those tissues are available. Then the beetles feed on pollen of alternate plant species (Cinereski & Chiang, 1968), frequently available only outside the maize habitat. In the field, *D. barberi* were commonly seen feeding on pollen in fields that would contain maize the following year. Oviposition by such females was judged to be the cause of damage to maize under annual rotation (for example, Hill & Mayo, 1980; Branson & Krysan, 1981; Gustin, 1984). This judgement was made despite the demonstration by Cinereski and Chiang (1968) that reproductively active *D. barberi* females collected in maize fields had clearly fed on pollen outside of those maize fields. This issue has generated confusion so a detailed review is merited; the conclusion is that *D. barberi* oviposits primarily among maize plants and occasionally elsewhere. Improved methods of egg sampling helped to clarify this issue.

Cinereski and Chiang (1968) related the pollen content of beetle guts to putative locations of pollen sources to track the movement of *D. barberi* (as *longicornis*) among maize and non-maize fields. Guts from some reproductively active beetles found in maize fields contained pollen only available outside that maize field. When maize pollen was readily available, beetles were rarely seen outside maize fields. When food quality of maize declined, beetles foraged elsewhere but few or no eggs were found in the fields where the beetles could have found the alternate pollen sources. When caged beetles (cages 2 by 4 meters) could oviposit in either of two plot types, Gustin (1984) found far more (83 to 88%) of the eggs in the plot of maize than in the plot of small grain stubble. In one year of the study, only the plants in the stubble were pollinating when beetles were in the cages, so the beetles fed among the stubble but oviposited mostly among the corn plants.

Siegfried and Mullin (1990) compared oviposition in plots planted to maize, sunflower, and squash by counting trapped emergent adults the following year when all plots were planted to maize. With linear plots only 3m wide, sunflower and squash were always near maize, and few differences were seen in numbers of emergent beetles. But, when the configuration was changed so emergence was monitored from the center of host plant blocks that were 81-m^2, significantly fewer beetles emerged from the plots that bore sunflower or squash during the ovipositional period.

Boetel *et al.* (1992) quantified the occurrence of both egg (using two sampling procedures) and adult (with sticky traps) stages of *D. barberi* in field plots of maize, foxtail, lambsquarters, pigweed, oat stubble, smartweed and soybean at three different locations. The numbers of beetles trapped in alternate plots often approached the numbers trapped in maize. Despite that distribution, highly significantly more eggs were found in maize plots compared with all other plots at all locations in both years of the study. Eggs in soil samples from plots of any one type of weed seldom exceeded 5% of the total in maize, although it reached 16% in one location.

Lance *et al.* (1989) and Naranjo (1991) addressed this issue by monitoring flight activity and ovarian stages of *D. barberi* in directional malaise traps placed at, or near the edges of maize (=source) fields. Few beetles were captured, compared with the numbers present, until the quality of food in the source fields declined. Then, the ovaries of beetles captured in traps facing the source field tended to have fewer eggs than those captured on the side facing away from the source field. This is consistent with beetle flights out of maize to feed, and back to maize for oviposition.

The pattern of cyclical movement of *D. barberi* out of maize (for feeding) and back to maize (for oviposition) as perceived by Cinereski and Chiang (1968) is upheld by the subsequent studies. The statistically greatest egg fraction is deposited in maize fields. But, those eggs that are laid in fields near plants other than maize may be consequential for species survival via the bet-hedging strategy.

6.4.2.2. *D. v. virgifera*

For many years all evidence indicated that eggs of *D. v. virgifera* were deposited in maize fields (Shaw *et al.*, 1978).

Study of oviposition by *D. v. virgifera* was provoked when damage to maize roots by D. *v. virgifera* occurred in a local area in east central Illinois in some fields with annual rotation of maize and soybeans. After several years of intense yet negative searches (Levine & Oloumi-Sadeghi, 1991;1996; Levine & Gray, 1996), eggs began to be found in soybeans; the numbers have increased with each passing year, and the geographic area affected has expanded (Levine & Gray, 1996; Sammons *et al.*, 1997; Spencer *et al.*, 1998; Onstad *et al.*, in press). Clearly, a population of *D. v. virgifera* has adapted to annual rotation of maize and soybean.

6.4.2.3. *D. v. zeae*

This subspecies occurs in Mexico, Texas and Oklahoma. It was effectively managed on maize in Texas by annual rotation for about a decade, but beginning in 1992 root damage caused by *D. v. zeae* has occurred in fields where maize followed sorghum (Stewart *et al.*, 1995). The hatching pattern of eggs from the area ruled out prolonged diapause as a causal factor, but eggs were found in soil samples from sorghum fields (Lingren, 1998). There were 58% as many eggs in the sorghum field compared with the maize field. The number of beetles captured in emergence cages over sorghum was trivial compared with those captured over maize so egg presence is not due to individuals completing their life cycle on sorghum. The reasonable conclusion from Lingren's work is that *D. v. zeae* that emerge from maize under annual rotation are progeny of females that oviposited in sorghum the preceding summer. Both subspecies of *D. virgifera* have adapted to annual rotation by oviposition near a non-host.

7. Flight and dispersal

For *Diabrotica*, dispersal by flight relates to feeding and oviposition and differs among species. Flight of *Diabrotica* in the field has been detected by finding beetles or their progeny in specific field locations where flight is the unquestioned mode of movement, and characterized by catch in traps monitored for height, temporal periodicity, flight direction, and reproductive status. Additionally, flight has been characterized in the laboratory with actographs, wind tunnels, and flight mills. Some flights related to oviposition might be thought of as "relocation" flights in the sense that the ecologically important outcome is placing the progeny in a different space, near or far.

7.1. Duration of flight

On flight mills, the duration of individual flights by *D. v. virgifera* females fell into two clearly disjunct modes (Coats *et al.,* 1986). The striking bimodality in flight duration was confirmed for females and extended to males by Naranjo (1990), who observed tethered flights with actographs. Flights of <20 (Naranjo, 1990) or <30 min (Coats *et al.*, 1986) are called "trivial" and those lasting longer are called "sustained". For comparing these data the 10 min difference in definitions is irrelevant (Naranjo, 1990). Sustained flights occurred in <15 (Coats *et al.*, 1986) or <24% (Naranjo, 1990) of the females tested and declined in incidence with age (Naranjo, 1990). Most flights by both *D. v. virgifera* (Coats et al., 1986, Naranjo, 1990) and *D. barberi* (Naranjo, 1990) are trivial, lasting 1 min or less. In response to maize silk in a flight tunnel most flights were only 10 to 20 cm in distance with stops apparently for reorientation (Prystupa *et al.*, 1988).

7.2. Onset and time of day

In a flight tunnel lacking known olfactory stimuli, *D. v. virgifera* were most inclined to initiate flight in still air and the tendency to fly decreased with increasing wind speed until no flights were initiated at 4.0 m/sec. In wind speeds of 0.0 and 0.5 m/sec flights were instantaneous and most were vertical, but as air speed increased pre-flight "preceptory" behaviors occurred and most flights were downwind; upwind flights were uncommon and never occurred at wind speeds of ≥1.5 m/sec (VanWoerkom *et al.*, 1983). Field observations fit this pattern. When malaise traps were monitored daily, few beetles were caught with higher wind speeds while most were caught when winds were moderate to low and temperatures were warm (Spencer *et al.*, 1998). Similarly, beetles tended to move up and out of the plant canopy when wind speeds and turbulence were low (Grant & Seevers, 1990). When a flight tunnel (air velocity 0.33 m/sec) had fresh maize ears or silks in the air stream, putative preceptory behaviors occurred before onset and the majority of flights were upwind (Prystupa *et al.*, 1988). Results with *D. barberi* in the latter study were similar to those for *D. v. virgifera.*

When tethered, most trivial flights for *D. v. virgifera* and *D. barberi* occurred during daylight hours with peaks in the morning, late afternoon and evening (Naranjo, 1990). On flight mills trivial flights occurred at any time of day with *D. v. virgifera* females (Coats *et al.*, 1986). Sustained flight occurred most often after sunrise and before sunset (Coats *et al.*, 1986; Naranjo, 1990). Trap catch of *D. v. virgifera* in the field revealed a diel pattern of flight. Most of the catch occurs in the morning and before sunset (Witkowski *et al.*, 1975; Grant & Seevers, 1990; Spencer *et al.*, 1998) including catch of males in traps baited with female sex pheromone (Bartelt & Chiang, 1977; Krysan *et al.,* 1980). Similar patterns were seen when locomotor activity was measured in the laboratory (VanWoerkom *et al.*, 1980). Optimum temperature for *D. v. virgifera* locomotion in the laboratory was 20 to 35 °C (VanWoerkom *et al.*, 1980), but flight out of the maize canopy occurred when the canopy temperature was between 18 and 26 °C (Grant & Seevers, 1990). However, the diel pattern of flight is modulated by cool temperatures (Dobson & Teal, 1986).

7.3. Species and sex differences in flight

Sustained flights occurred with *D. v. virgifera* males and females but with neither sex of *D. barberi.* Median trivial flight durations for *D. v. virgifera* were three and six times longer than those for *D. barberi* males and females, respectively. Females of *D. barberi* took more flights than did those of *D. v. virgifera,* but flight frequencies for males were similar. In 23 h assay periods *D. v. virgifera* logged a total of 11-12 minutes of trivial flight time whereas *D. barberi* flew some 3-4 minutes (Naranjo, 1990). In maize fields sticky traps tend to capture more male beetles below the plant canopy and more females above (Witkowski *et al.*, 1975; VanWoerkom *et al.*, 1983). Consistent with this is the finding that malaise traps capture more male than female beetles when trapping ≥ 1 m above ground, but traps capturing beetles flying vertically above the canopy caught more females, particularly with *D. v. virgifera* (Naranjo, 1991). Species and sex differences in flight are also related to duration (7.1) and age (paragraph 7.4.).

7.4. Age and reproductive status

On flight mills only preovipositional beetles had sustained flights and most occurred at age 5-6 days (Coats *et al.*, 1986). By contrast, using an actograph, more beetles, and a wider age span, Naranjo (1990) observed sustained flights by gravid females as old as 30 days, albeit with a drastic decline in frequency with age. Fewer males undertook sustained flight, but the frequency did not decline with age. Sustained flights occurred only once or twice during any 23 h period and individual flights lasted from 30 to 230 min. Age had no effect on trivial flight frequency for either *D. v. virgifera* or *D. barberi*. Coats *et al.* (1987) interpreted sustained flight in *D. v. virgifera* to reflect the "oogenesis-flight syndrome" common among insects. However, since some gravid females as well as males undergo sustained flight (Naranjo, 1990), the "oogenesis-flight syndrome" can only explain sustained flight by preovipositional females.

7.5. Short-range flight

7.5.1. Univoltine species

In large areas of north central U.S. cropland is a patchwork pattern of large, adjacent fields of maize, soybean, and sometimes small grain. *D. v. virgifera* and *D. barberi* differ markedly in the way flights between the different crops in this patchwork relate to feeding and oviposition behaviors. First consider adult feeding differences. Briefly, *D. v. virgifera* has been observed for many years to feed preferably on succulent reproductive tissues of maize but will feed on other maize tissues as well (Cinereski & Chiang, 1968; Ludwig & Hill, 1975; Elliott *et al.*, 1990). On maize, *D. barberi* prefers to feed on succulent reproductive tissues (Lance & Fisher, 1987 and references) but makes feeding sorties to adjacent habitats to satisfy food needs when maize tissues are not available. Movement patterns reflect these feeding tendencies. *D. v. virgifera* beetles tend to leave maize plantings when silks dry (Lance *et al.*, 1989; Naranjo, 1991) and appear in plantings where silks are fresh and maize is actively shedding pollen (Witkowski & Owens, 1979; Hill & Mayo, 1974). When flight activity at maize field borders was measured with sticky traps (Naranjo & Sawyer, 1988) or directional malaise traps (Lance *et al.*, 1989; Naranjo, 1991), capture was lowest when maize was at peak flower for both *D. barberi* and *D. v. virgifera*. Furthermore, beetles tend to leave maize plots where flowering is completed. However, the effect was far less pronounced for *D. v. virgifera* and those *D. barberi* females with developed ovaries. Additionally, some *D. barberi* left maize plots *before* flowering (Naranjo, 1991); those flights probably were feeding sorties. Those *D. v. virgifera* that leave maize early in the season (Witkowski & Owens, 1979; Naranjo, 1991) may reflect sustained flight (Coats *et al.*, 1986; Naranjo, 1990). Dispersal flights of *D. barberi* from maize are primarily to acquire food for continued egg production (Cinereski & Chiang, 1968; Lance *et al.*, 1989; Naranjo. 1991). See also paragraph 6.4. on oviposition as related to vegetation.

7.5.2. Regional differences in *D. v. virgifera* flight behavior

The view of flight outlined in the preceding paragraph no longer applies to a population of *D. v. virgifera* in eastern Illinois and northwestern Indiana, where a dramatic change in flight behavior has occurred. Using a variety of visual observations and trapping techniques, Spencer *et al.* (1997; 1998) showed that adults of *D. v. virgifera* fly between fields of maize and soybean. The beetles oviposit in soybean fields and then feed on soybean plants before flying again to maize. The propensity of beetles from this area to feed on soybean has been established (Sammons *et al.*, 1997). The incidence of beetles in soybean was highest near the border with maize suggesting that movement is largely between adjacent fields. Feeding on soybean appears to be incidental to oviposition because a diet of soybean foliage does not subtend reproduction (Spencer *et al.*, 1997); perhaps feeding on soybean replenishes water or energy for the return flight to maize.

7.6. *Multivoltine species*

The movement of *D. balteata* among plots with different host combinations varied in relation to feeding preferences in tests in Costa Rica (Risch, 1981). Briefly, when maize and squash leaves are <21 days old, *D. balteata* prefers maize as food, but as the plants age the preference reverses. Common bean plants were equally palatable at all ages. Thus, with older plants the preference was bean> squash > maize. Throughout the season the rates of emigration from bean were always lower than those from squash and maize; with plant age, emigration rates from squash decreased and those from maize increased. Early in the season there were no differences in emigration from bean monocultures compared with that from polycultures including bean; as plants aged, beetles left polycultures including maize and the effect was attributed largely to shade produced by the maize plant canopy. Measures of the amount of flight activity within plots showed that beetles flew about more in maize monocultures than in any of the following; bean or squash monocultures, maize bean diculture, or the triculture of maize, bean and squash. Movement in the maize-squash diculture exceeded that in all but the maize monoculture. Frequency of movement was greater in plots with non-hosts and they spent less time resting on non-hosts (Risch, 1981). Similarly, in a maize/cucurbit diculture in Mexico *D. balteata* was abundant early in the season and declined later; the beetles also preferred bean to non-bitter squash and maize foliage (Eben & Barbercheck, 1996). In field plots, flights by *D. u. howardi* occurred at a rate of 5.01 ± 0.63 times per hour per beetle and the mean distance moved was 2.24 ± 0.14 m (Lawrence & Bach, 1989). The generalist *D. u. howardi* flew seven times more frequently than the specialist *Acalymma vittatum* (Fabricius) which was interpreted to be a strategy allowing a generalist to adapt rapidly to temporally and spatially changing resource availability. Across all of the studies it is clear that flight patterns increase the chances of finding and ovipositing near hosts suitable for the larval stage.

7.7. *Long distance flights*

7.7.1. *D. v. virgifera*

Sustained flight seen with laboratory devices reveals a capacity for long distance flight that can account for the unquestioned eastward range expansion of *D. v. virgifera*. A rate of movement of about 62 km per generation can be inferred from the eastward extension of the range from the Nebraska-Iowa border in 1955 (Chiang, 1973) to New Jersey in 1983 (Krysan & Smith, 1987a). Interestingly, on mills sustained fliers covered a mean of 35 km per day (Coats *et al.*, 1986); if that reflects real distance, two such flights on windless days would account for the calculated distance covered each generation in the eastward range expansion.

Long range movement is further documented by washup on beaches of Lake Michigan as related to the passage of weather fronts, a pattern of appearance that has been interpreted to reflect both local and long range movements of *D. v. virgifera* (Grant & Seevers, 1989). The results are consistent with features of sustained flight as characterized by Coats *et al.*

(1987) and Naranjo (1990). While the data can be interpreted to mean that beetles were passively carried by storms, it is notable that descent from air systems was accounted for by active flight. Further, flight occurs mostly under calm conditions (paragraph 7.2). Whether *D. v. virgifera* disperses long distances by wind-assisted active flight or by active flight alone is not known, but active flight could account for the documented range expansion. Additionally, both sustained flights (Coats *et al.*, 1986) and upward flights from the canopy (Grant & Seevers, 1990) are common in the dawn hours, so long distance flights may actually occur during daylight hours.

Sustained flight by *D. v. virgifera* has been termed migratory but the adaptive value is not obvious in the corn belt of the United States, where maize is ubiquitous and ever present (Naranjo, 1990). However, in the highlands of Mexico, where the association with maize originated (Branson & Krysan, 1981), annual long distance migration by a portion of the population may adapt the species to intense drought conditions that can stifle maize production over large areas in some years. For example, the abandonment of major communities by the Anasazi in New Mexico some 700 years ago has been associated with an extended drought that eliminated maize production (Terrell, 1974). The extreme drought in the 1930s in the Great Plains of the United States reduced or eliminated *D. v. virgifera* as a pest of maize in Nebraska (Tate & Bare, 1946).

7.7.2. *D. barberi*

There is no clear evidence of a rapid range expansion by *D. barberi* (but see Webster, 1895). For *D. barberi,* lack of sustained flight in the laboratory and the existence of geographic variation (Krysan 1993; McDonald *et al.*, 1985; but see Krafsur *et al.*, 1993) suggests a lack of long-range dispersal ability comparable to that of *D. v. virgifera.*

7.7.3. *D. u. howardi*

For *D. u. howardi*, the existence of sustained flight is inferred from feeble cold-hardiness and the known range. *D. u. howardi* occurs throughout the year in northern Mexico and the southern United States from the Carolinas, west to Arizona. Each summer *D. u. howardi* occurs in large numbers as far as about 1200 km north of where it can reasonably survive winter (Krysan, 1986; Elsey, 1989). Northward migration unquestionably occurs annually; early studies on the migration are reported in Smith and Allen (1932). Some of their observations are difficult to interpret but particularly intriguing is their suggestion that a *southward* migration occurs in autumn. In Louisiana they observed high numbers of beetles in spring and autumn but few through the summer. They collected *D. u. howardi* beetles over Louisiana by airplane in spring and autumn at altitudes of up to 1000 feet. In Tennessee in October, "southerly moving" beetles were sighted atop a mountain and a 19 story building (Stanley, 1932). More recently, southerly flying beetles were observed in October to stop and imbibe condensed surface water on an oil platform in the Gulf of Mexico some 160 km SSE of Galveston, Texas (Baust *et al.*, 1981). Clearly, *D. u. howardi* undertakes long range flights in autumn but it is unlikely the oil platform sightings reflect a normal migration path because *D. u. howardi* is unknown from Yucatan or Central and South America. Perhaps these beetles overshot the wintering area along the gulf coast of the United States and in Mexico. Migration north in the spring and south in the autumn, as perceived by Smith and Allen, matches the seasonality of *D. u. howardi* in Charleston, SC; populations were high in spring and autumn and quite low during the summer months (Elsey, 1989). These observations denote hypotheses that merit testing.

8. Dormancy and overwintering

Knowledge of dormancy mechanisms applies to North American populations. Species differ in seasonal adaptation from *D. balteata*, which is multivoltine and is unable to be dormant, to *D. barberi* which can be dormant in the egg stage for up to four years. Cox (1994) recently reviewed diapause in *Diabrotica*; this review will place new findings in context and relate dormancy mechanisms to flight and oviposition behavior as a complex of adaptations to spatial and temporal changes in resources.

8. 1. Overwintering in multivoltine species

The polyphagous species *D. u. howardi* and *D. balteata* are multivoltine (Elsey, 1988a; 1988b; Rodriguez-del-Bosque & Magallanes-Estala, 1994). Neither species survives more than occasional periods of sub-freezing temperatures, but *D. u. howardi* survives better in such periods (Elsey, 1989). Charleston, S. C. appears to be the northernmost locality for *D. balteata,* while *D. u. howardi* overwinters as far north as North Carolina (Meinke & Gould, 1987) and perhaps Virginia and Tennessee (Isley, 1932).

8.1.1. *D. balteata*

Trap captures of adults in northern Mexico in every month of most (of 14) years (Rodriguez-del-Bosque & Magallanes-Estala, 1994) indicates multivoltinism and the species is multivoltine in Costa Rica (Risch, 1980) and South Carolina (Elsey, 1989). But in northern Mexico, trap captures were greatly reduced in spring following cold winters (extended periods at <0 °C) compared with normal winters (Rodriguez-del-Bosque & Magallanes-Estala, 1994). Biological (Saba, 1970; Elsey, 1989) and ecological (Elsey, 1988b) evidence shows that *D. balteata* lacks specific adaptation for a cold season. Therefore, passive quiescence in the adult stage appears to be the only means of survival when the temperature falls below thresholds for development and maintenance activities. The inability to survive subfreezing temperatures is reasonable given its range, that is, Venezuela and Columbia to California and South Carolina (Krysan, 1986). Furthermore, until early in the 20th Century, the northern limit of the range of *D. balteata* was the southernmost tip of Texas (Smith, 1966). The range expansion has been attributed to agricultural practices. Simply put, *D. balteata* is a tropical species whose northern range is limited by any but transitory periods of sub-freezing temperatures.

8.1.2. *D. u. howardi*

This taxon has reproductive diapause, photoperiodically induced in the adult stage, that halts egg production in those females eclosing in late fall or early winter; but this dormancy scheme is not rigid (Elsey, 1988a). Diapause duration is only about 20 to 60 days at

20 °C. At a given photoperiod, lower temperatures increase the incidence of diapause, suggesting that warm weather can delay diapause induction. The critical photoperiod is about LD 13:11. Adult feeding and locomotor activity continue normally during the overwintering phase (Meinke and Gould, 1987; Elsey, 1988a) and the females use basking behavior to elevate body temperature on cool, sunny days (Meinke and Gould, 1987). This non-rigid diapause scheme appears to well-adapt the subspecies to its wintertime range. The inability of the species to survive extended periods of sub-freezing temperatures limits the range to southern United States. There is no explanation why *D. u. howardi* occurs no further south than central Mexico. In wintertime it occupies the southern part of the temperate zone and the northern part of the tropical zone. The occurrence each summer in central and northern United States and Canada (Krysan, 1986) results not from overwintering, but from annual migration (paragraph 7.7.3). It is so reliably present each summer in northern United States that research projects depend on it without concern for its migration (Lawrence and Bach, 1989).

In the overwintering zone, where winter temperatures are fickle, opportunities to reproduce versus the risks of offspring mortality are balanced by at least four biological features; (1) brief photoperiodically induced diapause, (2) delayed diapause if the environment is warm when the short photoperiods of autumn ensue, (3) normal feeding and locomotion during diapause, and (4) basking behavior. A rigid, lengthy, diapause would reduce opportunities to reproduce in those years that beetles are located where autumn and/or winter and spring are warm. Similarly, diapause termination before the end of the cold of winter adapts for the variable onset of spring warmth, especially when combined with basking behavior and the ability to oviposit near frost-hardy larval hosts (Elsey, 1988b). Basking behavior, which often elevated body temperature to enhance feeding, mating, and preovipositional development (Meinke & Gould, 1987), is itself evidence that *D. u. howardi* is specifically adapted to exploit habitats characterized by periods of marginal warmth. *D. u. howardi* has two additional known adaptations to this climatic condition. Though not as cold hardy as "typical" temperate species, *D. u. howardi* beetles are more cold hardy than the clearly tropical relative, *D. balteata* (Elsey, 1989). Also, reproductive diapause does not inhibit mating; females mate *before* overwintering (Sweetman, 1926) and hence accommodate for poor winter survival of males (Elsey, 1989). Adaptation to this climatic zone provides *D. u. howardi* with an overwintering base from which first generation adults take part in the annual spring migration northward. A study of *D. u. howardi* flight with the methodology of Coats *et al.* (1987) or Naranjo (1990), combined with field studies of migration, would be illuminating.

8. 2. Overwintering in univoltine species

The seasonal cycle of *D. virgifera* s.l. that has been observed as prevalent through much of the 20th Century represents the general pattern for the *virgifera* group and may represent the phylogenetically primitive condition for this species group (Krysan, 1982; Krysan *et al.*, 1989). However, extrapolation to species in Mexico, other than *D. virgifera* s. l. may be inappropriate given that Eben and Barbercheck (1996) collected adult *D. viridula* (Fabricius) and *D. porracea* Harold in several regions of Veracruz, Mexico throughout the year. *D. virgifera* s. l. has one generation per year and passes the cold (temperate) or dry (tropical) season as the egg stage in soil. Survival through the dry season (Branson *et al.*, 1982) or cold season (temperature < the threshold of development, Chiang, 1973) is accommodated by diapause, then quiescence. At oviposition, soil temperature and moisture suffice for development, but diapause halts embryogenesis at a specific stage (Krysan, 1972). Subsequently, a decline in soil moisture (Krysan, 1978; Branson *et al.*, 1982), or soil temperature (Gustin, 1981) imposes quiescence. In the spring either soil warmth (temperate climate) or soil moisture (wet-dry tropics) ends quiescence in the eggs and concurrently promotes development of maize roots. With many temperate species diapause development occurs only under chill conditions (Tauber *et al.*, 1986), but that does not apply for *D. v. virgifera* (Chiang, 1974; Krysan, 1978) or *D. v. zeae* (Krysan & Branson, 1977). This physiological trait is consistent with the recent occupation of the temperate climate by *D. virgifera* s. l. (Krysan and Smith, 1987a).

Diapause in *D. barberi* and *D. longicornis* differs from that of *D. v. virgifera* in that most eggs in any population require exposure to chill for termination of diapause (Chiang *et al.,* 1972; Krysan, 1982; Golden & Meinke, 1991; Fisher *et al.*, 1994). The thermal optimum for diapause termination in *D. barberi* falls between 8 and 12 °C (Fisher *et al.*, 1994). The need for chill to terminate diapause in *D. barberi* and *D. longicornis* surely reflects their longer occupancy of the temperate climate compared with that of *D. v. virgifera* (Krysan *et al.*, 1989). Two other temperate, univoltine species in the *virgifera* group also diapause as eggs; *D. cristata* (Harris) and *D. lemniscata* LeConte (Krysan, 1982).

8. 3. Diapause in semivoltinism and beyond

Prolonged diapause yielding eggs that remain dormant for one additional year (=semivoltinism) occurs in some *D. barberi* (Chiang, 1965; Krysan *et al.*, 1984; 1986; Levine, Oloumi-Sadeghi & Fisher, 1992; Landis *et al.*, 1992), *D. v. virgifera* (Levine, Oloumi-Sadeghi & Ellis, 1992) and possibly *D. longicornis* (Golden & Meinke, 1991). For *D. barberi* some eggs can remain dormant for three or four years before hatching (Levine, Oloumi-Sadeghi & Fisher, 1992).

8.3.1. Prolonged diapause in *D. v. virgifera*

With *D. v. virgifera* prolonged diapause was detected in two populations, one from Illinois and another from Ontario, Canada, but the percentage with prolonged diapause was extremely low; 0.14 and 0.21%, respectively (Levine, Oloumi-Sadeghi & Ellis, 1992). The recent discovery that *D. v. virgifera* has adapted to annual rotation by a means other than prolonged diapause requires new analysis of the data related to prolonged diapause for this species. Krysan *et al.* (1986) argued that it would be highly unlikely for prolonged diapause to appear in *D. v. virgifera.* The argument was based on (1) the athermal nature of diapause termination and (2) the pattern of geographic variation in diapause duration. First, the only requirement for the termination of egg diapause is the passage of time. Secondly, as latitude becomes more northward, the duration of diapause shortens (Krysan, 1982). Shorter diapause at more northern latitudes can satisfy the adaptive role of diapause which is to maintain dormancy until the environment falls below the thermal threshold of development. In north central U.S. the more northerly the location the shorter is the time between oviposition and the onset of low temperatures that enforce quiescence. Most importantly for this argument, diapause duration shortened promptly as the species expanded its range into those northern areas (Krysan, 1982). Rapid appearance of shorter diapause implies that longer diapause carried genetic load so Krysan *et al.* (1986) speculated that it was unlikely that individuals would ever appear with diapause so long that dormancy would be maintained through an entire additional warm season. They erred by not conceiving a discontinuous pattern of diapause duration. *D. v. virgifera* eggs with prolonged diapause have a distinct bimodal hatch pattern (Levine, Oloumi-Sadeghi & Ellis, 1992). Eggs ceased hatching in June of the first year and did not begin hatching again until May of the following year. Therefore, prolonged diapause in *D. v. virgifera* is not simply the result of selection for longer diapause duration. If that were true, hatch would have continued through the first summer. The critical point is that eggs with prolonged diapause have not become prevalent under annual rotation, a condition that would obviously select for it (Steffey *et al.*, 1992; Levine & Oloumi-Sadeghi, 1996). Instead, that selection pattern has led to a change in the flight, oviposition, and feeding behaviors of the species. The undefined genetic load that caused rapid selection for short diapause (Krysan, 1982) may account for the apparent poor fitness of *D. v. virgifera* eggs with prolonged diapause. Indeed, an extensive survey in Illinois for prolonged diapause as a cause of rootworm damage in fields where maize and soybeans are planted in alternate seasons revealed that what damage occurred was most reasonably ascribable to *D. barberi* (Steffey *et al.*, 1992).

8.3.2. Prolonged diapause in *D. barberi*

Prolonged diapause in *D. barberi* was first detected in Minnesota (Chiang, 1965), but the incidence was low (0.3% of the eggs) and interest waned. Subsequently, Krysan *et al.* (1984) saw prolonged diapause in some 40% of the eggs in a South Dakota population and interest waxed. The trait has been detected or reasonably inferred in populations from Minnesota and Iowa (Krysan *et al.*, 1986; Ostlie, 1987), Illinois and North Dakota (Levine, Oloumi-Sadeghi & Fisher, 1992), and Michigan (Landis *et al.*, 1992). Furthermore, eggs have been found to diapause for up to 4 years in some populations (Levine, Oloumi-Sadeghi & Fisher, 1992). The proportion of eggs with prolonged diapause tends to reflect local crop sequences; where maize is generally planted each year on the same land, the fewest eggs have prolonged diapause (Krysan *et al.*, 1986; Ostlie, 1987; Levine, Oloumi-Sadeghi & Fisher, 1992). In an area with a history of annual rotation, 40 to 50% of the eggs had prolonged diapause (Krysan *et al.*, 1986; Levine, Oloumi-Sadeghi & Fisher, 1992). Eggs from a locality with a history of variable rotations (maize every second, third or fourth year) had hatches of 30, 13 and 13 % after one, two and three years, respectively (Levine, Oloumi-Sadeghi & Fisher, 1992).

The optimum temperature for diapause development of *D. barberi* falls between 8 and 12 °C and little diapause development occurs at temperatures <3 and >15 °C (Fisher *et al.*, 1994). An interesting question is whether diapause duration is continuously variable or terminates bi-, tri-, or tetramodally to account for one and two, three, or four years of diapause, respectively. While no experiment has been done to determine that, the detailed data in Tables 2 and 3 of Fisher *et al.* (1994) were interpreted to show that diapause duration varies continually, based on the percentages of hatch seen after chill periods ranging from 15 to 240 days. However, those data show that the percentage of hatch increased rapidly as days of chill increased from 60 to 105 and 120 days; after that, the percentage of hatch changed little, if any. There is no trend in the data to allow a prediction of the pattern of hatch that might have occurred had the experiments continued. The hatch pattern suggests some unknown event occurs to allow another round of hatch. Another issue is whether a distinct biotype with prolonged diapause is involved. A biotype is unlikely because Krafsur (1995) found virtually unrestricted gene flow between univoltine and semivoltine populations in the field, and concluded that semivoltinism will be favored where large areas practice annual rotation and will decline rapidly with two year rotations and slowly where maize is planted yearly on the same ground.

There is yet another important dimension in the prolonged diapause scheme of *D. barberi*. Eggs with prolonged diapause occur most commonly where the host is regularly planted every other year. This is compelling evidence that diapause in *D. barberi* is primarily and directly adaptive for regular temporal patterns in availability of the host plant, not for seasonal changes in temperature.

9. Dormancy, flight and oviposition related to host type and location

Host range and voltinism are closely related among the Nearctic *Diabrotica* species. Flight and ovipositional behavior patterns of the multivoltine species, which are polyphagous as larvae, favor exploitation of a medley of plants through the seasons of the year. Adults recognize some plants for oviposition and others for feeding. A migratory scheme allows a generation of *D. u. howardi* to be completed far north of its overwintering range.

Those univoltine species for which knowledge exists (*D. barberi*, *D. v. virgifera* and *D. v. zeae*) are essentially confined to roots of maize as larvae and are committed by obligatory egg diapause to no more than one generation per year. The system is simplest when maize is planted each year in the same fields; oviposition in maize fields bears little risk. The period of oviposition for *D. barberi* is longer than the period for which maize provides a diet suitable for reproduction, so females fly out of maize to forage but fly again to maize where most oviposition occurs. With corn planted continuously, *D. v. virgifera* usually dominates *D. barberi* (Hill & Mayo, 1980), at least in part because larvae have a competitive advantage (Piedrahita *et al.*, 1985; Woodson, 1993; 1994).

In arenas of intense and widespread annual rotation of host and non-host crops, *D. virgifera* and *D. barberi* have responded differently; with *D. barberi* the key response has been oviposition in maize coupled with extended diapause. With *D. virgifera* the response has been a single winter of dormancy coupled with oviposition in the alternate crop. These differences denote opposite flight-oviposition patterns between the adjacent host and non-host habitats. Female *D. v. virgifera* fly from maize to soybean to oviposit and then do some feeding; female *D. v. zeae* fly from maize to sorghum to oviposit. Conversely, female *D. barberi* fly away from maize to feed and lay a few eggs but fly again to maize where most oviposition occurs. The sequence and outcomes of the flight, oviposition and feeding steps are distinct between *D. barberi,* on one hand, and *D. virgifera,* on the other hand. Sammons *et al.* (1997) speculated that feeding on, and ovipositing in, soybeans by *D. v. virgifera* was acquired by hybridization with *D. barberi*. Hybridization seems an unlikely mechanism given these fundamental differences in behavior patterns, the existence of reproductive isolation mechanisms (Krysan & Guss, 1978), and the genetic distance between the taxa (Krysan *et al.*, 1989). Furthermore, *D. v. zeae* in Texas, a population geographically far removed from both Illinois/Indiana and *D. barberi*, has acquired a behavioral pattern analogous to that in *D. v. virgifera*.

10. Natural Enemies of *Diabrotica*

Given the economic importance of some *Diabrotica* species, many natural enemies have been noted, but their relationships to the host are sketchily known and useful biological control agents have yet to be recognized for any pest species. For each natural enemy/ species combination only key references are included in the Tables; reviews are used where appropriate.

10.1. Vertebrate Predators

Webster (1913) found *D. u. howardi* in the stomach contents of 24 species of birds, but few quantitative data were presented. Quantitative data are available for the redwing blackbird (*Agelaius phoeniceus*) and the big brown bat (*Eptesieus fuscus*). Redwing blackbirds visit maize fields where *D. barberi* populations are high and at times the beetles represented 50% of the gut contents of the birds. But when maize reached the milk stage the birds stopped eating *D. barberi* and began eating the grains of maize (Bollinger & Caslick, 1985).

In Indiana *D. u. howardi* is a staple in the diet of the big brown bat. Across several geographic locations *D. u. howardi* adults varied from 6.3 to 48.6% (mean 28.8%) of the volume of food consumed (Whitaker, 1995). Whitaker calculated that a typical colony with 150 bats would eat 600,000 *D. u. howardi* beetles during the period from April to October. A salient question is why *D. barberi* and *D. v. virgifera,* species that are almost surely more prevalent at times in Indiana than is *D. u. howardi,* were not detected in the droppings. The three species can be present together in, and over maize during the period when bats fed on *D. u. howardi*. Perhaps the bats do not eat *D. barberi* and *D. v. virgifera,* or they forage for *D. u. howardi* in habitats other than maize fields. Cucurbits are other plants near which *D. u. howardi* occurs, a habitat occupied also by species of *Acalymma*. Again, it is noteworthy that no *Acalymma* were found in the droppings. If species of *Acalymma* are absent due to bitter taste (since they develop only on cucurbits), it could be that *D. u. howardi* from cucurbit plantings would be bitter as well. My purpose is not to speculate, but to point out that the big brown bat may be useful for testing hypotheses on the role of cucurbitacins in protecting *Diabrotica* from predators. Gould and Massey (1984) found in laboratory experiments that four species of vertebrate predators, present in the habitat of *D. u. howardi* in North Carolina, were not deterred from preying on beetles which had fed on a cucurbitacin-containing cucumber variety.

10.2. Arthropod predators

Reports of arthropod predators of *Diabrotica* are listed in Table 2. Most reports either suggest that the impact of insect predators on *Diabrotica* is minimal or do not address the issue. In a laboratory experiment with carabids in a soil/maize system closely mimicking nature, and including radioactively labeled immature *D. barberi*, the carabid species attributed to Tyler and Ellis (1979) in Table 2 became labeled, suggesting they preyed on *D. barberi*. *Pterostichus melanarius* Ill. was concluded to be the most important carabid predator in the study. These results temper Kirk's (1982) view that it is impossible to determine whether a carabid might have eaten *Diabrotica*.

Table 2. Predators of *Diabrotica*

Species	Stage	Predator	Group	Reference
adelpha	E	*Solenopsis geminata*	Formicidae	Risch, 1981
adelpha	E	*Pheidole* sp.	Formicidae	Risch, 1981
balteata	A	*Labidura riparia*	Dermaptera	Waddill, 1978
balteata	A	Reduviidae sp.	Reduviidae	Eben & Barbercheck, 1996; Risch, 1981
balteata	A	*Oxyopes salticus*	Spider (Oxyopid)	Eben & Barbercheck, 1996
barberi	A	Red-winged blackbird	Aves	Bollinger & Caslick, 1985
barberi	**E/L**	Scaritini	Carabidae	Tyler & Ellis, 1979
barberi	E/L	Harpalini	Carabidae	Kirk, 1973
barberi		*Agonum* sp.	Carabidae	Tyler & Ellis, 1979
barberi	E/L	*Bembidion* sp.	Carabidae	Tyler & Ellis, 1979
barberi	A	Pentatomidae	Hemiptera	Quiring & Timmins, 1990
barberi	E/L	*Androlaelaps* sp.	Mite	Chiang, 1970
barberi	E/L	*Stratiolaelaps* sp.	Mite	Chiang, 1970
barberi	E/L/A	*Pterostichus* sp.	Carabidae	Kirk, 1971; Tyler & Ellis, 1979
u. howardi	A	*Chrysoperla carnea.* A/L	Chrysopidae [a]	Ellington *et al.*, 1997
u. howardi	A	*Arilus cristatus*	Reduviidae	Chittenden, 1910
u. howardi	A	*Collops vittatus.* A	Melyridae [a]	Ellington *et al.*, 1997
u. howardi	A	*Hippodamia convergens.* A	Coccinellidae [a]	Ellington *et al.*, 1997
u. howardi	A	*Geocoris* spp.	Lygaeidae [a]	Ellington *et al.*, 1997
u. howardi	E/L	*Tyrophagus putrascentiae*	Acarinae	Brust, 1991
u. howardi	A	*Epitesicus fuscus*	Bat	Whitaker, 1995
u. howardi	L	*Chauliognathus spp.* L	Cantharidae	Brust, 1991
u. howardi	L	*Harpalus caliginosus.* L	Carabidae	Brust, 1991
u. howardi	L	*Harpalus pennsylvanicus.* L	Carabidae	Brust, 1991
u. howardi	L	*Pterostichus spp.* L	Carabidae	Brust, 1991
u. howardi	E/L/A	*P. chalcites*	Carabidae	Heimpel & Hough-Goldstein, 1992
u. howardi	?	*Harpalus pennsylvanicus.* A	Carabidae	Brust, 1991
u. howardi	?	*Amara spp.* A	Carabidae	Brust, 1991
u. howardi	?	*Bembidion spp.* A	Carabidae	Brust, 1991
u. howardi	?	*Stenolophus spp.* A	Carabidae	Brust, 1991
u. howardi	?	Lithobiomorpha: Henicopidae	Chilopoda	Brust, 1991
u. howardi	?	*Cicindela* spp. L	Cicindelidae	Brust, 1991
u. howardi	?	Japygidae spp.	Diplura	Brust, 1991
u. howardi	E	*Lasius* spp.	Formicidae	Brust, 1991
u. howardi	E	*Pheidole* spp.	Formicidae	Brust, 1991
u. howardi	E/L	*Protolaelaps* spp.	Mesostigmata	Brust, 1991
u. howardi	E/L	*Alliphis* spp.	Mesostigmata	Brust, 1991
u. howardi	A	*Nabis* spp.	Nabidae [a]	Ellington *et al.*, 1997
u. howardi	A	*Phidippus audax*	Salticidae	Johnson, 1996
u. howardi	L	Staphilinidae spp. L	Staphilinidae	Brust, 1991
u. howardi	?	Staphilinidae spp. A	Staphilinidae	Brust, 1991
v. virgifera	E	*Myrmica*	Formicidae	Ballard & Mayo, 1979
v. virgifera	E	*Lasius neoniger*	Formicidae	Kirk, 1981b
v. virgifera	E/L	*Androlaelaps* sp.	Mite	Chiang, 1970
v. virgifera	E/L	*Stratiolaelaps* sp.	Mite	Chiang, 1970
v. zeae	L	*Chauliognathus* sp.	Cantharidae	Branson *et al.*, 1982

[a] Numerically associated with *Diabrotica* beyond chance alone; does not necessarily imply predation.

Ants are likely candidates as egg predators but experiments with three species that occur in maize fields in Nebraska provided little evidence of predation on eggs (Ballard &

Table 3. Parasites of *Diabrotica*

Diabrotica Species:	Locality	Parasitoid	Source
		Braconidae	
speciosa	Brasil	*Centistes gasseni* Shaw	Heineck-Leonel & Salles, 1997
		Nematoda	
balteata	South Carolina, USA	*Filipjevimermis leipsandra* Poinar & Welch	Creighton & Fassuloitis, 1980
balteata	South Carolina, USA	*Heterorhabditis* sp.	Creighton & Fassuliotis, 1985
balteata	Veracruz, Mexico	*Hexamermis* sp.	Eben & Barbercheck, 1996
speciosa	Brasil	*Hexamermis* sp.	Heineck-Leonel & Salles, 1997
speciosa vigens Erichson	Peru	*Hexamermis sp.*	Nickle *et al.*, 1984
u. howardi	Virginia, USA	*Diplogaster* sp.	Fronk, 1950
u. howardi	South Carolina, USA	*Heterorhabditis* sp.	Creighton & Fassuliotis, 1985
u. howardi	Va. & S. C., USA	*Howardula benigna* (Cobb)	Elsey, 1977, Fronk, 1950
		Tachinidae	
balteata	Veracruz, Mexico	*Celatoria compressa* Wulp	Eben & Barbercheck, 1996
barberi	Illinois, USA	*C. diabroticae* Shimer	Kuhlman, 1970
porracea	Veracruz, Mexico	*C. compressa*	Eben & Barbercheck, 1996
scutellata	Veracruz, Mexico	*C. compressa*	Eben & Barbercheck, 1996
sp.	Peru	*C.* sp. nr. *diabroticae*	Nickle *et al.*, 1984
speciosa	Brasil	C. *bosqi* Blanchard	Heineck-Leonel & Salles, 1997
tibialis	Veracruz, Mexico	*C. compressa*	Eben & Barbercheck, 1996
undecimpunctata howardi	New Mexico, USA	*C. diabroticae*	Gordon *et al.*, 1987
u. howardi	South Carolina, USA	*C. diabroticae*	Elsey, 1989
u. undecimpunctata	Western USA	*C. diabroticae*	Arnaud, 1978
v. virgifera	Illinois, USA	*C. diabroticae*	Kuhlman, 1970
viridula	Veracruz, Mexico	*C. compressa*	Eben & Barbercheck, 1996

Mayo, 1979). Risch (1981), working in Costa Rica with eggs of *D. balteata* and *D. adelpha* Harold, found that in the field, eggs were efficiently removed from petri dishes and that the predators were almost certainly ants (Table 3). In what may be the only finding of a natural enemy with impact on *Diabrotica*-as-pest, Kirk (1981b) reported that the presence of colonies of the ant *Lasius neoniger* Emery at the base of maize plants was correlated with lack of root damage by *D. v. virgifera*. Maize stalks with ant colonies present had no evidence of rootworm damage, while 75% of those without ants were damaged. Ten percent of the maize stalks had associated ant colonies.

The hypothesis that dietary cucurbitacins protect *Diabrotica* species from predation (Howe *et al.* 1976; Nishida & Fukami, 1990) was tested with Chinese praying mantids (Ferguson & Metcalf, 1985); the mantids rejected laboratory reared *D. u. howardi* that had fed on squash containing cucurbitacin, but rejected few field-collected beetles. Females of *D. u. howardi* sequester cucurbitacins in their eggs (Ferguson & Metcalf, 1985) but laboratory, greenhouse and field experiments in which up to 76% of eggs were consumed by generalist predators, no significant differences were noted between bitter (eggs contained cucurbitacin-c) and non-bitter eggs (Brust & Barbercheck, 1992). The authors noted that differences might be found had eggs contained cucurbitacin B or D, or if different predators had been used.

10.3. Parasitoids

Diabrotica have been found parasitized by tachinids, a braconid, and by nematodes in the genera *Filipjevimermis, Hexamermis, Heterorhabditis*, and *Howardula* (Table 3). An erroneous catalog entry (Thompson, 1943) listed five parasites of *Diabrotica melanocephala* (F.) (=*sinuata* Olivier) in the United States due to incorrect synonomy. The actual host is *Acalymma vitattum* (F); *D. sinuata* is strictly Neotropical. Arnaud (1978) provided an exhaustive catalog of literature on North American tachinid parasites of *Diabrotica*.

In a systematic seasonal survey, Heineck-Leonel and Salles (1997) found parasitism rates in *D. speciosa* ranged from 0.1 to 30.2%, 0 to 6.0%, and 0 to 2.0% for *Celatoria bosqi* Blanch., *Centistes gasseni* Shaw and *Hexamermis* sp., respectively. Parasitism was low from October through February and high from March through September. A survey by Eben and Barbercheck (1996) in Mexico revealed *Celatoria compressa* Wulp to be the most common parasitoid with the following percentages of parasitism; *D. balteata* (4), *D. scutellata* Harold (10.2) and *D. tibialis* (11.1). *Celatoria diabroticae* (Shimer) parasitized overwintering *D. u. howardi* in North and South Carolina; incidence in the area ranged from 3 to 15% (Meinke & Gould, 1987; Elsey, 1988b). In late summer in South Carolina the nematode *Filipjevimermis leipsandra* Poinar and Welch occurred in 50 to 100% of *D. balteata* (Creighton & Fassuloites, 1980). Nickle *et al.* (1984) found 5 to 90% infection rates for *Hexamermis* sp. in *D. speciosa* in Peru. Aside from these instances, the other reports of parasitoids in *Diabrotica* reveal very low incidences of parasitism.

There is little information on the biology of the tachinids and the only instance wherein an impact on populations has been suggested involves the related genus *Acalymma*; Elsey (1988b) found over 40% of *A. vittatum* (F.) parasitized by *Celatoria setosa* (Coquillet) and tentatively ascribed a concurrent decline in beetle population to the parasitoid. The many attempts to exploit nematodes as biological control agents for *Diabrotica* will not be reviewed here. Tests have not necessarily been with nematode species found naturally in *Diabrotica*.

10.4. Pathogens

Pathogens listed in Table 4 have been detected in *Diabrotica*. Attempts to use pathogens for management have been unsuccessful (for example, Krueger & Roberts, 1997). Diseases that occur in laboratory colonies have been discussed by Jackson (1986), Schalk (1986), and Oloumi-Sadeghi and Levine (1989a). Of these agents, it is interesting that the role in the biology and ecology of *Diabrotica* has been determined only for *Wolbachia* in *D. virgifera s.l.* Hardly a pathogen in the usual sense, the Rickettsia-like organism confers unidirectional reproductive incompatibility between populations of *D. v. virgifera* and *D. v. zeae* and hence may well function in hybrid zones in the process of speciation (Giordano *et al.*, 1997).

The attempts to establish that cucurbitacins protect *Diabrotica* from predators have yielded ambiguous results (see above). It seems ironic that the first demonstrated case of cucurbitacin-induced protection concerns the pathogenicity of a fungus. Tallamy et al. (1998) found that cucurbitacins sequestered in eggs and larvae of *D. u. howardi* reduced the pathogenicity of *Metarhizium anisopliae* toward those stages; significantly more cucurbitacin-containing eggs and larvae survived exposure to the fungus.

Table 4. Pathogens of *Diabrotica*

Diabrotica species	Pathogen	Group	Reference
balteata	*Beauveria bassiana*	Fungus	Humber, 1992
barberi	*Beauveria bassiana*	Fungus	Humber, 1992
barberi	Entomophthorlean (*Tarichium*)	Fungus	Naranjo & Steinkraus, 1988
barberi	*Metarhizium anisopliae*	Fungus	Charles, 1941 as *longicornis*
barberi	*Beauveria bassiana*	Fungus	Charles, 1941 as *longicornis*
paranaensis? Marques	*Beauveria bassiana*	Fungus	Humber, 1992
speciosa	*Beauveria bassiana*	Fungus	Heineck-Leonel & Salles, 1997; Humber, 1992
speciosa	*Metarhizium anisopliae*	Fungus	Heineck-Leonel & Salles, 1997; Humber, 1992
speciosa	*Paecilomyces fumosoroseus*	Fungus	Tigano-Milani et al., 1995
speciosa	*Paecilomyces lilacinus*	Fungus	Tigano-Milani et al., 1995
u. howardi	*Gregarina coronata*	Eugregarine	Clopton et al., 1992
u. howardi	Nonoccluded baculovirus	Virus	Kim & Kitajima, 1984
u. howardi	Mollicute	Virus	Kitajima & Kim, 1984
u. howardi	Mollicute (monotypic group XII spiroplasma)	Virus	Hackett & Lipa, 1996
u. howardi	*Spiroplasma diabroticae*	Virus?	Carle et al., 1997
u. undecimpunctata	*Beauveria bassiana*	Fungus	Charles, 1941 as *soror*; Humber, 1992
v. virgifera	*Beauveria bassiana*	Fungus	Humber, 1992
v. virgifera	*Cylindrocarpon destructans*	Fungus	Oloumi-Sadeghi & Levine, 1989a
v. virgifera	*Wolbachia*	Rickettsia-like organism	Degrugillier, 1996; Giordano *et al.*, 1997
viridula (Fabricius)	*Laboulbenia diabroticae*	Fungus	Eben & Barbercheck, 1996

Acknowledgements

I thank Larry Chandler, James Coppedge, Eli Levine and Doug Tallamy for providing reprints and preprints of critical publications. Carole Krysan provided assistance, support and toleration essential for this post-retirement activity.

References

Abou-Fakhr, E. M., Hibbard, B. E., Jewett, D. K. and Bjostad, L. B. 1996. Electroantennogram responses of western corn rootworm (Coleoptera: Chrysomelidae) adults in relation to maize silk morphology and phenology. Envir. Ent. 25: 430-435.

Arnaud, P. H., Jr. 1978. A host-parasite catalog of North American Tachinidae (Diptera). Misc. Pub. No. 131. U. S. Dept. Agric., Washington, D. C.

Baca, F. 1994. New member of the harmful entomofauna of Yugoslavia *Diabrotica virgifera virgifera* LeConte (Coleoptera: Chrysomelidae). Zašt. Bilja. 45: 125-131.

Ball, H. J. 1957. On the biology and egg-laying habits of the western corn rootworm. J. econ. Ent. 50: 126-128.

Ballard, J. B. and Mayo, Z B, 1979. Predatory potential of selected ant species on eggs of western corn rootworm. Envir. Ent. 8: 575-576.

Barbercheck, M. E., Herbert, D. A., Jr., and Warrick, W. C., Jr. 1995. Evaluation of Semiochemical baits for management of southern corn rootworm (Coleoptera: Chrysomelidae) in peanuts. J. econ. Ent. 88: 1754-1763.

Bartelt, R. J., and Chiang, H. C. 1977. Field studies involving the sex attractant pheromones of the western and northern corn rootworm beetles. Envir. Ent. 6: 853-861.

Baust, J. G., Benton, A. H. and Aumann, G. D. 1981. The influence of off-shore platforms on insect dispersal and migration. Bull. ent. Soc. Am. 27: 23-28.
Bechyné, J. 1997. Evaluacion de los datos sobre los Phytophaga daninos en lVenezuela (Coleoptera) Parte 11. In: Savini, V., (ed) Boletin de Entomologia Venezolana, Serie Monografias 1, November, 1997. Instituto de Zoologia Agricola, Marcay. pp 296-325.
Bechyné, J. and Springlova-Bechyné, B. 1970. Beitrage zur Kenntnis der Insektenfauna Boliviens XX. Coleoptera XV. Beitrage zur Galerucidenfauna Boliviens. Veröff. zool. StSamml. Münch. 14: 121-190.
Bechyné, J. and Springlova de Bechyné, B. 1976. Phytophages (Coleopteres) recoltes en Guyane Francaise par la misson du Museum National d'Histoire Naturalle. Ann. Soc. ent. Fr. 12: 527-556.
Boetel, M. A., Walgenbach, D. D., Hein, G. L., Fuller, B. W., and Gray, M. E. 1992. Oviposition and site selection of northern corn rootworm. J. econ. Ent. 85: 246-249.
Boetel, M. A. and Fuller, B. W. 1997. Seasonal emergence-time effects on adult longevity, fecundity, and egg viability of northern and western corn rootworms (Coleoptera: Chrysomelidae). Envir. Ent. 26: 1208-1212.
Bollinger, E. K. and J. W. Caslick. 1985. Red-winged blackbird predation on northern corn rootworm beetles in field corn. J. Appl. Ecol. 22: 39-48.
Branson, T. F. 1978. Optimum temperature for long-term storage of eggs of *Diabrotica virgifera* (Coleoptera: Chrysomelidae). Entomologia exp. appl. 24: 199-200.
Branson, T. F. 1987. The contribution of prehatch and post hatch development to protandry in the chrysomelid *Diabrotica virgifera virgifera*. Entomologia exp. appl. 43: 205-208.
Branson, T. F. 1989. Survival of starved neonate larvae of *Diabrotica virgifera virgifera* LeConte (Coleoptera: Chrysomelidae). J. Kans. ent. Soc. 62: 521-523.
Branson, T. F. and Johnson, R. D. 1973. Adult western corn rootworms: oviposition, fecundity, and longevity in the laboratory. J. econ. Ent. 66: 427-418.
Branson, T. F. and Krysan, J. L. 1981. Feeding and oviposition behavior and life cycle strategies of *Diabrotica*: an evolutionary view with implications for pest management. Envir. Ent. 10: 826-831.
Branson, T. F. and Krysan, J. L. 1987. Laboratory biology of *Diabrotica tibialis* and a closely related *Diabrotica* species (Coleoptera: Chrysomelidae). J. Kans. ent. Soc. 60: 446-450.
Branson, T. F., Guss, P. L. and Jackson, J. J. 1977. Mating frequency of the western corn rootworm. Ann. ent. Soc. Am. 70: 506-508.
Branson, T. F., Guss, P. L. and Krysan, J. L. 1978. Winter populations of some *Diabrotica* in central Mexico: voltinism and pheromone response. Ann. ent. Soc. Am. 71: 165-166.
Branson, T. F., Reyes R., J., and Valdes M., H. 1982. Field biology of the Mexican corn rootworm, *Diabrotica virgifera zeae* (Coleoptera: Chrysomelidae), in central Mexico. Envir. Ent. 11: 1078-83.
Branson, T. F., and Sutter, G. R. 1985. Influence of population density of immatures on size, longevity, and fecundity of adult *Diabrotica virgifera virgifera* (Coleoptera: Chrysomelidae). Envir. Ent. 14: 687-690.
Brust, G. E. 1991. Soil moisture, no tillage and predator effects on southern corn rootworm survival in peanut agroecosystems. Entomologia exp. appl. 58: 109-122.
Brust, G. E. and Barbercheck, M. E. 1992. Dietary cucurbitacins and southern corn rootworm survival. Envir. Ent. 21: 1466-1471.
Brust, G. E. and House, G. J. 1990a. Effects of soil moisture, texture and rate of soil drying on egg and larval survival of southern corn rootworm (Coleoptera: Chrysomelidae). Envir. Ent. 19: 697-703.
Brust, G. E. and House, G. J. 1990b. Influence of soil texture, soil moisture, organic cover, and weeds on oviposition preference of southern corn rootworm (Coleoptera: Chrysomelidae). Envir. Ent. 19: 966-971.
Carle, P., Whitcomb, R. F., Hackett, K. J., Tully, J. G., Rose, D. L., Bove, J. M., Henegar, R. B., Konai, M., and Williamson, D. L. 1997. *Spiroplasma diabroticae* sp. nov. from the southern corn rootworm beetle, *Diabrotica undecimpunctata.* (Coleoptera: Chrysomelidae). Int. J. Syst. Bact. 47: 78-80.
Charles, V. K. 1941. A preliminary check list of the entomogenous fungi of North America. Insect Pest Surv. Bull. 21: 707-785.
Chiang, H. C. 1965. Survival of northern corn rootworm eggs through one and two winters. J. econ. Ent. 58: 470-472.
Chiang, H. C. 1970. Effects of manure applications and mite predation on corn rootworm populations in Minnesota. J. econ. Ent. 63: 934-936.
Chiang, H. C. 1973. Bionomics of the northern and western corn rootworms. A. Rev. Ent. 18: 47-72.
Chiang, H. C. 1974. Temperature effects on hatching of eggs of the western corn rootworm: factual and theoretical interpretations. Entomologia exp. appl. 17: 149-156.
Chiang, H. C., Mihm, J. A. and Windels, M. B. 1972. Temperature effects on hatching of eggs of the western and northern corn rootworms. Proc. North Cent. Br. Entomol. Soc. Am. 27: 127-131.
Chittenden, F. H. 1910. Notes on cucumber beetles. Bull. U. S. Dept. Agric. Bur. Ent. 82: 67-75.
Chuman, T., Guss, P. L., Doolittle, R. E., McLaughlin, J. R., Krysan, J. L., Schalk, J. M. and Tumlinson, J. H. 1987. Identification of female-produced sex pheromone from banded cucumber beetle, *Diabrotica balteata* LeConte (Coleoptera: Chrysomelidae). J. Chem. Ecol. 13: 1601-1615.

Cinereski, J. E. and Chiang, H. C. 1968. The pattern of movements of adults of the northern corn rootworm inside and outside of corn fields. J. econ. Ent. 61: 1531-1536.

Clopton, R. E., T. J. Percival and J. Janovy, Jr. 1992. *Gregarina coronata* n. sp. (Apicomplexa: Eugregarinida) described from adults of the southern corn rootworm, *Diabrotica undecimpunctata howardi* (Coleoptera: Chrysomelidae). J. Protozool. 39: 417-420.

Coats, S. A., Mutchmor, J. A. and Tollefson, J. J. 1987. Regulation of migratory flight by juvenile hormone mimic and inhibitor in western corn rootworm (Coleoptera: Chrysomelidae). Ann. ent. Soc. Am. 80: 697-708.

Coats, S. A., Tollefson, J. J. and Mutchmor, J. A. 1986. Study of migratory flight in the western corn rootworm (Coleoptera: Chrysomelidae). Envir. Ent. 15: 620-625.

Cox, M. L. 1994. Diapause in Chrysomelidae. In: Jolivet, P. H., Cox, M. L., and Petitpierre, E. (eds) Novel aspects of the biology of Chrysomelidae. Kluwer Academic Publishers, Dordrecht, pp 419-467.

Creighton, C. S., and Fassuliotis, G. 1980. Seasonal population fluctuations of *Filipjevimermis leipsandra* and infectivity of juveniles in the banded cucumber beetle. J. econ. Ent. 73: 296-300.

Creighton, C. S., and Fassuliotis, G. 1985. *Heterorhabditis* sp. (Nematoda: Heterorhabditidae): a nematode parasite isolated from the banded cucumber beetle (*Diabrotica balteata*). J. Nemat. 117: 150-153.

Cuthbert, F. P., Creighton, C. S., and Cuthbert, R. B. 1968. Mass rearing banded cucumber beetles with notes on rearing spotted and striped cucumber beetles. J. econ. Ent. 61: 288-292.

Degrugillier, M. E., Degrugillier, S. S. and Jackson, J. J. 1991. Nonoccluded, cytoplasmic virus particles and rickettsia-like organisms in testes and spermathecae of *Diabrotica virgifera*. J. Invert. Path. 57: 50-58.

Dobson, I. and Teal, P. E. A. 1986. Field studies of the temporal response patterns of male *Diabrotica virgifera virgifera* LeConte and *D. barberi* Smith and Lawrence to 8R-methyl-2R-decyl propanoate. Physiol. Ent. 11: 405-410.

Dobson, I. D. and Teal, P. E. A. 1987. Analysis of long range reproductive behavior of male *Diabrotica virgifera virgifera* LeConte and D. *barberi* Smith and Lawrence to stereoisomers of 8-methyl-2-decyl propanoate. J. Chem. Ecol. 13: 1331-1341.

Dominique, C. R. M. and Yule, W. N. 1984. Bionomics of the northern corn rootworm, *Diabrotica longicornis* (Say) (Coleoptera: Chrysomelidae), in Quebec. Rev. Ent. Quebec 29: 12-16.

Eben, A. and Barbercheck, M. E. 1996. Field observations on host plant associations and natural enemies of diabroticite beetles (Chrysomelidae: Luperini) in Veracruz, Mexico. Acta zool. Mex. 67: 47-65.

Ellington, J., M. Southward, and Carillo, T. 1997. Association among cotton arthropods. Envir. Ent. 26: 1004-1008.

Elliott, N., Gustin, R. D. and Hanson, S. L. 1990. Influence of adult diet on the reproductive biology and survival of the western corn rootworm, *Diabrotica virgifera virgifera*. Entomologia exp. appl. 56: 115-122.

Ellsbury, M. M., Schumacher, T. E., Gustin, R. D., and Woodson, W. D. 1994. Soil compaction effect on corn rootworm populations in maize artificially infected with eggs of western corn rootworm (Coleoptera: Chrysomelidae). Envir. Ent. 23: 943-948.

Elsey, K. D. 1977. Parasitism of some economically important species of Chrysomelidae by nematodes of the genus *Howardula*. J. Invertebr. Path. 29: 384-385.

Elsey, K. 1988a. Reproductive diapause in the spotted cucumber beetle. Fla. Ent. 71: 78-83.

Elsey, K. 1988b. Cucumber beetle seasonality in coastal South Carolina. Envir. Ent. 17: 495-502.

Elsey, K. 1989. Cold tolerance of adult spotted and banded cucumber beetles (Coleoptera: Chrysomelidae). Envir. Ent. 18: 1112-1116.

Ferguson, J. E., and Metcalf, R. L. 1985. Cucurbitacins: Plant-derived defense compounds for the Diabroticina (Coleoptera: Chrysomelidae). J. Chem. Ecol. 11: 311-318.

Fisher, J. R. 1986. Development and survival of pupae of *Diabrotica virgifera* and *D. undecimpunctata howardi* at constant temperatures and humidities. Envir. Ent. 15: 626-630.

Fisher, J. R., Jackson, J. J. and Lew, A. C. 1994. Temperature and diapause development in the egg of *Diabrotica barberi* (Coleoptera: Chrysomelidae). Envir. Ent. 23: 464-471.

Fronk, W. D. 1950. Cultural and biological control of the southern corn rootworm in peanuts. J. econ. Ent. 43: 22-25.

George, B. W. and Hintz, A. M. 1966. Immature stages of the western corn rootworm. J. econ. Ent. 59: 1139-1142.

Giordano, R., J. J. Jackson, and H. M. Robertson. 1997. The role of *Wolbachia* bacteria in reproductive incompatibilities and hybrid zones of *Diabrotica* beetles and *Gryllus* crickets. Proc. Natl. Acad. Sci. USA. 94: 11439-44.

Godfrey, L. D. and Turpin, F. T. 1983. Comparison of western corn rootworm (Coleoptera: Chrysomelidae) adult populations and economic thresholds in first-year and continuous corn fields. J. econ. Ent. 76: 1028-1032.

Golden, K. L. and Meinke, L. J. 1991. Immature development, fecundity, longevity, and egg diapause of *Diabrotica longicornis* (Coleoptera: Chrysomelidae). J. Kans. ent. Soc. 64: 251-256.

Gordon, R., Ellington, J., Faubion, G. F. and Graham, H. 1987. A survey of the insect parasitoids from alfalfa and associated weeds in New Mexico. S.-west Entomologist 12: 335-350.

Gould, F. and Massey, A. 1984. Cucurbitacins and predation of the spotted cucumber beetle, *Diabrotica undecimpunctata howardi*. Entomologia exp. appl. 36: 273-278.

Grant, R. H. and Seevers, K. P. 1989. Local and long range movement of adult western corn rootworm as evidenced by washup on southern shore of Lake Michigan. Envir. Ent. 18: 266-272.

Grant, R. H. and Seevers, K. P. 1990. The vertical movement of adult western corn rootworms (*Diabrotica virgifera virgifera*) relative to the transport of momentum and heat. Agric. and Forest Meteorology 49: 191-203.

Gray, M. E. and Tollefson, J. J. 1987. Influence of tillage and western and northern Corn Rootworm (Coleoptera: Chrysomelidae) egg populations on larval populations and root damage. J. econ. Ent. 80: 911-915.

Gray, M. E., Hein, G. L., Boetel, M. A. and Walgenbach, D. D. 1992. Western and northern corn rootworm (Coleoptera: Chrysomelidae) egg densities at three soil depths: Implications for future ecological studies. J. Kans. ent. Soc. 65: 354-356.

Guss, P. L., Sonnet, P. E., Carney, R. L., Branson, T. F., and Tumlinson, J. H. 1984. Response of *Diabrotica virgifera virgifera*, *Diabrotica virgifera zeae*, and *Diabrotica porracea* to stereoisomers of 8-methyl-2-decyl propanoate. J. Chem. Ecol. 10: 1123-1131.

Guss, P. L., Tumlinson, J. H., Sonnet, P. E., and Proveaux, A. T. 1982. Identification of a female-produced pheromone of the western corn rootworm. J. Chem. Ecol. 8: 545-556.

Gustin, R. D. 1979. Effect of two moisture and population levels on oviposition of the western corn rootworm. Envir. Ent. 8: 406-407.

Gustin, R. D. 1981. Soil temperature environment of overwintering western corn rootworm eggs. Envir. Ent. 10: 483-487.

Gustin, R. D. 1983. *Diabrotica longicornis barberi* (Coleoptera: Chrysomelidae): Cold hardiness of eggs. Envir. Entomol: 12: 633-634.

Gustin, R. D. 1984. Effect of crop cover on oviposition of the northern corn rootworm, *Diabrotica longicornis barberi* Smith and Lawrence. J. Kans. ent. Soc. 57: 515-516.

Gustin, R. D. 1986. Effect of intermittent low temperatures on hatch of western corn rootworm eggs (Coleoptera: Chrysomelidae) J. Kans. ent. Soc. 59: 568-570.

Gustin, R. D., and Schumacher, T. E. 1989. Relationship of some soil pore parameters to movement of first -instar western corn rootworm (Coleoptera: Chrysomelidae). Envir. Ent. 18: 343-346.

Gustin, R. D., and Wilde, G. E. 1985. Cold hardiness of western corn rootworm (Coleoptera: Chrysomelidae) eggs from northern and southern zones of the corn belt. J. Kans. ent. Soc. 57: 722-725.

Hackett, K. J. and Lipa, J. J. 1996. Mollicutes associated with the Chrysomelidae. In : Jolivet, P., and Cox, M. L. (eds) Chrysomelidae Biology, vol. 2, Ecological Studies. SPB Academic Publishing, Amsterdam.

Hammack, L. 1995. Calling behavior in female western corn rootworm beetles (Coleoptera: Chrysomelidae). Ann. ent. Soc. Am. 88: 562-569,

Heimpel, G. E., and Hough-Goldstein, J. A. 1992. A survey of arthropod predators of *Leptinotarsa decemlineata* (Say) in Delaware potato fields. J. agric. Ent. 9: 137-142.

Heineck-Leonel, M. A. and Salles, A. B. 1997. Incidence of parasitoids and pathogens in adults of *Diabrotica speciosa* (Germar) (Coleoptera:Chrysomelidae) in Pelotas, R. S. Anais Soc. Ent. Bras. 26: 81-85.

Hill, R. E. 1975. Mating, oviposition patterns, fecundity and longevity of the western corn rootworm. J. econ. Ent. 68: 311-315.

Hill, R. E. and Mayo, Z B. 1974. Trap corn to control corn rootworms. J. econ. Ent. 67: 748-750.

Hill, R. E. and Mayo, Z B. 1980. Distribution and abundance of corn rootworm species as influenced by topography and crop rotation in eastern Nebraska. Envir. Ent. 9: 122-127.

Howe, W. L., Sandborn, J. R., and Rhodes, A. M. 1976. Western corn rootworms and spotted cucumber beetle associations with *Cucurbita* and cucurbitacin. Envir. Ent. 5: 1043-1048.

Humber, R. A. 1996. Fungal pathogens of the Chrysomelidae and prospects for their use in biological control. In: Jolivet, P. H. A. and Cox, M. L. (eds) Chrysomelidae Biology, vol. 3: General Studies. SPB Academic Publishing, Amsterdam. pp 95-115.

Isley, D. 1929. The southern corn rootworm. Bull. Ark. agric. Exp. Stn. 232, pp 1-31.

Jackson, J. J. 1986. Rearing and handling of *Diabrotica virgifera* and *Diabrotica undecimpunctata howardi*. In: Krysan, J. L., and Miller, T. A. (eds) Methods for the study of pest *Diabrotica*. Springer-Verlag, New York. pp. 25-47.

Jackson, J. J. and Elliott, N. C. 1988. Temperature dependent development of immature stages of the western corn rootworm, *Diabrotica virgifera virgifera* (Coleoptera: Chrysomelidae) Envir. Ent. 17: 166-171.

Jackson, J. J., Lew, A. C. and Woodson, W. D. 1995. Effect of egg age, storage temperature, and storage duration on egg hatch of *Diabrotica undecimpunctata howardi* (Coleoptera: Chrysomelidae). Ann. ent. Soc. Am. 88: 781-784.

Johnson, S. R. 1996. Use of coleopteran prey by *Phidippus audax* (Araneae, Salticidae) in tallgrass prairie wetlands. J. Arachnology 24: 39-42.

Kim, K. S., and Kitajima, E. W., 1984. Nonoccluded baculovirus-like and filamentous virus-like particles in the spotted cucumber beetle, *Diabrotica undecimpunctata* (Coleoptera: Chrysomelidae). J. Invertebr. Path. 43: 234-241.

Kirk, V. M. 1971. Ground beetles in cropland in South Dakota. Ann. ent. Soc. Am. 64: 238-241.

Kirk, V. M. 1973. Biology of a ground beetle, *Harpalus pennsylvanicus*. Ann. ent. Soc. Am. 66: 513-518.
Kirk, V. M. 1979. Drought cracks as oviposition sites for western and northern corn rootworms (*Diabrotica*: Coleoptera). J. Kans. ent. Soc. 52: 769-776.
Kirk, V. M. 1981a. Base of corn stalks as oviposition sites for western and northern corn rootworms (*Diabrotica*: Coleoptera). J. Kans. ent. Soc. 54: 255-262.
Kirk, V. M. 1981b. Corn rootworm: population reduction associated with the ant, *Lasius neoniger*. Envir. Ent. 10: 966-967.
Kirk, V. M., 1982. Carabids: minimal role in pest management of corn rootworms. Envir. Ent. 11: 5-8.
Kitajima, E. W., and Kim, K. S., 1984. A mollicute and associated virus-like particles in some tissues of the spotted cucumber beetle (*Diabrotica undecimpunctata*) (Coleoptera: Chrysomelidae). J. Submicrosc. Cytol. 16: 547-554.
Krafsur, E. S. 1995. Gene flow between univoltine and semivoltine northern corn rootworm (Coleoptera: Chrysomelidae) populations. Ann. ent. Soc. Am. 88: 699-704.
Krafsur, E. S., Nariboli, P. and Tollefson, J. J. 1993. Gene diversity in natural *Diabrotica barberi* populations (Coleoptera: Chrysomelidae). Ann. ent. Soc. Am. 86: 490-496.
Krysan, J. L. 1972. Embryonic stage of *Diabrotica virgifera* (Coleoptera: Chrysomelidae) at diapause. Ann. ent. Soc. Am. 65: 768-769.
Krysan, J. L. 1976. Moisture relationships of the egg of the southern corn rootworm, *Diabrotica undecimpunctata howardi* (Coleoptera: Chrysomelidae). Entomologia. exp. appl. 20: 154-162.
Krysan, J. L. 1977. Changes in water permeability of an insect egg in response to level of water in the environment. Experientia 33: 1181-1182.
Krysan, J. L. 1978. Diapause, quiescence, and moisture in the egg of the western corn rootworm, *Diabrotica virgifera*. J. Insect Physiol. 24: 535-540.
Krysan, J. L. 1982. Diapause in the Nearctic species of the *virgifera* group of *Diabrotica*: evidence for tropical origin and temperate adaptations. Ann. Ent. Soc. Am. 75: 136-142.
Krysan, J. L. 1986. Introduction: biology, distribution, and identification of pest *Diabrotica*. In: Krysan, J. L. and Miller, T. A. (eds) Methods for study of the pest *Diabrotica*. Springer-Verlag, New York. pp. 1-23.
Krysan, J. L. 1993. Adaptations of *Diabrotica* to habitat manipulations. In: Kim, K. C. and McPheron, B. A. (eds) Evolution of Insect Pests: Patterns of Variation. John Wiley, New York. pp 361-373.
Krysan, J. L. and Branson, T. F., 1977. Inheritance of diapause intensity in *Diabrotica virgifera*. J. Heredity 68: 415-417.
Krysan, J. L. and Branson, T. F., and Diaz Castro, G., 1977. Diapause in *Diabrotica virgifera* (Coleoptera: Chrysomelidae): a comparison of eggs from temperate and tropical climates. Entomologia exp. appl. 22: 81-89.
Krysan, J. L. and Guss, P. L. 1978. Barriers to hybridization between *Diabrotica virgifera* and *D. longicornis barberi* (Coleoptera: Chrysomelidae). Ann. Ent. Soc. Am. 71: 931-934.
Krysan, J. L. and Smith, R. F. 1987a. Systematics of the *virgifera* species group of *Diabrotica* (Coleoptera: Chrysomelidae: Galerucinae). Entomography 5: 375-484.
Krysan, J. L. and Smith, R. F. 1987b. Three new species of *Diabrotica* (Coleoptera: Chrysomelidae). Entomography 5: 523-530.
Krysan, J. L., Branson, T. F., Guss, P. L. and Gustin, R. D. 1981. Corn rootworm biology. Eight annual Illinois Crop Protection Workshop, March 1981, Cooperative Extension Service, Univ. Illinois, Urbana, Champaign. pp 59-64.
Krysan, J. L., Foster, D. E., Branson, T. F., Ostlie, K. R. and Cranshaw, W. S., 1986. Two years before the hatch: rootworms adapt to crop rotation. Bull. ent. Soc. Am. 32: 250-253.
Krysan, J. L., Jackson, J. J. and Lew, A. C. 1984. Field termination of egg diapause in *Diabrotica* with new evidence of prolonged diapause in *D. barberi* (Coleoptera: Chrysomelidae). Envir. Ent. 13: 1237-1240
Krysan, J. L., McDonald, I. C., and Tumlinson, J. H. 1989. Phenogram based on allozymes and its relationship to classical biosystematics and pheromone structure among eleven diabroticites (Coleoptera: Chrysomelidae). Ann. Ent. Soc. Am. 82: 574-581.
Krysan, J. L., Smith, R. F., Branson, T. F., and Guss, P. L. 1980. A new subspecies of *Diabrotica virgifera* (Coleoptera: Chrysomelidae): description, distribution, and sexual compatibility. Ann. ent. Soc. Am. 73: 123-130.
Krysan, J. L., Smith, R. F., and Guss, P. L. 1983. *Diabrotica barberi* (Coleoptera: Chrysomelidae) elevated to species rank based on behavior, habitat choice, morphometrics, and geographical variation of color. Ann. ent. Soc. Am. 76: 197-204.
Krysan, J. L., Wilkin, P. H., Tumlinson, J. H., Sonnet, P. E., Carney, R. L., and Guss, P. L. 1986. Responses of *Diabrotica lemniscata* and *D. longicornis* (Coleoptera: Chrysomelidae) to stereoisomers of 8-methyl-2-decyl-propanoate and studies on the pheromone of *D. longicornis*. Ann. ent. Soc. Am. 79: 742-746.
Krueger, S. R. and Roberts, D. W. 1997. Soil treatments with entomopathogenic fungi for corn rootworm (*Diabrotica* spp.) larval control. Biol. Cont. 9: 67-74.
Kuhlman, D. E. 1970. Bionomics of *Diabrotica longicornis* (Say) *and Diabrotica virgifera* LeConte. PhD thesis, University of Nebraska. 115 pp.

Lance, D. R. 1992. Odors influence choice of oviposition sites by *Diabrotica virgifera virgifera* (Coleoptera: Chrysomelidae). J. Chem. Ecol. 18: 1227-1237.

Lance, D., Elliott, N. and Hein, G. 1989. Flight activity of *Diabrotica* sp. at the borders of cornfields and its relation to ovarian stage in *D. barberi*. Entomologia exp. appl. 50: 61-67

Lance, D. R. and Fisher, J. R. 1987. Food quality of various plant tissues for adults of the northern corn rootworm (Coleoptera: Chrysomelidae). J. Kans. ent. Soc. 60: 462-466.

Landis, D. A., Levine, E., Haas, M. J. and Meints, V., 1992. Detection of prolonged diapause of northern corn rootworm in Michigan (Coleoptera: Chrysomelidae). Gt. Lakes Ent. 25: 215-222.

Lawrence, W. S. and Bach, C. E. 1989. Chrysomelid beetle movements in relation to host plant size and surrounding non-host vegetation. Ecology. 70: 1679-1690.

Levine, E., and Gray, M. 1996. First-year corn rootworm injury: East-central Illinois research progress to date and recommendations for 1996. Proceedings, 1996 Illinois Agricultural Pesticides Conference. Univ. of Illinois, Urbana-Champaign. pp 3-13.

Levine, E. and Oloumi-Sadeghi, H., 1991. Management of the Diabroticite rootworms in corn. A. Rev. Ent. 36: 229-255.

Levine, E., and Oloumi-Sadeghi, H. 1996. Western corn rootworm (Coleoptera: Chrysomelidae) larval injury to corn grown for seed production following soybeans grown for seed production. J. econ. Ent. 89: 1010-1016.

Levine, E., Oloumi-Sadeghi, H., and Fisher, J. R., 1992. Discovery of multiyear diapause in Illinois and South Dakota northern corn rootworm (Coleoptera: Chrysomelidae) eggs and incidence of the prolonged diapause trait in Illinois. J. econ. Ent. 85: 262-267.

Levine, E., Oloumi-Sadeghi, H. and Ellis, C. R. 1992. Thermal requirements, hatching patterns, and prolonged diapause in western corn rootworm (Coleoptera: Chrysomelidae) eggs. J. econ. Ent. 85: 2425-2432.

Lew, A., and Ball, H. C. 1978. The structure of the apparent pheromone-secreting cells in female *Diabrotica virgifera*. Ann. ent. Soc. Am. 71: 685-688.

Lew, A., and Ball, H. C. 1979. the mating behavior of the western corn rootworm *Diabrotica virgifera* (Coleoptera: Chrysomelidae). Ann. ent. Soc. Am. 72: 391-393.

Lew, A., and Ball, H. C. 1980. Effect of copulation time on spermatozoan transfer in *Diabrotica virgifera* (Coleoptera: Chrysomelidae). Ann. ent. Soc. Am. 73: 360-361.

Lingren, P. S. 1998. Areawide management of Mexican corn rootworm (Coleoptera: Chrysomelidae) through adult suppression and crop rotation. Ph. D. Thesis, Texas A. and M. University, College Station.

Ludwig, K. A. and Hill, R. E. 1975. Comparison of gut contents of adult western and northern corn rootworms in northeast Nebraska. Envir. Ent. 4: 435-438.

Lummus, P. F., Smith, J. C. and Powell, N. L. 1983. Soil moisture and texture effects on survival of immature southern corn rootworms, *Diabrotica undecimpunctata howardi* Barber (Coleoptera: Chrysomelidae). Envir. Ent. 12: 1529-1531.

MacDonald, P. J. , and Ellis, C. R. 1990. Survival time of unfed, first-instar western corn rootworm (Coleoptera: Chrysomelidae) and the effects of soil type, moisture, and compaction on their mobility in soil. Envir. Ent. 19: 666-671.

McDonald, I. C., Krysan, J. L. and Johnson, O. A. 1995. Genetic variation within and among geographic populations of *Diabrotica barberi* (Coleoptera: Chrysomelidae). Ann. ent. Soc. Am. 78: 271-278.

Meinke, L. J. and Gould, F. 1987. Thermoregulation by *Diabrotica undecimpunctata howardi* and potential effects on overwintering biology. Entomologia exp. appl. 45: 115-122.

Melhus, I. E., Painter, R. H., and Smith, F. O. 1954. A search for resistance to the injury caused by species of *Diabrotica* in the corns of Guatemala. Iowa St. Col. J. Sci. 29: 75-94.

Metcalf, R. L. 1994. Chemical ecology of the diabroticites. In Jolivet, P. H., Cox, M. L. and Petitpierre, E. (eds) Novel aspects of the biology of Chrysomelidae. Kluwer Academic Publishers, Dordrecht. pp 153-169.

Naranjo, S. 1990. Comparative flight behavior of *Diabrotica virgifera virgifera* and D. *barberi* in the laboratory. Entomologia exp. appl. 55: 79-90.

Naranjo, S. 1991. Movement of corn rootworm beetles, *Diabrotica* species, at cornfield boundaries related to sex, reproductive status and crop phenology. Envir. Ent. 20: 230-240.

Naranjo, S. and Sawyer, A. J. 1987. Reproductive biology and survival of *Diabrotica barberi* (Coleoptera: Chrysomelidae): effect of temperature, food, and seasonal time of emergence. Ann. ent. Soc. Am. 80: 841-848.

Naranjo, S. and Sawyer, A. J. 1988. Impact of host plant phenology on the population dynamics and oviposition of northern corn rootworms, *Diabrotica barberi* (Coleoptera: Chrysomelidae), in field corn. Envir. Ent. 17: 508-521.

Naranjo, S. and Steinkraus, D. C. 1988. Discovery of an entomophthorlean fungus (Zygomycetes: Entomophthorales) infecting northern corn rootworm, *Diabrotica barberi* (Coleoptera: Chrysomelidae). J. Invertebr. Path. 51: 298-300.

Nickle, W. R., Schroder, R. F. W. and Krysan, J. L. 1984. A new Peruvian Hexamermis sp. (Nematoda: Mermithidae) parasite of corn rootworms, *Diabrotica* spp. Proc. Helminthol. Soc. Wash. 51:212-216.

Nishida, R. and Fukami, H. 1990. Sequestration of distasteful compounds by some pharmacophagous insects. J. Chem. Ecol. 16: 151-164.

Olalquiaga, F. 1980. Aspectos fitosanitarios do la Isla de Pascua. Revta chil. Ent. 10: 101-102.

Oloumi-Sadeghi, H., and Levine, E. 1989a. Controlling fungi that colonize eggs of the western corn rootworm in the laboratory. Entomologia exp. appl. 50: 271-280.

Oloumi-Sadeghi, H., and Levine, E. 1989b. Effect of starvation and time of egg hatch on larval survival of western corn rootworm, *Diabrotica virgifera virgifera* (Coleoptera: Chrysomelidae), in the laboratory. J. Kans. ent. Soc. 62: 108-118.

Onstad, E. W., Joselyn, M. G., Isard, S. A., Levine, E., Spencer, J. L., Bledsoe, L. W., Edwards, C. R., Di Fonzo, C. E. and Willson, H. 1998. Modeling the spread of western corn rootworm (Coleoptera: Chrysomelidae) populations adapting to soybean-corn rotation. Envir. Ent. in press.

Ostlie, K. R. 1987. Extended diapause: northern corn rootworm adapts to crop rotation. Crops Soils Mag. 39: 23-25.

Palmer, D. F., French, L. K. and Chiang, H. C. 1977. Effects of chill temperatures on developing western corn rootworm eggs. Envir. Ent. 6: 862-864.

Piedrahita, O., Ellis, C. R., and Bogart, J. P. 1985. Interaction of northern and western corn rootworm larvae (Coleoptera: Chrysomelidae) in a controlled environment. Envir. Ent. 14: 138-141.

Prystupa, B., Ellis, C. R. and Teal, P. E. A. 1988. Attraction of adult *Diabrotica* (Coleoptera: Chrysomelidae) to corn silks and analysis of the host-finding response. J. Chem. Ecol. 14: 635-651.

Quiring, D. T. and Timmins, P. R. 1990. Influence of reproductive ecology on feasibility of mass trapping *Diabrotica virgifera virgifera* (Coleoptera: Chrysomelidae). J. Appl. Ecol. 27: 965-982.

Riedell, W. E. 1993. Advances in understanding corn rootworm damage effects on maize. Proceedings of the Forty-Eighth Annual Corn and Sorghum Industry Research Conference, 1993. Am. Seed Trade Assn., Washington. pp 76-90.

Riedell, W. E. and Kim, A. Y. 1990. Anatomical characterization of western corn rootworm damage to adventitious roots of maize. J. Iowa Acad. Sci. 97: 15-17.

Riedell, W. E., and Sutter, G. R. 1995. Soil moisture and survival of western corn rootworm larvae in field plots. J. Kans. ent. Soc. 68: 80-84.

Risch, S. 1980. The population dynamics of several herbivorous beetles in a tropical agroecosystem: the effect of intercropping corn, beans and squash in Costa Rica. J. Appl. Ecol. 17: 593-612.

Risch, S. 1981. Insect herbivore abundance in tropical monocultures and polycultures: an experimental test of 2 hypotheses. Ecology 62: 1325-1340.

Rodriguez-del-Bosque, L. A. and Magallanes-Estala, A. 1994: Seasonal abundance of *Diabrotica balteata* and other Diabroticina beetles (Coleoptera:Chrysomelidae) in Northeastern Mexico. Envir. Ent. 23: 1409-1415.

Ruesink, W. R. 1986. Egg sampling techniques. In: Krysan, J. L. and Miller, T. A. (eds) Methods for the study of pest *Diabrotica*. Springer-Verlag, New York. pp 83-99.

Saba, F. 1970. Host plant spectrum and temperature limitations of *Diabrotica balteata*. Can. Ent. 102: 648-691.

Sammons, A. E., Edwards, C. R., Bledsoe, L. W., Boeve, P. J., and Stuart, J. J. 1997. Behavorial and feeding assays reveal a western corn rootworm (Coleoptera: Chrysomelidae) variant that is attracted to soybean. Envir. Ent. 26: 1336-1342.

Schalk, J. M. 1986. Rearing and handling of *Diabrotica balteata*. In: Krysan, J. L. and Miller, T. A. (eds), Methods for the study of pest *Diabrotica*. Springer-Verlag, New York. pp 49-56.

Schalk, J. M. and Jones, A. 1982. Methods to evaluate sweet potatoes for resistance to the banded cucumber beetle in the field. J. econ. Ent. 75: 76-79.

Schenk, M. and Saunders, J. L. 1984. Vegetation management systems and insect responses in humid tropics of Costa Rica. Trop. Pest Mgmnt. 30: 186-193.

Shaw, J. T., Paullus, J. H., and Luckmann, W. H. 1978. Corn rootworm oviposition in soybean. J. econ. Ent. 71: 189-191.

Sherwood, D. R. and Levine, E. 1993. Copulation and its duration affects female weight, oviposition, hatching patterns, and ovarian development in the western corn rootworm (Coleoptera: Chrysomelidae). J. econ. Ent. 86: 1664-1671.

Siegfried, B. D. and Mullin, M. 1990. Effects of alternative host plants on longevity, oviposition and emergence of western and northern corn rootworms (Coleoptera: Chrysomelidae). Envir. Ent. 19: 474-480.

Silva-Werneck, J. O., de Faria, M. R., Abreu Neto, J. R. M. V., Magalhaes, B. P., and Schmidt, F. G. V. 1995. Rearing technique of *Diabrotica speciosa* (Germ.)(Coleoptera: Chrysomelidae) for bioassays with entomopathogenic bacteria and fungi. An. Soc. Entomol. Brasil 24: 45-52.

Smith, C. E. and Allen, N. 1932. The migratory habit of the spotted cucumber beetle. J. econ. Ent. 25: 53-57.

Smith, R. F. 1966. The distribution of diabroticites in western North America. Bull. ent. Soc. Am. 12: 108-110.

Spencer, J. L., Isard, S. A., and Levine, E. 1997. Corn rootworm injury to first-year corn: new research findings. Proceedings, 1997 Illinois Agricultural Pesticides Conference. University of Illinois, Urbana-Champaign. pp 73-81.

Spencer, J. L., Isard, S. A., and Levine, E. 1998. Western corn rootworms on the move: monitoring beetles in corn and soybeans. Proceedings, 1998 Illinois Agricultural Pesticides Conference. University of Illinois, Urbana-Champaign. pp 10-23.

Stamm, D. E., Mayo, Z B, Campbell, J. B., Witkowski, J. F., Andersen, and Kozub, R. 1985. Western corn rootworms (Coleoptera: Chrysomelidae) beetle counts as a means of making larval control recommendations in Nebraska. J. econ. Ent. 78: 794-798.
Stanley, W. W. 1932. Altitude records for *Diabrotica duodecimpunctata* (Fabricius) in Tennessee. J. econ. Ent. 25: 1311.
Steffey, K. L., Gray, M. and Kuhlman, D. 1992. Extent of corn rootworm damage in corn after soybeans: search for the expression of extended diapause trait in Illinois. J. econ. Ent. 85: 268-275.
Stewart, J. W., Cocke, J. Jr., Taylor, J. and Williams, R. 1995. Mexican corn rootworm emergence from corn fields where sorghum was grown the previous season. S.-west Entomologist 20: 229-30.
Strnad, S. P. and Bergman, M. K. 1987a. Movement of first instar western corn rootworms (Coleoptera: Chrysomelidae) in soil. Envir. Ent. 16: 975-978.
Strnad, S. P., and Bergman, M. K. 1987b. Distribution and orientation of western corn rootworm (Coleoptera: Chrysomelidae) larvae in corn roots. Envir. Ent. 16: 1193-98.
Sweetman, H. L. 1926. Results of life-history studies of *Diabrotica 12-punctata* Fabr. (Chrysomelidae, Coleoptera). J. econ. Ent. 19: 484-490.
Tallamy, D. W. and Halaweish, F. T. 1993. Effects of age, reproductive activity, sex and prior exposure on sensitivity to cucurbitacins in southern corn rootworm (Coleoptera: Chrysomelidae). Envir. Ent. 22: 925-932.
Tallamy, D. W., Whittington, D. P., Defurio, F., Fontaine, D. A., Gorski, P. M. and Gothro, P. W. 1998. Sequestered cucurbitacins and pathogenicity of *Metarhizium anisopliae* (Moniliales:Moniliaceae) on Spotted Cucumber Beetle Eggs and Larvae (Coleoptera:Chrysomelidae). Envir. Ent. 27: 366-372.
Tate, H. D. and Bare, O. S. 1946. Corn rootworms. Neb. St. Agric. Exp. Sta. Bull. 381: 1-12.
Tauber, M. J., Tauber, C. A. and Masaki, S. 1986. Seasonal Adaptations of Insects. Oxford, New York.
Teng, H-J., V. Waddill, V., Slansky, F. and Strager. F. 1984. Performance and host preference of adult banded cucumber beetles (*Diabrotica balteata*) when offered several crops. J. Agricult. Entomol. 1: 330-338.
Terrell, J. 1974. American Indian Almanac. Crowell, New York. 494 pp.
Thompson, W. R. 1943. A Catalogue of the Parasites and Predators of Insect Pests. Imperial Agricultural Bureaux, Institute of Entomology. Belleville, Ontario. p 40.
Tigano-Milani, M. S., Carneiro, R. G., De Faria, M. R., Frazao. H. S., and McCoy, C. W., 1995. Isozyme characteristics and pathogenicity of *Paecilomyces fumosoroseus* and *P. lilacinus* to *Diabrotica speciosa* (Coleoptera: Chrysomelidae) and *Meloidogyne javanica* (Nematoda: Tylenchidae). Biological Cont. 5: 378-382.
Turpin, F. T. and Peters, D. C. 1971. Survival of southern and western corn rootworm larvae in relation to soil texture. J. econ. Ent. 64: 1448-1451.
Tyler, B. M. J., and Ellis, C. R. 1979. Ground beetles in three tillage plots in Ontario and observations on their importance as predators of the northern corn rootworm, *Diabrotica longicornis* (Coleoptera: Chrysomelidae). Proc. ent. Soc. Ontario. 110: 65-73.
VanWoerkom, G. J., Turpin, F. T. and Barrett, J. R., Jr. 1980. Influence of constant and changing temperatures on locomotor activity of adult western corn rootworms (*Diabrotica virgifera*) in the laboratory. Envir. Ent. 9: 32-34.
VanWoerkom, G. J., Turpin, F. T. and Barrett, J. R., Jr. 1983. Wind effects on western corn rootworm (Coleoptera: Chrysomelidae) flight behavior. Envir. Ent. 12: 196-200.
Waddill, V. H., 1978. Sex differences in foraging on corn of adult *Labidura riparia* (Dermaptera: Labiduridae). Entomophaga 23: 337-342.
Webster, F. M. 1895. On the probable origin, development, and diffusion of the North American species of the genus *Diabrotica*. Jl. N. Y. ent. Soc. 3: 158-166.
Webster, F. M. 1913. The western corn rootworm. U.S.D.A. Bur. Ent. Bull. No. 8.
Whitaker, J. O. Jr. 1995. Food of big brown bat *Eptesicus fuscus* from maternity colonies in Indiana and Illinois. Am. Midland Naturalist. 134: 346-360.
Weiss, M. J., Mayo, Z B, and Newton, J. P. 1983. Influence of irrigation practices on the spatial distribution of corn rootworm (Coleoptera: Chrysomelidae) eggs in soil. Envir. Ent. 12: 1293-1295.
Wilcox, J. A. 1972: Coleopterorum Catalogus. Supplementa. (edita secunda). (Galerucinae, Luperini) Pars 78, Fasc. 2:296-431.
Witkowski, J. F. and Owens, J. C. 1979. Corn rootworm behavior in response to trap corn. Iowa St. J. Res. 53: 317-324.
Witkowski, J. F., Owens, J. C. and J. J. Tollefson, J. J. 1975. Diel activity and vertical flight distribution of adult western corn rootworms in Iowa cornfields. J. econ. Ent. 68: 351-352.
Woodson, W. D. 1993. Effect of species composition on the survival and development of western and northern corn rootworms (Coleoptera: Chrysomelidae). J. Kans. ent. Soc. 66: 377-382.
Woodson, W. D. 1994. Interspecific and intraspecific larval competition between *Diabrotica virgifera virgifera* and *Diabrotica barberi* (Coleoptera: Chrysomelidae). Envir. Ent. 23: 612-616.
Woodson, W. D. and Ellsbury, M. M. 1994. Low temperature effects on hatch of northern corn rootworm eggs (Coleoptera: Chrysomelidae). J. Kans. ent. Soc. 67: 102-106.

Woodson, W. D. and Gustin, R. D. 1993. Low temperature effects on hatch of western corn rootworm eggs (Coleoptera: Chrysomelidae). J. Kans. ent. Soc. 66: 104-107.
Woodson, W. D. and Jackson, J. J. 1996. Development rate as a function of temperature in northern corn rootworm (Coleoptera: Chrysomelidae). Ann. Entomol. Soc. Am. 89: 226-230.
Woodson, W. D., Jackson, J. J. and Ellsbury, M. M. 1996. Northern corn rootworm (Coleoptera: Chrysomelidae) temperature requirements for egg development. Ann. ent. Soc. Am. 89: 898-903.

THE BLACK *LONGITARSUS* SPECIES ASSOCIATED WITH BORAGINACEAE IN SOUTH AFRICA (COLEOPTERA, CHRYSOMELIDAE, ALTICINAE)

MAURIZIO BIONDI
Department of Environmental Sciences, University of L'Aquila, 67010 – Coppito (AQ), Italy

Abstract

In this contribution some taxonomic and ecological observations on the black species of the genus *Longitarsus* Berthold associated with Boraginaceae in South Africa are reported. *L. capensis* Baly, 1877 is re-described and six new species (*L. cedarbergensis*, *L. luctuosus*, *L. lugubris*, *L. melanicus*, *L. neseri*, *L. transvaalensis*) are described. Moreover the following new combinations are also proposed: *Longitarsus apicipes* Jacoby, 1903 = *Aphthona apicipes* (Jacoby, 1903); *Longitarsus barkeri* Jacoby, 1903 = *Aphthona barkeri* (Jacoby, 1903); *Aphthona barkeri* Jacoby, 1903 = *Gabonia barkeri* (Jacoby, 1903); *Longitarsus dunbrodensis* Jacoby, 1903 = *Phyllotreta dunbrodensis* (Jacoby, 1903).

Introduction

The botanical family Boraginaceae includes many host-plants for the genus *Longitarsus* Berthold.

In the Western Palaearctic area, where this flea beetle genus is better known also from the autoecological point of view, some species-groups, such as those of *anchusae*, *exoletus* and *echii*, are closely associated with these plants for their life-cycles.

On the basis of a recent study carried out in the Mediterranean area (Biondi, 1996), 50 species, equal to 26.4% of all known *Longitarsus* for this geographical region, are significantly associated with Boraginaceae. Therefore this botanical family is the most frequently used by these leaf beetles.

Moreover, in accordance with known data, the Palaearctic species of *Longitarsus* living on Boraginaceae are generally widespread in Central Asia and in the lands around the Mediterranean Basin (including the Macaronesian islands). On the contrary, these species rarely occur in the northern part of Asia and Europe. From an environmental point of view, the Palaearctic *Longitarsus* associated with Boraginaceae preferentially live both in steppic and parasteppic areas, also at high altitude, or in Mediterranean scrub.

As regards the Afrotropical Region, ecological data concerning the numerous *Longitarsus* is very scarce and practically lacking. Only one species, *Longitarsus punctifrons* Weise, 1895 (= *gossypii* Bryant, 1950), of the about 100 Afrotropical species of *Longitarsus* mentioned in the literature, is reported on Boraginaceae (*Heliotropium spp.*) (cf. Furth, 1985).

During my collecting trips in South Africa, I indicated some black species of *Longitarsus* feeding on leaves of Boraginaceae plants of the genus *Lobostemon* Lehm., endemic in the

Advances in Chysomelidae Biology 1, pp. 515–531
Edited by M.L. Cox

western area of South Africa and represented by about 30 species. In addition, other species of Boraginaceae, such as *Anchusa capensis*, or the alien invasive *Echium plantagineum* (= *lycopsis*), can be secondarily used as host-plants by these flea beetles.

Among the South African *Longitarsus* discussed in this note, only the species *capensis* Baly, 1877 from the Cape Province was already described, while six other species (*cedarbergensis*, *luctuosus*, *lugubris*, *melanicus*, *neseri*, and *transvaalensis*) are new to science. Moreover, it is significant that Jacoby (1903) described the species *L. apicipes*, *L. barkeri*, and *L. dunbrodensis* from South Africa (Cape Province and Natal). On the basis of the original descriptions, they seemed to belong to this group of species. However, by the examination of the relative types preserved at the Natural History Museum of London, I pointed out that these species really belong to the genus *Aphthona* Chevrolat (*apicipes* and *barkeri*) or to the genus *Phyllotreta* Stephens (*dunbrodensis*). Therefore I propose the following new combinations:

Longitarsus apicipes Jacoby, 1903 = *Aphthona apicipes* (Jacoby, 1903) comb. nov.
Longitarsus barkeri Jacoby, 1903 = *Aphthona barkeri* (Jacoby, 1903) comb. nov.[1]
Longitarsus dunbrodensis Jacoby, 1903 = *Phyllotreta dunbrodensis* (Jacoby, 1903) comb. nov.

In Jacoby's collection I found also the type of a further species from South Africa (Cape Province), *Longitarsus nigritus*. This species was never published by Jacoby, but it was quoted by Bryant (1959: 213) as "*Longitarsus nigritus* Jac.". Anyway this species also belongs to the genus *Aphthona*.

The South African *Longitarsus* species, the object of this note, show a close affinity with the Western-Palaearctic species of the *anchusae*-group, also associated with Boraginaceae.

These flea beetle species share many morphological and chromatic characters and similar aedeagal and spermathecal shape. The common presence so qualitatively and quantitatively significant of character-states between these two groups of species, also referred to no-adaptive structures, is indicative of their real and close relationship and it allows the exclusion of other hypotheses based on the scarcely probable phenomena of convergence for explaining the similarities that can be found between the Mediterranean and South African faunas. The scarce morphological differentiation existent between these two zoogeographical components suggests that their separation very probably happened recently. However some aspects regarding true or presumed affinities between the Western Palaearctic and South African fauna, will be discussed more widely and deeply in a future contribution (Biondi, in prep.).

The morphological and chromatic characters that the species of the *anchusae*-group share with the South African species associated with Boraginaceae, are as follows:

- integument black, often with more or less evident metallic reflection;
- elytra with subtruncate or independently rounded apices;
- elytral puncturation never with signs of striae, not even in the sutural area (this charac-

[1] The name *barkeri* was already used by Jacoby (1903) for a new species of *Aphthona* Chevrolat. I examined the type of this species preserved in the BM(NH) standing under the genus *Gabonia* Jacoby. Therefore the new nomenclatorial status is as follows: *Longitarsus barkeri* Jacoby, 1903 = *Aphthona barkeri* (Jacoby, 1903) comb. nov. and *Aphthona barkeri* Jacoby, 1903 = *Gabonia barkeri* (Jacoby, 1903) comb. nov.

teristic is important for distinguishing the species of this group from the black species associated with Labiatae);
– hind wings generally reduced;
– spermatheca big (normally LSP > 0.30 mm); spermathecal duct robust, widely arcuated, at most with one coil;
– aedeagus strongly sclerotized, generally little slender, with ventral sulcus always evident.

Abbreviations used

LAE: aedeagal length; LE: elytral length; LP: pronotal length; LSP: spermathecal length; WP: pronotal width; CB: collection of the author; MCZR: Civic Museum of Zoology of Rome, Italy; BM(NH): Natural History Museum of London, England; PPRI: Plant Protection Research Institute of Pretoria, Republic of South Africa.

List of species

***Longitarsus capensis* Baly, 1877**
TYPE MATERIAL. Lectotype (des. M. Biondi, 1997): "Cape of Good Hope", ♂ [BM(NH)]: ex Coll. Baly).
OTHER MATERIAL EXAMINED. South Africa: Western Cape Province, Cape Town, Table Mountain, Tafelberg road, 350 m, 33°55' S, 18°28' E, 25.x.1988, E. Colonnelli leg., 9 exx. (MCZR); ibidem, Tafelberg road, 300-500 m, 21.ix.1994, P. Audisio, M. Biondi & M. Bologna leg., 24 exx. (CB); Western Cape Province, Cape Town, Table Mountain, North-

Table 1. Morphometric data (see text for abbreviations).

♂♂	LE (mm)	LP (mm)	WP (mm)	LAE (mm)	LE/LP	WP/LP	LE/LAE
capensis	1.27 ± 0.11	0.47 ± 0.04	0.64 ± 0.04	0.82 ± 0.06	2.69 ± 0.11	1.35 ± 0.06	1.58 ± 0.07
cedarbergensis	1.15	0.43	0.56	0.74	2.58	1.30	1.54
luctuosus	1.38 ± 0.09	0.53 ± 0.05	0.70 ± 0.05	0.82 ± 0.04	2.60 ± 0.01	1.33 ± 0.05	1.64 ± 0.12
lugubris	1.20 ± 0.07	0.48 ± 0.03	0.65 ± 0.04	0.73 ± 0.03	2.53 ± 0.07	1.36 ± 0.05	1.67 ± 0.06
melanicus	1.49	0.49	0.66	0.84	3.06	1.36	1.77
neseri	1.37 ± 0.04	0.51 ± 0.02	0.69 ± 0.02	0.90 ± 0.03	2.66 ± 0.08	1.35 ± 0.01	1.52 ± 0.02
transvaalensis	1.18 ± 0.07	0.43 ± 0.03	0.58 ± 0.02	0.76 ± 0.01	2.69 ± 0.02	1.32 ± 0.04	1.55 ± 0.11
♀♀	LE (mm)	LP (mm)	WP (mm)	LSP (mm)	LE/LP	WP/LP	LE/LSP
capensis	1.53 ± 0.15	0.55 ± 0.05	0.75 ± 0.07	0.30 ± 0.01	2.76 ± 0.07	1.35 ± 0.04	5.03 ± 0.39
cedarbergensis	1.26 ± 0.06	0.45 ± 0.01	0.61 ± 0.01	0.29 ± 0.01	2.77 ± 0.07	1.36 ± 0.01	4.33 ± 0.34
luctuosus	1.58 ± 0.10	0.57 ± 0.05	0.77 ± 0.06	0.31 ± 0.02	2.78 ± 0.14	1.35 ± 0.04	5.15 ± 0.28
lugubris	1.38 ± 0.09	0.51 ± 0.03	0.72 ± 0.04	0.33 ± 0.01	2.68 ± 0.10	1.39 ± 0.05	4.10 ± 0.20
melanicus	1.63 ± 0.11	0.59 ± 0.04	0.78 ± 0.04	0.32 ± 0.02	2.77 ± 0.06	1.33 ± 0.04	5.10 ± 0.24
neseri	1.65 ± 0.07	0.58 ± 0.02	0.80 ± 0.03	0.30 ± 0.02	2.85 ± 0.09	1.38 ± 0.04	5.43 ± 0.30
transvaalensis	1.33 ± 0.03	0.47 ± 0.01	0.68 ± 0.02	0.30 ± 0.01	2.81 ± 0.06	1.43 ± 0.04	4.38 ± 0.04

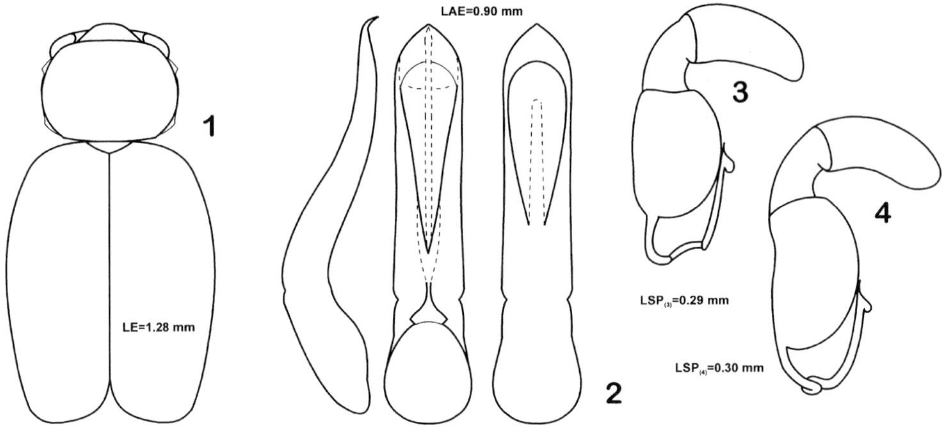

Figs. 1–4. Longitarsus capensis Baly. 1: dorsal habitus ♂ ; 2: median lobe of aedeagus in lateral, ventral and dorsal view; 3-4: spermatheca.

ern slope, 150-400 m, 23.x.1993, P. Audisio leg., 2 exx. (CB); ibidem, M. Bologna leg., 1 ex. (CB); Western Cape Province, 12 km ESE of Langabaan (R27), 30 m, 10.ix.1994 (fynbos and sandy grassland), P. Audisio, M. Biondi & M. Bologna leg., 8 exx. (CB); Western Cape Province, Cape Peninsula, Ocean View (1 km SW of M65), 120 m, 22.ix.1994 (damp fynbos), P. Audisio, M. Biondi & M. Bologna leg., 4 exx. (CB); Western Cape Province, Cape Peninsula, Cape of Good Hope Nature Reserve, 0-150 m, 22.ix.1994, P. Audisio, M. Biondi & M. Bologna leg., 10 exx. (CB).

DISTRIBUTION. South Africa: Western Cape Province.

REDESCRIPTION. Morphometric data are summarised in Tab. 1. Dorsal integument black, generally with weak bronze or greenish metallic reflection. Body shape (Fig. 1) elongate-oval, moderately convex. Maximum pronotal width in apical third; maximum elytral width in basal third.

Head with vertex and frons impunctate, but only with rather smooth or very slightly shagreened surface; interocular area with some large punctures near ocular margin; frontal tubercles scarcely delimited, elongate, oblique, with smooth surface; frontal grooves in distal part clearly divergent from ocular margin; frontal carina little raised, with rounded vertex; labrum piceous, triangular; palpi strongly blackened; eyes elliptical elongate; antennae generally with segment 1 partially obscured, segments 1-3 reddish and others gradually darker; length of each segment of right antenna generally proportional to following numerical sequences: 12:6:6:7:8:8:10:8:10:10:13 (♂) and 13:8:8:8:9:8:10:10:10:10:14 (♀).

Pronotum transverse, laterally rounded; basal and lateral margin finely bordered; puncturation clearly impressed, dense, homogeneously distributed; surface from smooth to clearly reticulate.

Elytra elongate, apically subtruncate, laterally rounded; pygidium in most part uncovered; puncturation very similar to that of pronotum but always impressed on rather smooth surface; humeral callus absent; subapterous wings in both sexes. Scutellum half-roundish, with clearly shagreened surface.

Legs generally entirely pale with exception of hind femora and ungual segments more or less widely blackened (sometimes also front and middle femora can be partially dark);

hind tibiae straight and clearly dentate on outer side from basal fourth; apical spur of hind tibiae relatively elongate, reddish.

Ventral part entirely black. Last tergite with clearly reticulate surface, weak and sparse points, and whitish pubescence.

Male: first front and middle tarsomeres weakly dilated. Last sternite without preapical impressions. Median lobe of aedeagus (Fig. 2) little elongate with smooth surface; in ventral view laterally subparallel and apically widely rounded with a small median slightly raised tooth; ventral sulcus wide in apical part, gradually narrower in basal part and interrupted in basal fifth; surface of ventral sulcus characterised by presence of a narrow, long, and rather deep longitudinal groove; dorsal sulcus wide, visible in distal half; median lobe in lateral view curved in basal third and apical fourth; apex clearly bent ventrally.

Female: generally larger than male; first front and middle tarsomeres not dilated; pygidium without preapical impressions; spermatheca (Figs. 3, 4) with basal part little elongate, sometimes subglobose, rarely reniform; distal part elongate, clearly distinct from basal part; spermathecal duct moderately elongate, generally with signs of a coil.

ECOLOGICAL NOTES. This species was collected in fynbos vegetation on plants of two *Lobostemon* species (one with pink flowers and the other with pale blue flowers with petals apically whitish) and on *Anchusa capensis*.

***Longitarsus cedarbergensis* n. sp.**

TYPE MATERIAL. Holotype (♂): South Africa, Western Cape Province, road Clanwilliam-Pakhuis Pass, m 200-900, 11.ix.1994 (fynbos and xeric grassland), P. Audisio, M. Biondi & M. Bologna leg. (CB). Paratypes: same locality, date and collectors of the holotype, 2 ♀♀ (with eggs) (CB).

DIAGNOSIS. This new species is externally very similar to *L. neseri* n.sp., but it is easily distinguishable by the different median lobe of aedeagus and spermathecal shape.

DISTRIBUTION. South Africa: Western Cape Province.

DESCRIPTION OF THE HOLOTYPE (♂). Dorsal integument piceous-brown with evident greenish metallic reflection. Body shape (Fig. 5) elongate-oval, scarcely convex.

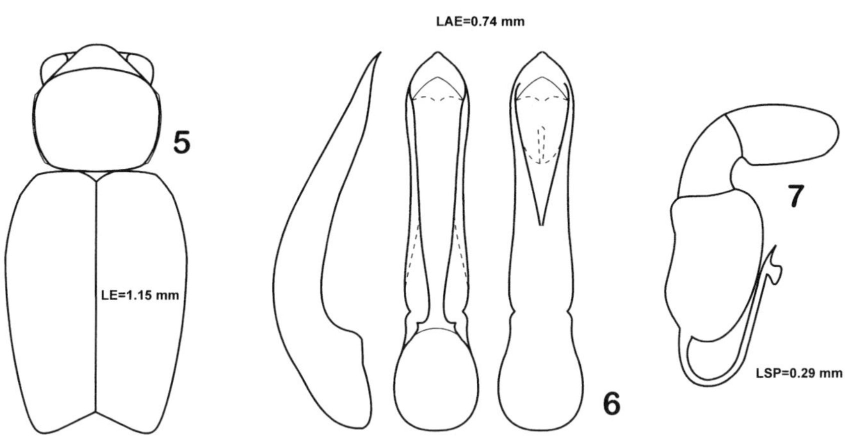

Figs. 5–7. Longitarsus cedarbergensis n.sp. 5: dorsal habitus ♂; 6: median lobe of aedeagus in lateral, ventral and dorsal view; 7: spermatheca.

Maximum pronotal width at middle: 0.58 mm; maximum elytral width at basal third: 0.80 mm.

Head with vertex and frons very finely shagreened; some points impressed along frontal grooves; frontal tubercles small, elongate, subhorizontal, scarcely delimited, with rather smooth surface; frontal grooves distally clearly divergent from ocular margin; frontal carina weakly raised; labrum black, distally broadly rounded; palpi dark brown; eyes roundish; antennae with segments 1-4 yellowish, 5-8 reddish, 9-11 brown; length of each segment of right antenna proportional to numerical sequence 12:7:6:8:9:8:10:10:10:10:14.

Pronotum scarcely transverse (LP = 0.43 mm; WP/LP = 1.30), laterally slightly rounded, narrower at base; lateral and basal margin finely bordered; puncturation clearly impressed, dense and homogeneously distributed; surface distinctly microreticulate, particularly laterally.

Elytra elongate (LE = 1.15 mm; LE/LP = 2.58), laterally moderately rounded, apically obliquely subtruncate; pygidium in most part uncovered; puncturation more strongly impressed than on pronotum, subrugose on sutural area; surface rather smooth; humeral calli absent; subapterous. Scutellum subtriangular, with finely shagreened surface.

Legs, including hind femora, entirely pale; hind tibiae dentate and slightly curved from basal third; apical spur of hind tibiae visible, reddish; first front and middle tarsomeres dilated.

Ventral part entirely black. Last sternite without preapical impressions; last tergite with distinctly microreticulate surface, covered by sparse whitish pubescence.

Median lobe of aedeagus (Fig. 6) little slender (LAE = 0.74 mm; LE/LAE = 1.54); in ventral view very weakly narrow in middle part; apical part moderately rounded, with a small obtuse median tooth; ventral sulcus long, rather wide; dorsal sulcus V-shaped, visible in apical third; median lobe in lateral view bent at basal third and then straight towards apex.

VARIABILITY OF THE PARATYPES. The two paratypes examined show some small differences with respect to the holotype, since hind femora are darker and pronotal puncturation is more strongly impressed, subrugose. Female bigger than male; first front and middle tarsomeres clearly narrower. Spermatheca (Fig. 7) with subreniform basal part; distal part little developed, distinct from basal part; ductus elongate, uncoiled. Morphometric data are summarised in Tab. 1.

ETYMOLOGY. This new species is named after the Cedarbergensis Mountains (South Africa, Western Cape Province).

ECOLOGICAL NOTES. *L. cedarbergensis* n.sp. was collected in fynbos vegetation on plants of *Lobostemon* sp. a shrub with oblong leaves and single pink flowers, together with specimens of *L. lugubris* n. sp.

***Longitarsus luctuosus* n. sp.**

TYPE MATERIAL. Holotype (♂): South Africa, Western Cape Province, between Akkedisberg Pass and Standford (R326), 60-160 m (fynbos), 20.ix.1994, P. Audisio, M. Biondi & M. Bologna leg. (CB). Paratypes: same locality, date and collectors of the holotype, 5 ♂♂ and 9 ♀♀ (CB). South Africa, Western Cape Province, NNW of Stellenbosch, 33°50' S, 18°46' E, 13.vii.1995, S. Neser leg.. (on *Echium plantagineum* L., leaves with holes), 15 ♂♂ and 20 ♀♀ (with eggs) (PPRI; CB).

DIAGNOSIS. In the aedeagal and spermathecal shape, this new species shows many similarities to *L. melanicus* n.sp., but it is distinguishable mainly by: different shape of the first

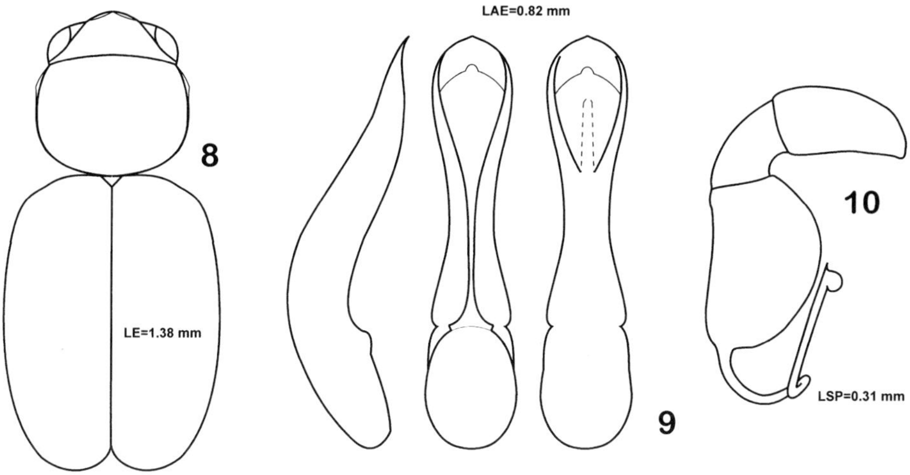

Figs. 8–10. Longitarsus luctuosus n.sp. 8: dorsal habitus ♂; 9: median lobe of aedeagus in lateral, ventral and dorsal view; 10: spermatheca.

front tarsomere in male; elytra apically less elongate and generally more strongly punctured.

DISTRIBUTION. South Africa: Western Cape Province.

DESCRIPTION OF THE HOLOTYPE (♂). Dorsal integument black, with weak greenish metallic reflection, particularly on elytra. Body shape (Fig. 8) elongate-oval, moderately convex. Maximum pronotal width at middle: 0.78 mm; maximum elytral width at middle: 1.22 mm.

Head with vertex and frons finely punctured, with rather smooth surface; some large points impressed along frontal grooves and laterally on frons; frontal tubercles with smooth surface, clearly delimited, elongate, subhorizontal; frontal grooves distally clearly divergent from ocular margin; frontal carina moderately wide, slightly raised, with broadly rounded vertex; labrum blackened, subrectangular; palpi piceous; eyes elliptical; antennae with segment 1 partially dark, segments 2-4 yellowish and others gradually darker towards apex; length of each segment of right antenna proportional to numerical sequence 14:9:8:8:10:10:10:10:10:12:14.

Pronotum scarcely transverse (LP = 0.59 mm; WP/LP = 1.32), laterally slightly rounded; basal and lateral margin finely bordered; puncturation moderately impressed, rather dense, homogeneously distributed; surface distinctly reticulate.

Elytra elongate (LE = 1.46 mm; LE/LP = 2.45), laterally rounded, apically subtruncate; pygidium in most part uncovered; puncturation more strongly impressed than on pronotum; surface dorsally nearly smooth and laterally clearly smooth; humeral calli absent; subapterous. Scutellum triangular, with finely shagreened surface.

Legs reddish, with exception of front and middle femora and two distal tarsomeres partially dark, and hind femora blackened; hind tibiae slightly curved from basal fourth and dentate on outer side from basal third to apex; apical spur of hind tibiae reddish, distinctly visible; first front and middle tarsomeres moderately dilated.

Ventral part entirely black. Last sternite without dimples or preapical impressions; last sternite with finely reticulate surface, covered by sparse whitish pubescence.

Median lobe of aedeagus (Fig. 9) moderately slender (LAE = 0.82 mm; LE/LAE = 1.77); in ventral view laterally clearly narrow in middle and apically lanceolate; ventral sulcus wide in apical third and strongly narrow at base; dorsal sulcus V-shaped, visible in apical fourth; median lobe in lateral view curved in basal 2/3 and almost straight in apical part.

VARIABILITY OF THE PARATYPES. From the chromatic and sculptural point of view, the paratypes are similar to the holotype. However some small difference can be pointed out such as: head with frons and vertex with almost obliterated puncturation; antennae basally more reddish; pronotal surface rather smooth; elytra apically more obliquely truncate; median lobe apically more elongate. Female normally larger than male; first front and middle tarsomeres narrow; last sternite without preapical impressions. Spermatheca (Fig. 10) with reniform basal part; distal part developed, scarcely distinct from basal one; ductus moderately elongate, normally with a narrow coil in middle part. Morphometric data are summarised in Tab. 1.

ETYMOLOGY. The name of this new species refers to the black coloration of the integument.

ECOLOGICAL NOTES. *L. luctuosus* n.sp. was collected in fynbos vegetation on plants of *Lobostemon* sp. and *Echium plantagineum*.

***Longitarsus lugubris* n. sp.**

TYPE MATERIAL. Holotype (♂): South Africa, Northern Cape Province, 3-8 km SW of Nieuwoudtville, 700-760 m, 15.ix.1994 (fynbos and grassland), P. Audisio, M. Biondi & M. Bologna leg. (CB). Paratypes: same locality, date and collectors of holotype, 7 ♂♂ and 16 ♀♀ [CB; BM(NH)]; South Africa, Western Cape Province, road Clanwilliam-Pakhuis Pass, 200-800 m, 11.ix.1994 (fynbos and xeric grassland), P. Audisio, M. Biondi & M. Bologna leg., 10 ♂♂ and 8 ♀♀ (CB; PPRI).

DIAGNOSIS. This new species shows many similarities to *L. cedarbergensis* n. sp. It is distinguishable mainly by: hind femora blackened; dorsal integument without evident greenish metallic reflection; median lobe of aedeagus with ventral sulcus interrupted at base; spermathecal duct more robust and widely arcuated.

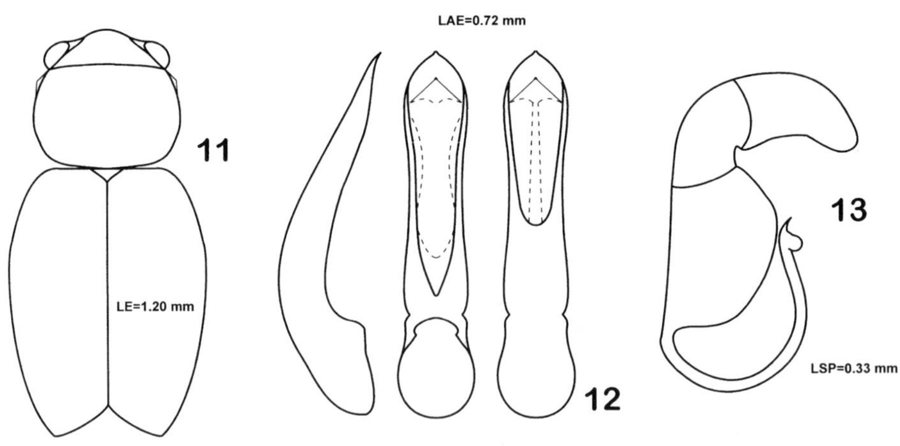

Figs. 11–13. Longitarsus lugubris n.sp. 11: dorsal habitus ♂; 12: median lobe of aedeagus in lateral, ventral and dorsal view; 13: spermatheca.

DISTRIBUTION. South Africa: Northern and Western Cape Province.
DESCRIPTION OF THE HOLOTYPE (♂). Dorsal integument black. Body shape (Fig. 11) elongate-oval, moderately convex. Maximum pronotal width at middle: 0.63 mm; maximum elytral width at basal third: 0.90 mm.

Head with vertex and frons impunctate, with rather smooth surface; some points impressed in interocular area along frontal grooves: frontal tubercles scarcely delimited, elongate, oblique, with smooth surface; frontal grooves distally clearly divergent from ocular margin; frontal carina little raised, with rounded vertex; labrum subrettangular, strongly blackened; palpi piceous-brown; eyes subelliptical elongate; antennae with segment 1 darkened, segments 2-4 reddish, others gradually darker; length of each segment of right antenna proportional to numerical sequence 12:7:6:8:10:8:10:9:9:10:14.

Pronotum little transverse (LP = 0.45 mm; WP/LP = 1.42), laterally regularly rounded; basal and lateral margin finely bordered; puncturation clearly impressed, rather dense and homogeneously distributed; surface finely reticulate.

Elytra elongate (LE = 1.19 mm; LE/LP = 2.67), laterally rounded, apically obliquely subtruncate; pygidium in most part uncovered; puncturation similar to pronotum but impressed on more smooth surface; humeral calli absent; subapterous. Scutellum semi-circular, with finely shagreened surface.

Legs entirely pale, with exception of hind femora darkened; hind tibiae slightly curved from basal fourth and dentate from basal third; apical spur of hind tibiae distinctly visible, reddish; first front and middle tarsomeres moderately dilated.

Ventral part entirely black. Last tergite with subrugose surface covered by dense, whitish pubescence; last sternite with a wide and shallow oval preapical dimple in middle.

Median lobe of aedeagus (Fig. 12) little slender (LAE = 0.72 mm; LE/LAE = 1.66); surface smooth with exception of ventral sulcus membranous; in ventral view weakly narrow in middle; apical part broadly rounded with a small median tooth; ventral sulcus wide in apical part, gradually narrower towards base and interrupted at about basal fifth; dorsal sulcus V-shaped, visible in apical third; in lateral view, median lobe bent in apical fourth in ventral direction and straight in basal ¾.
VARIABILITY OF THE PARATYPES. The paratypes are rather similar to the holotype. Small chromatic and sculptural differences regard mainly: dorsal integument sometimes with weak bronze metallic reflection particularly on elytra; first antennomere more or less browned, rarely entirely reddish; pronotal puncturation moderately or strongly impressed. Female larger than male; first front and middle tarsomeres narrow; last sternite without preapical impression. Spermatheca (Fig. 13) with subreniform basal part; distal part well developed, clearly distinct from basal one; duct widely arcuated and uncoiled. Morphometric data are summarised in Tab. 1.
ETYMOLOGY. The name of this new species refers to the black coloration of the integument.
ECOLOGICAL NOTES. *L. lugubris* n.sp. was collected in fynbos vegetation on plants of *Lobostemon* sp. a shrub with oblong leaves and single pink flowers.

***Longitarsus melanicus* n. sp.**
TYPE MATERIAL. Holotype (♂): South Africa, Western Cape Province, Danabaai, 6 km W of Mossel Bay, 0-60 m (sand dunes and fynbos), 19.ix.1994, P. Audisio, M. Biondi & M. Bologna leg. (CB). Paratypes: same locality, date and collectors of the holotype, 7 ♀♀ (with eggs) (CB).

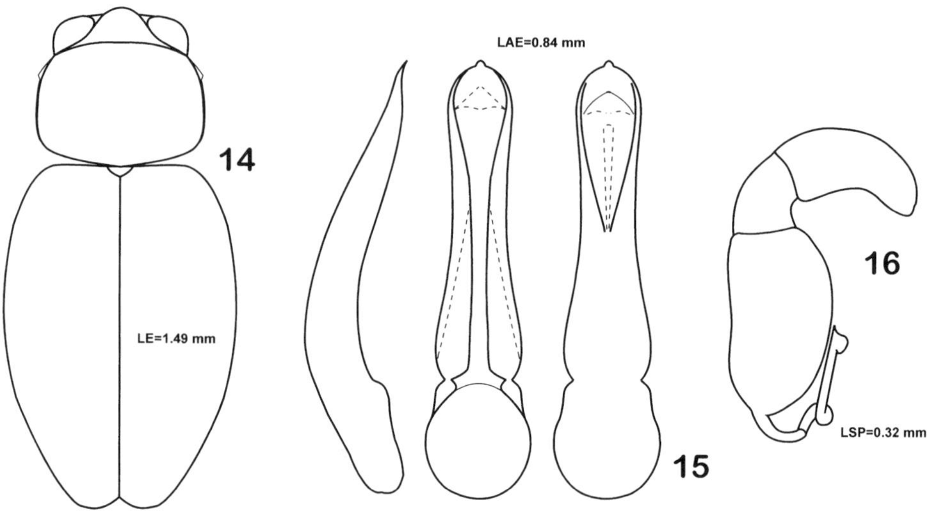

Figs. 14–16. Longitarsus melanicus n.sp. 14: dorsal habitus ♂; 15: median lobe of aedeagus in lateral, ventral and dorsal view; 16: spermatheca.

DIAGNOSIS. In the aedeagal and spermathecal shape, this new species shows many similarities with *L. luctuosus* n.sp., from which it is externally distinguishable mainly by: different shape of first front tarsomere in male; elytra apically more elongate and generally more finely punctured.

DISTRIBUTION. South Africa: Western Cape Province.

DESCRIPTION OF THE HOLOTYPE (♂). Dorsal integument black with weak bronze metallic reflection. Body shape (Fig. 14) elongate-oval, moderately convex. Maximum pronotal width in basal third: 0.66 mm; maximum elytral width in middle: 0.95 mm.

Head with vertex very finely shagreened; interocular area laterally with a single point impressed near frontal groove; frontal tubercles small, scarcely delimited, elongate, subhorizontal, with almost smooth surface; frontal grooves distally clearly divergent from ocular margin; frontal carina wide, little raised, with rounded vertex; labrum subrectangular, strongly blackened; palpi piceous-brown; eyes elliptical, little elongate; antennae with segments 1-3 pale and others gradually infuscate; length of each segment of right antenna proportional to numerical sequence 14:7:6:8:10:8:9:8:9:9:12.

Pronotum scarcely transverse (LP = 0.49 mm; WP/LP = 1.36), laterally slightly rounded; lateral and basal margin finely bordered; puncturation clearly impressed, rather dense and homogeneously distributed; surface very clearly microreticulate.

Elytra elongate (LE = 1.49 mm; LE/LP = 3.06), laterally rounded and apically not truncate but independently rounded; pygidium only apically uncovered; puncturation similar to pronotum, but impressed on smooth surface, slightly duller in sutural area; humeral calli absent; subapterous. Scutellum triangular, with finely shagreened surface.

Hind and middle legs with partially browned femora, tibiae and tarsi; hind legs with entirely blackened femora, reddish tibiae and tarsi, brown ungual segments; hind tibiae slightly curved from basal third and dentate from basal fourth; apical spur of hind tibiae distinctly visible, reddish; first front and middle tarsomeres clearly dilated, subcylindrical.

Ventral part entirely black. Last sternite without preapical impressions; last tergite with microreticulate surface and scarcely dense whitish pubescence.

Median lobe of aedeagus (Fig. 15) moderately slender (LAE = 0.84 mm; LE/LAE = 1.77); in ventral view slightly narrow in apical third; apex broadly rounded, with a median small rounded tooth; ventral sulcus basally not interrupted, wide in apical fifth and clearly and gradually narrower towards base; dorsal sulcus V-shaped, visible in apical third; in lateral view median lobe bent in basal fourth in ventral direction, then straight towards apex.

VARIABILITY OF THE PARATYPES. Compared to the holotype, some paratypes show small differences as bronze metallic reflection more evident and often not dark front tibiae. Female larger than male; hind and middle tarsi strongly narrow; last sternite without preapical impressions. Spermatheca (Fig. 16) with subreniform basal part; distal part moderately developed and distinct from basal one; duct little elongate, moderately robust, with a coil in middle part. Morphometric data are summarised in Tab. 1.

ETYMOLOGY. The name of this new species refers to the black coloration of the integument.

ECOLOGICAL NOTES. *L. melanicus* n.sp. was collected in fynbos vegetation on plants of *Lobostemon* sp.

Longitarsus neseri n. sp.

TYPE MATERIAL. Holotype (♂): South Africa, Western Cape Province, Piketberg, 12°44' S 29°25' E, 8.vii.1986, S. Neser leg. (on *Echium lycopsis*) (PPRI). Paratypes: same locality, date and collectors of the holotype, 5 ♂♂ and 9 ♀♀ (with eggs) (PPRI; CB).

DIAGNOSIS. This new species is externally very similar to *L. cedarbergensis* n.sp., from which it is easily distinguishable by the different aedeagal and spermathecal shape.

DISTRIBUTION. South Africa: Western Cape Province.

DESCRIPTION OF THE HOLOTYPE (♂). Dorsal integument black with evident greenish metallic reflection. Body shape (Fig. 17) elongate-oval, moderately convex. Maximum pronotal width at middle: 0.69 mm; maximum elytral width in basal third: 0.95 mm.

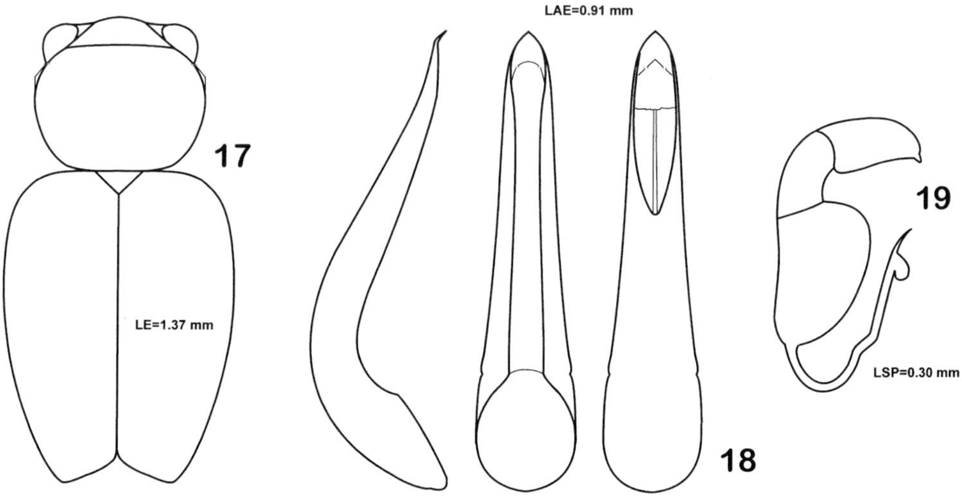

Figs. 17–19. Longitarsus neseri n.sp. 17: dorsal habitus ♂; 18: median lobe of aedeagus in lateral, ventral and dorsal view; 19: spermatheca.

Head with vertex and frons very finely shagreened; some large points impressed along frontal grooves; frontal tubercles elongate, oblique, subhorizontal, well delimited, with finely shagreened surface; frontal grooves distally clearly divergent from ocular margin; frontal carina narrow, moderately raised; labrum piceous-brown, subrectangular; palpi partially dark; eyes elliptical; antennae with segments 1-5 yellowish and 6-11 gradually darker; length of each segment of right antenna proportional to numerical sequence 14:8:8:10:10:10:11:11:10:10:14.

Pronotum scarcely transverse (LP = 0.52 mm; WP/LP = 1.35), laterally rounded, slightly narrower at base; lateral and basal margin finely bordered; puncturation clearly impressed, dense and homogeneously distributed; surface almost smooth.

Elytra elongate (LE = 1.33 mm; LE/LP = 2.56), laterally rounded, apically obliquely subtruncate; pygidium in most part uncovered; puncturation similar to pronotum, but impressed on clearly reticulate surface particularly on disc; humeral calli absent; subapterous. Scutellum subtriangular, with finely shagreened surface.

Legs, including hind femora, entirely pale; hind tibiae dentate and slightly curved from basal third; apical spur of hind tibiae visible, reddish; first front and middle tarsomeres dilated.

Ventral part entirely black. Last sternite with a narrow median preapical longitudinal incision; last tergite with distinctly microreticulate surface, covered by sparse whitish pubescence.

Median lobe of aedeagus (Fig. 18) elongate (LAE = 0.91 mm; LE/LAE = 1.50); in ventral view gradually narrowing from base to apex; apical part acute, with only a sign of small rounded median tooth; ventral sulcus wide, subparallel, not interrupted; dorsal sulcus U-shaped visible in apical half; in lateral view median lobe curved and with apex clearly bent in ventral direction..

VARIABILITY OF THE PARATYPES. The other specimens attributed to this new species are very similar to the holotype. Female larger than male; first front and middle tarsomeres clearly narrower. Spermatheca (Fig. 19) with subreniform basal part; distal part little developed, distinct from basal one; duct elongate, uncoiled, but clearly sinuous in ascendent tract. Morphometric data are summarised in Tab. 1.

ETYMOLOGY. The new species is named after its collector Stefan Neser (Republic of South Africa).

ECOLOGICAL NOTES. *L. neseri* n.sp. was collected on plants of the invasive alien *Echium plantagineum* (=*lycopsis*); the true host-plant is still unknown but, considering the affinities of this new species, also in this case it could be a *Lobostemon* sp.

***Longitarsus transvaalensis* n. sp.**

TYPE MATERIAL. Holotype (♂): South Africa, Eastern Transvaal, Nelshoogte Pass, 1300-1600 m, 25°49' S 30°48' E, 14.xii.1995 (forest edge), M. Biondi leg. (CB). Paratypes: same locality, date and collector of the holotype, 1 ♂ and 2 ♀♀ (CB).

DIAGNOSIS. In its median lobe shape, with apical part clearly lanceolate, and the elytral apices regularly rounded, *L. transvaalensis* n.sp. occupies a rather isolated position within this group of species.

DISTRIBUTION. South Africa: Eastern Transvaal.

DESCRIPTION OF THE HOLOTYPE (♂). Dorsal integument black with weak bronze metallic reflection. Body shape (Fig. 20) elongate-oval, moderately convex. Maximum

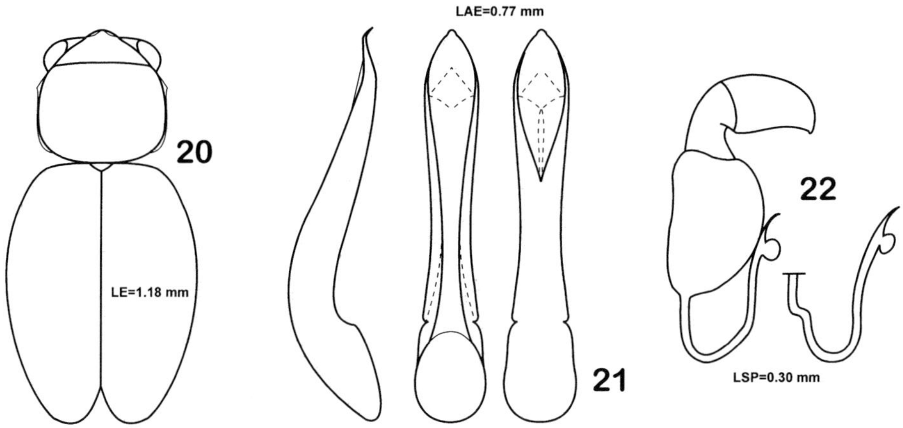

Figs. 20–22. Longitarsus transvaalensis n.sp. 20: dorsal habitus ♂; 21: median lobe of aedeagus in lateral, ventral and dorsal view; 22: spermatheca.

pronotal width near anterior angles: 0.57 mm; maximum elytral width in basal third: 0.81 mm.

Head with vertex and frons very finely shagreened; frontal tubercles small, elongate, slightly oblique, dorsally not well delimited; frontal grooves clearly marked and distally decisively divergent from ocular margin; frontal carina moderately raised, with rounded vertex; labrum black, subtrapezoidal; palpi piceous-brown; eyes elliptical; antennae blackened, with exception of segments 2-3 and apical half of segment 1; length of each segment of right antenna proportional to numerical sequence 14:8:8:10:12:10:12:12:12:12:15; left antenna with segments 3-11 lacking.

Pronotum transverse (LP = 0.42 mm; WP/LP = 1.35), laterally slightly rounded and narrower at base; lateral and basal margin finely bordered; anterior and posterior angles with long pubescence; puncturation weakly impressed on clearly reticulate surface.

Elytra elongate (LE = 1.13 mm; LE/LP = 2.71), laterally and apically rounded; pygidium in most part uncovered; puncturation more evident than on pronotum and more weakly impressed on apical declivity; surface rather smooth; humeral calli absent; subapterous. Scutellum semi-circular, with smooth surface.

Front and middle legs with blackened femora and tarsi, and weakly darkened tibiae; hind legs with entirely black femora and obscured tarsi; hind tibiae straight and dentate on outer side from basal fourth; apical spur of hind tibiae visible, reddish; first anterior and middle tarsomeres moderately dilated, subcylindrical.

Ventral part entirely black. Last sternite without preapical impressions; last tergite with clearly microreticulate surface with rather dense whitish pubescence.

Median lobe of aedeagus (Fig. 21) elongate (LAE = 0.77 mm; LE/LAE = 1.47), apically lanceolate; in ventral view laterally clearly narrow in middle part; ventral sulcus long and deeply excavated; apical part with a small rounded median tooth; dorsal sulcus V-shaped visible in distal half; in lateral view median lobe bent in basal third and clearly sinuous apically.

VARIABILITY OF THE PARATYPES. From the chromatic and sculptural point of view the paratypes are very similar to the holotype. Female slightly larger than male; first front

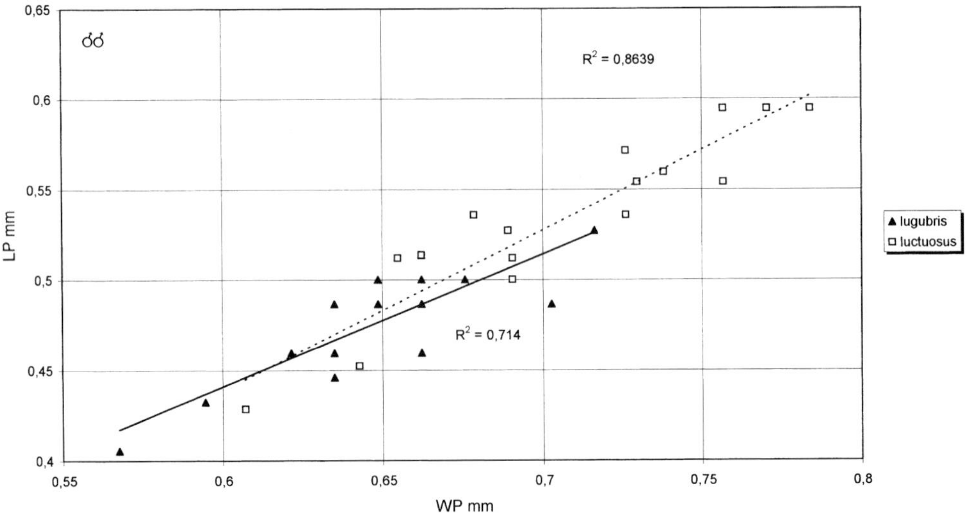

Fig. 23

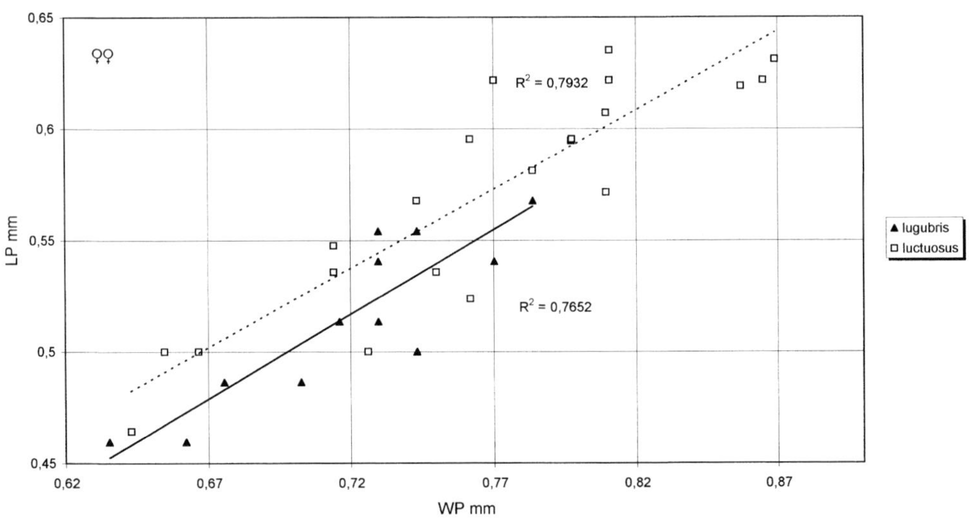

Fig. 24

Figs. 23–24. Width and length of pronotum in *Longitarsus lugubris* n.sp. and *L. luctuosus* n.sp. 23: males; 24: females.

and middle tarsomeres narrow and little elongate, subconical. Spermatheca (Fig. 22) with basal part subglobose; distal part well developed, thickset; ductus moderately elongate uncoiled or with only a slight indication of coil. Morphometric data are summarised in Table 1.

ETYMOLOGY. This new species is named after "Transvaal", which is a geographical region of northern-eastern South Africa.
ECOLOGICAL NOTES. *L. transvaalensis* n. sp. was collected in forest edge on leaves of a unidentified plant belonging to the Borage family.

Key to species

This key allows the identification of the known black South African species of *Longitarsus* associated with Boraginaceae. Since females sometimes do not have reliable diagnostic characters, this key could be not fully useful for them.

1. Hind femora pale. Dorsal integument with evident greenish metallic reflection. Spermathecal duct uncoiled 2
– Hind femora partially or completely blackened. Dorsal integument generally without evident greenish metallic reflection. Spermathecal duct often with a coil 3
2. Size smaller [LE+LP < 1.60 (♂♂) and < 1.80 (♀♀)]. Male with first front and middle tarsomeres little dilated. Median lobe of aedeagus (Fig. 5) in ventral view narrow in middle part and obtusely rounded at apically; ventral sulcus strongly narrow at base. Spermatheca (Fig. 7) comparatively less thickset, with less developed distal part; duct not undulate .... *L. cedarbergensis* n. sp.
– Size larger [LE+LP > 1.60 (♂♂) and > 1.80 (♀♀)]. Male with first front and middle tarsomeres more distinctly dilated. Median lobe of aedeagus (Fig. 18) in ventral view gradually narrow from base to apex and apically acute; ventral sulcus subparallel, not narrow at base. Spermatheca (Fig. 19) comparatively more thickset, with more developed distal part; duct clearly undulate in ascendent tract .... *L. neseri* n. sp.
3. Male with first front tarsomere clearly dilated, subcylindrical, as long as second and third tarsomeres together. Elytral puncturation generally formed by smaller and dense points, impressed on rather matt surface. Median lobe of aedeagus (Fig. 15) with apex broadly rounded, a median small rounded tooth, and ventral sulcus basally not interrupted. Spermathecal duct (Fig. 16) always with a narrow coil .... *L. melanicus* n. sp.
– Male with first front tarsomere less dilated, shorter than second and third tarsomeres together. Elytral puncturation generally formed by larger and sparse punctures, impressed on more smooth surface 4
4. Hind tibiae straight in male. Pronotal surface generally duller with finer puncturation. Median lobe of aedeagus (Figs. 2, 21) in lateral view with apex clearly bent ventrally 5
– Hind tibiae slightly but clearly curved in male. Pronotal surface generally more smooth with coarser puncturation. Median lobe of aedeagus (Figs. 9, 12) in ventral view with apex very slightly bent ventrally 6
5. Front and middle femora strongly blackened in basal half. Size smaller [generally LE+LP < 1.70 mm (♂♂) and < 1.85 mm (♀♀)]. Median lobe of aedeagus (Fig. 21) comparatively elongate, with apical part more clearly lanceolated. Spermathecal duct (Fig. 22) uncoiled or slightly undulate .... *L. transvaalensis* n. sp.
– Front and middle femora not blackened. Size larger [generally LE+LP > 1.70 mm (♂♂)

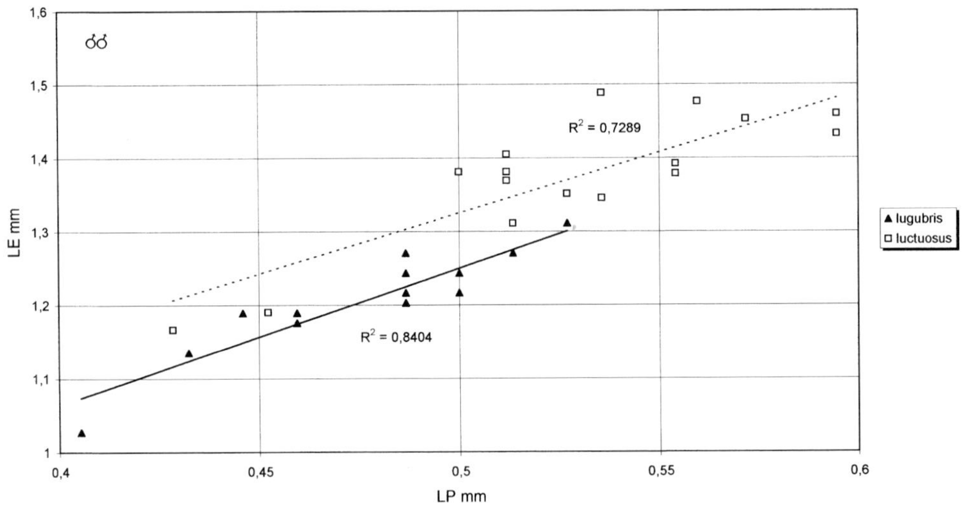

Fig. 25

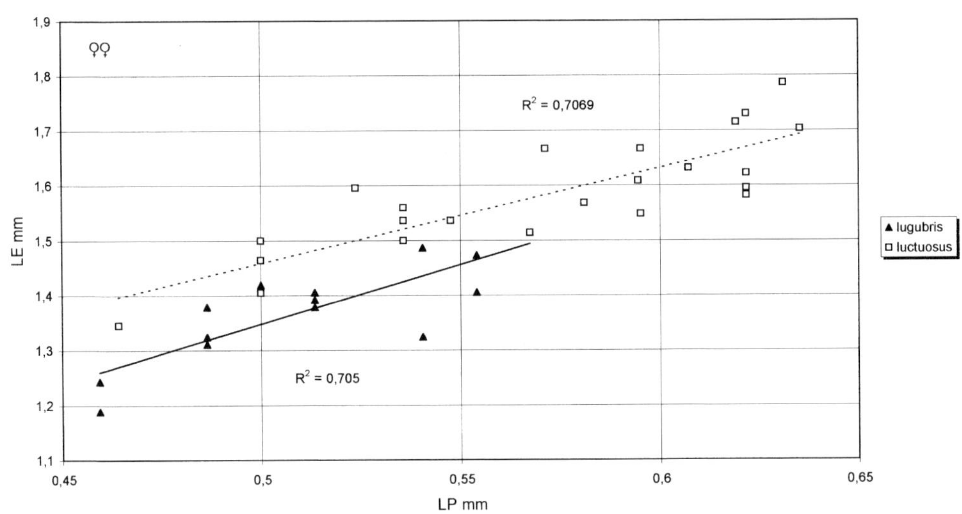

Fig. 26

Figs. 25–26. Elytral and pronotal length in *Longitarsus lugubris* n.sp. and *L. luctuosus* n.sp. 25: males; 26: females.

and > 1.85 mm (♀♀)]. Median lobe of aedeagus (Fig. 2) comparatively thickset, with apical part little lanceolate and with ventral sulcus characterised by a narrow middle longitudinal incision. Spermathecal duct (Figs. 3, 4) with a coil .... *L. capensis* Baly

6. Size generally larger (LE+LP: ♂♂ = 1.91 ± 0.13 mm; ♀♀ = 2.15 ± 0.15 mm) (Figs. 25, 26). Pronotum less transverse (WP/LP: ♂♂ = 1.33 ± 0.05; ♀♀ = 1.35 ± 0.05)

(Figs. 23, 24). Median lobe of aedeagus (Fig. 9) in ventral view strongly narrow in middle part; ventral sulcus not interrupted at base. Spermathecal duct (Fig. 10) with a coil .... *L. luctuosus* n. sp.

– Size generally smaller (LE+LP: ♂♂ = 1.68 ± 0.10 mm; ♀♀ = 1.90 ± 0.12 mm) (Figs. 25, 26). Pronotum more transverse (WP/LP: ♂♂ = 1.36 ± 0.05; ♀♀ = 1.39 ± 0.05) (Figs. 23, 24). Median lobe of aedeagus (Fig. 12) in ventral view very slightly narrow in middle part; ventral sulcus interrupted at base. Spermathecal duct (Fig. 13) uncoiled, widely arcuated .... *L. lugubris* n. sp.

Acknowledgements

I am indebted to following colleagues for providing me with material and data used in this study: Paolo Aldo Audisio, Marco Alberto Bologna, Enzo Colonnelli, Vincenzo Vomero, Sharon Lane Shute and Elizabeth Grobbelaar. Finally I would like thank Dr. Michael Cox for inviting me to contribute to this volume and for the linguistic improvements to the text.

References

Baly, J.S. 1877. Descriptions of new genera and of uncharacterized species of Halticinae. Trans. ent. Soc. London, 10 (4): 283-323.

Biondi, M. 1996. Proposal for an ecological and zoogeographical categorization of the Mediterranean species of the flea beetle genus *Longitarsus* Berthold. Chrysomelidae Biology. Vol. 3: General Studies. Jolivet P.H.A. & Cox M.L. (Eds), SPB Academic Publishing bv, Amsterdam, The Netherlands: 13-35

Bryant, G.E. 1959. Coleoptera: Chrysomelidae I. South African Animal Life, 6 (11): 194-226.

Furth, D.G. 1985. Some flea beetles and their foodplants from Kenya (Chrysomelidae: Alticinae). The Coleopterists Bulletin, 39 (3): 259-263.

Jacoby, M. 1903. A further contribution to our knowledge of African phytophagous Coleoptera. Part II. Trans. ent. Soc. London, 36: 1-38.

THE FLEA-BEETLE *ALTICA CORNI* WOODS IN NORTH AMERICA (COLEOPTERA: CHRYSOMELIDAE, ALTICINAE)

LAURENT LESAGE and JOCELYN DENIS
Agriculture Canada, Eastern Cereal and Oilseed Research Centre, Ottawa, Ontario K1A 0C6, Canada

Abstract

Altica corni Woods is redescribed including, for the first time, characters of male and female genitalia. Lectotypes are designated. Information on distribution, biology, host plants and parasites is also reported.

Introduction

Altica is a large genus of flea beetles with over 100 known species in North America. The costate species, a natural group of large species which are distinguished by a lateral costa on the elytra, have been recently revised (LeSage 1995). Non-costate species cannot yet be split into natural groups because information on their genitalic structures is not available. Consequently, the purpose of this paper is to describe in detail *A. corni*, including new information on male and female genitalia.

The materials, methods, and acronyms of collections mentioned here are the same as those used previously for the revision of the costate species (LeSage 1995).

Materials

Specimens studied here were obtained from:
JFLC : Jean-François Landry personal Collection. Aylmer, Québec.
UMC : University of Maine Collection, Orono, Maine 04469, U.S.A., S.E. Woods.

Systematic chorology

Altica corni Woods 1918:156; Blatchley 1921:24; Brown 1946:52; Wilcox 1954:447; Hatch 1971:215; Wilcox 1974:110; LeSage 1991:318; Downie & Arnett 1996:1372.
Haltica corni Woods: Leng 1920:300; Chagnon 1938:163; Heikertinger & Csiki 1940:238; Chagnon & Robert 1962:316.

Advances in Chysomelidae Biology 1, pp. 533–544
Edited by M.L. Cox

Fig. 1. Habitus of *Altica corni* Woods, dorsal view.

ETYMOLOGY. From *Cornus*, the generic Latin name of the dogwood host- plant.

DIAGNOSIS. Blackish species with bronzy-purplish metallic reflection and deep transverse pronotal groove (Fig. 1). In the male, median lobe of aedeagus with dorsal striations

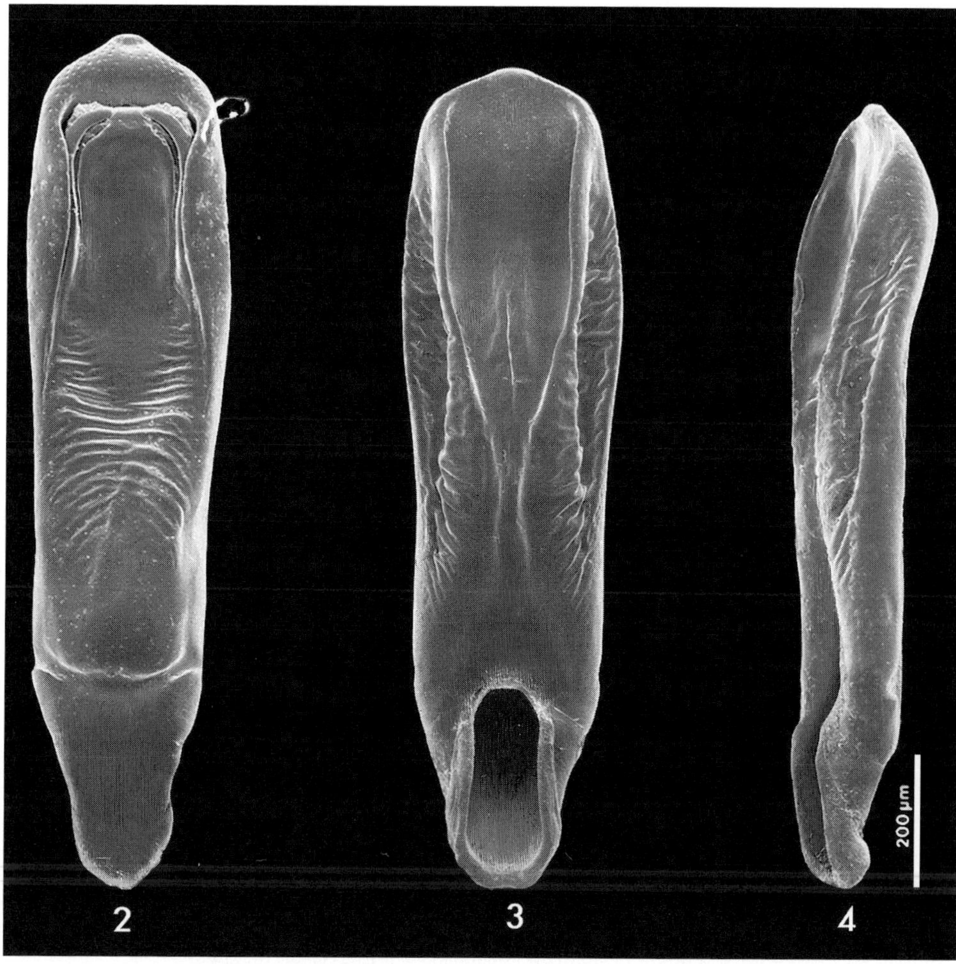

Figs. 2–4. Median lobe of the aedeagus of *Altica corni* Woods: 2, dorsal view; 3, ventral view; 4, lateral view.

(Fig. 2), and markedly high broadly separated ventral ridges (Fig. 4). In female, basal portion of the spermathecal duct elongate (Figs. 6, 7), and styli fused and obliquely divergent at apex (Fig. 8).

DESCRIPTION.

Body. Medium-sized species, 3-4 mm long (see Table 1 for measurements). Shape elongate, twice as long as wide. General coloration blackish with variable metallic reflection although bronzy-purplish color is most common (Table 2).

Head. Antennae proportionally longer in males than in females (Table 1); antennomeres 3 and 4 subequal, but distinctly longer than 2. Frontal carina moderately sharp, very finely punctulate, ending in middle of frontal tubercles. Frontal tubercles smooth, more or less well-defined posteriorly by interocular sulci. Coronal suture extending to frontal carina. Eyes not prominent, separated by 3 times their diameter. Only 1 pair of postocular

Table 1. Counts and measurements of structures found in the male (*n*= 10) and in the female (*n* =10) of *Altica corni*. Woods.

Character	Minimum	Maximum	Mean
Male			
Male body length (mm)	3.2	3.5	3.3
Male body width (mm)	1.6	1.8	1.7
Male length/body ratio	1.89	2.06	2.01
Male body/antenna ratio	1.44	1.7	1.54
Pronotum length (mm)	0.64	0.72	0.69
Pronotum width (mm)	1	1.12	1.05
Pronotum width/length ratio	1.43	1.63	1.51
Median lobe of aedeagus length (mm)	1.32	1.36	1.35
Median lobe of aedeagus width (mm)	0.32	0.36	0.34
Median lobe of aedeagus dorsal undulations	16	21	18
Median lobe of aedeagus latero-ventral wrinkles	14	25	19
Female			
Female body length (mm)	3.2	4	3.6
Female body width (mm)	1.7	2	1.8
Female length/width ratio	1.91	2.16	2
Female body/antenna ratio	1.52	1.8	1.63
Pronotum length (mm)	0.66	0.78	0.72
Pronotum width (mm)	1.06	1.2	1.13
Pronotum width/length ratio	1.51	1.7	1.57
Spermatheca length (um)	243	275	256
Style length (um)	288	358	336
Style width (um)	83	96	93
Style length/width ratio	3.2	4.31	3.61
Style apical setae	8	12	10
Style sensilla	8	14	11

Table 2. Color variations in adult *Altica corni* Woods (*n*= 100).

Coloration of the metallic reflections	Percentage
bronzy-purplish on pronotum and elytra	(65%)
bronze on pronotum and elytra	(23%)
purplish on pronotum and greenish on elytra	(7%)
bronze on pronotum and blueish on elytra	(3%)
greenish on pronotum and purplish on elytra	(1%)
greenish on pronotum and elytra	(1%)

macrochaetae present behind eyes. Vertex not punctate, feebly alutaceous, with 1-8 small punctures on each side of coronal suture. Labrum with 6 pairs of labral setae. Mandibles tridentate: outer and median teeth subequal in size, inner tooth curved mediad, cutting edge sinuous, basal indentation not visible (Fig. 5).

Thorax. Pronotum quadrate, widest in apical third, slightly narrower at apex than at base (Fig. 1). Anterior angles not prominent, obliquely truncate. Transverse groove deep, broadening in fovea before reaching lateral margins. Punctation fine and quite dense. Microsculpture variable: absent (50%), scarcely visible (45%), or weakly impressed (5%). Tarsal claws appendiculate, and moderately bent.

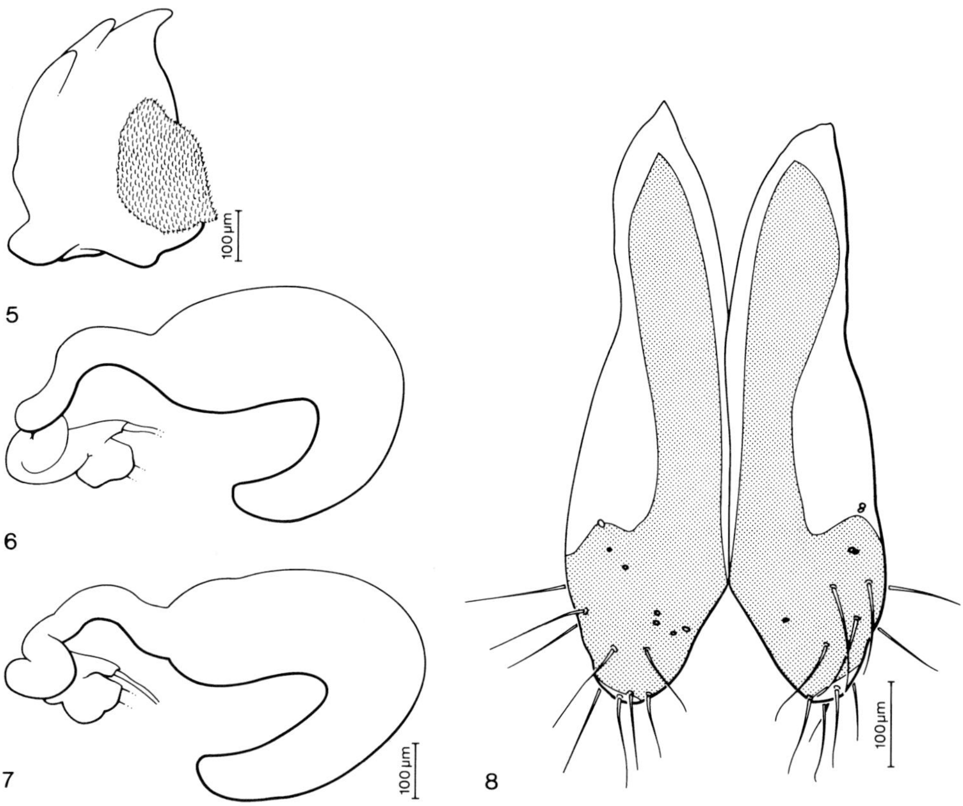

Figs. 5–8. Structures of the female genitalia of *Altica corni* Woods: 5, mandible, ventral view; 6, 7, variation in spermatheca; 8, styli..

Elytra. Humerus moderately prominent and defined on inner side by small depression. Elytral costa absent. Punctation moderately dense, coarser than that of pronotum, except at apex. Microsculpture in the form of numerous microreticulations, feeble but distinct.

Sexual dimorphism. Body slightly more elongate, antennae proportionally longer in male than in female (Table 1). First tarsomeres of front legs not distinctly broader in male than in female as usually found in most *Altica* species.

Male genitalia. Median lobe of aedeagus distinctly broadened in apical 1/4 in dorsal view, tip weakly nipple-shaped (Fig. 2); dorsal undulations distributed on 1/3 length of aedeagus, apical ones being interrupted medially; median lobe slightly bent in lateral view (Fig. 3); ventral longitudinal ridges markedly elevated and broadly separated (Fig. 4); latero-ventral wrinkles well-defined and numerous (Table1).

Female genitalia. Spermatheca slightly larger at base, with small ovoid receptacle; spermathecal pump cylindrical, not extending beyond base of receptacle, and without apical process (Figs. 6, 7); basal portion of spermathecal duct elongate and sinuous, followed by 2-2.5 loops. Styli largely fused together, and obliquely divergent at apex (Fig. 8); apical setae and sensilla moderately numerous (Table 1).

REMARKS. Adults of *Altica corni* are easily recognized externally by their dark metallic coloration with bronzy-purplish reflections and their deep pronotal groove (Fig. 1). The median lobe of the aedeagus is distinctly broadened before apex (Fig. 2), and the ventral ridges are broadly separated and high, a feature unique to this species (Fig. 3). Also, the first front tarsomeres are not broadened as they are usually in the males. In the female, styli are largely fused and tips are obliquely truncated (Fig. 8). These two characters separate females of *A. corni* from those of all other North American *Altica* species.

TYPE MATERIAL.

Lectotype. The selected specimen is a male which has been dissected and its aedeagus mounted on a plate. Typed on front side of label: "*Altica corni* Woods"; hand written on underside of label: "Orono/ Sept. 1917, Dogwood". The following designation label has been added: LECTOTYPE *Altica corni* Woods/ Des. LeSage 1997.

Paralelectotype. One female dissected and mounted on a plate with spermatheca and styli mounted separately on a transparent acetate plate. The first label reads: "Me Agr Exp Sta/ Orono Me/ 15 June 1917". The second label reads: "Lot 1791/ Sub 3/ Me Agr Exp Sta". The third label reads: "*Altica corni* Woods". The following designation label has been added: PARALECTOTYPE. "*Altica corni* Woods Des. LeSage 1997". Nine additional specimens have the same collection data except for the date: "July 17 (2), July 18 (4), July 23 (1), July 24 (1), July 27 (1)". Two specimens bear only two labels: a code "Lot 1781, Sub 13", and a determination label"*Altica corni* Woods". Finally, 6 specimens have only a determination label "*Altica corni* Woods". Appropriate designation labels have been added to all these specimens.

The primary type and 16 paralectotypes are preserved in the collection of the University of Maine, in Orono, and 2 paralectotypes are deposited in the Canadian National Collection, in Ottawa.

DISTRIBUTION. In Canada, coast to coast below the 49th parallel. In the United States, mostly east of the Great Lakes, but also with scattered locations westwards to Utah (Map 1).

Campbell & *al.* (1989) mentioned Northwest Territories, Canada, in the distribution of *Altica corni* but we have not seen any specimen from this area.

Locality records. Specimens examined: 1879.

For the meaning of acronyms see LeSage (1995). The new acronyms not listed in the previous paper are:

JFLC : Jean-François Landry personal Collection. Aylmer, Québec.
UMC : University of Maine Collection, Orono, Maine 04469, U.S.A., S.E. Woods.

– *CANADA.*

Alberta. SW Empress/ 4.2 km S Junction Hwys 41 & 555 (ISAC)

British Columbia. *Cariboo dist.*: Finlay Forks (UBC); *Coast co.*: Terrace (CNC); *Kootenay co.*: Atbara (CUIC, UBC), Bull River/ 6 miles from mouth (UBC), Creston (CMNC, CUIC, UBC), Robson (CNC), Wynndel (CNC, UBC, USNM); *New Westminster co.*: Hatzic (CNC), Lillooet river near John Sandy Creek (UBC), Rosedale (UBC); *Osoyoos co.*: Trinity Valley (UBC);*Yale co.*: Boundery Creek Provicial Park (ISAC), Manning Provincial Park/ Skagit River (CNC).

Manitoba. Brokenhead River/ crossing at Highway (CNC), / Hwy 15, 20 km E Anola

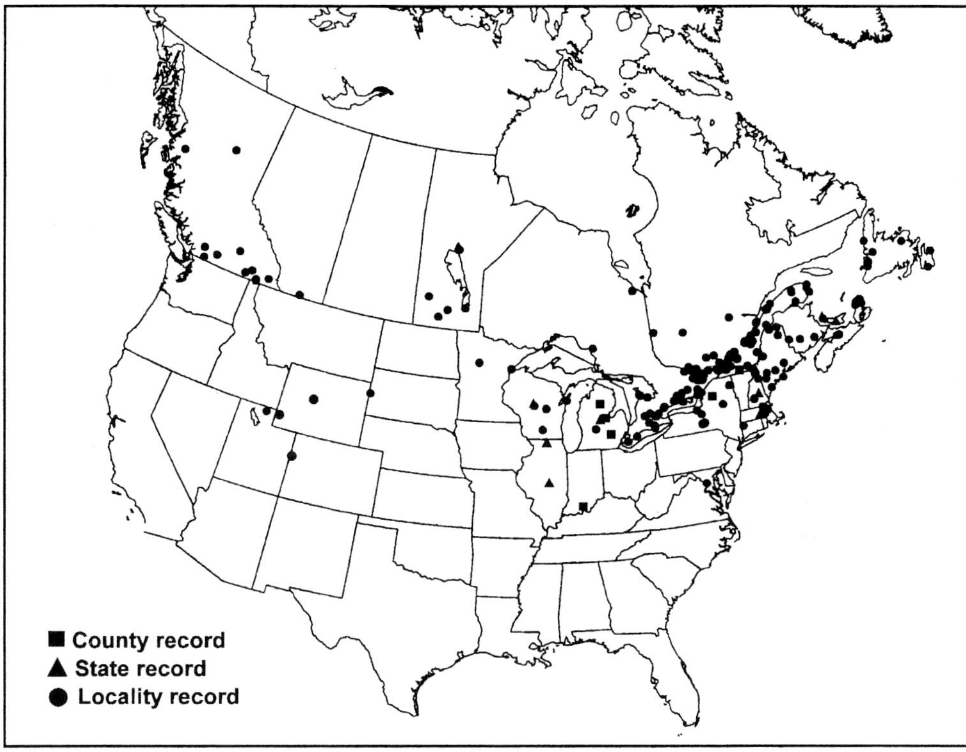

Map 1. Distribution of *Altica corni* Woods in North America north of Mexico.

(ISAC), \crossing Hwy 7 (ISAC), Delta Station (located at Delta Beach) (ISAC), Hadashville (ISC), Norway House (CNC), Riding Mountain National Park (CNC), Spruce Woods Provincial Park (CMNC).

New Brunswick. *Carleton co.*: Woodstock (CNC); *Kings co.*: Penobsquis (USNM); *York co.*: Fredericton (CNC).

Newfoundland. Badger (CNC), Deer Lake (CNC), / South Brook (CNC), Gambo (CNC), Gander (CNC), Halls Bay/ South Brook (CNC), Hampden (CNC), Hare Bay (CNC), Humber River/ Steady Brook (CNC), Parson's Pond (CNC), Peter's River (CNC), Rencontre West (CNC), South Branch (CNC), Spruce Brook (CNC), St.John's (CNC), Stephenville Crossing (CNC), /Harrys river (CNC).

Nova Scotia. *Colchester co.*: Masstown (NSDLF), Truro (CUIC); *Victoria co.*: Cape Breton/ White Point (CNC), Cape Breton Highlands National Park/ Aspy River (CNC), / French lake (CNC), / Freshwater lake (CNC), / 0.5 km N Grand Falaise (CNC), / 2 km NE Ingonish Beach, / Paquette lake (CNC), / Pleasant Bay (CNC), South Harbour Beach (CNC), Cape North (CNC), Dingwall (CNC).

Ontario. *Algoma dist.*: Michipicoten river/ south of Wawa (LLC); *Bruce co.*: Lions Head (CNC), Wiarton (CNC); *Carleton co.*: Bells Corners (CNC), Britannia Bay, W Ottawa (LLC), Britannia Heights (CNC), Constance Bay (CMNC), Dwyer Hill (CNC), Gloucester (CNC), Innis Point (CNC), Kanata (CNC), Leitrim (CNC), Marlborough forest /10 km W of North Gower (CNC, LLC), Marshall Bay (south of Arnprior) (LLC), Mer Bleue (SE Ottawa) (CNC), Nepean (CNC, LLC), Nepean/ National Capital Commission Log Farm (CNC), North Gower (LLC), / 10 km W (CNC), Ottawa (CDFA, CMNC, CNC), Ottawa/ Dow lake (CNC), Ottawa/ McCarthy Woods (LLC), Richmond (CNC); *Cochrane co.*: Moose Factory (CNC); *Dufferin co.*: bog 6 km E Wylde Lake/ near Arthur Wiau (LLC); *Durham co.*: Millbroook (CNC); *Essex co.*: Leamington (CNC); *Frontenac co.*: Kingston (CNC), St. Lawrence National Park/ Milton Island (CNC); *Grenville co.*: Kemptville (CNC); *Grey co.*: Meaford (CNC); *Halton co.*: Milton (CNC); *Hastings co.*: Marmora (CNC); *Kenora dist.*: Kenora (CNC); *Kent co.*: Chatham (CNC), Rondeau Provincial Park (CNC); *Lanark co.*: Carleton Place (CNC), / 7 km SW (CNC); *Leeds co.*: Chaffeys Locks (CNC), Chaffeys Locks Biological Station (CNC); St. Lawrence National Park/ Grenadier Island (CNC), / McDonald Island (CNC), / Mulcaster Island (CNC), / Thwartway Island (CNC); *Lennox & Addington co.*: Kaladar (CNC); *Norfolk co.*: Simcoe (CNC), Turkey Point (CNC); *Northumberland co.*: Brighton (CNC), Meyersburg (CNC), Presqu'Ile Provincial Park (CNC, LLC), Vittoria (CNC); *Perth co.*: Rostock (CNC); *Peterborough co.*: Warsaw (CNC); *Prince Edward co.*: county record only (CDFA, CNC); *Renfrew co.*: Arnprior (CNC), Pembrooke (CNC), White Lake (CNC); *Simcoe co.*: Springwater (CNC); *Toronto co.*: Toronto (CUIC, MCZ); *Waterloo co.*: Elmira/ Salem Creek (LLC), Wellesley (AVMC); *Wentworth co.*: Beverley Swamp (LLC), Hamilton (CNC); *York co.*: Markham (CNC).

Prince-Edward-Island. Province record only (CNC); *Kings co.*: Souris (CNC); *Prince co.*: West Devon (CNC).

Québec. *Abitibi co.*: Duparquet (CEUM), Paradis (CEUM); *Argenteuil co.*: Ile-de-Carillon (LLC); *Bellechasse co.*: Saint-Nérée (JFLC); *Berthier co.*: Berthierville (CEUM), Lanoraie (CTLC, LLC); *Bonaventure co.*: Carleton (LLC), Cascapédia (CNC); *Brome co.*: county record only (USNM), Knowlton (LLC); *Chambly co.*: Chambly (CCC); *Charlevoix-Est co.*: Lac à l'Ours near Port-au-Persil (LLC), Port-au-Persil (LLC), Port-au-Saumon (CNC), Saint-Fidèle (LLC); *Compton co.*: Johnville (LLC); *Deux-Montagnes co.*: La Trappe (CEUM), Saint-Placide (CEUM); *Dorchester co.*: Sainte-Hénédine (LLC); *Drummond co.*: Drummondville (LLC), Saint-Cyrille (= Saint-Cyrille-de-Wendover) (CNC, LLC); *Gaspé-Est co.*: Gaspé (CNC), Val-d'Espoir (CEUM); *Gaspé-Ouest co.*: Mont Jacques-Cartier (CNC), Parc de la Gaspésie/ ruisseau des Portes-de-l'Enfer (LLC); *Gatineau co.*: Aylmer (LLC), Lac Jean-Venne NW Sainte-Cécile-de-Masham (LLC), Parc de la Gatineau/ Notch road (CNC), / lac Brown (LLC), / lac Mulvihill (LLC), / lac Ramsay (LLC), Gatineau (CNC), Lucerne (CNC, LLC), Touraine (AVMC); *Hull co.:* Fairy Lake (= Lac des Fées) (CNC), Gatineau (LLC), Hull (CNC), Pointe-Gatineau (CNC); *Iles-de-la-Madeleine co.*: Anse-à-Fougère (as "Fougère") (AVMC), Étang-des-caps (CNC); *Ile-de-Montréal co.*: Dollard-des-Ormeaux (CNC), Montréal (CEUM, CNC, USNM); *Ile-Jésus co.*: Laval (CEUM), Laval/ rapides de (CEUM), Saint-Martin (CEUM); *Kamouraska co.*: Sully (CEUM); *Labelle co.*: Nominingue (CFLQ); *Lac-Saint-Jean co.*: Roberval (CEUM); *L'Assomption co.*: L'Assomption (CNC, LLC); *Lévis co.*: Lauzon (CEUM), Saint-Etienne (CCC), Saint-Nicholas (LLC), Saint-Romuald (JFLC); *L'Islet co.*: Saint-Aubert (LLC); *Lotbinière co.*: Dosquet (CCC), Villeroy (CCC); *Maskinongé co.*: Louiseville (LLC); *Montcalm co.*: Parc du Mont-Tremblant (CEUM); *Montmorency co.*: Cap Tourmente (JFLC); *Papineau co.*: Plaisance/ 4 k m N (CNC); *Pontiac co.:* Fort-Coulonge (LLC), Lac Davis (LLC); *Portneuf co.*: Saint-Augustin (= Saint-Augustin-de-des-Maures) (CNCC, CFLQ), Sainte-Catherine (USNM); *Québec co.*: Ancienne-Lorette (as "Lorette") (CFLQ), Cap-

Rouge (USNM), Loretteville (CFLQ, CNC, LLC), Québec (CCC), Sainte-Foy (CFLQ, LLC); *Rimouski co.*: Baie Hâtée/ 3 km NW du Bic (LLC), Le Bic (CEUM, CNC, LLC), / restaurant Charlot (LLL), Parc (de conservation) du Bic (CNC, LLC), Rimouski (CDFA, CNC, LLC), Sainte-Flavie (LLC), Saint-Valérien (LLC), / 9 km SE (LLC): *Rouville co.*: Richelieu (CNC), Saint-Hilaire (= Mont Saint-Hilaire) (USNM); *Saint-Maurice co.*: Trois-Rivières (CNC, LLC); *Terrebonne co.*: Saint-Hippolyte (LLC); *Vaudreuil co.*: Choisy (LLC), Pointe-Fortune (CNC), Rigaud (CEUM, CNC, LLC).

– *UNITED STATES.*

Connecticut. *Lichfield co.*: Cornwall (CUIC, USNM).

District of Columbia. Washington (USNM).

Florida. *Collier co.*: Naples/ 5 mi. NE (JEC); *Vollusia co.*: 3.5 mi. N Junction # 40 & # 17 (EGRC).

Idaho. *Clearwater co.*: Elk River/ 6 mi. N (JEC); *Fremont co.*: Ashton Riverside Camp/ 10 mi. N (DENH), Harriman State Park/ 20 mi. S Island Park (DENH); *Latah co.*: Moscow Mount (JEC).

Illinois. State record only (INHS); Northern Illinois (INHS).

Indiana. *Crawford co.*: county record only (PURC).

Maine. *Aroostock co.*: Allagash plantation (DENH), Ashland (DENH), Guerette (CMNC); *Cumberland co.*: Portland (CNC); *Kennebec co.*: Benton (DENH); *Knox co.*: Owls Head (DENH); *Oxford co.*: Paris (MCZ); *Penobscot co.*: Maine Agriculture Experimental Station (at Orono) (CUIC, PURC), Orono (MCZ); *Somerset co.*: (DENH), Jackman (DENH).

Massachusetts. State record only: (MCZ); *Middlesex co.*: Billerica (MCZ), Framingham (MCZ), Natick (CMNC), Sherborn (USNM), Tyngsboro (as "Tyng" or "Tings") (MCZ).

Michigan. *Benzie co.*: Beulah (UMMZ), county record only (CDFA, USNM), Beulah (CAS, UMMZ, USNM); *Cheboygan co.*: Indian river (CNC); St. Ignace (CNC); near Whitefish (DENH); *Delta co.*: Escanaba (CNC); *Emmet co.*: Mackinaw City Beach (NDSU); *Iona co.*: Iona (UMMZ); *Livingston co.*: county record only (UMMZ), E.S. George Reserve (as "G. Res.") (CDFA); *Marquette co.*: bay E of Marquette (ISAC); *Midland co.*: county record only (CAS); *Roscommon co.*: county record only (UMMZ). Not found: "Ag. Coll. " (MCZ).

Minnesota. *Cook co.*: Tofte (CNC); *Crow Wing co.*: Brainerd (CNC); *Hennepin co.*: Minneapolis (CNC); *Hubbard co.*: Camp Wilderness/ Bad Axe Lake (NDSU); *Itasca co.*: Deer Lake (EGRC); *St.Louis co.*: Duluth (MCZ).

Montana. *Gallatin co.*: Yellowstone/ 15 mi. NW route 287 (DENH).

New Hampshire. State record only: (INHS);*Coos co.*: Atkinson & Gilmanton Academy Grant (as " Academy Grant ") (CNC), Gorham (CNC), Pittsburg (DENH); *Hillborough co.*: Antrim (USNM); *Strafford co.*: Durham/ 1 mi. SW (DENH).

New York. *Clinton co.*: Peru (CUIC); *Essex co.*: top of Mount Whiteface (CUIC); *Hamilton co.*: Piseco Lake (MCZ); *Livingston co.*: Letchworth state Park (CUIC); *Niagara co.*: Olcott (CUIC); *Orleans co.*: Shelby (CUIC);*Tompkins co.*: Ithaca (NDSU), / vicinity (CUIC), / Saprucker West Road (CUIC), Lansing/ Monkey Run road (CUIC, NDSU), McLean/ 1 mi. S (CUIC), McLean Bog (CUIC), McLean Reserve/ Mud Pond (CUIC), Ringwood/ Dryden (CUIC); Ulysses/ N of Jacksonville (CUIC).

South Dakota. *Lawrence co.*: Roughlock Falls (NDSU).

Utah.*Cache co.*: Logan Canyon (NDSU); *Uintah co.*: Tony Canyon (CDFA, MCZ, USNM).

Washington. *Whitman co.*: Pullman (JEC).

Wisconsin. State record only: (INHS, MCZ); *Dane co.*: Madison (CDFA, MCZ, USNM); *Door co.*: Sturgeon Bay; *Waupaca co.*: Waupaca (USNM); *Wood co.*: Bethell (as "Bothell") (CNC).

HOST PLANTS. In the Northwest, *Altica corni* prefers the dogwood bush *Cornus pubescens* Nutt., which is no more than a variety of *C. stolonifera* according Scoggan (1979). We also found a record on *C. racemosa* Lam. from New York during the course of our work. Since we have no information on the circumstances of the capture, we consider the presence of *A. corni* on this dogwood species as accidental. In summary, our study shows that the host preferences appear considerably narrower than what has been reported before Wilcox (1979). In fact, we believe that *Altica corni* feeds exclusively on *Cornus stolonifera* Michx.

The following plant names were recorded on labels of specimens examined. They are all considered here as accidental occurrences: *Alnus,* flowers of *Barbarea vulgaris*, *Equisetum* (sweeping), *Euphorbia* (sweeping), *Rubus*, *Rumex*, *Solidago*, willows (=*Salix* sp.), weeds, and ex litter around *Salix petiolaris*.

BIOLOGY. *Altica corni* is common throughout its range. As its host plant, it prefers open or semi-open moist habitats, field depressions, edges of lakes, ponds and water courses. Adults hibernate underground under dogwood bushes. In early spring, adults hide for one or two weeks in the leaf litter before moving onto bushes, a period of low activity which corresponds to the appearance of buds on the host-plant (mid May in Ottawa area). Adults feed by biting holes through the leaves. After feeding for one or two weeks, they start mating. According to Woods (1918) copulation can last several hours.

Females deposit their eggs from mid-June to mid-July. The colour of eggs varies from whitish, to yellow or orange. These are laid in rows or in small clusters on the underside of leaves along the main veins. On average, females can lay about 400 eggs in a season. Eggs usually hatch within a week but more time is needed under cooler temperatures.

Larvae are gregarious. They feed only on the underside of leaves. Smaller larvae eat only the ventral parenchyma leaving the dorsal epidermis undamaged but which will turn brown later. Larger larvae eat the whole limb ignoring only the network of large veins. They go through three larval stages which last, roughly, one week each. Important variations occur in the period of time needed for their development according to local weather conditions. Examples of such variations are given by Woods (1918).

Pupation occurs underground from mid June to July. Full-grown larvae construct a pupal chamber by contortions of their body. Surrounding soil particles are cemented together by a mucous secretion of the maxillary glands (Woods 1918). Pupation may last one week or more. Pupae are very likely attacked by carabid larvae of the genus *Lebia* which are well known ectoparasites of *Altica* pupae (Capogreco 1989). In the Ottawa area, we have often collected adults of *Lebia viridis* Say in the company of adult *A. corni*, which leads us to believe that this carabid species parasitizes pupae of *A. corni*.

Adults of the new generation emerge from early July to August. According to Woods (1918), they do not feed after emerging but enter immediately into hibernation. We have not observed such haste in the course of our surveys, but rather believe that new adults feed for a few days before moving underground. Either way, their numbers steadily decline after mid-August. They are almost all gone by October.

PARASITES. Still unknown for eggs and larvae.

Pupae are very likely ectoparasitized by larvae of *Lebia viridis* Say which have been collected several times in the company of adult *Altica corni*.

Three parasites of the adults have been identified. The first instar larvae of the hymenopteran braconid *Microctonus loani* Mason has been obtained by Mason (1968) from adults of *Altica corni* collected in August and September. According to Mason, these parasites do not develop further in the autumn but rather hibernate inside their host which is still in first instar. This braconid is not specific to *A. corni* since it also parasitizes *A. ambiens alni* Harris, *A. chalybaea* Illiger, *A. ulmi* Woods, and *Calligrapha philadelphica* (Linné) (Mason l968, Krombein & *al.* 1979). The second known parasite is the tachinid fly *Celatoria spinosa* Coquillet reported by Woods (1918). The larvae of this fly develop inside the body of the adult flea beetle. The third parasite is the fungus *Beauveria globulifera* (Spegazzani) Piccard, reported as *Sporotrichum globuliferum* Speng. by Woods (1918).

PREDATORS. Unknown.

ECONOMIC IMPORTANCE. None. Native dogwoods are well adapted to the attacks of *Altica corni*. Although often severely defoliated, damaged bushes always fully recover after heavy infestations.

References

Blatchley, W. S. 1921. Notes on Indiana Halticini with characterization of new genus and descriptions of new species. *Jl N. Y. ent. Soc.* **29**: 16-27.

Brown, W. J. 1946. Some new Chrysomelidae, with notes on other species (Coleoptera). *Can. Ent.* **78**: 47-54.

Campbell, J. M., Sarazin, M. J. and Lyons, D. B. 1989. Canadian beetles (Coleoptera) injurious to crops, ornamentals, stored products, and buildings. Research Branch of Agriculture Canada, Publication 1826, Ottawa, Ontario. 491 pp.

Capogreco, J. V. 1989. Immature *Lebia viridis* Say (Coleoptera: Carabidae): bionomics, descriptions, and comparisons to other *Lebia* species. *Coleopts Bull.* **43**: 183-194.

Chagnon, G. 1938. Contribution à l'étude des Coléoptères de la province de Québec (suite). *Naturaliste can.* **65**: 157-166.

—, 1940. Contribution à l'étude des Coléoptères de la province de Québec. Département de Biologie de l'Université de Montréa, Montréal, Québec. 385 pp.

Chagon, G. and Robert, A. 1962. Principaux Coléoptères de la province de Québec. Les Presses de l'Université de Montréal, Montréal, Québec. 440 pp.

Downie, N. M. and Arnett, R. H. 1996. The beetles of Northeastern North America. The Sandhill Crane Press, Gainesville, Florida. 1721 pp. (2 volumes).

Hatch, M. H. 1971. The beetles of the Pacific Northwest. Part V: Rhipiceroidea, Sternoxi, Phytophaga, Rhynchophora, and Lamellicornia. *Univ. Wash. Publ. Biol.* **16**: i-xiv + 622 pp.

Heikertinger, F. and Csiki, E. 1940. Pars 166 and 169. Chrysomelidae: Halticinae. *In* Junk, W. and S. Schenkling (Eds.), *Coleopterorum Catalogus*. W. Junk, 's-Gravenhage. 635 pp.

Krombein, K. V., Hurd, P. D., Smith, D. R. and Burks, B. D. 1979. Catalog of Hymenoptera in America North of Mexico. Smithsonian Institution. Washington, D.C. 2735 pages (3 volumes).

Leng, C. W. 1920. Catalogue of the Coleoptera of America, north of Mexico. J. D. Sherman, Jr., Mount Vernon, New York. 470 pp.

LeSage, L. 1991. Family Chrysomelidae – leaf beetles. pp. 301-323 *in* Bousquet, Y. (Ed.). Checklist of Beetles of Canada and Alaska. Agriculture Canada Publication 1861/E, Ottawa, Ontario. 430 pp.

—, 1995. Revision of the costate species of *Altica* Müller of North America north of Mexico (Coleoptera: Chrysomelidae). *Can. Ent.* **127**: 295-411.

Mason, W. R. M. 1968. New Canadian Braconidae (Hymenoptera). *Can. Ent.* **100**: 715-728.

Scoggan, H. J. 1979. The flora of Canada. Natural Museum of Natural Sciences, *Publs Bot. natn. Mus. nat. Sci. Can.* **7**: 1-89 (part 1); **7**: 93-545 (part 2); **7**: 547-1115 (part 3), **7**: 1117-1711 (part 4).

Wilcox, J. A. 1954. Leaf beetles of Ohio. *Bull. Ohio biol. Surv.* **43**: 353-506.

—, 1974. Checklist of the Chrysomelidae of Canada, United States, Mexico, Central America, and the West Indies. Family No. 104. (Red version). Biological Research Institute of America Inc. Latham, New York. 166 pp.

—, 1979. Leaf beetles host plants in Northeastern North America (Coleoptera, Chrysomelidae). Published for the Biological Research Institute of America, Inc. by World Natural History Publications, Kinderhook, New York, NY. 30 pp.

Woods, W.C. 1918. The biology of Maine species of *Altica*. *Bull. Me agric. Exp. Stn, Orono* **265**: 149-204.

BIOLOGY AND ECOLOGY OF THE GENUS *PROCALUS* CLARK, 1865, ENDEMIC TO THE ANDINOPATAGONIAN REGION (ALTICINAE)

VIVIANE JEREZ
Departamento de Zoología, Facultad de Ciencias Naturales y Oceanográficas, Casilla 160-C, Universidad de Concepción, Concepción, Chile. e-mail:vijerez@udec.cl

Abstract

Procalus is a small genus of flea beetles, endemic to the Andinopatagonian region, and associated with Anacardiaceae, which in Chile are represented by the genera *Lithrea* and *Schinus*. The distribution of *Procalus* species agrees with the distribution of the host plants, between 30°50' lat. S and 40°50' lat. S. In this study, information about life cycle, seasonality, and host plant preferences of *Procalus* species are given. All *Procalus* species are univoltine. One group of taxa, associated principally with *Lithrea caustica* has a reproductive period occurring in spring-summer and another group of species, associated with *Schinus latifolius*, *S. velutinus*, *S. patagonicus*, *S. montanus* and *S. polygamus*, reproducing in autumn-winter. The genus *Procalus* has similarities in the life cycle and host plants with *Blepharida* of the New World, which support the hypothesis of the existence of phylogenetic connections between these two genera.

Introduction

Procalus Clark, 1865 is a small genus of flea beetles, endemic to the Andinopatagonian region and associated with Anarcardiaceae (Jerez, 1992). The distribution of *Procalus* species agrees with the distribution of the host plants between 30° 50' lat. S and 40° 50' lat. S (Fig. 5). In Chile, the Anarcardiaceae form part of the shrub region and sclerophyllous forest, that extends throughout the central zone, following the arid vegetational belt that crosses the continent diagonally from SE to NW (Donoso, 1993; Villagrán & Hinojosa, 1997). Following a revision of the genus by Jerez (1992), nine *Procalus* species were recognised, but in this study only the following six species were studied: *P. mutans* (Blanchard, 1851); *P. viridis* (Philippi & Philippi, 1864); *P. lenzi* (Harold, 1876); *P. reduplicatus* Bechyne, 1951; *P. malaisei* Bechyne, 1951, and *P. silvai* Jerez, 1995.

Two genera of Anarcardiaceae are the host plants of *Procalus*: *Lithrea* and *Schinus*, with disjunct species between both slopes of the Cordillera de los Andes. *Lithrea caustica*, *Schinus latifolius*, *S. velutinus* and *S. montanus* are endemics, only found in the Mediterranean region of Chile. These four plant species, together with *Schinus polygamus* and *S. patagonicus*, taxa also present on the eastern slope, are all host plants of *Procalus* species.

According to the locality records of the conserved material in collections and personal collections of the author, the genus *Procalus* is distributed between 30° 50' lat. S (south of the Limarí River) and 40° 50' lat. S (Malleco River). Notwithstanding, in the Museum National d'Histoire Naturelle de Paris there is material of *Procalus* collected at the beginning of the 1900's in the south of Argentina, but without host plant data. The host plant was

Advances in Chysomelidae Biology 1, pp. 545–555
Edited by M.L. Cox

probably *Schinus patagonicus*, because this is a relatively abundant species in this area and its distribution range reaches Magallanes (Barrera & Meza, 1982).

The aim of this study is to produce a synthesis of biological and ecological information on the life cycles and the host plants of the described *Procalus* species.

This paper is based upon the study of specimens of all the developmental stages of species described in the genus *Procalus*. Pairs of beetles were collected in the field and confined in plastic containers in order to obtain eggs and larvae. The immature stages were described using scanning electron microscope (SEM) and compound microscopes. All ontogenetic stages were fixed in 90% ethanol. The micromorphology of eggs and the morphology of the first, second and third instar larvae are described, using the nomenclature of Rowley and Peters (1972), Lawrence (1991) and Cox (1997).

Life cycle

All *Procalus* species are univoltine, with a moderate overlap of the different developmental stages and in some geographical areas, it is possible to find two species coexisting on the same host plant. The ontogenetic cycle of *Procalus* species, includes the egg, three larval instars, a prolonged period of aestival diapause or hibernation in the third larval instar within a pupal chamber, a pupal stage, and finally the imago or adult stage (Fig. 1).

Description of immature stages

Eggs

The eggs are cylindrical, tapering at both ends, with the basal pole often covered by a secretion that the females use to attach the eggs to the leaves. The egg shell or chorion is sculptured into polygonal cells, uniformly distributed over the entire surface and with the micropyle clearly differentiated in a circular area with a pore at the centre. The micropyle is slightly invaginated and rounded, with a texture that varies between the *Procalus* species. The peripheral micropyles are covered with aeropyles of variable height, number, and diameter between the species (Fig. 2A). Measurements of the eggs of *Procalus* species are given in Table 1.

Larvae

Three larval stages are recognized in the genus *Procalus* and they are clearly differentiated by the width of the head and some morphological characteristics. However, the three instars have a certain degree of similarity between the different *Procalus* species.

General description (Based on the morphology of *P. mutans* and *P. viridis*).

The larvae are recognized by their soft body, which is moderately dorsally convex, ventrally flat and green colour. The head is strongly sclerotized, shiny dark brown or black.

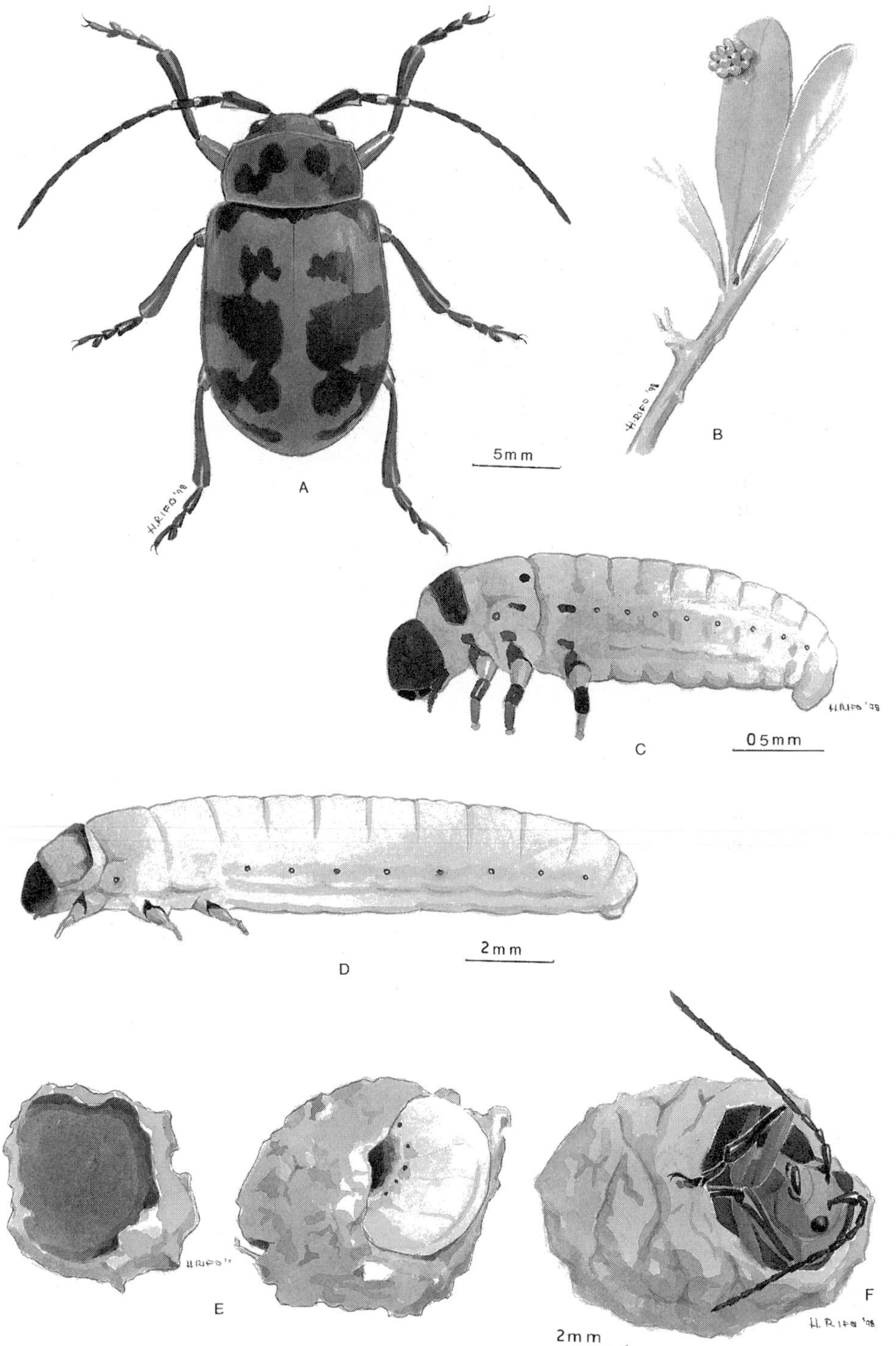

Fig. 1. Procalus mutans. A) Dorsal habitus male. B) Eggs on leaves of host plant. C) First instar larva, lateral view. D) Third instar larva, lateral view. E) Pupal chamber and prepupal stage. F) Pupal chamber and newly emerged adult.

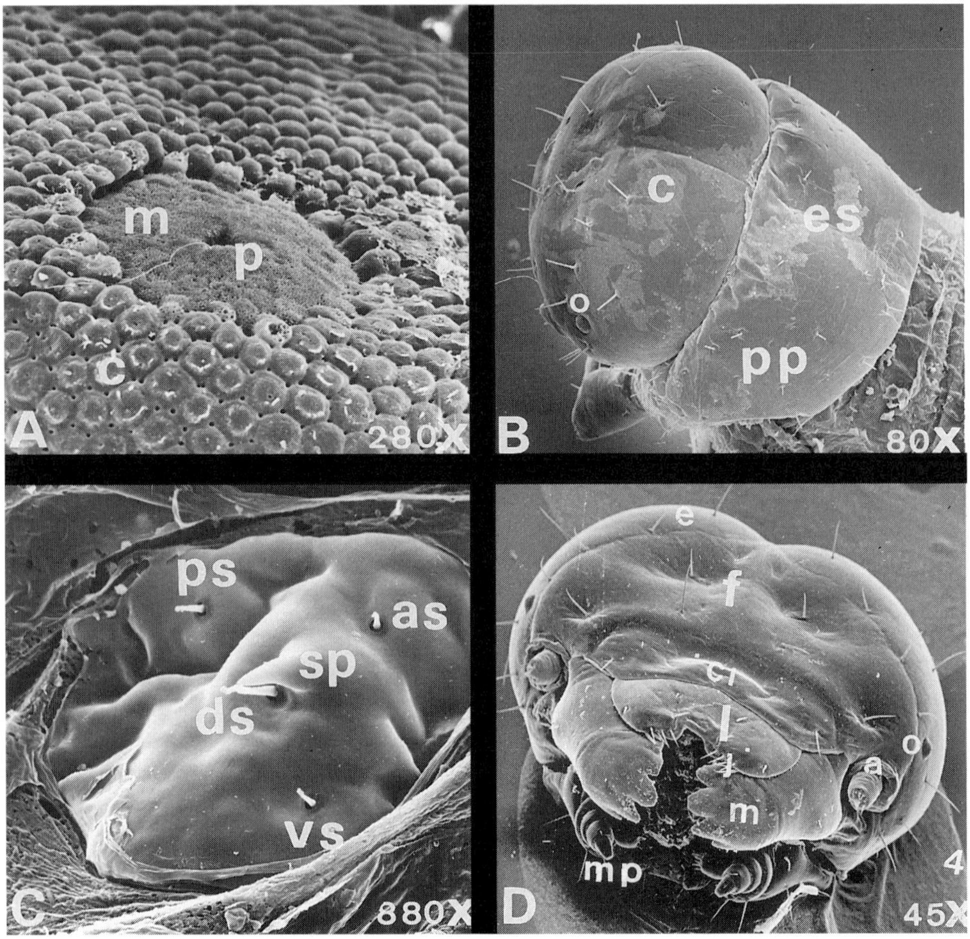

Fig. 2. Procalus and scanning electron micrograph of immature stages. A) *Procalus reduplicatus*, egg, apical view; abbreviations c, chorion, m, micropyle, p, pore. B) *P. viridis*, first instar larva; head and prothorax dorsal view; abbreviations c, head, es, ecdysial suture, o, ocellus, pp, pronotal plate. C) *P. silvai*, tubercle bearing egg burster; abbreviations as, antero-dorsal setae, ds dorsal setae, ps, posterior setae, sp, spine, vs, ventral setae. D) *P. mutans*, third instar larva; head frontal view; abbreviations a, antenna, cl, clypeus, e, epicranium, f, frons, l, labrum, m, mandible; mp, maxillary palpus, o, ocellus.

Table 1. Measurements of eggs of *Procalus* species

Species	Length $\bar{x}$ (mm)	Range (mm)	Width $\bar{x}$ (mm)	Range (mm)	n
P. mutans	1.94	1.8-2.3	1.01	1.0-1.1	7
P. viridis	2.16	2.0-2.1	1.01	0.9-1.2	10
P. lenzi	2.02	1.8-2.2	0.98	0.9-1.1	5
P. reduplicatus	1.86	1.8-1.9	0.96	0.9-1.0	5
P. silvai	1.78	1.7-1.8	0.95	0.9-1.1	7

There is an anal foot on the ninth abdominal segment. The three pairs of thoracic legs are black. The cuticle is smooth, lacking tubercles and short setae (Figs 1C, 1D). Measurements of the three larval instars of *P. mutans* are given in Table 2.

First instar

The first instar is recognized by the general green body coloration and the shiny black head and pronotal plate (Fig. 2B). Egg bursters are present on the mesothoracic sclerites. Each hatching spine is associated with several setae. The spiracles on the mesothorax and abdominal segments 1-8 are annular uniforous, with black peritremes. The meso- and metathoracic pleura have black sclerites bearing microsetae. Thoracic legs green with the femora and tibiae black (Fig. 1C).

Second instar

Similar to the first instar, but differing in body length and the absence of egg bursters and mesothoracic sclerites; and also by the presence of a less strongly sclerotized pronotal plate and by the paler, yellowish green colour.

Third instar

Diagnosis

The third instar larva of all *Procalus* species is characterized by its relatively large size, by the sclerotized pronotal plate, and by the rounded abdominal segments in cross section. Its external morphology is similar to that of Galerucinae larvae, with sclerotized tergites absent on the tegument and microsetae visible only under a compound microscope. Larva with body curved, with head black, well sclerotized. Pronotal plate sclerotized, but slightly paler green than the body; in *P. mutans*, the posterior pronotal margin has a black border, whilst it is completely green in *P. viridis*. Meso- and metanotum with a pair of black pleural spots. Thoracic legs black (Fig. 1D).

Detailed description (based on *P. mutans*)

Head. Hypognathous, rounded, heavily sclerotized; epicranium smooth, bearing 5 pairs of setae; epicranial suture distinct, Y-shaped; frontal arms broadly divergent, weakly curved,

Table 2. Measurements of larval instars of *Procalus mutans*

Instar	Length $\bar{x}$ (mm)	Range (mm)	Head width $\bar{x}$ (mm)	Range (mm)	Pronotal width $\bar{x}$ (mm)	Range (mm)	n
L1	3.67	1.7-6.4	0.75	0.6-0.9	0.95	0.9-1.2	14
L2	5.18	4.3-5.6	0.91	0.9-1.0	1.21	1.1-1.3	7
L3	14.8	12.6-16.8	1.78	1.7-1.8	3	2.9-3.1	6

3 pairs of frontal setae; fronto-clypeal suture distinct. Clypeus bearing 2 dorsal and 2 lateral setae. Labrum rectangular, with apical edge strongly emarginate medially; bearing 2 dorsal, 2 lateral, 3 pairs marginal, 3 pairs ventral setae of subequal width and length. Antennae 2-segmented. One ocellus on each side laterodorsal to antennae. Mandibles palmate, apex rounded 5-dentate, lacking penicillus. Maxillary palpi 3-segmented. Labial palpi 2-segmented (Fig. 2D).

Thorax. Rounded in lateral view; cuticle smooth. Pronotal plate strongly sclerotized, bearing 8 pairs microsetae on anterior margin, 3 pairs on posterior margin; ecdysial suture distinct. Prothorax bearing 3 pairs of pleural setae on oval sclerite. Microsetae of meso- and metathorax located on sclerites of diverse sizes. Mesothoracic spiracle annular uniforous. Legs short, similar in size and shape; tarsungulus with curved claw; pulvilli broad with adhesive function; claw shape intraspecifically variable. Ventrally, each thoracic segment with pair of sclerotized areas in the mid-line.

Abdomen. Curved, with 8 rounded segments in lateral view; cuticle smooth, indistinct sclerites bearing microsetae, visible when using a compound microscope. Segment 9 rounded, apically bearing numerous more or less erect setae. Segment 10 forming an anal pseudopod. Anal opening vertical. Ventral ambulatory lobes present on all segments. Spiracles annular uniforous, with peritremes well defined. In *P. mutans* and *P. viridis*, the spiracles have the peritremes with straight areola, but in the other species the areola is longer and darker. These two species also have black spots on the pleural region which distinguishes them from other *Procalus* species (Fig. 1D).

Pupa (based on P. viridis and P. mutans)

Exarate, body pale green; cuticle smooth, without tubercles.

Head. Strongly deflexed, not visible from above. Eyes pigmented dark brown. Labrum emarginate on apical margin. Mandibles 4-dentate.

Thorax. Pronotal disc curved dorsally, bearing microsetae in each half; 5 setae on anterior margin, 2 setae on lateral margin, 3 setae on posterior margin. Pterothecae extending to abdominal segment 4. Front, mid- and hind podothecae extending to abdominal segment 4. Podothecae only partially visible in dorsal view. Metanotum quadrangular, disc smooth, with 4 pairs setae near posterior margin.

Abdomen. With 7 segments visible dorsally. Spiracles distinct, annular uniforous, visible dorsally on abdominal segments 1-8.

Bionomics and life cycle

In the wild, *Procalus* eggs were found on the upper or lower surface of leaves of the host plant (Fig. 1B). The chorion is yellowish and shiny when newly laid, but subsequently becomes yellow or dark brown and opaque. In *P. viridis*, *P. mutans* and *P. reduplicatus*, the first instar larvae hatch 8 to 10 days after oviposition. They feed at first on the chorion and can even eat unhatched eggs so that there is some degree of cannibalism. They later feed by rasping the epidermis of the new leaves of the host plant.

The first moult occurs 4 to 5 days after hatching. During this period, the larva attaches to the foliar lamina by the pygopod and remains immobile for approximately two days until the emergence of the second instar. The moult that produces the third instar larva occurs 6 to 8 days later, the second instar larva attaching itself by the pygopod to the foliar lamina in a similar manner. The third instar is characterized by the growth of the larvae and when fully grown they attain a length of approximately 15 mm, considerably increasing in volume. Their presence on the plant is evidenced by the abundant faeces which are left in the apical parts of the branches. The green larvae are well camouflaged against the similarly coloured foliage of the host plant.

When the larvae are fully grown, they abandon the foliage of the host plant and reach the ground where they construct a pupal chamber composed of grains of soil covering smooth, shiny walls. The pupal chamber is oval in shape and its external texture depends upon the particular structure of the soil (sandy or pebbley) and no greater intraspecific differences are seen from an external morphological point of view (Fig. 2E). Within the pupal cell, the third instar larva or prepupal stage remains in diapause about 9 months, spanning three seasons (summer, autumn and winter), according to the type of life cycle that the species undergoes. Pupation takes place within the pupal chamber, two weeks before the emergence of adults the following spring or autumn, depending on the life cycle of the species (Fig. 1F).

Reproductive cycles and seasonality

In *Procalus*, all of the species are univoltine. The species associated with *Lithrea caustica* initiate their reproductive cycles at the beginning of spring, with the emergence of the adults after a long larval diapause period. In contrast, the species associated with *Schinus*, preferably start their reproductive cycles at the beginning of autumn, with the exception of *P. silvai*, whose reproductive period occurs in spring-summer.

a) Spring-summer cycle

Procalus lenzi, *P. malaisei*, *P. mutans*, and *P. reduplicatus* are spring-breeding species. They oviposit after emerging in spring, and the larvae remain on the foliage until the end of December. The larval diapause period extends from approximately mid-December until the first days of September of the following year (Fig. 3).

b) Autumn-winter cycle

P. viridis is autumn-breeding and aestivates in the summer. It does not oviposit until the beginning of autumn of the following year. It is possible to find eggs and larvae until the middle of September. The construction of pupal chambers begins in mid-July (Jerez, 1988)(Fig. 4).

Coexistence

It has been determined that throughout the geographic distribution of the species of *Procalus* associated with *Lithrea caustica*, in some sectors, cases of sympatry and syntopy can be

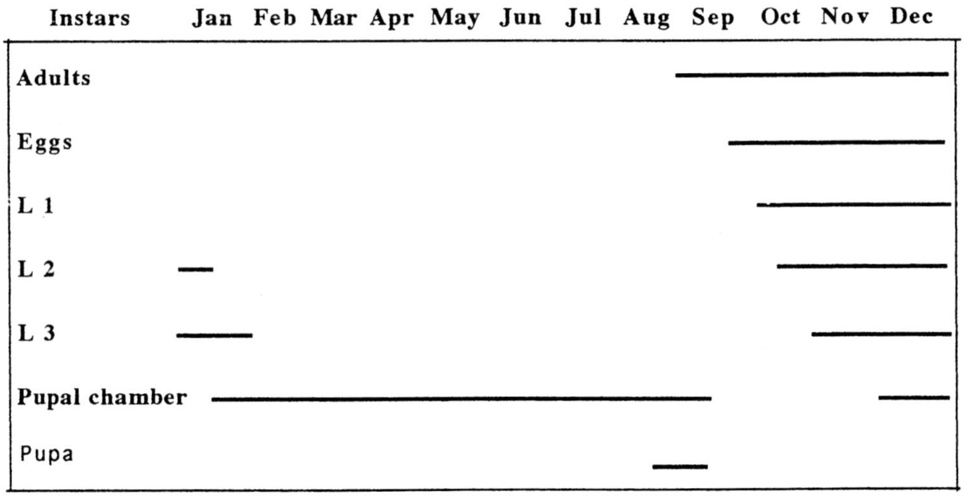

Fig. 3. Phenology of *Procalus mutans* with spring-breeding.

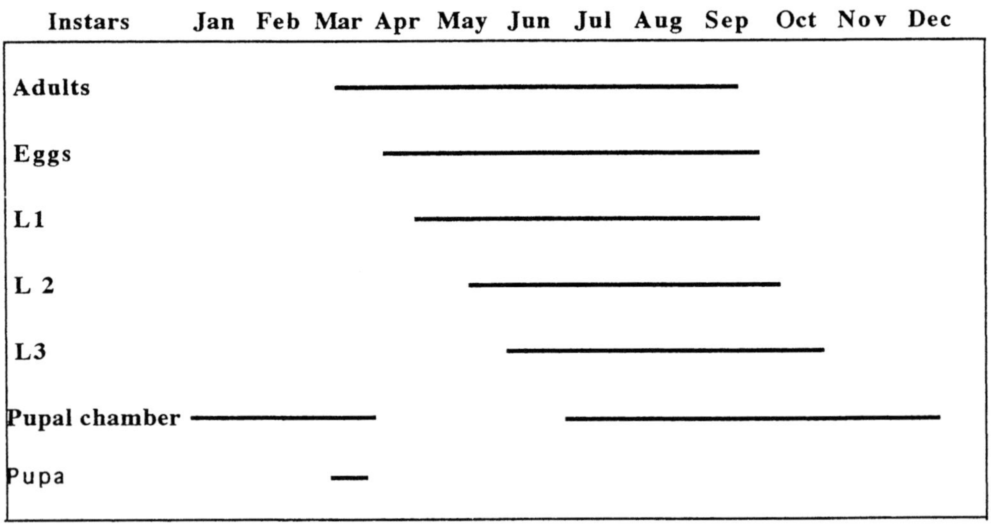

Fig. 4. Phenology of *Procalus viridis* with autumn-breeding.

found. This is true with *P. lenzi* and *P. malaisei*, in the foothill sectors of the central zone of Chile (Grez, 1988; Poiani, 1989) and *P. mutans* and *P. reduplicatus* in the coastal zones of the VIII Region (Cortes et al., 1993).

Studies aimed at explaining the coexistence of *P. mutans* and *P. reduplicatus* on *Lithrea caustica* in the Península de Hualpen (36° 47' S – 73° 10' S), showed that this is because of the low population densities of both species on *L. caustica*, their only food resource, which is relatively abundant in the area, so that interspecific competition is unlikely (Centella, 1994).

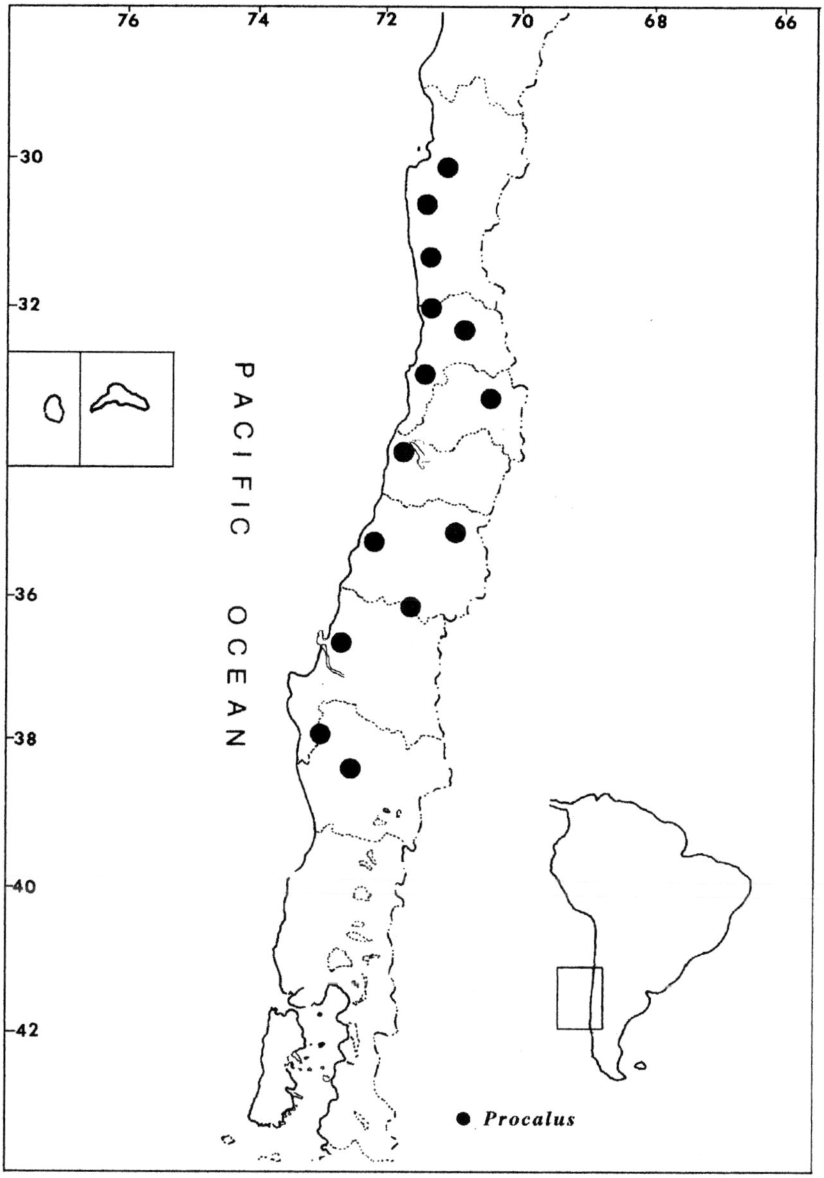

Fig. 5. Geographical distribution of *Procalus*.

Parasitoids

A parasitic wasp of the family Ichneumonidae emerged from a pupal chamber; also a Tachinidae. Recently Jerez and Centella (1996) obtained Mermithidae Nematodes from *P. mutans* and *P. reduplicatus* adults maintained in rearing boxes.

Concluding remarks

The life cycle pattern of *Procalus* species is very similar to that of *Blepharida sacra* Weise, described by Furth (1982) as are also their larval morphology, pupal chamber, diapause period, and association with Anarcardiaceae (Furth & Young, 1988).

The existence of an obligate larval diapause period within a pupal chamber is characteristic of species of temperate zones. These species are subject to extreme climatic conditions and have only one generation per year (Cox, 1994; Scriber, 1994). In the case of *B. sacra*, the diapause period would be an adaptation to arid or desert zones (Furth, 1988). For *Procalus*, its host plants, which belong to the genera *Lithrea* and *Schinus*, form part of a continuous climate and vegetational strip which crosses the South American continent from SE to NW. This strip represents a disconnected zone between the original subtropical vegetation and the southern forests of South America and would have permitted the diversification of *Procalus* and its host plants, parting from Neotropical ancestors, most likely due to the climatic changes which occurred during the Pliocene (Villagrán & Hinojosa, 1997).

The genus *Procalus*, presumably of Neotropical origin, has similarities to the life cycle and host plants of *Blepharida* of the New World, all of which support the hypothesis of the existence of phylogenetic connections between both genera.

Acknowledgements

I am grateful to Dr Michael Cox for inviting me to contribute to this volume and for his critical reading of the manuscript. The drawings in this paper were prepared by Hipolito Riffo. The photographs were obtained from the Electronic Microscopy Laboratory of the Dirección de Investigación, Universidad de Concepción and Thomas Ogden translated the paper into English. This work was supported by a grant DIUC 92.38.07-6 (Thesis Doctoral Project) and by FONDECYT 1940995.

References

Bechyné, J., 1951. Chrysomeloidea américains nouveaux ou peu connus. Revta chil. Ent., 1: 75-112.

Blanchard, E., 1851. Fauna Chilena. Coleópteros. In Gay, Historia Física y Política de Chile, 5: 522-558.

Barrera, E. & Meza, I., 1982. Análisis de la cutícula foliar de las especies chilenas del género *Schinus* L. (Anarcardiaceae). Bol. Mus. Nac. Chile, 39: 29-36.

Centella, C., 1994. Aspectos ecológicos de dos especies congenéricas, *Procalus mutans* y *Procalus reduplicatus* (Coleoptera: Chrysomelidae), especialistas de *Lithrea caustica*. Un caso de coexistencia. Tesis de grado. Licenciatura en Biología. Pontificia Universidad Católica, sede Talcahuano.

Clark, H., 1865. An examination of the Dejeanian Genus *Coelomera* (Coleoptera-Phytophaga) and its affinities. Ann. Mag. nat. Hist., 16 (3 ème series): 256-268.

Cortés, X., Jerez, V., & Galleguillos, R., 1993. Estudio electroforético en dos especies simpátricas y sintópicas del genero *Procalus* Clark, 1865 (Coleoptera: Chrysomelidae). Rev. Chilena Ent., 20: 43-48.

Cox, M., 1994. Diapause in Chrysomelidae. In Jolivet, P.H., Cox, M.L., & Petitpierre, E. (eds). Novel aspects of the biology of Chrysomelidae. Kluwer Academic Publishers, Dordrecht, The Netherlands. pp 469-502.

Cox, M.L., 1997. *Homichloda barkeri* (Jacoby), (Coleoptera: Chrysomelidae: Alticinae), a candidate agent for the biocontrol of priockly acacia, *Acacia nilotica* (Mimosaceae) in Australia. J. nat. Hist., 31: 935-964.

Donoso, C., 1993. Bosques templados de Chile y Argentina. Variación, estructura y dinámica. Ecologia Forestal. Ed. Universitaria, pp 1-483.

Furth, D., 1982. *Blepharida* biology as demonstrated by the sacred sumac flea beetle (*B. sacra* Weise) (Coleoptera: Chrysomelidae: Alticinae). Spixiana, suppl., 7: 43-52.

Furth, D., 1988. Relations of herbivore feeding and plant flavonoids (Coleoptera: Chrysomelidae and Anarcardiaceae: *Rhus*). Oecologia, 74: 496-500.

Furth, D. & Young, D.A., 1988. Relationships of herbivore feeding and plant flavonoids (Coleoptera: Chrysomelidae and Anacardiaceae: *Rhus*). Oecologia, 74(4): 496-500.

Grez, A., 1988. *Procalus lenzi* y *Procalus malaisei* (Coleoptera: Chrysomelidae): dos especialistas del matorral. Revta chil. Ent., 16: 65-67.

Harold, E., 1876. Diagnosen neuer Arten. *Elithia lenzi*. Coleopterologische Heften, 15: 11.

Jerez, V., 1988. Ciclo de vida y biologia de *Procalus viridis* (Phil. y Phil., 1864) (Coleoptera Chrysomelidae). Comn. Mus. Reg. Concepción, 2: 7-11.

Jerez, V., 1992. Revisión taxonómica del género *Procalus* Clark, 1865 (Chrysomelidae-Alticinae). Gayana Zoologia, 56(3-4): 109-125.

Jerez, V., 1995. *Procalus silvai*, n. sp. Descripción e interacción con *Schinus patagonicus* (Chrysomelidae-Alticinae). Gayana Zoologia, 59(2): 161-165.

Jerez, V. & Centella, C., 1996. Primer registro de Nemátodos Mermithidae, parásitos de *Procalus mutans* y *Procalus reduplicatus* (Chrysomelidae: Alticinae). Acta ent. Chil., 20: 107-110.

Lawrence, J.F., 1991. Order Coleoptera. In Stehr, F.W. (ed.). Immature Insects. Kendall/Hunt, Dubuque. Vol. 2: 144-658.

Philippi, R.A. & Philippi, F., 1864. Beschreibung einiger neuen Chilenischen Kafer. Stettin. ent. Ztg., 25(1-12): 382-4019.

Poiani, A., 1989. Interacciones entre *Procalus* (Coleoptera, Chrysomelidae) y *Lithrea caustica* (Sapindales, Anacardiaceae). Un caso de monofagia en el matorral de Chile central. Orsis, 4: 99-112.

Rowley, W.A. & Peters, D.C., 1972. Scanning electron microscopy of the eggshell of four species of *Diabrotica* (Coleoptera: Chrysomelidae). Ann. ent. Soc. Am., 65(5): 1188-1191.

Scriber, J.M., 1994. Climatic legacies and sex chromosomes: latitudinal patterns of voltinism, diapause, size, and host-plant selection in two species of swallowtail butterflies at their hybrid zone. In Danks, H.V. (ed.) Insect life-cycle polymorphism. Theory, evolution and ecological consequences for seasonality and diapause control. Kluwer Academic Publishers, Dordrecht, The Netherlands. pp 133-171.

Villagrán, C. & Hinojosa, L.F., 1997. Historia de los bosques del sur de Sudamérica, II: Análisis fitogeográfico. Revta chil. Hist. nat., 70: 241-267.

BIOLOGY AND ECOLOGY OF THE LEAF-MINING HISPINAE, *OCTUROPLATA OCTOPUSTULATA* (BALY)

CLAUDIO R. TEIXEIRA, MARGARETE V. DE MACÊDO
and RICARDO F. MONTEIRO
Laboratório de Ecologia de Insetos, Depto de Ecologia CP 68020, IB, Universidade Federal do Rio de Janeiro, Ilha do Fundão, CEP 21941-590, Rio de Janeiro, Brazil.

Abstract

This study was developed at "Restinga" of Barra de Maricá, Rio de Janeiro State, Brazil. Leaves from eight *Senna australis* (Leguminosae) individuals containing larval mines of *Octuroplata octopustulata* were collected monthly from December 1990 to December 1991. This paper examines the pattern of female oviposition site preference with time, the population fluctuation, parasitism and some aspects of host preference. *O. octopustulata* lays single eggs on the upper surface of *S. australis* leaflets the larva constructing a blotch mine. Mines were collected throughout the year but were most abundant during the autumn and winter. Females oviposited mainly on the first and largest of the four leaflet pairs of the leaf, suggesting that leaf size is important for female oviposition site selection. The species was also found mining leaves of *Eugenia ovalifolia* (Myrtaceae) and *Ouratea cuspidata* (Ochnaceae). Larval development on *E. ovalifolia* leaves was faster than on *S. australis*. However, the parasitism rate of *O. octopustulata* larvae found in *E. ovalifolia* was much higher than in those found in *S. australis* and this may be one of the factors making *O. octopustulata* more abundant on the last host plant.

1. Introduction

Mines are feeding channels inside the parenchyma or epidermal tissues of plants, in which the epidermis or at least its outer wall remains undamaged, thus shutting off the mine cavity from outside (Hering, 1951). A mine represents a source of food and also a shelter (Damman, 1994). However, living within a mine places limits on the quality and, especially, the quantity of food to which the leaf miner has access (West, 1985). The mine cavity can be considered a kind of "hot house" which creates particularly favourable conditions of temperature and humidity for the larvae, and also for their parasitoids (Hering, 1951).

Leaf-mining insects include holometabolous species in the orders Lepidoptera, Diptera, Coleoptera and Hymenoptera. Needham *et al.* (1928), Hering (1951) and Hespenheide (1991) completed general reviews of the taxa and bionomics of leaf-mining insects. Despite the fact that in all leaf-mining taxa only the larvae feed within leaf mines, in many taxa, adults and pupae can sometimes be found inside the mines because pupation frequently takes place within the mine and the adult must leave it through the epidermis of the leaf (Connor & Tavener, 1997). In the Coleoptera the leaf-mining habit occurs in six families but is most extensively developed in the Chrysomelidae (Connor & Tavener, 1997). Within the Chrysomelidae the mining habit is very common in the Hispinae (Jolivet, 1988) whose larvae are usually leaf miners or free feeders among appressed leaves (Jolivet, 1989).

Advances in Chysomelidae Biology 1, pp. 557–563
Edited by M.L. Cox

However, despite this, very little is known about their biology and ecology.

This paper examines the pattern of *Octuroplata octopustulata* (Baly), female oviposition site preference with time, the population fluctuation, parasitism and some aspects of host species preference.

2. Biology and behaviour

Octuroplata octopustulata females lay their eggs on the upper surface of *Senna australis* (Leguminosae), *Eugenia ovalifolia* (Myrtaceae) and *Ouratea cuspidata* (Ochnaceae) mature leaves in the Restinga (coastal sand dunes) of Barra de Maricá (22° 57' S and 12° 52' W), Maricá County, Rio de Janeiro State, Brazil. These are the first host records for this tiny (6 X 2 mm) mining hispine (Fig. 1) (C.L. Staines, pers. com.). Having three host species belonging to three different plant families seems an uncommon pattern among leaf-miners, since most of them tend to be monophagous or associated with more than one congeneric or confamilial plant species (Hespenheide, 1991). In the UK, the leaf-mining alticine, *Apteropeda orbiculata* (Marsham) is also known to mine leaves of species in different plant families, *Plantago lanceolata* (Plantaginaceae), *Teucrium scorodonia* (Labiatae), *Centaurea nigra* (Asteraceae) and *Digitalis purpurea* (Scrophulariaceae) (Cox, 1997). *O. octopustulata* larvae construct blotch mines (Fig. 2) and pupate within them just leaving the mine as an adult. The adults feed by rasping the leaves of the same host plant species, producing characteristic linear scars similar to those of other leaf-mining hispines (for example, Winder *et al.*, 1984).

This Hispinae clearly preferred to oviposit on *S. australis* leaflets, where it was abundant most of the year, being rare on the two other host plants. In *S. australis* leaflets 94.4% of the 1521 *O. octopustulata* collected mines were solitary, but a maximum of five mines occurred.

In the warmest months, December to February, *O. octopustulata* developmental time in the field ranged from 43 to 44 days (n = 4) in *S. australis*, and 37 to 39 days (n = 3) in *E. ovalifolia*. At lower mean temperatures, from September to November, the observed developmental time was 56 to 68 days (n = 4) in the preferred host plant and 46 to 49 days (n = 3) in *E. ovalifolia*. *O. octopustulata* adults were observed feeding, mating and ovipositing during the day throughout the year. No diapause was observed in this species, contrasting with Temperate leaf-miners, most of which undergo a winter season diapause (Hespenheide, 1991)

3. Population fluctuation and resource use

Eight individual plants of *S. australis* were marked and checked monthly from December 1990 to December 1991. Leaves with mature mines were collected and taken to the laboratory for rearing out of adults. *O octopustulata* mines were found throughout the year on leaflets of the eight plants but more abundantly during the months of May to September (Fig. 3). Despite this greater abundance in these months, *O. octopustulata* was never rare in any month. Another Hispinae species, *Mecistomela marginata* (Thunberg), studied at the same place was also reproductively active throughout the year (Macêdo *et al.*, 1994).

Fig. 1. Octuroplata octopustulata adult on a *Senna australis* leaflet.

Fig. 2. O. octopustulata mature mine on a *S. australis* leaflet.

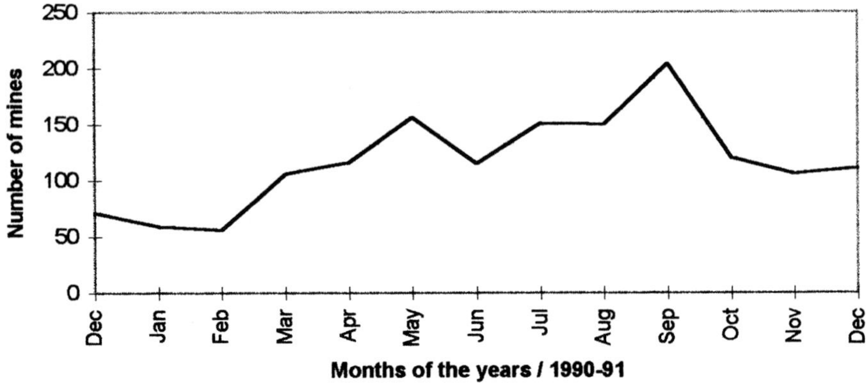

Fig. 3. Number of *O. octopustulata* mines found on eight *S. australis* individuals from December 1990 to December 1991.

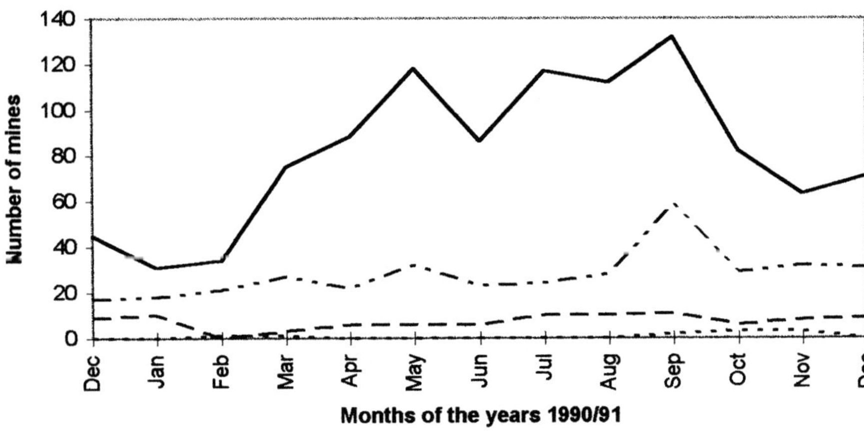

Fig. 4. Number of *O. octopustulata* mines found on the first (——), second (——-—), third (- - -) and fourth (·····) pairs of *S. australis* leaflets from December 1990 to December 1991.

In both cases the feeding resources were available all year long. Still at the "Restinga" of Barra de Maricá, a third species of Chrysomelidae, *Echoma marginata* (L.) (Cassidinae), could only be found as both larvae and adults during the months of April and May (Windsor *et al.*, 1995). This last species feeds only on flower buds of an annual plant, thus its resource is quite ephemeral. Although abiotic factors have already been proved to be important in population fluctuation for Chrysomelidae species (for example, Medeiros & Vasconcellos-Neto, 1994), it is clear that this is not a general pattern (for example, Nakamura *et al.*, 1989). Where abiotic factors do not constrain population occurrence throughout the year, as seems to occur in the tropics, resource availability is certainly of great importance in determining the general pattern of population fluctuation.

S. australis leaves have four pairs of leaflets, which get smaller from the first or distal to the fourth basal pair (Table 1). Although *O. octopustulata* can develop in either of the four pairs of leaflets it showed a clear preference for ovipositing on the more distal, larger pair

Table 1. Average size (± sd) of the four pairs of leaflets of *S. australis* leaves.

Leaflet pair (n = 50)	Leaflet area (in cm^2)
1°	22.83 ± 4.85
2°	17.30 ± 5.60
3°	12.14 ± 3.66
4°	7.94 ± 2.62

of leaflets in all months of the year (Fig. 4). It is probable that the size of the leaflet blade is the most important factor in determining this preference. Preference for larger leaves has already been shown for leaf-mining moth species (for example, Hileman & Lieto, 1981; Simberloff & Stiling, 1987). As in most of the miners *O. octopustulata* develops in a single mine and once the female has chosen a leaflet in which to oviposit the larva has no chance of changing its resource. In these cases, adult females are responsible for the living conditions of their progeny by their choice of oviposition sites (Futuyma & Moreno, 1988). This determines a strong pressure over females to choose good sites for their brood to develop, which in the case of *O. octopustulata* seems to be big leaflets. However, this does not mean that in these sites the larvae will do best. Empirical evidence from herbivorous insects has shown that sometimes they do (for example, Rausher, 1979; Preszler & Price, 1988) but sometimes they do not (for example, Auerbach & Simberloff, 1989; Rausher, 1979; Valladares & Lawton, 1991).

4. Parasitism

O. octopustulata larvae on *S. australis* leaflets were parasitized by a parasitoid, *Zagrammosoma* sp (Hymenoptera: Eulophidae), only with 1.3% of the hispine larvae parasitized by it (n = 1521). *O. octopustulata* larvae on *E. ovalifolia* leaves harboured three Eulophidae species: *Closterocerus* sp, *Cirrospilus* sp and *Diaulomorpha* sp. All three species were responsible for an overall parasitism rate of 34.5% (n = 64). This chalcid wasp family was the richest one found as parasitoids of *Uroplata lantanae* Buzzi & Winder, another leaf-mining Hispinae in Brazil (Winder *et al.*, 1984). Although we cannot say anything about *E. ovalifolia* secondary chemistry as a factor influencing host-plant selection, which is clearly an important consideration (Denno *et al.*, 1990), it is possible that the difference in parasitism rates plays a role in *O. octopustulata* host species preference. Despite its faster development time in *E. ovalifolia* leaves, a measure of larval performance, which may mean that this is a higher quality host, this was the less utilised plant. Denno *et al.* (1990) found that the oviposition preferences of a chrysomelid beetle did not correspond well to the patterns of larval performance. The higher parasitism rate of *O. octopustulata* larvae on *E. ovalifolia* may limits its use in the area. Natural enemies have already been suggested as an influential force in herbivore distribution and niche (for example, Atsatt, 1981 a, b; Lawton, 1986; Rausher, 1988). In some cases a weak correspondence between patterns of oviposition and larval performance may be the result of the importance of predation and other non-host quality factors on larval fitness (Courtney, 1988; Thompson, 1988).

Acknowledgements

We would like to thank C.L. Staines for the Hispinae identification and M.A.P. de Azevedo and J. La Salle for the parasitoid identifications. Thanks also go to M.L. Cox for critically reading the manuscript. Financial support came from CNPq, FAPERJ and CEPG/UFRJ.

References

Atsatt, P.R., 1981a Ant-dependent food plant selection by the mistletoe butterfly, *Ogyris amaryllis* (Lycaenidae). Oecologia 48: 60-63.

Atsatt, P.R., 1981b Lycaenid butterflies and ants: selection for enemy-free space. Am. Nat. 118: 638-654.

Auerbach, M. and Simberloff, D., 1989 Oviposition site preference and larval mortality in a leaf-mining moth. Ecol. Ent. 14: 131-140.

Connor, E.F. and Tavener, M.P., 1997 The evolution and adaptive significance of the leaf-mining habit. Oikos 70: 6-25.

Courtney, S.P., 1988 If it's not coevolution it must be predation? Ecology 69: 910-911.

Cox, M.L., 1997 The larva of the flea beetle, *Mniophila muscorum* (Koch, 1803) (Coleoptera: Chrysomelidae, Alticinae), not a leaf-miner. Entomologist's Gaz. 48: 275-283.

Damman, H., 1994 Defence and development in a gregarious leaf-mining beetle. Ecol. Ent. 19: 335-343.

Denno, R.F., Larsson, S. and Olmstead, K.L., 1990 Role of enemy-free space and plant quality in host-plant selection by willow beetles. Ecology 71: 124-137.

Futuyma, D.J. and Moreno, G., 1988 The evolution of ecological specialization. A. Rev. Ecol. Syst. 19: 207-233.

Hering, E.M., 1951 Biology of the leaf miners. Junk, Gravenhage.

Hespenheide, H.A., 1991 Bionomics of leaf-mining insects. A. Rev. Ent. 36: 535-560.

Hileman, D.R. and Lieto, L.F., 1981 Mortality and area reduction in leaves of the bog shrub *Chamaedaphne calyculata* (Ericaceae) caused by the leaf miner *Coptodisca kalmiella* (Lepidoptera: Heliozelidae). Am. Midl. Nat. 106: 180-188.

Jolivet, P., 1988 Food habits and food selection of Chrysomelidae, bionomic and evolutionary perspective. In Jolivet, P., Petitpierre, E. and Hsiao, T.H. (eds.), Biology of the Chrysomelidae. Kluwer Academic Publishers, Dordrecht, The Netherlands.

Jolivet, P., 1989 Selection trophique chez les Hispinae (Coleoptera: Chrysomelidae: Cryptostoma). Bull. mens. Soc. linn. Lyon 58: 297-317.

Lawton, J.H., 1986 The effect of parasitoids on phytophagous insect communities. In: Waage, J. and Greathead, D. (eds.), Insect parasitoids. Academic Press, London.

Macêdo, M.V., Monteiro, R.F. and Lewinsohn, T.M., 1994 Biology and ecology of *Mecistomela marginata* (Thunberg, 1821) (Hispinae: Alurnini) in Brazil. In Jolivet, P.H., Cox, M. and Petitpierre, E. (eds.), Novel Aspects of the Biology of Chrysomelidae. Kluwer Academic Publishers, Dordrecht, The Netherlands.

Medeiros, L. and Vasconcelos-Neto, J., 1994 Host plants and seasonal abundance patterns of some Brazilian Chrysomelidae. In Jolivet, P.H., Cox, M. Petitpierre, E. (eds.), Novel Aspects of the Biology of Chrysomelidae. Kluwer Academic Publishers, Dordrecht, The Netherlands.

Nakamura, K., Abbas, I. and Hasyim, A., 1989 Survivorship and fertility schedules of two Sumatran tortoise beetles, *Aspidomorpha miliaris* and *A. sanctaecrucis* (Coleoptera: Chrysomelidae) under laboratory conditions. Res. Popul. Ecol. 31: 21-44.

Needham, J.G., Frost, S.W. and Tothill, B.H., 1928 Leaf-mining insects. Williams & Wilkins, Baltimore, MD.

Preszler, R.W. and Price, P.W., 1988 Host quality and sawfly populations: a new approach to life-table analysis. Ecology 69: 2012-2020.

Rausher, M.D., 1979 Larval habitat suitability and oviposition preference in three related butterflies. Ecology 60: 503-511.

Rausher, M.D., 1988 Is coevolution dead? Ecology 69: 898-901.

Simberloff, D. and Stiling, P., 1987 Larval dispersion and survivorship in a leaf-mining moth. Ecology 68: 1647-1657.

Thompson, J.N., 1988 Evolutionary ecology of the relationship between oviposition preference and performance of offspring in phytophagous insects. Entomologia exp. appl. 47: 3-14.

Valladares, G. and Lawton, J.H., 1991 Host plant selection in the holly leaf-miner: does mother know best? J. anim. Ecol. 60: 227-240.

West, C., 1985 Factors underlying the late seasonal appearance of the lepidopterous leaf-mining guild on oak. Ecol. Ent. 10: 111-120.

Winder, J.A., Harley, K.L.S. and Kassulke, R.C., 1984 *Uroplata lantanae* Buzzi and Winder (Coleoptera:

Chrysomelidae: Hispinae), a potential biological control agent of *Lantana camara* in Australia. Bull. ent. Res. 74: 327-340.

Windsor, D.M., Macêdo, M.V. and Siqueira-Campos, A.T. 1995 Flower feeding by species of *Echoma* Chevrolat (Coleoptera: Chrysomelidae: Cassidinae) on *Mikania* (Asteraceae) in Panama and Brazil. Coleopts. Bull. 49: 101-108.

THE BIOLOGY OF *CONCHYLOCTENIA PUNCTATA* (FABRICIUS) – A CYCLOALEXIC CASSID (*CHRYSOMELIDAE*: *CASSIDINAE*)

HUGH D.C. HERON
P.O. Box 39042, Escombe, Queensburgh, 4070, South Africa

Abstract

Conchyloctenia punctata is a common and widespread species in eastern, central and southern Africa and, in the Durban-Queensburgh region of Natal, South Africa, has been recorded from four host plants of the Convolvulaceae. The imago is briefly described, as also is the life cycle. Ootheca, found on lower leaf surfaces, contain up to circa 40 ova and are composed of stacked papery scales. They are somewhat similar to the complex ootheca of *Aspidimorpha puncticosta* Boheman. The larvae are gregarious and, particularly during instars 1 to 3, exhibit strong cycloalexic behaviour. This weakens during the 4th instar and some larvae may become solitary in instar 5. Only cast exuviae are retained in the supra-anal processes (faeces are scattered), but all are shed during ecdysis into instar 5 for most larvae. They scatter to pupate and pupae are seldom encountered on the host plants. Several parasitoids and predators are known, as also is an entomopathogenic fungus. The beetle is triple-brooded and overwinters as imagines in coastal Natal. The damage done to host plants during instar 5 and during the intensive feeding phase may, on occasions, render this species a potential minor garden pest. Comparisons are drawn with the brief life-cycle study of Paterson (1941:6-8). Her specimens were smaller and the life cycle of longer duration; there being two brood per season. The differences are probably the result of environmental factors.

Introduction

Conchyloctenia punctata was originally described by Fabricius (1787) in the genus *Cassida* (syntypes preserved in the Zoological Museum, Copenhagen), later transferred to the genus *Aspidomorpha* (Weise, 1895) and finally to the genus *Conchyloctenia* (Spaeth, 1902). It is considered to be the most abundant species in the genus (Borowiec, 1994:101) and this fact, together with its cycloalexic behaviour (larvae and young imagines may completely defoliate host plants on occasions) renders it a potential pest of certain convolvulous plants. Apart from taxonomic descriptions, the only publications on this species are the papers by Paterson (1941), briefly dealing with the life cycle, Heron (1992), dealing with cycloalexic behaviour, and Heron & Borowiec (1997), dealing with host plants and feeding patterns.

Distribution

This common tortoise beetle is found throughout Africa south of the Equator (particularly in the eastern and southern regions) and extends northwards into Ethiopia. It is rarer in the western portions of the continent, being reported from isolated localities in southwestern Angola, western Congo Democratic Republic (formerly Zaire), Gabon and Cameroon. The

Advances in Chysomelidae Biology 1, pp. 565–580
Edited by M.L. Cox

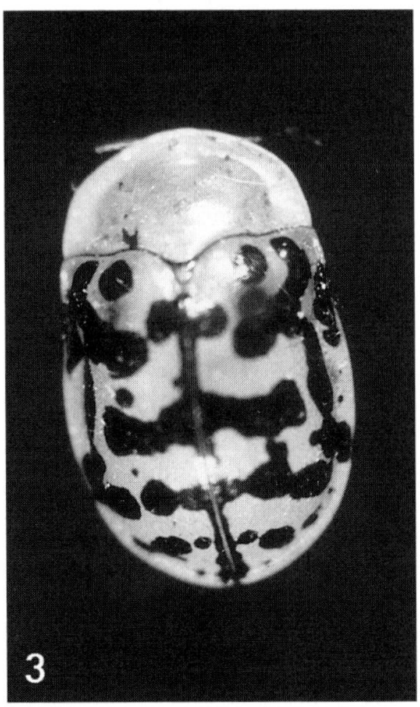

Fig. 1. *Conchyloctenia punctata* var *parummaculata*. This is the dominant variety in the Durban-Queensburgh area of Natal. Malvern, Queensburgh. Specimen: Durban Natural Science Museum.

Fig. 2. *Conchyloctenia punctata* var *parummaculata*. Heavier maculation. Malvern, Queensburgh. Specimen: Durban Natural Science Museum.

Fig. 3. *Conchyloctenia punctata* var *maculipennis*. From near Morija, Lesotho. Collected 15/01/1955 by L. Devis. Specimen: Durban Natural Science Museum.

Fig. 4. *Conchyloctenia punctata* var *luteicollis*. Malvern, Queensburgh. Specimen: Durban Natural Science Museum.

records from Cote d'Ivoire and Guinea are considered to be of introduced populations (Borowiec, 1994:101). In South Africa, the beetle has been reported from all except the drier northern and north-western Cape regions and, in addition to the Natal localities listed by Borowiec (1994:102-103), has been found at Queensburgh, near Durban. The observations in this study were made upon specimens raised in captivity, and recorded from host plants, in the Durban-Queensburgh region of Natal.

Description

This species is very variably patterned and exhibits a moderately convex elongated form. The size varies from 9×5 mm to 9×6 mm for males, and from 10×6 to 11×7 mm for females. The pronotal disc is reddish-brown with the explanate margin of a sub-metallic coppery-brown colouring. It has a matt surface and is devoid of maculation. The explanate margin exhibits micro-reticulation and has a slightly upturned, gutter-like, lip. The triangular scutellum is black. The elytral bases and posterior portions of the pronotal disc are about equally wide. The elytra are reddish-brown with a sub-metallic coppery lustre which exhibits a hint of green iridescence. The elytral suture appears as a black line. Maculation is extremely variable: some specimens are devoid of any, whilst others have entirely black elytra (var. *luteicollis* Boheman). The latter is scarce in the Durban-Queensburgh area although, on one occasion, a small colony was found to comprise up to 36% of that form. Maculation is generally sparse and seldom symmetrical, and maculae are not connected: the specimens studied conform to the variety *parummaculata* Boheman. The elytra exhibit evenly distributed micro puncturation; larger along the costa (marginal row). The legs and underside are vitreous black (explanate margin reddish-brown except where maculae present). Antennae bicoloured: yellowish basally and black apically. A full taxonomic description of this beetle appears in Borowiec (1994:96-105).

Metamorphosis

1. Oothecae

Although somewhat simplified and less ornate, these resemble the oothecae of *Aspidimorpha puncticosta* Boheman, as described by Muir & Sharp (1904:2-6). They are found attached to the undersides of host plant leaves, generally those that are more mature, and are initially pale pink, rapidly fading to pale beige and straw-white. They vary in size and shape, and are attached to the leaf at one end (there is no well-defined "petiole" or "stalk" as in *A. puncticosta*). They curve sharply so that the complete ootheca lies parallel to the leaf surface and give the illusion of being attached along its length. In *A. puncticosta*, the basal petiole area is partially masked by overlapping sterile papery membranes (Muir & Sharp, 1904: plate 1, fig. 1), and this is followed by the body of the ootheca: rectangular, curved, almost square in cross-section, and consisting of a succession of recurved membranes containing ova in compartments. *C. punctata* is very similar but appears less ornate and well formed. The sterile papery scale-like membranes appear as a slightly enlarged pleated spongy aggregate around the base of the ootheca: not clearly defined scales as seen in *A.*

puncticosta. The body structure of the ootheca appears less "tidy" than that of *A. puncticosta*, the ridging of the folded membranes appearing more loose and spongy. The terminal "face" is distinctly concave. The number of segments varies and the ova, measuring 2×1 mm (Paterson, 1941:7 gave 1.5×0.7-0.8 mm) and of a whitish colour, appear to be deposited in four rows along the length of the ootheca. Each ovum is in its own compartment arranged in offset fashion from the ova preceding and following. The ootheca is widest at its base (enlarged due to the presence of spongy membranes) and of variable length. Specimens measured varied from 6×4×3 mm to 9.5×5×5 mm, with up to 40 ova present. Paterson (1941:7) gave 5.25×4.0 mm with ova varying from 28 to 40 per ootheca. They take 14 to 16 days to hatch (Paterson (1941) gave 14 to 15 days) and, late in the season, may be heavily parasitised.

2. Larvae

Instar 1: Upon emerging, larvae measure circa 1 mm in length and are straw-white, becoming darker and suffused with greyish olive-green after feeding. They reach between 1.5×0.7 mm and 2×1 mm (Paterson (1941) gave 1.3×0.6 mm) (excluding lateral spines) by the end of the instar which lasts 3 to 4 days (Paterson (1941): 8 days). The lateral spines, supra-anal process, head, pronotal shields, and legs are black. The supra-anal process is somewhat filamentous and oriented almost vertically. The lateral spines appear nearly equal in length and are curved upwards at about 40°-50° from the horizontal. The larvae are gregarious and exhibit strong cycloalexic behaviour: packed together with heads inwards when at rest, and keeping in a group when feeding. Feeding takes place from the underside of the leaves from the ootheca, systematically stripping away the leaf tissue and leaving the transparent upper epidermal cuticle intact. The thread-like faeces are scattered. Towards the end of the instar the body takes on a translucent greenish-brown colour with a darker median line of internal organs.

Instar 2: Begins at approximately 2×1 mm and reaches 3×2 mm after 2 to 3 days (Paterson (1941) gave 2.1×1.16 after 7 days). They appear translucent pale chocolate- to olive-chocolate-brown with a darker median line. The pronotal shields are dark brown to black with a sub-vitreous lustre, as also are the head and legs. Lateral spines are black. The larvae are oval in shape, appearing distinctly parallel-sided, with the first instar exuviae retained in the supra-anal processes which are held vertically or slightly forward over the abdomen. They still exhibit strong cycloalexic behaviour.

Instar 3: These measure 3×2 mm (3.5×2.5 mm with lateral spines) and reach between 5×2 mm (6×3.5 mm) and 5×2.3 mm (6.7×4 mm) by the end of the instar (Paterson (1941) gave 2.8×1.6 mm after 6 days). The colour is identical to that of instar 2 and the scattered thick filamentous faeces may become conspicuous on leaves below the feeding site. Cycloalexic behaviour continues and, on softer leaves such as those of *Convolvulus farinosus*, holes may be left where feeding penetrates the upper leaf surface.

Instar 4: At this stage the cycloalexic bond begins to weaken (Heron, 1992) and, toward the end of the instar, larvae start to divide into smaller groups. When disturbed, the lateral spines are raised and if the disturbance persists, the larvae drop to the undergrowth. Beginning at 3×4 mm (4×5.5 mm with lateral spines) at the lower end of the size range, they grow to circa 7.5×4.5 mm (9×6 mm) by the end of the instar (Paterson (1941) gave 4.6×2.4 mm after 11 days). They appear chocolate-brown except for black pronotal shields and

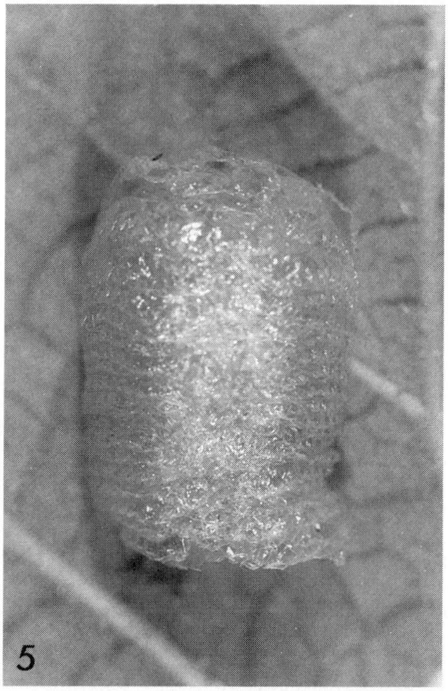

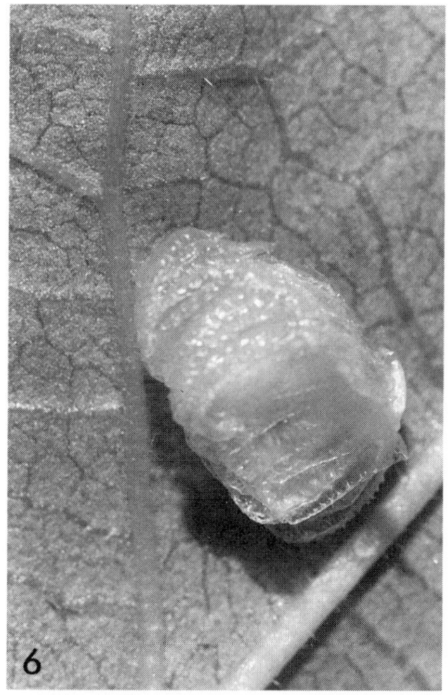

Fig. 5. Ootheca on underside of *Hewittia sublobata* leaf. Escombe, Queensburgh.
Fig. 6. Ootheca on underside of *Hewittia sublobata* leaf: exhibits a less compact structure than that of Fig. 5.
Fig. 7. Instar 3 larvae exhibiting cycloalexic clustering on underside of *Ipomoea batatas* leaf. Escombe, Queensburgh.
Fig. 8. Early instar 5 larva on *Hewittia sublobata*. Note short supra-anal process. Escombe, Queensburgh.

maculae, as described for instar 5, below. They may be suffused with pale greyish. The lateral spines are pale ash-grey to blackish. Spiracles are white. The cast exuviae are retained on the supra-anal processes but some may be lost before the end of the instar. Feeding, from the lower leaf surface, is from the leaf margin although, on tougher leaves, scrapings may still be produced early in the instar.

Instar 5: Size varies from 8×4 mm (10×6 mm with lateral spines) to 10×5 mm (11×7 mm) but there is considerable variation, depending upon gender and living conditions. Paterson (1941) gave 8.75×4.25-5.25 mm after 10 days. The larvae are pale brown to translucent whitish-brown with an orange tinge, and are surrounded by 16 pairs of black lateral spines held horizontally but which can be raised if the larvae are disturbed. The shape is elongate oval but not markedly tapering posteriorly. Body maculae, lateral spines and head are black to brownish-black. The relative spinal lengths are approximately as follows: $3 \approx 4 \approx 6 \approx 8 \approx 9 \approx 10 \approx 11 \approx 12 \approx 13 \approx 14 \approx 15 \approx 16 > 2 \approx 5 \approx 7 > 1$. Cast exuviae are only retained on the supra-anal processes until the end of the 4th instar and all are usually discarded during ecdysis into instar 5. The supra-anal process is short (circa 2 mm) and blackish: about two-thirds the length of lateral spine 4. The actual length varies depending upon the size of the larva (Paterson (1941) gave 2.5 mm). Their limited length is responsible for the loss, as early as during the 4th instar in some cases, of the cast exuviae. Filamentous faeces are scattered and never retained on the exuvial pile. The prothorax has four lateral spines. The pronotum has two small oval maculae, and a paler greyish patch, composed of minute grey speckles, is present along the median line. The mesothorax has two small maculae positioned anteriorly, a continuation of the greyish median line, and a large oval greyish patch situated posteriorly on each side. A large conspicuous white spiracle is present, as also are two lateral spines. Metathorax as for mesothorax but without the spiracle. Abdominal segments 1 to 7 host a pair each of small black maculae positioned anteriorly on the segmental margin, linked posteriorly by a pale greyish line composed of minute black speckles. The base of the supra-anal process is pale translucent orange. The head is black. Underside colour as for upperside laterally, but is otherwise distinctly orange. A black lateral band extends from the prothorax to the 16th lateral spine, and two small black maculae are present dorsally on each abdominal segment alongside the lateral greyish band. Black transverse bands extend between the segments from the greyish band laterally. Legs are black, except for the third segment which is pale whitish-brown. The larvae scatter to pupate and, for some days prior to pupation, may remain motionless: usually on the lower leaf surface.

3. *Pupae*

Pupation takes place after the larvae have remained immobile for a period of 48 to 72 hours (Paterson (1941) gave 6 days for this pre-pupal phase). The pupal body is flattened and oval, varying from 8×5 mm (8×6 mm including lateral spines) to 10×6 mm (10×7 mm) depending upon gender and the conditions under which the larva developed. Paterson (1941) gave 9.25×5.0 mm. Occasionally, larvae may pupate gregariously. The pupal colour is pale brownish-cream with a pinkish tinge and variable black markings, particularly on the abdominal segments. Pronotum vitreous with short setae. A large brownish-cream broadly cruciform marking with a pinkish tinge is flanked posterolaterally by dark greyish oval markings. A pair of black sub-triangular, elongated markings are present at the posterior

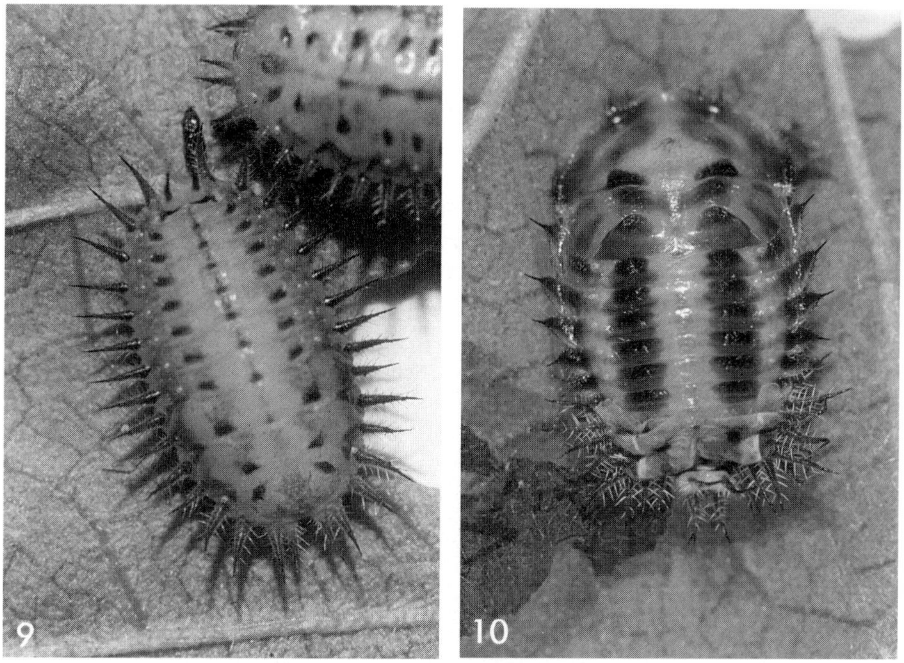

Fig. 9. Instar 5 larvae on *Hewitta sublobata*. Cast exuviae are lost during ecdysis at the end of instar 4. Escombe, Queensburgh.
Fig. 10. Pupa on underside of *Hewittia sublobata* leaf. Larvae generally abandon their host plant to pupate. Fifth instar exuvium is retained. Escombe, Queensburgh.
Fig. 11. Newly emerged imago during intensive feeding phase and prior to the development of mature colouration. Elytra partially transparent. Escombe, Queensburgh.

Fig. 12. Damage done to *Hewittia sublobata* creeper following mass emergence of imagines. Such extensive defoliation is rare but could result in the beetle being regarded as a minor pest were it to occur on garden subjects or crop species. Umbilo, Durban. Photo by the author.

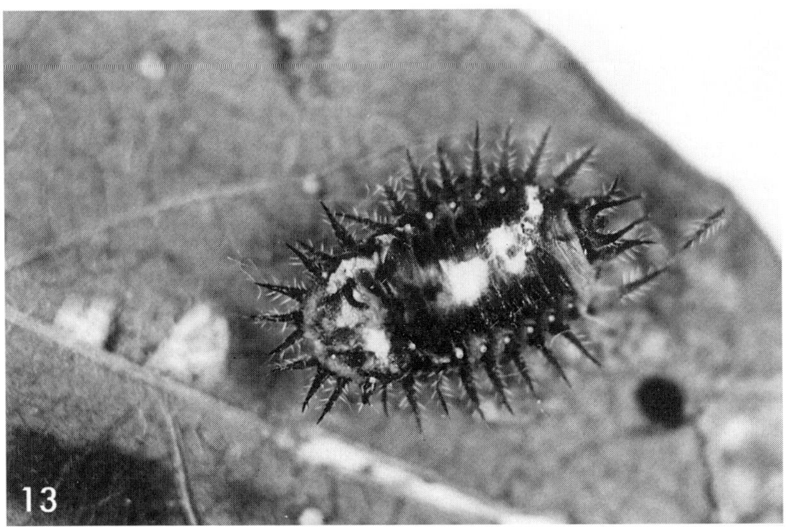

Fig. 13. The entomopathogenic fungus *Beauveria bassiana* erupting from 4th instar larva on underside of *Hewittia sublobata* leaf. Specimen: collection of the Plant Protection Research Institute, Pretoria.

edge of the pronotum. A small diffuse grey spot is located in the cruciform region, just anterior to the centre of the pronotum. The explanate margin has diffuse grey-black lateral markings, a pale creamy-brown patch along a portion of the margin, as well as along the median line, and a patch midway between the diffuse grey markings and a similar marking on the thorax. Approximately 80 minute spinules border the pronotal explanate margin. Two pairs on the anterior margin are larger, stouter, and strongly thorn-like (twice the length of the others) and black in colour. The spinules are colourless, except on those portions of a diffuse greyish colour where they assume the same hue. The anterior edge of the explanate margin has a sharp median indentation. The posterolateral margin is not strongly incurved and the explanate margin is depressed anterolaterally. The mesothorax and elytral area are pale creamy-brown with a diffuse smokey-grey chevron band anteriorly, and in the basal zone. The abdominal segments are glabrous with five pairs of leaf-like lateral projections, each terminating in a lateral spine circa 0.5 to 0.3 of the projection's length. The colour is smokey-grey: the spine black. The spiracles are white and, on the anterior segments, are obscured by the cast exuvium of the 5th instar which is retained by the pupa. The abdominal markings are variable: essentially pale creamy-brown with a variably wide dark smokey-grey to blackish band between the pale dorsal line and the spiracles. In some pupae the entire abdomen may be suffused with smokey-grey (darker in the vicinity of the band) and then the white spiracles are conspicuous. The supra-anal process is entirely obscured by the exuvium. On the underside, the pronotal explanate margin appears smokey-grey and the two largest spinules on the anterior margin are especially conspicuous. The legs, lateral abdominal projections, and the margin of the body are all blackish-grey. The remainder is pale brownish-orange suffused with pale grey. The pupal period lasts 8 to 10 days (Paterson (1941) gave 6 to 7 days).

4. Newly emerged imagines

At first the newly emerged imago is whitish with the pronotum and elytra, including epipleura, a transparent orange colour. The maculae, if any, begin to appear first and the epipleura darken to orange-brown. The pronotal area becomes pale greenish-grey and the elytra a uniform pale grey. This immature colouring is retained for 10 to 14 days, gradually changing as the mature sub-metallic colouring develops. Dispersal of imagines takes place at this stage.

Host plants

Only three of the fifteen species of *Conchyloctenia* in Africa have had their host plants recorded (Heron & Borowiec, 1997). *C. punctata* utilises four members of the Convolvulaceae, viz. *Convolvulus farinosus*, *Hewittia sublobata*, *Ipomoea batatas* and *I. cairica*. Of these, *Convolvulus farinosus* is relatively infrequently used. There is a single record of *Ipomoea ficifolia* being used by an imago over 2 to 3 days but the beetle has not been observed to breed upon it. Paterson (1941:6) gave *Ipomoea purpurea*. Bearing in mind the beetle's widespread distribution, many additional host plants are to be anticipated.

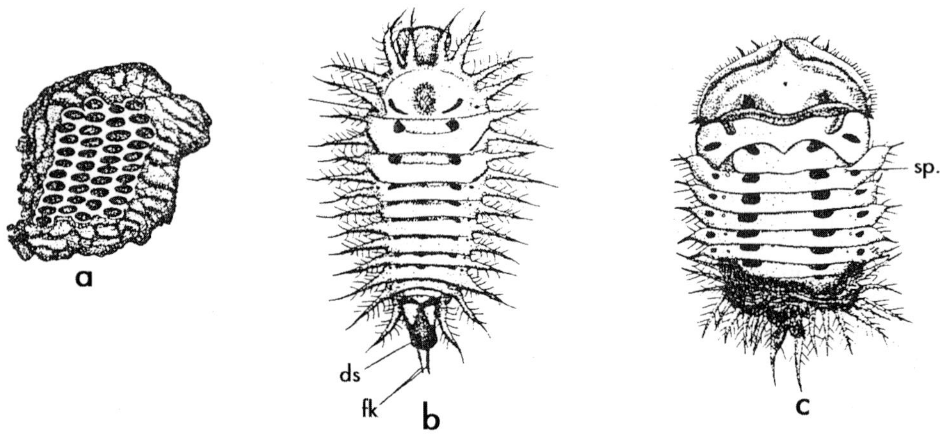

Fig. 14. Sectioned ootheca, larva and pupa of *Conchyloctenia punctata* (Paterson, 1941:8). (Reproduced with permission).

Behaviour

The larvae are gregarious, forming roughly circular to oval cycloalexic groupings, heads inwards, when at rest. The significance of circular defensive clustering in insect larvae was first recognised and named in a joint publication by Vasconcellos-Neto and Jolivet in 1988 (see Jolivet *et al.*, 1990). In *Conchyloctenia punctata*, the bond appears to weaken after the 3rd instar when groups begin to divide (sometimes as small as 3 or 4 individuals) and, by instar 5, some larvae may become solitary. Once the cycloalexic bond weakens, larvae drop to the undergrowth when disturbed: particularly during the 5th instar when even the observer's shadow may alarm them.

Imagines are generally located on the undersides of leaves and their presence betrayed by their feeding patterns. They are best located in late afternoon when they may be frequently observed clambering about their host plants. They are shy: readily dropping to the undergrowth when disturbed. This is particularly true during the intensive feeding phase (10 to 14 days following eclosion) when they may be found in considerable numbers (sometimes defoliating the host plant). Defoliation of host plants is not regarded as a problem and has only been noted in the case of *Hewittia sublobata* on one occasion. Were this to occur on a garden subject such as the *Ipomoea purpurea* recorded by Paterson (1941), such defoliation might render the beetle a minor pest. The imagines disperse once their metallic colouring begins to appear. They are very rarely noted in flight but individuals are suspected to range widely: this is an area for further study. The species overwinters as imagines and, in coastal Natal, appears suddenly a few weeks before the Spring rains to deposit oothecae (generally about mid-August). There are three generations per season (September-October, January-February, and March-April) and the beetle may all but vanish at the end of each cycle. Paterson (1941) confirmed the overwintering as imagines but recorded that Highveld populations were double-brooded.

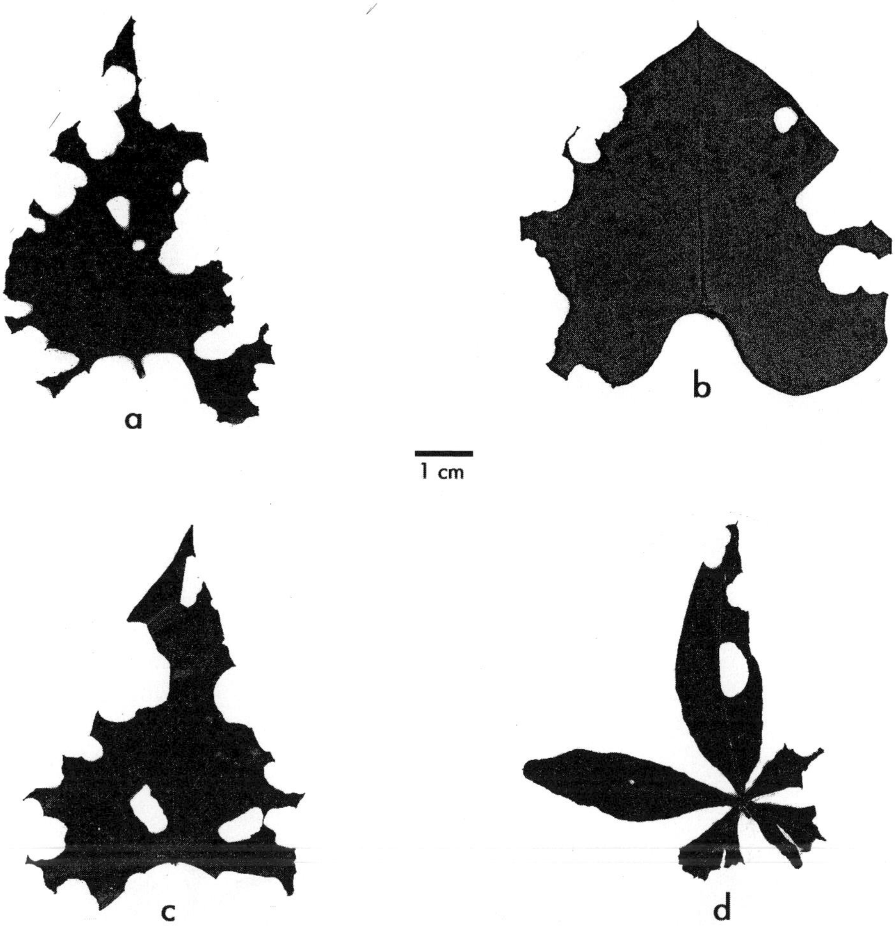

Fig. 15. Imaginal and larval feeding patterns. Imagines: large irregular marginal indentations are produced (holes, where present, generally penetrate the leaf margin). The irregular to ragged marginal feeding patterns are distinct from those produced by other cassid species known to use the same plants in the study area of coastal Natal. The pattern on *Ipomoea cairica* may be ambiguous since the deeply lobed structure of the leaf allows little room and the feeding patterns of other cassid species may penetrate the leaf margin and thus resemble marginal feeding. *a. Convolvulus farinosus*. Queensburgh, Natal. No: 893/53; *b. Hewittia sublobata*. Durban, Natal. No. 1302/70; *c. Ipomoea batatas*. Queensburgh, Natal. No. 167/9; *d. Ipomoea cairica*. Queensburgh, Natal. No. 561/26.

Feeding patterns

The feeding patterns or traces left by cassid beetles studied in coastal Natal appear to be characteristic of the species for a particular host plant (Heron & Borowiec, 1997:637). This is especially true of imagines: larvae are more likely to prove ambiguous. The imagines of *C. punctata* are marginal feeders, that is, large irregular indentations are made from the leaf margin. On *Convolvulus farinosus* the pattern may be confused with that made by *Aspidimorpha icterica* Boheman, a common cassid, but that species also produces large irregularly rounded holes and its larvae are almost always present. The irregular to ragged

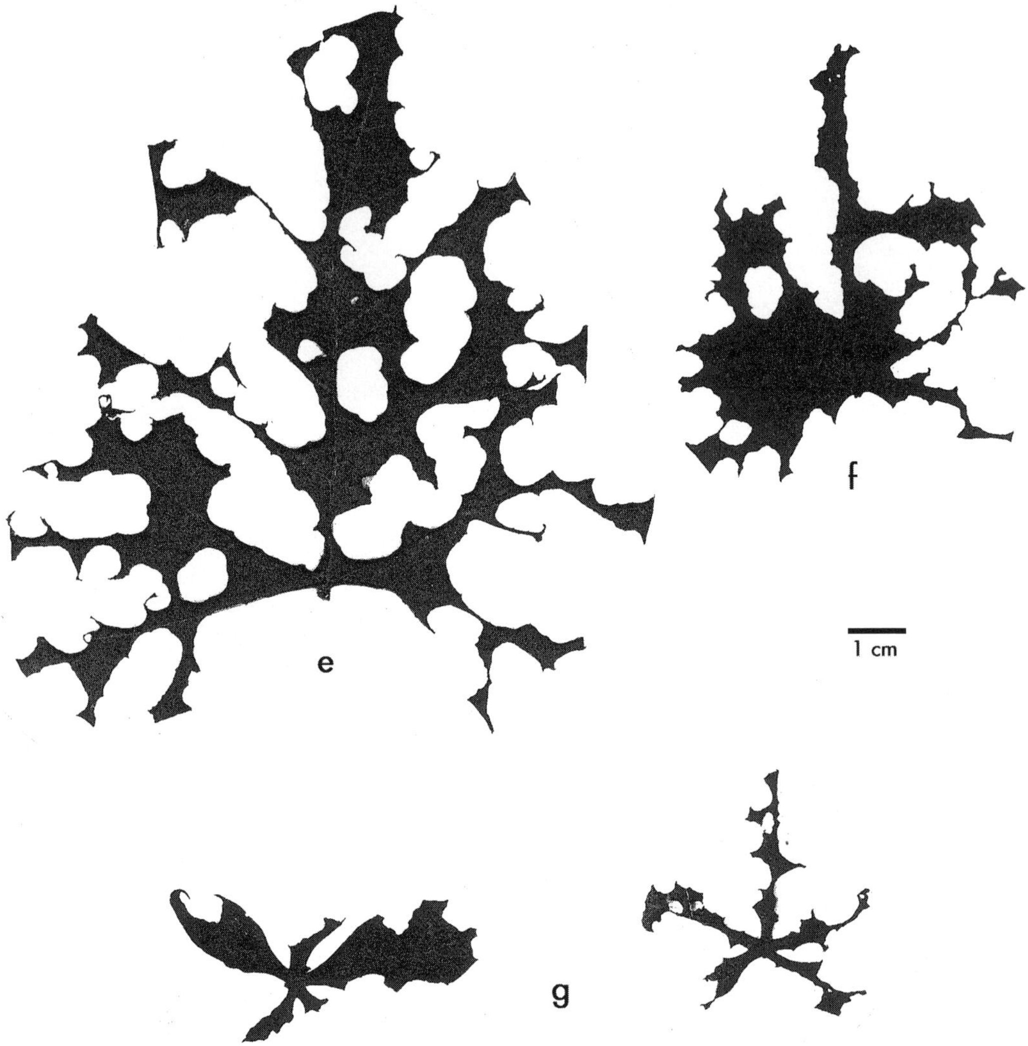

Fig. 15e–g. Intensive feeding phase: Following eclosion, large numbers of imagines may suddenly appear and, on occasions, have been known to defoliate small areas (see Fig. 12). Severe damage is generally limited to small areas and portions of individual plants. This phase lasts 10 to 14 days but imagines begin to scatter within a few days and may vanish completely until the next cycle. Despite the damage, this beetle is unlikely to be viewed as more than a very minor pest. *e. Hewittia sublobata*. Queensburgh, Natal. No: 1308/74; *f. Ipomoea batatas*. Queensburgh, Natal. No: 934/67; *g. Ipomoea cairica*. Durban, Natal. Nos: 836/41 & 838/42.

marginal indentations on *Hewittia sublobata* and *Ipomoea batatas* are distinct from other cassids on the same plants. Where irregular holes are produced on *Hewittia*, these almost always penetrate the leaf margin. The deeply lobed leaves of *Ipomoea cairica* are a problem since their limited areas result in the holes created by many cassid species penetrating the leaf margins and resembling marginal feeding. Parts of the leaves may be completely consumed. During the intensive feeding phase, 10 to 14 days following eclosion, the imagines may largely or completely consume the host plant leaves.

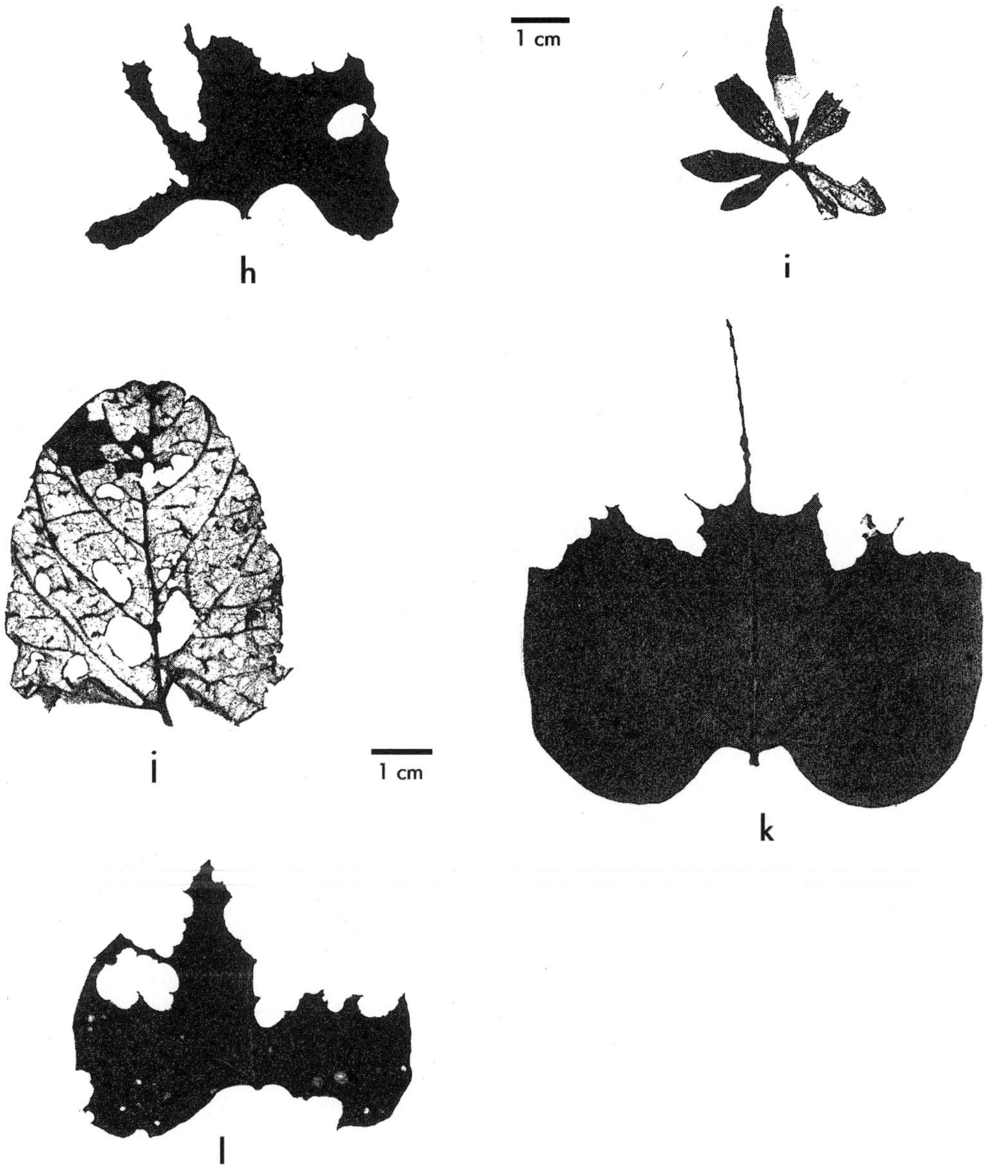

Fig. 15h–l. Larvae: During instars 1 and 2 (sometimes extending into instar 3 for tough leaves) the cycloalexic larvae strip away the lower leaf surface, leaving the upper epidermal cuticle intact. This is followed by systematic marginal feeding and the patterns thus produced may be confused with those made by other insects such as hawk moth larvae (Sphingidae); *h. Convulvulus farinosus*. Instars 3 to 4. Queensburgh, Natal. No: 903/59; *i. Ipomoea cairica*. Instar 1 in proximity to ootheca. Durban, Natal. No: 685/35. *j. Hewittia sublobata*. Instars 1-2, upper cuticle intact. Holes produced earlier by larvae of *Aspidimorpha icterica*. Queensburgh, Natal. No: 498/25; *k. Hewittia sublobata*. Instar 5. Queensburgh, Natal. No: 619/33; *l. Ipomoea batatas*. Instars 4 to 5. Queensburgh, Natal. No: 910/61.

During instars 1 and 2, sometimes extending into 3 in the case of older, tougher, leaves, the larvae feed by stripping away the lower leaf surface, leaving the upper epidermal cuticle intact. From instar 3 onwards, they feed systematically from the leaf margin. Leaf stripping is conspicuous: such leaves tend to brown and curl, and the scattered filamentous faeces on vegetation below the feeding site may attract attention. The feeding pattern on *Ipomoea purpurea* has not been recorded but would not be expected to differ from the above.

Parasites and predators

Parasitoid wasps (Chalcididae and Eulophidae) appear to be the principal controlling agents of early stages, with social wasps (Vespidae) and ambushing spiders (Thomisidae and Salticidae) putting additional pressure upon larvae. The following observations have been made:

a. *Oothecae*: A damaged ootheca was attacked by a lacewing larva (Neuroptera: Chrysopidae). Numerous small parasitoid wasps of the genus *Tetrastichus* (Eulophidae) emerged from an ootheca and small wasps which failed to escape from another ootheca were identified as *Pediobius sp.* (Eulophidae).

b. *Larvae*: Several Hemiptera have been recorded taking larvae. Imagines and nymphs of assassin bugs (Reduviidae) were noted taking larvae, particularly during the later instars. A small scutellerid-type bug (Scutelleridae: possibly near *Callidea*) was observed with a 4th instar larva. This record requires confirmation since the Scutelleridae are described as being phytophagous (Scholz & Holm, 1985:144). A patterned bug, possibly near *Bagrada sp.* (Pentatomidae) was also recorded taking larvae. Larvae and imagines of carabid beetles (Carabidae) took larvae. Social wasps (Vespidae), notably *Polistes sp.* and *Belonogaster sp.* are active predators and were observed taking larvae (instars unspecified) on numerous occasions. Two spider families have been recorded, viz. Salticidae (genus not identified) and Thomisidae (*Synaema sp.*). The latter was observed to take circa twenty instar 1 and 2 larvae, despite their cycloalexic behaviour, over a week or so (Heron, 1996:6). A mitosporic (no sexual reproductive state known) entomopathogenic fungus, *Beauveria bassiana*, was identified as having killed a 4th instar larva.

c. *Pupae*: Two parasitoid wasp genera have been recorded as emerging during the pupal period, viz. *Brachymeria sp.* (Chalcididae) and *Pediobius sp.*, near *P. aspidomorphae* (Girault) (Eulophidae). Eighteen of the latter emerged from a single pupa.

d. *Imagines*: An unidentified mantis (Order Mantodea) took an imago during the intensive feeding phase (10 to 14 days after eclosion).

Future research

Since this is such a widespread species, many more host plants are anticipated. Parasitoids and predators are imperfectly known and require further study. Genetic studies to determine the origin of the *luteicollis* form would prove interesting.

Acknowledgements

Sincere thanks are extended to Miss R. Williams and Mr A. Ngwenya of the Natal Herbarium, Durban, for their identification of host plants, to Dr G. Prinsloo of the Plant Protection Research Institute, Pretoria, for the identification of the parasitoid wasps, to Dr I. Rong, also of the P.P.R.I., Pretoria, for identifying the entomopathogenic fungus, to Peter Croeser of the Natal Museum, Pietermaritzburg, for assistance with the spiders, to Dr T. Gijswijt of the Instituut voor Taxonomische Zoologie, Amsterdam, for identifying the *Tetrastichus* wasps, and to Willem Hogenes, keeper of the collection at the Instituut, for preparing and mounting the wasps (now in the collection of the P.P.R.I., Pretoria). Special thanks go to Prof. S.A. Hanrahan, President of the Entomological Society of South Africa, for permission to use the figures from Paterson (1941), and to Andrew Carter of the Durban Natural Science Museum, for taking the photographs. Dr Pierre Jolivet very kindly supplied a copy of his paper on cycloalexy. The assistance and encouragement given by Prof. Lech Borowiec of Wrocław University, Poland, is gratefully acknowledged.

References

Borowiec, L. 1994. A Monograph of the Afrotropical *Cassidinae* (Coleoptera: Chrysomelidae). Part 1. Introduction, morphology, key to the genera, and reviews of the tribes *Epistictinini*, *Basiprionotini* and *Aspidimorphini* (except the genus *Aspidimorpha*). Genus (suppl.), Biologica Silesiae, Wrocław, 176pp.

Fabricius, J.Ch. 1787. Mantissa Insectorum sistens eorum species nuper detectas adiectis characteribus genericus, differentiis specificus, emendationibus, observationibus. I. Hafniae, XX + 348pp.

Heron, H.D.C. 1992. Cycloalexy in two South African tortoise beetles (Chrysomelidae, Cassidinae). Chrysomela, *27*, 3-4.

Heron, H.D.C. 1996. Some arachnids preying upon tortoise beetles. Spider Club News, *11*(1):3-9 (Official organ of the Spider Club of Southern Africa).

Heron, H. & Borowiec, L. 1997. Host plants and feeding patterns of some South African tortoise beetles. (*Coleoptera*: *Chrysomelidae*: cassidoid *Hispinae*). Genus, *8*(3):625-658.

Jolivet, P., Vasconcellos-Neto, J. & Weinstein, P. 1990. Cycloalexy: a new concept in the larval defence of insects. Insecta Mundi, *4*:133-142.

Muir, F. & Sharp, D. 1904. On the egg-cases and early stages of some *Cassididae*. Trans. ent. Soc. London, *1904*:1-23.

Paterson, N.F. 1941. The early stages of some South African Chrysomelidae (Coleoptera). J. ent. Soc. Sth. Afr., IV:1-15.

Scholtz, C.H. & Holm, E. 1985. *Insects of Southern Africa*. Butterworth Publishers: Durban.

Spaeth, F. 1902. Beschreibung neuer Centralafrikanischer Cassiden aus dem Museum zu Brüssel. Annis. Soc. ent. Belg. *46*:446-461.

Weise, J. 1895. In: G.A. Graf von Götzen, Durch Afrika von Ost nach West. Resultate und Begebenheiten einer Reise von der Deutsch-Ostafrikanischen Küste bis zur Kongomündung in den Jahren 1893/94. Geographische Verlagshandlung Dietrich Reimer, Berlin, XI + 417pp.

DESCRIPTION OF IMMATURES AND LIFE CYCLE OF *STOLAS LACORDAIREI* (BOHEMAN) (COLEOPTERA, CHRYSOMELIDAE, CASSIDINAE)*

ZUNDIR JOSÉ BUZZI[1] and ROSINA DJUNKO MIYAZAKI[2]
[1]*Zoology Department, Paraná Federal University, P.O. Box 19020, 81531-990 Curitiba, PR, Brazil. E-mail: zbuzzi@bio.ufpr.br;* [2]*Zoology-Biology Department, Mato Grosso Federal University, MT, Brazil*

Abstract

Immatures and life cycle of *Stolas lacordairei* (Boheman,1850) are described. The incubation period, duration and number of larval instars, the duration of the pupal period and some biological data on the adult including oviposition period, fecundity, longevity, diapause, and sex ratio are presented and compared with other Stolaini. Immatures were reared at a temperature of 25°C and fed with leaves of *Mikania micrantha* (Asteraceae). Pairs of adults were kept under laboratory conditions at 23.8°C and fed with leaves from the same plant.

Introduction

Stolas lacordairei occurs in Buenos Aires and Tucuman (Argentina) and in the south of Brazil (Spaeth, 1914). Blackwelder (1946) also recorded it from Brazil. Specimens from Brazil, labelled São Paulo (São Paulo), Jaguariaíva and Curitiba (Paraná) in the Entomology collection. "Pe. Jesus S. Moure" are standing in the Zoology Department of Paraná Federal University. A series of specimens from Uruguay and also specimens from Rio Grande do Sul State, Brazil and La Plata, Argentina are present in the BM (NH), London (M. Cox pers. comm.).

Adults and larvae of *S. lacordairei* feed on leaves of *Mikania micrantha* (Asteraceae). Specimens of this plant in the UPCB herbarium of the Botany Department indicated the following geographical distribution in Paraná state: Antonina, Campo Largo, Curitiba, Mel and Farol Islands (Paranaguá), Lapa, Maringá, Morretes, São José dos Pinhais and São Mateus do Sul.

Material and methods

All the experiments were perfomed in the Zoology Department laboratory of Paraná Federal University, in Paraná State, Brazil, in the course of the years 1996 and 1997.

*Contribution 1068 from Zoology Department of Paraná Federal University.

Advances in Chysomelidae Biology 1, pp. 581–597
Edited by M.L. Cox

The length of the incubation period, the number of larval instars and their duration, as well as the length of the pupal stage were determined inside a Biometron greenhouse, at a temperature of 25°C ± 1, under a 16:8 hour L:D photoperiod. The eggs obtained from adults collected in the field were placed in petri dishes; larvae and pupae were kept in transparent cylindrical plastic containers, 70 mm in height by 40 mm in diameter. The larvae were fed on leaves of *M. micrantha* which were changed daily.

Eggs, larvae and pupae used for the descriptions and photographs were fixed in Kahle-dietrich (300 parts distilled water: 150 parts 95% alcohol: 10 parts glacial acetic acid).

Pairs of adults were kept under laboratory conditions at an average temperature of 23.8°C, in 10 x 12 cm transparent plastic containers and fed with *M. micrantha* leaves were re-newed daily.

Descriptions of immature stages

Eggs (Fig. 3)

The eggs are long, fusiform, pale pink, about 2.4 x longer than the greatest external diameter. The average length is 1.36 mm, ranging from 1.26 to 1.42; the peduncle average length is 0.26 mm, varying from 0.12 to 0.52 mm; and the average hood length is 0.20 mm, ranging from 0.14 to 0.24 mm. The surface of the chorion is almost even and both ends have a micropyle. At the lower end which is attached to the substratum, the micropyle is distinct. Laterally there is a short peduncle which sticks the egg to the substratum via a small plate. Usually, this plate fuses with others that are nearby, forming a large plate on the substratum. At the upper end a brown substance covers the micropyle.

The eggs obtained in the laboratory comprised compact masses of on average 57.65 eggs, but the number in a mass varied from 10 to 140. The eggs are closely-arranged, stuck together at the mid-point, and usually making elongate masses. Fig. 3 shows a mass of 172 eggs collected in the field. The largest field-collected mass comprised 187 eggs.

The incubation period lasts on average about 6.08 days, with a range of 5 do 9 days.

Fifth or final instar larva (Figs. 8-9)

The larva has a long body two times longer than the width. They are orange, except fot the scoli, anal fork and the dorsal part of the legs and spiracular peritreme, which are black. The chalazae which are distributed dorsally over the body almost always have a very dark, short, apical bristle. The subparallel sides slightly narrow posteriorly from the sixth segment and before that very slightly narrowed from the mesothorax. The anterior margin of the prothorax is broadly curved. The body laterally has 14 pairs of scoli, subequal in length, with the following measurements in mm taken from a fifth instar larva from the anterior posteriorly: 1.28, 1.20, 2.00, 1.20, 1.60, 1.44, 1.20, 1.20, 1.20, 1.20, 1.12, 1.52, 1.60 and 1.60. They are arranged on the segments as follows: three pairs on the prothorax with the last posterior scoli very close to the base; two on the mesothorax; one on the metathorax; eight on the abdomen, from the first to the eighth segment. The eighth segment carries dorsally an anal fork approximately 2mm in length, whose tip when bent forwards reaches the fifth tergum.

Figs. 1-3. Stolas lacordairei. 1. Female; 2. Male; 3. Eggs. (Scale line: 3.7 mm).

Table 1. Length, in mm, of eggs (E) and pupae (P), number of eggs in egg masses (EM) of Cassidinae (Stolaini).

	E	EM	P	Source
Acromis nebulosa	1.19	32	——	Buzzi (1980)
A. sparsa	1.50	48	——	Windsor (1987)
A. spinifex	1.5	30	——	Fiebrig (1910)
Chelymorpha alternans	1.55	49	8.0	(unpublished data)
C. cribraria	1.5	8	8.0	Marques (1932)
C. indigesta	1.75	38-48	——	Frers (1922)
C. inflata	1.47	52	——	(unpublished data)

Table 1. Cont.

	E	EM	P	Source
C. variabilis	1.25	20-40	7.8	Zolessi (1968)
C. puncticollis	1.50	18	8.0	Marques (1932)
C. rufipennis	1.50	28	8.0	Marques (1932)
C. constellata	1.50	38	8.0	Marques (1932)
Echoma sp.nr. *bonfilsi*	——	4.7	——	Windsor *et al.* (1995)
Stolas (Anacassis) dubia	1.82	9.17	7.84	Buzzi (1996)
S. (A.) fuscata	1.68	12.54	7.89	Buzzi (1996)
S. (A.) languida	2.03	6.7	9.52	Buzzi & Garcia (1983)
S. (A.) cribrum	1.92	23.9	8.14	Buzzi (1975a)
S. (A.) phaeopoda	1.78	1-3	8.23	Buzzi (1996)
S. (A.) punctulata	2.24	6-7	10.83	Buzzi (1996)
Stolas (Botanochara) impressa	3.40	6	10.9	Buzzi (1977)
Stolas (Cyrtonota) sexpustulata	2.5	3-20	12.0	Marques (1932)
Stolas lacordairei	**1.36**	**57.65**	**7.6**	**(present work)**
Zatrephina meticulosa	1.8	1	7.6	Buzzi (1976b)

Table 2. Stolas lacordairei. Mean width (m), in mm, of the cephalic capsules of the larval instars (I–V). Standard deviation (s), Range (< >), Coefficient of variation (CV). Number of cephalic capsules recorded (N).

	I	II	III	IV	V
N	22	20	18	15	15
m	0.43	0.62	0.85	1.24	1.65
s	0. 0129	0.0179	0.0331	0.0553	0.0821
<	0.40	0.60	0.78	1.16	1.50
>	0.46	0.64	0.92	1.34	1.76
CV	3.00	2.89	3.89	4.50	4.98

Table 3. Stolas lacordairei. Mean duration (m), in days, incubation period (E), larval instars (I – V) and pupal period (P). Standard deviation (s), Range (< >), Coefficient of variation (CV). Number of incubation periods recorded (N).

	E	I	II	III	IV	V	P
N	15	18	18	18	17	16	14
m	6.08	3.72	2.56	2.61	2.94	5.69	5.43
s	0.8997	0.7519	0.6157	0.5016	0.5557	0.6021	0.8516
<	4	2	2	2	2	5	5
>	7	5	4	3	4	7	8
CV	15.78	20.21	24.05	19.22	18.90	10.58	15.68

The head is subcircular in frontal view (Fig. 10). It is somewhat flat in side view. The parietals in frontal view are slightly longer than the greatest width in the middle ocellar height and it has a remarkable depression between the epicranial suture and the middle ocelli (stemmata). The surface is sparsely covered with bristles. The clypeus is about 2.6 x wider that its length at the mid-point. The anterior margin is almost straight and the lateral margins converge anteriorly. The labrum (Fig. 11) is approximately 2.8 x wider that its length at the mid-point; bearing 4 setae in a transverse row; anterior margin is slightly chamfered medially with short bristles laterally; posterior margin sublinear. There are six

Figs. 4-9. Instars of the larva of *Stolas lacordairei*. 4. First; 5. Second; 6.Third; 7. Fourth; 8-9. Fifth. (Scale line: 2.4 mm).

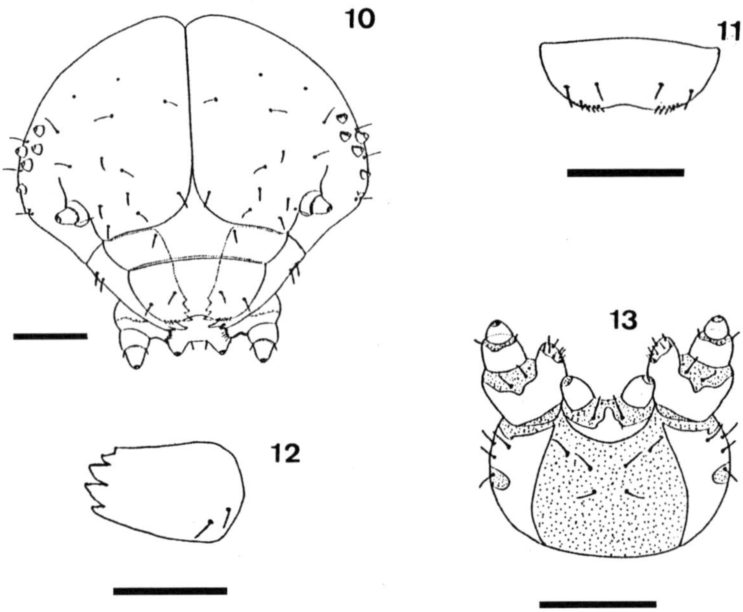

Figs. 10-13. Head capsule of the fifth instar of the larva of *Stolas lacordairei*. 10. Head; 11. Labrum; 12. Mandible; 13. Maxilla and labium (Scale line: 0.5 mm)

ocelli or stemmata, in pairs of which the mosto anterior one is on the external part of the antennae. The antennae are 2-segmented, with the first segment very short and ring-shaped; the second segment is sub-conical with some apical sensorial processes of which one is distinctly larger. The mandibles (Fig. 12) are about 1.5 x longer that their basal width; apically 5-dentate, two teeth distinctly larger than others; externo-basally bearing two setae. Maxillae with cardo long, bearing some lateral bristles; stipes almost as long as maxillary palpus (Fig. 13). The stipes is continuous with the mala along its inner margin; mala bearing some apical bristles. The maxillary palpi are 2-segmented; basal segment cylindrical, width slightly greater than length; distal segment subconical, almost as long as its basal width. Labium (Fig. 13) with part of mentum and submentum membranous, broadening posteriorly, bearing a few setae. The stipe has a membranous back and its front shows a narrow, transverse sclerotized band with a slightly median anterior projection. The labial palpi are one-segmented and subconical.

The pronotal plate is very wide, covering almost the entire prothorax, and is divided at the middle by a narrow, whitish band; bearing furrows and sparse chalazae. The mesonotum is about four x wider length at the mid-point, excluding the scoli and laterally there is a short dark spot. The pre-mesotergite also has transverse lines and the posterior part has chalazae, some small ones more posteriorly forming a second line. Laterally there are a few irregularly arranged chalazae. At the base of the fourth pair of scoli, the first pair of spiracles project since their peritreme and the ostium is better developed than in the other spiracles. The metanotum is similar to the mesonotum. However, the former is wider and has slightly larger lateral dark spots. The pre-tergite and the post-metatergite have only a transverse chalazal line. The abdominal tergum is divided into pre- and post-tergites, each one with a transverse chalazal line. Laterally, next to the base of the scoli, from the first to the seventh segment, there are lines of spiracles, one pair per segment. The anterior spiracles have a larger peritreme and ostium which gradually decrease in size posteriorly.

The legs are black, heavily sclerotized, bearing many setae. The thoracic and abdominal sterna have many chalazae, usually with an apical bristle. The abdominal sternum is very wide and short. The apical sternites are narrower, the ninth and tenth forming an anal tube; The tenth is strongly ventrally scletorotized.

The larva before pupating becomes inactive, stops feeding and attaches to the substratum by the anterior abdominal sterna. It remains in this pre-pupal stage for about two days and its average length is 9.64 mm, ranging from 7.5 to 10.5 mm.

Larva of earlier instars (Figs 4-7)

The first four larval instars basically have the same colour as the fifth instar. The length of the anal fork of the first two instars is very long compared with the total larval length and they reach the first tergum if bent anteriorly over the larva. This is not true for the anal fork of the other instars. The neonate larva has an average length of 1.23 mm, ranging from 1.20 to 1.32 mm.

The average head width of the different larval instars, with the range and standard deviation in parenthesis, are respectiverly from the first to the fifth instar: 0.43 (0.40 – 0.46; 0.0129); 0.62 (0.60 – 0.64; 0.0179); 0.85 (0.78 – 0.92; 0.0331); 1.24 (1.16 – 1.34; 0.0553); 1.65 (1.50 – 1.76; 0.0821); (Table 2). This shows that no overlap in measurements occurs, and this allows the recognition of the instar to which a larva belongs.

Fig. 17, shows the regression line for the geometric growth of head width as a function of larval instars whose line equation is $y = 0{,}04 + 0.31x$ and the linear correlation coefficient is 0.99, indicating a good fit to linear models throughout.

Fig. 18, shows the line of regression of larval length as a function of age, whose line equation is: $y = 0.78 + 0{,}51\ x$ and the linear correlation coefficient is 0.99, indicating a good fit to linear models throughout.

The growth ratio between the variou instars is: 1.44 – 1.37 – 1.46 and 1.33.

The average duration of the larval instars in days, the range and the standard deviation, in parenthesis, is from the first instar, respectively: 3.72(2 – 5; 0.7519); 2.56(2 – 4; 0.6157); 2.61 (2 – 3; 0.5016); 2.94 (2 – 4; 0.5557); 5.69 (5 – 7; 0.6021); 5.43 (5 – 8; 0.8516);

Figs 19 – 23 show the larval feeding damage to the leaves of the host plant *Mikania micrantha* and the corresponding area of consumption during 24 hours. Fig. 19 shows the area consumed during the first day of larval life. Fig. 20, shows the area on the third day (second instar); Fig. 21, the area on the sixth day (third instar); Fig. 22, the area on the ninth day (fourth instar) and Fig. 23, the area on the twelfth day (fifth instar). In the first instars, the larva just grazes the upper epidermis of the leaf until it reaches the lower epidermis. At the end of third instar and even more so in the fourth, perforations appear, showing that the larva can cut through the upper and lower leaf epidermis. In the fifth instar, the larva cuts the leaf thoroughly and there are no more grazed parts. This often happens at the end of the fourth instar.

The Pupa (Figs 14 – 16)

The pupa is elongate, about 1.7 x longer than the width of the second abdominal tergum, excluding the scoli. Its sides, except for the pronotum are strongly curved, converging anteriorly, weakly from the second tergum and more strongly backwards from the fifth one. In lateral view it is strongly dorsally convex, but flat ventrally. It is orange with black or brown spots. The body is covered with very short, sparse and indistinct whitish setae. The average length of a male pupa is 7.38 mm and for that of the female is 7.93 mm. This difference is significant at the 1% level (Table 4).

The head is furrowed medially, from the clypeus to the vertex and is bilobed. The antennae are very dark brown and are bent sideways and backwards reaching the anterior femoral-tibial articulation. The labrum is salient, brown and trapezoidal. The maxillary palpi are very dark and digitiform, a little longer than the labial palpi.

The sub-retangular pronotum, which is about twice as wide as its average length, is surrounded by a black band. The anterior margin is bordered. The posterior margin medially bearing a strong projection. On the disc there is a long transverse brown spot and laterally it is wider, divided in the middle by the longitudinal median furrow. It extends from the anterior margin to the apex of the median projection.

The mesonotum is about 3.3 x wider than its average length, with three subcircular spots next to the superior edge: a median one which is browner and two lateral ones which are black. Laterally and towards the ventral part, there are the elytra. They show three subparallel bands: the posterior one which is dark, follows the respective edge and becomes browner on the inner margin; the two remaining bands are brown, shortened, the anterior one on the next to the median line more so.

The metanotum has an almost straight posterior margin and is about 4.6 x wider than the average length, taking into consideration the posterior margin.

The prothoracic and mesothoracic legs are visible, with flat femora and tibiae. The metathoracic legs are hidden beneath the wings with only part of the femoro-tibial articulation and tarsi visible.

The abdominal terga 1 and 2 are brown; 3-6 have a transverse dark brown, sometimes almost black band next to the posterior margin; terga 1-4 are approximately six times wider than their average length, excluding the scoli; the remaining terga are gradually narrowed to the apex. The pupa bearing 8 pairs of scoli, one pair per segment; the scoli from segment 1-5 are black with short setae and are turned forwards, the last (7tn and 8th pairs) more markedly. The others are yellowish brown, glabrous, backwards which are only visible when the last larval exuvia with which the pupa attaches itself to the substratum is removed. The same is true of the anal fork that is fixed to the abdominal apex, with short, glabrous, yellowish brown branches. There are paired spiracles with black peritreme on the first five abdominal terga; the ostium of the first pair of spiracles is larger than that of the following spiracles which decrease in size posteriorly.

Adult (Fig. 1-2)

The adult are easily identified by their colour. The female and males are similarly coloured, but the latter are significantly smaller (Table 4). Entirely blue males and females occur in low numbers, suggesting that it is a recessive characteristic.

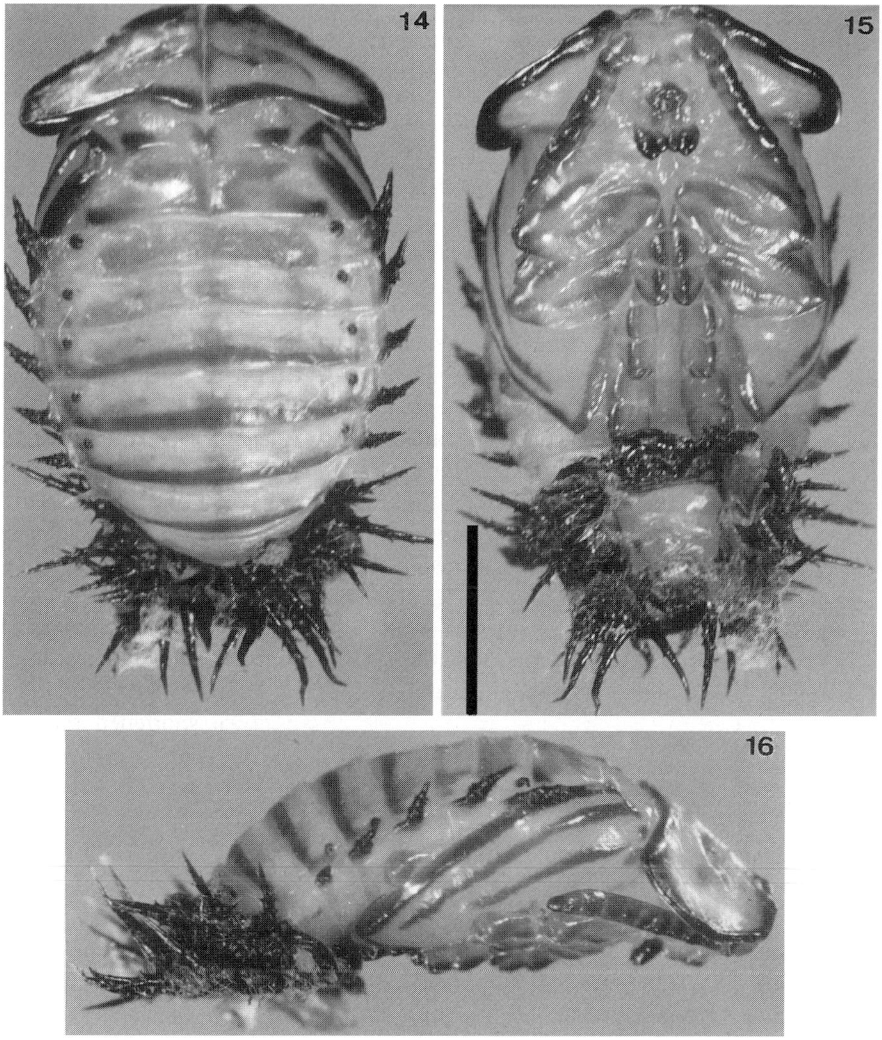

Figs. 14-16. Pupa of *Stolas lacordairei.* 14. Dorsal, 15. Ventral and 16. Lateral view (Scale line: 2.5 mm).

Immediately after emergence from the pupa, the elytra and the abdomen become whitish; the antennae, pronotum, legs and sides of the abdomen blacken. After approximately three hours, the adults achieve their definitive colour.

From total of 21 field collected pupae kept in the laboratory, 8 females and 13 males emerged, that is 38% females and 62% males, representing a ratio of 1.6 males to 1 female. This sex ratio was also confirmed by random collection of adults.

Many adults enter diapause in the laboratoy, from the end of May until the end of August. That is a total of 94 days. During this period which coincides with the decrease in photoperiod, they did not eat, remaining together, sometimes overlaping one another and only rarely making slight movements. At the end of this period, they began to eat and reproduce again.

Table 4. Mean length (m), in mm, of the male and female pupa (MP, FP) and male and female adult (MA, FA). Standard deviation (s), Range (< >), Coefficient of variation (CV). Number of specimens recorded (N).

	MP	FP	MA	FA
N	14	14	14	13
m	7.36	8.11	9.70	11.15
s	0.34843	0.30429	0.72327	0.80878
<	6.50	7.50	8.50	9.33
>	8.00	8.42	10.33	11.83
CV	4.73	3.73	7.70	7.54

Adults survived in the laboratory an average of 109.5 days, with a maximum of 189 days and a minimum of 65 days. The females live longer on average than the males.

In the laboratory, the average fecundity per female was 443, with the maximum number of 737 eggs and the minimum 125 eggs (Table 6 and Fig. 24). The percentage of viable eggs is very high and around 98% hatch occurred and this was also the case in the field.

From females raised in the laboratory the highest number of egg-masses produced was 10 and the lowest 3 per female. Thus, the average number is 7.17 egg-masses. On average, the interval between egg-laying was 6,33 days, with a maximum of 12 and a minimum of 2.

The pre-oviposition period of females reared in the laboratory was about 40 days. However, during this period copulation occurred quite regularly. The average oviposition period lasted 95.3 days, with a maximum of 140 days and a minimum of 30.

Discussion

Eggs

The eggs of *Stolas lacordairei* are easily identified by their beige colour and disposition in a large mass containing many eggs. Comparing the eggs of this species with those of other Stolaini (Table 1) it can be seen that this cassid has relatively small eggs and only those os *Acromis nebulosa* (Boheman) and *Chelymorpha variabilis* Boheman are smaller. The biggest eggs are those of *Stolas (Botanochara) impressa* (Panzer), followed by *Stolas (Anacassis) punctulata* (Klug), *Stolas (A.) languida* (Boheman) and *Stolas (A.) cribrum* (Klug). Apparently, there is a possible relationship between egg length and the number of eggs in each egg mass.

Table 1 also shows the average number of eggs in the egg masses of different cassid species and as can be seen, no species produced more eggs than *Stolas lacordairei* and in the laboratory, the average number of eggs per egg mass was 57.65. Some other Stolaini species produced egg masses with a relatively large average number of eggs, for example *Acromis sparsa* Boheman with 48; *Chelymorpha alternans* Boheman with 49; *C. indigesta* Boheman with 45; *C. variabilis* with 43 and *C. constellata* (Klug) with 38 eggs. *Chelymorpha inflata* Boheman with 52, produced almost as many eggs per batch as *S. lacordairei*. However, the former species lays the eggs in a small pile and they are differently coloured from those of the latter species. *Stolas (Cyrtonota) sexpustulata* (F.) arranges its eggs in a bunch, each egg being linked to the next by the peduncle. In *Stolas (Anacassis) phaeopoda* Buzzi,

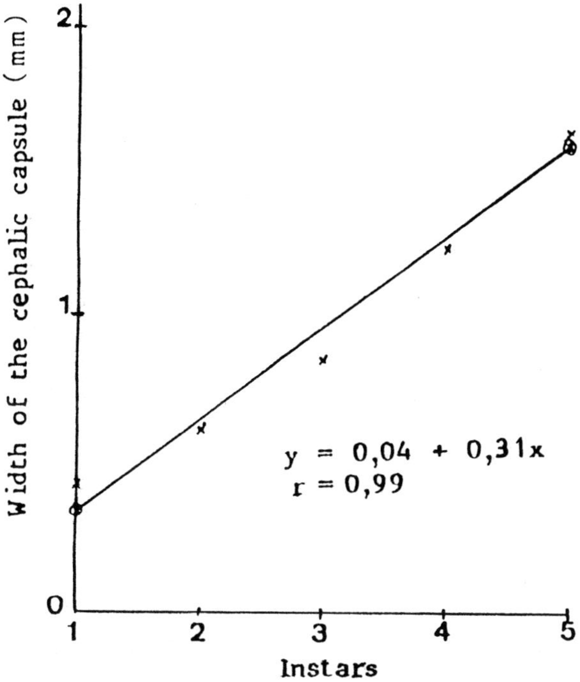

Fig. 17. Stolas lacordairei. Line of regression between geometric growth of the cephalic width as a function of the larval instars (y = equation line; r = coefficient of correlation).

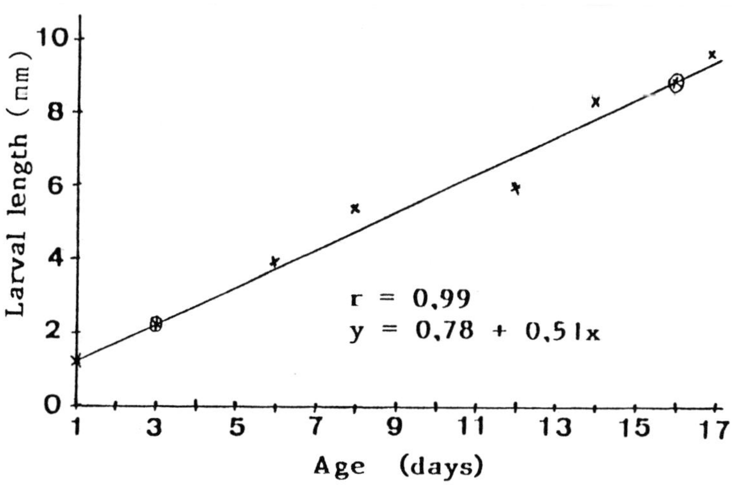

Fig. 18. Stolas lacordairei. Line of regression between the larval length as a function of the age (y = equation line; r = coefficient of correlation).

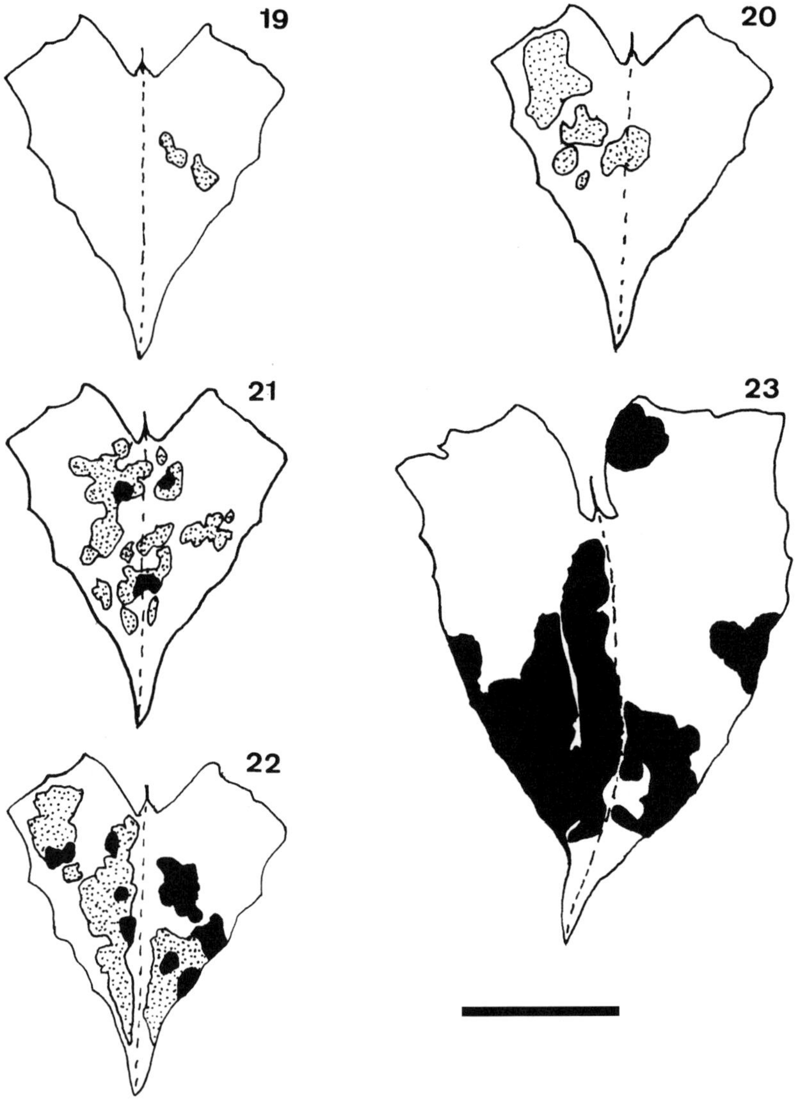

Figs. 19-23. Leaves of *Mikania micrantha* (Asteraceae) attacked by the larva of *Stolas lacordairei,* during 24 hours, respectively in the 1st (Fig.19), 3rd (Fig.20), 6th (Fig.21), 9th (Fig. 22) and 12th (Fig.23) day of life. In black totaly perforated; spotted, only grazed. (Scale line: 20 mm).

the majority of eggs are laid singly (Buzzi, 1996), as is also true for the majority of eggs of *Stolas (Botanochara) impressa* (Buzzi, 1977).

Table 5 shows the development time in days for the immature stages of various cassids. The length of the incubation period in *S. lacordairei* at 6.08 days, is somewhat shorter than in the majority of other species listed. However, *Echoma marginata* (L.) and *Echoma* sp. near *bonfilsi* (Boheman) had shorter incubation periods. However, published incubation times for other species may not be directly comparable because of differing temperature, RH, and photoperiod under which the eggs were kept and also the different feeding regimes of the females.

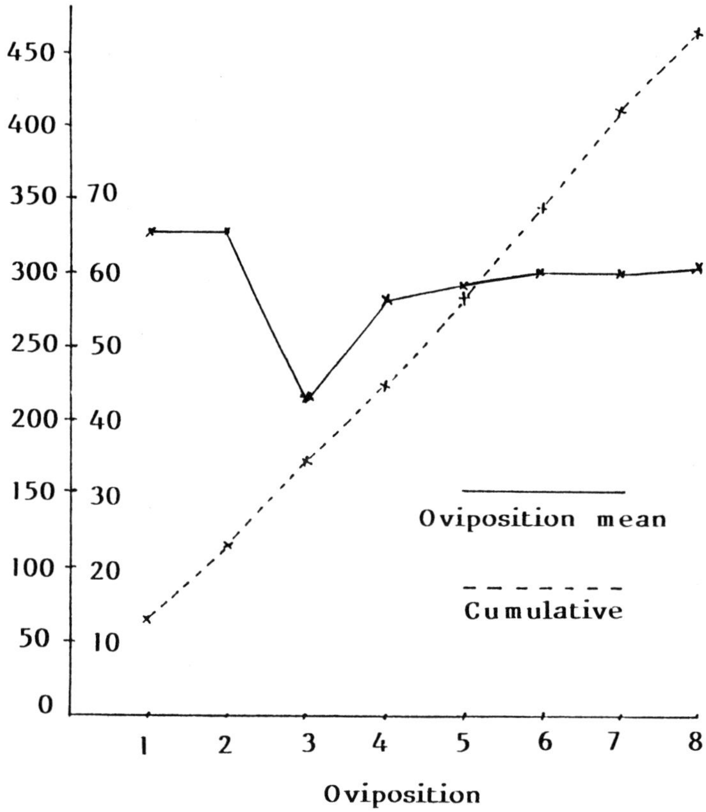

Fig. 24. Stolas lacordairei. Oviposition mean and cumulative of six females, reared in laboratory.

Table 5. Development times (days) for Cassidinae (Stolaini) and their host plants.

	incubation / developmental period				
	egg	larva	pupa	host plants	source
Acromis sparsa	12.0	18.0	14.0	*Merremia umbellata*[1]	Windsor (1987)
Chelymorpha alternans	10.5	13.2	4.5	—	Windsor *et al.* (1995)
C. inflata	9.0	16.8	5.5	*Ipomoea cairica*[2]	(unpublished data)
C. variabilis	8-15	16	5-6	*Convolvulus arvensis*[2]	Zolessi (1968)
C. cribraria, C.puncticollis, C. rufipennis	8-10	13-18 21-28	8-10** 12-14*	*Ipomoea batatas*[2]	Marques (1932)
C. constellata	8-10	10-15 21-28	8-18** 12-14*	*Ipomoea batatas*[2]	Marques (1932)
Stolas (Cyrtonota) sexpustulata	8-10 12-14	16-21 24-34	10-12** 14-16*	*Ipomoea batatas*[2]	Marques (1932)
Echoma sp.nr. *bonfilsi*	5.8	—	5.6	*Mikania guaco*[1]	Windsor *et. al.* (1995)
E. dichroa	7.0	11.0	9.0	*Calonyction preciosum*[2]	Silva *et al.* (1968)
E. marginata	5.0	11.0	5.0	*Mikania stipulacea*[1]	Windsor *et al.* (1995)

Table 5. Cont.

	incubation / developmental period				
	egg	larva	pupa	host plants	source
Stolas (Anacassis) cribrum	11.2	21.1	8.1**	*Baccharidastrum argutum*[1]	Buzzi (1975a)
	—	26.7	11.9*		
S. (A.) dubia	11.11	30.36	7.21	*Baccharis retusa*[1]	Buzzi (1996)
S. (A.) fuscata	6.22	22.7	6.0	*Baccharidastrum triplinerve*[1]	Buzzi (1975b)
S. (A.) languida	13.5	31.6	11.5	*Baccharis pauciflosculosa*[1]	
				B. retusa[1]	Buzzi & Garcia (1983)
S. (A) phaeopoda	7.96	21.80	6.20	*Baccharis gaudichaudiana*[1]	Buzzi (1996)
Stolas (Botanochara)					
impressa	6.9	13.63	5.2	*Ipomoea acuminata*[2]	Habib & Vascon-
	—	17.94	—		celos-Neto (1979)
Stolas pictilis	10.6	19.1	5.5	—	Windsor *et. al.* (1995)
S. festiva	6.0	—	5.0	*Mikania scandens*[1]	Frers (1925)
S. lacordairei	**6.08**	**17.5**	**5.4**	***Mikania micrantha***[1]	**(present work)**

* = winter; ** = summer. 1 = Asteraceae; 2 = Convolvulaceae.

Stolas lacordairei females oviposited on *Mikania micrantha*, whilst *Echoma marginata* and *Echoma* sp. near *bonfilsi* laid eggs on leaves of *Mikania stipulacea* and *M. guaco*, respectively, according to Windsor *et al.* (1995). The other Stolaini species feed on other genera of Asteraceae and also on species of Convulvulaceae (Table 5).

Larva

The larva of *Stolas lacordairei* is very easily distinguished from those of other Stolaini by the almost uniforme orange integumental colour and the disposition of the dorsal chalazae. Larvae of other Stolaini are recognizable by their colour and the number and size of the scoli. According to Buzzi (1975a) the larva of *Stolas (Anacassis) cribrum, Stolas (A.) dubia* (Boheman), *Stolas (A.) fuscata* (Klug), *Stolas (A.) languida, Stolas (A.) phaeopoda* and *Stolas (A.) punctulata* (Buzzi, 1996) have 14 pairs of scoli as in the larva of *Stolas lacordairei* and their number and position on the different body segments is the same. However the scoli differ significantly in colour as reported by Buzzi (1975a, b, 1976b, 1983, 1986). The larvae of *Acromis nebulosa* (Boheman) and *A. spinifex* (L.) bear 10 pairs of very short scoli: 8 on the abdomen; one on the pronotum and one on the mesonotum (Buzzi, 1980; Fiebrig, 1910). The larva of *Echoma decipiens* (Boheman) has 15 pairs of scoli and the first pair is divided into three ramifications (Maulik, 1948).

The larva of *S. lacordairei* retains the exuviae and faeces on the anal fork (Buzzi, 1988). This is also true for the larvae of *Stolas (Anacassis)* sp. (Buzzi 1975a, b, 1983, 1988, 1996) and of *Chelymorpha* sp. (Marques, 1932; Zolessi, 1968), but in the larva of *Stolas (Cyrtonota) sexpustulata* and *Stolas (Botanochara)* sp. only exuviae are retained (Marques, 1932; Buzzi, 1977).

Table 6. Oviposition period, in days, total production of eggs, Mean (m), Confidence limits (CL) and Cumulative of six females, reared in laboratory.

	Female nº								
	1	2	3	4	5	6	m	±CL	Cum.
Oviposition period	129	35	140	30	130	108			
Oviposition									
1	66	30	140	15	19	110	66.50	54.47	63.33
2	68	30	130	56	17	98	66.50	44.31	129.83
3	43	14	38	54	18	90	42.83	3.98	172.66
4	47	70	44		17	105	56.60	4.78	229.26
5	110	24	87		16	54	58.20	50.04	287.46
6	92		48		15	85	60.00	56.72	347.46
7	99		54		14	74	60.25	57.10	407.71
8	44		72		13	115	61.00	68.88	468.71
9	32		120		11				
10	5		85		10				
Total	660	168	818	125	150	737	443		

Pupa

The pupa of *Stolas lacordairei* is readily distinguished by its orange colour with brown and black spots as is shown in Figs 14-16. What is more notable are the dark brown transverse band on tergites 3 to 6, the three bands on the fore wings and the black border on the pronotum. Table 4 shows the morphometrics of male and female pupae and it can be seen that there is a significant difference between the pupae of the two sexes. This has been also shown by Buzzi (1996) for several other cassids, for example, *Stolas (Anacassis) dubia,* the male pupa of which is 7.34 mm in length as compared with that of the female which is 8.34 mm in length; in *Stolas (A). fuscata* the male pupa is 7.57 mm in length, whilst the female is 8.43 mm in length; and in *Stolas (A.) phaeopoda* the male pupa is 7.80 mm in length as compared with 8.50 mm for the male.

Adult

In the laboratory, the longevity of adults of *S. lacordairei* was 75 days for the males and 100.9 days for the females. This difference also ocurred in laboratory reared *Chelymorpha inflata* where the males lived for 305 days and the females 510 days fed on *Ipomoea batatas* (unpublished data). According to Marques (1932) adults of *Chelymorpha cribraria* (F.), *C. puncticollis* (Boheman), *C. rufipennis* (Boheman), *C. constellata* and *Stolas (Cyrtonota) sexpustulata* lived on average 180 days. *Stolas (Botanochara) impressa* males lived for 175.9 days, whilst the females only survived for 162.9 days (Habib & Vasconcelos-Neto, 1979). Laboratory reared adults of *Stolas (Anacassis) cribrum* lived for 2.5 – 5 months (Buzzi, 1975a).

Stolas lacordairei adults under laboratory conditions produced on average 443 eggs, which is lower than the fecundity of *Chelymorpha cribraria, C. puncticollis, C. rufipennis* and *C. constellata* which according to Marques (1932) each had an average fecundity of 1500 eggs, and *Stolas (Cyrtonota) sexpustulata,* 1000 eggs. Laboratory reared adults of *Chelymorpha inflata* produced 1650 eggs (unpublished data), whilst those of *Stolas (Anacassis) cribrum* produced on average 161.8 eggs (Buzzi, 1975a). More datailed studies on the fecundity of *S. lacordairei* are necessary because of the relatively low number of females reared in this study.

Maternal care of the eggs, larvae and pupae was never observed in *Stolas lacordairei.* However, such care has been confirmed in *Acromis sparsa* (Windsor, 1987), *A. nebulosa* (Buzzi, 1980), *Omaspides pallidipennis* (Boheman) by Lima (1914) who related that the female walking over the pupae seemed to watch over them. Maternal care has also been observed in *Echoma dichroa* (Germar), *E. flava* (L.), *E. solieri* (Boheman) and *Omaspides nigrolineata* (Boheman) by Monte (1932) who noted that the females have the habit of protecting the eggs under her body.

Adults of *Stolas lacordairei* enter diapause during the months when the photoperiod is very short and this coincides with the coldest months of lthe year. During this period they remain immobile, aggregating together in places protected from light, and some remain on top of each other for long periods. Adult diapause was also observed in *Stolas (Botanochara) impressa* (Buzzi, 1977), in *Stolas (Anacassis) cribrum* (Buzzi, 1975a) and *Chelymorpha inflata* (unpublished data).

Host plants

The larvae and adults of *Stolas lacordairei* feed on the leaves of the Asteraceae, *Mikania micrantha.* The host plants of various other Stolaini are listed in Table 5. The leaves of eleven Asteraceae and 7 Convolvulaceae are consumed by those species listed. The typical damage by the various larval instars of *S. lacordairei* to the leaves of *Mikania micrantha* is shown in Figs. 19-23. The early instars graze the leaves, leaving the lower epidermis intact, whilst the late third to fifth instars completely perforate the leaves with holes. Similar leaf damages is caused by *Stolas (Anacassis) cribrum* (Buzzi, 1975a). Windsor *et al.* (1995) observed adults and larvae of *Echoma* sp. nr *bonfilsi* feeding on the flowers of *Mikania guaco,* whilst Rodriguez (pers. com.) observed the same species feeding on the inflorescences of *Mikania micrantha* at the La Selva field station in the Atlantic lowlands of Costa Rica.

References

Blackwelder, R. E., 1946. Checklist of the Coleopterous insects of Mexico, Central America, The West Indies and South America. Bull. U. S. natn. Mus., 185(4): 551-763.

Buzzi, Z.J., 1975a. Redescrição e bionomia de *Anacassis cribrum* (Klug, 1829)(Coleoptera, Chrysomelidae). Acta Biol. paran., 4(1/2):59-89.

Buzzi, Z. J., 1975b. Contribuição ao conhecimento da biologia de *Anacassis fuscata* (Klug, 1829) (Coleoptera, Chrysomelidae). Rev. Brasil. Biol., 35(4):767-774.

Buzzi, Z. J., 1976a. Uma nova espécie de *Anacassis* do sul do Brasil (Coleoptera, Chrysomelidae). Dusenia, 9(3):107-111.

Buzzi, Z. J., 1976b. Contribuição ao conhecimento de *Zatrephina meticulosa* Spaeth, 1909 (Coleoptera, Chrysomelidae). Rev. Brasil. Biol., 36(2): 381-385.

Buzzi, Z. J., 1977. Nota sobre a ontogenia de *Botanochara impressa* (Panzer, 1798)(Coleoptera, Chrysomelidae, Cassidinae). Dusenia, 10(4): 221- 228.

Buzzi, Z. J., 1980. Estágios imaturos de *Acromis nebulosa* (Boheman, 1854) (Coleoptera, Chrysomelidae, Cassidinae). Dusenia, 12(2): 63-67.

Buzzi, Z. J., 1988. Biology of neotropical Cassidinae. *In* Jolivet, P; Petitpierre, E. & Hsiao,T.H. (eds). Biology of Chrysomelidae. Kluwer Academic Publishers, Dordrecht, The Netherlands. pp. 559-580.

Buzzi, Z. J., 1996. Morfologia dos imaturos e ciclo evolutivo de *Anacassis dubia* (Boheman). *A. fuscata* (Klug), *A. languida* (Boheman), *A. phaeopoda* Buzzi e *A. punctulata* (Klug) (Coleoptera, Chrysomelidae, Cassidinae). Revta. bras. zool., 13(1):215-289.

Buzzi, Z. J. & Garcia, C., 1983. Immature stages and life cycle of *Anacassis languida* (Boheman, 1854)(Coleoptera, Chrysomelidae, Cassidinae). Coleopts Bull., 37(2):193-198.

Fiebrig, K.,1910. Cassiden und Cryptocephaliden Paraguays. Zool. Jahrb. Suppl., 12:161-264.

Frers, A. G. 1922. Metamorfosis de coleopteros argentinos. Physis, B. Aires, 5:245-262.

Frers, A. G., 1925. Metamorfosis de coleopteros argentinos. Physis, B. Aires, 8:82-90.

Habib, M. E. M. & Vasconcelos-neto, J., 1979. Biological studies on *Botanochara impressa*. Panzer, 1789 (Coleoptera, Chrysomelidae). Rev. Biol.Trop. , 27(1): 103-110.

Lima, A. da Costa, 1914. Nota relativa ao cassidideo *Omoplata pallidipennis* (Dejean). Mems. Inst. Oswaldo Cruz, 6:112-145.

Marques, L. A. de A., 1932. Insetos damninhos à batata dôce, seus habitos e os meios de combatel-os. Bol. Inst. Biol. Def. Agric., 9, 81 pp.

Maulik, S., 1948. Larva of *Echoma decipiens* (Boheman, 1854) (Cassidinae, Chrysomelidae, Coleoptera). Ann. Mag. nat. Hist., 1(12): 821-825.

Monte, O., 1932. Alguns cassidideos pragas da batata doce. O Campo, 3(4):62-64.

Silva, A. G. d'A.; Gonçalves, C. R.; Galvão, D. M.; Gonçalves A. J.; Gomes J.; Silva M. do N.; & Simoni L. de, 1968. Quarto catálogo dos insetos que vivem nas plantas do Brasil. Seus parasitos e predadores. Ministério da Agricultura. Laboratório central de patologia vegetal. Rio de Janeiro. Parte II, 1° Tomo, 622 p.

Spaeth, F., 1914. Coleopterorum Catalogus 62; Chrysomelidae: 16. Cassidinae. W. Junk, Berlin, 182 pp.

Windsor, D.M., 1987. Natural history os a subsocial tortoise beetle, *Acromis sparsa* Boheman (Chrysomelidae, Cassidinae) in Panama. Psyche, 94(1-2): 127-150.

Windsor, D. M.; Macedo, M.V. de & SIQUEIRA-CAMPOS A. de T., 1995. Flower feeding by species of *Echoma* Chevrolat (Coleoptera: Chrysomelidae: Cassidinae) on *Mikania* (Asteraceae) in Panama and Brazil. Coleopts Bull., 49(2):101-108.

Zolessi, L.C.de, 1968. Estudio genetico de las variaciones cromaticas y estudio anatomico de *Chelymorpha variabilis* Boh. (= *Ch. Crucifera* Boh.), 1854, (Coleoptera, Chrysomelidae, Cassidinae). Rev. Soc. Uruguaya Ent., 7: 20-44.

AUTHOR INDEX

SUBJECT INDEX

TAXONOMIC INDEX – ANIMALS

TAXONOMIC INDEX – PLANTS